Video Demystified

Video Demystified

A Handbook
for the
Digital Engineer
Fourth Edition

by Keith Jack

AMSTERDAM • BOSTON • HEIDELBERG • LONDON
NEW YORK • OXFORD • PARIS • SAN DIEGO
SAN FRANCISCO • SINGAPORE • SYDNEY • TOKYO

Newnes is an imprint of Elsevier

Newnes is an imprint of Elsevier
200 Wheeler Road, Burlington, MA 01803, USA
Linacre House, Jordan Hill, Oxford OX2 8DP, UK

 Recognizing the importance of preserving what has been written, Elsevier prints its books on acid-free paper whenever possible.

Library of Congress Cataloging-in-Publication Data

(Application submitted.)

British Library Cataloguing-in-Publication Data
A catalogue record for this book is available from the British Library.

ISBN: 0-7506-7822-4

For information on all Newnes publications
visit our Web site at www.books.elsevier.com

04 05 06 07 08 09 10 9 8 7 6 5 4 3 2 1

Printed in the United States of America

Contents

Chapter 5 · **Analog Video Interfaces** 66

Chapter 6 · **Digital Video Interfaces   100**

Chapter 7 · **Digital Video Processing   202**

Chapter 8 • NTSC, PAL, and SECAM Overview 265

Chapter 9 · NTSC and PAL Digital Encoding and Decoding 394

Chapter 12 · **MPEG-1** **543**

Chapter 13 · **MPEG-2   581**

Chapter 14 · MPEG-4 and H.264   736

Chapter 15 • **ATSC Digital Television**   **760**

Chapter 16 • **OpenCable™ Digital Television**   **778**

Chapter 17 · **DVB Digital Television**   **796**

About the Author

Keith Jack is Director of Product Marketing at Sigma Designs, a leading supplier of Digital Media Processors that provide high-quality processing of MPEG-4, MPEG-2, MPEG-1 and Microsoft® Windows Media® 9 content.

Prior to joining Sigma Designs, Mr. Jack held various marketing and chip design positions at Harris Semiconductor, Brooktree and Rockwell International. He has been involved in over 30 multimedia chips for the consumer market.

What's on the CD-ROM?

The included CD-ROM contains documents and tools to assist in designing, testing and evaluating various video subsystems.

- A fully searchable eBook version of the text in Adobe PDF format.

- Test images to enable the evaluation of a video subsystem. They are primarily used to test color accuracy and filter design. The sharper the transitions are without ringing, the better the filters. Since the images are computer-generated, there may be flicker along horizontal edges when viewed on an interlaced display. Unless the video signal is RF modulated, these images should pass through a system with no problems.

- Links to associations, licensing authorities and standards organizations. Before starting any design, and several times during the design process, verify the latest specifications are being used since they are continually updated. The list of licensing authorities should be consulted before starting any chip or system design, since several licenses are required prior to purchasing chips.

Chapter 1

Introduction

A few short years ago, the applications for video were somewhat confined—analog was used for broadcast and cable television, VCRs, set-top boxes, televisions and camcorders. Since then, there has been a tremendous and rapid conversion to digital video, mostly based on the MPEG-2 video compression standard.

Today, in addition to the legacy DV, MPEG-1 and MPEG-2 audio and video compression standards, there are three new high-performance video compression standards. These new video codecs offer much higher video compression for a given level of video quality.

- *MPEG-4.* This video codec typically offers a 1.5–2× improvement in compression ratio over MPEG-2. Able to address a wide variety of markets, MPEG-4 never really achieved widespread acceptance due to its complexity. Also, many simply decided to wait for the new H.264 video codec to become available.

- *H.264.* Also known as MPEG-4 Part 10, this video codec typically offers a 2–3× improvement in compression ratio over MPEG-2. Additional improvements in compression ratios and quality are expected as the encoders become better and use more of the available tools that H.264 offers. Learning a lesson from MPEG-4, H.264 is optimized for implementing on low-cost single-chip solutions.

- *SMPTE VC-9 (Microsoft® Windows Media® Video 9 or WMV9).* A competitor to H.264, this video codec also typically offers a 2–3× improvement in compression ratios over MPEG-2. Again, additional improvements in compression ratios and quality are expected as the encoders become better.

Many more audio codecs are also available as a result of the interest in 6.1- and 7.1-channel audio, multi-channel lossless compression, lower bit rates for the same level of audio quality, and finally, higher bit rates for applications needing the highest audio quality at a reasonable bit rate.

In addition to decoding audio, real-time high-quality audio encoding is needed for recordable DVD and digital video recorder (DVR) applications. Combining all these audio requirements mandates that any single-chip solution for the consumer market incorporate a DSP for audio processing.

Equipment for the consumer has also become more sophisticated, supporting a much wider variety of content and interconnectivity. Today we have:

- *Networked DVD Players.* In addition to playing normal CDs and DVDs, these advanced DVD players also support the playback of MPEG-4, H.264, Microsoft® Windows Media® 9 (WM9) and JPEG (for photos) content. An Ethernet or 802.11 connection enables PC-based content to be enjoyed easily on any television. Web radio and viewing of on-line movies may also be supported.

- *Digital Media Adapters.* These small, low-cost boxes use an Ethernet or 802.11 connection to enable PC-based content to be enjoyed easily on any television. Playback of MPEG-2, MPEG-4, H.264, WM9 and JPEG content is supported.

- *Digital Set-top Boxes.* Cable, satellite and terrestrial set-top boxes are now including digital video recorder (DVR) capabilities, allowing viewers to enjoy content at their convenience. Many are looking at H.264 and/or WMV9 to enable system operators to offer more channels of content and reduce the chance of early obsolescence.

- *Advanced Digital Televisions (DTV).* In addition to the tuners and decoders being incorporated inside the television, some also include an advanced DVD player, surround sound processor, wireless networking (802.11 or UWB), etc.

- *IP Video Set-top Boxes.* Also known as "IPTV" and "video over IP", these low-cost set-top boxes are gaining popularity in regions that have high-speed DSL and FTTH (fiber to the home) available. Many are also moving to H.264 or WMV9 to be able to offer HDTV content.

- *Portable Media Players.* Using an internal hard disc drive (HDD), these players connect to the PC via USB or 802.11 network for downloading a wide variety of content. Playback of MPEG-2, MPEG-4, H.264, WM9 and JPEG content is supported.

- *Mobile Video Receivers.* Being incorporated into cell phones, H.264 is used to transmit a high-quality video signal. Example applications are the DMB and DVB-H standards.

Of course, to make these advanced consumer products requires more than just supporting an audio and video codec. There is also the need to support:

- *Closed Captioning, Subtitles, Teletext, and V-Chip.* These standards were updated to support digital broadcasts.

- *Advanced Video Processing.* Due to the wide range of resolutions and aspect ratios for both content and displays, sophisticated high-quality scaling is usually required. Since the standard-definition (SD) and high-definition (HD) standards use different colorimetry standards, this should be corrected when viewing SD content on a HDTV or HD content on a SDTV.

- *Sophisticated image composition.* The ability to render a sophisticated image composed of a variety of video, OSD (on-screen display), subtitle/captioning/subpicture, text and graphics elements.

- *ARIB and DVB over IP.* The complexity of supporting IP Video is increasing, with deployments now incorporating ARIB and DVB over IP.

- *Digital Rights Management (DRM).* The protection of content from unauthorized copying or viewing.

This fourth edition of *Video Demystified* has been updated to reflect these changing times. Implementing "real-world" solutions is not easy, and many engineers have little knowledge or experience in this area. This book is a guide for those engineers charged with the task of understanding and implementing video features into next-generation designs.

This book can be used by engineers who need or desire to learn about video, VLSI design engineers working on new video products, or anyone who wants to evaluate or simply know more about video systems.

Contents

The book is organized as follows:

Chapter 2, an *Introduction to Video*, discusses the various video formats and signals, where they are used, and the differences between interlaced and progressive video. Block diagrams of DVD players and digital set-top boxes are provided.

Chapter 3 reviews the common *Color Spaces*, how they are mathematically related, and when a specific color space is used. Color spaces reviewed include RGB, YUV, YIQ, YCbCr, HSI, HSV and HLS. Considerations for converting from a non-RGB to a RGB color space and gamma correction are also discussed.

Chapter 4 is a *Video Signals Overview* that reviews the video timing and the analog and digital representations of various video formats, including 480i, 480p, 576i, 576p, 720p, 1080i, and 1080p.

Chapter 5 discusses the *Analog Video Interfaces*, including the analog RGB, YPbPr, S-Video and SCART interfaces for consumer and pro-video applications.

Chapter 6 discusses the various *Digital Video Interfaces* for semiconductors, pro-video equipment and consumer equipment. It reviews the BT.601 and BT.656 semiconductor interfaces, the SDI, SDTI and HD-SDTI pro-video interfaces, and the DVI, HDMI and IEEE 1394 consumer interfaces. Also reviewed are the formats for digital audio, timecode, error correction, and so on for transmission over various digital interfaces.

Chapter 7 covers several *Digital Video Processing* requirements such as 4:4:4 to 4:2:2 YCbCr, YCbCr digital filter templates, scaling, interlaced/noninterlaced conversion, frame rate conversion, alpha mixing, flicker filtering and chroma keying. Brightness, contrast, saturation, hue, and sharpness controls are also discussed.

Chapter 8 provides an *NTSC, PAL and SECAM Overview.* The various composite analog video signal formats are reviewed, along with video test signals. VBI data discussed includes timecode, closed captioning and extended data services (XDS), widescreen signaling and teletext. In addition, PALplus, RF modulation, BTSC and Zweiton analog stereo audio and NICAM 728 digital stereo audio are reviewed.

Chapter 9 covers digital techniques used for the *Encoding and Decoding of NTSC and PAL* color video signals. Also reviewed are various luma/chroma (Y/C) separation techniques and their trade-offs.

Chapter 10 discusses the *H.261 and H.263* video compression standards used for video teleconferencing.

Chapter 11 discusses the *Consumer DV* video compression standards used by digital camcorders.

Chapter 12 reviews the *MPEG-1* video compression standard.

Chapter 13 discusses the *MPEG-2* video compression standard.

Chapter 14 discusses the *MPEG-4* video compression standard, including H.264.

Chapter 15 discusses the *ATSC Digital Television* standard used in the United States.

Chapter 16 discusses the *OpenCable™ Digital Television* standard used in the United States.

Chapter 17 discusses the *DVB Digital Television* standard used in Europe and Asia.

Chapter 18 discuses the *ISDB Digital Television* standard used in Japan.

Chapter 19 discusses *IPTV.* This technology sends compressed video over broadband networks such as Internet, DSL, FTTH (Fiber To The Home), etc.

Finally, chapter 20 is a glossary of over 400 video terms which has been included for reference. If you encounter an unfamiliar term, it likely will be defined in the glossary.

Organization Addresses

Many standards organizations, some of which are listed below, are involved in specifying video standards.

Advanced Television Systems Committee (ATSC)

1750 K Street NW, Suite 1200
Washington, DC 20006
Tel: (202) 872-9160
Fax: (202) 872-9161
www.atsc.org

Cable Television Laboratories

858 Coal Creek Circle
Louisville, CO 80027
Tel: (303) 661-9100
Fax: (303) 661-9199
www.cablelabs.com

Digital Video Broadcasting (DVB)

17a Ancienne Route
CH-1218 Grand Sacconnex, Geneva
Switzerland
Tel: +41 (0)22 717 27 14
Fax: +41 (0)22 717 27 27
www.dvb.org

European Broadcasting Union (EBU)
17A, Ancienne Route
CH-1218 Grand-Saconnex
Switzerland
Tel: +41 (0)22 717 2111
Fax: +41 (0)22 747 4000
www.ebu.ch

Electronic Industries Alliance (EIA)
2500 Wilson Boulevard
Arlington, Virginia 22201
Tel: (703) 907-7500
Fax: (703) 907-7501
www.eia.org

European Telecommunications Standards Institute (ETSI)
650, route des Lucioles
06921 Sophia Antipolis Cedex
France
Tel: +33 (0)4 92 94 42 00
Fax: +33 (0)4 93 65 47 16
www.etsi.org

International Electrotechnical Commission (IEC)
3, rue de Varembé
P.O. Box 131
CH - 1211 GENEVA 20
Switzerland
Tel: +41 (0)22 919 02 11
Fax: +41 (0)22 919 03 00
www.iec.ch

Institute of Electrical and Electronics Engineers (IEEE)
1828 L Street, N.W., Suite 1202
Washington, D.C. 20036
Tel: (202) 785-0017
Fax: (202) 785-0835
www.ieee.org

International Telecommunication Union (ITU)
Place des Nations
CH-1211 Geneva 20
Switzerland
Tel: +41 (0)22 730 5111
Fax: +41 (0)22 733 7256
www.itu.int

Society of Cable Telecommunications Engineers (SCTE)
140 Philips Road
Exton, PA 19341
Tel: (610) 363-6888
Fax: (610) 363-5898
www.scte.org

Society of Motion Picture and Television Engineers (SMPTE)
595 West Hartsdale Avenue
White Plains, NY 10607
Tel: (914) 761-1100
Fax: (914) 761-3115
www.smpte.org

Video Electronics Standards Association (VESA)
920 Hillview Ct., Suite 140
Milpitas, CA 95035
Tel: (408) 957-9270
www.vesa.org

Video Demystified Web Site

At the Video Demystified web site, you'll find links to video-related newsgroups, standards, standards organizations and associations.

www.video-demystified.com

Chapter 2

Introduction to Video

Although there are many variations and implementation techniques, video signals are just a way of transferring visual information from one point to another. The information may be from a VCR, DVD player, a channel on the local broadcast, cable television, or satellite system, the internet or one of many other sources.

Invariably, the video information must be transferred from one device to another. It could be from a satellite set-top box or DVD player to a television. Or it could be from one chip to another inside the satellite set-top box or television. Although it seems simple, there are many different requirements, and therefore many different ways of doing it.

Analog vs. Digital

Until a few years ago, most video equipment was designed primarily for analog video. Digital video was confined to professional applications, such as video editing.

The average consumer now uses digital video every day thanks to continuing falling costs. This trend has led to the development of DVD players and recorders, digital set-top boxes, digital television (DTV), portable video players and the ability to use the Internet for transferring video data.

Video Data

Initially, video contained only gray-scale (also called black-and-white) information.

While color broadcasts were being developed, attempts were made to transmit color video using analog RGB (red, green, blue) data. However, this technique occupied 3× more bandwidth than the current gray-scale solution, so alternate methods were developed that led to using Y, R–Y and G–Y data to represent color information. A technique was then developed to transmit this Y, R–Y and G–Y information using one signal, instead of three separate signals, and in the same bandwidth as the original gray-scale video signal. This *com-*

posite video signal is what the NTSC, PAL, and SECAM video standards are still based on today. This technique is discussed in more detail in Chapters 8 and 9.

Today, even though there are many ways of representing video, they are still all related mathematically to RGB. These variations are discussed in more detail in Chapter 3.

S-Video was developed for connecting consumer equipment together (it is not used for broadcast purposes). It is a set of two analog signals, one gray-scale (Y) and one that carries the analog R–Y and B–Y color information in a specific format (also called C or chroma). Once available only for S-VHS, it is now supported on most consumer video products. This is discussed in more detail in Chapter 9.

Although always used by the professional video market, analog RGB video data has made a temporary come-back for connecting high-end consumer equipment together. Like S-Video, it is not used for broadcast purposes.

A variation of the Y, R–Y and G–Y video signals, called *YPbPr*, is now commonly used for connecting consumer video products together. Its primary advantage is the ability to transfer high-definition video between consumer products. Some manufacturers incorrectly label the YPbPr connectors YUV, YCbCr, or Y(B-Y)(R-Y).

Chapter 5 discusses the various analog interconnect schemes in detail.

Digital Video

The most common digital signals used are *RGB* and *YCbCr*. RGB is simply the digitized version of the analog RGB video signals. YCbCr is basically the digitized version of the analog YPbPr video signals, and is the format used by DVD and digital television.

Chapter 6 further discusses the various digital interconnect schemes.

Best Connection Method

There is always the question of "what is the best connection method for equipment?". For DVD players and digital cable/satellite/terrestrial set-top boxes, the typical order of decreasing video quality is:

1. HDMI (digital YCbCr)
2. DVI (digital RGB)
3. Analog YPbPr
4. Analog RGB
5. Analog S-Video
6. Analog Composite

Some will disagree about the order. However, most consumer products do digital video processing in the YCbCr color space. Therefore, using YCbCr as the interconnect for equipment reduces the number of color space conversions required. Color space conversion of digital signals is still preferable to D/A (digital-to-analog) conversion followed by A/D (analog-to-digital) conversion, hence the positioning of DVI above analog YPbPr.

The computer industry has standardized on analog and digital RGB for connecting to the computer monitor.

Video Timing

Although it looks like video is continuous motion, it is actually a series of still images, changing fast enough that it looks like continuous motion, as shown in Figure 2.1. This typically occurs 50 or 60 times per second for consumer video, and 70–90 times per second for computer displays. Special timing information, called *vertical sync*, is used to indicate when a new image is starting.

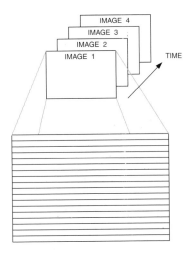

Figure 2.1. Video Is Composed of a Series of Still Images. Each image is composed of individual lines of data.

Each still image is also composed of *scan lines*, lines of data that occur sequentially one after another down the display, as shown in Figure 2.1. Additional timing information, called *horizontal sync*, is used to indicate when a new scan line is starting.

The vertical and horizontal sync information is usually transferred in one of three ways:

1. Separate horizontal and vertical sync signals

2. Separate composite sync signal

3. Composite sync signal embedded within the video signal

The composite sync signal is a combination of both vertical and horizontal sync.

Computer and consumer equipment that uses analog RGB video usually uses technique 1 or 2. Consumer equipment that supports composite video or analog YPbPr video usually uses technique 3.

For digital video, either technique 1 is commonly used or timing code words are embedded within the digital video stream. This is discussed in Chapter 6.

Interlaced vs. Progressive

Since video is a series of still images, it makes sense to simply display each full image consecutively, one after the another.

This is the basic technique of progressive, or non-interlaced, displays. For progressive displays that "paint" an image on the screen, such as a CRT, each image is displayed starting at the top left corner of the display, moving to the right edge of the display. Then scanning then moves down one line, and repeats scanning left-to-right. This process is repeated until the entire screen is refreshed, as seen in Figure 2.2.

In the early days of television, a technique called "interlacing" was used to reduce the amount of information sent for each image. By transferring the odd-numbered lines, followed by the even-numbered lines (as shown in Figure 2.3), the amount of information sent for each image was halved.

Given this advantage of interlacing, why bother to use progressive?

With interlace, each scan line is refreshed half as often as it would be if it were a progressive display. Therefore, to avoid line flicker on sharp edges due to a too-low refresh rate, the line-to-line changes are limited, essentially by vertically lowpass filtering the image. A progressive display has no limit on the line-to-line changes, so is capable of providing a higher-resolution image (vertically) without flicker.

Today, most broadcasts (including HDTV) are still transmitted as interlaced. Most CRT-based displays are still interlaced while LCD, plasma and computer displays are progressive.

Video Resolution

Video resolution is one of those "fuzzy" things in life. It is common to see video resolutions of 720 × 480 or 1920 × 1080. However, those are just the number of horizontal samples and vertical scan lines, and do not necessarily convey the amount of useful information.

For example, an analog video signal can be sampled at 13.5 MHz to generate 720 samples per line. Sampling the same signal at 27 MHz would generate 1440 samples per line. However, only the number of samples per line has changed, not the resolution of the content.

Therefore, video is usually measured using *"lines of resolution"*. In essence, how many distinct black and white vertical lines can be seen across the display? This number is then normalized to a 1:1 display aspect ratio (dividing the number by 3/4 for a 4:3 display, or by 9/16 for a 16:9 display). Of course, this results in a lower value for widescreen (16:9) displays, which goes against intuition.

Standard Definition

Standard definition video is usually defined as having 480 or 576 interlaced active scan lines, and is commonly called "480i" and "576i" respectively.

For a fixed-pixel (non-CRT) consumer display with a 4:3 aspect ratio, this translates into an active resolution of 720 × 480i or 720 × 576i. For a 16:9 aspect ratio, this translates into an active resolution of 960 × 480i or 960 × 576i.

Enhanced Definition

Enhanced definition video is usually defined as having 480 or 576 progressive active scan lines, and is commonly called "480p" and "576p" respectively.

For a fixed-pixel (non-CRT) consumer display with a 4:3 aspect ratio, this translates into an active resolution of 720 × 480p or 720 × 576p. For a 16:9 aspect ratio, this translates into an active resolution of 960 × 480p or 960 × 576p.

The difference between standard and enhanced definition is that standard definition is interlaced, while enhanced definition is progressive.

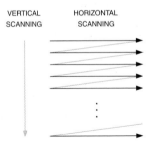

Figure 2.2. Progressive Displays "Paint" the Lines of An Image Consecutively, One After Another.

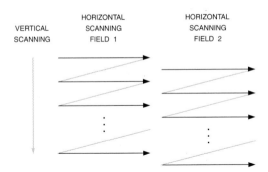

Figure 2.3. Interlaced Displays "Paint" First One-Half of the Image (Odd Lines), Then the Other Half (Even Lines).

High Definition

High definition video is usually defined as having 720 progressive (720p) or 1080 interlaced (1080i) active scan lines. For a fixed-pixel (non-CRT) consumer display with a 16:9 aspect ratio, this translates into an active resolution of 1280 × 720p or 1920 × 1080i, respectively.

However, HDTV displays are technically defined as being capable of displaying a minimum of 720p or 1080i active scan lines. They also must be capable of displaying 16:9 content using a minimum of 540 progressive (540p) or 810 interlaced (810i) active scan lines. This enables the manufacturing of CRT-based HDTVs with a 4:3 aspect ratio and LCD/plasma 16:9 aspect ratio displays with resolutions of 1024 × 1024p, 1280 × 768p, 1024 × 768p, and so on, lowering costs.

Audio and Video Compression

The recent advances in consumer electronics, such as digital television, DVD players and recorders, digital video recorders, and so on, were made possible due to audio and video compression based largely on MPEG-2 video with Dolby® Digital, DTS®, MPEG-1 or MPEG-2 audio.

New audio and video codecs, such as MPEG-4 HE-AAC, H.264 and SMPTE VC-9, offer better compression than previous codecs for the same quality. These advances are enabling new ways of distributing content (both to consumers and within the home), new consumer products (such as portable video players and mobile video/cell phones) and more cable/satellite channels.

Application Block Diagrams

Looking at a few simplified block diagrams helps envision how video flows through its various operations.

DVD Players

Figure 2.4 is a simplified block diagram for a basic DVD-Video player, showing the common blocks. Today, all of this is on a single low-cost chip.

In addition to playing DVDs (which are based on MPEG-2 video compression), DVD players are now expected to handle MP3 and WMA audio, MPEG-4 video (for DivX Video), JPEG images, and so on. Special playback modes such as slow/fast forward/reverse at various speeds are also expected. Support for DVD-Audio and SACD is also increasing.

A recent enhancement to DVD players is the ability to connect to a home network for playing content (music, video, pictures, etc.) residing on the PC. These "networked DVD players" may also include the ability to play movies from the Internet and download content onto an internal hard disc drive (HDD) for later viewing. Support for playing audio, video and pictures from a variety of flash-memory cards is also growing.

In an attempt to "look different" to quickly grab buyers attention, some DVD player manufacturers "tweak" the video frequency response. Since this "feature" is usually irritating over the long term, it should be defeated or properly adjusted. For the "film look" many video enthusiasts strive for, the frequency response should be as flat as possible.

Another issue is the output levels of the analog video signals. Although it is easy to generate very accurate video levels, they vary con-

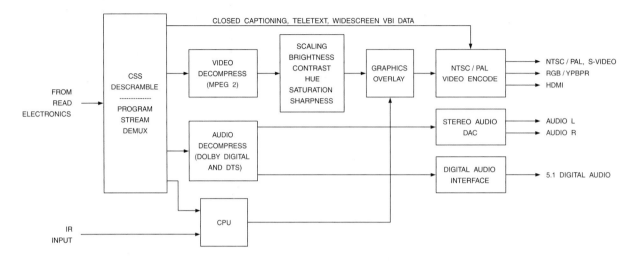

Figure 2.4. Simplified Block Diagram of a Basic DVD Player.

siderably. Reviews are now pointing out this issue since switching between sources may mean changing brightness or black levels, defeating any television calibration or personal adjustments that may have been done by the user.

Digital Media Adapters

Digital media adapters connect to a home network for playing content (music, video, pictures, and so on) residing on the PC. These small, low-cost boxes enable PC-based content to be enjoyed on any or all televisions in the home. Many support optional wireless networking, simplifying installation. Except for DVD playback, they have capabilities similar to networked DVD players.

Figure 2.5 is a simplified block diagram for a basic digital media adapter, showing the common blocks. Today, all of this is on a single low-cost chip.

Digital Television Set-top Boxes

The digital television standards fall into six major categories:

ATSC (Advanced Television Systems Committee)

DVB (Digital Video Broadcast)

ARIB (Association of Radio Industries and Businesses)

Open digital cable standards, such as OpenCable

Proprietary digital cable standards

Proprietary digital satellite standards

Currently based on MPEG-2 video compression, with Dolby® Digital or MPEG audio compression, work is progressing on supporting the new advanced audio and video codecs, such as HE-AAC, H.264 and VC-9.

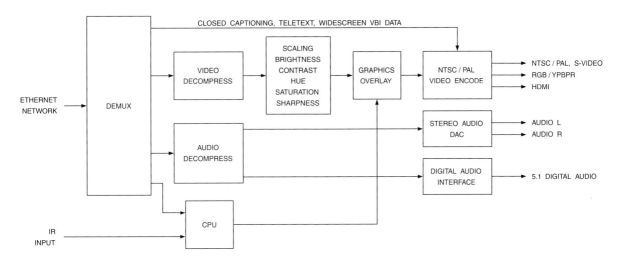

Figure 2.5. Simplified Block Diagram of a Digital Media Adapter.

Figure 2.6 is a simplified block diagram for a digital television set-top box, showing the common audio and video processing blocks. It is used to receive digital television broadcasts, from either terrestrial (over-the-air), cable, or satellite. A digital television may include this circuitry inside the television.

Many set-top boxes now include two tuners and digital video recorder (DVR) capability. This enables recording one program onto an internal HDD while watching another. Two tuners are also common in digital television receivers to support a picture-in-picture (PIP) feature.

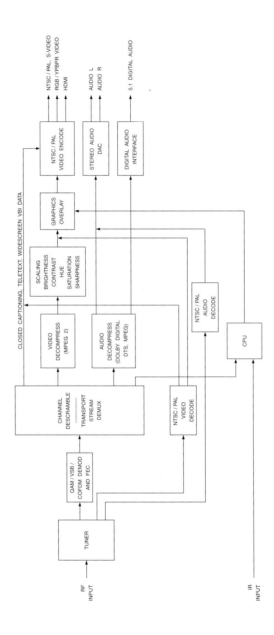

Figure 2.6. Simplified Block Diagram of a Digital Television Set-top Box.

Chapter 3

Color Spaces

A color space is a mathematical representation of a set of colors. The three most popular color models are RGB (used in computer graphics); YIQ, YUV, or YCbCr (used in video systems); and CMYK (used in color printing). However, none of these color spaces are directly related to the intuitive notions of hue, saturation, and brightness. This resulted in the temporary pursuit of other models, such as HSI and HSV, to simplify programming, processing, and end-user manipulation.

All of the color spaces can be derived from the RGB information supplied by devices such as cameras and scanners.

RGB Color Space

The red, green, and blue (RGB) color space is widely used throughout computer graphics. Red, green, and blue are three primary additive colors (individual components are added together to form a desired color) and are represented by a three-dimensional, Cartesian coordinate system (Figure 3.1). The indicated diagonal of the cube, with equal amounts of each primary component, represents various gray levels. Table 3.1 contains the RGB values for 100% amplitude, 100% saturated color bars, a common video test signal.

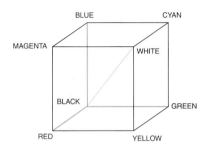

Figure 3.1. The RGB Color Cube.

	Nominal Range	White	Yellow	Cyan	Green	Magenta	Red	Blue	Black
R	0 to 255	255	255	0	0	255	255	0	0
G	0 to 255	255	255	255	255	0	0	0	0
B	0 to 255	255	0	255	0	255	0	255	0

Table 3.1. 100% RGB Color Bars.

The RGB color space is the most prevalent choice for computer graphics because color displays use red, green and blue to create the desired color. Therefore, the choice of the RGB color space simplifies the architecture and design of the system. Also, a system that is designed using the RGB color space can take advantage of a large number of existing software routines, since this color space has been around for a number of years.

However, RGB is not very efficient when dealing with "real-world" images. All three RGB components need to be of equal bandwidth to generate any color within the RGB color cube. The result of this is a frame buffer that has the same pixel depth and display resolution for each RGB component. Also, processing an image in the RGB color space is usually not the most efficient method. For example, to modify the intensity or color of a given pixel, the three RGB values must be read from the frame buffer, the intensity or color calculated, the desired modifications performed, and the new RGB values calculated and written back to the frame buffer. If the system had access to an image stored directly in the intensity and color format, some processing steps would be faster.

For these and other reasons, many video standards use luma and two color difference signals. The most common are the YUV, YIQ,

and YCbCr color spaces. Although all are related, there are some differences.

YUV Color Space

The YUV color space is used by the PAL (Phase Alternation Line), NTSC (National Television System Committee), and SECAM (Sequentiel Couleur Avec Mémoire or Sequential Color with Memory) composite color video standards. The black-and-white system used only luma (Y) information; color information (U and V) was added in such a way that a black-and-white receiver would still display a normal black-and-white picture. Color receivers decoded the additional color information to display a color picture.

The basic equations to convert between gamma-corrected RGB (notated as R′G′B′ and discussed later in this chapter) and YUV are:

$$Y = 0.299R' + 0.587G' + 0.114B'$$

$$U = -0.147R' - 0.289G' + 0.436B'$$
$$= 0.492\,(B' - Y)$$

$$V = 0.615R' - 0.515G' - 0.100B'$$
$$= 0.877\,(R' - Y)$$

R′ = Y + 1.140V

G′ = Y − 0.395U − 0.581V

B′ = Y + 2.032U

For digital R′G′B′ values with a range of 0–255, Y has a range of 0–255, U a range of 0 to ±112, and V a range of 0 to ±157. These equations are usually scaled to simplify the implementation in an actual NTSC or PAL digital encoder or decoder.

Note that for digital data, 8-bit YUV and R′G′B′ data should be saturated at the 0 and 255 levels to avoid underflow and overflow wrap-around problems.

If the full range of (B′ − Y) and (R′ − Y) had been used, the composite NTSC and PAL levels would have exceeded what the (then current) black-and-white television transmitters and receivers were capable of supporting. Experimentation determined that modulated subcarrier excursions of 20% of the luma (Y) signal excursion could be permitted above white and below black. The scaling factors were then selected so that the maximum level of 75% amplitude, 100% saturation yellow and cyan color bars would be at the white level (100 IRE).

YIQ Color Space

The YIQ color space, further discussed in Chapter 8, is derived from the YUV color space and is optionally used by the NTSC composite color video standard. (The "I" stands for "in-phase" and the "Q" for "quadrature," which is the modulation method used to transmit the color information.) The basic equations to convert between R′G′B′ and YIQ are:

Y = 0.299R′ + 0.587G′ + 0.114B′

$$
\begin{aligned}
I &= 0.596R′ − 0.275G′ − 0.321B′ \\
&= V\cos 33° − U\sin 33° \\
&= 0.736(R′ − Y) − 0.268(B′ − Y)
\end{aligned}
$$

$$
\begin{aligned}
Q &= 0.212R′ − 0.523G′ + 0.311B′ \\
&= V\sin 33° + U\cos 33° \\
&= 0.478(R′ − Y) + 0.413(B′ − Y)
\end{aligned}
$$

or, using matrix notation:

$$
\begin{bmatrix} I \\ Q \end{bmatrix} = \begin{bmatrix} 0 & 1 \\ 1 & 0 \end{bmatrix} \begin{bmatrix} \cos(33) & \sin(33) \\ -\sin(33) & \cos(33) \end{bmatrix} \begin{bmatrix} U \\ V \end{bmatrix}
$$

R′ = Y + 0.956I + 0.621Q

G′ = Y − 0.272I − 0.647Q

B′ = Y − 1.107I + 1.704Q

For digital R′G′B′ values with a range of 0–255, Y has a range of 0–255, I has a range of 0 to ±152, and Q has a range of 0 to ±134. I and Q are obtained by rotating the U and V axes 33°. These equations are usually scaled to simplify the implementation in an actual NTSC digital encoder or decoder.

Note that for digital data, 8-bit YIQ and R′G′B′ data should be saturated at the 0 and 255 levels to avoid underflow and overflow wrap-around problems.

YCbCr Color Space

The YCbCr color space was developed as part of ITU-R BT.601 during the development of a world-wide digital component video standard (discussed in Chapter 4). YCbCr is a scaled and offset version of the YUV color space. Y is defined to have a nominal 8-bit range of 16–235; Cb and Cr are defined to have a nominal range of 16–240. There are several YCbCr sampling formats, such as 4:4:4, 4:2:2, 4:1:1, and 4:2:0 that are also described.

RGB - YCbCr Equations: SDTV

The basic equations to convert between 8-bit digital R′G′B′ data with a 16–235 nominal range (Studio RGB) and YCbCr are:

$$Y_{601} = 0.299R' + 0.587G' + 0.114B'$$

$$Cb = -0.172R' - 0.339G' + 0.511B' + 128$$

$$Cr = 0.511R' - 0.428G' - 0.083B' + 128$$

$$R' = Y_{601} + 1.371(Cr - 128)$$

$$G' = Y_{601} - 0.698(Cr - 128) - 0.336(Cb - 128)$$

$$B' = Y_{601} + 1.732(Cb - 128)$$

When performing YCbCr to R′G′B′ conversion, the resulting R′G′B′ values have a nominal range of 16–235, with possible occasional excursions into the 0–15 and 236–255 values. This is due to Y and CbCr occasionally going outside the 16–235 and 16–240 ranges, respectively, due to video processing and noise. Note that 8-bit YCbCr and R′G′B′ data should be saturated at the 0 and 255 levels to avoid underflow and overflow wrap-around problems.

Table 3.2 lists the YCbCr values for 75% amplitude, 100% saturated color bars, a common video test signal.

Computer Systems Considerations

If the R′G′B′ data has a range of 0–255, as is commonly found in computer systems, the following equations may be more convenient to use:

$$Y_{601} = 0.257R' + 0.504G' + 0.098B' + 16$$

$$Cb = -0.148R' - 0.291G' + 0.439B' + 128$$

$$Cr = 0.439R' - 0.368G' - 0.071B' + 128$$

	Nominal Range	White	Yellow	Cyan	Green	Magenta	Red	Blue	Black
SDTV									
Y	16 to 235	180	162	131	112	84	65	35	16
Cb	16 to 240	128	44	156	72	184	100	212	128
Cr	16 to 240	128	142	44	58	198	212	114	128
HDTV									
Y	16 to 235	180	168	145	133	63	51	28	16
Cb	16 to 240	128	44	147	63	193	109	212	128
Cr	16 to 240	128	136	44	52	204	212	120	128

Table 3.2. 75% YCbCr Color Bars.

$R' = 1.164(Y_{601} - 16) + 1.596(Cr - 128)$

$G' = 1.164(Y_{601} - 16) - 0.813(Cr - 128) - 0.391(Cb - 128)$

$B' = 1.164(Y_{601} - 16) + 2.018(Cb - 128)$

Note that 8-bit YCbCr and R'G'B' data should be saturated at the 0 and 255 levels to avoid underflow and overflow wrap-around problems.

RGB - YCbCr Equations: HDTV

The basic equations to convert between 8-bit digital R'G'B' data with a 16–235 nominal range (Studio RGB) and YCbCr are:

$Y_{709} = 0.213R' + 0.715G' + 0.072B'$

$Cb = -0.117R' - 0.394G' + 0.511B' + 128$

$Cr = 0.511R' - 0.464G' - 0.047B' + 128$

$R' = Y_{709} + 1.540(Cr - 128)$

$G' = Y_{709} - 0.459(Cr - 128) - 0.183(Cb - 128)$

$B' = Y_{709} + 1.816(Cb - 128)$

When performing YCbCr to R'G'B' conversion, the resulting R'G'B' values have a nominal range of 16–235, with possible occasional excursions into the 0–15 and 236–255 values. This is due to Y and CbCr occasionally going outside the 16–235 and 16–240 ranges, respectively, due to video processing and noise. Note that 8-bit YCbCr and R'G'B' data should be saturated at the 0 and 255 levels to avoid underflow and overflow wrap-around problems.

Table 3.2 lists the YCbCr values for 75% amplitude, 100% saturated color bars, a common video test signal.

Computer Systems Considerations

If the R'G'B' data has a range of 0–255, as is commonly found in computer systems, the following equations may be more convenient to use:

$Y_{709} = 0.183R' + 0.614G' + 0.062B' + 16$

$Cb = -0.101R' - 0.338G' + 0.439B' + 128$

$Cr = 0.439R' - 0.399G' - 0.040B' + 128$

$R' = 1.164(Y_{709} - 16) + 1.793(Cr - 128)$

$G' = 1.164(Y_{709} - 16) - 0.534(Cr - 128) - 0.213(Cb - 128)$

$B' = 1.164(Y_{709} - 16) + 2.115(Cb - 128)$

Note that 8-bit YCbCr and R'G'B' data should be saturated at the 0 and 255 levels to avoid underflow and overflow wrap-around problems.

4:4:4 YCbCr Format

Figure 3.2 illustrates the positioning of YCbCr samples for the 4:4:4 format. Each sample has a Y, a Cb and a Cr value. Each sample is typically 8 bits (consumer applications) or 10 bits (pro-video applications) per component. Each sample therefore requires 24 bits (or 30 bits for pro-video applications).

4:2:2 YCbCr Format

Figure 3.3 illustrates the positioning of YCbCr samples for the 4:2:2 format. For every two horizontal Y samples, there is one Cb and Cr sample. Each sample is typically 8 bits (consumer applications) or 10 bits (pro-video applications) per component. Each sample therefore requires 16 bits (or 20 bits for pro-video applications), usually formatted as shown in Figure 3.4.

To display 4:2:2 YCbCr data, it is first converted to 4:4:4 YCbCr data, using interpolation to generate the missing Cb and Cr samples.

4:1:1 YCbCr Format

Figure 3.5 illustrates the positioning of YCbCr samples for the 4:1:1 format (also known as YUV12), used in some consumer video and DV video compression applications. For every four horizontal Y samples, there is one Cb and Cr value. Each component is typically 8 bits. Each sample therefore requires 12 bits, usually formatted as shown in Figure 3.6.

To display 4:1:1 YCbCr data, it is first converted to 4:4:4 YCbCr data, using interpolation to generate the missing Cb and Cr samples.

4:2:0 YCbCr Format

Rather than the horizontal-only 2:1 reduction of Cb and Cr used by 4:2:2, 4:2:0 YCbCr implements a 2:1 reduction of Cb and Cr in both the vertical and horizontal directions. It is commonly used for video compression.

As shown in Figures 3.7 through 3.11, there are several 4:2:0 sampling formats. Table 3.3 lists the YCbCr formats for various DV applications.

To display 4:2:0 YCbCr data, it is first converted to 4:4:4 YCbCr data, using interpolation to generate the new Cb and Cr samples. Note that some MPEG decoders do not properly convert the 4:2:0 YCbCr data to the 4:4:4 format, resulting in a "chroma bug."

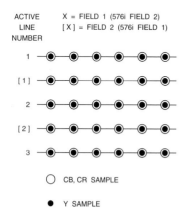

Figure 3.2. 4:4:4 Co-Sited Sampling. The sampling positions on the active scan lines of an interlaced picture.

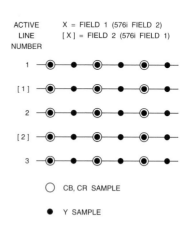

Figure 3.3. 4:2:2 Co-Sited Sampling. The sampling positions on the active scan lines of an interlaced picture.

SAMPLE 0	SAMPLE 1	SAMPLE 2	SAMPLE 3	SAMPLE 4	SAMPLE 5	
Y7 - 0	Y7 - 1	Y7 - 2	Y7 - 3	Y7 - 4	Y7 - 5	
Y6 - 0	Y6 - 1	Y6 - 2	Y6 - 3	Y6 - 4	Y6 - 5	
Y5 - 0	Y5 - 1	Y5 - 2	Y5 - 3	Y5 - 4	Y5 - 5	
Y4 - 0	Y4 - 1	Y4 - 2	Y4 - 3	Y4 - 4	Y4 - 5	
Y3 - 0	Y3 - 1	Y3 - 2	Y3 - 3	Y3 - 4	Y3 - 5	
Y2 - 0	Y2 - 1	Y2 - 2	Y2 - 3	Y2 - 4	Y2 - 5	
Y1 - 0	Y1 - 1	Y1 - 2	Y1 - 3	Y1 - 4	Y1 - 5	
Y0 - 0	Y0 - 1	Y0 - 2	Y0 - 3	Y0 - 4	Y0 - 5	16 BITS PER SAMPLE
CB7 - 0	CR7 - 0	CB7 - 2	CR7 - 2	CB7 - 4	CR7 - 4	
CB6 - 0	CR6 - 0	CB6 - 2	CR6 - 2	CB6 - 4	CR6 - 4	
CB5 - 0	CR5 - 0	CB5 - 2	CR5 - 2	CB5 - 4	CR5 - 4	
CB4 - 0	CR4 - 0	CB4 - 2	CR4 - 2	CB4 - 4	CR4 - 4	
CB3 - 0	CR3 - 0	CB3 - 2	CR3 - 2	CB3 - 4	CR3 - 4	
CB2 - 0	CR2 - 0	CB2 - 2	CR2 - 2	CB2 - 4	CR2 - 4	
CB1 - 0	CR1 - 0	CB1 - 2	CR1 - 2	CB1 - 4	CR1 - 4	
CB0 - 0	CR0 - 0	CB0 - 2	CR0 - 2	CB0 - 4	CR0 - 4	

- 0 = SAMPLE 0 DATA
- 1 = SAMPLE 1 DATA
- 2 = SAMPLE 2 DATA
- 3 = SAMPLE 3 DATA
- 4 = SAMPLE 4 DATA

Figure 3.4. 4:2:2 Frame Buffer Formatting.

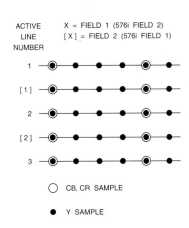

ACTIVE LINE NUMBER

X = FIELD 1 (576i FIELD 2)
[X] = FIELD 2 (576i FIELD 1)

○ CB, CR SAMPLE

● Y SAMPLE

Figure 3.5. 4:1:1 Co-Sited Sampling. The sampling positions on the active scan lines of an interlaced picture.

SAMPLE 0	SAMPLE 1	SAMPLE 2	SAMPLE 3	SAMPLE 4	SAMPLE 5	
Y7 - 0	Y7 - 1	Y7 - 2	Y7 - 3	Y7 - 4	Y7 - 5	
Y6 - 0	Y6 - 1	Y6 - 2	Y6 - 3	Y6 - 4	Y6 - 5	
Y5 - 0	Y5 - 1	Y5 - 2	Y5 - 3	Y5 - 4	Y5 - 5	
Y4 - 0	Y4 - 1	Y4 - 2	Y4 - 3	Y4 - 4	Y4 - 5	
Y3 - 0	Y3 - 1	Y3 - 2	Y3 - 3	Y3 - 4	Y3 - 5	
Y2 - 0	Y2 - 1	Y2 - 2	Y2 - 3	Y2 - 4	Y2 - 5	12 BITS PER SAMPLE
Y1 - 0	Y1 - 1	Y1 - 2	Y1 - 3	Y1 - 4	Y1 - 5	
Y0 - 0	Y0 - 1	Y0 - 2	Y0 - 3	Y0 - 4	Y0 - 5	
CB7 - 0	CB5 - 0	CB3 - 0	CB1 - 0	CB7 - 4	CB5 - 4	
CB6 - 0	CB4 - 0	CB2 - 0	CB0 - 0	CB6 - 4	CB4 - 4	
CR7 - 0	CR5 - 0	CR3 - 0	CR1 - 0	CR7 - 4	CR5 - 4	
CR6 - 0	CR4 - 0	CR2 - 0	CR0 - 0	CR6 - 4	CR4 - 4	

- 0 = SAMPLE 0 DATA
- 1 = SAMPLE 1 DATA
- 2 = SAMPLE 2 DATA
- 3 = SAMPLE 3 DATA
- 4 = SAMPLE 4 DATA

Figure 3.6. 4:1:1 Frame Buffer Formatting.

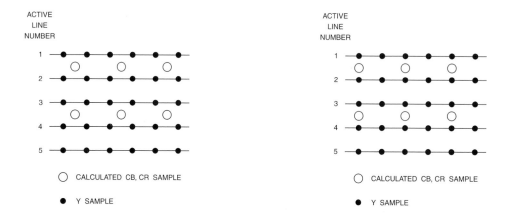

Figure 3.7. 4:2:0 Sampling for H.261, H.263, and MPEG-1. The sampling positions on the active scan lines of a progressive or noninterlaced picture.

Figure 3.8. 4:2:0 Sampling for MPEG-2, MPEG-4 Part 2 and H.264. The sampling positions on the active scan lines of a progressive or noninterlaced picture.

| YCbCr Format | 25 Mbps DV | | | | | 50 Mbps DV | | | 100 Mbps DV | | | | |
	480-Line DV	576-Line DV	480-Line DVCAM	576-Line DVCAM	D-7 (DVCPRO)	DVCPRO 50	Digital Betacam	D-9 (Digital S)	DVCPRO HD	D-9 HD	MPEG-1	MPEG-2, -4 Part 2, H.264	H.261, H.263
4:4:4 Co-Sited												×	
4:2:2 Co-Sited						×	×	×	×	×		×	
4:1:1 Co-Sited	×		×		×								
4:2:0											×	×	×
4:2:0 Co-Sited		×		×									

Table 3.3. YCbCr Formats for Various DV Applications.

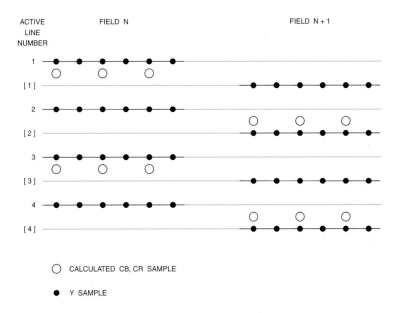

Figure 3.9. 4:2:0 Sampling for MPEG-2, MPEG-4 Part 2 and H.264. The sampling positions on the active scan lines of an interlaced picture (top_field_first = 1).

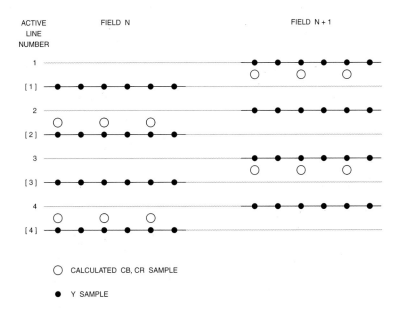

Figure 3.10. 4:2:0 Sampling for MPEG-2, MPEG-4 Part 2 and H.264. The sampling positions on the active scan lines of an interlaced picture (top_field_first = 0).

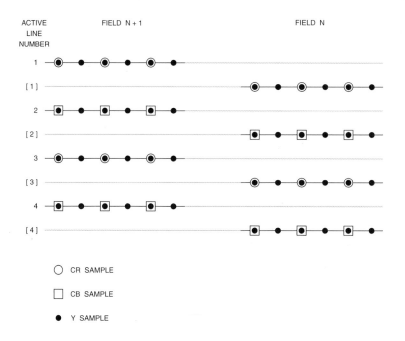

Figure 3.11. 4:2:0 Co-Sited Sampling for 576i DV and DVCAM. The sampling positions on the active scan lines of an interlaced picture.

PhotoYCC Color Space

PhotoYCC (a trademark of Eastman Kodak Company) was developed to encode Photo CD image data. The goal was to develop a display-device-independent color space. For maximum video display efficiency, the color space is based upon ITU-R BT.601 and BT.709.

The encoding process (RGB to PhotoYCC) assumes CIE Standard Illuminant D_{65} and that the spectral sensitivities of the image capture system are proportional to the color-matching functions of the BT.709 reference primaries. The RGB values, unlike those for a computer graphics system, may be negative. PhotoYCC includes colors outside the BT.709 color gamut; these are encoded using negative values.

RGB to PhotoYCC

Linear RGB data (normalized to have values of 0 to 1) is nonlinearly transformed to PhotoYCC as follows:

for R, G, B ≥ 0.018

$$R' = 1.099\, R^{0.45} - 0.099$$

$$G' = 1.099\, G^{0.45} - 0.099$$

$$B' = 1.099\, B^{0.45} - 0.099$$

for $-0.018 < R, G, B < 0.018$

$$R' = 4.5\, R$$

$$G' = 4.5\, G$$

$$B' = 4.5\, B$$

for R, G, B ≤ –0.018

$$R' = -1.099 |R|^{0.45} - 0.099$$

$$G' = -1.099 |G|^{0.45} - 0.099$$

$$B' = -1.099 |B|^{0.45} - 0.099$$

From R′G′B′ with a 0–255 range, a luma and two chrominance signals (C1 and C2) are generated:

$$Y = 0.213R' + 0.419G' + 0.081B'$$

$$C1 = -0.131R' - 0.256G' + 0.387B' + 156$$

$$C2 = 0.373R' - 0.312G' - 0.061B' + 137$$

As an example, a 20% gray value (R, G, and B = 0.2) would be recorded on the PhotoCD disc using the following values:

$$Y = 79$$

$$C1 = 156$$

$$C2 = 137$$

PhotoYCC to RGB

Since PhotoYCC attempts to preserve the dynamic range of film, decoding PhotoYCC images requires the selection of a color space and range appropriate for the output device. Thus, the decoding equations are not always the exact inverse of the encoding equations. The following equations are suitable for generating RGB values for driving a CRT display, and assume a unity relationship between the luma in the encoded image and the displayed image.

$$R' = 0.981Y + 1.315(C2 - 137)$$

$$G' = 0.981Y - 0.311(C1 - 156) - 0.669(C2 - 137)$$

$$B' = 0.981Y + 1.601(C1 - 156)$$

The R′G′B′ values should be saturated to a range of 0 to 255. The equations above assume the display uses phosphor chromaticities that are the same as the BT.709 reference primaries, and that the video signal luma (V) and the display luminance (L) have the relationship:

for V ≥ 0.0812

$$L = ((V + 0.099) / 1.099)^{1/0.45}$$

for V < 0.0812

$$L = V / 4.5$$

HSI, HLS and HSV Color Spaces

The HSI (hue, saturation, intensity) and HSV (hue, saturation, value) color spaces were developed to be more "intuitive" in manipulating color and were designed to approximate the way humans perceive and interpret color. They were developed when colors had to be specified manually, and are rarely used now that users can select colors visually or specify Pantone colors. These color spaces are discussed for "historic" interest. HLS (hue, lightness, saturation) is similar to HSI; the term lightness is used rather than intensity.

The difference between HSI and HSV is the computation of the brightness component (I or V), which determines the distribution and dynamic range of both the brightness (I or V) and saturation (S). The HSI color space is best for traditional image processing functions such as convolution, equalization, histograms, and so on, which operate by manipulation of the brightness values since I is equally dependent on R, G, and B. The HSV color space is preferred for manipulation of hue and saturation (to shift colors or adjust the amount of color) since it yields a greater dynamic range of saturation.

Figure 3.12 illustrates the single hexcone HSV color model. The top of the hexcone corresponds to V = 1, or the maximum intensity colors. The point at the base of the hexcone is black and here V = 0. Complementary colors are 180° opposite one another as measured by H, the angle around the vertical axis (V), with red at 0°. The value of S is a ratio, ranging from 0 on the center line vertical axis (V) to 1 on the sides of the hexcone. Any value of S between 0 and 1 may be associated with the point V = 0. The point S = 0, V = 1 is white. Intermediate values of V for S = 0 are the grays. Note that when S = 0, the value of H is irrelevant. From an artist's viewpoint, any color with V = 1, S = 1 is a pure pigment (whose color is defined by H). Adding white corresponds to decreasing S (without changing V); adding black corresponds to decreasing V (without changing S). Tones are created by decreasing both S and V. Table 3.4 lists the 75% amplitude, 100% saturated HSV color bars.

Figure 3.13 illustrates the double hexcone HSI color model. The top of the hexcone corresponds to I = 1, or white. The point at the base of the hexcone is black and here I = 0. Complementary colors are 180° opposite one another as measured by H, the angle around the vertical axis (I), with red at 0° (for consistency with the HSV model, we have changed from the Tektronix convention of blue at 0°). The value of S ranges from 0 on the vertical axis (I) to 1 on the surfaces of the hexcone. The grays all have S = 0, but maximum saturation of hues is at S = 1, I = 0.5. Table 3.5 lists the 75% amplitude, 100% saturated HSI color bars.

Chromaticity Diagram

The color gamut perceived by a person with normal vision (the 1931 CIE Standard Observer) is shown in Figure 3.14. The diagram and underlying mathematics were updated in 1960 and 1976; however, the NTSC television system is based on the 1931 specifications.

Color perception was measured by viewing combinations of the three standard CIE (International Commission on Illumination or Commission Internationale de l'Eclairage) primary colors: red with a 700-nm wavelength, green at 546.1 nm, and blue at 435.8 nm. These primary colors, and the other spectrally pure colors resulting from mixing of the primary colors, are located along the curved outer boundary line (called the *spectrum locus*), shown in Figure 3.14.

The ends of the spectrum locus (at red and blue) are connected by a straight line that represents the purples, which are combinations of red and blue. The area within this closed boundary contains all the colors that can be generated by mixing light of different colors. The closer a color is to the boundary, the more saturated it is. Colors within the boundary are perceived as becoming more pastel as the center of the diagram (white) is approached. Each point on the diagram, representing a unique color, may be identified by its x and y coordinates.

In the CIE system, the intensities of red, green, and blue are transformed into what are called the *tristimulus values*, which are represented by the capital letters X, Y, and Z. These values represent the relative quantities of the primary colors.

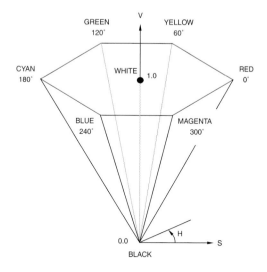

Figure 3.12. Single Hexcone HSV Color Model.

	Nominal Range	White	Yellow	Cyan	Green	Magenta	Red	Blue	Black
H	0° to 360°	–	60°	180°	120°	300°	0°	240°	–
S	0 to 1	0	1	1	1	1	1	1	0
V	0 to 1	0.75	0.75	0.75	0.75	0.75	0.75	0.75	0

Table 3.4. 75% HSV Color Bars.

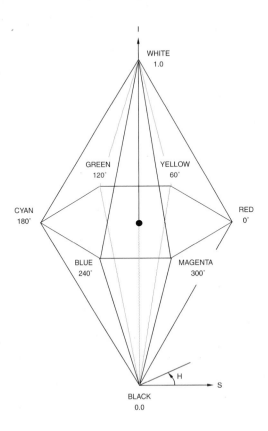

Figure 3.13. Double Hexcone HSI Color Model. For consistency with the HSV model, we have changed from the Tektronix convention of blue at 0° and depict the model as a double hexcone rather than as a double cone.

	Nominal Range	White	Yellow	Cyan	Green	Magenta	Red	Blue	Black
H	0° to 360°	–	60°	180°	120°	300°	0°	240°	–
S	0 to 1	0	1	1	1	1	1	1	0
I	0 to 1	0.75	0.375	0.375	0.375	0.375	0.375	0.375	0

Table 3.5. 75% HSI Color Bars. For consistency with the HSV model, we have changed from the Tektronix convention of blue at 0°.

The coordinate axes of Figure 3.14 are derived from the tristimulus values:

$$x = X/(X + Y + Z)$$
$$= red/(red + green + blue)$$
$$y = Y/(X + Y + Z)$$
$$= green/(red + green + blue)$$
$$z = Z/(X + Y + Z)$$
$$= blue/(red + green + blue)$$

The coordinates x, y, and z are called *chromaticity coordinates*, and they always add up to 1. As a result, z can always be expressed in terms of x and y, which means that only x and y are required to specify any color, and the diagram can be two-dimensional.

Typically, a source or display specifies three (x, y) coordinates to define the three primary colors it uses. The triangle formed by the three (x, y) coordinates encloses the gamut of colors that the source or display can reproduce. This is shown in Figure 3.15, which compares the color gamuts of NTSC, PAL, and typical inks and dyes. Note that no set of three colors can generate all possible colors, which is why television pictures are never completely accurate.

In addition, a source or display usually specifies the (x, y) coordinate of the white color used, since pure white is not usually captured or reproduced. White is defined as the color captured or produced when all three primary signals are equal, and it has a subtle shade of color to it. Note that luminance, or brightness information, is not included in the standard CIE 1931 chromaticity diagram, but is an axis that is orthogonal to the (x, y) plane. The lighter a color is, the more restricted the chromaticity range is.

The chromaticities and reference white (CIE illuminate C) for the 1953 NTSC standard are:

R:	$x_r = 0.67$	$y_r = 0.33$
G:	$x_g = 0.21$	$y_g = 0.71$
B:	$x_b = 0.14$	$y_b = 0.08$
white:	$x_w = 0.3101$	$y_w = 0.3162$

Modern NTSC, 480i and 480p video systems use a different set of RGB phosphors, resulting in slightly different chromaticities of the RGB primaries and reference white (CIE illuminate D_{65}):

R:	$x_r = 0.630$	$y_r = 0.340$
G:	$x_g = 0.310$	$y_g = 0.595$
B:	$x_b = 0.155$	$y_b = 0.070$
white:	$x_w = 0.3127$	$y_w = 0.3290$

The chromaticities and reference white (CIE illuminate D_{65}) for PAL, SECAM, 576i and 576p video systems are:

R:	$x_r = 0.64$	$y_r = 0.33$
G:	$x_g = 0.29$	$y_g = 0.60$
B:	$x_b = 0.15$	$y_b = 0.06$
white:	$x_w = 0.3127$	$y_w = 0.3290$

The chromaticities and reference white (CIE illuminate D_{65}) for HDTV are based on BT.709:

R:	$x_r = 0.64$	$y_r = 0.33$
G:	$x_g = 0.30$	$y_g = 0.60$
B:	$x_b = 0.15$	$y_b = 0.06$
white:	$x_w = 0.3127$	$y_w = 0.3290$

Since different chromaticity and reference white values are used for various video standards, minor color errors may occur when the source and display values do not match; for example, displaying a 480i or 480p program on a HDTV, or displaying a HDTV program on a NTSC television. These minor color errors can easily be corrected at the display by using a 3 × 3 matrix multiplier, as discussed in Chapter 7.

Non-RGB Color Space Considerations

When processing information in a non-RGB color space (such as YIQ, YUV, or YCbCr), care must be taken that combinations of values are not created that result in the generation of invalid RGB colors. The term invalid refers to RGB components outside the normalized RGB limits of (1, 1, 1).

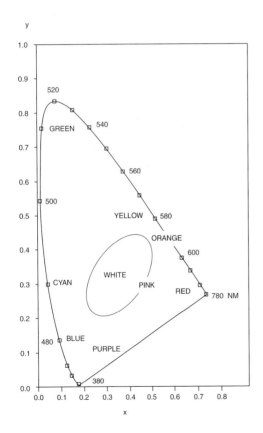

Figure 3.14. CIE 1931 Chromaticity Diagram Showing Various Color Regions.

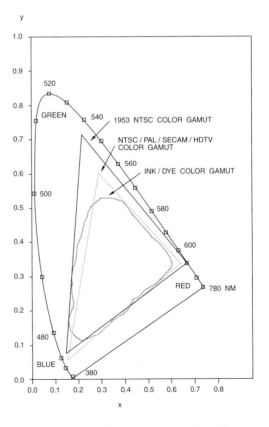

Figure 3.15. CIE 1931 Chromaticity Diagram Showing Various Color Gamuts.

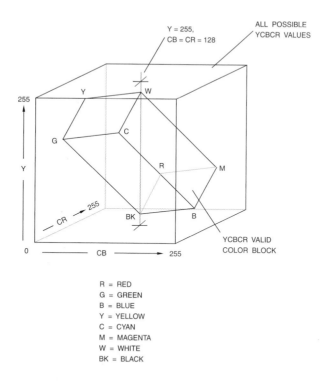

R = RED
G = GREEN
B = BLUE
Y = YELLOW
C = CYAN
M = MAGENTA
W = WHITE
BK = BLACK

Figure 3.16. RGB Limits Transformed into 3-D YCbCr Space.

For example, given that RGB has a normalized value of (1, 1, 1), the resulting YCbCr value is (235, 128, 128). If Cb and Cr are manipulated to generate a YCbCr value of (235, 64, 73), the corresponding RGB normalized value becomes (0.6, 1.29, 0.56)—note that the green value exceeds the normalized value of 1.

From this illustration it is obvious that there are many combinations of Y, Cb, and Cr that result in invalid RGB values; these YCbCr values must be processed so as to generate valid RGB values. Figure 3.16 shows the RGB normalized limits transformed into the YCbCr color space.

Best results are obtained using a constant luma and constant hue approach—Y is not altered while Cb and Cr are limited to the maximum valid values having the same hue as the invalid color prior to limiting. The constant hue principle corresponds to moving invalid CbCr combinations directly towards the CbCr origin (128, 128), until they lie on the surface of the valid YCbCr color block.

When converting to the RGB color space from a non-RGB color space, care must be taken to include saturation logic to ensure overflow and underflow wrap-around conditions do not occur due to the finite precision of digital circuitry. 8-bit RGB values less than 0 must be set to 0, and values greater than 255 must be set to 255.

Gamma Correction

The transfer function of most CRT displays produces an intensity that is proportional to some power (referred to as gamma) of the signal amplitude. As a result, high-intensity ranges are expanded and low-intensity ranges are compressed (see Figure 3.17). This is an advantage in combatting noise, as the eye is approximately equally sensitive to equally relative intensity changes. By "gamma correcting" the video signals before transmission, the intensity output of the display is roughly linear (the gray line in Figure 3.17), and transmission-induced noise is reduced.

To minimize noise in the darker areas of the image, modern video systems limit the gain of the curve in the black region. This technique limits the gain close to black and stretches the remainder of the curve to maintain function and tangent continuity.

Although video standards assume a display gamma of about 2.2, a gamma of about 2.5 is more realistic for CRT displays. However, this difference improves the viewing in a dimly lit environment. More accurate viewing in a brightly lit environment may be accomplished by applying another gamma factor of about 1.14 (2.5/2.2). It is also common to tweak the gamma curve in the display to get closer to the "film look."

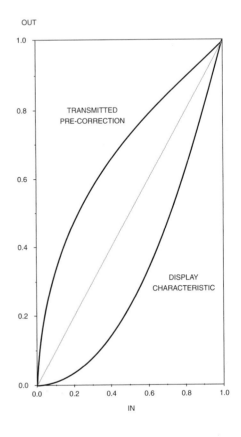

Figure 3.17. Effect of Gamma.

Early NTSC Systems

Early NTSC systems assumed a simple transform at the display, with a gamma of 2.2. RGB values are normalized to have a range of 0 to 1:

$$R = R'^{2.2}$$

$$G = G'^{2.2}$$

$$B = B'^{2.2}$$

To compensate for the nonlinear display, linear RGB data was "gamma-corrected" prior to transmission by the inverse transform. RGB values are normalized to have a range of 0 to 1:

$$R' = R^{1/2.2}$$

$$G' = G^{1/2.2}$$

$$B' = B^{1/2.2}$$

Early PAL and SECAM Systems

Most early PAL and SECAM systems assumed a simple transform at the display, with a gamma of 2.8. RGB values are normalized to have a range of 0 to 1:

$$R = R'^{2.8}$$

$$G = G'^{2.8}$$

$$B = B'^{2.8}$$

To compensate for the nonlinear display, linear RGB data was "gamma-corrected" prior to transmission by the inverse transform. RGB values are normalized to have a range of 0 to 1:

$$R' = R^{1/2.8}$$

$$G' = G^{1/2.8}$$

$$B' = B^{1/2.8}$$

Current Systems

Current NTSC, 480i, 480p and HDTV video systems assume the following transform at the display, with a gamma of [1/0.45]. RGB values are normalized to have a range of 0 to 1:

for $(R', G', B') < 0.0812$

$$R = R' / 4.5$$

$$G = G' / 4.5$$

$$B = B' / 4.5$$

for $(R', G', B') \geq 0.0812$

$$R = ((R' + 0.099) / 1.099)^{1/0.45}$$

$$G = ((G' + 0.099) / 1.099)^{1/0.45}$$

$$B = ((B' + 0.099) / 1.099)^{1/0.45}$$

To compensate for the nonlinear display, linear RGB data is "gamma-corrected" prior to transmission by the inverse transform. RGB values are normalized to have a range of 0 to 1:

for $R, G, B < 0.018$

$$R' = 4.5 R$$

$$G' = 4.5 G$$

$$B' = 4.5 B$$

for $R, G, B \geq 0.018$

$$R' = 1.099 R^{0.45} - 0.099$$

$$G' = 1.099 G^{0.45} - 0.099$$

$$B' = 1.099 B^{0.45} - 0.099$$

Although most PAL and SECAM standards specify a gamma of 2.8, a value of [1/0.45] is now commonly used. Thus, these equations are also now used for PAL, SECAM, 576i and 576p video systems.

Non-CRT Displays

Since they are not based on CRTs, the LCD, LCOS, DLP and plasma displays have different display transforms. To simplify interfacing to these displays, their electronics are designed to accept standard gamma-corrected video and then compensate for the actual transform of the display panel.

Constant Luminance Problem

Due to the wrong order of the gamma and matrix operations, the U and V (or Cb and Cr) signals also contribute to the luminance (Y) signal. This causes an error in the perceived luminance when the amplitude of U and V is not correct. This may be due to bandwidth-limiting U and V or a non-nominal setting of the U and V gain (color saturation).

For low color frequencies, there is no problem. For high color frequencies, U and V disappear and consequently R′, G′, and B′ degrade to be equal to (only) Y.

References

1. Benson, K. Blair, *Television Engineering Handbook*. McGraw-Hill, Inc., 1986.
2. Clarke, C.K.P., 1986, *Colour Encoding and Decoding Techniques for Line-Locked Sampled PAL and NTSC Television Signals*, BBC Research Department Report BBC RD1986/2.
3. Devereux, V. G., 1987, *Limiting of YUV digital video signals*, BBC Research Department Report BBC RD1987 22.
4. EIA Standard EIA–189–A, July 1976, *Encoded Color Bar Signal*.
5. Faroudja, Yves Charles, *NTSC and Beyond. IEEE Transactions on Consumer Electronics*, Vol. 34, No. 1, February 1988.
6. ITU-R BT.470–6, 1998, *Conventional Television Systems*.
7. ITU-R BT.601–5, 1995, *Studio Encoding Parameters of Digital Television for Standard 4:3 and Widescreen 16:9 Aspect Ratios*.
8. ITU-R BT.709–5, 2002, *Parameter Values for the HDTV Standards for Production and International Programme Exchange*.
9. Photo CD Information Bulletin, *Fully Utilizing Photo CD Images–PhotoYCC Color Encoding and Compression Schemes*, May 1994, Eastman Kodak Company.

Chapter 4

Video Signals Overview

Video signals come in a wide variety of options—number of scan lines, interlaced vs. progressive, analog vs. digital, and so on. This chapter provides an overview of the common video signal formats and their timing.

Digital Component Video Background

In digital component video, the video signals are in digital form (YCbCr or R′G′B′), being encoded to composite NTSC, PAL or SECAM only when it is necessary for broadcasting or recording purposes.

The European Broadcasting Union (EBU) became interested in a standard for digital component video due to the difficulties of exchanging video material between the 576i PAL and SECAM systems. The format held the promise that the digital video signals would be identical whether sourced in a PAL or SECAM country, allowing subsequent encoding to the appropriate composite form for broadcasting. Consultations with the Society of Motion Pic-ture and Television Engineers (SMPTE) resulted in the development of an approach to support international program exchange, including 480i systems.

A series of demonstrations was carried out to determine the quality and suitability for signal processing of various methods. From these investigations, the main parameters of the digital component coding, filtering, and timing were chosen and incorporated into ITU-R BT.601. BT.601 has since served as the starting point for other digital component video standards.

Coding Ranges

The selection of the coding ranges balanced the requirements of adequate capacity for signals beyond the normal range and minimizing quantizing distortion. Although the black level of a video signal is reasonably well defined, the white level can be subject to variations due to video signal and equipment tolerances. Noise, gain variations, and transients produced by filtering can produce signal levels outside the nominal ranges.

8 or 10 bits per sample are used for each of the YCbCr or R′G′B′ components. Although 8-bit coding introduces some quantizing distortion, it was originally felt that most video sources contained sufficient noise to mask most of the quantizing distortion. However, if the video source is virtually noise-free, the quantizing distortion is noticeable as contouring in areas where the signal brightness gradually changes. In addition, at least two additional bits of fractional YCbCr or R′G′B′ data were desirable to reduce rounding effects when transmitting between equipment in the studio editing environment. For these reasons, most pro-video equipment uses 10-bit YCbCr or R′G′B′, allowing 2 bits of fractional YCbCr or R′G′B′ data to be maintained.

Initial proposals had equal coding ranges for all three YCbCr components. However, this was changed so that Y had a greater margin for overloads at the white levels, as white level limiting is more visible than black. Thus, the nominal 8-bit Y levels are 16–235, while the nominal 8-bit CbCr levels are 16–240 (with 128 corresponding to no color). Occasional excursions into the other levels are permissible, but never at the 0 and 255 levels.

For 8-bit systems, the values of 00_H and FF_H are reserved for timing information. For 10-bit systems, the values of 000_H–003_H and $3FC_H$–$3FF_H$ are reserved for timing information, to maintain compatibility with 8-bit systems.

The YCbCr or R′G′B′ levels to generate 75% color bars are discussed in Chapter 3. Digital R′G′B′ signals are defined to have the same nominal levels as Y to provide processing margin and simplify the digital matrix conversions between R′G′B′ and YCbCr.

SDTV Sample Rate Selection

Line-locked sampling of analog R′G′B′ or YUV video signals is done. This technique produces a static orthogonal sampling grid in which samples on the current scan line fall directly beneath those on previous scan lines and fields, as shown Figures 3.2 through 3.11.

Another important feature is that the sampling is locked in phase so that one sample is coincident with the 50% amplitude point of the falling edge of analog horizontal sync (0_H). This ensures that different sources produce samples at nominally the same positions in the picture. Making this feature common simplifies conversion from one standard to another.

For 480i and 576i video systems, several Y sampling frequencies were initially examined, including four times F_{sc}. However, the four-times F_{sc} sampling rates did not support the requirement of simplifying international exchange of programs, so they were dropped in favor of a single common sampling rate. Because the lowest sample rate possible (while still supporting quality video) was a goal, a 12-MHz sample rate was preferred for a long time, but eventually was considered to be too close to the Nyquist limit, complicating the filtering requirements. When the frequencies between 12 MHz and 14.3 MHz were examined, it became evident that a 13.5-MHz sample rate for Y provided some commonality between 480i and 576i systems. Cb and Cr, being color difference signals, do not require the same bandwidth as the Y, so may be sampled at one-half the Y sample rate, or 6.75 MHz.

The "4:2:2" notation now commonly used originally applied to NTSC and PAL video, implying that Y, U and V were sampled at 4×, 2× and 2× the color subcarrier frequency, respectively. The "4:2:2" notation was then adapted to BT.601 digital component video, implying that the sampling frequencies of Y, Cb and Cr were 4×, 2× and 2× 3.375 MHz, respectively. "4:2:2" now commonly means that the sample rate of Cb and Cr is one-half that of Y, regardless of the actual sample rates used.

With 13.5-MHz sampling, each scan line contains 858 samples (480i systems) or 864 samples (576i systems) and consists of a digital blanking interval followed by an active line period. Both the 480i and 576i systems use 720 samples during the active line period. Having a common number of samples for the active line period simplifies the design of multistandard equipment and standards conversion. With a sample rate of 6.75 MHz for Cb and Cr (4:2:2 sampling), each active line period contains 360 Cr samples and 360 Cb samples.

With analog systems, problems may arise with repeated processing, causing an extension of the blanking intervals and softening of the blanking edges. Using 720 digital samples for the active line period accommodates the range of analog blanking tolerances of both the 480i and 576i systems. Therefore, repeated processing may be done without affecting the digital blanking interval. Blanking to define the analog picture width need only be done once, preferably at the display or upon conversion to analog video.

Initially, BT.601 supported only 480i and 576i systems with a 4:3 aspect ratio (720 × 480i and 720 × 576i active resolutions). Support for a 16:9 aspect ratio was then added (960 × 480i and 960 × 576i active resolutions) using an 18 MHz sample rate.

EDTV Sample Rate Selection

ITU BT.1358 defines the progressive SDTV video signals, also known as 480p or 576p, or Enhanced Digital Television (EDTV). The sample rate is doubled to 27 MHz (4:3 aspect ratio) or 36 MHz (16:9 aspect ratio) in order to keep the same static orthogonal sampling grid as that used by BT.601.

HDTV Sample Rate Selection

ITU BT.709 defines the 720p, 1080i and 1080p video signals, respectively. With HDTV, a different technique was used—the number of active samples per line and the number of active lines per frame is constant regardless of the frame rate. Thus, in order to keep a static orthogonal sampling grid, each frame rate uses a different sample clock rate.

480i and 480p Systems

Interlaced Analog Composite Video

(M) NTSC and (M) PAL are analog composite video signals that carry all timing and color information within a single signal. These analog interfaces use 525 lines per frame and are discussed in detail in Chapter 8.

Interlaced Analog Component Video

Analog component signals are comprised of three signals, analog R′G′B′ or YPbPr. Referred to as 480i (since there are typically 480 active scan lines per frame and they are interlaced), the frame rate is usually 29.97 Hz (30/1.001) for compatibility with (M) NTSC timing. The analog interface uses 525 lines per frame, with active video present on lines 23–262 and 286–525, as shown in Figure 4.1.

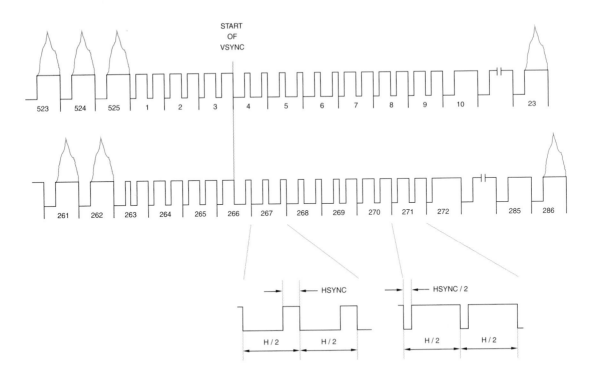

Figure 4.1. 480i Vertical Interval Timing.

Figure 4.2. 480p Vertical Interval Timing.

For the 29.97 Hz frame rate, each scan line time (H) is about 63.556 μs. Detailed horizontal timing is dependent on the specific video interface used, as discussed in Chapter 5.

Progressive Analog Component Video

Analog component signals are comprised of three signals, analog R′G′B′ or YPbPr. Referred to as 480p (since there are typically 480 active scan lines per frame and they are progressive), the frame rate is usually 59.94 Hz (60/1.001) for easier compatibility with (M) NTSC timing. The analog interface uses 525 lines per frame, with active video present on lines 45–524, as shown in Figure 4.2.

For the 59.94 Hz frame rate, each scan line time (H) is about 31.776 μs. Detailed horizontal timing is dependent on the specific video interface used, as discussed in Chapter 5.

Interlaced Digital Component Video

BT.601 and SMPTE 267M specify the representation for 480i digital R′G′B′ or YCbCr video signals. Active resolutions defined within BT.601 and SMPTE 267M, their 1× Y and R′G′B′ sample rates (F_s), and frame rates, are:

960 × 480i	18.0 MHz	29.97 Hz
720 × 480i	13.5 MHz	29.97 Hz

Other common active resolutions, their 1× sample rates (F_s), and frame rates, are:

864 × 480i	16.38 MHz	29.97 Hz
704 × 480i	13.50 MHz	29.97 Hz
640 × 480i	12.27 MHz	29.97 Hz
544 × 480i	10.12 MHz	29.97 Hz
528 × 480i	9.900 MHz	29.97 Hz
480 × 480i	9.000 MHz	29.97 Hz
352 × 480i	6.750 MHz	29.97 Hz

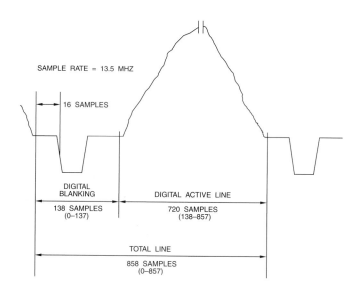

**Figure 4.3. 480i Analog - Digital Relationship
(4:3 Aspect Ratio, 29.97 Hz Refresh, 13.5 MHz Sample Clock).**

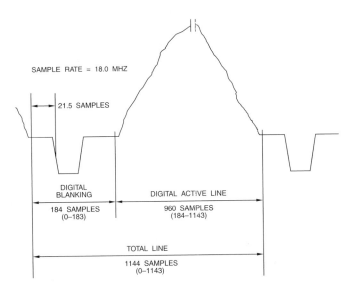

**Figure 4.4. 480i Analog - Digital Relationship
(16:9 Aspect Ratio, 29.97 Hz Refresh, 18 MHz Sample Clock).**

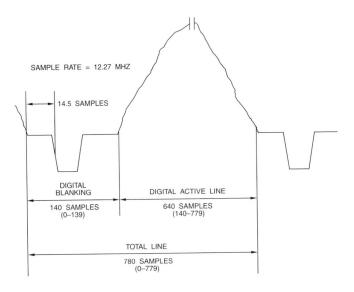

**Figure 4.5. 480i Analog - Digital Relationship
(4:3 Aspect Ratio, 29.97 Hz Refresh, 12.27 MHz Sample Clock).**

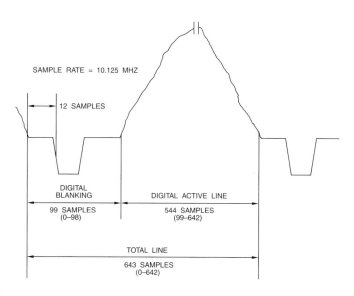

**Figure 4.6. 480i Analog - Digital Relationship
(4:3 Aspect Ratio, 29.97 Hz Refresh, 10.125 MHz Sample Clock).**

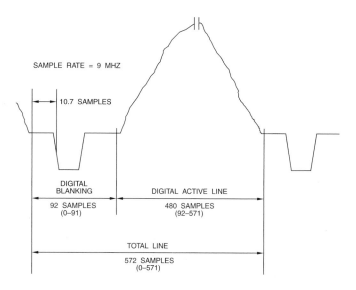

**Figure 4.7. 480i Analog - Digital Relationship
(4:3 Aspect Ratio, 29.97 Hz Refresh, 9 MHz Sample Clock).**

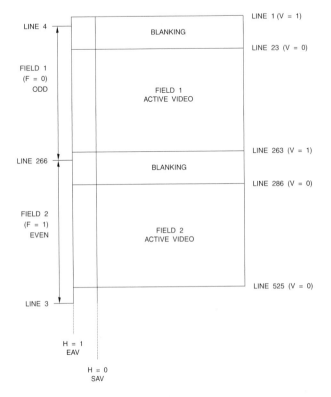

LINE NUMBER	F	V
1–3	1	1
4–22	0	1
23–262	0	0
263–265	0	1
266–285	1	1
286–525	1	0

Figure 4.8. 480i Digital Vertical Timing (480 Active Lines). F and V change state at the EAV sequence at the beginning of the digital line. Note that the digital line number changes state prior to start of horizontal sync, as shown in Figures 4.3 through 4.7.

These active lines are used by the SMPTE RP-202, ATSC A/54a and ARIB STD-B32 standards. EIA–861B (DVI and HDMI timing) specifies lines 22–261 and 285–524 for active video. IEC 61834-2, ITU-R BT.1618 and SMPTE 314M (DV formats) specify lines 23–262 and 285–524 for active video.

ITU-R BT.656 specifies lines 20–263 and 283–525 for active video, resulting in 487 total active lines per frame.

864 × 480i is a 16:9 square pixel format, while 640 × 480i is a 4:3 square pixel format. Although the ideal 16:9 resolution is 854 × 480i, 864 × 480i supports the MPEG 16 × 16 block structure. The 704 × 480i format is done by using the 720 × 480i format, and blanking the first eight and last eight samples each active scan line. Example relationships between the analog and digital signals are shown in Figures 4.3 through 4.7.

The H (horizontal blanking), V (vertical blanking), and F (field) signals are defined in Figure 4.8. The H, V and F timing indicated is compatible with video compression standards rather than BT.656 discussed in Chapter 6.

Progressive Digital Component Video

BT.1358 and SMPTE 293M specify the representation for 480p digital R′G′B′ or YCbCr video signals. Active resolutions defined within BT.1358 and SMPTE 293M, their 1× sample rates (F_s), and frame rates, are:

960 × 480p	36.0 MHz	59.94 Hz
720 × 480p	27.0 MHz	59.94 Hz

Other common active resolutions, their 1× Y and R′G′B′ sample rates (F_s), and frame rates, are:

864 × 480p	32.75 MHz	59.94 Hz
704 × 480p	27.00 MHz	59.94 Hz
640 × 480p	24.54 MHz	59.94 Hz
544 × 480p	20.25 MHz	59.94 Hz
528 × 480p	19.80 MHz	59.94 Hz
480 × 480p	18.00 MHz	59.94 Hz
352 × 480p	13.50 MHz	59.94 Hz

864 × 480p is a 16:9 square pixel format, while 640 × 480p is a 4:3 square pixel format. Although the ideal 16:9 resolution is 854 × 480p, 864 × 480p supports the MPEG 16 × 16 block structure. The 704 × 480p format is done

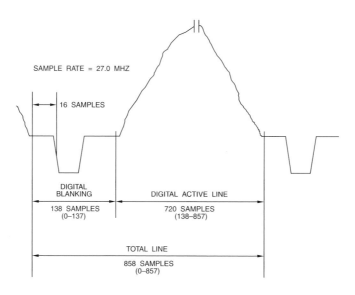

Figure 4.9. 480p Analog - Digital Relationship (4:3 Aspect Ratio, 59.94 Hz Refresh, 27 MHz Sample Clock).

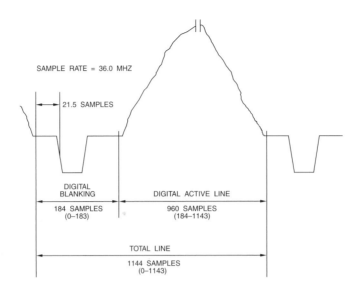

SAMPLE RATE = 36.0 MHZ

21.5 SAMPLES

DIGITAL
BLANKING

DIGITAL ACTIVE LINE

184 SAMPLES
(0–183)

960 SAMPLES
(184–1143)

TOTAL LINE

1144 SAMPLES
(0–1143)

**Figure 4.10. 480p Analog - Digital Relationship
(16:9 Aspect Ratio, 59.94 Hz Refresh, 36 MHz Sample Clock).**

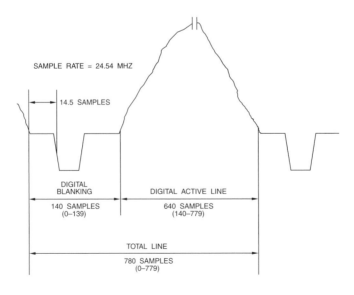

SAMPLE RATE = 24.54 MHZ

14.5 SAMPLES

DIGITAL
BLANKING

DIGITAL ACTIVE LINE

140 SAMPLES
(0–139)

640 SAMPLES
(140–779)

TOTAL LINE

780 SAMPLES
(0–779)

**Figure 4.11. 480p Analog - Digital Relationship
(4:3 Aspect Ratio, 59.94 Hz Refresh, 24.54 MHz Sample Clock).**

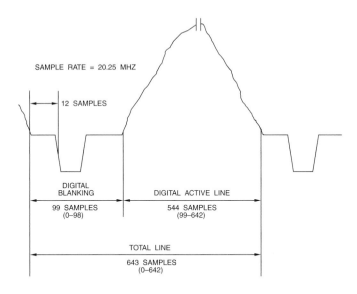

SAMPLE RATE = 20.25 MHZ

12 SAMPLES

DIGITAL
BLANKING

99 SAMPLES
(0–98)

DIGITAL ACTIVE LINE

544 SAMPLES
(99–642)

TOTAL LINE

643 SAMPLES
(0–642)

**Figure 4.12. 480p Analog - Digital Relationship
(4:3 Aspect Ratio, 59.94 Hz Refresh, 20.25 MHz Sample Clock).**

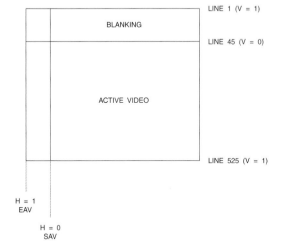

LINE 1 (V = 1)

BLANKING

LINE 45 (V = 0)

ACTIVE VIDEO

LINE 525 (V = 1)

H = 1
EAV

H = 0
SAV

LINE NUMBER	F	V
1–44	0	1
45–524	0	0
525	0	1

Figure 4.13. 480p Digital Vertical Timing (480 Active Lines). V changes state at the EAV sequence at the beginning of the digital line. Note that the digital line number changes state prior to start of horizontal sync, as shown in Figures 4.9 through 4.12.

These active lines are used by the SMPTE RP-202, ATSC A/54a and ARIB STD-B32 standards. However, EIA–861B (DVI and HDMI timing) specifies lines 43–522 for active video.

by using the 720 × 480p format, and blanking the first eight and last eight samples each active scan line. Example relationships between the analog and digital signals are shown in Figures 4.9 through 4.12.

The H (horizontal blanking), V (vertical blanking), and F (field) signals are defined in Figure 4.13. The H, V and F timing indicated is compatible with video compression standards rather than BT.656 discussed in Chapter 6.

SIF and QSIF

SIF is defined to have an active resolution of 352 × 240p. Square pixel SIF is defined to have an active resolution of 320 × 240p.

QSIF is defined to have an active resolution of 176 × 120p. Square pixel QSIF is defined to have an active resolution of 160 × 120p.

576i and 576p Systems

Interlaced Analog Composite Video

(B, D, G, H, I, N, N_C) PAL are analog composite video signals that carry all timing and color information within a single signal. These analog interfaces use 625 lines per frame and are discussed in detail in Chapter 8.

Interlaced Analog Component Video

Analog component signals are comprised of three signals, analog R′G′B′ or YPbPr. Referred to as 576i (since there are typically 576 active scan lines per frame and they are interlaced), the frame rate is usually 25 Hz for compatibility with PAL timing. The analog interface uses 625 lines per frame, with active video present on lines 23–310 and 336–623, as shown in Figure 4.14.

For the 25 Hz frame rate, each scan line time (H) is 64 µs. Detailed horizontal timing is dependent on the specific video interface used, as discussed in Chapter 5.

Progressive Analog Component Video

Analog component signals are comprised of three signals, analog R′G′B′ or YPbPr. Referred to as 576p (since there are typically 576 active scan lines per frame and they are progressive), the frame rate is usually 50 Hz for compatibility with PAL timing. The analog interface uses 625 lines per frame, with active video present on lines 45–620, as shown in Figure 4.15.

For the 50 Hz frame rate, each scan line time (H) is 32 µs. Detailed horizontal timing is dependent on the specific video interface used, as discussed in Chapter 5.

Interlaced Digital Component Video

BT.601 specifies the representation for 576i digital R′G′B′ or YCbCr video signals. Active resolutions defined within BT.601, their 1× Y and R′G′B′ sample rates (F_s), and frame rates, are:

960 × 576i	18.0 MHz	25 Hz
720 × 576i	13.5 MHz	25 Hz

Other common active resolutions, their 1× Y and R′G′B′ sample rates (F_s), and frame rates, are:

1024 × 576i	19.67 MHz	25 Hz
768 × 576i	14.75 MHz	25 Hz
704 × 576i	13.50 MHz	25 Hz
544 × 576i	10.12 MHz	25 Hz
480 × 576i	9.000 MHz	25 Hz

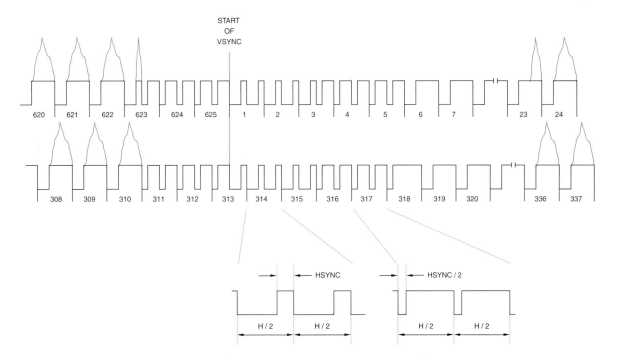

Figure 4.14. 576i Vertical Interval Timing.

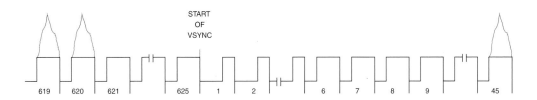

Figure 4.15. 576p Vertical Interval Timing.

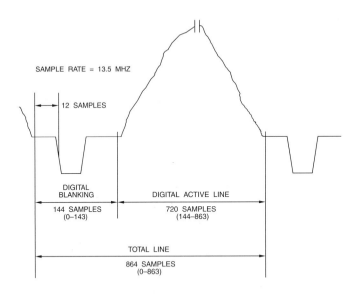

**Figure 4.16. 576i Analog - Digital Relationship
(4:3 Aspect Ratio, 25 Hz Refresh, 13.5 MHz Sample Clock).**

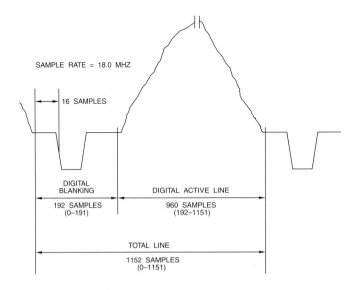

**Figure 4.17. 576i Analog - Digital Relationship
(16:9 Aspect Ratio, 25 Hz Refresh, 18 MHz Sample Clock).**

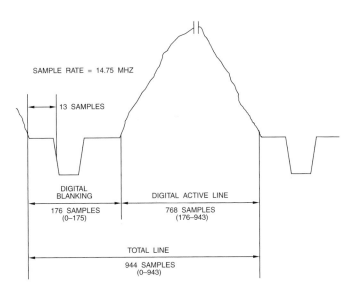

Figure 4.18. 576i Analog - Digital Relationship
(4:3 Aspect Ratio, 25 Hz Refresh, 14.75 MHz Sample Clock).

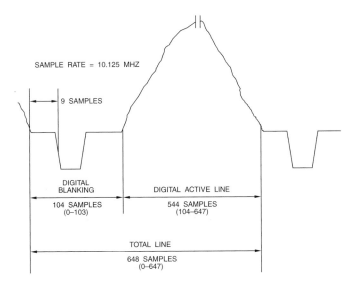

Figure 4.19. 576i Analog - Digital Relationship
(4:3 Aspect Ratio, 25 Hz Refresh, 10.125 MHz Sample Clock).

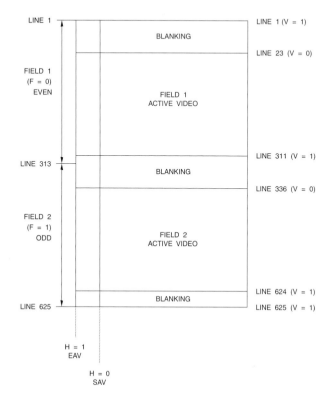

LINE NUMBER	F	V
1–22	0	1
23–310	0	0
311–312	0	1
313–335	1	1
336–623	1	0
624–625	1	1

Figure 4.20. 576i Digital Vertical Timing (576 Active Lines). F and V change state at the EAV sequence at the beginning of the digital line. Note that the digital line number changes state prior to start of horizontal sync, as shown in Figures 4.16 through 4.19.

IEC 61834-2, ITU-R BT.1618 and SMPTE 314M (DV formats) specify lines 23–310 and 335–622 for active video.

1024 × 576i is a 16:9 square pixel format, while 768 × 576i is a 4:3 square pixel format. The 704 × 576i format is done by using the 720 × 576i format, and blanking the first eight and last eight samples each active scan line. Example relationships between the analog and digital signals are shown in Figures 4.16 through 4.19.

The H (horizontal blanking), V (vertical blanking), and F (field) signals are defined in Figure 4.20. The H, V and F timing indicated is compatible with video compression standards rather than BT.656 discussed in Chapter 6.

Progressive Digital Component Video

BT.1358 specifies the representation for 576p digital R′G′B′ or YCbCr signals. Active resolutions defined within BT.1358, their 1× Y and R′G′B′ sample rates (F_s), and frame rates, are:

960 × 576p	36.0 MHz	50 Hz
720 × 576p	27.0 MHz	50 Hz

Other common active resolutions, their 1× Y and R′G′B′ sample rates (F_s), and frame rates, are:

1024 × 576p	39.33 MHz	50 Hz
768 × 576p	29.50 MHz	50 Hz
704 × 576p	27.00 MHz	50 Hz
544 × 576p	20.25 MHz	50 Hz
480 × 576p	18.00 MHz	50 Hz

1024 × 576p is a 16:9 square pixel format, while 768 × 576p is a 4:3 square pixel format. The 704 × 576p format is done by using the 720 × 576p format, and blanking the first eight and last eight samples each active scan line. Example relationships between the analog and digital signals are shown in Figures 4.21 through 4.24.

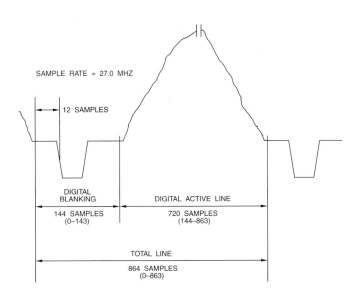

Figure 4.21. 576p Analog - Digital Relationship (4:3 Aspect Ratio, 50 Hz Refresh, 27 MHz Sample Clock).

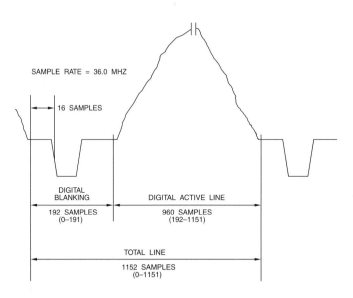

**Figure 4.22. 576p Analog - Digital Relationship
(16:9 Aspect Ratio, 50 Hz Refresh, 36 MHz Sample Clock).**

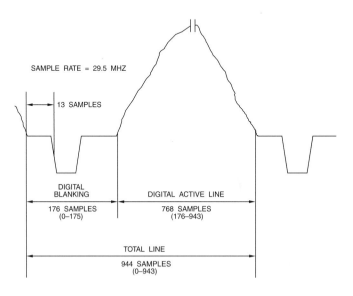

**Figure 4.23. 576p Analog - Digital Relationship
(4:3 Aspect Ratio, 50 Hz Refresh, 29.5 MHz Sample Clock).**

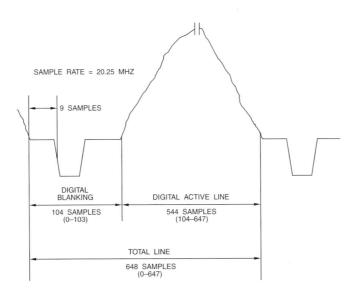

SAMPLE RATE = 20.25 MHZ

9 SAMPLES

DIGITAL
BLANKING

DIGITAL ACTIVE LINE

104 SAMPLES
(0–103)

544 SAMPLES
(104–647)

TOTAL LINE

648 SAMPLES
(0–647)

**Figure 4.24. 576p Analog - Digital Relationship
(4:3 Aspect Ratio, 50 Hz Refresh, 20.25 MHz Sample Clock).**

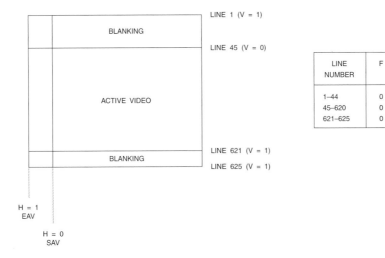

LINE 1 (V = 1)

BLANKING

LINE 45 (V = 0)

ACTIVE VIDEO

LINE 621 (V = 1)

BLANKING

LINE 625 (V = 1)

H = 1
EAV

H = 0
SAV

LINE NUMBER	F	V
1–44	0	1
45–620	0	0
621–625	0	1

**Figure 4.25. 576p Digital Vertical Timing (576 Active Lines). V changes state at the EAV
sequence at the beginning of the digital line. Note that the digital line number changes state prior
to start of horizontal sync, as shown in Figures 4.21 through 4.24.**

The H (horizontal blanking), V (vertical blanking), and F (field) signals are defined in Figure 4.25. The H, V and F timing indicated is compatible with video compression standards rather than BT.656 discussed in Chapter 6.

720p Systems

Progressive Analog Component Video

Analog component signals are comprised of three signals, analog R′G′B′ or YPbPr. Referred to as 720p (since there are typically 720 active scan lines per frame and they are progressive), the frame rate is usually 59.94 Hz (60/1.001) to simplify the generation of (M) NTSC video. The analog interface uses 750 lines per frame, with active video present on lines 26–745, as shown in Figure 4.26.

For the 59.94 Hz frame rate, each scan line time (H) is about 22.24 μs. Detailed horizontal timing is dependent on the specific video interface used, as discussed in Chapter 5.

Progressive Digital Component Video

SMPTE 296M specifies the representation for 720p digital R′G′B′ or YCbCr signals. Active resolutions defined within SMPTE 296M, their 1× Y and R′G′B′ sample rates (F_S), and frame rates, are:

1280 × 720p	74.176 MHz	23.976 Hz
1280 × 720p	74.250 MHz	24.000 Hz
1280 × 720p	74.250 MHz	25.000 Hz
1280 × 720p	74.176 MHz	29.970 Hz
1280 × 720p	74.250 MHz	30.000 Hz
1280 × 720p	74.250 MHz	50.000 Hz
1280 × 720p	74.176 MHz	59.940 Hz
1280 × 720p	74.250 MHz	60.000 Hz

Note that square pixels and a 16:9 aspect ratio are used. Example relationships between the analog and digital signals are shown in Figures 4.27 and 4.28, and Table 4.1. The H (horizontal blanking), V (vertical blanking), and F (field) signals are as defined in Figure 4.29.

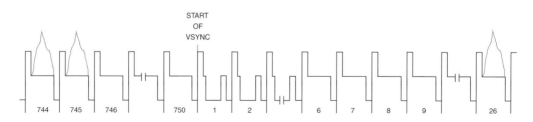

Figure 4.26. 720p Vertical Interval Timing.

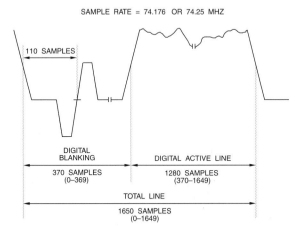

SAMPLE RATE = 74.176 OR 74.25 MHZ

110 SAMPLES

DIGITAL
BLANKING

DIGITAL ACTIVE LINE

370 SAMPLES
(0–369)

1280 SAMPLES
(370–1649)

TOTAL LINE

1650 SAMPLES
(0–1649)

Figure 4.27. 720p Analog - Digital Relationship (16:9 Aspect Ratio, 59.94 Hz Refresh, 74.176 MHz Sample Clock and 60 Hz Refresh, 74.25 MHz Sample Clock).

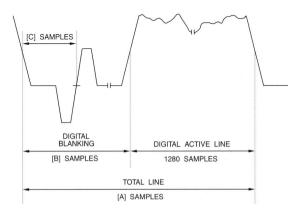

[C] SAMPLES

DIGITAL
BLANKING

DIGITAL ACTIVE LINE

[B] SAMPLES

1280 SAMPLES

TOTAL LINE

[A] SAMPLES

Figure 4.28. General 720p Analog - Digital Relationship.

Active Horizontal Samples	Frame Rate (Hz)	1× Y Sample Rate (MHz)	Total Horizontal Samples (A)	Horizontal Blanking Samples (B)	C Samples
1280	24/1.001	74.25/1.001	4125	2845	2585
	24	74.25	4125	2845	2585
	25[1]	48	1536	256	21
	25[1]	49.5	1584	304	25
	25	74.25	3960	2680	2420
	30/1.001	74.25/1.001	3300	2020	1760
	30	74.25	3300	2020	1760
	50	74.25	1980	700	440
	60/1.001	74.25/1.001	1650	370	110
	60	74.25	1650	370	110

Notes:

1. Useful for CRT-based 50 Hz HDTVs based on a 31.250 kHz horizontal frequency. Sync pulses are –300 mV bi-level, rather than ±300 mV tri-level. 720p content scaled vertically to 1152i active scan lines; 1250i total scan lines instead of 750p.

Table 4.1. Various 720p Analog - Digital Parameters for Figure 4.28.

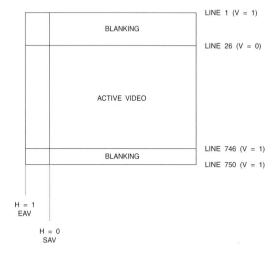

LINE NUMBER	F	V
1–25	0	1
26–745	0	0
746–750	0	1

Figure 4.29. 720p Digital Vertical Timing (720 Active Lines). V changes state at the EAV sequence at the beginning of the digital line. Note that the digital line number changes state prior to start of horizontal sync, as shown in Figures 4.27 and 4.28.

1080i and 1080p Systems

Interlaced Analog Component Video

Analog component signals are comprised of three signals, analog R′G′B′ or YPbPr. Referred to as 1080i (since there are typically 1080 active scan lines per frame and they are interlaced), the frame rate is usually 25 or 29.97 Hz (30/1.001) to simplify the generation of (B, D, G, H, I) PAL or (M) NTSC video. The analog interface uses 1125 lines per frame, with active video present on lines 21–560 and 584–1123, as shown in Figure 4.30.

MPEG-2 and MPEG-4 systems use 1088 lines, rather than 1080, in order to have a multiple of 32 scan lines per frame. In this case, an additional 4 lines per field after the active video are used.

For the 25 Hz frame rate, each scan line time is about 35.56 μs. For the 29.97 Hz frame rate, each scan line time is about 29.66 μs. Detailed horizontal timing is dependent on the specific video interface used, as discussed in Chapter 5.

1152i Format

The 1152i active (1250 total) line format is not a broadcast transmission format. However, it is being used as an analog interconnection standard from HD set-top boxes and DVD players to 50 Hz CRT-based HDTVs. This enables 50 Hz HDTVs to use a fixed 31.25 kHz horizontal frequency, reducing their cost. Other HDTV display technologies, such as DLP, LCD and plasma, are capable of handling the native timing of 720p50 (750p50 with VBI) and 1080i25 (1125i25 with VBI) analog signals.

The set-top box or DVD player converts 720p50 and 1080i25 content to the 1152i25 format. 1280 × 720p50 content is scaled to 1280 × 1152i25; 1920 × 1080i25 content is presented letterboxed in a 1920 × 1152i25 format. HDTVs will have a nominal vertical zoom mode for correcting the geometry of 1080i25, which can be recognized by the vertical synchronizing signal.

Progressive Analog Component Video

Analog component signals are comprised of three signals, analog R′G′B′ or YPbPr. Referred to as 1080p (since there are typically 1080 active scan lines per frame and they are progressive), the frame rate is usually 50 or 59.94 Hz (60/1.001) to simplify the generation of (B, D, G, H, I) PAL or (M) NTSC video. The analog interface uses 1125 lines per frame, with active video present on lines 42–1121, as shown in Figure 4.31.

MPEG-2 and MPEG-4 systems use 1088 lines, rather than 1080, in order to have a multiple of 16 scan lines per frame. In this case, an additional 8 lines per frame after the active video are used.

For the 50 Hz frame rate, each scan line time is about 17.78 μs. For the 59.94 Hz frame rate, each scan line time is about 14.83 μs. Detailed horizontal timing is dependent on the specific video interface used, as discussed in Chapter 5.

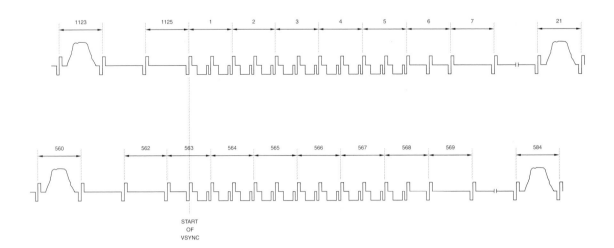

Figure 4.30. 1080i Vertical Interval Timing.

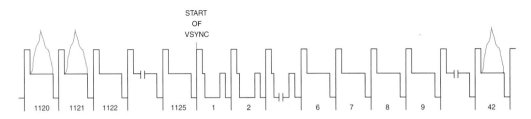

Figure 4.31. 1080p Vertical Interval Timing.

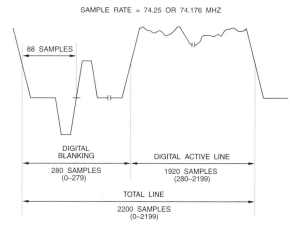

SAMPLE RATE = 74.25 OR 74.176 MHZ

88 SAMPLES

DIGITAL
BLANKING

DIGITAL ACTIVE LINE

280 SAMPLES
(0–279)

1920 SAMPLES
(280–2199)

TOTAL LINE

2200 SAMPLES
(0–2199)

Figure 4.32. 1080i Analog - Digital Relationship (16:9 Aspect Ratio, 29.97 Hz Refresh, 74.176 MHz Sample Clock and 30 Hz Refresh, 74.25 MHz Sample Clock).

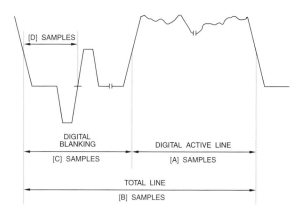

[D] SAMPLES

DIGITAL
BLANKING

DIGITAL ACTIVE LINE

[C] SAMPLES

[A] SAMPLES

TOTAL LINE

[B] SAMPLES

Figure 4.33. General 1080i Analog - Digital Relationship.

Active Horizontal Samples (A)	Frame Rate (Hz)	1× Y Sample Rate (MHz)	Total Horizontal Samples (B)	Horizontal Blanking Samples (C)	D Samples
1920	25[1]	72	2304	384	32
	25[1]	74.25	2376	456	38
	25	74.25	2640	720	528
	30/1.001	74.25/1.001	2200	280	88
	30	74.25	2200	280	88
1440	25[1]	54	1728	288	24
	25	55.6875	1980	540	396
	30/1.001	55.6875/1.001	1650	210	66
	30	55.6875	1650	210	66
1280	25[1]	48	1536	256	21
	25	49.5	1760	480	352
	30/1.001	49.5/1.001	1466.7	186.7	58.7
	30	49.5	1466.7	186.7	58.7

Notes:
1. Useful for CRT-based 50 Hz HDTVs based on a 31.250 kHz horizontal frequency. Sync pulses are –300 mV bi-level, rather than ±300 mV tri-level. 1080i content letterboxed in 1152i active scan lines; 1250i total scan lines instead of 1125i.

Table 4.2. Various 1080i Analog - Digital Parameters for Figure 4.33.

Interlaced Digital Component Video

ITU-R BT.709 and SMPTE 274M specify the digital component format for the 1080i digital R′G′B′ or YCbCr signal. Active resolutions defined within BT.709 and SMPTE 274M, their 1× Y and R′G′B′ sample rates (F_s), and frame rates, are:

1920 × 1080i	74.250 MHz	25.00 Hz
1920 × 1080i	74.176 MHz	29.97 Hz
1920 × 1080i	74.250 MHz	30.00 Hz

Note that square pixels and a 16:9 aspect ratio are used. Other common active resolutions, their 1× Y and R′G′B′ sample rates (F_s), and frame rates, are:

1280 × 1080i	49.500 MHz	25.00 Hz
1280 × 1080i	49.451 MHz	29.97 Hz
1280 × 1080i	49.500 MHz	30.00 Hz
1440 × 1080i	55.688 MHz	25.00 Hz
1440 × 1080i	55.632 MHz	29.97 Hz
1440 × 1080i	55.688 MHz	30.00 Hz

Example relationships between the analog and digital signals are shown in Figures 4.32 and 4.33, and Table 4.2. The H (horizontal blanking) and V (vertical blanking) signals are as defined in Figure 4.34.

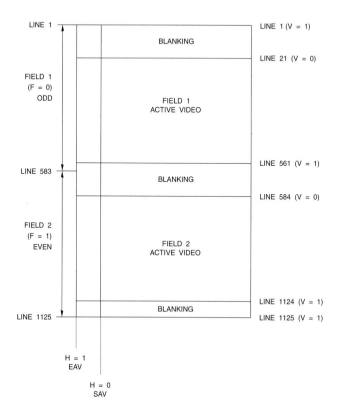

LINE 1

LINE 1 (V = 1)

BLANKING

LINE 21 (V = 0)

FIELD 1
(F = 0)
ODD

FIELD 1
ACTIVE VIDEO

LINE 583

LINE 561 (V = 1)

BLANKING

LINE 584 (V = 0)

FIELD 2
(F = 1)
EVEN

FIELD 2
ACTIVE VIDEO

LINE 1124 (V = 1)
BLANKING
LINE 1125

LINE 1125 (V = 1)

H = 1
EAV

H = 0
SAV

LINE NUMBER	F	V
1–20	0	1
21–560	0	0
561–562	0	1
563–583	1	1
584–1123	1	0
1124–1125	1	1

Figure 4.34. 1080i Digital Vertical Timing (1080 Active Lines). F and V change state at the EAV sequence at the beginning of the digital line. Note that the digital line number changes state prior to start of horizontal sync, as shown in Figures 4.32 and 4.33.

Progressive Digital Component Video

ITU-R BT.709 and SMPTE 274M specify the digital component format for the 1080p digital R′G′B′ or YCbCr signal. Active resolutions defined within BT.709 and SMPTE 274M, their 1× Y and R′G′B′ sample rates (F_s), and frame rates, are:

1920 × 1080p	74.176 MHz	23.976 Hz
1920 × 1080p	74.250 MHz	24.000 Hz
1920 × 1080p	74.250 MHz	25.000 Hz
1920 × 1080p	74.176 MHz	29.970 Hz
1920 × 1080p	74.250 MHz	30.000 Hz
1920 × 1080p	148.50 MHz	50.000 Hz
1920 × 1080p	148.35 MHz	59.940 Hz
1920 × 1080p	148.50 MHz	60.000 Hz

Note that square pixels and a 16:9 aspect ratio are used. Other common active resolutions, their 1× Y and R′G′B′ sample rates (F_s), and frame rates, are:

1280 × 1080p	49.451 MHz	23.976 Hz
1280 × 1080p	49.500 MHz	24.000 Hz
1280 × 1080p	49.500 MHz	25.000 Hz
1280 × 1080p	49.451 MHz	29.970 Hz
1280 × 1080p	49.500 MHz	30.000 Hz
1280 × 1080p	99.000 MHz	50.000 Hz
1280 × 1080p	98.901 MHz	59.940 Hz
1280 × 1080p	99.000 MHz	60.000 Hz
1440 × 1080p	55.632 MHz	23.976 Hz
1440 × 1080p	55.688 MHz	24.000 Hz
1440 × 1080p	55.688 MHz	25.000 Hz
1440 × 1080p	55.632 MHz	29.970 Hz
1440 × 1080p	55.688 MHz	30.000 Hz
1440 × 1080p	111.38 MHz	50.000 Hz
1440 × 1080p	111.26 MHz	59.940 Hz
1440 × 1080p	111.38 MHz	60.000 Hz

Example relationships between the analog and digital signals are shown in Figures 4.35 and 4.36, and Table 4.3. The H (horizontal blanking), V (vertical blanking), and F (field) signals are as defined in Figure 4.37.

Other Video Systems

Some consumer displays, such as those based on LCD and plasma technologies, have adapted other resolutions as their native resolution. Common active resolutions and their names are:

640 × 400	VGA
640 × 480	VGA
854 × 480	WVGA
800 × 600	SVGA
1024 × 768	XGA
1280 × 768	WXGA
1366 × 768	WXGA
1024 × 1024	XGA
1280 × 1024	SXGA
1600 × 1024	WSXGA
1600 × 1200	UXGA
1920 × 1200	WUXGA

These resolutions, and their timings, are defined for computer monitors by the Video Electronics Standards Association (VESA). Displays based on one of these native resolutions are usually capable of accepting many input resolutions, scaling the source to match the display resolution.

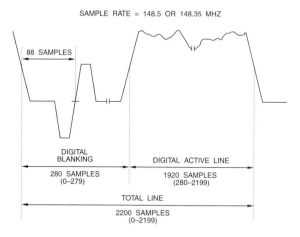

Figure 4.35. 1080p Analog - Digital Relationship (16:9 Aspect Ratio, 59.94 Hz Refresh, 148.35 MHz Sample Clock and 60 Hz Refresh, 148.5 MHz Sample Clock).

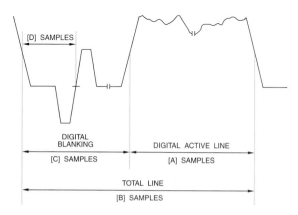

Figure 4.36. General 1080p Analog - Digital Relationship.

Active Horizontal Samples (A)	Frame Rate (Hz)	1× Y Sample Rate (MHz)	Total Horizontal Samples (B)	Horizontal Blanking Samples (C)	D Samples
1920	24/1.001	74.25/1.001	2750	830	638
	24	74.25	2750	830	638
	25	74.25	2640	720	528
	30/1.001	74.25/1.001	2200	280	88
	30	74.25	2200	280	88
	50	148.5	2640	720	528
	60/1.001	148.5/1.001	2200	280	88
	60	148.5	2200	280	88
1440	24/1.001	55.6875/1.001	2062.5	622.5	478.5
	24	55.6875	2062.5	622.5	478.5
	25	55.6875	1980	540	396
	30/1.001	55.6875/1.001	1650	210	66
	30	55.6875	1650	210	66
	50	111.375	1980	540	396
	60/1.001	111.375/1.001	1650	210	66
	60	111.375	1650	210	66
1280	24/1.001	49.5/1.001	1833.3	553.3	425.3
	24	49.5	1833.3	553.3	425.3
	25	49.5	1760	480	352
	30/1.001	49.5/1.001	1466.7	186.7	58.7
	30	49.5	1466.7	186.7	58.7
	50	99	1760	480	352
	60/1.001	99/1.001	1466.7	186.7	58.7
	60	99	1466.7	186.7	58.7

Table 4.3. Various 1080p Analog - Digital Parameters for Figure 4.36.

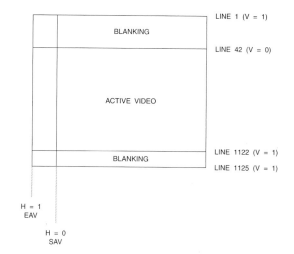

Figure 4.37. 1080p Digital Vertical Timing (1080 Active Lines). V changes state at the EAV sequence at the beginning of the digital line. Note that the digital line number changes state prior to start of horizontal sync, as shown in Figures 4.35 and 4.36.

References

1. EIA–770.1, *Analog 525-Line Component Video Interface—Three Channels*, November 2001.
2. EIA–770.2, *Standard Definition TV Analog Component Video Interface*, November 2001.
3. EIA–770.3, *High Definition TV Analog Component Video Interface*, November 2001.
4. EIA/CEA–861B, *A DTV Profile for Uncompressed High Speed Digital Interfaces*, May 2002.
5. ITU-R BT.601–5, 1995, *Studio Encoding Parameters of Digital Television for Standard 4:3 and Widescreen 16:9 Aspect Ratios.*
6. ITU-R BT.709–5, 2002, *Parameter Values for the HDTV Standards for Production and International Programme Exchange.*
7. ITU-R BT.1358, 1998, *Studio Parameters of 625 and 525 Line Progressive Scan Television Systems.*
8. SMPTE 267M–1995, *Television—Bit-Parallel Digital Interface—Component Video Signal 4:2:2 16 × 9 Aspect Ratio.*
9. SMPTE 274M–2003, *Television—1920 × 1080 Image Sample Structure, Digital Representation and Digital Timing Reference Sequences for Multiple Picture Rates.*
10. SMPTE 293M–2003, *Television—720 × 483 Active Line at 59.94-Hz Progressive Scan Production—Digital Representation.*
11. SMPTE 296M–2001, *Television—1280 × 720 Progressive Image Sample Structure, Analog and Digital Representation and Analog Interface.*

Chapter 5

Analog Video
Interfaces

For years, the primary video signal used by the consumer market has been composite NTSC or PAL video (Figures 8.2 and 8.13). Attempts have been made to support S-Video, but, until recently, it has been largely limited to S-VHS VCRs and high-end televisions.

With the introduction of DVD players, digital set-top boxes, and DTV, there has been renewed interest in providing high-quality video to the consumer market. This equipment not only supports very high quality composite and S-Video signals, but many devices also allow the option of using analog R′G′B′ or YPbPr video.

Using analog R′G′B′ or YPbPr video eliminates NTSC/PAL encoding and decoding artifacts. As a result, the picture is sharper and has less noise. More color bandwidth is also available, increasing the horizontal detail.

S-Video Interface

The RCA phono connector (consumer market) or BNC connector (pro-video market) transfers a composite NTSC or PAL video signal, made by adding the intensity (Y) and color (C) video signals together. The television then has to separate these Y and C video signals in order to display the picture. The problem is that the Y/C separation process is never perfect, as discussed in Chapter 9.

Many video components now support a 4-pin "S1" S-Video connector, illustrated in Figure 5.1 (the female connector viewpoint). This connector keeps the intensity (Y) and color (C) video signals separate, eliminating the Y/C separation process in the TV. As a result, the picture is sharper and has less noise. Figures 9.2 and 9.3 illustrate the Y signal, and Figures 9.10 and 9.11 illustrate the C signal.

NTSC and PAL VBI (vertical blanking interval) data, discussed in Chapter 8, may be present on the 480i or 576i Y video signal.

The "S2" version adds a +5V DC offset to the C signal when a widescreen (16:9) anamorphic program (horizontally squeezed by 25%) is present. A 16:9 TV detects the DC offset and horizontally expands the 4:3 image to fill the screen, restoring the correct aspect ratio of the program. The "S3" version also supports using a +2.3V offset when a program is letterboxed.

The IEC 60933-5 standard specifies the S-Video connector, including signal levels.

Extended S-Video Interface

The PC market also uses an extended S-Video interface. This interface has 7 pins, as shown in Figure 5.1, and is backwards compatible with the 4-pin interface.

The use of the three additional pins varies by manufacturer. They may be used to support an I^2C interface (SDA bi-directional data pin and SCL clock pin), +12V power, a composite NTSC/PAL video signal (CVBS), or analog R'G'B' or YPbPr video signals.

SCART Interface

Most consumer video components in Europe support one or two 21-pin SCART connectors (also known as Peritel, Peritelevision, and Euroconnector). This connection allows analog R'G'B' video or S-Video, composite video, and analog stereo audio to be transmitted between equipment using a single cable. The composite video signal must always be present, as it provides the basic video timing for the analog R'G'B' video signals. Note that the 700 mV R'G'B' signals do not have a blanking pedestal or sync information, as illustrated in Figure 5.4.

PAL VBI (vertical blanking interval) data, discussed in Chapter 8, may be present on the 576i composite video signal.

There are now several types of SCART pinouts, depending on the specific functions implemented, as shown in Tables 5.1 through 5.3. Pinout details are shown in Figure 5.2.

The CENELEC EN 50049–1 and IEC 60933 standards specify the basic SCART connector, including signal levels.

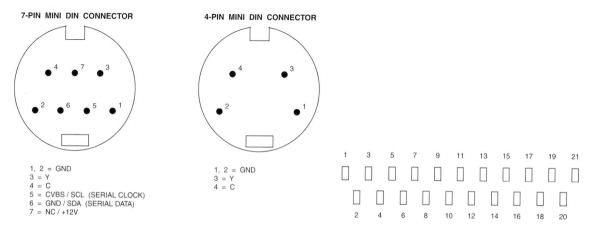

Figure 5.1. S-Video Connector and Signal Names.

Figure 5.2. SCART Connector.

Pin	Function	Signal Level	Impedance
1	right audio out	0.5V rms	< 1K ohm
2	right audio in	0.5V rms	> 10 K ohm
3	left / mono audio out	0.5V rms	< 1K ohm
4	ground - for pins 1, 2, 3, 6		
5	ground - for pin 7		
6	left / mono audio in	0.5V rms	> 10K ohm
7	blue (or C) video in / out	0.7V (or 0.3V burst)	75 ohms
8	status and aspect ratio in / out	9.5V–12V = 4:3 source	> 10K ohm
		4.5V–7V = 16:9 source	
		0V–2V = inactive source	
9	ground - for pin 11		
10	data 2		
11	green video in / out	0.7V	75 ohms
12	data 1		
13	ground - for pin 15		
14	ground - for pin 16		
15	red (or C) video in / out	0.7V (or 0.3V burst)	75 ohms
16	RGB control in / out	1–3V = RGB,	75 ohms
		0–0.4V = composite	
17	ground - for pin 19		
18	ground - for pin 20		
19	composite (or Y) video out	1V	75 ohms
20	composite (or Y) video in	1V	75 ohms
21	ground - for pins 8, 10, 12, shield		

Note:

Often, the SCART 1 connector supports composite video and RGB, the SCART 2 connector supports composite video and S-Video, and the SCART 3 connector supports only composite video. SCART connections may also be used to add external decoders or descramblers to the video path – the video signal goes out and comes back in.

The RGB control signal controls the TV switch between the composite and RGB inputs, enabling the overlaying of text onto the video, even the internal TV program. This enables an external teletext or closed captioning decoder to add information over the current program. If pin 16 is held high, signifying RGB signals are present, the sync is still carried on the Composite Video pin. Some devices (such DVD players) may provide RGB on a SCART and hold pin 16 permanently high.

When a source becomes active, it sets a 12V level on pin 8. This causes the TV to automatically switch to that SCART input. When the source stops, the signal returns to 0V and TV viewing is resumed. If an anamorphic 16:9 program is present, the source raises the signal on pin 8 to only 6V. This causes the TV to switch to that SCART input and at the same time enable the video processing for anamorphic 16:9 programs.

Table 5.1.   SCART Connector Signals.

SDTV RGB Interface

Some SDTV consumer video equipment supports an analog R'G'B' video interface. NTSC and PAL VBI (vertical blanking interval) data, discussed in Chapter 8, may be present on 480i or 576i R'G'B' video signals. Three separate RCA phono connectors (consumer market) or BNC connectors (pro-video and PC market) are used.

The horizontal and vertical video timing are dependent on the video standard, as discussed in Chapter 4. For sources, the video signal at the connector should have a source impedance of 75Ω ±5%. For receivers, video inputs should be AC-coupled and have a 75-Ω ±5% input impedance. The three signals must be coincident with respect to each other within ±5 ns.

Sync information may be present on just the green channel, all three channels, as a separate composite sync signal, or as separate horizontal and vertical sync signals. A gamma of 1/0.45 is used.

7.5 IRE Blanking Pedestal

As shown in Figure 5.3, the nominal active video amplitude is 714 mV, including a 7.5 ±2 IRE blanking pedestal. A 286 ±6 mV composite sync signal may be present on just the green channel (consumer market), or all three channels (pro-video market). DC offsets up to ±1V may be present.

Analog R'G'B' Generation

Assuming 10-bit D/A converters (DACs) with an output range of 0–1.305V (to match the video DACs used by the NTSC/PAL encoder in Chapter 9), the 10-bit YCbCr to R'G'B' equations are:

$$R' = 0.591(Y_{601} - 64) + 0.810(Cr - 512)$$

$$G' = 0.591(Y_{601} - 64) - 0.413(Cr - 512) - 0.199(Cb - 512)$$

$$B' = 0.591(Y_{601} - 64) + 1.025(Cb - 512)$$

R'G'B' has a nominal 10-bit range of 0–518 to match the active video levels used by the NTSC/PAL encoder in Chapter 9. Note that negative values of R'G'B' should be supported at this point.

To implement the 7.5 IRE blanking pedestal, a value of 42 is added to the digital R'G'B' data during active video. 0 is added during the blanking time.

After the blanking pedestal is added, the R'G'B' data is clamped by a blanking signal that has a raised cosine distribution to slow the slew rate of the start and end of the video signal. For 480i and 576i systems, blank rise and fall times are 140 ±20 ns. For 480p and 576p systems, blank rise and fall times are 70 ±10 ns.

Composite sync information may be added to the R'G'B' data after the blank processing has been performed. Values of 16 (sync present) or 240 (no sync) are assigned. The sync rise and fall times should be processed to generate a raised cosine distribution (between 16 and 240) to slow the slew rate of the sync signal. For 480i and 576i systems, sync rise and fall times are 140 ±20 ns, and horizontal sync width at the 50%-point is 4.7 ±0.1 μs. For 480p and 576p systems, sync rise and fall times are 70 ±10 ns, and horizontal sync width at the 50%-point is 2.33 ±0.05 μs.

At this point, we have digital R'G'B' with sync and blanking information, as shown in Figure 5.3 and Table 5.2. The numbers in parentheses in Figure 5.3 indicate the data value for a 10-bit DAC with a full-scale output value of 1.305V. The digital R'G'B' data drive

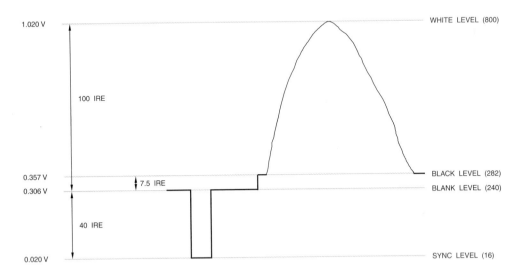

GREEN, BLUE, OR RED CHANNEL, SYNC PRESENT

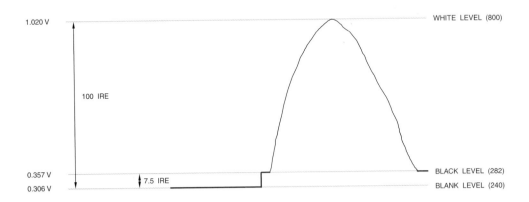

GREEN, BLUE, OR RED CHANNEL, NO SYNC PRESENT

Figure 5.3. SDTV Analog RGB Levels. 7.5 IRE blanking level.

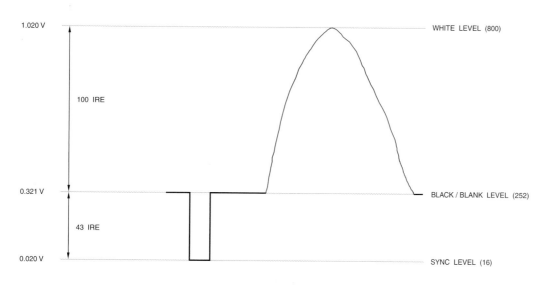

GREEN, BLUE, OR RED CHANNEL, SYNC PRESENT

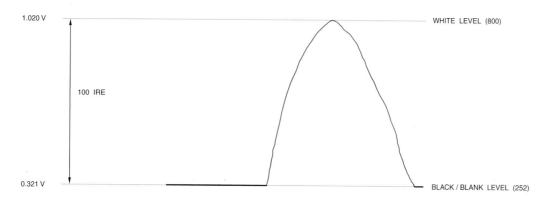

GREEN, BLUE, OR RED CHANNEL, NO SYNC PRESENT

Figure 5.4. SDTV Analog RGB Levels. 0 IRE blanking level.

three 10-bit DACs to generate the analog R′G′B′ video signals.

As the sample-and-hold action of the DAC introduces a (sin x)/x characteristic, the video data may be digitally filtered by a $[(\sin x)/x]^{-1}$ filter to compensate. Alternately, as an analog lowpass filter is usually present after each DAC, the correction may take place in the analog filter.

Video Level	7.5 IRE Blanking Pedestal	0 IRE Blanking Pedestal
white	800	800
black	282	252
blank	240	252
sync	16	16

Table 5.2. SDTV 10-Bit R′G′B′ Values.

Analog R′G′B′ Digitization

Assuming 10-bit A/D converters (ADCs) with an input range of 0–1.305V (to match the video ADCs used by the NTSC/PAL decoder in Chapter 9), the 10-bit R′G′B′ to YCbCr equations are:

$$Y_{601} = 0.506(R' - 282) + 0.992(G' - 282) + 0.193(B' - 282) + 64$$

$$Cb = -0.291(R' - 282) - 0.573(G' - 282) + 0.864(B' - 282) + 512$$

$$Cr = 0.864(R' - 282) - 0.724(G' - 282) - 0.140(B' - 282) + 512$$

R′G′B′ has a nominal 10-bit range of 282–800 to match the active video levels used by the NTSC/PAL decoder in Chapter 9. Table 5.2 and Figure 5.3 illustrate the 10-bit R′G′B′ values for the white, black, blank, and (optional) sync levels.

0 IRE Blanking Pedestal

As shown in Figure 5.4, the nominal active video amplitude is 700 mV, with no blanking pedestal. A 300 ±6 mV composite sync signal may be present on just the green channel (consumer market), or all three channels (pro-video market). DC offsets up to ±1V may be present.

Analog R′G′B′ Generation

Assuming 10-bit DACs with an output range of 0–1.305V (to match the video DACs used by the NTSC/PAL encoder in Chapter 9), the 10-bit YCbCr to R′G′B′ equations are:

$$R' = 0.625(Y_{601} - 64) + 0.857(Cr - 512)$$

$$G' = 0.625(Y_{601} - 64) - 0.437(Cr - 512) - 0.210(Cb - 512)$$

$$B' = 0.625(Y_{601} - 64) + 1.084(Cb - 512)$$

R′G′B′ has a nominal 10-bit range of 0–548 to match the active video levels used by the NTSC/PAL encoder in Chapter 9. Note that negative values of R′G′B′ should be supported at this point.

The R′G′B′ data is processed as discussed when using a 7.5 IRE blanking pedestal. However, no blanking pedestal is added during active video, and the sync values are 16–252 instead of 16–240.

At this point, we have digital R′G′B′ with sync and blanking information, as shown in Figure 5.4 and Table 5.2. The numbers in parentheses in Figure 5.4 indicate the data value for a 10-bit DAC with a full-scale output value of 1.305V. The digital R′G′B′ data drive three 10-bit DACs to generate the analog R′G′B′ video signals.

Analog R'G'B' Digitization

Assuming 10-bit ADCs with an input range of 0–1.305V (to match the video ADCs used by the NTSC/PAL decoder in Chapter 9), the 10-bit R'G'B' to YCbCr equations are:

$$Y_{601} = 0.478(R' - 252) + 0.938(G' - 252) + 0.182(B' - 252) + 64$$

$$Cb = -0.275(R' - 252) - 0.542(G' - 252) + 0.817(B' - 252) + 512$$

$$Cr = 0.817(R' - 252) - 0.685(G' - 252) - 0.132(B' - 252) + 512$$

R'G'B' has a nominal 10-bit range of 252–800 to match the active video levels used by the NTSC/PAL decoder in Chapter 9. Table 5.2 and Figure 5.4 illustrate the 10-bit R'G'B' values for the white, black, blank, and (optional) sync levels.

HDTV RGB Interface

Some HDTV consumer video equipment supports an analog R'G'B' video interface. Three separate RCA phono connectors (consumer market) or BNC connectors (pro-video and PC market) are used.

The horizontal and vertical video timing are dependent on the video standard, as discussed in Chapter 4. For sources, the video signal at the connector should have a source impedance of 75Ω ±5%. For receivers, video inputs should be AC-coupled and have a 75-Ω ±5% input impedance. The three signals must be coincident with respect to each other within ±5 ns.

Sync information may be present on just the green channel, all three channels, as a separate composite sync signal, or as separate horizontal and vertical sync signals. A gamma of 1/0.45 is used.

As shown in Figure 5.5, the nominal active video amplitude is 700 mV, and has no blanking pedestal. A ±300 ±6 mV tri-level composite sync signal may be present on just the green channel (consumer market), or all three channels (pro-video market). DC offsets up to ±1V may be present.

Analog R'G'B' Generation

Assuming 10-bit DACs with an output range of 0–1.305V (to match the video DACs used by the NTSC/PAL encoder in Chapter 9), the 10-bit YCbCr to R'G'B' equations are:

$$R' = 0.625(Y_{709} - 64) + 0.963(Cr - 512)$$

$$G' = 0.625(Y_{709} - 64) - 0.287(Cr - 512) - 0.114(Cb - 512)$$

$$B' = 0.625(Y_{709} - 64) + 1.136(Cb - 512)$$

R'G'B' has a nominal 10-bit range of 0–548 to match the active video levels used by the NTSC/PAL encoder in Chapter 9. Note that negative values of R'G'B' should be supported at this point.

The R'G'B' data is clamped by a blanking signal that has a raised cosine distribution to slow the slew rate of the start and end of the video signal. For 1080i and 720p systems, blank rise and fall times are 54 ±20 ns. For 1080p systems, blank rise and fall times are 27 ±10 ns.

Composite sync information may be added to the R'G'B' data after the blank processing has been performed. Values of 16 (sync low), 488 (high sync), or 252 (no sync) are assigned. The sync rise and fall times should be processed to generate a raised cosine distribution to slow the slew rate of the sync signal. For 1080i systems, sync rise and fall times are 54 ±20 ns, and the horizontal sync low and high widths at the 50%-points are 593 ±40 ns. For

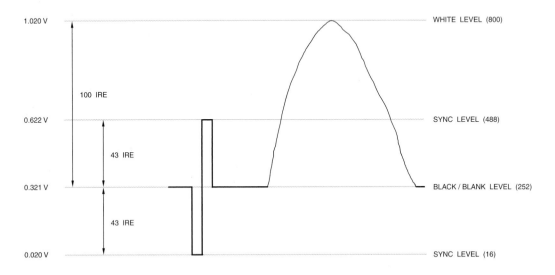

GREEN, BLUE, OR RED CHANNEL, SYNC PRESENT

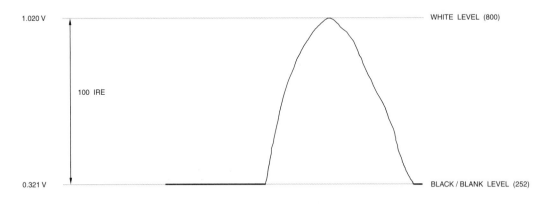

GREEN, BLUE, OR RED CHANNEL, NO SYNC PRESENT

Figure 5.5. HDTV Analog RGB Levels. 0 IRE blanking level.

720p systems, sync rise and fall times are 54 ±20 ns, and the horizontal sync low and high widths at the 50%-points are 539 ±40 ns. For 1080p systems, sync rise and fall times are 27 ±10 ns, and the horizontal sync low and high widths at the 50%-points are 296 ±20 ns.

At this point, we have digital R′G′B′ with sync and blanking information, as shown in Figure 5.5 and Table 5.3. The numbers in parentheses in Figure 5.5 indicate the data value for a 10-bit DAC with a full-scale output value of 1.305V. The digital R′G′B′ data drive three 10-bit DACs to generate the analog R′G′B′ video signals.

Video Level	0 IRE Blanking Pedestal
white	800
sync - high	488
black	252
blank	252
sync - low	16

Table 5.3. HDTV 10-Bit R′G′B′ Values.

Analog R′G′B′ Digitization

Assuming 10-bit ADCs with an input range of 0–1.305V (to match the video ADCs used by the NTSC/PAL decoder in Chapter 9), the 10-bit R′G′B′ to YCbCr equations are:

$$Y_{709} = 0.341(R' - 252) + 1.143(G' - 252) + 0.115(B' - 252) + 64$$

$$Cb = -0.188(R' - 252) - 0.629(G' - 252) + 0.817(B' - 252) + 512$$

$$Cr = 0.817(R' - 252) - 0.743(G' - 252) - 0.074(B' - 252) + 512$$

R′G′B′ has a nominal 10-bit range of 252–800 to match the active video levels used by the NTSC/PAL decoder in Chapter 9. Table 5.3 and Figure 5.5 illustrate the 10-bit R′G′B′ values for the white, black, blank, and (optional) sync levels.

Copy Protection

Currently, there is no approved copy protection technology for this high-definition analog interface. For this reason, some standards and DRM implementations disable these video outputs or only allow a "constrained image" to be output. A "constrained image" has an effective maximum resolution of 960 × 540p, although the total number of video samples and the video timing remain unchanged (for example, 1280 × 720p or 1920 × 1080i).

SDTV YPbPr Interface

Some SDTV consumer video equipment supports an analog YPbPr video interface. NTSC and PAL VBI (vertical blanking interval) data, discussed in Chapter 8, may be present on 480i or 576i Y video signals. Three separate RCA phono connectors (consumer market) or BNC connectors (pro-video market) are used.

The horizontal and vertical video timing are dependent on the video standard, as discussed in Chapter 4. For sources, the video signal at the connector should have a source impedance of 75Ω ±5%. For receivers, video inputs should be AC-coupled and have a 75-Ω ±5% input impedance. The three signals must be coincident with respect to each other within ±5 ns.

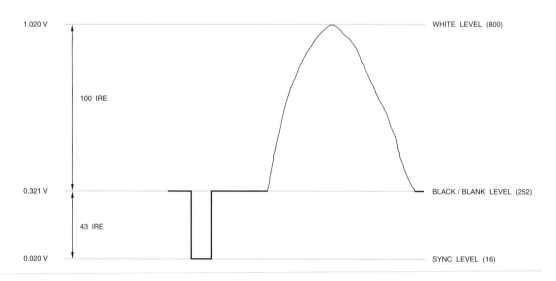

Y CHANNEL, SYNC PRESENT

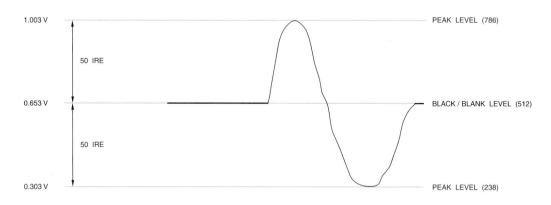

PB OR PR CHANNEL, NO SYNC PRESENT

Figure 5.6. EIA-770.2 SDTV Analog YPbPr Levels. Sync on Y.

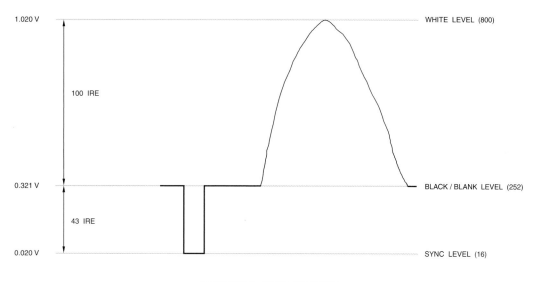

1.020 V — WHITE LEVEL (800)

100 IRE

0.321 V — BLACK / BLANK LEVEL (252)

43 IRE

0.020 V — SYNC LEVEL (16)

Y CHANNEL, SYNC PRESENT

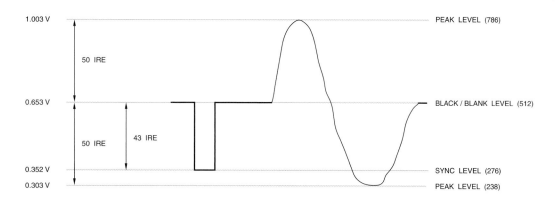

1.003 V — PEAK LEVEL (786)

50 IRE

0.653 V — BLACK / BLANK LEVEL (512)

50 IRE   43 IRE

0.352 V — SYNC LEVEL (276)
0.303 V — PEAK LEVEL (238)

PB OR PR CHANNEL, SYNC PRESENT

Figure 5.7. SDTV Analog YPbPr Levels. Sync on YPbPr.

		White	Yellow	Cyan	Green	Magenta	Red	Blue	Black
Y	IRE	100	88.6	70.1	58.7	41.3	29.9	11.4	0
	mV	700	620	491	411	289	209	80	0
Pb	IRE	0	−50	16.8	−33.1	33.1	−16.8	50	0
	mV	0	−350	118	−232	232	−118	350	0
Pr	IRE	0	8.1	−50	−41.8	41.8	50	−8.1	0
	mV	0	57	−350	−293	293	350	−57	0
Y	64 to 940	940	840	678	578	426	326	164	64
Cb	64 to 960	512	64	663	215	809	361	960	512
Cr	64 to 960	512	585	64	137	887	960	439	512

Table 5.4. EIA-770.2 SDTV YPbPr and YCbCr 100% Color Bars. YPbPr values relative to the blanking level.

		White	Yellow	Cyan	Green	Magenta	Red	Blue	Black
Y	IRE	75	66.5	52.6	44	31	22.4	8.6	0
	mV	525	465	368	308	217	157	60	0
Pb	IRE	0	−37.5	12.6	−24.9	24.9	−12.6	37.5	0
	mV	0	−262	88	−174	174	−88	262	0
Pr	IRE	0	6.1	−37.5	−31.4	31.4	37.5	−6.1	0
	mV	0	43	−262	−220	220	262	−43	0
Y	64 to 940	721	646	525	450	335	260	139	64
Cb	64 to 960	512	176	625	289	735	399	848	512
Cr	64 to 960	512	567	176	231	793	848	457	512

Table 5.5. EIA-770.2 SDTV YPbPr and YCbCr 75% Color Bars. YPbPr values relative to the blanking level.

For consumer products, composite sync is present on only the Y channel. For pro-video applications, composite sync is present on all three channels. A gamma of 1/0.45 is specified.

As shown in Figures 5.6 and 5.7, the Y signal consists of 700 mV of active video (with no blanking pedestal). Pb and Pr have a peak-to-peak amplitude of 700 mV. A 300 ±6 mV composite sync signal is present on just the Y channel (consumer market), or all three channels (pro-video market). DC offsets up to ±1V may be present. The 100% and 75% YPbPr color bar values are shown in Tables 5.4 and 5.5.

Analog YPbPr Generation

Assuming 10-bit DACs with an output range of 0–1.305V (to match the video DACs used by the NTSC/PAL encoder in Chapter 9), the 10-bit YCbCr to YPbPr equations are:

$$Y = ((800 - 252)/(940 - 64))(Y_{601} - 64)$$

$$Pb = ((800 - 252)/(960 - 64))(Cb - 512)$$

$$Pr = ((800 - 252)/(960 - 64))(Cr - 512)$$

Y has a nominal 10-bit range of 0–548 to match the active video levels used by the NTSC/PAL encoder in Chapter 9. Pb and Pr have a nominal 10-bit range of 0 to ±274. Note that negative values of Y should be supported at this point.

The YPbPr data is clamped by a blanking signal that has a raised cosine distribution to slow the slew rate of the start and end of the video signal. For 480i and 576i systems, blank rise and fall times are 140 ±20 ns. For 480p and 576p systems, blank rise and fall times are 70 ±10 ns.

Composite sync information is added to the Y data after the blank processing has been performed. Values of 16 (sync present) or 252 (no sync) are assigned. The sync rise and fall times should be processed to generate a raised cosine distribution (between 16 and 252) to slow the slew rate of the sync signal.

Composite sync information may also be added to the PbPr data after the blank processing has been performed. Values of 276 (sync present) or 512 (no sync) are assigned. The sync rise and fall times should be processed to generate a raised cosine distribution (between 276 and 512) to slow the slew rate of the sync signal.

For 480i and 576i systems, sync rise and fall times are 140 ±20 ns, and horizontal sync width at the 50%-point is 4.7 ±0.1 μs. For 480p and 576p systems, sync rise and fall times are 70 ±10 ns, and horizontal sync width at the 50%-point is 2.33 ±0.05 μs.

At this point, we have digital YPbPr with sync and blanking information, as shown in Figures 5.6 and 5.7 and Table 5.6. The numbers in parentheses in Figures 5.6 and 5.7 indicate the data value for a 10-bit DAC with a full-scale output value of 1.305V. The digital YPbPr data drive three 10-bit DACs to generate the analog YPbPr video signals.

Video Level	Y	PbPr
white	800	512
black	252	512
blank	252	512
sync	16	276

Table 5.6. SDTV 10-Bit YPbPr Values.

Analog YPbPr Digitization

Assuming 10-bit ADCs with an input range of 0–1.305V (to match the video ADCs used by the NTSC/PAL decoder in Chapter 9), the 10-bit YPbPr to YCbCr equations are:

$$Y_{601} = 1.5985\,(Y - 252) + 64$$

$$Cb = 1.635\,(Pb - 512) + 512$$

$$Cr = 1.635\,(Pr - 512) + 512$$

Y has a nominal 10-bit range of 252–800 to match the active video levels used by the NTSC/PAL decoder in Chapter 9. Table 5.6 and Figures 5.6 and 5.7 illustrate the 10-bit YPbPr values for the white, black, blank, and (optional) sync levels.

VBI Data for 480p Systems

CGMS

EIA/CEA-805, IEC 61880–2 and EIA-J CPR–1204–1 define the format of CGMS (Copy Generation Management System) data on line 41 for 480p systems. The waveform is illustrated in Figure 5.8.

A sample clock rate of 27 MHz (59.94 Hz frame rate) or 27.027 MHz (60 Hz frame rate) is used. Each data bit is 26 clock cycles, or 963 ±30 ns, wide with a maximum rise and fall time of 50 ns. A logical "1" has an amplitude of 70 ±10 IRE; a logical "0" has an amplitude of 0 ±5 IRE.

The 2-bit start symbol begins 156 clock cycles, or about 5.778 μs, after 0_H. It consists of a "1" followed by a "0."

The 6-bit header follows the start symbol, and defines the nature of the payload data as shown in Table 5.7. The End of Message immediately follows the last packet of any data service that uses more than one packet. It has an associated payload consisting of all zeros. ECCI is a data service that may use more than one packet, thus requiring the use of the End of Message.

The 14-bit payload for CGMS data is shown in Table 5.8. The 14-bit payload for ECCI data is currently "reserved", consisting of all ones.

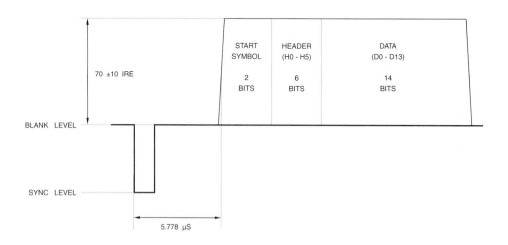

Figure 5.8. EIA-805, IEC 61880–2 and EIA-J CPR–1204–1 480p Line 41 CGMS Timing.

H0	H1	Aspect Ratio	Picture Display Format
0	0	4:3	normal
0	1	16:9	normal
1	0	4:3	letter box
1	1	reserved	reserved

H2	H3	H4	H5	Service Name
0	0	0	0	CGMS
0	0	0	1	Extended Copy Control Information (ECCI)
0	0	1	0	
:				reserved
1	1	1	0	
1	1	1	1	End of Message (default if no copyright information)

Table 5.7. EIA-805, IEC 61880–2 and EIA-J CPR–1204–1 Header Format.

D0	D1	D2	D3	D4	D5	D6	D7	D8	D9	D10	D11	D12	D13
CGMS	CGMS	APS	APS	ASB	0	0	0						

D0	D1	D2	D3	D4	D5	D6	D7	D8	D9	D10	D11	D12	D13
CGMS	CGMS	APS	APS	ASB	0	0	0	$CRC = x^6 + x + 1$					

D0–D1: CGMS Definition

    00   copying permitted without restriction
    01   no more copies (one copy has already been made)
    10   one copy allowed
    11   no copying permitted

D2–D3: Analog Protection Services (valid only if both D0 and D1 are "1")

    00   no Macrovision pseudo-sync pulse
    01   Macrovision pseudo-sync pulse on; color striping off
    10   Macrovision pseudo-sync pulse on; 2-line color striping on
    11   Macrovision pseudo-sync pulse on; 4-line color striping on

D4: Analog Source Bit

    0   non-analog pre-recorded package medium
    1   analog pre-recorded package medium

Table 5.8. EIA-805, IEC 61880–2 and EIA-J CPR–1204–1 CGMS Payload Format.

VBI Data for 576p Systems

CGMS

IEC 62375 defines the format of CGMS (Copy Generation Management System) and widescreen signalling (WSS) data on line 43 for 576p systems. The waveform is illustrated in Figure 5.9. This standard allows a WSS-enhanced 16:9 TV to display programs in their correct aspect ratio.

Data Timing

CGMS and WSS data is normally on line 43, as shown in Figure 5.9. However, due to video editing, the data may reside on any line between 43–47.

The clock frequency is 10 MHz (±1 kHz). The signal waveform should be a sine-squared pulse, with a half-amplitude duration of 100 ±10 ns. The signal amplitude is 500 mV ±5%.

The NRZ data bits are processed by a bi-phase code modulator, such that one data period equals 6 elements at 10 MHz.

Data Content

The WSS consists of a run-in code, a start code, and 14 bits of data, as shown in Table 5.9.

Run-In

The run-in consists of 29 elements of a specific sequence at 10 MHz, shown in Table 5.9.

Start Code

The start code consists of 24 elements of a specific sequence at 10 MHz, shown in Table 5.9.

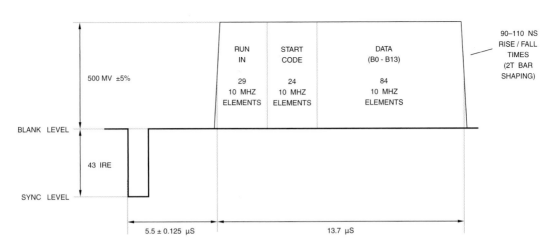

Figure 5.9. IEC 62375 576p Line 43 CGMS Timing.

Group 1 Data

The group 1 data consists of 4 data bits that specify the aspect ratio. Each data bit generates 6 elements at 10 MHz. b0 is the LSB.

Table 5.9 lists the data bit assignments and usage. The number of active lines listed in Table 5.10 are for the exact aspect ratio (a = 1.33, 1.56, or 1.78).

The aspect ratio label indicates a range of possible aspect ratios (a) and number of active lines:

4:3	$a \leq 1.46$	527–576
14:9	$1.46 < a \leq 1.66$	463–526
16:9	$1.66 < a \leq 1.90$	405–462
>16:9	$a > 1.90$	< 405

To allow automatic selection of the display mode, a 16:9 receiver should support the following minimum requirements:

Case 1: The 4:3 aspect ratio picture should be centered on the display, with black bars on the left and right sides.

Case 2: The 14:9 aspect ratio picture should be centered on the display, with black bars on the left and right sides. Alternately, the picture may be displayed using the full display width by using a small (typically 8%) horizontal geometrical error.

Case 3: The 16:9 aspect ratio picture should be displayed using the full width of the display.

Case 4: The >16:9 aspect ratio picture should be displayed as in Case 3 or use the full height of the display by zooming in.

Group 3 Data

The group 3 data consists of three data bits that specify subtitles. Each data bit generates six elements at 10 MHz. Data bit b8 is the LSB.

b10, b9: open subtitles
00	no
01	inside active picture
10	outside active picture
11	reserved

Group 4 Data

The group 4 data consists of three data bits that specify surround sound and copy protection. Each data bit generates six elements at 10 MHz. Data bit b11 is the LSB.

b11: surround sound
0	no
1	yes

b12: copyright
0	no copyright asserted or unknown
1	copyright asserted

b13: copy protection
0	copying not restricted
1	copying restricted

run-in	29 elements at 10 MHz	1 1111 0001 1100 0111 0001 1100 0111 (1F1C 71C7$_H$)
start code	24 elements at 10 MHz	0001 1110 0011 1100 0001 1111 (1E 3C1F$_H$)
group 1 (aspect ratio)	24 elements at 10 MHz "0" = 000 111 "1" = 111 000	b0, b1, b2, b3
group 2 (enhanced services)	24 elements at 10 MHz "0" = 000 111 "1" = 111 000	b4, b5, b6, b7 (b4, b5, b6 and b7 = "0" since reserved)
group 3 (subtitles)	18 elements at 10 MHz "0" = 000 111 "1" = 111 000	b8, b9, b10 (b8 = "0" since reserved)
group 4 (reserved)	18 elements at 10 MHz "0" = 000 111 "1" = 111 000	b11, b12, b13

Table 5.9. IEC 62375 576p Line 43 WSS Information.

b3, b2, b1, b0	Aspect Ratio Label	Format	Position On 4:3 Display	Active Lines	Minimum Requirements
1000	4:3	full format	–	576	case 1
0001	14:9	letterbox	center	504	case 2
0010	14:9	letterbox	top	504	case 2
1011	16:9	letterbox	center	430	case 3
0100	16:9	letterbox	top	430	case 3
1101	> 16:9	letterbox	center	–	case 4
1110	14:9	full format	center	576	–
0111	16:9	full format (anamorphic)	–	576	–

Table 5.10. IEC 62375 Group 1 (Aspect Ratio) Data Bit Assignments and Usage.

HDTV YPbPr Interface

Some HDTV consumer video equipment supports an analog YPbPr video interface. Three separate RCA phono connectors (consumer market) or BNC connectors (pro-video market) are used.

The horizontal and vertical video timing are dependent on the video standard, as discussed in Chapter 4. For sources, the video signal at the connector should have a source impedance of 75Ω ±5%. For receivers, video inputs should be AC-coupled and have a 75-Ω ±5% input impedance. The three signals must be coincident with respect to each other within ±5 ns.

For consumer products, composite sync is present on only the Y channel. For pro-video applications, composite sync is present on all three channels. A gamma of 1/0.45 is specified.

As shown in Figures 5.10 and 5.11, the Y signal consists of 700 mV of active video (with no blanking pedestal). Pb and Pr have a peak-to-peak amplitude of 700 mV. A ±300 ±6 mV composite sync signal is present on just the Y channel (consumer market), or all three channels (pro-video market). DC offsets up to ±1V may be present. The 100% and 75% YPbPr color bar values are shown in Tables 5.11 and 5.12.

Analog YPbPr Generation

Assuming 10-bit DACs with an output range of 0–1.305V (to match the video DACs used by the NTSC/PAL encoder in Chapter 9), the 10-bit YCbCr to YPbPr equations are:

$$Y = ((800 - 252) / (940 - 64)) (Y_{709} - 64)$$

$$Pb = ((800 - 252) / (960 - 64)) (Cb - 512)$$

$$Pr = ((800 - 252) / (960 - 64)) (Cr - 512)$$

Y has a nominal 10-bit range of 0–548 to match the active video levels used by the NTSC/PAL encoder in Chapter 9. Pb and Pr have a nominal 10-bit range of 0 to ±274. Note that negative values of Y should be supported at this point.

The YPbPr data is clamped by a blanking signal that has a raised cosine distribution to slow the slew rate of the start and end of the video signal. For 1080i and 720p systems, blank rise and fall times are 54 ±20 ns. For 1080p systems, blank rise and fall times are 27 ±10 ns.

Composite sync information is added to the Y data after the blank processing has been performed. Values of 16 (sync low), 488 (high sync), or 252 (no sync) are assigned. The sync rise and fall times should be processed to generate a raised cosine distribution to slow the slew rate of the sync signal.

Composite sync information may be added to the PbPr data after the blank processing has been performed. Values of 276 (sync low), 748 (high sync), or 512 (no sync) are assigned. The sync rise and fall times should be processed to generate a raised cosine distribution to slow the slew rate of the sync signal.

For 1080i systems, sync rise and fall times are 54 ±20 ns, and the horizontal sync low and high widths at the 50%-points are 593 ±40 ns. For 720p systems, sync rise and fall times are 54 ±20 ns, and the horizontal sync low and high widths at the 50%-points are 539 ±40 ns. For 1080p systems, sync rise and fall times are 27 ±10 ns, and the horizontal sync low and high widths at the 50%-points are 296 ±20 ns.

At this point, we have digital YPbPr with sync and blanking information, as shown in Figures 5.10 and 5.11 and Table 5.13. The numbers in parentheses in Figures 5.10 and 5.11 indicate the data value for a 10-bit DAC with a full-scale output value of 1.305V. The digital

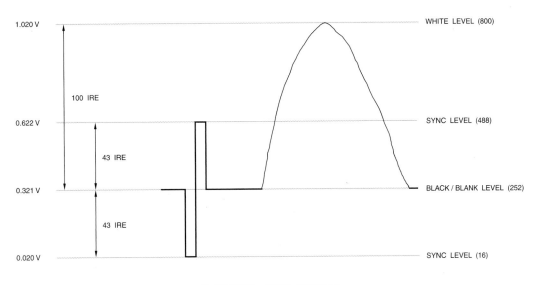

Y CHANNEL, SYNC PRESENT

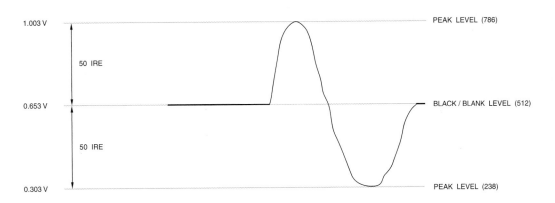

PB OR PR CHANNEL, NO SYNC PRESENT

Figure 5.10. EIA-770.3 HDTV Analog YPbPr Levels. Sync on Y.

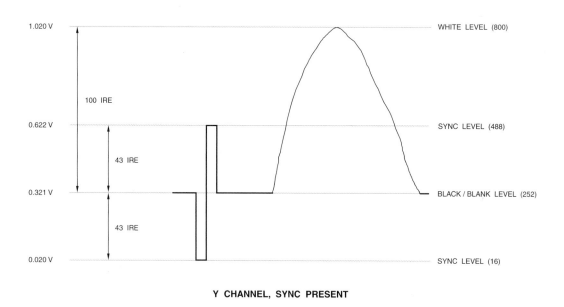

Y CHANNEL, SYNC PRESENT

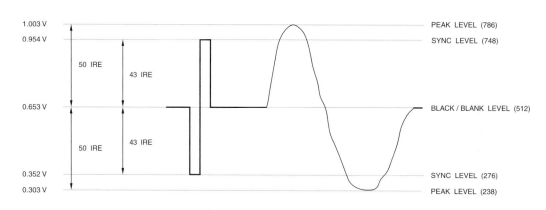

PB OR PR CHANNEL, SYNC PRESENT

Figure 5.11. SMPTE 274M and 296M HDTV Analog YPbPr Levels. Sync on YPbPr.

		White	Yellow	Cyan	Green	Magenta	Red	Blue	Black
Y	IRE	100	92.8	78.7	71.5	28.5	21.3	7.2	0
	mV	700	650	551	501	200	149	50	0
Pb	IRE	0	−50	11.4	−38.5	38.5	−11.4	50	0
	mV	0	−350	80	−270	270	−80	350	0
Pr	IRE	0	4.6	−50	−45.4	45.4	50	−4.6	0
	mV	0	32	−350	−318	318	350	−32	0
Y	64 to 940	940	877	753	690	314	251	127	64
Cb	64 to 960	512	64	614	167	857	410	960	512
Cr	64 to 960	512	553	64	106	918	960	471	512

Table 5.11. EIA-770.3 HDTV YPbPr and YCbCr 100% Color Bars. YPbPr values relative to the blanking level.

		White	Yellow	Cyan	Green	Magenta	Red	Blue	Black
Y	IRE	75	69.6	59	53.7	21.3	16	5.4	0
	mV	525	487	413	376	149	112	38	0
Pb	IRE	0	−37.5	8.6	−28.9	28.9	−8.6	37.5	0
	mV	0	−263	60	−202	202	−60	263	0
Pr	IRE	0	3.5	−37.5	−34	34	37.5	−3.5	0
	mV	0	24	−263	−238	238	263	−24	0
Y	64 to 940	721	674	581	534	251	204	111	64
Cb	64 to 960	512	176	589	253	771	435	848	512
Cr	64 to 960	512	543	176	207	817	848	481	512

Table 5.12. EIA-770.3 HDTV YPbPr and YCbCr 75% Color Bars. YPbPr values relative to the blanking level.

YPbPr data drive three 10-bit DACs to generate the analog YPbPr video signals.

Video Level	Y	PbPr
white	800	512
sync - high	488	748
black	252	512
blank	252	512
sync - low	16	276

Table 5.13. HDTV 10-Bit YPbPr Values.

Analog YPbPr Digitization

Assuming 10-bit ADCs with an input range of 0–1.305V (to match the video ADCs used by the NTSC/PAL decoder in Chapter 9), the 10-bit YPbPr to YCbCr equations are:

$$Y_{709} = 1.5985(Y - 252) + 64$$

$$Cb = 1.635(Pb - 512) + 512$$

$$Cr = 1.635(Pr - 512) + 512$$

Y has a nominal 10-bit range of 252–800 to match the active video levels used by the NTSC/PAL decoder in Chapter 9. Table 5.13 and Figures 5.10 and 5.11 illustrate the 10-bit YPbPr values for the white, black, blank, and (optional) sync levels.

VBI Data for 720p Systems

CGMS

EIA/CEA-805 and EIA-J CPR–1204–2 define the format of CGMS (Copy Generation Management System) data on line 24 for 720p systems. The waveform is illustrated in Figure 5.12.

A sample clock rate of 74.176 MHz (59.94 Hz frame rate) or 74.25 MHz (60 Hz frame rate) is used. Each data bit is 58 clock cycles, or 782 ±30 ns, wide with a maximum rise and fall time of 50 ns. A logical "1" has an amplitude of 70 ±10 IRE; a logical "0" has an amplitude of 0 ±5 IRE.

The 2-bit start symbol begins 232 clock cycles, or about 3.128 μs, after 0_H. It consists of a "1" followed by a "0."

The 6-bit header and 14-bit CGMS payload data format is the same as for 480p systems discussed earlier in this chapter.

VBI Data for 1080i Systems

CGMS

EIA/CEA-805 and EIA-J CPR–1204–2 define the format of CGMS (Copy Generation Management System) data on lines 19 and 582 for 1080i systems. The waveform is illustrated in Figure 5.13.

A sample clock rate of 74.176 MHz (59.94 Hz field rate) or 74.25 MHz (60 Hz field rate) is used. Each data bit is 77 clock cycles, or 1038 ±30 ns, wide with a maximum rise and fall time of 50 ns. A logical "1" has an amplitude of 70 ±10 IRE; a logical "0" has an amplitude of 0 ±5 IRE.

The 2-bit start symbol begins 308 clock cycles, or about 4.152 μs, after 0_H. It consists of a "1" followed by a "0."

The 6-bit header and 14-bit CGMS payload data format is the same as for 480p systems discussed earlier in this chapter.

Copy Protection

Currently, there is no approved copy protection technology for this high-definition analog interface. For this reason, some standards and DRM implementations disable these video outputs or only allow a "constrained image" to be output. A "constrained image" has an effective maximum resolution of 960 × 540p, although the total number of video samples and the video timing remain unchanged (for example, 1280 × 720p or 1920 × 1080i).

D-Connector Interface

A 14-pin female D-Connector (EIA-J CP–4120 standard, EIA-J RC–5237 connector) is optionally used on some high-end consumer equipment in Japan, Hong Kong and Singapore. It is used to transfer EIA 770.2 or EIA 770.3 interlaced or progressive analog YPbPr video.

There are five flavors of the D-Connector, referred to as D1, D2, D3, D4 and D5, each used to indicate supported video formats, as indicated in Table 5.14. Figure 5.14 illustrates the connector and Table 5.15 lists the pin names.

Three Line signals (Line 1, Line 2 and Line 3) indicate the resolution and refresh rate of the YPbPr source video, as indicated in Table 5.16.

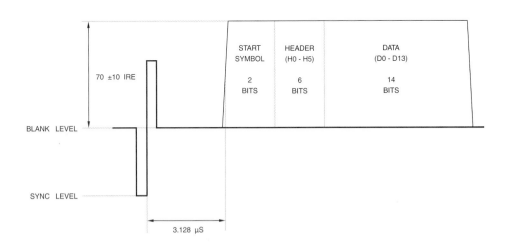

Figure 5.12. EIA-805 and EIA-J CPR–1204–2 720p Line 24 CGMS Timing.

Other Pro-Video Analog Interfaces

Tables 5.17 and 5.18 list some other common component analog video formats. The horizontal and vertical timing is the same as for 525-line (M) NTSC and 625-line (B, D, G, H, I) PAL. The 100% and 75% color bar values are shown in Tables 5.19 through 5.22. The SMPTE, EBU N10, 625-line Betacam, and 625-line MII values are the same as for SDTV YPbPr.

VGA Interface

Table 5.23 and Figure 5.15 illustrate the 15-pin VGA connector used by computer equipment, and some consumer equipment, to transfer analog RGB signals. The analog RGB signals do not contain sync information and have no blanking pedestal, as shown in Figure 5.4.

References

1. EIA–770.1, *Analog 525-Line Component Video Interface—Three Channels*, November 2001.
2. EIA–770.2, *Standard Definition TV Analog Component Video Interface*, November 2001.
3. EIA–770.3, *High Definition TV Analog Component Video Interface*, November 2001.
4. EIA/CEA–805, *Data Services on the Component Video Interfaces*, October 2000.
5. EIA-J CPR–1204–1, *Transfer Method of Video ID Information using Vertical Blanking Interval (525P System)*, March 1998.
6. EIA-J CPR–1204–2, *Transfer Method of Video ID Information using Vertical Blanking Interval (720P, 1125I System)*, January 2000.
7. EIA-J CP–4120, *Interface Between Digital Tuner and Television Receiver Using D-Connector*, January 2000.

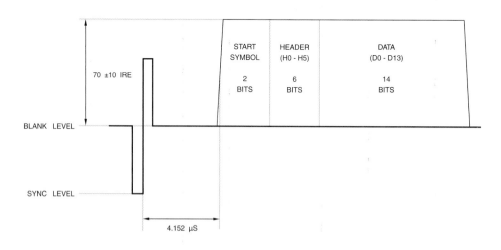

Figure 5.13. EIA-805 and EIA-J CPR–1204–2 1080i Lines 19 and 582 CGMS Timing.

	480i	480p	720p	1080i	1080p
D1	×				
D2	×	×			
D3	×	×		×	
D4	×	×	×	×	
D5	×	×	×	×	×

Table 5.14. D-Connector Supported Video Formats.

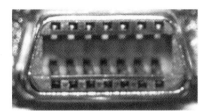

Figure 5.14. D-Connector.

Pin	Function	Signal Level	Impedance
1	Y	0.700V + sync	75 ohms
2	ground - Y		
3	Pb	±0.350V	75 ohms
4	ground - Pb		
5	Pr	±0.350V	75 ohms
6	ground - Pr		
7	reserved 1		
8	line 1	0V, 2.2V, or 5V[1]	10K ±3K ohm
9	line 2	0V, 2.2V, or 5V[1]	10K ±3K ohm
10	reserved 2		
11	line 3	0V, 2.2V, or 5V[1]	10K ±3K ohm
12	ground - detect plugged		
13	reserved 3		
14	detect plugged	0V = plugged in[2]	> 100K ohm

Notes:
1. 2.2V has range of 2.2V ±0.8V. 5V has a range of 5V ±1.5V.
2. Inside equipment, pin 12 is connected to ground and pin 14 is pulled to 5V through a resistor. Inside each D-Connector plug, pins 12 and 14 are shorted together.

Table 5.15. D-Connector Pin Descriptions.

Resolution		Refresh Rate	Line 1 Scan Lines	Line 2 Refresh Rate	Line 3 Aspect Ratio	Chromaticity and Reference White	Color Space Equations	Gamma Correction	Sync Amplitude on Y
1920x1080		60i	5V	0V	5V	EIA-770.3	EIA-770.3	EIA-770.3	$\pm 0.300V^3$
		50i[2]	5V	2.2V	5V				
		30p	5V	2.2V	5V				
		25p[2]	5V	2.2V	5V				
		24p[2]	5V	2.2V	5V				
		24sF[2]	5V	2.2V	5V				
1280x720		60p	2.2V	5V	5V				
		50p[2]	2.2V	2.2V	5V				
		30p	2.2V	2.2V	5V				
		25p[2]	2.2V	2.2V	5V				
		24p[2]	2.2V	2.2V	5V				
640x480		60p[2]	0V	5V	0V	EIA-770.2	EIA-770.2	EIA-770.2	$-0.300V^3$
720x480	16:9 Squeeze	60p	0V	5V	5V				
	16:9 Squeeze	60i	0V	0V	5V				
	16:9 Letterbox	60i	0V	0V	2.2V				
	4:3	60i	0V	0V	0V				

Notes:
1. 60p, 60i, 30p, and 24p refresh rates also include the 59.94p, 59.94i, 29.97p and 23.976p refresh rates.
2. Not part of EIAJ CP-4120 specification, but commonly supported by equipment.
3. Relative to the blanking level.

Table 5.16. Voltage Levels of Line Signals for Various Video Formats for D-Connector.

Format	Output Signal	Signal Amplitudes (volts)	Notes
SMPTE, EBU N10	Y	+0.700	0% setup on Y 100% saturation three wire = (Y + sync), (R′−Y), (B′−Y)
	sync	−0.300	
	R′−Y, B′−Y	±0.350	
525-Line Betacam[1]	Y	+0.714	7.5% setup on Y only 100% saturation three wire = (Y + sync), (R′−Y), (B′−Y)
	sync	−0.286	
	R′−Y, B′−Y	±0.467	
625-Line Betacam[1]	Y	+0.700	0% setup on Y 100% saturation three wire = (Y + sync), (R′−Y), (B′−Y)
	sync	−0.300	
	R′−Y, B′−Y	±0.350	
525-Line MII[2]	Y	+0.700	7.5% setup on Y only 100% saturation three wire = (Y + sync), (R′−Y), (B′−Y)
	sync	−0.300	
	R′−Y, B′−Y	±0.324	
625-Line MII[2]	Y	+0.700	0% setup on Y 100% saturation three wire = (Y + sync), (R′−Y), (B′−Y)
	sync	−0.300	
	R′−Y, B′−Y	±0.350	

Notes:
1. Trademark of Sony Corporation.
2. Trademark of Matsushita Corporation.

Table 5.17. Common Pro-Video Component Analog Video Formats.

Format	Output Signal	Signal Amplitudes (volts)	Notes
SMPTE, EBU N10	G′, B′, R′	+0.700	0% setup on G′, B′, and R′ 100% saturation three wire = (G′ + sync), B′, R′
	sync	−0.300	
NTSC (setup)	G′, B′, R′	+0.714	7.5% setup on G′, B′, and R′ 100% saturation three wire = (G′ + sync), B′, R′
	sync	−0.286	
NTSC (no setup)	G′, B′, R′	+0.714	0% setup on G′, B′, and R′ 100% saturation three wire = (G′ + sync), B′, R′
	sync	−0.286	
MII[1]	G′, B′, R′	+0.700	7.5% setup on G′, B′, and R′ 100% saturation three wire = (G′ + sync), B′, R′
	sync	−0.300	

Notes:
1. Trademark of Matsushita Corporation.

Table 5.18. Common Pro-Video RGB Analog Video Formats.

			White	Yellow	Cyan	Green	Magenta	Red	Blue	Black
Y		IRE	100	89.5	72.3	61.8	45.7	35.2	18.0	7.5
		mV	714	639	517	441	326	251	129	54
B′–Y		IRE	0	−65.3	22.0	−43.3	43.3	−22.0	65.3	0
		mV	0	−466	157	−309	309	−157	466	0
R′–Y		IRE	0	10.6	−65.3	−54.7	54.7	65.3	−10.6	0
		mV	0	76	−466	−391	391	466	−76	0

Table 5.19. 525-Line Betacam 100% Color Bars. Values are relative to the blanking level.

			White	Yellow	Cyan	Green	Magenta	Red	Blue	Black
Y		IRE	76.9	69.0	56.1	48.2	36.2	28.2	15.4	7.5
		mV	549	492	401	344	258	202	110	54
B′–Y		IRE	0	−49.0	16.5	−32.5	32.5	−16.5	49.0	0
		mV	0	−350	118	−232	232	−118	350	0
R′–Y		IRE	0	8.0	−49.0	−41.0	41.0	49.0	−8.0	0
		mV	0	57	−350	−293	293	350	−57	0

Table 5.20. 525-Line Betacam 75% Color Bars. Values are relative to the blanking level.

			White	Yellow	Cyan	Green	Magenta	Red	Blue	Black
Y		IRE	100	89.5	72.3	61.8	45.7	35.2	18.0	7.5
		mV	700	626	506	433	320	246	126	53
B′–Y		IRE	0	–46.3	15.6	–30.6	30.6	–15.6	46.3	0
		mV	0	–324	109	–214	214	–109	324	0
R′–Y		IRE	0	7.5	–46.3	–38.7	38.7	46.3	–7.5	0
		mV	0	53	–324	–271	271	324	–53	0

Table 5.21. 525-Line MII 100% Color Bars. Values are relative to the blanking level.

			White	Yellow	Cyan	Green	Magenta	Red	Blue	Black
Y		IRE	76.9	69.0	56.1	48.2	36.2	28.2	15.4	7.5
		mV	538	483	393	338	253	198	108	53
B′–Y		IRE	0	–34.7	11.7	–23.0	23.0	–11.7	34.7	0
		mV	0	–243	82	–161	161	–82	243	0
R′–Y		IRE	0	5.6	–34.7	–29.0	29.0	34.7	–5.6	0
		mV	0	39	–243	–203	203	243	–39	0

Table 5.22. 525-Line MII 75% Color Bars. Values are relative to the blanking level.

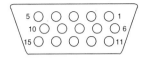

Figure 5.15. VGA 15-Pin D-SUB Female Connector.

Pin	Function	Signal Level	Impedance
1	red	0.7v	75 ohms
2	green	0.7v	75 ohms
3	blue	0.7v	75 ohms
4	ground		
5	ground		
6	ground - red		
7	ground - green		
8	ground - blue		
9	+5V DC		
10	ground - HSYNC		
11	ground - VSYNC		
12	DDC SDA (data)	≥ 2.4v	
13	HSYNC (horizontal sync)	≥ 2.4v	
14	VSYNC (vertical sync)	≥ 2.4v	
15	DDC SCL (clock)	≥ 2.4v	

Notes:
1. DDC = Display Data Channel.

Table 5.23. VGA Connector Signals.

8. IEC 60993–1, *Audio, Video and Audiovisual Systems—Interconnections and Matching Values—Part 1: 21-pin Connector for Video Systems, Application No. 1*, April 1988.

9. IEC 61880–2, *Video Systems (525/60)—Video and Accompanied Data Using the Vertical Blanking Interval—Part 2: 525 Progressive Scan System*, September 2002.

10. IEC 62375, *Video Systems (625/50 Progressive)—Video and Accompanied Data Using the Vertical Blanking Interval—Analog Interface*, February 2004.

11. ITU-R BT.709–5, 2002, *Parameter Values for the HDTV Standards for Production and International Programme Exchange.*

12. SMPTE 253M–1998, *Television—Three-Channel RGB Analog Video Interface.*

13. SMPTE 274M–2003, *Television—1920 × 1080 Image Sample Structure, Digital Representation and Digital Timing Reference Sequences for Multiple Picture Rates.*

14. SMPTE 293M–2003, *Television—720 × 483 Active Line at 59.94-Hz Progressive Scan Production—Digital Representation.*

15. SMPTE RP-160–1997, *Three-Channel Parallel Analog Component High-Definition Video Interface.*

Chapter 6

Digital Video Interfaces

Pro-Video Component Interfaces

Pro-video equipment, such as that used within studios, has unique requirements and therefore its own set of digital video interconnect standards. Table 6.1 lists the various pro-video parallel and serial digital interface standards.

Video Timing

Rather than digitize and transmit the blanking intervals, special sequences are inserted into the digital video stream to indicate the start of active video (SAV) and end of active video (EAV). These EAV and SAV sequences indicate when horizontal and vertical blanking are present and which field is being transmitted. They also enable the transmission of ancillary data such as digital audio, teletext, captioning, etc. during the blanking intervals.

The EAV and SAV sequences must have priority over active video data or ancillary data to ensure that correct video timing is always maintained at the receiver. The receiver decodes the EAV and SAV sequences to recover the video timing.

The video timing sequence of the encoder is controlled by three timing signals discussed in Chapter 4: H (horizontal blanking), V (vertical blanking), and F (Field 1 or Field 2). A zero-to-one transition of H triggers an EAV sequence while a one-to-zero transition triggers an SAV sequence. F and V are allowed to change only at EAV sequences.

Usually, both 8-bit and 10-bit interfaces are supported, with the 10-bit interface used to transmit 2 bits of fractional video data to minimize cumulative processing errors and to support 10-bit ancillary data.

YCbCr or $R'G'B'$ data may not use the 10-bit values of 000_H–003_H and $3FC_H$–$3FF_H$, or the 8-bit values of 00_H and FF_H, since they are used for timing information.

Active Resolution (H × V)	Total Resolution[1] (H × V)	Display Aspect Ratio	Frame Rate (Hz)	1× Y Sample Rate (MHz)	SDTV or HDTV	Digital Parallel Standard	Digital Serial Standard
720 × 480i	858 × 525i	4:3	29.97	13.5	SDTV	BT.656 BT.799 SMPTE 125M	BT.656 BT.799
720 × 480p	858 × 525p	4:3	59.94	27	SDTV	–	BT.1362 SMPTE 294M
720 × 576i	864 × 625i	4:3	25	13.5	SDTV	BT.656 BT.799	BT.656 BT.799
720 × 576p	864 × 625p	4:3	50	27	SDTV	–	BT.1362
960 × 480i	1144 × 525i	16:9	29.97	18	SDTV	BT.1302 BT.1303 SMPTE 267M	BT.1302 BT.1303
960 × 576i	1152 × 625i	16:9	25	18	SDTV	BT.1302 BT.1303	BT.1302 BT.1303
1280 × 720p	1650 × 750p	16:9	59.94	74.176	HDTV	SMPTE 274M	–
1280 × 720p	1650 × 750p	16:9	60	74.25	HDTV	SMPTE 274M	–
1920 × 1080i	2200 × 1125i	16:9	29.97	74.176	HDTV	BT.1120 SMPTE 274M	BT.1120 SMPTE 292M
1920 × 1080i	2200 × 1125i	16:9	30	74.25	HDTV	BT.1120 SMPTE 274M	BT.1120 SMPTE 292M
1920 × 1080p	2200 × 1125p	16:9	59.94	148.35	HDTV	BT.1120 SMPTE 274M	–
1920 × 1080p	2200 × 1125p	16:9	60	148.5	HDTV	BT.1120 SMPTE 274M	–
1920 × 1080i	2376 × 1250i	16:9	25	74.25	HDTV	BT.1120	BT.1120
1920 × 1080p	2376 × 1250p	16:9	50	148.5	HDTV	BT.1120	–

Table 6.1. Pro-Video Parallel and Serial Digital Interface Standards for Various Component Video Formats. [1]i = interlaced, p = progressive.

The EAV and SAV sequences are shown in Table 6.2. The status word is defined as:

F = "0" for Field 1   F = "1" for Field 2
V = "1" during vertical blanking
H = "0" at SAV   H = "1" at EAV
P3–P0 = protection bits

> P3 = V ⊕ H
> P2 = F ⊕ H
> P1 = F ⊕ V
> P0 = F ⊕ V ⊕ H

where ⊕ represents the exclusive-OR function. These protection bits enable one- and two-bit errors to be detected and one-bit errors to be corrected at the receiver. For most progressive video systems, F is usually a "0" since there is no field information.

For 4:2:2 YCbCr data, after each SAV sequence, the stream of active data words always begins with a Cb sample, as shown in Figure 6.1. In the multiplexed sequence, the co-sited samples (those that correspond to the same point on the picture) are grouped as Cb, Y, Cr. During blanking intervals, unless ancillary data is present, 10-bit Y or R′G′B′ values should be set to 040_H and 10-bit CbCr values should be set to 200_H.

The receiver detects the EAV and SAV sequences by looking for the 8-bit FF_H 00_H 00_H preamble. The status word (optionally error corrected at the receiver, see Table 6.3) is used to recover the H, V, and F timing signals.

Ancillary Data

General Format

Ancillary data packets are used to transmit information (such as digital audio, closed captioning, and teletext data) during the blanking intervals. ITU-R BT.1364 and SMPTE 291M describe the ancillary data formats.

During horizontal blanking, ancillary data may be transmitted in the interval between the EAV and SAV sequences. During vertical blanking, ancillary data may be transmitted in the interval between the SAV and EAV sequences. Multiple ancillary packets may be present in a horizontal or vertical blanking interval, but they must be contiguous with each other. Ancillary data should not be present where indicated in Table 6.4 as these regions may be affected by video switching.

There are two types of ancillary data formats. The older Type 1 format uses a single data ID word to indicate the type of ancillary data; the newer Type 2 format uses two words for the data ID. The general packet format is shown in Table 6.5.

	8-bit Data								10-bit Data	
	D9 (MSB)	D8	D7	D6	D5	D4	D3	D2	D1	D0
preamble	1	1	1	1	1	1	1	1	1	1
	0	0	0	0	0	0	0	0	0	0
	0	0	0	0	0	0	0	0	0	0
status word	1	F	V	H	P3	P2	P1	P0	0	0

Table 6.2. EAV and SAV Sequence.

Received D5–D2	Received F, V, H (Bits D8–D6)							
	000	001	010	011	100	101	110	111
0000	000	000	000	*	000	*	*	111
0001	000	*	*	111	*	111	111	111
0010	000	*	*	011	*	101	*	*
0011	*	*	010	*	100	*	*	111
0100	000	*	*	011	*	*	110	*
0101	*	001	*	*	100	*	*	111
0110	*	011	011	011	100	*	*	011
0111	100	*	*	011	100	100	100	*
1000	000	*	*	*	*	101	110	*
1001	*	001	010	*	*	*	*	111
1010	*	101	010	*	101	101	*	101
1011	010	*	010	010	*	101	010	*
1100	*	001	110	*	110	*	110	110
1101	001	001	*	001	*	001	110	*
1110	*	*	*	011	*	101	110	*
1111	*	001	010	*	100	*	*	*

Notes:
* = uncorrectable error.

Table 6.3. SAV and EAV Error Correction at Decoder.

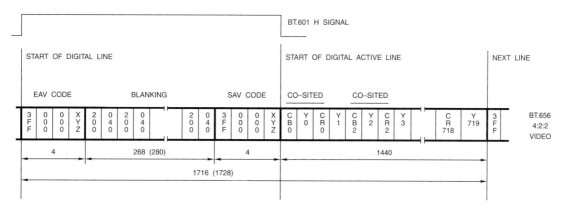

Figure 6.1. BT.656 Parallel Interface Data For One Scan Line. 480i; 4:2:2 YCbCr; 720 active samples per line; 27 MHz clock; 10-bit system. The values for 576i systems are shown in parentheses.

Video Standard	Line Numbers Affected
480i (525i)	10, 273
480p (525p)	10
576i (625i)	6, 319
576p (625ip	6
720p (750p)	7
1080i (1125i)	7, 569
1080p (1125p)	7

Table 6.4. Lines Affected by Switching.

Data ID (DID)

DID indicates the type of data being sent. The assignment of most of the DID values is controlled by the ITU and SMPTE to ensure equipment compatibility. A few DID values are available that don't require registration. Some DID values are listed in Table 6.6.

Secondary ID (SDID, Type 2 Only)

SDID is also part of the data ID for Type 2 ancillary formats. The assignment of most of the SDID values is also controlled by the ITU and SMPTE to ensure equipment compatibility. A few SDID values are available that don't require registration. Some SDID values are listed in Table 6.6.

Data Block Number (DBN, Type 1 Only)

DBN is used to allow multiple ancillary packets (sharing the same DID) to be put back together at the receiver. This is the case when there are more than 255 user data words required to be transmitted, thus requiring

more than one ancillary packet to be used. The DBN value increments by one for each consecutive ancillary packet.

Data Count (DC)

DC specifies the number of user data words in the packet. In 8-bit applications, it specifies the six MSBs of an 8-bit value, so the number of user data words must be an integral number of four.

User Data Words (UDW)

Up to 255 user data words may be present in the packet. In 8-bit applications, the number of user data words must be an integral number of four. Padding words may be added to ensure an integral number of four user data words are present.

User data may not use the 10-bit values of 000_H–003_H and $3FC_H$–$3FF_H$, or the 8-bit values of 00_H and FF_H, since they are used for timing information.

	8-bit Data								10-bit Data	
	D9 (MSB)	D8	D7	D6	D5	D4	D3	D2	D1	D0
ancillary data flag (ADF)	0	0	0	0	0	0	0	0	0	0
	1	1	1	1	1	1	1	1	1	1
	1	1	1	1	1	1	1	1	1	1
data ID (DID)	$\overline{D8}$	even parity	Value of 0000 0000 to 1111 1111							
data block number or SDID	$\overline{D8}$	even parity	Value of 0000 0000 to 1111 1111							
data count (DC)	$\overline{D8}$	even parity	Value of 0000 0000 to 1111 1111							
user data word 0	Value of 00 0000 0100 to 11 1111 1011									
	:									
user data word N	Value of 00 0000 0100 to 11 1111 1011									
check sum	$\overline{D8}$	Sum of D0–D8 of data ID through last user data word. Preset to all zeros; carry is ignored.								

Table 6.5. Ancillary Data Packet General Format.

8-bit DID Type 2	Function	8-bit DID Type 1	Function
00_H	undefined	80_H	marked for deletion
01_H–03_H	reserved	81_H–83_H	reserved
04_H, 08_H, $0C_H$	8-bit applications	84_H	end marker
10_H–$3F_H$	reserved	85_H–BF_H	reserved
40_H–$5F_H$	user application	$C0_H$–DF_H	user application
60_H	timecode	$E0_H$–EB_H	registered
61_H	closed captioning	EC_H	AES control packet, group 4
62_H–$7F_H$	registered	ED_H	AES control packet, group 3
		EE_H	AES control packet, group 2
		EF_H	AES control packet, group 1

8-bit SDID Type 2	Function	8-bit DID Type 1	Function
		$F4_H$	error detection
		$F5_H$	longitudinal timecode
00_H	undefined format	$F8_H$	AES extended packet, group 4
$x0_H$	8-bit applications	$F9_H$	AES audio data, group 4
$x4_H$	8-bit applications	FA_H	AES extended packet, group 3
$x8_H$	8-bit applications	FB_H	AES audio data, group 3
xC_H	8-bit applications	FC_H	AES extended packet, group 2
all others	unassigned	FD_H	AES audio data, group 2
		FE_H	AES extended packet, group 1
		FF_H	AES audio data, group 1

Table 6.6. DID and SDID Assignments.

Audio Sampling Rate (kHz)	Samples per Frame: 29.97 Hz Video					Samples per Frame: 25 Hz Video
	Samples per Frame	Samples per Field 1	Samples per Field 2	Exceptions: Frame Number	Exceptions: Number of Samples	
48.0	8008 / 5	1602	1601	–	–	1920
44.1	147147 / 100	1472	1471	23 47 71	1471 1471 1471	1764
32	16016 / 15	1068	1067	4 8 12	1068 1068 1068	1280

Table 6.7. Isochronous Audio Sample Rates.

Audio Format

ITU-R BT.1305 and SMPTE 272M describe the transmission of digital audio as ancillary data. 2–16 channels of up to 24-bit digital audio are supported, with sample rates of 32–48 kHz. Table 6.7 lists the number of audio samples per video frame for various audio sample rates.

Audio data of up to 20 bits per sample is transferred using the format in Table 6.8. "V" is the AES/EBU sample valid bit, "U" is the AES/EBU user bit, and "C" is the AES/EBU audio channel status bit. "P" is an even parity bit for the 26 previous bits in the sample (excluding D9 in the first and second words of the audio sample). Audio is represented as two's complement linear PCM data.

To support 24-bit audio samples, extended data packets may be used to transfer the four auxiliary bits of the AES/EBU audio stream.

Audio data is formatted as 1–4 groups, defined by [gr 1] and [gr 0], with each group having 1–4 channels of audio data, defined by [ch 1] and [ch 0].

Optional control packets may be used on lines 12 and 275 (480i systems) or lines 8 and 320 (576i systems) to specify the sample rate, delay relative to the video, etc. If present, it must be transmitted prior to any audio packets. If not transmitted, a default condition of 48 kHz isochronous audio is assumed.

Timecode Format

ITU-R BT.1366 and SMPTE RP-188 define the transmission of timecode using ancillary data for 480i, 576i, and 1080i systems. The ancillary packet format is shown in Table 6.9, and is used to convey longitudinal (LTC) or vertical interval timecode (VITC) information. For additional information on the timecode format, and the meaning of the flags in Table 6.9, see the timecode discussion in Chapter 8.

VITC 1 and VITC 2 timecode packets for 480p systems use lines 27 and 28, respectively. LTC, VITC 1 and VITC 2 timecode packets for 1080i systems use lines 10, 9 and 571, respectively.

	8-bit Data								10-bit Data	
	D9 (MSB)	D8	D7	D6	D5	D4	D3	D2	D1	D0
ancillary data flag (ADF)	0	0	0	0	0	0	0	0	0	0
	1	1	1	1	1	1	1	1	1	1
	1	1	1	1	1	1	1	1	1	1
data ID (DID)	$\overline{D8}$	even parity	1	1	1	1	1	gr 1	gr 0	1
data block number (DBN)	$\overline{D8}$	even parity	Value of 0000 0000 to 1111 1111							
data count (DC)	$\overline{D8}$	even parity	Value of 0000 0000 to 1111 1111							
audio sample 0	$\overline{D8}$	A5	A4	A3	A2	A1	A0	ch 1	ch 0	Z
	$\overline{D8}$	A14	A13	A12	A11	A10	A9	A8	A7	A6
	$\overline{D8}$	P	C	U	V	A19	A18	A17	A16	A15
					:					
audio sample N	$\overline{D8}$	A5	A4	A3	A2	A1	A0	ch 1	ch 0	Z
	$\overline{D8}$	A14	A13	A12	A11	A10	A9	A8	A7	A6
	$\overline{D8}$	P	C	U	V	A19	A18	A17	A16	A15
check sum	$\overline{D8}$	Sum of D0–D8 of data ID through last audio sample word. Preset to all zeros; carry is ignored.								

Table 6.8. Digital Audio Ancillary Data Packet Format.

	8-bit Data								10-bit Data	
	D9 (MSB)	D8	D7	D6	D5	D4	D3	D2	D1	D0
ancillary data flag (ADF)	0	0	0	0	0	0	0	0	0	0
	1	1	1	1	1	1	1	1	1	1
	1	1	1	1	1	1	1	1	1	1
data ID (DID)	$\overline{D8}$	EP	0	1	1	0	0	0	0	0
SDID	$\overline{D8}$	EP	0	1	1	0	0	0	0	0
data count (DC)	$\overline{D8}$	EP	0	0	0	1	0	0	0	0
timecode data	$\overline{D8}$	EP	units of frames				DBB10	0	0	0
	$\overline{D8}$	EP	user group 1				DBB11	0	0	0
	$\overline{D8}$	EP	flag 2	flag 1	tens of frames		DBB12	0	0	0
	$\overline{D8}$	EP	user group 2				DBB13	0	0	0
	$\overline{D8}$	EP	units of seconds				DBB14	0	0	0
	$\overline{D8}$	EP	user group 3				DBB15	0	0	0
	$\overline{D8}$	EP	flag 3	tens of seconds			DBB16	0	0	0
	$\overline{D8}$	EP	user group 4				DBB17	0	0	0
	$\overline{D8}$	EP	units of minutes				DBB20	0	0	0
	$\overline{D8}$	EP	user group 5				DBB21	0	0	0
	$\overline{D8}$	EP	flag 4	tens of minutes			DBB22	0	0	0
	$\overline{D8}$	EP	user group 6				DBB23	0	0	0
	$\overline{D8}$	EP	units of hours				DBB24	0	0	0
	$\overline{D8}$	EP	user group 7				DBB25	0	0	0
	$\overline{D8}$	EP	flag 6	flag 5	tens of hours		DBB26	0	0	0
	$\overline{D8}$	EP	user group 8				DBB27	0	0	0
check sum	$\overline{D8}$	Sum of D0–D8 of data ID through last timecode data word. Preset to all zeros; carry is ignored.								

Notes:
EP = even parity for D0–D7.

Table 6.9. Timecode Ancillary Data Packet Format.

DBB17	DBB16	DBB15	DBB14	DBB13	DBB12	DBB11	DBB10	Definition
0	0	0	0	0	0	0	0	LTC
0	0	0	0	0	0	0	1	VITC 1
0	0	0	0	0	0	1	0	VITC 2
0	0	0	0	0	0	1	1	user defined
				:				
0	0	0	0	0	1	0	1	
0	0	0	0	0	1	1	0	film data block
0	0	0	0	0	1	1	1	production data block
0	0	0	0	1	0	0	0	locally generated time address and user data
				:				
0	1	1	1	1	1	0	0	
0	1	1	1	1	1	0	1	video tape data block
0	1	1	1	1	1	1	0	film data block
0	1	1	1	1	1	1	1	production data block
1	0	0	0	0	0	0	0	reserved
				:				
1	1	1	1	1	1	1	1	

Table 6.10. Binary Bit Group 1 Definitions.

Binary Bit Group 1

The eight bits that comprise binary bit group 1 (DBB10–DBB17) specify the type of timecode and user data, as shown in Table 6.10.

Binary Bit Group 2

The eight bits that comprise binary bit group 2 (DBB20–DBB27) specify line numbering and status information.

DBB20–DBB24 specify the VITC line select as shown in Table 6.11. These convey the VITC line number location.

If DBB25 is a "1," when the timecode information is converted into an analog VITC signal on line N, it must also be repeated on line N + 2.

If DBB26 is a "1," a timecode error was received, and the transmitted timecode has been interpolated from a previous timecode.

DBB24	DBB23	DBB22	DBB21	DBB20	480i Systems		576i Systems	
					VITC on Line N	VITC on Line N + 2	VITC on Line N	VITC on Line N + 2
0	0	1	1	0	–	–	6, 319	8, 321
0	0	1	1	1	–	–	7, 320	9, 322
0	1	0	0	0	–	–	8, 321	10, 323
0	1	0	0	1	–	–	9, 322	11, 324
0	1	0	1	0	10, 273	12, 275	10, 323	12, 325
0	1	0	1	1	11, 274	13, 276	11, 324	13, 326
0	1	1	0	0	12. 275	14, 277	12, 325	14, 327
0	1	1	0	1	13, 276	15, 278	13, 326	15, 328
0	1	1	1	0	14, 277	16, 279	14, 327	16, 329
0	1	1	1	1	15, 278	17, 280	15, 328	17, 330
1	0	0	0	0	16, 279	18, 281	16, 329	18, 331
1	0	0	0	1	17, 280	19, 282	17, 330	19, 332
1	0	0	1	0	18, 281	20, 283	18, 331	20, 333
1	0	0	1	1	19, 282	–	19, 332	21, 334
1	0	1	0	0	20, 283	–	20, 333	22, 335
1	0	1	0	1	–	–	21, 334	–
1	0	1	1	0	–	–	22, 335	–

Table 6.11. VITC Line Select Definitions for 480i and 576i Systems.

If DBB27 is a "0," the user group bits are processed to compensate for any latency. If a "1," the user bits are retransmitted with no delay compensation.

User Group Bits

32 bits of user data may be transferred with each timecode packet. User data is organized as eight groups of four bits each, with the D7 bit being the MSB. For additional information on user bits, see the timecode discussion in Chapter 8.

SMPTE 266M

SMPTE 266M also defines a digital vertical interval timecode (DVITC) for 480i and 576i systems. It is an 8-bit digital representation of the analog VITC signal, transferred using the 8 MSBs. If the VITC is present, it is carried on the Y data channel in the active portion of lines 14 and 277 (and optionally lines 16 and 279) for 480i systems; lines 19 and 332 (and optionally lines 21 and 334) are used for 576i systems.

The 90 bits of VITC information are carried by 675 consecutive Y samples. A 10-bit value of 040_H represents a "0;" a 10-bit value of 300_H represents a "1." Unused Y samples have a value of 040_H.

Closed Captioning Format

ITU-R BT.1619 and SMPTE 334M define the ancillary packet format for EIA-608 closed captioning, as shown in Table 6.12.

The field bit is a "0" for Field 2 and a "1" for Field 1.

The offset value is a 5-bit unsigned integer which represents the offset (in lines) of the data insertion line, relative to line 9 or 272 for 480i systems and line 5 or 318 for 576i systems.

ITU-R BT.1619 and SMPTE 334M also define the ancillary packet format for EIA-708 digital closed captioning, as shown in Table 6.13. The payload is the EIA-708 caption distribution packet (CDP), which has a variable length.

Error Detection Checksum Format

ITU-R BT.1304 and SMPTE RP-165 define a checksum for error detection. The ancillary packet format is shown in Table 6.14.

For 13.5 MHz 480i systems, the ancillary packet occupies sample words 1689–1711 on lines 9 and 272 (sample words 2261–2283 for 18 MHz 480i systems). For 13.5 MHz 576i systems, the ancillary packet occupies sample words 1701–1723 on lines 5 and 318 (sample words 2277–2299 for 18 MHz 576i systems). Note that these locations are immediately prior to the SAV code words.

Checksums

Two checksums are provided: one for a field of active video data and one for a full field of data. Each checksum is a 16-bit value calculated as follows:

$$CRC = x^{16} + x^{12} + x^5 + x^1$$

For the active CRC, the starting and ending samples for 13.5 MHz 480i systems are sample word 0 on lines 21 and 284 (start) and sample word 1439 on lines 262 and 525 (end). The starting and ending samples for 13.5 MHz 576i systems are sample word 0 on lines 24 and 336 (start) and sample word 1439 on lines 310 and 622 (end).

For the field CRC, the starting and ending samples for 13.5 MHz 480i systems are sample word 1444 on lines 12 and 275 (start) and sam-

	8-bit Data								10-bit Data	
	D9 (MSB)	D8	D7	D6	D5	D4	D3	D2	D1	D0
ancillary data flag (ADF)	0	0	0	0	0	0	0	0	0	0
	1	1	1	1	1	1	1	1	1	1
	1	1	1	1	1	1	1	1	1	1
data ID (DID)	$\overline{D8}$	EP	0	1	1	0	0	0	0	1
SDID	$\overline{D8}$	EP	0	0	0	0	0	0	1	0
data count (DC)	$\overline{D8}$	EP	0	0	0	0	0	0	1	1
line	$\overline{D8}$	EP	field	0	0	offset				
caption word 0	$\overline{D8}$	EP	D07	D06	D05	D04	D03	D02	D01	D00
caption word 1	$\overline{D8}$	EP	D17	D16	D15	D14	D13	D12	D11	D10
check sum	$\overline{D8}$	Sum of D0–D8 of data ID through last caption word. Preset to all zeros; carry is ignored.								

Notes:
EP = even parity for D0–D7.

Table 6.12. EIA-608 Closed Captioning Ancillary Data Packet Format.

ple word 1439 on lines 8 and 271 (end). The starting and ending samples for 13.5 MHz 576i systems are sample word 1444 on lines 8 and 321 (start) and sample word 1439 on lines 4 and 317 (end).

Error Flags

Error flags indicate the status of the previous field.

edh (error detected here): A "1" indicates that a transmission error was detected since one or more ancillary packets did not match its checksum.

eda (error detected already): A "1" indicates a transmission error was detected at a prior point in the data path. A device that receives data with this flag set should forward the data with the flag set and the *edh* flag reset to "0" if no further errors are detected.

idh (internal error detected here): A "1" indicates that an error unrelated to the transmission has been detected.

ida (internal error status): A "1" indicates data was received from a device that does not support this error detection method.

	8-bit Data								10-bit Data	
	D9 (MSB)	D8	D7	D6	D5	D4	D3	D2	D1	D0
ancillary data flag (ADF)	0	0	0	0	0	0	0	0	0	0
	1	1	1	1	1	1	1	1	1	1
	1	1	1	1	1	1	1	1	1	1
data ID (DID)	$\overline{D8}$	EP	0	1	1	0	0	0	0	1
SDID	$\overline{D8}$	EP	0	0	0	0	0	0	0	1
data count (DC)	$\overline{D8}$	EP	Value of 0000 0000 to 1111 1111							
data word 0	$\overline{D8}$	EP	Value of 0000 0000 to 1111 1111							
	:									
data word N	$\overline{D8}$	EP	Value of 0000 0000 to 1111 1111							
check sum	$\overline{D8}$	Sum of D0–D8 of data ID through last data word. Preset to all zeros; carry is ignored.								

Notes:
EP = even parity for D0–D7.

Table 6.13. EIA-708 Digital Closed Captioning Ancillary Data Packet Format.

	8-bit Data								10-bit Data	
	D9 (MSB)	D8	D7	D6	D5	D4	D3	D2	D1	D0
ancillary data flag (ADF)	0	0	0	0	0	0	0	0	0	0
	1	1	1	1	1	1	1	1	1	1
	1	1	1	1	1	1	1	1	1	1
data ID (DID)	$\overline{D8}$	EP	1	1	1	1	0	1	0	0
SDID	$\overline{D8}$	EP	0	0	0	0	0	0	0	0
data count (DC)	$\overline{D8}$	EP	0	0	0	1	0	0	0	0
active CRC	$\overline{D8}$	EP	crc5	crc4	crc3	crc2	crc1	crc0	0	0
	$\overline{D8}$	EP	crc11	crc10	crc9	crc8	crc7	crc6	0	0
	$\overline{D8}$	EP	V	0	crc15	crc14	crc13	crc12	0	0
field CRC	$\overline{D8}$	EP	crc5	crc4	crc3	crc2	crc1	crc0	0	0
	$\overline{D8}$	EP	crc11	crc10	crc9	crc8	crc7	crc6	0	0
	$\overline{D8}$	EP	V	0	crc15	crc14	crc13	crc12	0	0
ancillary flags	$\overline{D8}$	EP	0	ues	ida	idh	eda	edh	0	0
active flags	$\overline{D8}$	EP	0	ues	ida	idh	eda	edh	0	0
field flags	$\overline{D8}$	EP	0	ues	ida	idh	eda	edh	0	0
reserved	1	0	0	0	0	0	0	0	0	0
	1	0	0	0	0	0	0	0	0	0
	1	0	0	0	0	0	0	0	0	0
	1	0	0	0	0	0	0	0	0	0
	1	0	0	0	0	0	0	0	0	0
	1	0	0	0	0	0	0	0	0	0
	1	0	0	0	0	0	0	0	0	0
check sum	$\overline{D8}$	Sum of D0–D8 of data ID through last reserved word. Preset to all zeros; carry is ignored.								

Notes:
EP = even parity for D0–D7.

Table 6.14. Error Detection Ancillary Data Packet Format.

VBI Data Service

The VBI data service (SMPTE RP-208) is intended for use in reconstructing VBI data in a standard-definition analog video signal produced from the digital video program. Ancillary data packets contain the sampled VBI data, as seen in Table 6.15.

The *field* bit is "0" for Field 2 and "1" for Field 1.

The *offset* value is a 5-bit unsigned integer which represents the offset (in lines) of the data insertion line, relative to line 9 or 272 for 480i systems and line 5 or 318 for 576i systems.

The *type* value identifies the VBI data content, with EIA-516 (DC = 36_D; 34 data bytes starting with byte sync value), Guide Plus + (DC = 6_D; 4 data bytes), AMOL (DC = 8_D; 6 data bytes) and AMOL II (DC = 14_D; 12 data bytes) already defined.

	8-bit Data								10-bit Data	
	D9 (MSB)	D8	D7	D6	D5	D4	D3	D2	D1	D0
ancillary data flag (ADF)	0	0	0	0	0	0	0	0	0	0
	1	1	1	1	1	1	1	1	1	1
	1	1	1	1	1	1	1	1	1	1
data ID (DID)	$\overline{D8}$	EP	0	1	1	0	0	0	0	1
SDID	$\overline{D8}$	EP	0	0	0	0	0	0	1	0
data count (DC)	$\overline{D8}$	EP	0	0	0	0	0	0	1	1
line	$\overline{D8}$	EP	field	0	0	offset				
type	$\overline{D8}$	EP	type							
data word 0	$\overline{D8}$	EP	Value of 0000 0000 to 1111 1111							
:										
data word N	$\overline{D8}$	EP	Value of 0000 0000 to 1111 1111							
check sum	$\overline{D8}$	Sum of D0–D8 of data ID through last data word. Preset to all zeros; carry is ignored.									

Notes:
EP = even parity for D0–D7.

Table 6.15. VBI Data Service Ancillary Packet Format.

Video Index Format

If the video index for 480i or 480p systems (SMPTE RP-186) is present, it is carried on the CbCr data channels in the active portion of lines 14 and 277 for 480i systems and lines 27 and 28 for 480p systems.

A total of 90 8-bit data words are transferred serially by D2 of the 720 CbCr samples of the active portion of the lines. A 10-bit value of 200_H represents a "0;" a 10-bit value of 204_H represents a "1." Unused CbCr samples have a 10-bit value of 200_H.

Video Payload Identification Format

ITU-R BT.1614 and SMPTE 352M define a 4-byte payload identifier which may be used for identifying the video payload.

For 480i digital interfaces, lines 13 and 276 are used to carry the ancillary data packet. For 480p digital interfaces, line 13 is used to carry the ancillary data packet.

For 576i digital interfaces, lines 9 and 322 are used to carry the ancillary data packet. For 576p digital interfaces, line 9 is used to carry the ancillary data packet.

	8-bit Data								10-bit Data	
	D9 (MSB)	D8	D7	D6	D5	D4	D3	D2	D1	D0
ancillary data flag (ADF)	0	0	0	0	0	0	0	0	0	0
	1	1	1	1	1	1	1	1	1	1
	1	1	1	1	1	1	1	1	1	1
data ID (DID)	$\overline{D8}$	EP	0	1	0	0	0	0	0	1
SDID	$\overline{D8}$	EP	0	0	0	0	0	0	0	1
data count (DC)	$\overline{D8}$	EP	0	0	0	0	0	1	0	0
video payload	$\overline{D8}$	EP	version ID	payload ID						
picture rate	$\overline{D8}$	EP	I/P transport	I/P picture	0	0	picture rate			
sampling structure	$\overline{D8}$	EP	16:9	0	0	0	sampling structure			
other info	$\overline{D8}$	EP	channel		0	dynamic range		0	bit depth	
check sum	$\overline{D8}$	Sum of D0–D8 of data ID through other info word. Preset to all zeros; carry is ignored.								

Notes:
EP = even parity for D0–D7.

Table 6.16. Video Payload ID Ancillary Packet Format.

For 720p digital interfaces, line 10 is used to carry the ancillary data packet.

For 1080i digital interfaces, lines 10 and 572 are used to carry the ancillary data packet. For 1080p digital interfaces, line 10 is used to carry the ancillary data packet.

The *version ID* bit is "1" if based on SMPTE 352M-2002; "0" if based on SMPTE 352M as trial published in the July, 2001 issue of the SMPTE Journal.

The *I/P transport* bit is provided for rapid and reliable detection of the transport scanning structure. The scanning structure can be determined through the "F" bit, which has a static "0" value for progressive transports, and toggles between "0" (field 1) and "1" (field 2) for interlaced transports. However, this detection may take several transport frames to ensure accuracy. The *I/P transport* bit can provide the information on a per-frame basis.

The *I/P picture* bit is used to identify whether the picture has been scanned as progressive ("1") or interlaced ("0").

The *picture rate* bits indicate the picture frame rate in Hz, as specified in Table 6.17.

The *16:9* bit is used to identify whether the picture aspect ratio is 4:3 ("0") or 16:9 ("1").

The *sampling structure* bits indicate the YCbCr or RGB sampling structure of the picture, as specified in Table 6.18.

The *channel* bits indicate channel identification information:

"00" = single channel, or channel 1 of a multi-channel, video payload

"01" = channel 2 of a multi-channel video payload

"10" = channel 3 of a multi-channel video payload

"11" = channel 4 of a multi-channel video payload

The *dynamic range* bits identify the dynamic range of the sample quantization:

"00" = quantization value range normal

"01" = quantization value range extended to 200% of normal (1 overhead bit in the MSB location)

"10" = quantization value range extended to 400% of normal (2 overhead bits in the MSB locations)

"11" = reserved

The *bit depth* bits identify the bit depth of the sample quantization:

"00" = 8 bits per sample

"01" = 10 bits per sample

"10" = 12 bits per sample

"11" = reserved

Picture Rate Code (D3-D0)	Picture Rate Value	Picture Rate Code (D3-D0)	Picture Rate Value	Picture Rate Code (D3-D0)	Picture Rate Value	Picture Rate Code (D3-D0)	Picture Rate Value
0000	not defined	0100	reserved	1000	reserved	1100	reserved
0001	reserved	0101	25	1001	50	1101	reserved
0010	24/1.001	0110	30/1.001	1010	60/1.001	1110	reserved
0011	24	0111	30	1011	60	1111	reserved

Table 6.17. *Picture Rate* **Values.**

Sampling Structure Code (D3-D0)	Sampling Structure Format	Sampling Structure Code (D3-D0)	Sampling Structure Format	Sampling Structure Code (D3-D0)	Sampling Structure Format	Sampling Structure Code (D3-D0)	Sampling Structure Format
0000	4:2:2 YCbCr	0100	4:2:2:4 YCbCrA	1000	4:2:2:4 YCbCrD	1100	reserved
0001	4:4:4 YCbCr	0101	4:4:4:4 YCbCrA	1001	4:4:4:4 YCbCrD	1101	reserved
0010	4:4:4 GBR	0110	4:4:4:4 GBRA	1010	4:4:4:4 GBRD	1110	reserved
0011	4:2:0 YCbCr	0111	reserved	1011	reserved	1111	reserved

Notes:
"A" notation refers to picture channel, "D" notation refers to a non-picture (i.e. data) channel.

Table 6.18. *Sampling Structure* **Formats.**

Pin	Signal	Pin	Signal
1	clock	14	clock–
2	system ground A	15	system ground B
3	D9	16	D9–
4	D8	17	D8–
5	D7	18	D7–
6	D6	19	D6–
7	D5	20	D5–
8	D4	21	D4–
9	D3	22	D3–
10	D2	23	D2–
11	D1	24	D1–
12	D0	25	D0–
13	cable shield		

Table 6.19. 25-Pin Parallel Interface Connector Pin Assignments. For 8-bit interfaces, D9–D2 are used.

25-pin Parallel Interface

This interface is used to transfer SDTV resolution 4:2:2 YCbCr data. 8-bit or 10-bit data and a clock are transferred. The individual bits are labeled D0–D9, with D9 being the most significant bit. The pin allocations for the signals are shown in Table 6.19.

Y has a nominal 10-bit range of 040_H–$3AC_H$. Values less than 040_H or greater than $3AC_H$ may be present due to processing. During blanking, Y data should have a value of 040_H, unless other information is present.

Cb and Cr have a nominal 10-bit range of 040_H–$3C0_H$. Values less than 040_H or greater than $3C0_H$ may be present due to processing. During blanking, CbCr data should have a value of 200_H, unless other data is present.

Signal levels are compatible with ECL-compatible balanced drivers and receivers. The generator must have a balanced output with a maximum source impedance of 110 Ω; the signal must be 0.8–2.0V peak-to-peak measured across a 110-Ω load. At the receiver, the transmission line is terminated by 110 ±10 Ω.

27 MHz Parallel Interface

This BT.656 and SMPTE 125M interface is used for 480i and 576i systems with an aspect ratio of 4:3. Y and multiplexed CbCr information at a sample rate of 13.5 MHz are multiplexed into a single 8-bit or 10-bit data stream, at a clock rate of 27 MHz.

The 27 MHz clock signal has a clock pulse width of 18.5 ±3 ns. The positive transition of the clock signal occurs midway between data transitions with a tolerance of ±3 ns (as shown in Figure 6.2).

To permit reliable operation at interconnect lengths of 50–200 meters, the receiver must use frequency equalization, with typical

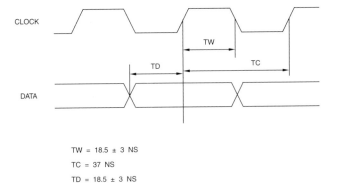

CLOCK

DATA

TW = 18.5 ± 3 NS

TC = 37 NS

TD = 18.5 ± 3 NS

Figure 6.2. 25-Pin 27 MHz Parallel Interface Waveforms.

RELATIVE GAIN (DB)

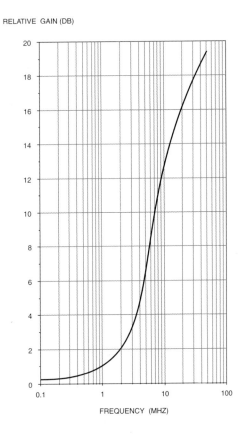

FREQUENCY (MHZ)

Figure 6.3. Example Line Receiver Equalization Characteristics for Small Signals.

characteristics shown in Figure 6.3. This example enables operation with a range of cable lengths down to zero.

36 MHz Parallel Interface

This BT.1302 and SMPTE 267M interface is used for 480i and 576i systems with an aspect ratio of 16:9. Y and multiplexed CbCr information at a sample rate of 18 MHz are multiplexed into a single 8-bit or 10-bit data stream, at a clock rate of 36 MHz.

The 36 MHz clock signal has a clock pulse width of 13.9 ±2 ns. The positive transition of the clock signal occurs midway between data transitions with a tolerance of ±2 ns (as shown in Figure 6.4.

To permit reliable operation at interconnect lengths of 40–160 meters, the receiver must use frequency equalization, with typical characteristics shown in Figure 6.3.

93-pin Parallel Interface

This interface is used to transfer 16:9 HDTV resolution R′G′B′ data, 4:2:2 YCbCr data, or 4:2:2:4 YCbCrK data. The pin allocations for the signals are shown in Table 6.20. The most significant bits are R9, G9, and B9.

When transferring 4:2:2 YCbCr data, the green channel carries Y information and the red channel carries multiplexed CbCr information.

When transferring 4:2:2:4 YCbCrK data, the green channel carries Y information, the red channel carries multiplexed CbCr information, and the blue channel carries K (alpha keying) information.

Y has a nominal 10-bit range of 040_H–$3AC_H$. Values less than 040_H or greater than $3AC_H$ may be present due to processing. During blanking, Y data should have a value of 040_H, unless other information is present.

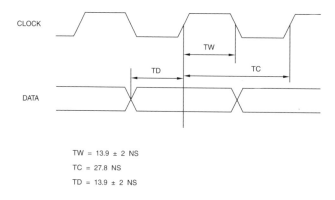

TW = 13.9 ± 2 NS
TC = 27.8 NS
TD = 13.9 ± 2 NS

Figure 6.4. 25-Pin 36 MHz Parallel Interface Waveforms.

Pin	Signal	Pin	Signal	Pin	Signal	Pin	Signal
1	clock	26	GND	51	B2	76	GND
2	G9	27	GND	52	B1	77	GND
3	G8	28	GND	53	B0	78	GND
4	G7	29	GND	54	R9	79	B4–
5	G6	30	GND	55	R8	80	B3–
6	G5	31	GND	56	R7	81	B2–
7	G4	32	GND	57	R6	82	B1–
8	G3	33	clock–	58	R5	83	B0–
9	G2	34	G9–	59	R4	84	R9–
10	G1	35	G8–	60	R3	85	R8–
11	G0	36	G7–	61	R2	86	R7–
12	B9	37	G6–	62	R1	87	R6–
13	B8	38	G5–	63	R0	88	R5–
14	B7	39	G4–	64	GND	89	R4–
15	B6	40	G3–	65	GND	90	R3–
16	B5	41	G2–	66	GND	91	R2–
17	GND	42	G1–	67	GND	92	R1–
18	GND	43	G0–	68	GND	93	R0–
19	GND	44	B9–	69	GND		
20	GND	45	B8–	70	GND		
21	GND	46	B7–	71	GND		
22	GND	47	B6–	72	GND		
23	GND	48	B5–	73	GND		
24	GND	49	B4	74	GND		
25	GND	50	B3	75	GND		

Table 6.20. 93-Pin Parallel Interface Connector Pin Assignments. For 8-bit interfaces, bits 9–2 are used.

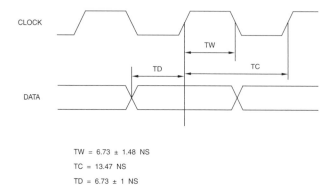

TW = 6.73 ± 1.48 NS

TC = 13.47 NS

TD = 6.73 ± 1 NS

Figure 6.5. 93-Pin 74.25 MHz Parallel Interface Waveforms.

Cb and Cr have a nominal 10-bit range of 040_H–$3C0_H$. Values less than 040_H or greater than $3C0_H$ may be present due to processing. During blanking, CbCr data should have a value of 200_H, unless other information is present.

R'G'B' and K have a nominal 10-bit range of 040_H–$3AC_H$. Values less than 040_H or greater than $3AC_H$ may be present due to processing. During blanking, R'G'B' data should have a value of 040_H, unless other information is present.

Signal levels are compatible with ECL-compatible balanced drivers and receivers. The generator must have a balanced output with a maximum source impedance of 110 Ω; the signal must be 0.6–2.0V peak-to-peak measured across a 110-Ω load. At the receiver, the transmission line must be terminated by 110 ±10 Ω

74.25 MHz Parallel Interface

This ITU-R BT.1120 and SMPTE 274M interface is primarily used for 16:9 HDTV systems.

The 74.25 MHz clock signal has a clock pulse width of 6.73 ±1.48 ns. The positive transition of the clock signal occurs midway between data transitions with a tolerance of ±1 ns (as shown in Figure 6.5).

To permit reliable operation at interconnect lengths greater than 20 meters, the receiver must use frequency equalization.

74.176 MHz Parallel Interface

This BT.1120 and SMPTE 274M interface is primarily used for 16:9 HDTV systems.

The 74.176 MHz (74.25/1.001) clock signal has a clock pulse width of 6.74 ±1.48 ns. The positive transition of the clock signal occurs midway between data transitions with a tolerance of ±1 ns (similar to Figure 6.5).

To permit reliable operation at interconnect lengths greater than 20 meters, the receiver must use frequency equalization.

148.5 MHz Parallel Interface

This BT.1120 and SMPTE 274M interface is used for 16:9 HDTV systems.

The 148.5 MHz clock signal has a clock pulse width of 3.37 ±0.74 ns. The positive transition of the clock signal occurs midway between data transitions with a tolerance of ±0.5 ns (similar to Figure 6.5).

To permit reliable operation at interconnect lengths greater than 14 meters, the receiver must use frequency equalization.

148.35 MHz Parallel Interface

This BT.1120 and SMPTE 274M interface is used for 16:9 HDTV systems.

The 148.35 MHz (148.5/1.001) clock signal has a clock pulse width of 3.37 ±0.74 ns. The positive transition of the clock signal occurs midway between data transitions with a tolerance of ±0.5 ns (similar to Figure 6.5).

To permit reliable operation at interconnect lengths greater than 14 meters, the receiver must use frequency equalization.

Serial Interfaces

The parallel formats can be converted to a serial format (Figure 6.6), allowing data to be transmitted using a 75-Ω coaxial cable (or optical fiber). Equipment inputs and outputs both use BNC connectors so that interconnect cables can be used in either direction.

For cable interconnect, the generator has an unbalanced output with a source impedance of 75Ω; the signal must be 0.8V ±10% peak-to-peak measured across a 75-Ω load. The receiver has an input impedance of 75Ω.

In an 8-bit environment, before serialization, the 00_H and FF_H codes during EAV and SAV are expanded to 10-bit values of 000_H and $3FF_H$, respectively. All other 8-bit data is appended with two least significant "0" bits before serialization.

The 10 bits of data are serialized (LSB first) and processed using a scrambled and polarity-free NRZI algorithm:

$$G(x) = (x^9 + x^4 + 1)(x + 1)$$

The input signal to the scrambler (Figure 6.7) uses positive logic (the highest voltage represents a logical one; lowest voltage represents a logical zero).

The formatted serial data is output at the 10× sample clock rate. Since the parallel clock may contain large amounts of jitter, deriving the 10× sample clock directly from an unfiltered parallel clock may result in excessive signal jitter.

At the receiver, phase-lock synchronization is done by detecting the EAV and SAV sequences. The PLL is continuously adjusted slightly each scan line to ensure that these patterns are detected and to avoid bit slippage. The recovered 10× sample clock is divided by ten to generate the sample clock, although care must be taken not to mask word-related jitter components. The serial data is low- and high-frequency equalized, inverse scrambling performed (Figure 6.8), and deserialized.

270 Mbps Serial Interface

This BT.656 and SMPTE 259M interface (also called SDI) converts a 27 MHz parallel stream into a 270 Mbps serial stream. The 10× PLL generates a 270 MHz clock from the 27 MHz clock signal. This interface is primarily used for 480i and 576i 4:3 systems.

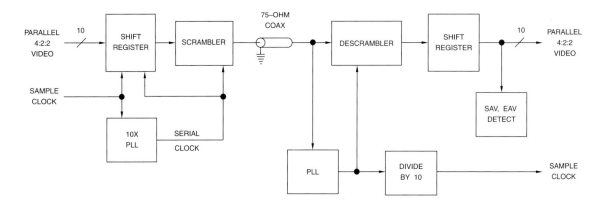

Figure 6.6. Serial Interface Block Diagram.

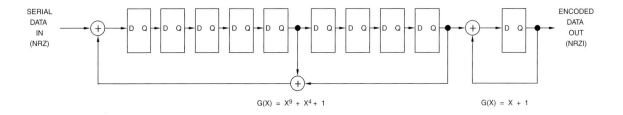

$$G(X) = X^9 + X^4 + 1 \qquad\qquad G(X) = X + 1$$

Figure 6.7. Typical Scrambler Circuit.

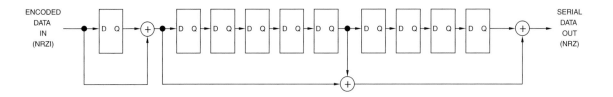

Figure 6.8. Typical Descrambler Circuit.

360 Mbps Serial Interface

This BT.1302 and SMPTE 259M interface converts a 36 MHz parallel stream into a 360 Mbps serial stream. The 10× PLL generates a 360 MHz clock from the 36 MHz clock signal. This interface is primarily used for 480i and 576i 16:9 systems.

540 Mbps Serial Interface

This SMPTE 344M interface converts a 54 MHz parallel stream, or two 27 MHz parallel streams, into a 540 Mbps serial stream. The 10× PLL generates a 540 MHz clock from the 54 MHz clock signal. This interface is primarily used for 480p and 576p 4:3 systems.

1.485 Gbps Serial Interface

This BT.1120 and SMPTE 292M interface multiplexes two 74.25 MHz parallel streams (Y and CbCr) into a single 1.485 Gbps serial stream. A 20× PLL generates a 1.485 GHz clock from the 74.25 MHz clock signal. This interface is used for 16:9 HDTV systems.

Before multiplexing the two parallel streams together, line number and CRC information (Table 6.21) is added to each stream after each EAV sequence. The CRC is used to detect errors in the active video and EAV. It consists of two words generated by the polynomial:

$$CRC = x^{18} + x^5 + x^4 + 1$$

The initial value is set to zero. The calculation starts with the first active line word and ends at the last word of the line number (LN1).

1.4835 Gbps Serial Interface

This BT.1120 and SMPTE 292M interface multiplexes two 74.176 (74.25/1.001) MHz parallel streams (Y and CbCr) into a single 1.4835 (1.485/1.001) Gbps serial stream. A 20× PLL generates a 1.4835 GHz clock from the 74.176 MHz clock signal. This interface is used for 16:9 HDTV systems.

Line number and CRC information is added as described for the 1.485 Gbps serial interface.

	D9 (MSB)	D8	D7	D6	D5	D4	D3	D2	D1	D0
LN0	$\overline{D8}$	L6	L5	L4	L3	L2	L1	L0	0	0
LN1	$\overline{D8}$	0	0	0	L10	L9	L8	L7	0	0
CRC0	$\overline{D8}$	crc8	crc7	crc6	crc5	crc4	crc3	crc2	crc1	crc0
CRC1	$\overline{D8}$	crc17	crc16	crc15	crc14	crc13	crc12	crc11	crc10	crc9

Table 6.21. Line Number and CRC Data.

SDTV—Interlaced

Supported active resolutions, with their corresponding aspect ratios and frame refresh rates, are:

720 × 480i	4:3	29.97 Hz
720 × 576i	4:3	25.00 Hz
960 × 480i	16:9	29.97 Hz
960 × 576i	16:9	25.00 Hz

4:2:2 YCbCr Parallel Interface

The ITU-R BT.656 and BT.1302 parallel interfaces were developed to transfer BT.601 4:2:2 YCbCr digital video between equipment. SMPTE 125M and 267M further clarify the operation for 480i systems.

Figure 6.9 illustrates the timing for one scan line for the 4:3 aspect ratio, using a 27 MHz sample clock. Figure 6.10 shows the timing for one scan line for the 16:9 aspect ratio, using a 36 MHz sample clock. The 25-pin parallel interface is used.

4:2:2 YCbCr Serial Interface

BT.656 and BT.1302 also define a YCbCr serial interface. The 10-bit 4:2:2 YCbCr parallel streams shown in Figures 6.9 and 6.10 are serialized using the 270 or 360 Mbps serial interface.

4:4:4:4 YCbCrK Parallel Interface

The ITU-R BT.799 and BT.1303 parallel interfaces were developed to transfer BT.601 4:4:4:4 YCbCrK digital video between equipment. K is an alpha keying signal, used to mix two video sources, discussed in Chapter 7. SMPTE RP-175 further clarifies the operation for 480i systems.

Multiplexing Structure

Two transmission links are used. Link A contains all the Y samples plus those Cb and Cr samples located at even-numbered sample points. Link B contains samples from the keying channel and the Cb and Cr samples from the odd-numbered sampled points. Although it may be common to refer to Link A as 4:2:2 and Link B as 2:2:4, Link A is not a true 4:2:2 signal since the CbCr data was sampled at 13.5 MHz, rather than 6.75 MHz.

Figure 6.11 shows the contents of links A and B when transmitting 4:4:4:4 YCbCrK video data. Figure 6.12 illustrates the contents when transmitting R′G′B′K video data. If the keying signal (K) is not present, the K sample values should have a 10-bit value of $3AC_H$.

Figure 6.13 illustrates the YCbCrK timing for one scan line for the 4:3 aspect ratio, using a 27 MHz sample clock. Figure 6.14 shows the YCbCrK timing for one scan line for the 16:9 aspect ratio, using a 36 MHz sample clock. Two 25-pin parallel interfaces are used.

4:4:4:4 YCbCrK Serial Interface

BT.799 and BT.1303 also define a YCbCr serial interface. The two 10-bit 4:2:2 YCbCr parallel streams shown in Figure 6.13 or 6.14 are serialized using two 270 or 360 Mbps serial interfaces. SMPTE RP-175 further clarifies the operation for 480i systems.

RGBK Parallel Interface

BT.799 and BT.1303 also support transferring BT.601 R′G′B′K digital video between equipment. For additional information, see the 4:4:4:4 YCbCrK parallel interface. SMPTE RP-175 further clarifies the operation for 480i systems. The G′ samples are sent in the Y locations, the R′ samples are sent in the Cr locations, and the B′ samples are sent in the Cb locations.

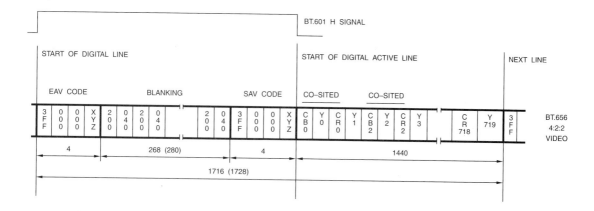

Figure 6.9. BT.656 and SMPTE 125M Parallel Interface Data For One Scan Line.
480i; 4:2:2 YCbCr; 720 active samples per line; 27 MHz clock; 10-bit system. The
values for 576i systems are shown in parentheses.

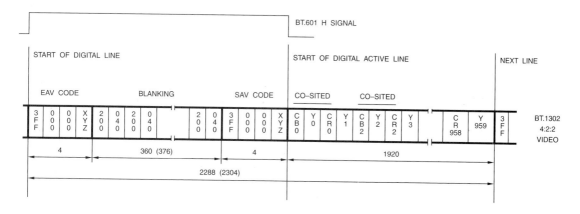

Figure 6.10. BT.1302 and SMPTE 267M Parallel Interface Data For One Scan Line.
480i; 4:2:2 YCbCr; 960 active samples per line; 36 MHz clock; 10-bit system. The
values for 576i systems are shown in parentheses.

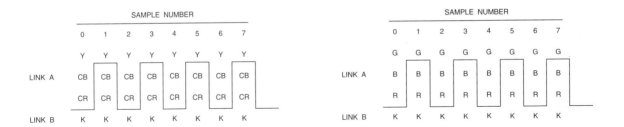

Figure 6.11. Link Content Representation for YCbCrK Video Signals.

Figure 6.12. Link Content Representation for R´G´B´K Video Signals.

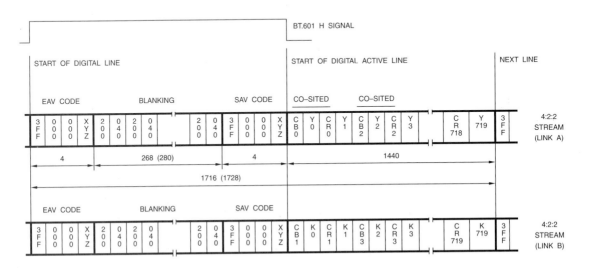

Figure 6.13. BT.799 and SMPTE RP-175 Parallel Interface Data For One Scan Line. 480i; 4:4:4:4 YCbCrK; 720 active samples per line; 27 MHz clock; 10-bit system. The values for 576i systems are shown in parentheses.

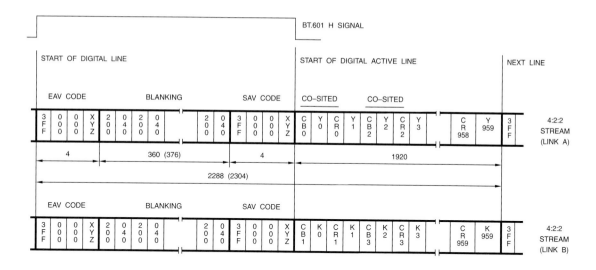

Figure 6.14. BT.1303 Parallel Interface Data For One Scan Line. 480i; 4:4:4:4 YCbCrK; 960 active samples per line; 36 MHz clock; 10-bit system. The values for 576i systems are shown in parentheses.

RGBK Serial Interface

BT.799 and BT.1303 also define a R′G′B′K serial interface. The two 10-bit R′G′B′K parallel streams are serialized using two 270 or 360 Mbps serial interfaces.

SDTV—Progressive

Supported active resolutions, with their corresponding aspect ratios and frame refresh rates, are:

720 × 480p	4:3	59.94 Hz
720 × 576p	4:3	50.00 Hz

4:2:2 YCbCr Serial Interface

ITU-R BT.1362 defines two 10-bit 4:2:2 YCbCr data streams (Figure 6.15), using a 27 MHz sample clock. SMPTE 294M further clarifies the operation for 480p systems.

What stream is used for which scan line is shown in Table 6.22. The two 10-bit parallel streams shown in Figure 6.15 are serialized using two 270 Mbps serial interfaces.

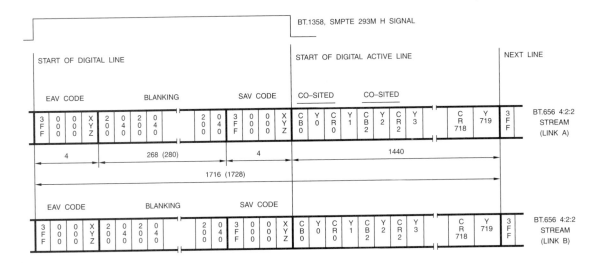

Figure 6.15. BT.1362 and SMPTE 294M Parallel Data For Two Scan Lines. 480p; 4:2:2 YCbCr; 720 active samples per line; 27 MHz clock; 10-bit system. The values for 576p systems are shown in parentheses.

480p (525p) System				576p (625p) System			
Link A	Link B	Link A	Link B	Link A	Link B	Link A	Link B
7	8	6	7	1	2	4	5
9	10	:	:	3	4	6	7
:	:	522	523	:	:	8	9
523	524	524	525	621	622	:	:
525	1	1	2	623	624	620	621
2	3	3	4	625	1	622	623
4	5	5	6	2	3	624	625

Table 6.22. BT.1362 and SMPTE 294M Scan Line Numbering and Link Assignment.

HDTV—Interlaced

Supported active resolutions, with their corresponding aspect ratios and frame refresh rates, are:

1920 × 1080i	16:9	25.00 Hz
1920 × 1080i	16:9	29.97 Hz
1920 × 1080i	16:9	30.00 Hz

4:2:2 YCbCr Parallel Interface

The ITU-R BT.1120 parallel interface was developed to transfer interlaced HDTV 4:2:2 YCbCr digital video between equipment. SMPTE 274M further clarifies the operation for 29.97 and 30 Hz systems.

Figure 6.16 illustrates the timing for one scan line for the 1920 × 1080i active resolutions. The 93-pin parallel interface is used with a sample clock rate of 74.25 MHz (25 or 30 Hz refresh) or 74.176 MHz (29.97 Hz refresh).

4:2:2 YCbCr Serial Interface

BT.1120 also defines a YCbCr serial interface. SMPTE 292M further clarifies the operation for 29.97 and 30 Hz systems. The two 10-bit 4:2:2 YCbCr parallel streams shown in Figure 6.16 are multiplexed together, then serialized using a 1.485 or 1.4835 Gbps serial interface.

4:2:2:4 YCbCrK Parallel Interface

BT.1120 also supports transferring HDTV 4:2:2:4 YCbCrK digital video between equipment. SMPTE 274M further clarifies the operation for 29.97 and 30 Hz systems.

Figure 6.17 illustrates the timing for one scan line for the 1920 × 1080i active resolutions. The 93-pin parallel interface is used with a sample clock rate of 74.25 MHz (25 or 30 Hz refresh) or 74.176 MHz (29.97 Hz refresh).

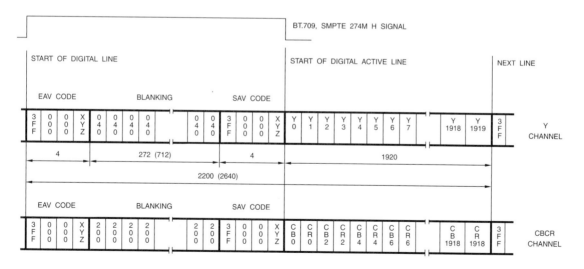

Figure 6.16. BT.1120 and SMPTE 274M Parallel Interface Data For One Scan Line. 1080i; 29.97-, 30-, 59.94-, and 60-Hz systems; 4:2:2 YCbCr; 1920 active samples per line; 74.176, 74.25, 148.35, or 148.5 MHz clock; 10-bit system. The values for 25- and 50-Hz systems are shown in parentheses.

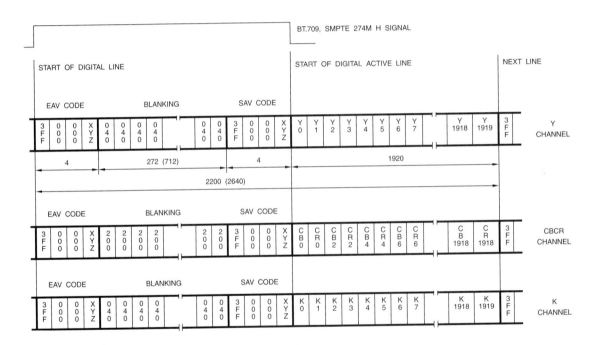

Figure 6.17. BT.1120 and SMPTE 274M Parallel Interface Data For One Scan Line. 1080i; 29.97-, 30-, 59.94-, and 60-Hz systems; 4:2:2:4 YCbCrK; 1920 active samples per line; 74.176, 74.25, 148.35, or 148.5 MHz clock; 10-bit system. The values for 25- and 50-Hz systems are shown in parentheses.

RGB Parallel Interface

BT.1120 also supports transferring HDTV R′G′B′ digital video between equipment. SMPTE 274M further clarifies the operation for 29.97 and 30 Hz systems.

Figure 6.18 illustrates the timing for one scan line for the 1920 × 1080i active resolutions. The 93-pin parallel interface is used with a sample clock rate of 74.25 MHz (25 or 30 Hz refresh) or 74.176 MHz (29.97 Hz refresh).

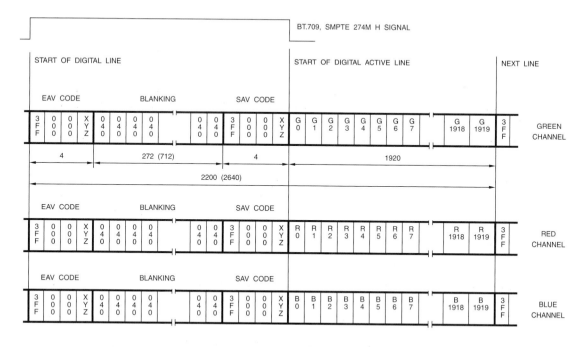

Figure 6.18. BT.1120 and SMPTE 274M Parallel Interface Data For One Scan Line. 1080i; 29.97-, 30-, 59.94-, and 60-Hz systems; R′G′B′; 1920 active samples per line; 74.176, 74.25, 148.35, or 148.5 MHz clock; 10-bit system. The values for 25- and 50-Hz systems are shown in parentheses.

HDTV—Progressive

Supported active resolutions, with their corresponding aspect ratios and frame refresh rates, are:

1280 × 720p	16:9	23.98 Hz
1280 × 720p	16:9	24.00 Hz
1280 × 720p	16:9	25.00 Hz
1280 × 720p	16:9	29.97 Hz
1280 × 720p	16:9	30.00 Hz
1280 × 720p	16:9	50.00 Hz
1280 × 720p	16:9	59.94 Hz
1280 × 720p	16:9	60.00 Hz
1920 × 1080p	16:9	23.98 Hz
1920 × 1080p	16:9	24.00 Hz
1920 × 1080p	16:9	25.00 Hz
1920 × 1080p	16:9	29.97 Hz
1920 × 1080p	16:9	30.00 Hz
1920 × 1080p	16:9	50.00 Hz
1920 × 1080p	16:9	59.94 Hz
1920 × 1080p	16:9	60.00 Hz

4:2:2 YCbCr Parallel Interface

The ITU-R BT.1120 and SMPTE 274M parallel interfaces were developed to transfer progressive HDTV 4:2:2 YCbCr digital video between equipment.

Figure 6.16 illustrates the timing for one scan line for the 1920 × 1080p active resolutions. The 93-pin parallel interface is used with a sample clock rate of 148.5 MHz (24, 25, 30, 50 or 60 Hz refresh) or 148.35 MHz (23.98, 29.97 or 59.94 Hz refresh).

Figure 6.19 illustrates the timing for one scan line for the 1280 × 720p active resolutions. The 93-pin parallel interface is used with a sample clock rate of 74.25 MHz (24, 25, 30, 50 or 60 Hz refresh) or 74.176 MHz (23.98, 29.97, or 59.94 Hz refresh).

4:2:2:4 YCbCrK Parallel Interface

BT.1120 and SMPTE 274M also support transferring HDTV 4:2:2:4 YCbCrK digital video between equipment.

Figure 6.17 illustrates the timing for one scan line for the 1920 × 1080p active resolutions. The 93-pin parallel interface is used with a sample clock rate of 148.5 MHz (24, 25, 30, 50 or 60 Hz refresh) or 148.35 MHz (23.98, 29.97 or 59.94 Hz refresh).

Figure 6.20 illustrates the timing for one scan line for the 1280 × 720p active resolutions. The 93-pin parallel interface is used with a sample clock rate of 74.25 MHz (24, 25, 30, 50 or 60 Hz refresh) or 74.176 MHz (23.98, 29.97 or 59.94 Hz refresh).

RGB Parallel Interface

BT.1120 and SMPTE 274M also support transferring HDTV R′G′B′ digital video between equipment.

Figure 6.18 illustrates the timing for one scan line for the 1920 × 1080p active resolutions. The 93-pin parallel interface is used with a sample clock rate of 148.5 MHz (24, 25, 30, 50 or 60 Hz refresh) or 148.35 MHz (23.98, 29.97 or 59.94 Hz refresh).

Figure 6.21 illustrates the timing for one scan line for the 1280 × 720p active resolutions. The 93-pin parallel interface is used with a sample clock rate of 74.25 MHz (24, 25, 30, 50 or 60 Hz refresh) or 74.176 MHz (23.98, 29.97 or 59.94 Hz refresh).

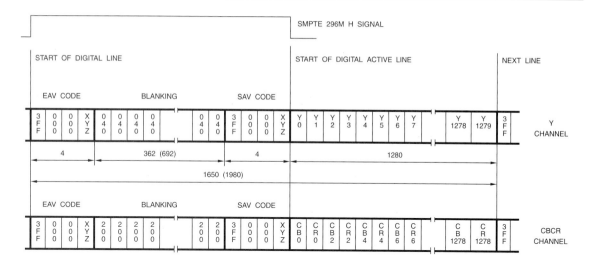

Figure 6.19. SMPTE 274M Parallel Interface Data For One Scan Line. 720p; 59.94- and 60-Hz systems; 4:2:2 YCbCr; 1280 active samples per line; 74.176 or 74.25 MHz clock; 10-bit system. The values for 50-Hz systems are shown in parentheses.

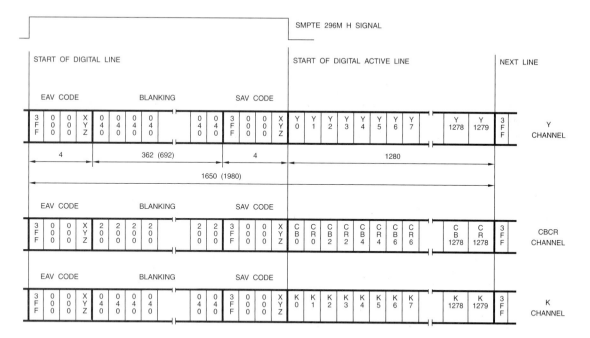

Figure 6.20. SMPTE 274M Parallel Interface Data For One Scan Line. 720p; 59.94- and 60-Hz systems; 4:2:2:4 YCbCrK; 1280 active samples per line; 74.176 or 74.25 MHz clock; 10-bit system. The values for 50-Hz systems are shown in parentheses.

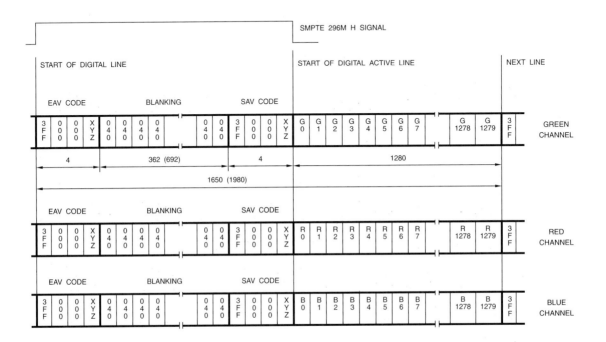

Figure 6.21. SMPTE 274M Parallel Interface Data For One Scan Line. 720p; 59.94- and 60-Hz systems; R´G´B´; 1280 active samples per line; 74.176 or 74.25 MHz clock; 10-bit system. The values for 50-Hz systems are shown in parentheses.

Pro-Video Composite Interfaces

Digital composite video is essentially a digital version of a composite analog (M) NTSC or (B, D, G, H, I) PAL video signal. The sample clock rate is four times F_{SC}: about 14.32 MHz for (M) NTSC and about 17.73 MHz for (B, D, G, H, I) PAL.

Usually, both 8-bit and 10-bit interfaces are supported, with the 10-bit interface used to transmit 2 bits of fractional video data to minimize cumulative processing errors and to support 10-bit ancillary data.

Table 6.23 lists the digital composite levels. Video data may not use the 10-bit values of 000_H–003_H and $3FC_H$–$3FF_H$, or the 8-bit values of 00_H and FF_H, since they are used for timing information.

NTSC Video Timing

There are 910 total samples per scan line, as shown in Figure 6.22. Horizontal count 0 corresponds to the start of active video, and a horizontal count of 768 corresponds to the start of horizontal blanking.

Sampling is along the ±I and ±Q axes (33°, 123°, 213°, and 303°). The sampling phase at horizontal count 0 of line 10, Field 1 is on the +I axis (123°).

The sync edge values, and the horizontal counts at which they occur, are defined as shown in Figure 6.23 and Tables 6.24–6.26. 8-bit values for one color burst cycle are 45, 83, 75, and 37. The burst envelope starts at horizontal count 857, and lasts for 43 clock cycles, as shown in Table 6.24. Note that the peak amplitudes of the burst are not sampled.

Video Level	(M) NTSC	(B, D, G, H, I) PAL
peak chroma	972	1040 (limited to 1023)
white	800	844
peak burst	352	380
black	280	256
blank	240	256
peak burst	128	128
peak chroma	104	128
sync	16	4

Table 6.23. 10-Bit Video Levels for Digital Composite Video Signals.

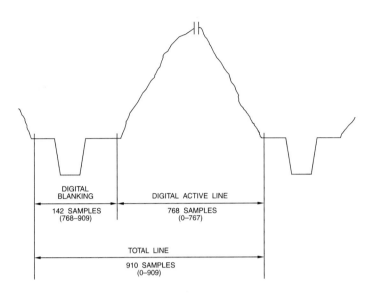

Figure 6.22. Digital Composite (M) NTSC Analog and Digital Timing Relationship.

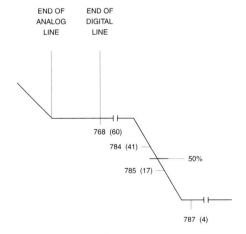

Figure 6.23. Digital Composite (M) NTSC Sync Timing. The horizontal counts are shown with the corresponding 8-bit sample values in parentheses.

Sample	8-bit Hex Value		10-bit Hex Value	
	Fields 1, 3	Fields 2, 4	Fields 1, 3	Fields 2, 4
768–782	3C	3C	0F0	0F0
783	3A	3A	0E9	0E9
784	29	29	0A4	0A4
785	11	11	044	044
786	04	04	011	011
787–849	04	04	010	010
850	06	06	017	017
851	17	17	05C	05C
852	2F	2F	0BC	0BC
853	3C	3C	0EF	0EF
854–856	3C	3C	0F0	0F0
857	3C	3C	0F0	0F0
858	3D	3B	0F4	0EC
859	37	41	0DC	104
860	36	42	0D6	10A
861	4B	2D	12C	0B4
862	49	2F	123	0BD
863	25	53	096	14A
864	2D	4B	0B3	12D
865	53	25	14E	092
866	4B	2D	12D	0B3
867	25	53	092	14E
868	2D	4B	0B3	12D
869	53	25	14E	092
870	4B	2D	12D	0B3
871	25	53	092	14E
872	2D	4B	0B3	12D
873	53	25	14E	092

Table 6.24a. Digital Values During the Horizontal Blanking Intervals for Digital Composite (M) NTSC Video Signals.

Sample	8-bit Hex Value		10-bit Hex Value	
	Fields 1, 3	Fields 2, 4	Fields 1, 3	Fields 2, 4
874	4B	2D	12D	0B3
875	25	53	092	14E
876	2D	4B	0B3	12D
877	53	25	14E	092
878	4B	2D	12D	0B3
879	25	53	092	14E
880	2D	4B	0B3	12D
881	53	25	14E	092
882	4B	2D	12D	0B3
883	25	53	092	14E
884	2D	4B	0B3	12D
885	53	25	14E	092
886	4B	2D	12D	0B3
887	25	53	092	14E
888	2D	4B	0B3	12D
889	53	25	14E	092
890	4B	2D	12D	0B3
891	25	53	092	14E
892	2D	4B	0B3	12D
893	53	25	14E	092
894	4A	2E	129	0B7
895	2A	4E	0A6	13A
896	33	45	0CD	113
897	44	34	112	0CE
898	3F	39	0FA	0E6
899	3B	3D	0EC	0F4
900–909	3C	3C	0F0	0F0

Table 6.24b. Digital Values During the Horizontal Blanking Intervals for Digital Composite (M) NTSC Video Signals.

Fields 1, 3			Fields 2, 4		
Sample	8-bit Hex Value	10-bit Hex Value	Sample	8-bit Hex Value	10-bit Hex Value
768–782	3C	0F0	313–327	3C	0F0
783	3A	0E9	328	3A	0E9
784	29	0A4	329	29	0A4
785	11	044	330	11	044
786	04	011	331	04	011
787–815	04	010	332–360	04	010
816	06	017	361	06	017
817	17	05C	362	17	05C
818	2F	0BC	363	2F	0BC
819	3C	0EF	364	3C	0EF
820–327	3C	0F0	365–782	3C	0F0
328	3A	0E9	783	3A	0E9
329	29	0A4	784	29	0A4
330	11	044	785	11	044
331	04	011	786	04	011
332–360	04	010	787–815	04	010
361	06	017	816	06	017
362	17	05C	817	17	05C
363	2F	0BC	818	2F	0BC
364	3C	0EF	819	3C	0EF
365–782	3C	0F0	820–327	3C	0F0

Table 6.25. Equalizing Pulse Values During the Vertical Blanking Intervals for Digital Composite (M) NTSC Video Signals.

Fields 1, 3			Fields 2, 4		
Sample	8-bit Hex Value	10-bit Hex Value	Sample	8-bit Hex Value	10-bit Hex Value
782	3C	0F0	327	3C	0F0
783	3A	0E9	328	3A	0E9
784	29	0A4	329	29	0A4
785	11	044	330	11	044
786	04	011	331	04	011
787–260	04	010	332–715	04	010
261	06	017	716	06	017
262	17	05C	717	17	05C
263	2F	0BC	718	2F	0BC
264	3C	0EF	719	3C	0EF
265–327	3C	0F0	720–782	3C	0F0
328	3A	0E9	783	3A	0E9
329	29	0A4	784	29	0A4
330	11	044	785	11	044
331	04	011	786	04	011
332–715	04	010	787–260	04	010
716	06	017	261	06	017
717	17	05C	262	17	05C
718	2F	0BC	263	2F	0BC
719	3C	0EF	264	3C	0EF
720–782	3C	0F0	265–327	3C	0F0

Table 6.26. Serration Pulse Values During the Vertical Blanking Intervals for Digital Composite (M) NTSC Video Signals.

To maintain zero SCH phase, horizontal count 784 occurs 25.6 ns (33° of the subcarrier phase) before the 50% point of the falling edge of horizontal sync, and horizontal count 785 occurs 44.2 ns (57° of the subcarrier phase) after the 50% point of the falling edge of horizontal sync.

PAL Video Timing

There are 1135 total samples per line, except for two lines per frame which have 1137 samples per line, making a total of 709,379 samples per frame. Figure 6.24 illustrates the typical line timing. Horizontal count 0 corresponds to the start of active video, and a horizontal count of 948 corresponds to the start of horizontal blanking.

Sampling is along the ±U and ±V axes (0°, 90°, 180°, and 270°), with the sampling phase at horizontal count 0 of line 1, Field 1 on the +V axis (90°).

8-bit color burst values are 95, 64, 32, and 64, continuously repeated. The swinging burst causes the peak burst (32 and 95) and zero burst (64) samples to change places. The burst envelope starts at horizontal count 1058, and lasts for 40 clock cycles.

Sampling is not H-coherent as with (M) NTSC, so the position of the sync pulses change from line to line. Zero SCH phase is defined when alternate burst samples have a value of 64.

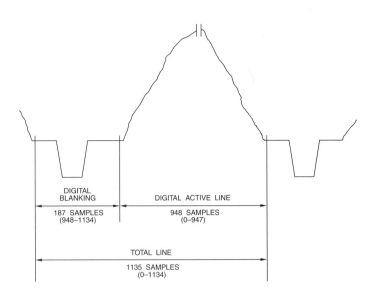

DIGITAL BLANKING

187 SAMPLES (948–1134)

DIGITAL ACTIVE LINE

948 SAMPLES (0–947)

TOTAL LINE

1135 SAMPLES (0–1134)

Figure 6.24. Digital Composite (B, D, G, H, I) PAL Analog and Digital Timing Relationship.

Ancillary Data

Ancillary data packets are used to transmit information (such as digital audio, closed captioning, and teletext data) during the blanking intervals. ITU-R BT.1364 and SMPTE 291M describe the ancillary data formats.

The ancillary data formats are the same as for digital component video, discussed earlier in this chapter. However, instead of a 3-word preamble, a one-word ancillary data flag is used, with a 10-bit value of $3FC_H$. There may be multiple ancillary data flags following the TRS-ID, with each flag identifying the beginning of another ancillary packet.

Ancillary data may be present within the following word number boundaries (see Figures 6.25 through 6.30).

NTSC	PAL	
795–849	972–1035	horizontal sync period
795–815	972–994	equalizing pulse periods
340–360	404–426	
795–260	972–302	vertical sync periods
340–715	404–869	

User data may not use the 10-bit values of 000_H–003_H and $3FC_H$–$3FF_H$, or the 8-bit values of 00_H and FF_H, since they are used for timing information.

25-pin Parallel Interface

The SMPTE 244M parallel interface is based on that used for 27 MHz 4:2:2 digital component video (Table 6.19), except for the timing differences. This interface is used to transfer SDTV resolution digital composite

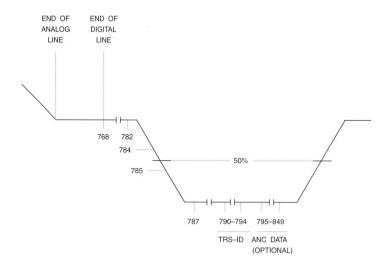

Figure 6.25. (M) NTSC TRS-ID and Ancillary Data Locations During Horizontal Sync Intervals.

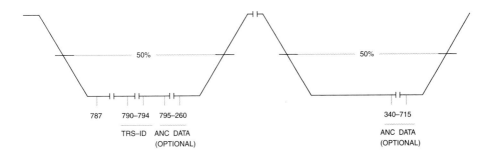

Figure 6.26. (M) NTSC TRS-ID and Ancillary Data Locations During Vertical Sync Intervals.

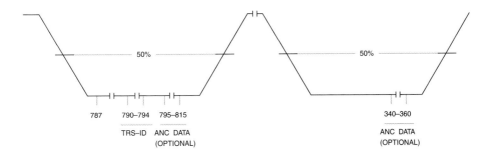

Figure 6.27. (M) NTSC TRS-ID and Ancillary Data Locations During Equalizing Pulse Intervals.

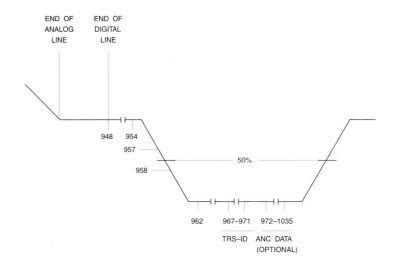

Figure 6.28. (B, D, G, H, I) PAL TRS-ID and Ancillary Data Locations During Horizontal Sync Intervals.

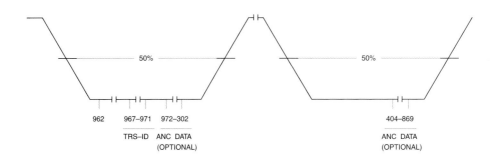

Figure 6.29. (B, D, G, H, I) PAL TRS-ID and Ancillary Data Locations During Vertical Sync Intervals.

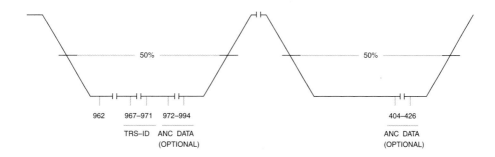

Figure 6.30. (B, D, G, H, I) PAL TRS-ID and Ancillary Data Locations During Equalizing Pulse Intervals.

data. 8-bit or 10-bit data and a $4\times F_{SC}$ clock are transferred.

Signal levels are compatible with ECL-compatible balanced drivers and receivers. The generator must have a balanced output with a maximum source impedance of 110 Ω; the signal must be 0.8–2.0V peak-to-peak measured across a 110-Ω load. At the receiver, the transmission line must be terminated by 110 ±10 Ω.

The clock signal is a $4\times F_{SC}$ square wave, with a clock pulse width of 35 ±5 ns for (M) NTSC or 28 ±5 ns for (B, D, G, H, I) PAL. The positive transition of the clock signal occurs midway between data transitions with a tolerance of ±5 ns (as shown in Figure 6.31).

To permit reliable operation at interconnect lengths of 50–200 meters, the receiver must use frequency equalization, with typical characteristics shown in Figure 6.3. This example enables operation with a range of cable lengths down to zero.

Serial Interface

The parallel format can be converted to a SMPTE 259M serial format (Figure 6.32), allowing data to be transmitted using a 75-Ω coaxial cable (or optical fiber). This interface converts the 14.32 or 17.73 MHz parallel stream into a 143 or 177 Mbps serial stream. The 10× PLL generates the 143 or 177 MHz clock from the 14.32 or 17.73 MHz clock signal.

For cable interconnect, the generator has an unbalanced output with a source impedance of 75Ω; the signal must be 0.8V ±10% peak-to-peak measured across a 75-Ω load. The receiver has an input impedance of 75Ω.

The 10 bits of data are serialized (LSB first) and processed using a scrambled and polarity-free NRZI algorithm:

$$G(x) = (x^9 + x^4 + 1)(x + 1)$$

This algorithm is the same as used for digital component video discussed earlier. In an 8-bit

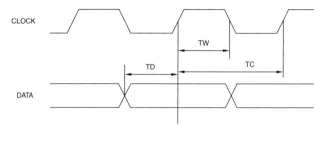

TW = 35 ± 5 NS (M) NTSC; 28 ± 5 NS (B, D, G, H, I) PAL

TC = 69.84 NS (M) NTSC; 56.39 NS (B, D, G, H, I) PAL

TD = 35 ± 5 NS (M) NTSC; 28 ± 5 NS (B, D, G, H, I) PAL

Figure 6.31. Digital Composite Video Parallel Interface Waveforms.

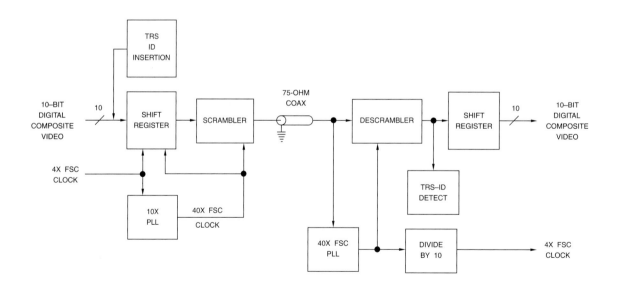

Figure 6.32. Serial Interface Block Diagram.

environment, 8-bit data is appended with two least significant "0" bits before serialization.

The input signal to the scrambler (Figure 6.7) uses positive logic (the highest voltage represents a logical one; lowest voltage represents a logical zero). The formatted serial data is output at the $40\times F_{SC}$ rate.

At the receiver, phase-lock synchronization is done by detecting the TRS-ID sequences. The PLL is continuously adjusted slightly each scan line to ensure that these patterns are detected and to avoid bit slippage. The recovered $10\times$ clock is divided by ten to generate the $4\times F_{SC}$ sample clock. The serial data is low- and high-frequency equalized, inverse scrambling performed (Figure 6.8), and deserialized.

TRS-ID

When using the serial interface, a special five-word sequence, known as the TRS-ID, must be inserted into the digital video stream during the horizontal sync time. The TRS-ID is present only following sync leading edges which identify a horizontal transition, and

occupies horizontal counts 790–794, inclusive (NTSC) or 967–971, inclusive (PAL). Table 6.27 shows the TRS-ID format; Figures 6.25 through 6.30 show the TRS-ID locations for digital composite (M) NTSC and (B, D, G, H, I) PAL video signals.

The line number ID word at horizontal count 794 (NTSC) or 971 (PAL) is defined as shown in Table 6.28.

PAL requires the reset of the TRS-ID position relative to horizontal sync once per field on only one of lines 625–4 and 313–317 due to the 25-Hz offset. All lines have 1135 samples except the two lines used for reset, which have 1137 samples. The two additional samples are numbered 1135 and 1136, and occur just prior to the first active picture sample (sample 0).

Due to the 25-Hz offset, the samples occur slightly earlier each line. Initial determination of the TRS-ID position should be done on line 1, Field 1, or a nearby line. The TRS-ID location always starts at sample 967, but the distance from the leading edge of sync varies due to the 25-Hz offset.

	D9 (MSB)	D8	D7	D6	D5	D4	D3	D2	D1	D0
TRS word 0	1	1	1	1	1	1	1	1	1	1
TRS word 1	0	0	0	0	0	0	0	0	0	0
TRS word 2	0	0	0	0	0	0	0	0	0	0
TRS word 3	0	0	0	0	0	0	0	0	0	0
line number ID	$\overline{D8}$	EP	line number ID							

Notes:
EP = even parity for D0–D7.

Table 6.27. TRS-ID Format.

D2	D1	D0	(M) NTSC	(B, D, G, H, I) PAL
0	0	0	line 1–263 field 1	line 1–313 field 1
0	0	1	line 264–525 field 2	line 314–625 field 2
0	1	0	line 1–263 field 3	line 1–313 field 3
0	1	1	line 264–525 field 4	line 314–625 field 4
1	0	0	not used	line 1–313 field 5
1	0	1	not used	line 314–625 field 6
1	1	0	not used	line 1–313 field 7
1	1	1	not used	line 314–625 field 8

D7–D3	(M) NTSC	(B, D, G, H, I) PAL
$1 \le x \le 30$	line number 1–30 [264–293]	line number 1–30 [314–343]
$x = 31$	line number $\ge$ 31 [294]	line number $\ge$ 31 [344]
$x = 0$	not used	not used

Table 6.28. Line Number ID Word at Horizontal Count 794 (NTSC) or 971 (PAL).

Pro-Video Transport Interfaces

Serial Data Transport Interface (SDTI)

SMPTE 305M and ITU-R BT.1381 define a Serial Data Transport Interface (SDTI) that enables transferring data between equipment. The physical layer uses the 270 or 360 Mbps BT.656, BT.1302, and SMPTE 259M digital component video serial interface. Figure 6.33 illustrates the signal format.

A 53-word header is inserted immediately after the EAV sequence, specifying the source, destination, and data format. Table 6.29 illustrates the header contents.

The payload data is defined within BT.1381 and by other application-specific standards such as SMPTE 326M. It may consist of MPEG-2 program or transport streams, DV streams, etc., and uses either 8-bit words plus even parity and $\overline{D8}$, or 9-bit words plus $\overline{D8}$.

Line Number

The line number specifies a value of 1–525 (480i systems) or 1–625 (576i systems). L0 is the least significant bit.

Line Number CRC

The line number CRC applies to the data ID through the line number, for the entire 10 bits. C0 is the least significant bit. It is an 18-bit value, with an initial value set to all ones:

$$CRC = x^{18} + x^5 + x^4 + x^1$$

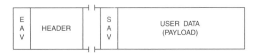

Figure 6.33. SDTI Signal Format.

Code and AAI

The 4-bit code value (CD3–CD0) specifies the length of the payload (the user data contained between the SAV and EAV sequences):

0000	4:2:2 YCbCr video data
0001	1440 word payload (uses 270 Mbps interface)
0010	1920 word payload (uses 360 Mbps interface)
1000	143 Mbps digital composite video

The 4-bit authorized address identifier (AAI) value, AAI3–AAI0, specifies the format of the destination and source addresses:

0000	unspecified format
0001	IPv6 address

Destination and Source Addresses

These specify the address of the source and destination devices. A universal address is indicated when all address bits are zero and AAI3–AAI0 = 0000.

Block Type

The block type value specifies the segmentation of the payload. BL7–BL6 indicate the payload block structure:

00	fixed block size without ECC
01	fixed block size with ECC
10	unassigned
11	variable block size

BL5–BL0 indicate the segmentation for fixed block sizes. Variable block sizes are indicated by BL7–BL0 having a value of 11000001. The ECC format is application-dependent.

Payload CRC Flag

The CRCF bit indicates whether or not the payload CRC is present at the end of the payload:

0	no CRC
1	CRC present

	10-bit Data									
	D9 (MSB)	**D8**	**D7**	**D6**	**D5**	**D4**	**D3**	**D2**	**D1**	**D0**
ancillary data flag (ADF)	0	0	0	0	0	0	0	0	0	0
	1	1	1	1	1	1	1	1	1	1
	1	1	1	1	1	1	1	1	1	1
data ID (DID)	$\overline{D8}$	EP	0	1	0	0	0	0	0	0
SDID	$\overline{D8}$	EP	0	0	0	0	0	0	0	1
data count (DC)	$\overline{D8}$	EP	0	0	1	0	1	1	1	0
line number	$\overline{D8}$	EP	L7	L6	L5	L4	L3	L2	L1	L0
	$\overline{D8}$	EP	0	0	0	0	0	0	L9	L8
line number CRC	$\overline{D8}$	C8	C7	C6	C5	C4	C3	C2	C1	C0
	$\overline{D8}$	C17	C16	C15	C14	C13	C12	C11	C10	C9
code and AAI	$\overline{D8}$	EP	AAI3	AAI2	AAI1	AAI0	CD3	CD2	CD1	CD0
destination address	$\overline{D8}$	EP	DA7	DA6	DA5	DA4	DA3	DA2	DA1	DA0
	$\overline{D8}$	EP	DA15	DA14	DA13	DA12	DA11	DA10	DA9	DA8
					:					
	$\overline{D8}$	EP	DA127	DA126	DA125	DA124	DA123	DA122	DA121	DA120
source address	$\overline{D8}$	EP	SA7	SA6	SA5	SA4	SA3	SA2	SA1	SA0
	$\overline{D8}$	EP	SA15	SA14	SA13	SA12	SA11	SA10	SA9	SA8
					:					
	$\overline{D8}$	EP	SA127	SA126	SA125	SA124	SA123	SA122	SA121	SA120

Notes:
EP = even parity for D0–D7.

Table 6.29a. SDTI Header Structure.

	10-bit Data									
	D9 (MSB)	D8	D7	D6	D5	D4	D3	D2	D1	D0
block type	$\overline{D8}$	EP	BL7	BL6	BL5	BL4	BL3	BL2	BL1	BL0
payload CRC flag	$\overline{D8}$	EP	0	0	0	0	0	0	0	CRCF
reserved	$\overline{D8}$	EP	0	0	0	0	0	0	0	0
reserved	$\overline{D8}$	EP	0	0	0	0	0	0	0	0
reserved	$\overline{D8}$	EP	0	0	0	0	0	0	0	0
reserved	$\overline{D8}$	EP	0	0	0	0	0	0	0	0
reserved	$\overline{D8}$	EP	0	0	0	0	0	0	0	0
header CRC	$\overline{D8}$	C8	C7	C6	C5	C4	C3	C2	C1	C0
	$\overline{D8}$	C17	C16	C15	C14	C13	C12	C11	C10	C9
check sum	$\overline{D8}$	Sum of D0–D8 of data ID through last header CRC word. Preset to all zeros; carry is ignored.								

Notes:
EP = even parity for D0–D7.

Table 6.29b. SDTI Header Structure (continued).

Header CRC

The header CRC applies to the code and AAI word through the last reserved data word, for the entire 10 bits. C0 is the least significant bit. It is an 18-bit value, with an initial value set to all ones:

$$CRC = x^{18} + x^5 + x^4 + x^1$$

High Data-Rate Serial Data Transport Interface (HD-SDTI)

SMPTE 348M and ITU-R BT.1577 define a High Data-Rate Serial Data Transport Interface (HD-SDTI) that enables transferring data between equipment. The physical layer uses the 1.485 (or 1.485/1.001) Gbps SMPTE 292M digital component video serial interface.

Figure 6.34 illustrates the signal format. Two data channels are multiplexed onto the single HD-SDTI stream such that one 74.25 (or 74.25/1.001) MHz data stream occupies the Y data space and the other 74.25 (or 74.25/1.001) MHz data stream occupies the CbCr data space.

A 49-word header is inserted immediately after the line number CRC data, specifying the source, destination, and data format. Table 6.30 illustrates the header contents.

The payload data is defined by other application-specific standards. It may consist of MPEG-2 program or transport streams, DV streams, etc., and uses either 8-bit words plus even parity and $\overline{D8}$, or 9-bit words plus $\overline{D8}$.

Code and AAI

The 4-bit code value (CD3–CD0) specifies the length of the payload (the user data contained between the SAV and EAV sequences):

0000	4:2:2 YCbCr video data
0001	1440 word payload
0010	1920 word payload
0011	1280 word payload
1000	143 Mbps digital composite video
1001	2304 word payload (extended mode)
1010	2400 word payload (extended mode)
1011	1440 word payload (extended mode)
1100	1728 word payload (extended mode)
1101	2880 word payload (extended mode)
1110	3456 word payload (extended mode)
1111	3600 word payload (extended mode)

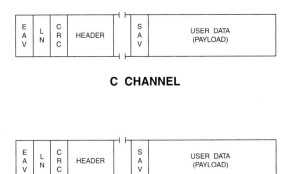

Figure 6.34. HD-SDTI Signal Format. LN = line number (two 10-bit words), CRC = Line Number CRC (two 10-bit words).

	10-bit Data									
	D9 (MSB)	D8	D7	D6	D5	D4	D3	D2	D1	D0
ancillary data flag (ADF)	0	0	0	0	0	0	0	0	0	0
	1	1	1	1	1	1	1	1	1	1
	1	1	1	1	1	1	1	1	1	1
data ID (DID)	$\overline{D8}$	EP	0	1	0	0	0	0	0	0
SDID	$\overline{D8}$	EP	0	0	0	0	0	0	1	0
data count (DC)	$\overline{D8}$	EP	0	0	1	0	1	0	1	0
code and AAI	$\overline{D8}$	EP	AAI3	AAI2	AAI1	AAI0	CD3	CD2	CD1	CD0
destination address	$\overline{D8}$	EP	DA7	DA6	DA5	DA4	DA3	DA2	DA1	DA0
	$\overline{D8}$	EP	DA15	DA14	DA13	DA12	DA11	DA10	DA9	DA8
					:					
	$\overline{D8}$	EP	DA127	DA126	DA125	DA124	DA123	DA122	DA121	DA120
source address	$\overline{D8}$	EP	SA7	SA6	SA5	SA4	SA3	SA2	SA1	SA0
	$\overline{D8}$	EP	SA15	SA14	SA13	SA12	SA11	SA10	SA9	SA8
					:					
	$\overline{D8}$	EP	SA127	SA126	SA125	SA124	SA123	SA122	SA121	SA120
block type	$\overline{D8}$	EP	BL7	BL6	BL5	BL4	BL3	BL2	BL1	BL0
payload CRC flag	$\overline{D8}$	EP	0	0	0	0	0	0	0	0
reserved	$\overline{D8}$	EP	0	0	0	0	0	0	0	0

Notes:
EP = even parity for D0–D7.

Table 6.30a. HD-SDTI Header Structure.

	10-bit Data									
	D9 (MSB)	**D8**	**D7**	**D6**	**D5**	**D4**	**D3**	**D2**	**D1**	**D0**
reserved	$\overline{D8}$	EP	0	0	0	0	0	0	0	0
reserved	$\overline{D8}$	EP	0	0	0	0	0	0	0	0
reserved	$\overline{D8}$	EP	0	0	0	0	0	0	0	0
reserved	$\overline{D8}$	EP	0	0	0	0	0	0	0	0
header CRC	$\overline{D8}$	C8	C7	C6	C5	C4	C3	C2	C1	C0
	$\overline{D8}$	C17	C16	C15	C14	C13	C12	C11	C10	C9
check sum	$\overline{D8}$	Sum of D0–D8 of data ID through last header CRC word. Preset to all zeros; carry is ignored.								

Notes:
EP = even parity for D0–D7.

Table 6.30b. HD-SDTI Header Structure (continued).

The extended mode advances the timing of the SAV sequence, shortening the blanking interval, so that the payload data rate remains a constant 129.6 (or 129.6/1.001) MBps.

The 4-bit authorized address identifier (AAI) format is the same as for SDTI.

Destination and Source Addresses

The source and destination address formats are the same as for SDTI.

Block Type

The block type format is the same as for SDTI.

Header CRC

The header CRC applies to the DID through the last reserved data word, for the entire 10 bits. C0 is the least significant bit. It is an 18-bit value, with an initial value set to all ones:

$$CRC = x^{18} + x^5 + x^4 + x^1$$

IC Component Interfaces

Many solutions for transferring digital video between chips are derived from the pro-video interconnect standards. Chips for the pro-video market typically support 10 or 12 bits of data per video component, while chips for the consumer market typically use 8 bits of data per video component. "BT.601" and "BT.656" are the most popular interfaces for chips.

YCbCr Values: 8-bit Data

Y has a nominal range of 10_H–EB_H. Values less than 10_H or greater than EB_H may be present due to processing. Cb and Cr have a nominal range of 10_H–$F0_H$. Values less than 10_H or greater than $F0_H$ may be present due to processing. YCbCr data may not use the values of 00_H and FF_H since those values may be used for timing information.

During blanking, Y data should have a value of 10_H and CbCr data should have a value of 80_H, unless other information is present.

YCbCr Values: 10-bit Data

For higher accuracy, pro-video solutions typically use 10-bit YCbCr data. Y has a nominal range of 040_H–$3AC_H$. Values less than 040_H or greater than $3AC_H$ may be present due to processing. Cb and Cr have a nominal range of 040_H–$3C0_H$. Values less than 040_H or greater than $3C0_H$ may be present due to processing. The values 000_H–003_H and $3FC_H$–$3FF_H$ may not be used to avoid timing contention with 8-bit systems.

During blanking, Y data should have a value of 040_H and CbCr data should have a value of 200_H, unless other information is present.

RGB Values: 8-bit Data

Consumer solutions typically use 8-bit R′G′B′ data, with a range of 10_H–EB_H (note that PCs typically use a range of 00_H–FF_H). Values less than 10_H or greater than EB_H may be present due to processing.

During blanking, R′G′B′ data should have a value of 10_H, unless other information is present.

RGB Values: 10-bit Data

For higher accuracy, pro-video solutions typically use 10-bit R′G′B′ data, with a nominal range of 040_H–$3AC_H$. Values less than 040_H or greater than $3AC_H$ may be present due to processing. The values 000_H–003_H and $3FC_H$–$3FF_H$ may not be used to avoid timing contention with 8-bit systems.

During blanking, R′G′B′ data should have a value of 040_H, unless other data is present.

"BT.601" Video Interface

The "BT.601" video interface has been used for years, with the control signal names and timing reflecting the video standard. Supported active resolutions and sample clock rates are dependent on the video standard and aspect ratio.

Devices usually support multiple data formats to simplify using them in a wide variety of applications.

Video Data Formats

The 24-bit 4:4:4 YCbCr data format is shown in Figure 6.35. Y, Cb, and Cr are each 8 bits, and all are sampled at the same rate, resulting in 24 bits of data per sample clock. Pro-video solutions typically use a 30-bit interface, with the Y, Cb, and Cr streams each being

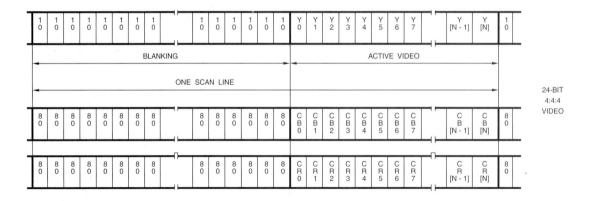

Figure 6.35. 24-Bit 4:4:4 YCbCr Data Format.

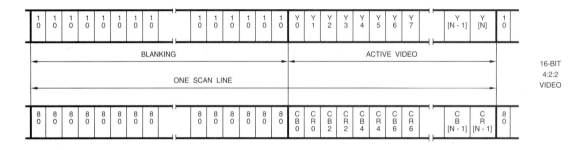

Figure 6.36. 16-Bit 4:2:2 YCbCr Data Format.

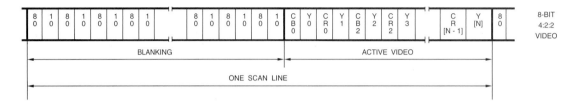

Figure 6.37. 8-Bit 4:2:2 YCbCr Data Format.

10 bits. Y0, Cb0, and Cr0 are the least significant bits.

The 16-bit 4:2:2 YCbCr data format is shown in Figure 6.36. Cb and Cr are sampled at one-half the Y sample rate, then multiplexed together. The CbCr stream of active data words always begins with a Cb sample. Pro-video solutions typically use a 20-bit interface, with the Y and CbCr streams each being 10 bits.

The 8-bit 4:2:2 YCbCr data format is shown in Figure 6.37. The Y and CbCr streams from the 16-bit 4:2:2 YCbCr format are simply multiplexed at 2× the sample clock rate. The YCbCr stream of active data words always begins with a Cb sample. Pro-video solutions typically use a 10-bit interface.

Tables 6.31 and 6.32 illustrate the 15-bit RGB, 16-bit RGB, and 24-bit RGB formats. For the 15-bit RGB format, the unused bit is sometimes used for keying (alpha) information. R0, G0, and B0 are the least significant bits.

Control Signals

In addition to the video data, there are four control signals:

HSYNC#	horizontal sync
VSYNC#	vertical sync
BLANK#	blanking
CLK	1× or 2× sample clock

For the 8-bit and 10-bit 4:2:2 YCbCr data formats, CLK is a 2× sample clock. For the other data formats, CLK is a 1× sample clock. For sources, the control signals and video data are output following the rising edge of CLK. For receivers, the control signals and video data are sampled on the rising edge of CLK.

While BLANK# is negated, active R′G′B′ or YCbCr video data is present.

HSYNC# is asserted during the horizontal sync time each scan line, with the leading edge indicating the start of a new line. The amount of time that HSYNC# is asserted is usually the same as that specified by the video standard.

VSYNC# is asserted during the vertical sync time each field or frame, with the leading edge indicating the start of a new field or frame. The number of scan lines that VSYNC# is asserted is usually same as that specified by the video standard.

For interlaced video, if the leading edges of VSYNC# and HSYNC# are coincident, the field is Field 1. If the leading edge of VSYNC# occurs mid-line, the field is Field 2. For noninterlaced video, the leading edge of VSYNC# indicates the start of a new frame. Figure 6.38 illustrates the typical HSYNC# and VSYNC# relationships.

The 8-/10-bit interface is typically limited to SDTV resolutions. To support HDTV resolutions, some designs transfer data and control information using both edges of the clock.

Receiver Considerations

Assumptions should not be made about the number of samples per line or horizontal blanking interval. Otherwise, the implementation may not work with all sources.

To ensure compatibility between various sources, horizontal counters should be reset by the leading edge of HSYNC#, not by the trailing edge of BLANK#.

To handle real-world sources, a receiver should use a "window" for detecting whether Field 1 or Field 2 is present. For example, if the leading edge of VSYNC# occurs within ±64 1× clock cycles of the leading edge of HSYNC#, the field is Field 1. Otherwise, the field is Field 2.

Some video sources indicate sync timing by having Y data be an 8-bit value less than 10_H. However, most video ICs do not do this. In addition, to allow real-world video and test signals to be passed through with minimum disruption, many ICs now allow the Y data to

24-bit RGB	16-bit RGB (5,6,5)	15-bit RGB (5,5,5)	24-bit 4:4:4 YCbCr	16-bit 4:2:2 YCbCr	8-bit 4:2:2 YCbCr
R7			Cr7		
R6			Cr6		
R5			Cr5		
R4			Cr4		
R3			Cr3		
R2			Cr2		
R1			Cr1		
R0			Cr0		
G7	R4	–	Y7	Y7	Cb7, Y7, Cr7
G6	R3	R4	Y6	Y6	Cb6, Y6, Cr6
G5	R2	R3	Y5	Y5	Cb5, Y5, Cr5
G4	R1	R2	Y4	Y4	Cb4, Y4, Cr4
G3	R0	R1	Y3	Y3	Cb3, Y3, Cr3
G2	G5	R0	Y2	Y2	Cb2, Y2, Cr2
G1	G4	G4	Y1	Y1	Cb1, Y1, Cr1
G0	G3	G3	Y0	Y0	Cb0, Y0, Cr0
B7	G2	G2	Cb7	Cb7, Cr7	
B6	G1	G1	Cb6	Cb6, Cr6	
B5	G0	G0	Cb5	Cb5, Cr5	
B4	B4	B4	Cb4	Cb4, Cr4	
B3	B3	B3	Cb3	Cb3, Cr3	
B2	B2	B2	Cb2	Cb2, Cr2	
B1	B1	B1	Cb1	Cb1, Cr1	
B0	B0	B0	Cb0	Cb0, Cr0	

Table 6.31. Transferring YCbCr and RGB Data over a 16-bit or 24-bit Interface.

24-bit RGB	16-bit RGB (5,6,5)	15-bit RGB (5,5,5)	24-bit 4:4:4 YCbCr	16-bit 4:2:2 YCbCr	8-bit 4:2:2 YCbCr
	R4	–		Y7	
	R3	R4		Y6	
	R2	R3		Y5	
	R1	R2		Y4	
	R0	R1		Y3	
	G5	R0		Y2	
	G4	G4		Y1	
	G3	G3		Y0	
R7	G2	G2	Cr7	Cb7, Cr7	
R6	G1	G1	Cr6	Cb6, Cr6	
R5	G0	G0	Cr5	Cb5, Cr5	
R4	B4	B4	Cr4	Cb4, Cr4	
R3	B3	B3	Cr3	Cb3, Cr3	
R2	B2	B2	Cr2	Cb2, Cr2	
R1	B1	B1	Cr1	Cb1, Cr1	
R0	B0	B0	Cr0	Cb0, Cr0	
G7	R4	–	Y7	Y7	Cb7, Y7, Cr7
G6	R3	R4	Y6	Y6	Cb6, Y6, Cr6
G5	R2	R3	Y5	Y5	Cb5, Y5, Cr5
G4	R1	R2	Y4	Y4	Cb4, Y4, Cr4
G3	R0	R1	Y3	Y3	Cb3, Y3, Cr3
G2	G5	R0	Y2	Y2	Cb2, Y2, Cr2
G1	G4	G4	Y1	Y1	Cb1, Y1, Cr1
G0	G3	G3	Y0	Y0	Cb0, Y0, Cr0
B7	G2	G2	Cb7	Cb7, Cr7	
B6	G1	G1	Cb6	Cb6, Cr6	
B5	G0	G0	Cb5	Cb5, Cr5	
B4	B4	B4	Cb4	Cb4, Cr4	
B3	B3	B3	Cb3	Cb3, Cr3	
B2	B2	B2	Cb2	Cb2, Cr2	
B1	B1	B1	Cb1	Cb1, Cr1	
B0	B0	B0	Cb0	Cb0, Cr0	

Table 6.32. Transferring YCbCr and RGB Data over a 32-bit Interface.

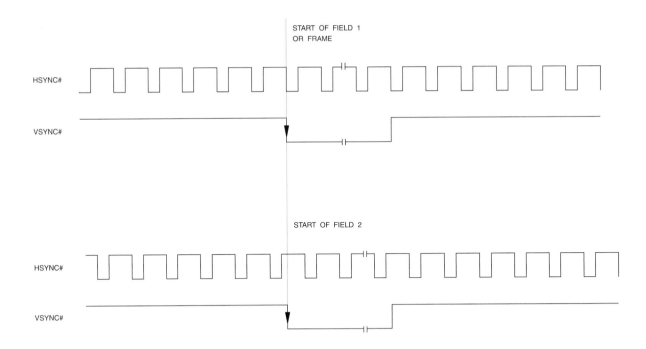

Figure 6.38. Typical HSYNC# and VSYNC# Relationships (Not to Scale).

have a value less than 10_H during active video. Thus, receiver designs assuming sync timing is present on the Y channel may no longer work.

Video Module Interface (VMI)

VMI (Video Module Interface) was developed in cooperation with several multimedia IC manufacturers. The goal was to standardize the video interfaces between devices such as MPEG decoders, NTSC/PAL decoders, and graphics chips.

Video Data Formats

The VMI specification specifies an 8-bit 4:2:2 YCbCr data format as shown in Figure 6.39. Many devices also support the other YCbCr and R′G′B′ formats discussed in the "BT.601 Video Interface" section.

Control Signals

In addition to the video data, there are four control signals:

HREF	horizontal blanking
VREF	vertical sync
VACTIVE	active video
PIXCLK	2× sample clock

For the 8-bit and 10-bit 4:2:2 YCbCr data formats, PIXCLK is a 2× sample clock. For the other data formats, PIXCLK is a 1× sample clock. For sources, the control signals and video data are output following the rising edge of PIXCLK. For receivers, the control signals and video data are sampled on the rising edge of PIXCLK.

While VACTIVE is asserted, active R′G′B′ or YCbCr video data is present. Although transitions in VACTIVE are allowed, it is intended to allow a hardware mechanism for cropping video data. For systems that do not support a VACTIVE signal, HREF can generally be connected to VACTIVE with minimal loss of function.

To support video sources that do not generate a line-locked clock, a DVALID# (data valid) signal may also be used. While DVALID# is asserted, valid data is present.

HREF is asserted during the active video time each scan line, including during the vertical blanking interval.

VREF is asserted for 6 scan line times, starting one-half scan line after the start of vertical sync.

For interlaced video, the trailing edge of VREF is used to sample HREF. If HREF is asserted, the field is Field 1. If HREF is negated, the field is Field 2. For noninterlaced video, the leading edge of VREF indicates the start of a new frame. Figure 6.40 illustrates the typical HREF and VREF relationships.

Receiver Considerations

Assumptions should not be made about the number of samples per line or horizontal blanking interval. Otherwise, the implementation may not work with all sources.

Video data has input setup and hold times, relative to the rising edge of PIXCLK, of 5 and 0 ns, respectively.

VACTIVE has input setup and hold times, relative to the rising edge of PIXCLK, of 5 and 0 ns, respectively.

HREF and VREF both have input setup and hold times, relative to the rising edge of PIXCLK, of 5 and 5 ns, respectively.

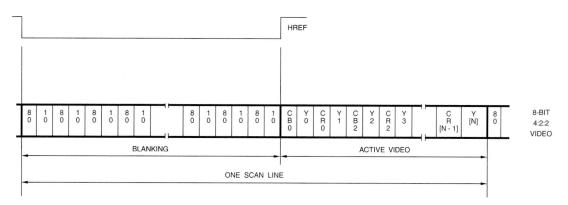

Figure 6.39. VMI 8-bit 4:2:2 YCbCr Data for One Scan Line.

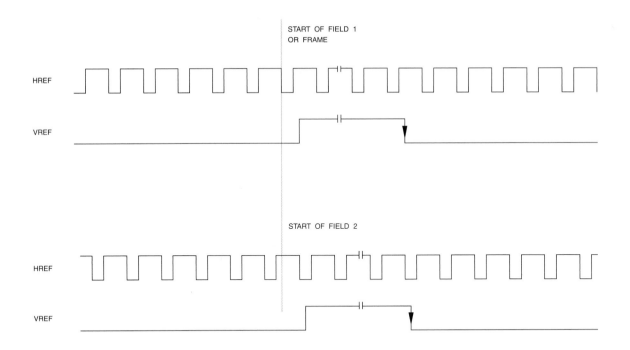

Figure 6.40. VMI Typical HREF and VREF Relationships (Not to Scale).

"BT.656" Interface

The BT.656 interface for ICs is based on the pro-video BT.656-type parallel interfaces, discussed earlier in this chapter (Figures 6.1 and 6.9). Using EAV and SAV sequences to indicate video timing reduces the number of pins required. The timing of the H, V, and F signals for common video formats is illustrated in Chapter 4.

Standard IC signal levels and timing are used, and any resolution can be supported.

Video Data Formats

8-bit or 10-bit 4:2:2 YCbCr data is used, as shown in Figures 6.1 and 6.9. Although sources should generate the four protection bits in the EAV and SAV sequences, receivers may choose to ignore them due to the reliability of point-to-point transfers between chips.

Control Signals

CLK is a 2× sample clock. For sources, the video data is output following the rising edge of CLK. For receivers, the video data is sampled on the rising edge of CLK.

This interface is typically limited to SDTV resolutions. To support HDTV resolutions, some designs transfer data using both edges of the clock.

Zoomed Video Port (ZV Port)

Used on laptops, the ZV Port is a point-to-point uni-directional bus between the PC Card host adaptor and the graphics controller. It enables video data to be transferred real-time directly from the PC Card into the graphics frame buffer.

The PC Card host adaptor has a special multimedia mode configuration. If a non-ZV PC Card is plugged into the slot, the host adaptor is not switched into the multimedia mode, and the PC Card behaves as expected. Once a ZV card has been plugged in and the host adaptor has been switched to the multimedia mode, the pin assignments change. As shown in Table 6.33, the PC Card signals A6–A25, SPKR#, INPACK#, and IOIS16# are replaced by ZV Port video signals (Y0–Y7, CbCr0–CbCr7, HREF, VREF, and PCLK) and 4-channel audio signals (MCLK, SCLK, LRCK, and SDATA).

Video Data Formats

16-bit 4:2:2 YCbCr data is used, as shown in Figure 6.36.

Control Signals

In addition to the video data, there are four control signals:

HREF	horizontal reference
VREF	vertical sync
PCLK	1× sample clock

HREF, VREF, and PCLK have the same timing as the VMI interface discussed earlier in this chapter.

PC Card Signal	ZV Port Signal	PC Card Signal	ZV Port Signal	PC Card Signal	ZV Port Signal
A25	CbCr7	A17	Y1	A9	Y0
A24	CbCr5	A16	CbCr2	A8	Y2
A23	CbCr3	A15	CbCr4	A7	SCLK
A22	CbCr1	A14	Y6	A6	MCLK
A21	CbCr0	A13	Y4	SPKR#	SDATA
A20	Y7	A12	CbCr6	IOIS16#	PCLK
A19	Y5	A11	VREF	INPACK#	LRCK
A18	Y3	A10	HREF		

Table 6.33. PC Card vs. ZV Port Signal Assignments.

Video Interface Port (VIP)

The VESA VIP specification is an enhancement to the "BT.656" interface for ICs, previously discussed. The primary application is to interface up to four devices to a graphics controller chip, although the concept can easily be applied to other applications.

There are three sections to the interface:

Host Interface:

VIPCLK	host clock
HAD0–HAD7	host address/data bus
HCTL	host control

Video Interface:

PIXCLK	video sample clock
VID0–VID7	lower video data bus
VIDA, VIDB	10-bit data extension
XPIXCLK	video sample clock
XVID0–XVID7	upper video data bus
XVIDA, XVIDB	10-bit data extension

System Interface:

VRST#	reset
VIRQ#	interrupt request

The host interface signals are provided by the graphics controller. Essentially, a 2-, 4-, or 8-bit version of the PCI interface is used. VIPCLK has a frequency range of 25–33 MHz. PIXCLK and XPIXCLK have a maximum frequency of 75 and 80 MHz, respectively.

Video Interface

As with the "BT.656" interface, special four-word sequences are inserted into the 8-bit or 10-bit 4:2:2 YCbCr video stream to indicate the start of active video (SAV) and end of active video (EAV). These sequences also indicate when horizontal and vertical blanking are present and which field is being transmitted.

VIP modifies the BT.656 EAV and SAV sequences as shown in Table 6.34. BT.656 uses four protection bits (P0–P3) in the status word since it was designed for long cable connections between equipment. With chip-to-chip interconnect, this protection isn't required, so the bits are used for other purposes. The timing of the H, V, and F signals for common video formats are illustrated in Chapter 4. The status word for VIP is defined as:

T = "0" for task B	T = "1" for task A
F = "0" for Field 1	F = "1" for Field 2
V = "1" during vertical blanking	
H = "0" at SAV	H = "1" at EAV

The task bit, T, is programmable. If BT.656 compatibility is required, it should always be a "1." Otherwise, it may be used to indicate which one of two data streams are present: stream A = "1" and stream B = "0." Alternately, T may be a "0" when raw 2× oversampled VBI data is present, and a "1" otherwise.

The noninterlaced bit, N, indicates whether the source is progressive ("1") or interlaced ("0").

The repeat bit, R, is a "1" if the current field is a repeat field. This occurs only during 3:2 pull-down. The repeat bit (R), in conjunction with the noninterlaced bit (N), enables the graphics controller to handle Bob and Weave, as well as 3:2 pull-down (further discussed in Chapter 7), in hardware.

The extra flag bit, E, is a "1" if another byte follows the EAV. Table 6.35 illustrates the extra flag byte. This bit is valid only during EAV sequences. If the E bit in the extra byte is "1," another extra byte immediately follows. This allows chaining any number of extra bytes together as needed.

	8-bit Data							
	D7 (MSB)	D6	D5	D4	D3	D2	D1	D0
preamble	1	1	1	1	1	1	1	1
	0	0	0	0	0	0	0	0
	0	0	0	0	0	0	0	0
status word	T	F	V	H	N	R	0	E

Table 6.34. VIP EAV and SAV Sequence.

	8-bit Data							
	D7 (MSB)	D6	D5	D4	D3	D2	D1	D0
extra byte	$\overline{D0}$			user defined				E

Table 6.35. VIP EAV Extra Byte.

Unlike pro-video interfaces, code 00_H may be used during active video data to indicate an invalid video sample. This is used to accommodate scaled video and square pixel timing.

Video Data Formats

In the 8-bit mode (Figure 6.41), the video interface is similar to BT.656, except for the differences mentioned. XVID0–XVID7 are not used.

In the 16-bit mode (Figure 6.42), SAV sequences, EAV sequences, Y video data, ancillary packet headers, and even-numbered ancillary data values are transferred across the lower 8 bits (VID0–VID7). CbCr video data and odd-numbered ancillary data values are transferred across the upper 8 bits (XVID0–XVID7).

Note that "skip data" (value 00_H) during active video must also appear in 16-bit format to preserve the 16-bit data alignment.

10-bit video data is supported by the VIDA, VIDB, XVIDA, and XVIDB signals. VIDA and XVIDA are the least significant bits.

Ancillary Data

Ancillary data packets are used to transmit information (such as digital audio, closed captioning, and teletext data) during the blanking intervals, as shown in Table 6.36. Unlike pro-video interfaces, the 00_H and FF_H values may be used by the ancillary data. Note that the ancillary data formats were defined prior to many of the pro-video ancillary data formats, and therefore may not match.

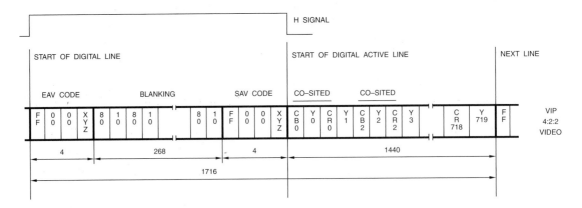

Figure 6.41. VIP 8-Bit Interface Data for One Scan Line. 480i; 720 active samples per line; 27 MHz clock.

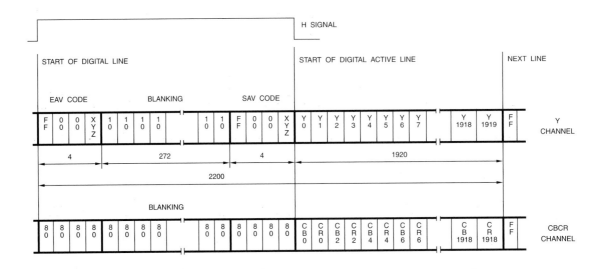

Figure 6.42. VIP 16-Bit Interface Data for One Scan Line. 1080i; 1920 active samples per line; 74.176 or 74.25 MHz clock.

	8-bit Data							
	D7 (MSB)	**D6**	**D5**	**D4**	**D3**	**D2**	**D1**	**D0**
ancillary data flag (ADF)	0	0	0	0	0	0	0	0
	1	1	1	1	1	1	1	1
	1	1	1	1	1	1	1	1
data ID (DID)	$\overline{D6}$	EP	0	1	0	DID2	DID1	DID0
SDID	$\overline{D6}$	EP	user defined value					
data count (DC)	$\overline{D6}$	EP	DC5	DC4	DC3	DC2	DC1	DC0
internal data ID 0	user defined value							
internal data ID 1	user defined value							
data word 0	D7	D6	D5	D4	D3	D2	D1	D0
:				:				
data word N	D7	D6	D5	D4	D3	D2	D1	D0
check sum	$\overline{D6}$	EP	CS5	CS4	CS3	CS2	CS1	CS0
optional fill data	$\overline{D6}$	EP	0	0	0	0	0	0

Notes:
EP = even parity for D0–D5.

Table 6.36. VIP Ancillary Data Packet General Format.

DID2 of the DID field indicates whether Field 1 or Field 2 ancillary data is present:

0   Field 1
1   Field 2

DID1–DID0 of the DID field indicate the type of ancillary data present:

00   start of field
01   sliced VBI data, lines 1–23
10   end of field VBI data, line 23
11   sliced VBI data, line 24 to end of field

The data count value (DC) specifies the number of D-words (4-byte blocks) of ancillary data present. Thus, the number of data words in the ancillary packet after the DID must be a multiple of four. 1–3 optional fill bytes may be added after the check sum data to meet this requirement.

When DID1–DID0 are "00" or "10," no ancillary data or check sum is present. The data count (DC) value is "00000," and is the last field present in the packet.

Consumer Component Interfaces

Many solutions for transferring digital video between equipment has been developed over the years. The DVI and HDMI standards are the most popular digital video interfaces for consumer equipment.

Digital Visual Interface (DVI)

In 1998, the Digital Display Working Group (DDWG) was formed to address the need for a standardized digital video interface between a PC and VGA monitor, as illustrated in Figure 6.43. The DVI 1.0 specification was released in April 1999.

Designed to transfer uncompressed real-time digital video, DVI supports PC graphics resolutions beyond 1600 × 1200 and HDTV resolutions, including 720p, 1080i, and 1080p.

In 2003, the consumer electronics industry started adding DVI outputs to DVD players and cable/satellite settop boxes. DVI inputs also started appearing on digital televisions and LCD/plasma monitors.

Technology

DVI is based on the Digital Flat Panel (DFP) Interface, enhancing it by supporting more formats and timings. It also includes support for the High-bandwidth Digital Content Protection (HDCP) specification to deter unauthorized copying of content.

DVI also supports VESA's Extended Display Identification Data (EDID) standard, Display Data Channel (DDC) standard (used to read the EDID), and Monitor Timing Specification (DMT).

DDC and EDID enable automatic display detection and configuration. Extended Display Identification Data (EDID) was created to enable plug and play capabilities of displays. Data is stored in the display, describing the supported video formats This information is supplied to the source device, over DVI, at the request of the source device. The source device then chooses its output format, taking into account the format of the original video stream and the formats supported by the display. The source device is responsible for the format conversions necessary to supply video in an understandable form to the display.

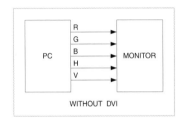 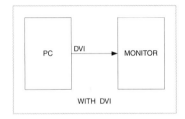

Figure 6.43. Using DVI to Connect a VGA Monitor to a PC.

In addition, the EIA-861 standard specifies mandatory and optionally supported resolutions and timings, and how to include data such as aspect ratio and format information.

TMDS Links

DVI uses transition-minimized differential signaling (TMDS). Eight bits of video data are converted to a 10-bit transition-minimized, DC-balanced value, which is then serialized. The receiver deserializes the data, and converts it back to eight bits. Thus, to transfer digital R′G′B′ data requires three TMDS signals that comprise one TMDS link.

"TFT data mapping" is supported as the minimum requirement: one pixel per clock, eight bits per channel, MSB justified.

Either one or two TMDS links may be used, as shown in Figures 6.44 and 6.45, depending on the formats and timing required. A system supporting two TMDS links must be able to switch dynamically between formats requiring a single link and formats requiring a dual link. A single DVI connector can handle two TMDS links.

A single TMDS link supports resolutions and timings using a video sample rate of 25–165 MHz. Resolutions and timings using a video sample rate of 165–330 MHz are implemented using two TMDS links, with each TMDS link operating at one-half the frequency. Thus, the two TMDS links share the same clock and the bandwidth is shared evenly between the two links.

Video Data Formats

Typically, 24-bit R′G′B′ data is transferred over a link. For applications requiring more than eight bits per color component, the second TMDS link may be used for the additional least significant bits.

For PC applications, R′G′B′ data typically has a range of 00_H–FF_H. For consumer applications, R′G′B′ data typically has a range of 10_H–EB_H (values less than 10_H or greater than EB_H may be occasionally present due to processing).

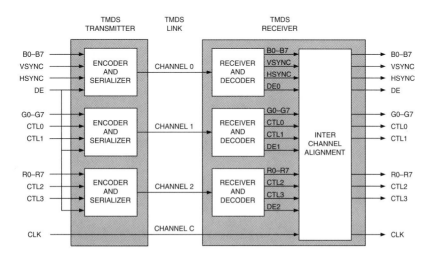

Figure 6.44. DVI Single TMDS Link.

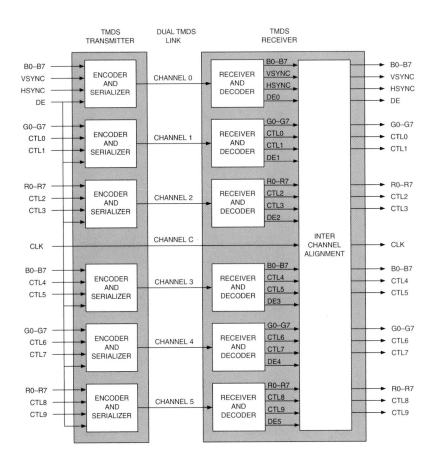

Figure 6.45. DVI Dual TMDS Link.

Control Signals

In addition to the video data, DVI transmitter and receiver chips typically use up to 14 control signals for interfacing to other chips in the system:

HSYNC	horizontal sync
VSYNC	vertical sync
DE	data enable
CTL0–CTL3	reserved (link 0)
CTL4–CTL9	reserved (link 1)
CLK	1× sample clock

While DE is a "1," active video is processed. While DE is a "0," the HSYNC, VSYNC and CTL0–CTL9 signals are processed. HSYNC and VSYNC may be either polarity.

One issue is that some HDTVs use the falling edge of the YPbPr tri-level sync, rather than the center (rising edge), for horizontal timing. When displaying content from DVI, this results in the image shifting by 2.3%. Providing the ability to adjust the DVI embedded sync timing relative to the YPbPr tri-level sync timing is a useful capability in this case. Many fixed-pixel displays, such as DLP, LCD and plasma, instead use the DE signal as a timing reference, avoiding the issue.

Digital-Only (DVI-D) Connector

The digital-only connector, which supports dual link operation, contains 24 contacts arranged as three rows of eight contacts, as shown in Figure 6.46. Table 6.37 lists the pin assignments.

Digital-Analog (DVI-I) Connector

In addition to the 24 contacts used by the digital-only connector, the 29-contact digital-analog connector adds five additional contacts to support analog video as shown in Figure 6.47. Table 6.38 lists the pin assignments.

HSYNC	horizontal sync
VSYNC	vertical sync
RED	analog red video
GREEN	analog green video
BLUE	analog blue video

The operation of the analog signals is the same as for a standard VGA connector.

DVI-A is available as a plug (male) connector only and mates to the analog-only pins of a DVI-I connector. DVI-A is only used in adapter cables, where there is the need to convert to or from a traditional analog VGA signal.

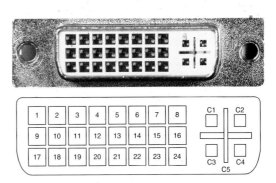

Figure 6.46. DVI-D Connector.   **Figure 6.47. DVI-I Connector.**

Pin	Signal	Pin	Signal	Pin	Signal
1	D2–	9	D1–	17	D0–
2	D2	10	D1	18	D0
3	shield	11	shield	19	shield
4	D4–	12	D3–	20	D5–
5	D4	13	D3	21	D5
6	DDC SCL	14	+5V	22	shield
7	DDC SDA	15	ground	23	CLK
8	reserved	16	Hot Plug Detect	24	CLK–

Table 6.37. DVI-D Connector Signal Assignments.

Pin	Signal	Pin	Signal	Pin	Signal
1	D2–	9	D1–	17	D0–
2	D2	10	D1	18	D0
3	shield	11	shield	19	shield
4	D4–	12	D3–	20	D5–
5	D4	13	D3	21	D5
6	DDC SCL	14	+5V	22	shield
7	DDC SDA	15	ground	23	CLK
8	VSYNC	16	Hot Plug Detect	24	CLK–
C1	RED	C2	GREEN	C3	BLUE
C4	HSYNC	C5	ground		

Table 6.38. DVI-I Connector Signal Assignments.

High Definition Multimedia Interface (HDMI)

Although DVI handles transferring uncompressed real-time digital RGB video to a display, the consumer electronics industry preferred a smaller, more flexible solution, based on DVI technology. In April 2002, the HDMI working group was formed by Hitachi, Matsushita Electric (Panasonic), Philips, Silicon Image, Sony, Thomson and Toshiba.

HDMI is capable of replacing up to eight audio cables (7.1 channels) and up to three video cables with a single cable, as illustrated in Figure 6.48. In 2004, the consumer electronics industry started adding HDMI outputs to DVD players and cable/satellite set-top boxes. HDMI inputs also started appearing on digital televisions and LCD/plasma monitors, with wide usage expected by 2005.

Through the use of an adaptor cable, HDMI is backwards compatible with equipment using DVI and the EIA-861 DTV profile. However, the advanced features of HDMI, such as digital audio and Consumer Electronics Control (used to enable passing control commands between equipment), are not available.

Technology

HDMI, based on DVI, supports VESA's Extended Display Identification Data (EDID) standard and Display Data Channel (DDC) standard (used to read the EDID).

In addition, the EIA-861 standard specifies mandatory and optionally supported resolutions and timings, and how to include data such as aspect ratio and format information

HDMI also supports the High-bandwidth Digital Content Protection (HDCP) specification to deter unauthorized copying of content. A common problem is sources not polling the TV often enough (twice per second) to see if its HDCP circuit is active. This results in "snow" if the TV's HDMI input is deselected, then later selected again.

The 19-pin "Type A" connector uses a single TMDS link and can therefore carry video signals with a 25–165 MHz sample rate. Video with sample rates below 25 MHz (i.e. 13.5 MHz 480i and 576i) are transmitted using a pixel-repetition scheme. This smaller connector is well-suited for consumer devices due to its small size.

To support video signals sampled at greater than 165 MHz, the dual-link capability of the 29-pin "Type B" connector designed for the PC market must be used.

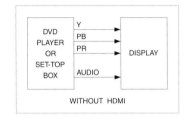
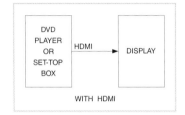

Figure 6.48. Using HDMI Eliminates Confusing Cable Connections For Consumers.

Video Data Formats

HDMI supports RGB, 4:4:4 YCbCr and 4:2:2 YCbCr. Up to 24 bits per pixel can be transferred.

For the 640 × 480 resolution, the R′G′B′ data has a range of 00_H–FF_H. For YCbCr data and all other RGB resolutions, data has a range of 10_H–EB_H (values less than 10_H or greater than EB_H may be occasionally present due to processing).

Audio Data Formats

Driven by the DVD-Audio standard, audio support consists of 1–8 uncompressed audio streams with a sample rate of up to 48, 96 or 192 kHz, depending on the video format. It can alternately carry a compressed multi-channel audio stream at sample rates up to 192 kHz.

Digital Flat Panel (DFP) Interface

The VESA DFP interface was developed for transferring uncompressed digital video from a computer to a digital flat panel display. It supports VESA's Plug and Display (P&D) standard, Extended Display Identification Data (EDID) standard, Display Data Channel (DDC) standard, and Monitor Timing Specification (DMT). DDC and EDID enable automatic display detection and configuration. Only "TFT data mapping" is supported: one pixel per clock, eight bits per channel, MSB justified.

Like DVI, DFP uses transition-minimized differential signaling (TMDS). Eight bits of video data are converted to a 10-bit transition-minimized, DC-balanced value, which is then serialized. The receiver deserializes the data, and converts it back to eight bits. Thus, to transfer digital R′G′B′ data requires three TMDS signals that comprise one TMDS link. Cable lengths may be up to 5 meters.

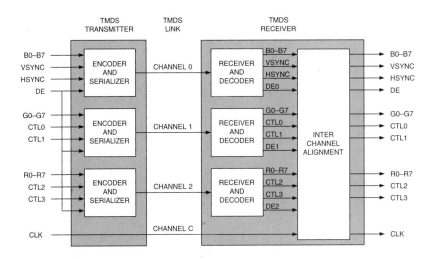

Figure 6.49. DFP TMDS Link.

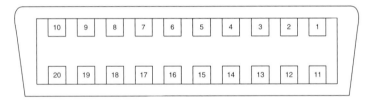

Figure 6.50. DFP Connector.

TMDS Links

A single TMDS link, as shown in Figure 6.49, supports formats and timings requiring a clock rate of 22.5–160 MHz.

Video Data Formats

24-bit R′G′B′ data is transferred over the link, as shown in Figure 6.49.

Control Signals

In addition to the video data, DFP transmitter and receiver chips typically use up to 8 control signals for interfacing to other chips in the system:

HSYNC	horizontal sync
VSYNC	vertical sync
DE	data enable
CTL0–CTL3	reserved
CLK	1× sample clock

While DE is a "1," active video is processed. While DE is a "0," the HSYNC, VSYNC and CTL0–CTL3 signals are processed. HSYNC and VSYNC may be either polarity.

Connector

The 20-pin mini-D ribbon (MDR) connector contains 20 contacts arranged as two rows of ten contacts, as shown in Figure 6.50. Table 6.39 lists the pin assignments.

Pin	Signal	Pin	Signal
1	D1	11	D2
2	D1–	12	D2–
3	shield	13	shield
4	shield	14	shield
5	CLK	15	D0
6	CLK–	16	D0–
7	ground	17	no connect
8	+5V	18	Hot Plug Detect
9	no connect	19	DDC SDA
10	no connect	20	DDC SCL

Table 6.39. DFP Connector Signal Assignments.

Open LVDS Display Interface (OpenLDI)

OpenLDI was developed for transferring uncompressed digital video from a computer to a digital flat panel display. It enhances the FPD-Link standard used to drive the displays of laptop computers, and adds support for VESA's Plug and Display (P&D) standard, Extended Display Identification Data (EDID) standard, and Display Data Channel (DDC) standard. DDC and EDID enable automatic display detection and configuration.

Unlike DVI and DFP, OpenLDI uses low-voltage differential signaling (LVDS). Cable lengths may be up to 10 meters.

LVDS Link

The LVDS link, as shown in Figure 6.51, supports formats and timings requiring a clock rate of 32.5–160 MHz.

Eight serial data lines (A0–A7) and two sample clock lines (CLK1 and CLK2) are used. The number of serial data lines actually used is dependent on the pixel format, with the serial data rate being 7× the sample clock rate. The CLK2 signal is used in the dual pixel modes for backwards compatibility with FPD-Link receivers.

Video Data Formats

18-bit single pixel, 24-bit single pixel, 18-bit dual pixel, or 24-bit dual pixel R′G′B′ data is transferred over the link. Table 6.40 illustrates the mapping between the pixel data bit number and the OpenLDI bit number.

The 18-bit single pixel R′G′B′ format uses three 6-bit R′G′B′ values: R0–R5, G0–G5, and B0–B5. OpenLDI serial data lines A0–A2 are used to transfer the data.

The 24-bit single pixel R′G′B′ format uses three 8-bit R′G′B′ values: R0–R7, G0–G7, and B0–B7. OpenLDI serial data lines A0–A3 are used to transfer the data.

The 18-bit dual pixel R′G′B′ format represents two pixels as three upper/lower pairs of 6-bit R′G′B′ values: RU0–RU5, GU0–GU5, BU0–BU5, RL0–RL5, GL0–GL5, BL0–BL5. Each upper/lower pair represents two pixels. OpenLDI serial data lines A0–A2 and A4–A6 are used to transfer the data.

The 24-bit dual pixel R′G′B′ format represents two pixels as three upper/lower pairs of 8-bit R′G′B′ values: RU0–RU7, GU0–GU7, BU0–BU7, RL0–RL7, GL0–GL7, BL0–BL7. Each upper/lower pair represents two pixels. OpenLDI serial data lines A0–A7 are used to transfer the data.

Control Signals

In addition to the video data, OpenLDI transmitter and receiver chips typically use up to seven control signals for interfacing to other chips in the system:

HSYNC	horizontal sync
VSYNC	vertical sync
DE	data enable
CNTLE	reserved
CNTLF	reserved
CLK1	1× sample clock
CLK2	1× sample clock

During unbalanced operation, the DE, HSYNC, VSYNC, CNTLE, and CNTLF levels are sent as unencoded bits within the A2 and A6 bitstreams.

During balanced operation (used to minimize short- and long-term DC bias), a DC Balance bit is sent within each of the A0–A7 bitstreams to indicate whether the data is unmodified or inverted. Since there is no room left for the control signals to be sent directly, the DE level is sent by slightly modifying the

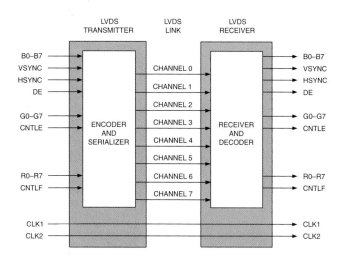

Figure 6.51. OpenLDI LVDS Link.

18 Bits per Pixel Bit Number	24 Bits per Pixel Bit Number	OpenLDI Bit Number
5	7	5
4	6	4
3	5	3
2	4	2
1	3	1
0	2	0
	1	7
	0	6

Table 6.40. OpenLDI Bit Number Mappings.

timing of the falling edge of the CLK1 and CLK2 signals. The HSYNC, VSYNC, CNTLE and CNTLF levels are sent during the blanking intervals using 7-bit code words on the A0, A1, A5, and A4 signals, respectively.

Connector

The 36-pin mini-D ribbon (MDR) connector is similar to the one shown in Figure 6.50, except that there are two rows of eighteen contacts. Table 6.41 lists the pin assignments.

Gigabit Video Interface (GVIF)

The Sony GVIF was developed for transferring uncompressed digital video using a single differential signal, instead of the multiple signals that DVI, DFP, and OpenLDI use. Cable lengths may be up to 10 meters.

GVIF Link

The GVIF link, as shown in Figure 6.52, supports formats and timings requiring a clock rate of 20–80 MHz. For applications requiring higher clock rates, more than one GVIF link may be used.

The serial data rate is 24× the sample clock rate for 18-bit R′G′B′ data, or 30× the sample clock rate for 24-bit R′G′B′ data.

Video Data Formats

18-bit or 24-bit R′G′B′ data, plus timing, is transferred over the link. The 18-bit R′G′B′ format uses three 6-bit R′G′B′ values: R0–R5, G0–G5, and B0–B5. The 24-bit R′G′B′ format uses three 8-bit R′G′B′ values: R0–R7, G0–G7, and B0–B7.

Pin	Signal	Pin	Signal	Pin	Signal
1	A0–	13	+5V	25	reserved
2	A1–	14	A4–	26	reserved
3	A2–	15	A5–	27	ground
4	CLK1–	16	A6–	28	DDC SDA
5	A3–	17	A7–	29	ground
6	ground	18	CLK2–	30	USB–
7	reserved	19	A0	31	ground
8	reserved	20	A1	32	A4
9	reserved	21	A2	33	A5
10	DDC SCL	22	CLK1	34	A6
11	+5V	23	A3	35	A7
12	USB	24	reserved	36	CLK2

Table 6.41. OpenLDI Connector Signal Assignments.

18-bit R′G′B′ data is converted to 24-bit data by slicing the R′G′B data into six 3-bit values that are in turn transformed into six 4-bit codes. This ensures rich transitions for receiver PLL locking and good DC balance.

24-bit R′G′B′ data is converted to 30-bit data by slicing the R′G′B data into six 4-bit values that are in turn transformed into six 5-bit codes.

Control Signals

In addition to the video data, there are six control signals:

HSYNC	horizontal sync
VSYNC	vertical sync
DE	data enable
CTL0	reserved
CTL1	reserved
CLK	1× sample clock

If any of the HSYNC, VSYNC, DE, CTL0, or CTL1 signals change, during the next CLK cycle a special 30-bit format is used. The first six bits are header data indicating the new levels of HSYNC, VSYNC, DE, CTL0, or CTL1. This is followed by 24 bits of R′G′B data (unencoded except for inverting the odd bits).

Note that during the blanking periods, non-video data, such as digital audio, may be transferred. The CTL signals may be used to indicate when non-video data is present.

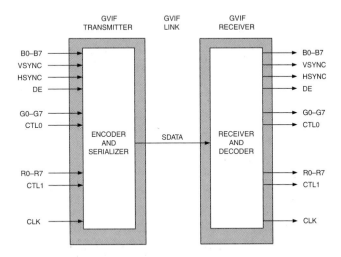

Figure 6.52. GVIF Link.

Consumer Transport Interfaces

Several transport interfaces, such as USB 2.0, Ethernet and IEEE 1394, are available for consumer products. Of course, each standard has its own advantages and disadvantages.

USB 2.0

Well known in the PC market for connecting peripherals to a PC, there is growing interest in using USB (Universal Serial Bus) 2.0 to transfer compressed audio/video data between products.

USB 2.0 is capable of operating up to 480 Mbps and supports an isochronous mode to guarantee data delivery timing. Thus, it can easily transfer compressed real-time audio/video data from a cable/satellite set-top box or DVD player to a digital television. DTCP (Digital Transmission Copy Protection) may be used to encrypt the audio and video content over USB.

Due to USB's lower cost and widespread usage, many companies are interested in using USB 2.0 instead of IEEE 1394 to transfer compressed audio/video data between products. However, some still prefer IEEE 1394 since the methods for transferring various types of data are much better defined.

USB On-The-Go

With portable devices increasing in popularity, there was a growing desire for them to communicate directly with each other without requiring a PC or other USB host.

"On-The-Go" addresses this desire by allowing a USB device to communicate directly with other "On-The-Go" products. It also features a smaller USB connector and low power features to preserve battery life.

Ethernet

With the widespread adoption of home networks, DSL and FTTH (Fiber-To-The-Home), Ethernet has become a common interface for transporting digital audio and video data. Initially used for file transfers, streaming of real-time compressed video over wired or wireless Ethernet networks is now becoming common.

Ethernet commonly supports up to 100 Mbps, with use of 1 Gbps starting to increase. DTCP (Digital Transmission Copy Protection) may be used to encrypt the audio and video content over wired or wireless networks.

IEEE 1394

IEEE 1394 was originally developed by Apple Computer as Firewire. Designed to be a generic interface between devices, 1394 specifies the physical characteristics; separate application-specific specifications describe how to transfer data over the 1394 network. The Open-Cable™, SCTE DVS-194, EIA-775, and ITU-T J.117 specifications for compatibility between digital televisions and settop boxes specifically include IEEE 1394 support.

1394 is a transaction-based packet technology, using a bi-directional serial interconnect that features hot plug-and-play. This enables devices to be connected and disconnected without affecting the operation of other devices connected to the network.

Guaranteed delivery of time-sensitive data is supported, enabling digital audio and video to be transferred in real time. In addition, multiple independent streams of digital audio and video can be carried.

Specifications

The original 1394-1995 specification supports bit rates of 98.304, 196.608, and 393.216 Mbps.

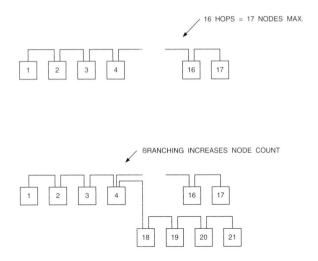

Figure 6.53. IEEE 1394 Network Topology Examples.

The 1394A-2000 specification clarifies areas that were vague and led to system interoperability issues. It also reduces the overhead lost to bus control, arbitration, bus reset duration, and concatenation of packets. 1394A-2000 also introduces advanced power-savings features. The electrical signalling method is also common between 1394-1995 and 1394A-2000, using data-strobe (DS) encoding and analog-speed signaling.

The 1394B-2002 specification adds support for bit rates of 786.432, 1572.864, and 3145.728 Mbps. It also includes

- 8B/10B encoding technique used by Giga-bit Ethernet

- Continuous dual simplex operation

- Longer distance (up to 100 meters over Cat5)

- Changes the speed signalling to a more digital method

- Three types of ports: Legacy (1395A compatible), Beta and Bilingual (supports both Legacy and Beta). Connector keying ensures that incompatible connections cannot physically be made.

Endian Issues

1394 uses a big-endian architecture, defining the most significant bit as bit 0. However, many processors are based on the little endian architecture which defines the most significant bit as bit 31 (assuming a 32-bit word).

Network Topology

Like many networks, there is no designated bus master. The tree-like network structure has a root node, branching out to logical nodes in other devices (Figure 6.53). The root is responsible for certain control functions, and is chosen during initialization. Once chosen, it retains that function for as long as it remains powered-on and connected to the network.

A network can include up to 63 nodes, with each node (or device) specified by a 6-bit physical identification number. Multiple networks may be connected by bridges, up to a system maximum of 1,023 networks, with each network represented by a separate 10-bit bus ID. Combined, the 16-bit address allows up to 64,449 nodes in a system. Since device addresses are 64 bits, and 16 of these bits are used to specify nodes and networks, 48 bits remain for memory addresses, allowing up to 256TB of memory space per node.

Node Types

Nodes on a 1394 bus may vary in complexity and capability (listed simplest to most complex):

Transaction nodes respond to asynchronous communication, implement the minimal set of control status registers (CSR), and implement a minimal configuration ROM.

Isochronous nodes add a 24.576 MHz clock used to increment a cycle timer register that is updated by cycle start packets.

Cycle master nodes add the ability to generate the 8 kHz cycle start event, generate cycle start packets, and implement a bus timer register.

Isochronous resource manager (IRM) nodes add the ability to detect bad self-ID packets, determine the node ID of the chosen IRM, and implement the channels available, bandwidth available, and bus manager ID registers. At least one node must be capable of acting as an IRM to support isochronous communication.

Bus manager (BM) nodes are the most complex. This level adds responsibility for storing every self-ID packet in a topology map and analyzing that map to produce a speed map of the entire bus. These two maps are used to manage the bus. Finally, the BM must be able to activate the cycle master node, write configuration packets to allow optimization of the bus, and act as the power manager.

Node Ports

In the network topology, a one-port device is known as a "leaf" device since it is at the end of a network branch. They can be connected to the network, but cannot expand the network.

Two-port devices can be used to form daisy-chained topologies. They can be connected to and continue the network, as shown in Figure 6.53. Devices with three or more ports are able to branch the network to the full 63-node capability.

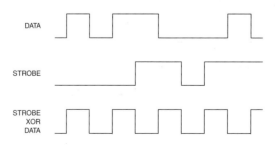

Figure 6.54. IEEE 1394 Data and Strobe Signal Timing.

It is important to note that no loops or parallel connections are allowed within the network. Also, there are no reserved connectors—any connector may be used to add a new device to the network.

Since 1394-1995 mandates a maximum of 16 cable "hops" between any two nodes, a maximum of 17 peripherals can be included in a network if only two-port peripherals are used. Later specifications implement a "ping" packet to measure the round-trip delay to any node, removing the 16 "hop" limitation.

For 1394-1995 and 1394A-2000, a 4- or 6-pin connector is used. The 6-pin connector can provide power to peripherals. For 1394B-2002, the 9-pin Beta and Bilingual connector includes power, two extra pins for signal integrity and one pin for reserved for future use.

Figure 6.54 illustrates the 1394-1995 and 1394A-2000 data and strobe timing. The strobe signal changes state on every bit period for which the data signal does not. Therefore, by exclusive-ORing the data and strobe signals, the clock is recovered.

Physical Layer

The typical hardware topology of a 1394 network consists of a physical layer (PHY) and link layer (LINK), as shown in Figure 6.55. The 1394-1995 standard also defined two software layers, the transaction layer and the bus management layer, parts of which may be implemented in hardware.

The PHY transforms the point-to-point network into a logical physical bus. Each node is also essentially a data repeater since data is

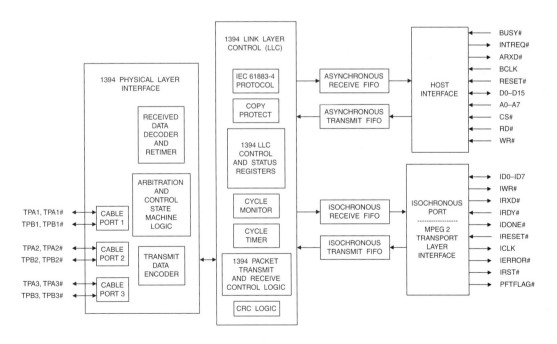

Figure 6.55. IEEE 1394 Typical Physical and Link Layer Block Diagrams.

reclocked at each node. The PHY also defines the electrical and mechanical connection to the network. Physical signaling circuits and logic responsible for power-up initialization, arbitration, bus-reset sensing, and data signaling are also included.

Link Layer

The LINK provides interfacing between the physical layer and application layer, formatting data into packets for transmission over the network. It supports both asynchronous and isochronous data.

Asynchronous Data

Asynchronous packets are guaranteed delivery since after an asynchronous packet is received, the receiver transmits an acknowledgment to the sender, as shown in Figure 6.56. However, there is no guaranteed bandwidth. This type of communication is useful for commands, non-real-time data, and error-free transfers.

The delivery latency of asynchronous packets is not guaranteed and depends upon the network traffic. However, the sender may continually retry until an acknowledgment is received.

Asynchronous packets are targeted to one node on the network or can be sent to all nodes, but can not be broadcast to a subset of nodes on the bus.

The maximum asynchronous packet size is:

512 * (n / 100) bytes

n = network speed in Mbps

Isochronous Data

Isochronous communications have a guaranteed bandwidth, with up to 80% of the network bandwidth available for isochronous use. Up to 63 independent isochronous channels are available, although the 1394 Open Host Controller Interface (OHCI) currently only supports 4–32 channels. This type of communi-

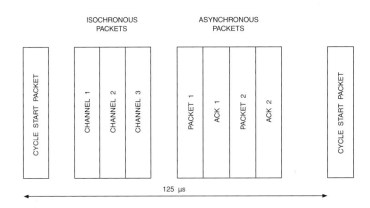

Figure 6.56. IEEE 1394 Isochronous and Asynchronous Packets.

cation is useful for real-time audio and video transfers since the maximum delivery latency of isochronous packets is calculable and may be targeted to multiple destinations. However, the sender may not retry sending a packet.

The maximum isochronous packet size is:

1024 * (n / 100) bytes

n = network speed in Mbps

Isochronous operation guarantees a time slice each 125 µs. Since time slots are guaranteed, and isochronous communication takes priority over asynchronous, isochronous bandwidth is assured.

Once an isochronous channel is established, the sending device is guaranteed to have the requested amount of bus time for that channel every isochronous cycle. Only one device may send data on a particular channel, but any number of devices may receive data on a channel. A device may use multiple isochronous channels as long as capacity is available.

Transaction Layer

The transaction layer supports asynchronous write, read, and lock commands. A lock combines a write with a read by producing a round trip routing of data between the sender and receiver, including processing by the receiver.

Bus Management Layer

The bus management layer control functions of the network at the physical, link, and transaction layers.

Digital Transmission Content Protection (DTCP)

To prevent unauthorized copying of content, the DTCP system was developed. Although originally designed for 1394, it is applicable to any digital network that supports bi-directional communications, such as USB and Ethernet.

Device authentication, content encryption, and renewability (should a device ever be compromised) are supported by DTCP. The Digital Transmission Licensing Administrator (DTLA) licenses the content protection system and distributes cipher keys and device certificates.

DTCP outlines four elements of content protection:

1. Copy control information (CCI)

2. Authentication and key exchange (AKE)

3. Content encryption

4. System renewability

Copy Control Information (CCI)

CCI allows content owners to specify how their content can be used, such as "copy-never," "copy-one-generation," "no-more-copies," and "copy-free." DTCP is capable of securely communicating copy control information between devices. Two different CCI mechanisms are supported: *embedded* and *encryption mode indicator.*

Embedded CCI is carried within the content stream. Tampering with the content stream results in incorrect decryption, maintaining the integrity of the embedded CCI.

The *encryption mode indicator* (EMI) provides a secure, yet easily accessible, transmission of CCI by using the two most significant bits of the sync field of the isochronous packet header. Devices can immediately determine the CCI of the content stream without decoding the content. If the two EMI bits are tampered with, the encryption and decryption modes do not match, resulting in incorrect content decryption.

Authentication and Key Exchange (AKE)

Before sharing content, a device must first verify that the other device is authentic. DTCP includes a choice of two authentication levels: *full* and *restricted*. Full authentication can be used with all content protected by the system. Restricted authentication enables the protection of "copy-one-generation" and "no-more-copies" content only.

Full authentication

Compliant devices are assigned a unique public/private key pair and a device certificate by the DTLA, both stored within the device so as to prevent their disclosure. In addition, devices store other necessary constants and keys.

Full authentication uses the public key-based digital signature standard (DSS) and Diffie-Hellman (DH) key exchange algorithms. DSS is a method for digitally signing and verifying the signatures of digital documents to verify the integrity of the data. DH key exchange is used to establish control-channel symmetric cipher keys, which allows two or more devices to generate a shared key.

Initially, the receiver sends a request to the source to exchange device certificates and random challenges. Then, each device calculates a DH key exchange first-phase value. The devices then exchange signed messages that contain the following elements:

1. The other device's random challenge

2. The DH key-exchange first-phase value

3. The renewability message version number of the newest system renewability message (SRM) stored by the device

The devices check the message signatures using the other device's public key to verify that the message has not been tampered with and also verify the integrity of the other device's certificate. Each device also examines the certificate revocation list (CRL) embedded in its system renewability message (SRM) to verify that the other device's certificate has not been revoked due to its security having been compromised. If no errors have occurred, the two devices have successfully authenticated each other and established an authorization key.

Restricted authentication

Restricted authentication may be used between sources and receivers for the exchange of "copy-one-generation" and "no-more-copies" contents. It relies on the use of a shared secret to respond to a random challenge.

The source initiates a request to the receiver, requests its device ID, and sends a random challenge. After receiving the challenge back from the source, the receiver computes a response and sends it to the source.

The source compares this response with similar information generated by the source using its service key and the ID of the receiver. If the comparison matches its own calculation, the receiver has been verified and authenticated. The source and receiver then each calculate an authorization key.

Content Encryption

To ensure interoperability, all compliant devices must support the 56-bit M6 baseline cipher. Additional content protection may be supported by using additional, optional ciphers.

System Renewability

Devices that support full authentication can receive and process SRMs that are created by the DTLA and distributed with content. System renewability is used to ensure the long-term system integrity by revoking the device IDs of compromised devices.

SRMs can be updated from other compliant devices that have a newer list, from media with prerecorded content, or via compliant devices with external communication capability (Internet, phone, cable, network, and so on).

Example Operation

For this example, the source has been instructed to transmit a copy protected system stream of content.

The source initiates the transmission of content marked with the copy protection status: "copy-one-generation," "copy-never," "no-more-copies," or "copy-free."

Upon receiving the content stream, the receiver determines the copy protection status. If marked "copy never," the receiver requests that the source initiate full authentication. If the content is marked "copy once" or "no more copies," the receiver will request full authentication if supported, or restricted authentication if it isn't.

When the source receives the authentication request, it proceeds with the requested type of authentication. If full authentication is requested but the source can only support restricted authentication, then restricted authentication is used.

Once the devices have completed the authentication procedure, a content-channel encryption key (content key) is exchanged between them. This key is used to encrypt the content at the source device and decrypt the content at the receiver.

1394 Open Host Controller Interface (OHCI)

The 1394 Open Host Controller Interface (OHCI) specification is an implementation of the 1394 link layer, with additional features to support the transaction and bus management layers. It provides a standardized way of interacting with the 1394 network.

Home AV Interoperability (HAVi)

Home AV Interoperability (HAVi) is another layer of protocols for 1394. HAVi is directed at making 1394 devices plug-and-play interoperable in a 1394 network whether or not a PC host is present.

Serial Bus Protocol (SBP-2)

The ANSI Serial Bus Protocol 2 (SBP-2) defines standard way of delivering command and status packets over 1394 for devices such as DVD players, printers, scanners, hard drives, and other devices.

IEC 61883 Specifications

Certain types of isochronous signals, such as MPEG-2 and the IEC 61834, SMPTE 314M and ITU-R BT.1618 digital video (DV) standards, use specific data transport protocols and formats. When this data is sent isochronously over a 1394 network, special packetization techniques are used.

The IEC 61883 series of specifications define the details for transferring various application-specific data over 1394:

IEC 61883-1 = General specification

IEC 61883-2 = SD-DVCR data transmission 25 Mbps continuous bit rate

IEC 61883-3 = HD-DVCR data transmission

IEC 61883-4 = MPEG-2 TS data transmission bit rate bursts up to 44 Mbps

IEC 61883-5 = SDL-DVCR data transmission

IEC 61883-6 = Audio and music data transmission

IEC 61883-7 = Transmission of ITU-R BO.1294 System B

IEC 61883-1

IEC 61883-1 defines the general structure for transferring digital audio and video data over 1394. It describes the general packet format, data flow management, and connection management for digital audio and video data, and also the general transmission rules for control commands.

A common isochronous packet (CIP) header is placed at the beginning of the data field of isochronous data packets, as shown in Figure 6.57. It specifies the source node, data block size, data block count, time stamp, type of real-time data contained in the data field, etc.

A connection management procedure (CMP) is also defined for making isochronous connections between devices.

In addition, a functional control protocol (FCP) is defined for exchanging control commands over 1394 using asynchronous data.

IEC 61883-2

IEC 61883-2 and SMPTE 396M define the CIP header, data packet format, and transmission timing for IEC 61834, SMPTE 314M and ITU-R BT.1618 digital video (DV) standards over 1394. Active resolutions of 720 × 480 (at 29.97 frames per second) and 720 × 576 (at 25 frames per second) are supported.

DV data packets are 488 bytes long, made up of 8 bytes of CIP header and 480 bytes of DV data, as shown in Figure 6.57. Figure 6.58 illustrates the frame data structure.

Each of the 720 × 480 4:1:1 YCbCr frames are compressed to 103,950 bytes, resulting in a 4.9:1 compression ratio. Including overhead and audio increases the amount of data to 120,000 bytes.

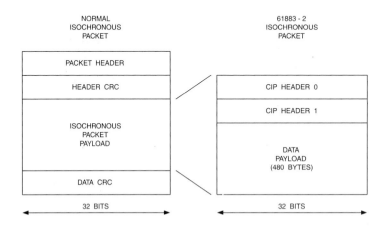

Figure 6.57. 61883-2 Isochronous Packet Formatting.

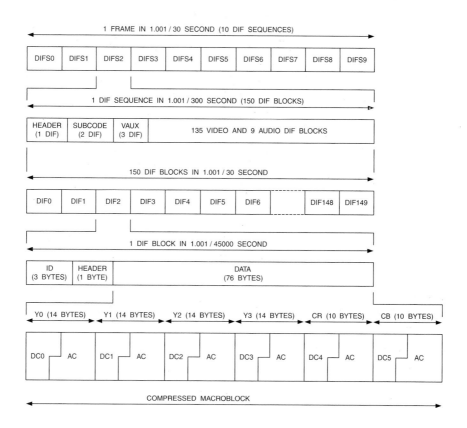

Figure 6.58. IEC 61834, SMPTE 314M and ITU-R BT.1618 Packet Formatting for 720 × 480 Systems (4:1:1 YCbCr).

The compressed 720 × 480 frame is divided into 10 DIF (data in frame) sequences. Each DIF sequence contains 150 DIF blocks of 80 bytes each, used as follows:

135 DIF blocks for video

9 DIF blocks for audio

6 DIF blocks used for Header, Subcode, and Video Auxiliary (VAUX) information

Figure 6.59 illustrates the DIF sequence structure in detail. The audio DIF blocks contain both audio data and audio auxiliary data (AAUX). IEC 61834 supports four 32-kHz, 12-bit nonlinear audio signals or two 48-, 44.1-, or 32-kHz, 16-bit audio signals. SMPTE 314M and ITU-R BT.1618 at 25 Mbps support two 48-kHz 16-bit audio signals, while the 50 Mbps version supports four. Video auxiliary data (VAUX) DIF blocks include recording date and time, lens aperture, shutter speed, color balance, and other camera setting data. The subcode DIF blocks store a variety of information, the most important of which is timecode.

Each video DIF block contains 80 bytes of compressed macroblock data:

3 bytes for DIF block ID information

1 byte for the header that includes the quantization number (QNO) and block status (STA)

14 bytes each for Y0, Y1, Y2, and Y3

10 bytes each for Cb and Cr

As the 488-byte packets come across the 1394 network, the start of a video frame is determined. Once the start of a frame is detected, 250 valid packets of data are collected to have a complete DV frame; each packet contains 6 DIF blocks of data. Every 15th packet is a null packet and should be discarded. Once 250 valid packets of data are in the buffer, discard the CIP headers. If all went well, you have a frame buffer with a 120,000 byte compressed DV frame in it.

720 × 576 frames may use either the 4:2:0 YCbCr format (IEC 61834) or the 4:1:1 YCbCr format (SMPTE 314M and ITU-R BT.1618), and require 12 DIF sequences. Each 720 × 576 frame is compressed to 124,740 bytes. Including overhead and audio increases the amount of data to 144,000 bytes, requiring 300 packets to transfer.

Note that the organization of data transferred over 1394 differs from the actual DV recording format since error correction is not required for digital transmission. In addition, although the video blocks are numbered in sequence in Figure 6.59, the sequence does not correspond to the left-to-right, top-to-bottom transmission of blocks of video data. Compressed macroblocks are shuffled to minimize the effect of errors and aid in error concealment. Audio data also is shuffled. Data is transmitted in the same shuffled order as recorded.

To illustrate the video data shuffling, DV video frames are organized as 50 super blocks, with each super block being composed of 27 compressed macroblocks, as shown in Figure 6.60. A group of 5 super blocks (one from each super block column) make up one DIF sequence. Table 6.42 illustrates the transmission order of the DIF blocks. Additional information on the DV data structure is available in Chapter 11.

IEC 61883-4

IEC 61883-4 defines the CIP header, data packet format, and transmission timing for MPEG-2 transport streams over 1394.

It is most efficient to carry an integer number of 192 bytes (188 bytes of MPEG-2 data plus 4 bytes of time stamp) per isochronous packet, as shown in Figure 6.61. However, MPEG data rates are rarely integer multiples of the isochronous data rate. Thus, it is more

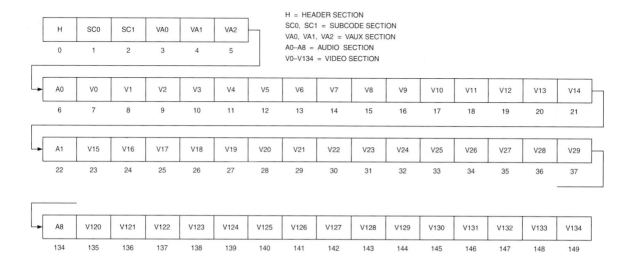

Figure 6.59. IEC 61834, SMPTE 314M and ITU-R BT.1618 DIF Sequence Detail (25 Mbps).

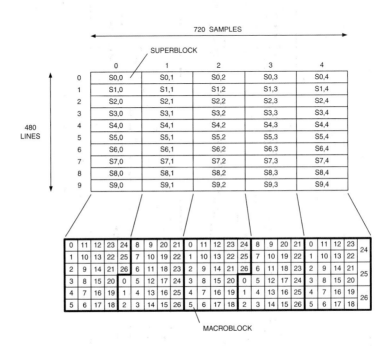

Figure 6.60. Relationship Between Super Blocks and Macroblocks (720 × 480, 4:1:1 YCbCr).

DIF Sequence Number	Video DIF Block Number	Compressed Macroblock		DIF Sequence Number	Video DIF Block Number	Compressed Macroblock	
		Superblock Number	Macroblock Number			Superblock Number	Macroblock Number
0	0	2, 2	0		:		
	1	6, 1	0	n–1	0	1, 2	0
	2	8, 3	0		1	5, 1	0
	3	0, 0	0		2	7, 3	0
	4	4, 4	0		3	n–1, 0	0
	:				4	3, 4	0
	133	0, 0	26		:		
	134	4, 4	26		133	n–1, 0	26
1	0	3, 2	0		134	3, 4	26
	1	7, 1	0				
	2	9, 3	0				
	3	1, 0	0				
	4	5, 4	0				
	:						
	133	1, 0	26				
	134	5, 4	26				

Notes:
1. n = 10 for 480-line systems, n = 12 for 576-line systems.

Table 6.42. Video DIF Blocks and Compressed Macroblocks for 25 Mbps.

efficient to divide the MPEG packets into smaller components of 24 bytes each to maximize available bandwidth. The transmitter then uses an integer number of data blocks (restricted multiples of 0, 1, 2, 4, or 8) placing them in an isochronous packet and adding the 8-byte CIP header.

50 Mbps DV

Like the 25 Mbps DV format, the 50 Mbps DV format supports 720 × 480i30 and 720 × 576i25 sources. However, the 50 Mbps DV format uses 4:2:2 YCbCr rather than 4:1:1 YCbCr.

As previously discussed, the source packet size for the 25 Mbps DV streams is 480 bytes (consisting of 6 DIF blocks). The 250 packets (300 packets for 576i25 systems) of 480-byte data are transferred over a 25 Mbps channel.

The source packet size for the 50 Mbps DV streams is 960 bytes (consisting of 12 DIF blocks). The first 125 packets (150 packets for 576i25 systems) of 960-byte data are sent over one 25 Mbps channel and the next 125 packets (150 packets for 576i25 systems) of 960-byte data are sent over a second 25 Mbps channel.

100 Mbps DV

100 Mbps DV streams support 1920 × 1080i30, 1920 × 1080i25 and 1280 × 720p60 sources. 1920 × 1080i30 sources are horizontally scaled to 1280 × 1080i30. 1920 × 1080i25 sources are horizontally scaled to 1440 × 1080i25. 1280 × 720p60 sources are horizontally scaled to 960 × 720p60. The 4:2:2 YCbCr format is used.

The source packet size for the 100 Mbps DV streams is 1920 bytes (consisting of 24 DIF blocks). The first 63 packets (75 packets for 1080i25 systems) of 1920-byte data are sent over one 25 Mbps channel, the next 62 packets (75 packets for 1080i25 systems) of 1920-byte data are sent over a second 25 Mbps channel, the next 63 packets (75 packets for 1080i25 systems) of 1920-byte data are sent over a third 25 Mbps channel and the last 62 packets (75 packets for 1080i25 systems) of 1920-byte data are sent over a fourth 25 Mbps channel.

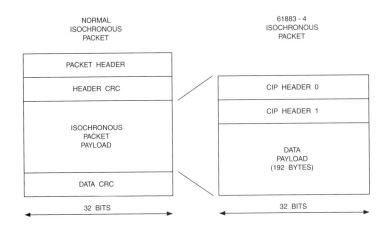

Figure 6.61. 61883-4 Isochronous Packet Formatting.

Digital Camera Specification

The 1394 Trade Association has written a specification for 1394-based digital video cameras. This was done to avoid the silicon and software cost of implementing the full IEC 61883 specification.

Seven resolutions are defined, with a wide range of format support:

160 × 120	4:4:4 YCbCr
320 × 240	4:2:2 YCbCr
640 × 480	4:1:1, 4:2:2 YCbCr, 24-bit RGB
800 × 600	4:2:2 YCbCr, 24-bit RGB
1024 × 768	4:2:2 YCbCr, 24-bit RGB
1280 × 960	4:2:2 YCbCr, 24-bit RGB
1600 × 1200	4:2:2 YCbCr, 24-bit RGB

Supported frame rates are 1.875, 3.75, 7.5, 15, 30, and 60 frames per second.

Isochronous packets are used to transfer the uncompressed digital video data over the 1394 network.

References

1. 1394-based Digital Camera Specification, Version 1.20, July 23, 1998.
2. Digital Transmission Content Protection Specification, Volume 1 (Informational Version), July 25, 2000.
3. Digital Visual Interface (DVI), April 2, 1999.
4. EBU Tech. 3267-E, 1992, *EBU Interfaces for 625-Line Digital Video Signals at the 4:2:2 Level of CCIR Recommendation 601*, European Broadcasting Union, June, 1991.
5. IEC 61883–1, 2003, *Consumer Audio/Video Equipment—Digital Interface—Part 1: General.*
6. IEC 61883–2, 1998, *Consumer Audio/Video Equipment—Digital Interface—Part 2: SD-DVCR Data Transmission.*
7. IEC 61883–3, 1998, *Consumer Audio/Video Equipment—Digital Interface—Part 3: HD-DVCR Data Transmission.*
8. IEC 61883–4, 1998, *Consumer Audio/Video Equipment—Digital Interface—Part 4: MPEG-2 TS Data Transmission.*
9. IEC 61883–5, 1998, *Consumer Audio/Video Equipment—Digital Interface—Part 5: SDL-DVCR Data Transmission.*
10. ITU-R BT.656–4, 1998, *Interfaces for Digital Component Video Signals in 525-Line and 625-Line Television Systems Operating at the 4:2:2 Level of Recommendation ITU-R BT.601.*
11. ITU-R BT.799–3, 1998, *Interfaces For Digital Component Video Signals in 525-Line and 625-Line Television Systems Operating at the 4:4:4 Level of Recommendation ITU-R BT.601 (Part A).*
12. ITU-R BT.1302, 1997, *Interfaces for Digital Component Video Signals in 525-Line and 625-Line Television Systems Operating at the 4:2:2 Level of ITU-R BT.601.*
13. ITU-R BT.1303, 1997, *Interfaces For Digital Component Video Signals in 525-Line and 625-Line Television Systems Operating at the 4:4:4 Level of Recommendation ITU-R BT.601 (Part B).*
14. ITU-R BT.1304, 1997, *Checksum for Error Detection and Status Information in Interfaces Conforming to ITU-R BT.656 and ITU-R BT.799.*
15. ITU-R BT.1305, 1997, *Digital Audio and Auxiliary Data as Ancillary Data Signals in Interfaces Conforming to ITU-R BT.656 and ITU-R BT.799.*
16. ITU-R BT.1362, 1998, *Interfaces For Digital Component Video Signals in 525-Line and 625-Line Progressive Scan Television Systems.*

17. ITU-R BT.1364, 1998, *Format of Ancillary Data Signals Carried in Digital Component Studio Interfaces*.

18. ITU-R BT.1365, 1998, *24-Bit Digital Audio Format as Ancillary Data Signals in HDTV Serial Interfaces*.

19. ITU-R BT.1366, 1998, *Transmission of Time Code and Control Code in the Ancillary Data Space of a Digital Television Stream According to ITU-R BT.656, ITU-R BT.799, and ITU-R BT.1120*.

20. ITU-R BT.1381–1, 2001, *Serial Digital Interface-based Transport Interface for Compressed Television Signals In Networked Television Production Based On Recommendations ITU-R BT.656 and ITU-R BT.1302*.

21. ITU-R BT.1577, 2002, *Serial digital interface-based transport interface for compressed television signals in networked television production based on Recommendation ITU-R BT.1120*.

22. ITU-R BT.1616, 2003, *Data stream format for the exchange of DV-based audio, data and compressed video over interfaces complying with Recommendation ITU-R BT.1381*.

23. ITU-R BT.1617, 2003, *Format for transmission of DV compressed video, audio and data over interfaces complying with Recommendation ITU-R BT.1381*.

24. ITU-R BT.1618, 2003, *Data structure for DV-based audio, data and compressed video at data rates of 25 and 50 Mbit/s*.

25. ITU-R BT.1619, 2003, *Vertical ancillary data mapping for serial digital interface*.

26. ITU-R BT.1620, 2003, *Data structure for DV-based audio, data and compressed video at a data rate of 100 Mbit/s*.

27. Kikuchi, Hidekazu et. al., *A 1-bit Serial Interface Chip Set for Full-Color XGA Pictures*, Society for Information Display, 1999.

28. Kikuchi, Hidekazu et. al., *Gigabit Video Interface: A Fully Serialized Data Transmission System for Digital Moving Pictures*, International Conference on Consumer Electronics, 1998.

29. Open LVDS Display Interface (OpenLDI) Specification, v0.95, May 13, 1999.

30. SMPTE 125M–1995, *Television—Component Video Signal 4:2:2—Bit-Parallel Digital Interface*.

31. SMPTE 240M–1999, *Television—Signal Parameters—1125-Line High-Definition Production Systems*.

32. SMPTE 244M–2003, *Television—System M/NTSC Composite Video Signals—Bit-Parallel Digital Interface*.

33. SMPTE 259M–1997, *Television—10-Bit 4:2:2 Component and $4F_{SC}$ Composite Digital Signals—Serial Digital Interface*.

34. SMPTE 260M–1999, *Television—1125/60 High-Definition Production System—Digital Representation and Bit-Parallel Interface*.

35. SMPTE 266M–2002, *Television—4:2:2 Digital Component Systems—Digital Vertical Interval Time Code*.

36. SMPTE 267M–1995, *Television—Bit-Parallel Digital Interface—Component Video Signal 4:2:2 16 × 9 Aspect Ratio*.

37. SMPTE 272M–1994, *Television—Formatting AES/EBU Audio and Auxiliary Data into Digital Video Ancillary Data Space*.

38. SMPTE 274M–2003, *Television—1920 × 1080 Image Sample Structure, Digital Representation and Digital Timing Reference Sequences for Multiple Picture Rates*.

39. SMPTE 291M–1998, *Television—Ancillary Data Packet and Space Formatting*.

40. SMPTE 292M–1998, *Television—Bit-Serial Digital Interface for High-Definition Television Systems*.

41. SMPTE 293M–2003, *Television—720 × 483 Active Line at 59.94-Hz Progressive Scan Production—Digital Representation*.

42. SMPTE 294M–2001, *Television—720 × 483 Active Line at 59.94 Hz Progressive Scan Production—Bit-Serial Interfaces*.

43. SMPTE 296M–2001, *Television—1280 × 720 Progressive Image Sample Structure, Analog and Digital Representation and Analog Interface*.

44. SMPTE 305.2M–2000, *Television—Serial Data Transport Interface (SDTI)*.

45. SMPTE 314M–1999, *Television—Data Structure for DV-Based Audio, Data and Compressed Video—25 and 50 Mb/s*.

46. SMPTE 326M–2000, *Television—SDTI Content Package Format (SDTI-CP)*.

47. SMPTE 334M–2000, *Television—Vertical Ancillary Data Mapping for Bit-Serial Interface*.

48. SMPTE 344M–2000, *Television—540 Mbps Serial Digital Interface*.

49. SMPTE 348M–2000, *Television—High Data-Rate Serial Data Transport Interface (HD-SDTI)*.

50. SMPTE 370M–2002, *Television—Data Structure for DV-Based Audio, Data and Compressed Video at 100 Mb/s 1080/60i, 1080/50i, 720-60p*.

51. SMPTE 372M–2002, *Television—Dual Link 292M Interface for 1920 × 1080 Picture Raster*.

52. SMPTE 396M–2003, *Television—Packet Format and Transmission Timing of DV-Based Data Streams over IEEE 1394*.

53. SMPTE RP-165–1994, *Error Detection Checkwords and Status Flags for Use in Bit-Serial Digital Interfaces for Television*.

54. SMPTE RP-174–1993, *Bit-Parallel Digital Interface for 4:4:4:4 Component Video Signal (Single Link)*.

55. SMPTE RP-175–1997, *Digital Interface for 4:4:4:4 Component Video Signal (Dual Link)*.

56. SMPTE RP-168–2002, *Definition of Vertical Interval Switching Point for Synchronous Video Switching*.

57. SMPTE RP-188–1999, *Transmission of Time Code and Control Code in the Ancillary Data Space of a Digital Television Data Stream*.

58. SMPTE RP-208–2002, *Transport of VBI Packet Data in Ancillary Data Packets*.

59. Teener, Michael D. Johas, *IEEE 1394-1995 High Performance Serial Bus*, 1394 Developer's Conference, 1997.

60. VESA DFP 1.0: Digital Flat Panel (DFP) Standard.

61. VESA Video Interface Port (VIP), Version 2, October 21, 1998.

62. VMI Specification, v1.4, January 30, 1996.

63. Wickelgren, Ingrid J., *The Facts about FireWire*, IEEE Spectrum, April 1997.

Chapter 7

Digital Video Processing

In addition to encoding and decoding MPEG, NTSC/PAL and many other types of video, a typical system usually requires considerable additional video processing.

Since many consumer displays, and most computer displays, are progressive (noninterlaced), interlaced video must be converted to progressive ("deinterlaced"). Progressive video must be converted to interlaced to drive a conventional analog VCR or TV, requiring noninterlaced-to-interlaced conversion.

Many computer displays have a vertical refresh rate of about 75 Hz, whereas consumer video has a vertical refresh rate of 25 or 29.97 (30/1.001) frames per second. For DVD and DTV, source material may only be 24 frames per second. Thus, some form of frame rate conversion must be done.

Another not-so-subtle problem includes video scaling. SDTV and HDTV support multiple resolutions, yet the display may be a single, fixed resolution.

Alpha mixing and chroma keying are used to mix multiple video signals or video with computer-generated text and graphics. Alpha mixing ensures a smooth crossover between sources, allows subpixel positioning of text, and limits source transition bandwidths to simplify eventual encoding to composite video signals.

Since no source is perfect, even digital sources, user controls for adjustable brightness, contrast, saturation, and hue are always desirable.

Rounding Considerations

When two 8-bit values are multiplied together, a 16-bit result is generated. At some point, a result must be rounded to some lower precision (for example, 16 bits to 8 bits or 32 bits to 16 bits) in order to realize a cost-effective hardware implementation. There are several rounding techniques: truncation, conventional rounding, error feedback rounding, and dynamic rounding.

Truncation

Truncation drops any fractional data during each rounding operation. As a result, after only a few operations, a significant error may be introduced. This may result in contours being visible in areas of solid colors.

Conventional Rounding

Conventional rounding uses the fractional data bits to determine whether to round up or round down. If the fractional data is 0.5 or greater, rounding up should be performed—positive numbers should be made more positive and negative numbers should be made more negative. If the fractional data is less than 0.5, rounding down should be performed—positive numbers should be made less positive and negative numbers should be made less negative.

Error Feedback Rounding

Error feedback rounding follows the principle of "never throw anything away." This is accomplished by storing the residue of a truncation and adding it to the next video sample. This approach substitutes less visible noise-like quantizing errors in place of contouring effects caused by simple truncation. An example of an error feedback rounding implementation is shown in Figure 7.1. In this example, 16 bits are reduced to 8 bits using error feedback.

Dynamic Rounding

This technique (a licensable Quantel patent) dithers the LSB according to the weighting of the discarded fractional bits. The original data word is divided into two parts, one representing the resolution of the final output word and one dealing with the remaining fractional data. The fractional data is compared to the output of a random number generator equal in resolution to the fractional data. The output of the comparator is a 1-bit random pattern weighted by the value of the fractional

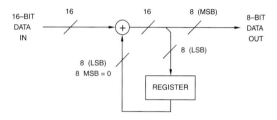

Figure 7.1. Error Feedback Rounding.

data, and serves as a carry-in to the adder. In all instances, only one LSB of the output word is changed, in a random fashion. An example of a dynamic rounding implementation is shown in Figure 7.2.

SDTV - HDTV YCbCr Transforms

SDTV and HDTV applications have different colorimetric characteristics, as discussed in Chapter 3. Thus, when SDTV (HDTV) data is displayed on a HDTV (SDTV) display, the YCbCr data should be processed to compensate for the different colorimetric characteristics.

SDTV to HDTV

A 3×3 matrix can be used to convert from $Y_{601}CbCr$ (SDTV) to $Y_{709}CbCr$ (HDTV):

$$\begin{bmatrix} 1 & -0.11554975 & -0.20793764 \\ 0 & 1.01863972 & 0.11461795 \\ 0 & 0.07504945 & 1.02532707 \end{bmatrix}$$

Note that before processing, the 8-bit DC offset (16 for Y and 128 for CbCr) must be removed, then added back in after processing.

HDTV to SDTV

A 3×3 matrix can be used to convert from $Y_{709}CbCr$ (HDTV) to $Y_{601}CbCr$ (SDTV):

$$\begin{bmatrix} 1 & 0.09931166 & 0.19169955 \\ 0 & 0.98985381 & -0.11065251 \\ 0 & -0.07245296 & 0.98339782 \end{bmatrix}$$

Note that before processing, the 8-bit DC offset (16 for Y and 128 for CbCr) must be removed, then added back in after processing.

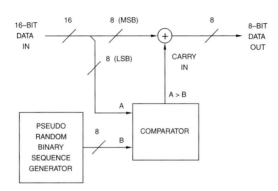

Figure 7.2. Dynamic Rounding.

4:4:4 to 4:2:2 YCbCr Conversion

Converting 4:4:4 YCbCr to 4:2:2 YCbCr (Figure 7.3) is a common function in digital video. 4:2:2 YCbCr is the basis for many digital video interfaces, and requires fewer connections to implement than 4:4:4.

Saturation logic should be included in the Y, Cb, and Cr data paths to limit the 8-bit range to 1–254. The 16 and 128 values shown in Figure 7.3 are used to generate the proper levels during blanking intervals.

Y Filtering

A template for the Y lowpass filter is shown in Figure 7.4 and Table 7.1.

Because there may be many cascaded conversions (up to 10 were envisioned), the filters were designed to adhere to very tight toler-ances to avoid a buildup of visual artifacts. Departure from flat amplitude and group delay response due to filtering is amplified through successive stages. For example, if filters exhib-iting –1 dB at 1 MHz and –3 dB at 1.3 MHz were employed, the overall response would be –8 dB (at 1 MHz) and –24 dB (at 1.3 MHz) after four conversion stages (assuming two fil-ters per stage).

Although the sharp cut-off results in ring-ing on Y edges, the visual effect should be min-imal provided that group-delay performance is adequate. When cascading multiple filtering operations, the passband flatness and group-delay characteristics are very important. The passband tolerances, coupled with the sharp cut-off, make the template very difficult (some say impossible) to match. As a result, there is usually a temptation to relax passband accu-racy, but the best approach is to reduce the rate of cut-off and keep the passband as flat as possible.

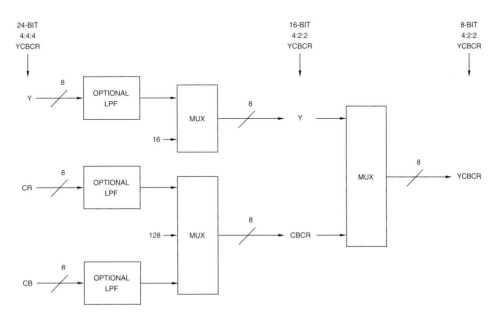

Figure 7.3. 4:4:4 to 4:2:2 YCbCr Conversion.

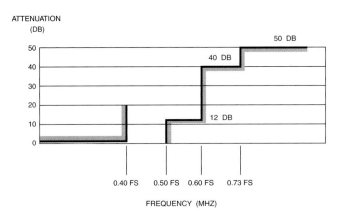

Figure 7.4. Y Filter Template. F$_s$ = Y 1× sample rate.

Frequency Range	Typical SDTV Tolerances	Typical HDTV Tolerances
Passband Ripple Tolerance		
0 to 0.40F$_s$	±0.01 dB increasing to ±0.05 dB	±0.05 dB
Passband Group Delay Tolerance		
0 to 0.27F$_s$	0 increasing to ±1.35 ns	±0.075T
0.27F$_s$ to 0.40F$_s$	±1.35 ns increasing to ±2 ns	±0.110T

Table 7.1. Y Filter Ripple and Group Delay Tolerances. F$_s$ = Y 1× sample rate. T = 1 / F$_s$.

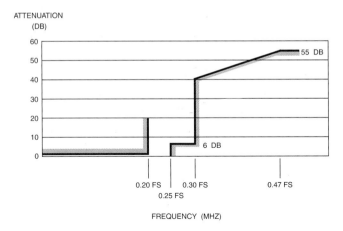

Figure 7.5. Cb and Cr Filter Template for Digital Filter for Sample Rate Conversion from 4:4:4 to 4:2:2. F$_s$ = Y 1× sample rate.

Frequency Range	Typical SDTV Tolerances	Typical HDTV Tolerances
Passband Ripple Tolerance		
0 to 0.20F$_s$	0 dB increasing to ±0.05 dB	±0.05 dB
Passband Group Delay Tolerance		
0 to 0.20F$_s$	delay distortion is zero by design	

Table 7.2. CbCr Filter Ripple and Group Delay Tolerances. F$_s$ = Y 1× sample rate. T = 1 / F$_s$.

CbCr Filtering

Cb and Cr are lowpass filtered and decimated. In a standard design, the lowpass and decimation filters may be combined into a single filter, and a single filter may be used for both Cb and Cr by multiplexing.

As with Y filtering, the Cb and Cr lowpass filtering requires a sharp cut-off to prevent repeated conversions from producing a cumulative resolution loss. However, due to the low cut-off frequency, the sharp cut-off produces ringing that is more noticeable than for Y.

A template for the Cb and Cr filters is shown in Figure 7.5 and Table 7.2.

Since aliasing is less noticeable in color difference signals, the attenuation at half the sampling frequency is only 6 dB. There is an advantage in using a skew-symmetric response passing through the –6 dB point at half the sampling frequency—this makes alternate coefficients in the digital filter zero, almost halving the number of taps, and also allows using a single digital filter for both the Cb and Cr signals. Use of a transversal digital filter has the advantage of providing perfect linear phase response, eliminating the need for group-delay correction.

As with the Y filter, the passband flatness and group-delay characteristics are very important, and the best approach again is to reduce the rate of cut-off and keep the passband as flat as possible.

Display Enhancement

Brightness, Contrast, Saturation (Color) and Hue (Tint)

Working in the YCbCr color space simplifies the implementation of brightness, contrast, saturation and hue controls, as shown in Figure 7.6. Also illustrated are multiplexers to allow the output of black screen, blue screen and color bars.

The design should ensure that no overflow or underflow wrap-around errors occur, effectively saturating results to the 0 and 255 values.

Y Processing

16 is subtracted from the Y data to position the black level at zero. This removes the DC offset so adjusting the contrast does not vary the black level. Since the Y input data may have values below 16, negative Y values should be supported at this point.

The contrast (or "picture" or "white level") control is implemented by multiplying the YCbCr data by a constant. If Cb and Cr are not adjusted, a color shift will result whenever the contrast is changed. A typical 8-bit contrast adjustment range is 0–1.992×.

The brightness (or "black level") control is implemented by adding or subtracting from the Y data. Brightness is done after the contrast to avoid introducing a varying DC offset due to adjusting the contrast. A typical 6-bit brightness adjustment range is –32 to +31.

Finally, 16 is added to position the black level at 16.

CbCr Processing

128 is subtracted from Cb and Cr to position the range about zero.

The hue (or "tint") control is implemented by mixing the Cb and Cr data:

$$Cb' = Cb \cos \theta + Cr \sin \theta$$

$$Cr' = Cr \cos \theta - Cb \sin \theta$$

where θ is the desired hue angle. A typical 8-bit hue adjustment range is –30° to +30°.

The saturation (or "color") control is implemented by multiplying both Cb and Cr by a constant. A typical 8-bit saturation adjustment

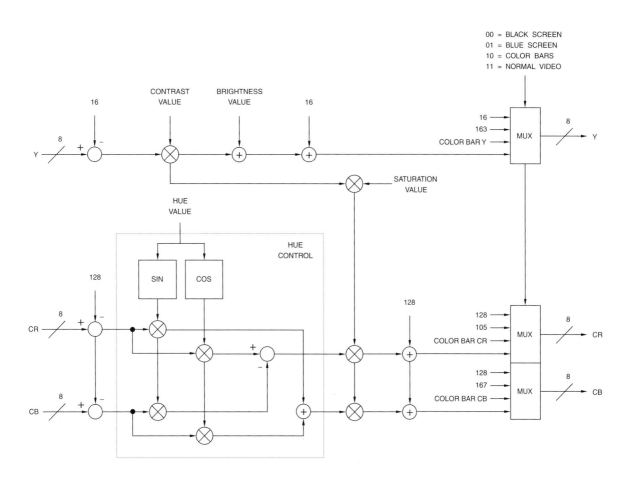

Figure 7.6. Hue, Saturation, Contrast, and Brightness Controls.

range is 0–1.992×. In the example shown in Figure 7.6, the contrast and saturation values are multiplied together to reduce the number of multipliers in the CbCr datapath.

Finally, 128 is added to both Cb and Cr.

Color Transient Improvement

YCbCr transitions should be aligned. However, the Cb and Cr transitions are usually degraded due to the narrow bandwidth of color difference information.

By monitoring coincident Y transitions, faster transitions may be synthesized for Cb and Cr. These edges are then aligned with the Y edge, as shown in Figure 7.7.

Alternately, Cb and Cr transitions may be differentiated, and the results added to the original Cb and Cr signals. Small amplitudes in the differentiation signals should be suppressed by coring. To eliminate "wrong colors" due to overshoots and undershoots, the enhanced CbCr signals should also be limited to the proper range.

In some cases, the Y rise and fall times are also shortened to artificially sharpen the image.

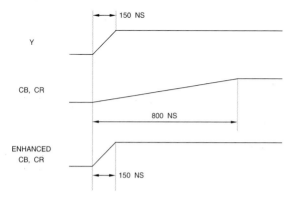

Figure 7.7. Color Transient Improvement.

Since this technique artificially increases the high-frequency component of video signals, it should not be used if the video will be compressed, as the compression ratio will be reduced.

Sharpness

The apparent sharpness of a picture may be increased by increasing the amplitude of high-frequency luminance information.

As shown in Figure 7.8, a simple bandpass filter with selectable gain (also called a peaking filter) may be used. The frequency where maximum gain occurs is usually selectable to be either at the color subcarrier frequency or at about 2.6 MHz. A coring circuit is typically used after the filter to reduce low-level noise.

Figure 7.9 illustrates a more complex sharpness control circuit. The high-frequency luminance is increased using a variable bandpass filter, with adjustable gain. The coring function (typically ±1 LSB) removes low-level noise. The modified luminance is then added to the original luminance signal.

Since this technique artificially increases the high-frequency component of the video signals, it should not be used if the video will be compressed, as the compression ratio will be reduced.

In addition to selectable gain, selectable attenuation of high frequencies should also be supported. Many televisions boost high-frequency gain to improve the apparent sharpness of the picture. If this is applied to a compressed (such as MPEG) source, the picture quality can be substantially degraded. Although the sharpness control on the television may be turned down, this affects the picture quality of analog broadcasts. Therefore, many MPEG sources have the option of attenuating high frequencies to negate the sharpness control on the television.

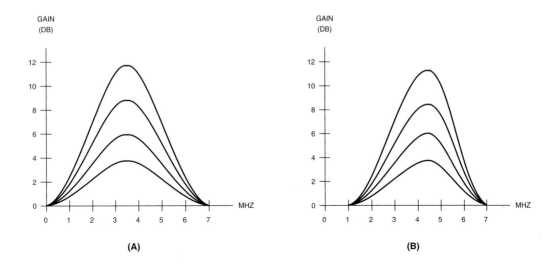

Figure 7.8. Simple Adjustable Sharpness Control. (a) NTSC. (b) PAL.

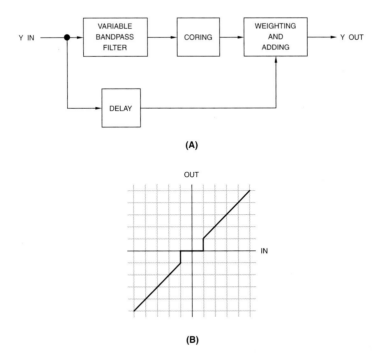

Figure 7.9. More Complex Sharpness Control. (a) Typical implementation. (b) Coring function.

Video Mixing and Graphics Overlay

Mixing video signals may be as simple as switching between two video sources. This is adequate if the resulting video is to be displayed on a computer monitor.

For most other applications, a technique known as alpha mixing should be used. Alpha mixing may also be used to fade to or from a specific color (such as black) or to overlay computer-generated text and graphics onto a video signal.

Alpha mixing must be used if the video is to be encoded to composite video. Otherwise, ringing and blurring may appear at the source switching points, such as around the edges of computer-generated text and graphics. This is due to the color information being lowpass filtered within the NTSC/PAL encoder. If the filters have a sharp cut-off, a fast color transition will produce ringing. In addition, the intensity information may be bandwidth-limited to about 4–5 MHz somewhere along the video path, slowing down intensity transitions.

Mathematically, with alpha normalized to have values of 0–1, alpha mixing is implemented as:

$$out = (alpha\_0)(in\_0) + (alpha\_1)(in\_1) + ...$$

In this instance, each video source has its own alpha information. The alpha information may not total to one (unity gain).

Figure 7.10 shows mixing of two YCbCr video signals, each with its own alpha information. As YCbCr uses an offset binary notation, the offset (16 for Y and 128 for Cb and Cr) is removed prior to mixing the video signals. After mixing, the offset is added back in. Note that two 4:2:2 YCbCr streams may also be processed directly; there is no need to convert

them to 4:4:4 YCbCr, mix, then convert the result back to 4:2:2 YCbCr.

When only two video sources are mixed and alpha_0 + alpha_1 = 1 (implementing a crossfader), a single alpha value may be used, mathematically shown as:

$$out = (alpha)(in\_0) + (1 - alpha)(in\_1)$$

When alpha = 0, the output is equal to the in_1 video signal; when alpha = 1, the output is equal to the in_0 video signal. When alpha is between 0 and 1, the two video signals are proportionally multiplied, and added together.

Expanding and rearranging the previous equation shows how a two-channel mixer may be implemented using a single multiplier:

$$out = (alpha)(in\_0 - in\_1) + in\_1$$

Fading to and from a specific color is done by setting one of the input sources to a constant color.

Figure 7.11 illustrates mixing two YCbCr sources using a single alpha channel. Figures 7.12 and 7.13 illustrate mixing two R′G′B′ video sources (R′G′B′ has a range of 0–255). Figures 7.14 and 7.15 show mixing two digital composite video signals.

A common problem in computer graphics systems that use alpha is that the frame buffer may contain preprocessed R′G′B′ or YCbCr data; that is, the R′G′B′ or YCbCr data in the frame buffer has already been multiplied by alpha. Assuming an alpha value of 0.5, non-processed R′G′B′A values for white are (255, 255, 255, 128); preprocessed R′G′B′A values for white are (128, 128, 128, 128). Therefore, any mixing circuit that accepts R′G′B′ or YCbCr data from a frame buffer should be able to handle either format.

By adjusting the alpha values, slow to fast crossfades are possible, as shown in Figure

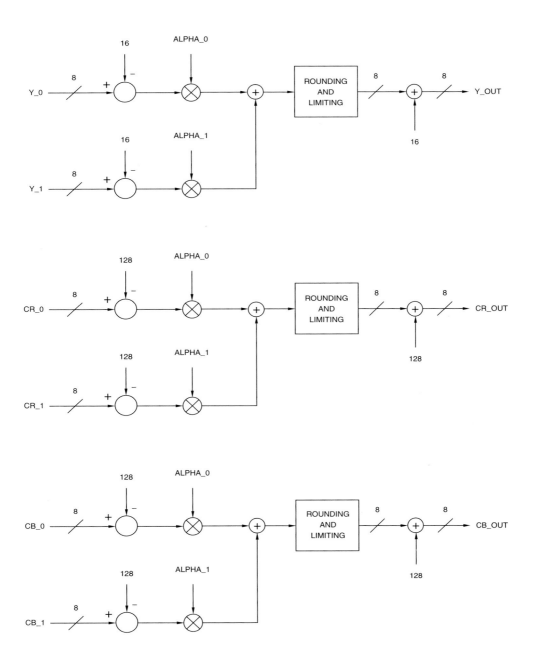

Figure 7.10. Mixing Two YCbCr Video Signals, Each With Its Own Alpha Channel.

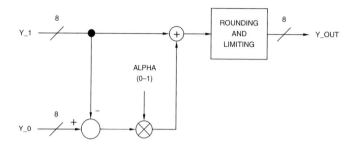

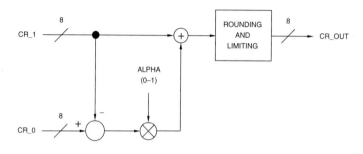

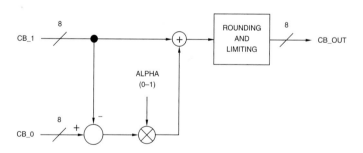

Figure 7.11. Simplified Mixing (Crossfading) of Two YCbCr Video Signals Using a Single Alpha Channel.

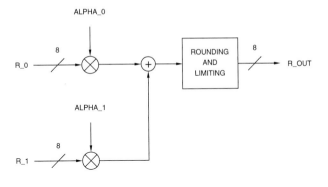

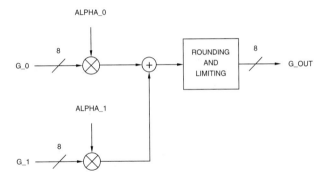

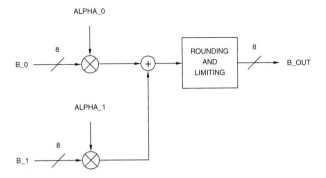

Figure 7.12. Mixing Two RGB Video Signals (RGB has a Range of 0–255), Each With Its Own Alpha Channel.

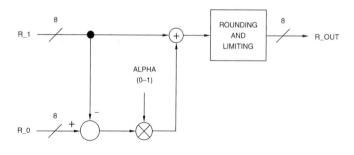

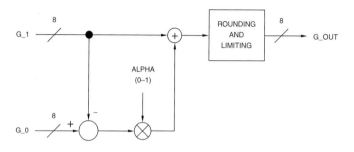

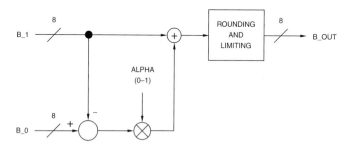

Figure 7.13. Simplified Mixing (Crossfading) of Two RGB Video Signals (RGB has a Range of 0–255) Using a Single Alpha Channel.

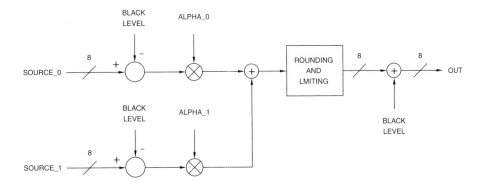

Figure 7.14. Mixing Two Digital Composite Video Signals, Each With Its Own Alpha Channel.

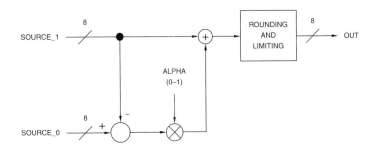

Figure 7.15. Simplified Mixing (Crossfading) of Two Digital Composite Video Signals Using a Single Alpha Channel.

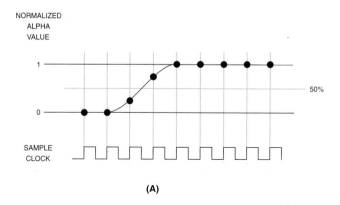

(A)

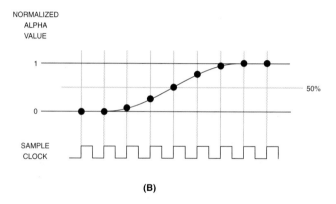

(B)

Figure 7.16. Controlling Alpha Values to Implement (a) Fast or (b) Slow Keying. In (a), the effective switching point lies between two samples. In (b), the transition is wider and is aligned at a sample instant.

7.16. Large differences in alpha between samples result in a fast crossfade; smaller differences result in a slow crossfade. If using alpha mixing for special effects, such as wipes, the switching point (where 50% of each video source is used) must be able to be adjusted to an accuracy of less than one sample to ensure smooth movement. By controlling the alpha values, the switching point can be effectively positioned anywhere, as shown in Figure 7.16a.

Text can be overlaid onto video by having a character generator control the alpha inputs. By setting one of the input sources to a constant color, the text will assume that color.

Note that for those designs that subtract 16 (the black level) from the Y channel before processing, negative Y values should be supported after the subtraction. This allows the design to pass through real-world and test video signals with minimum artifacts.

Luma and Chroma Keying

Keying involves specifying a desired foreground color; areas containing this color are replaced with a background image. Alternately, an area of any size or shape may be specified; foreground areas inside (or outside) this area are replaced with a background image.

Luminance Keying

Luminance keying involves specifying a desired foreground luminance level; foreground areas containing luminance levels above (or below) the keying level are replaced with the background image.

Alternately, this hard keying implementation may be replaced with soft keying by specifying two luminance values of the foreground image: Y_H and Y_L ($Y_L < Y_H$). For keying the background into "white" foreground areas, foreground luminance values (Y_{FG}) above Y_H are replaced with the background image; Y_{FG} values below Y_L contain the foreground image. For Y_{FG} values between Y_L and Y_H, linear mixing is done between the foreground and background images. This operation may be expressed as:

if $Y_{FG} > Y_H$
$K = 1$ = background only

if $Y_{FG} < Y_L$
$K = 0$ = foreground only

if $Y_H \geq Y_{FG} \geq Y_L$
$K = (Y_{FG} - Y_L)/(Y_H - Y_L)$ = mix

By subtracting K from 1, the new luminance keying signal for keying into "black" foreground areas can be generated.

Figure 7.17 illustrates luminance keying for two YCbCr sources. Although chroma keying typically uses a suppression technique to remove information from the foreground image, this is not done when luminance keying as the magnitudes of Cb and Cr are usually not related to the luminance level.

Figure 7.18 illustrates luminance keying for R'G'B' sources, which is more applicable for computer graphics. Y_{FG} may be obtained by the equation:

$$Y_{FG} = 0.299R' + 0.587G' + 0.114B'$$

In some applications, the red and blue data is ignored, resulting in Y_{FG} being equal to only the green data.

Figure 7.19 illustrates one technique of luminance keying between two digital composite video sources.

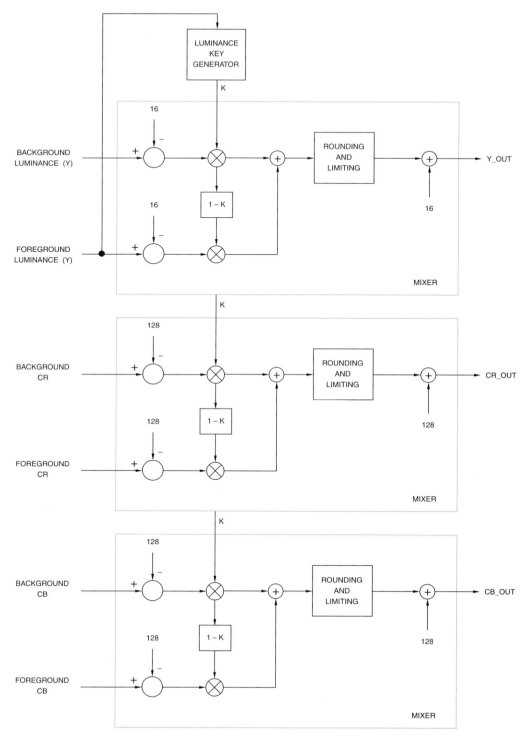

Figure 7.17. Luminance Keying of Two YCbCr Video Signals.

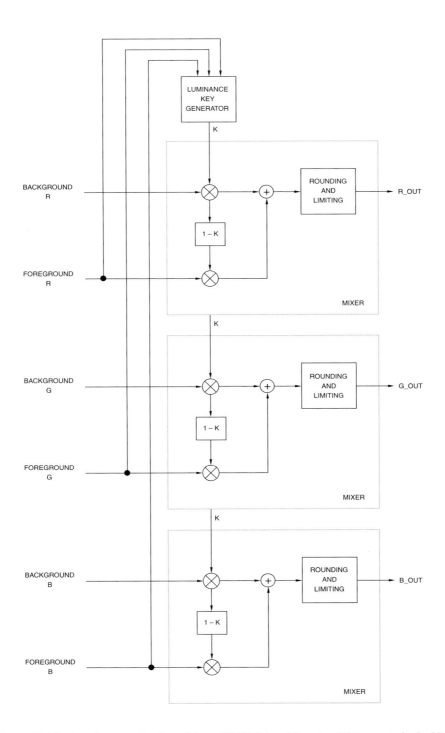

Figure 7.18. Luminance Keying of Two RGB Video Signals. RGB range is 0–255.

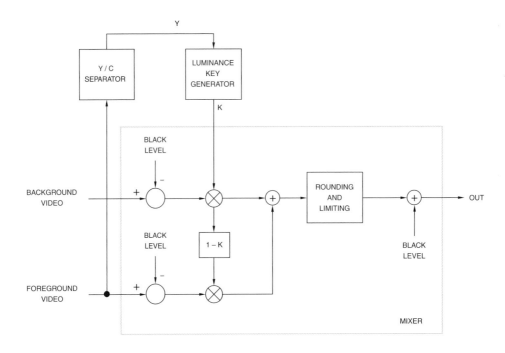

Figure 7.19. Luminance Keying of Two Digital Composite Video Signals.

Chroma Keying

Chroma keying involves specifying a desired foreground key color; foreground areas containing the key color are replaced with the background image. Cb and Cr are used to specify the key color; luminance information may be used to increase the realism of the chroma keying function. The actual mixing of the two video sources may be done in the component or composite domain, although component mixing reduces artifacts.

Early chroma keying circuits simply performed a hard or soft switch between the foreground and background sources. In addition to limiting the amount of fine detail maintained in the foreground image, the background was not visible through transparent or translucent foreground objects, and shadows from the foreground were not present in areas containing the background image.

Linear keyers were developed that combine the foreground and background images in a proportion determined by the key level, resulting in the foreground image being attenuated in areas containing the background image. Although allowing foreground objects to appear transparent, there is a limit on the fineness of detail maintained in the foreground. Shadows from the foreground are not present in areas containing the background image unless additional processing is done—the luminance levels of specific areas of the background image must be reduced to create the effect of shadows cast by foreground objects.

If the blue or green backing used with the foreground scene is evenly lit except for shadows cast by the foreground objects, the effect on the background will be that of shadows cast by the foreground objects. This process, referred to as shadow chroma keying, or luminance modulation, enables the background luminance levels to be adjusted in proportion to the brightness of the blue or green backing in the foreground scene. This results in more realistic keying of transparent or translucent foreground objects by preserving the spectral highlights.

Note that green backgrounds are now more commonly used due to lower chroma noise.

Chroma keyers are also limited in their ability to handle foreground colors that are close to the key color without switching to the background image. Another problem may be a bluish tint to the foreground objects as a result of blue light reflecting off the blue backing or being diffused in the camera lens. Chroma spill is difficult to remove since the spill color is not the original key color; some mixing occurs, changing the original key color slightly.

One solution to many of the chroma keying problems is to process the foreground and background images individually before combining them, as shown in Figure 7.20. Rather than choosing between the foreground and background, each is processed individually and then combined. Figure 7.21 illustrates the major processing steps for both the foreground and background images during the chroma key process. Not shown in Figure 7.20 is the circuitry to initially subtract 16 (Y) or

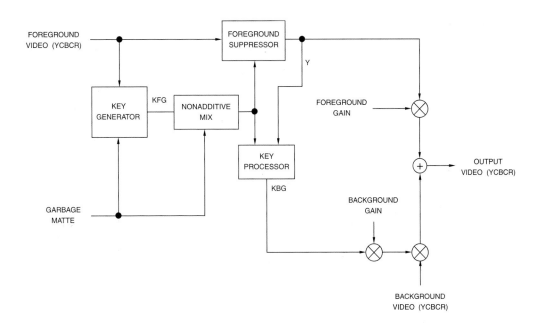

Figure 7.20. Typical Component Chroma Key Circuit.

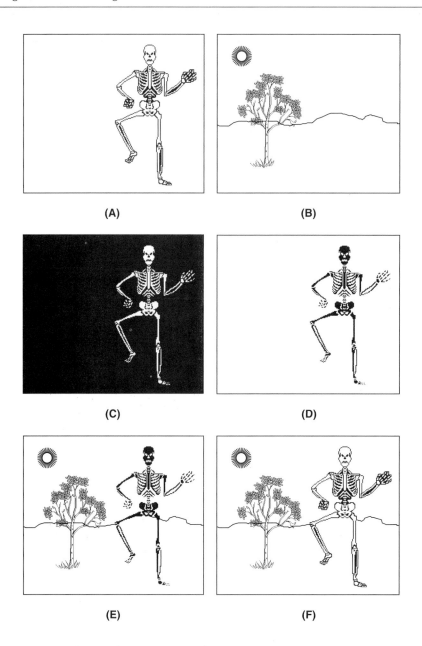

Figure 7.21. Major Processing Steps During Chroma Keying. (a) Original foreground scene. (b) Original background scene. (c) Suppressed foreground scene. (d) Background keying signal. (e) Background scene after multiplication by background key. (f) Composite scene generated by adding (c) and (e).

128 (Cb and Cr) from the foreground and background video signals and the addition of 16 (Y) or 128 (Cb and Cr) after the final output adder. Any DC offset not removed will be amplified or attenuated by the foreground and background gain factors, shifting the black level.

The foreground key (K_{FG}) and background key (K_{BG}) signals have a range of 0 to 1. The garbage matte key signal (the term matte comes from the film industry) forces the mixer to output the foreground source in one of two ways.

The first method is to reduce K_{BG} in proportion to increasing K_{FG}. This provides the advantage of minimizing black edges around the inserted foreground.

The second method is to force the background to black for all nonzero values of the matte key, and insert the foreground into the background "hole." This requires a cleanup function to remove noise around the black level, as this noise affects the background picture due to the straight addition process.

The garbage matte is added to the foreground key signal (K_{FG}) using a non-additive mixer (NAM). A nonadditive mixer takes the brighter of the two pictures, on a sample-by-sample basis, to generate the key signal. Matting is ideal for any source that generates its own keying signal, such as character generators, and so on.

The key generator monitors the foreground Cb and Cr data, generating the foreground keying signal, K_{FG}. A desired key color is selected, as shown in Figure 7.22. The foreground Cb and Cr data are normalized (generating Cb′ and Cr′) and rotated θ degrees to generate the X and Z data, such that the positive X axis passes as close as possible to the desired key color. Typically, θ may be varied in 1° increments, and optimum chroma keying

occurs when the X axis passes through the key color.

X and Z are derived from Cb and Cr using the equations:

$$X = Cb' \cos \theta + Cr' \sin \theta$$

$$Z = Cr' \cos \theta - Cb' \sin \theta$$

Since Cb′ and Cr′ are normalized to have a range of ±1, X and Z have a range of ±1.

The foreground keying signal (K_{FG}) is generated from X and Z and has a range of 0–1:

$$K_{FG} = X - (|Z| / (\tan (\alpha/2)))$$

$$K_{FG} = 0 \text{ if } X < (|Z| / (\tan (\alpha/2)))$$

where α is the acceptance angle, symmetrically centered about the positive X axis, as shown in Figure 7.23. Outside the acceptance angle, K_{FG} is always set to zero. Inside the acceptance angle, the magnitude of K_{FG} linearly increases the closer the foreground color approaches the key color and as its saturation increases. Colors inside the acceptance angle are further processed by the foreground suppressor.

The foreground suppressor reduces foreground color information by implementing X = X – K_{FG}, with the key color being clamped to the black level. To avoid processing Cb and Cr when K_{FG} = 0, the foreground suppressor performs the operations:

$$Cb_{FG} = Cb - K_{FG} \cos \theta$$

$$Cr_{FG} = Cr - K_{FG} \sin \theta$$

where Cb_{FG} and Cr_{FG} are the foreground Cb and Cr values after key color suppression. Early implementations suppressed foreground information by multiplying Cb and Cr by a clipped version of the K_{FG} signal. This, however, generated in-band alias components due

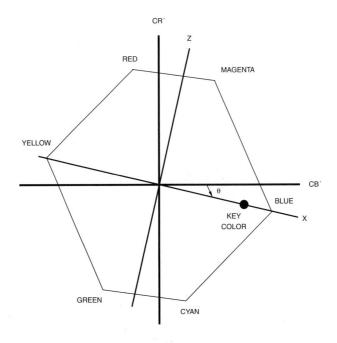

Figure 7.22. Rotating the Normalized Cb and Cr (Cb′ and Cr′) Axes by θ to Obtain the X and Z Axes, Such That the X Axis Passes Through the Desired Key Color (Blue in This Example).

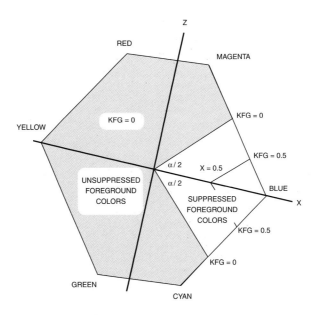

Figure 7.23. Foreground Key Values and Acceptance Angle.

to the multiplication and clipping process and produced a hard edge at key color boundaries.

Unless additional processing is done, the Cb_{FG} and Cr_{FG} components are set to zero only if they are exactly on the X axis. Hue variations due to noise or lighting will result in areas of the foreground not being entirely suppressed. Therefore, a suppression angle is set, symmetrically centered about the positive X axis. The suppression angle (β) is typically configurable from a minimum of zero degrees, to a maximum of about one-third the acceptance angle (α). Any CbCr components that fall within this suppression angle are set to zero. Figure 7.24 illustrates the use of the suppression angle.

Foreground luminance, after being normalized to have a range of 0–1, is suppressed by:

$$Y_{FG} = Y' - y_S K_{FG}$$

$$Y_{FG} = 0 \text{ if } y_S K_{FG} > Y'$$

Here, y_S is a programmable value and used to adjust Y_{FG} so that it is clipped at the black level in the key color areas.

The foreground suppressor also removes key-color fringes on wanted foreground areas caused by chroma spill, the overspill of the key color, by removing discolorations of the wanted foreground objects.

Ultimatte® improves on this process by measuring the difference between the blue and green colors, as the blue backing is never pure blue and there may be high levels of blue in the foreground objects. Pure blue is rarely found in nature, and most natural blues have a higher content of green than red. For this reason, the red, green, and blue levels are monitored to differentiate between the blue backing and blue in wanted foreground objects.

If the difference between blue and green is great enough, all three colors are set to zero to produce black; this is what happens in areas of the foreground containing the blue backing.

If the difference between blue and green is not large, the blue is set to the green level unless the green exceeds red. This technique allows the removal of the bluish tint caused by the blue backing while being able to reproduce natural blues in the foreground. As an example, a white foreground area normally would consist of equal levels of red, green, and blue. If the white area is affected by the key color (blue in this instance), it will have a bluish tint—the blue levels will be greater than the red or green levels. Since the green does not exceed the red, the blue level is made equal to the green, removing the bluish tint.

There is a price to pay, however. Magenta in the foreground is changed to red. A green backing can be used, but in this case, yellow in the foreground is modified. Usually, the clamping is released gradually to increase the blue content of magenta areas.

The key processor generates the initial background key signal (K'_{BG}) used to remove areas of the background image where the foreground is to be visible. K'_{BG} is adjusted to be zero in desired foreground areas and unity in background areas with no attenuation. It is generated from the foreground key signal (K_{FG}) by applying lift (k_L) and gain (k_G) adjustments followed by clipping at zero and unity values:

$$K'_{BG} = (K_{FG} - k_L)k_G$$

Figure 7.25 illustrates the operation of the background key signal generation. The transition between $K'_{BG} = 0$ and $K'_{BG} = 1$ should be made as wide as possible to minimize discontinuities in the transitions between foreground and background areas.

For foreground areas containing the same CbCr values, but different luminance (Y) val-

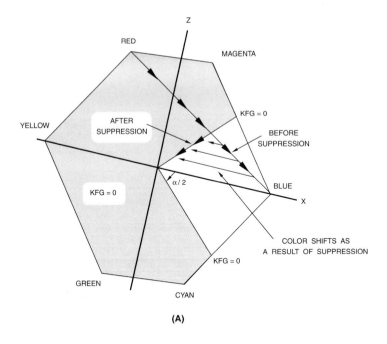

(A)

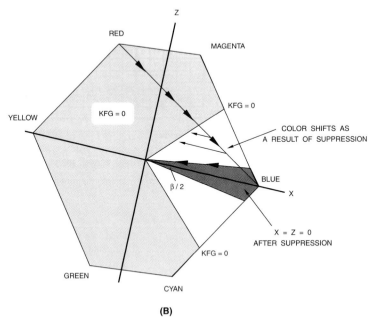

(B)

Figure 7.24. Suppression Angle Operation for a Gradual Change from a Red Foreground Object to the Blue Key Color. (a) Simple suppression. (b) Improved suppression using a suppression angle.

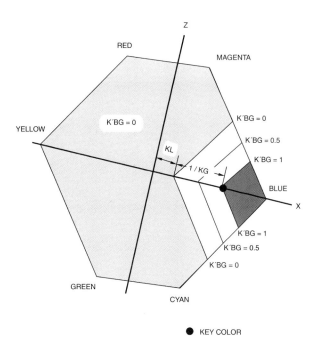

Figure 7.25. Background Key Generation.

ues, as the key color, the key processor may also reduce the background key value as the foreground luminance level increases, allowing turning off the background in foreground areas containing a "lighter" key color, such as light blue. This is done by:

$$K_{BG} = K'_{BG} - y_c Y_{FG}$$

$$K_{BG} = 0 \text{ if } y_c Y_{FG} > K_{FG}$$

To handle shadows cast by foreground objects, and opaque or translucent foreground objects, the luminance level of the blue backing of the foreground image is monitored. Where the luminance of the blue backing is reduced, the luminance of the background image also is reduced. The amount of background luminance reduction must be controlled so that defects in the blue backing

(such as seams or footprints) are not interpreted as foreground shadows.

Additional controls may be implemented to enable the foreground and background signals to be controlled independently. Examples are adjusting the contrast of the foreground so it matches the background or fading the foreground in various ways (such as fading to the background to make a foreground object vanish or fading to black to generate a silhouette).

In the computer environment, there may be relatively slow, smooth edges—especially edges involving smooth shading. As smooth edges are easily distorted during the chroma keying process, a wide keying process is usually used in these circumstances. During wide keying, the keying signal starts before the edge of the graphic object.

Composite Chroma Keying

In some instances, the component signals (such as YCbCr) are not directly available. For these situations, composite chroma keying may be implemented, as shown in Figure 7.26.

To detect the chroma key color, the foreground video source must be decoded to produce the Cb and Cr color difference signals. The keying signal, K_{FG}, is then used to mix between the two composite video sources. The garbage matte key signal forces the mixer to output the background source by reducing K_{FG}.

Chroma keying using composite video signals usually results in unrealistic keying, since there is inadequate color bandwidth. As a result, there is a lack of fine detail, and halos may be present on edges.

Superblack Keying

Video editing systems also may make use of superblack keying. In this application, areas of the foreground composite video signal that have a level of 0 to –5 IRE are replaced with the background video information.

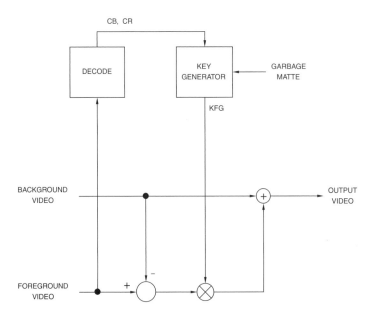

Figure 7.26. Typical Composite Chroma Key Circuit.

Video Scaling

With all the various video resolutions (Table 7.3), scaling is usually needed in almost every solution.

When generating objects that will be displayed on SDTV, computer users must be concerned with such things as text size, line thickness, and so forth. For example, text readable on a 1280 × 1024 computer display may not be readable on a SDTV display due to the large amount of downscaling involved. Thin horizontal lines may either disappear completely or flicker at a 25- or 29.97-Hz rate when converted to interlaced SDTV.

Note that scaling must be performed on component video signals (such as R′G′B′ or YCbCr). Composite color video signals cannot be scaled directly due to the color subcarrier phase information present, which would be meaningless after scaling.

In general, the spacing between output samples can be defined by a Target Increment (tarinc) value:

$$\text{tarinc} = I \, / \, O$$

where I and O are the number of input (I) and output (O) samples, either horizontally or vertically.

The first and last output samples may be aligned with the first and last input samples by adjusting the equation to be:

$$\text{tarinc} = (I - 1) \, / \, (O - 1)$$

Displays		SDTV Sources		HDTV Sources
704 × 480	640 × 480	704 × 360[1]	704 × 432[1]	1280 × 720
854 × 480	800 × 600	480 × 480	480 × 576	1440 × 816[2]
704 × 576	1024 × 768	528 × 480		1440 × 1040[3]
854 × 576	1280 × 768	544 × 480	544 × 576	1280 × 1080
1280 × 720	1366 × 768	640 × 480		1440 × 1080
1280 × 768	1024 × 1024	704 × 480	704 × 576	1920 × 1080
1920 × 1080	1280 × 1024		768 × 576	

Table 7.3. Common Active Resolutions for Consumer Displays and Broadcast Sources. [1]16:9 letterbox on a 4:3 display. [2]2.35:1 anamorphic for a 16:9 1920x1080 display. [3]1.85:1 anamorphic for a 16:9 1920x1080 display.

Pixel Dropping and Duplication

This is also called "nearest neighbor" scaling since only the input sample closest to the output sample is used.

The simplest form of scaling down is pixel dropping, where (m) out of every (n) samples are thrown away both horizontally and vertically. A modified version of the Bresenham line-drawing algorithm (described in most computer graphics books) is typically used to determine which samples not to discard.

Simple upscaling can be accomplished by pixel duplication, where (m) out of every (n) samples are duplicated both horizontally and vertically. Again, a modified version of the Bresenham line-drawing algorithm can be used to determine which samples to duplicate.

Scaling using pixel dropping or duplication is not recommended due to the visual artifacts and the introduction of aliasing components.

Linear Interpolation

An improvement in video quality of scaled images is possible using linear interpolation. When an output sample falls between two input samples (horizontally or vertically), the output sample is computed by linearly interpolating between the two input samples. However, scaling to images smaller than one-half of the original still results in deleted samples.

Figure 7.27 illustrates the vertical scaling of a 16:9 image to fit on a 4:3 display. A simple bi-linear vertical filter is commonly used, as shown in Figure 7.28a. Two source samples, L_n and L_{n+1}, are weighted and added together to form a destination sample, D_m.

$$D_0 = 0.75L_0 + 0.25L_1$$
$$D_1 = 0.5L_1 + 0.5L_2$$
$$D_2 = 0.25L_2 + 0.75L_3$$

However, as seen in Figure 7.28a, this results in uneven line spacing, which may result in visual artifacts. Figure 7.28b illustrates vertical filtering that results in the output lines being more evenly spaced:

$$D_0 = L_0$$
$$D_1 = (2/3)L_1 + (1/3)L_2$$
$$D_2 = (1/3)L_2 + (2/3)L_3$$

The linear interpolator is a poor bandwidth-limiting filter. Excess high-frequency detail is removed unnecessarily and too much energy above the Nyquist limit is still present, resulting in aliasing.

Anti-Aliased Resampling

The most desirable approach is to ensure the frequency content scales proportionally with the image size, both horizontally and vertically.

Figure 7.29 illustrates the fundamentals of an anti-aliased resampling process. The input data is upsampled by A and lowpass filtered to remove image frequencies created by the interpolation process. Filter B bandwidth-limits the signal to remove frequencies that will alias in the resampling process B. The ratio of B/A determines the scaling factor.

Filters A and B are usually combined into a single filter. The response of the filter largely determines the quality of the interpolation. The ideal lowpass filter would have a very flat passband, a sharp cutoff at half of the lowest sampling frequency (either input or output), and very high attenuation in the stopband. However, since such a filter generates ringing on sharp edges, it is usually desirable to roll off the top of the passband. This makes for slightly softer pictures, but with less pronounced ringing.

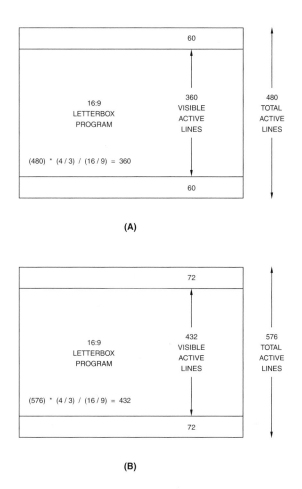

(A)

(B)

Figure 7.27. Vertical Scaling of 16:9 Images to Fit on a 4:3 Display. (a) 480-line systems. (b) 576-line systems.

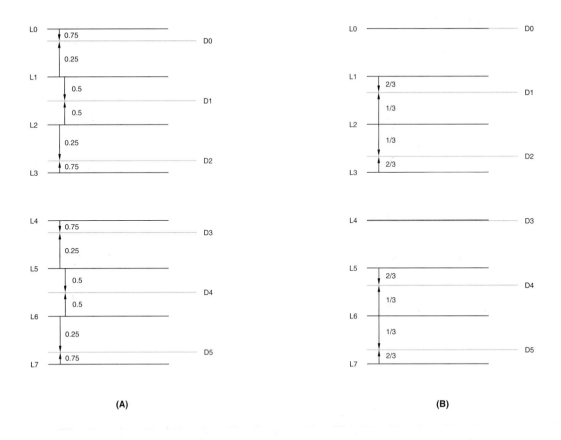

**Figure 7.28. 75% Vertical Scaling of 16:9 Images to Fit on a 4:3 Display.
(a) Unevenly spaced results. (b) Evenly spaced results.**

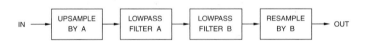

Figure 7.29. General Anti-Aliased Resampling Structure.

Passband ripple and stopband attenuation of the filter provide some measure of scaling quality, but the subjective effect of ringing means a flat passband might not be as good as one might think. Lots of stopband attenuation is almost always a good thing.

There are essentially three variations of the general resampling structure. Each combines the elements of Figure 7.29 in various ways.

One approach is a variable-bandwidth anti-aliasing filter followed by a combined interpolator/resampler. In this case, the filter needs new coefficients for each scale factor—as the scale factor is changed, the quality of the image may vary. In addition, the overall response is poor if linear interpolation is used. However, the filter coefficients are time-invariant and there are no gain problems.

A second approach is a combined filter/interpolator followed by a resampler. Generally, the higher the order of interpolation, n, the better the overall response. The center of the filter transfer function is always aligned over the new output sample. With each scaling factor, the filter transfer function is stretched or compressed to remain aligned over n output samples. Thus, the filter coefficients, and the number of input samples used, change with each new output sample and scaling factor. Dynamic gain normalization is required to ensure the sum of the filter coefficients is always equal to one.

A third approach is an interpolator followed by a combined filter/resampler. The input data is interpolated up to a common multiple of the input and output rates by the insertion of zero samples. This is filtered with a low-pass finite-impulse-response (FIR) filter to interpolate samples in the zero-filled gaps, then re-sampled at the required locations. This type of design is usually achieved with a "polyphase" filter which switches its coefficients as the relative position of input and output samples change.

Display Scaling Examples

Figures 7.30 through 7.38 illustrate various scaling examples for displaying 16:9 and 4:3 pictures on 4:3 and 16:9 displays, respectively.

How content is displayed is a combination of user preferences and content aspect ratio. For example, when displaying 16:9 content on a 4:3 display, many users prefer to have the entire display filled with the cropped picture (Figure 7.31) rather than seeing black or gray bars with the letterbox solution (Figure 7.32).

In addition, some displays incorrectly assume any progressive video signal on their YPbPr inputs is from an "anamorphic" source. As a result, they horizontally upscale progressive 16:9 programs by 25% when no scaling should be applied. Therefore, for set-top boxes it is useful to include a "16:9 (Compressed)" mode, which horizontally downscales the progressive 16:9 program by 25% to pre-compensate for the horizontally upscaling being done by the 16:9 display.

Scan Rate Conversion

In many cases, some form of scan rate conversion (also called temporal rate conversion, frame rate conversion, or field rate conversion) is needed. Multi-standard analog VCRs and scan converters use scan rate conversion to

1920 Samples

1080
Scan
Lines

Figure 7.30. 16:9 Source Example.

720 Samples

480
Scan
Lines

Figure 7.31. Scaling 16:9 Content for a 4:3 Display: "Normal" or pan-and-scan mode. Results in some of the 16:9 content being ignored (indicated by gray regions).

720 Samples

360 Scan Lines

480 Scan Lines

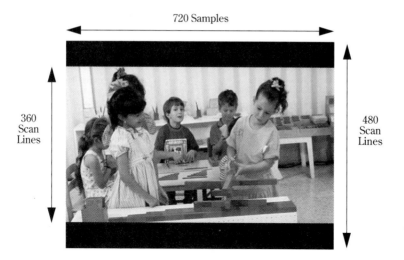

Figure 7.32. Scaling 16:9 Content for a 4:3 Display: "Letterbox" mode. Entire 16:9 program visible, with black bars at top and bottom of display.

720 Samples

480 Scan Lines

Figure 7.33. Scaling 16:9 Content for a 4:3 Display: "Squeezed" mode. Entire 16:9 program horizontally squeezed to fit 4:3 display, resulting in a distorted picture.

720 Samples

480
Scan
Lines

Figure 7.34. 4:3 Source Example.

1920 Samples

1080
Scan
Lines

Figure 7.35. Scaling 4:3 Content for a 16:9 Display: "Normal" mode. Left and right portions of 16:9 display not used, so made black or gray.

1920 Samples

1080
Scan
Lines

Figure 7.36. Scaling 4:3 Content for a 16:9 Display: "Wide" mode. Entire picture linearly scaled horizontally to fill 16:9 display, resulting in distorted picture unless used with anamorphic content.

1920 Samples

1080
Scan
Lines

Figure 7.37. Scaling 4:3 Content for a 16:9 Display: "Zoom" mode. Top and bottom portion of 4:3 picture deleted, then scaled to fill 16:9 display.

1920 Samples

1080
Scan
Lines

Figure 7.38. Scaling 4:3 Content for a 16:9 Display: "Panorama" mode. Left and right 25% edges of picture are nonlinearly scaled horizontally to fill 16:9 display, distorted picture on left and right sides.

convert between various video standards. Computers usually operate the display at about 75 Hz noninterlaced, yet need to display 50- and 60-Hz interlaced video. With digital television, multiple refresh rates can be supported.

Note that processing must be performed on component video signals (such as R'G'B' or YCbCr). Composite color video signals cannot be processed directly due to the color subcarrier phase information present, which would be meaningless after processing.

Frame or Field Dropping and Duplicating

Simple scan-rate conversion may be done by dropping or duplicating one out of every N fields. For example, the conversion of 60-Hz to 50-Hz interlaced operation may drop one out of

every six fields, as shown in Figure 7.39, using a single field store.

The disadvantage of this technique is that the viewer may be see jerky motion, or motion "judder." In addition, some MPEG decoders use top-field only to convert from 60Hz to 50Hz, degrading the vertical resolution.

The worst artifacts are present when a non-integer scan rate conversion is done—for example, when some frames are displayed three times, while others are displayed twice. In this instance, the viewer will observe double or blurred objects. As the human brain tracks an object in successive frames, it expects to see a regular sequence of positions, and has trouble reconciling the apparent stop-start motion of objects. As a result, it incorrectly concludes that there are two objects moving in parallel.

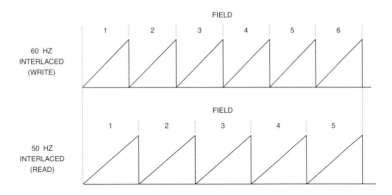

Figure 7.39. 60-Hz to 50-Hz Conversion Using a Single Field Store by Dropping One out of Every Six Fields.

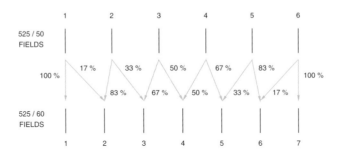

Figure 7.40. 50-Hz to 60-Hz Conversion Using Temporal Interpolation with No Motion Compensation.

Temporal Interpolation

This technique generates new frames from the original frames as needed to generate the desired frame rate. Information from both past and future input frames should be used to optimally handle objects appearing and disappearing.

Conversion of 50-Hz to 60-Hz operation using temporal interpolation is illustrated in Figure 7.40. For every five fields of 50-Hz video, there are six fields of 60-Hz video.

After both sources are aligned, two adjacent 50-Hz fields are mixed together to generate a new 60-Hz field. This technique is used in some inexpensive standards converters to convert between 625/50 and 525/60 standards. Note that no motion analysis is done. Therefore, if the camera operating at 625/50 pans horizontally past a narrow vertical object, you see one object once every six 525/60 fields, and for the five fields in between, you see two objects, one fading in while the other fades out.

625/50 to 525/60 Examples

Figure 7.41 illustrates a scan rate converter that implements vertical, followed by temporal, interpolation. Figure 7.42 illustrates the spectral representation of the design in Figure 7.41.

Many designs now combine the vertical and temporal interpolation into a single design, as shown in Figure 7.43, with the corresponding spectral representation shown in Figure 7.44. This example uses vertical, followed by temporal, interpolation. If temporal, followed by vertical, interpolation were implemented, the field stores would be half the size. However, the number of line stores would increase from four to eight.

In either case, the first interpolation process must produce an intermediate, higher-resolution progressive format to avoid interlace components that would interfere with the second interpolation process. It is insufficient to interpolate, either vertically or temporally, using a mixture of lines from both fields, due to the interpolation process not being able to compensate for the temporal offset of interlaced lines.

Motion Compensation

Higher-quality scan rate converters using temporal interpolation incorporate motion compensation to minimize motion artifacts. This results in extremely smooth and natural motion, and images appear sharper and do not suffer from motion "judder."

Motion estimation for scan rate conversion differs from that used by MPEG. In MPEG, the goal is to minimize the displaced frame difference (error) by searching for a high correlation between areas in subsequent frames. The resulting motion vectors do not necessarily correspond to true motion vectors.

For scan rate conversion, it is important to determine true motion information to perform correct temporal interpolation. The interpolation should be tolerant of incorrect motion vectors to avoid introducing artifacts as unpleasant as those the technique is attempting to remove. Motion vectors could be incorrect for several reasons, such as insufficient time to track the motion, out-of-range motion vectors, and estimation difficulties due to aliasing.

100 Hz Interlaced Television Example

A standard PAL television shows 50 fields per second. The images flicker, especially when you look at large areas of highly-saturated color. A much improved picture can be achieved using a 100 Hz interlaced refresh (also called double scan).

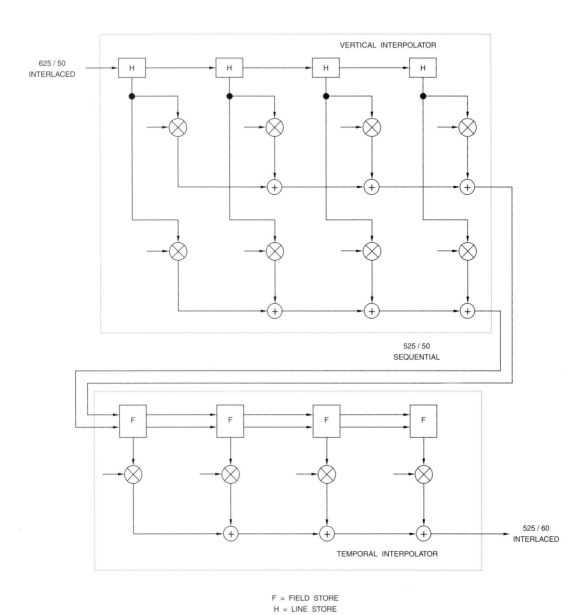

Figure 7.41. Typical 625/50 to 525/60 Conversion Using Vertical, Followed by Temporal, Interpolation.

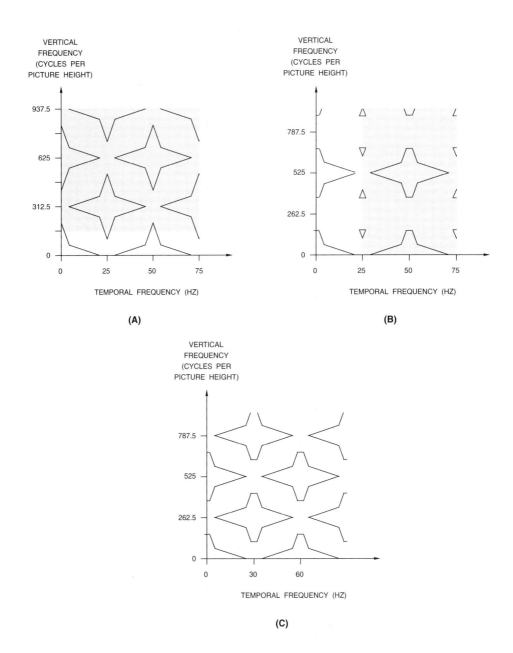

Figure 7.42. Spectral Representation of Vertical, Followed by Temporal, Interpolation. (a) Vertical lowpass filtering. (b) Resampling to intermediate sequential format and temporal lowpass filtering. (c) Resampling to final standard.

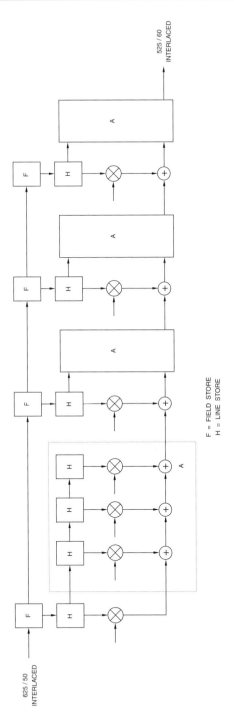

Figure 7.43. Typical 625/50 to 525/60 Conversion Using Combined Vertical and Temporal Interpolation.

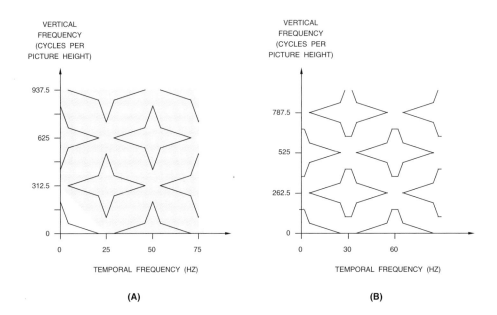

Figure 7.44. Spectral Representation of Combined Vertical and Temporal Interpolation. (a) Two-dimensional lowpass filtering. (b) Resampling to final standard.

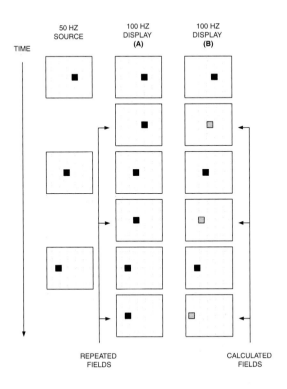

Figure 7.45. 50 Hz to 100 Hz (Double Scan Interlaced) Techniques.

Early 100 Hz televisions simply repeated fields ($F_1F_1F_2F_2F_3F_3F_4F_4$...), as shown in Figure 7.45a. However, they still had line flicker, where horizontal lines constantly jumped between the odd and even lines. This disturbance occurred once every twenty-fifth of a second.

The field sequence $F_1F_2F_1F_2F_3F_4F_3F_4$... can be used, which solves the line flicker problem. Unfortunately, this gives rise to the problem of judder in moving images. This can be compensated for by using the $F_1F_2F_1F_2F_3F_4F_3F_4$... sequence for static images, and the $F_1F_1F_2F_2F_3F_3F_4F_4$... sequence for moving images.

An ideal picture is still not obtained when viewing programs created for film. They are subject to judder, owing to the fact that each film frame is transmitted twice. Instead of the field sequence $F_1F_1F_2F_2F_3F_3F_4F_4$..., the situation calls for the sequence $F_1F_1{'}F_2F_2{'}F_3F_3{'}F_4F_4{'}$... (Figure 7.45b), where $F_{n'}$ is a motion-compensated generated image between F_n and F_{n+1}.

2:2 Pulldown

This technique is used with some film-based MPEG content for 50 Hz regions. Film is usually recorded at 24 frames per second.

During MPEG encoding, the telecine machine is sped up from 24 to 25 frames per second, making the content 25 frames per second progressive. During MPEG decoding, each film frame is simply mapped into two video fields.

This technique provides higher video quality and avoids motion judder artifacts. However, it shortens the duration of the program by about 4 percent, cutting the duration of a two-hour movie by ~5 minutes.

To compensate the audio changing pitch due to the telecine speedup, it may be resampled during decoding to restore the original pitch (costly to do in a low-cost consumer product) or resampling may be done during the program authoring. One of these two solutions must be used since many audio decoders cannot handle the 4% faster audio data via S/PDIF (IEC 60958).

3:2 Pulldown

When converting 24 frames per second content to NTSC (59.94-Hz field rate), 3:2 pulldown is commonly used, as shown in Figure 7.46. The film speed is slowed down by 0.1% to 23.976 (24/1.001) frames per second. Two film frames generate five video fields. In scenes of high-speed motion of objects, the specific film frame used for a particular video field may be manually adjusted to minimize motion artifacts.

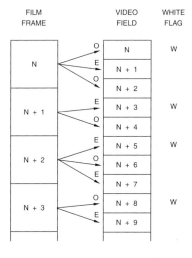

O = ODD LINES OF FILM FRAME
E = EVEN LINES OF FILM FRAME

Figure 7.46. Typical 3:2 Pulldown for Transferring Film to NTSC Video.

3:2 pulldown may also be used by MPEG decoders simply to increase the frame rate from 23.976 (24/1.001) to 59.94 (60/1.001) frames per second, avoiding the deinterlacing issue.

Varispeed is commonly used to cover up problems such as defects, splicing, censorship cuts, or to change the running time of a program. Rather than repeating film frames and causing a "stutter," the 3:2 relationship between the film and video is disrupted long enough to ensure a smooth temporal rate.

Analog laserdiscs use a white flag signal to indicate the start of another sequence of related fields for optimum still-frame performance. During still-frame mode, the white flag signal tells the system to back up two fields (to use two fields that have no motion between them) to re-display the current frame.

3:3 Pulldown

This technique is used in some displays that support 72 Hz refresh. The 24 frames per second film-based content is converted to 72 Hz progressive by simply duplicating each film frame three times.

24:1 Pulldown

This technique, also called "12:1 pull-down," can also be used to convert 24 frames/second content to 50 fields per second.

Two video fields are generated from every film frame, except every 12th film frame generates 3 video fields. Although the audio pitch is correct, motion judder is present every one-half second when smooth motion is present.

Noninterlaced-to-Interlaced Conversion

In some applications, it is necessary to display a noninterlaced video signal on an interlaced display. Thus, some form of "noninterlaced-to-interlaced conversion" may be required.

Noninterlaced to interlaced conversion must be performed on component video signals (such as R′G′B′ or YCbCr). Composite color video signals (such as NTSC or PAL) cannot be processed directly due to the presence of color subcarrier phase information, which would be meaningless after processing. These signals must be decoded into component color signals, such as R′G′B′ or YCbCr, prior to conversion.

There are essentially two techniques: scan line decimation and vertical filtering.

Scan Line Decimation

The easiest approach is to throw away every other active scan line in each noninterlaced frame, as shown in Figure 7.47. Although the cost is minimal, there are problems with this approach, especially with the top and bottom of objects.

If there is a sharp vertical transition of color or intensity, it will flicker at one-half the refresh rate. The reason is that it is only displayed every other field as a result of the decimation. For example, a horizontal line that is one noninterlaced scan line wide will flicker on and off. Horizontal lines that are two noninterlaced scan lines wide will oscillate up and down.

Simple decimation may also add aliasing artifacts. While not necessarily visible, they will affect any future processing of the picture.

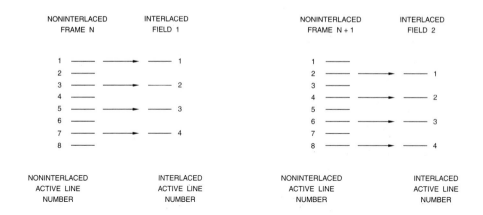

Figure 7.47. Noninterlaced-to-Interlaced Conversion Using Scan Line Decimation.

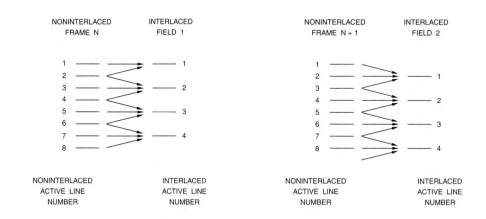

Figure 7.48. Noninterlaced-to-Interlaced Conversion Using 3-Line Vertical Filtering.

Vertical Filtering

A better solution is to use two or more lines of noninterlaced data to generate one line of interlaced data. Fast vertical transitions are smoothed out over several interlaced lines.

For a 3-line filter, such as shown in Figure 7.48, typical coefficients are [0.25, 0.5, 0.25]. Using more than three lines usually results in excessive blurring, making small text difficult to read.

An alternate implementation uses IIR rather than FIR filtering. In addition to averaging, this technique produces a reduction in brightness around objects, further reducing flicker.

Note that care must be taken at the beginning and end of each frame in the event that fewer scan lines are available for filtering.

Interlaced-to-Noninterlaced Conversion

In some applications, it is necessary to display an interlaced video signal on a noninterlaced display. Thus, some form of "deinterlacing" or "progressive scan conversion" may be required.

Note that deinterlacing must be performed on component video signals (such as R′G′B′ or YCbCr). Composite color video signals (such as NTSC or PAL) cannot be deinterlaced directly due to the presence of color subcarrier phase information, which would be meaningless after processing. These signals must be decoded into component color signals, such as R′G′B′ or YCbCr, prior to deinterlacing.

There are two fundamental deinterlacing algorithms: video mode and film mode. Video mode deinterlacing can be further broken down into interfield and intrafield processing.

The goal of a good deinterlacer is to correctly choose the best algorithm needed at a particular moment.

In systems where the vertical resolution of the source and display do not match (due to, for example, displaying SDTV content on a HDTV), the deinterlacing and vertical scaling can be merged into a single process.

Video Mode: Intrafield Processing

This is the simplest method for generating additional scan lines using only information in the original field. The computer industry has coined this technique as "bob."

Although there are two common techniques for implementing intrafield processing, scan line duplication and scan line interpolation, the resulting vertical resolution is always limited by the content of the original field.

Scan Line Duplication

Scan line duplication (Figure 7.49) simply duplicates the previous active scan line. Although the number of active scan lines is doubled, there is no increase in the vertical resolution.

Scan Line Interpolation

Scan line interpolation generates interpolated scan lines between the original active scan lines. Although the number of active scan lines is doubled, the vertical resolution is not.

The simplest implementation, shown in Figure 7.50, uses linear interpolation to generate a new scan line between two input scan lines:

$$\text{out}_n = (\text{in}_{n-1} + \text{in}_{n+1}) / 2$$

Better results, at additional cost, may be achieved by using a FIR filter:

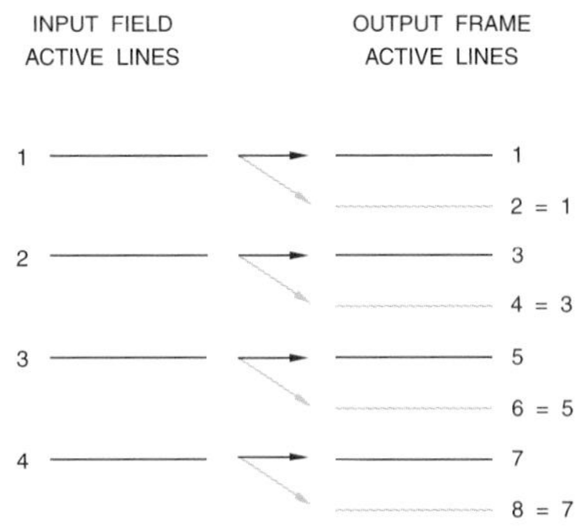

INPUT FIELD
ACTIVE LINES

OUTPUT FRAME
ACTIVE LINES

1 ——————→ ——————— 1

———————— 2 = 1

2 ——————→ ——————— 3

———————— 4 = 3

3 ——————→ ——————— 5

———————— 6 = 5

4 ——————→ ——————— 7

———————— 8 = 7

Figure 7.49. Deinterlacing Using Scan Line Duplication. New scan lines are generated by duplicating the active scan line above it.

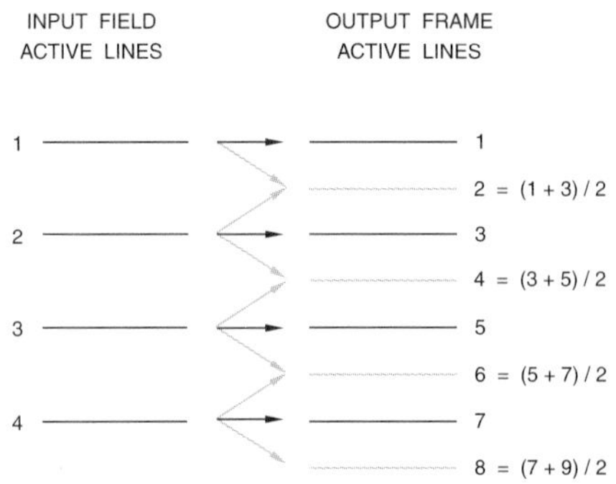

INPUT FIELD
ACTIVE LINES

OUTPUT FRAME
ACTIVE LINES

1 ——————→ ——————— 1

———————— $2 = (1 + 3)/2$

2 ——————→ ——————— 3

———————— $4 = (3 + 5)/2$

3 ——————→ ——————— 5

———————— $6 = (5 + 7)/2$

4 ——————→ ——————— 7

———————— $8 = (7 + 9)/2$

Figure 7.50. Deinterlacing Using Scan Line Interpolation. New scan lines are generated by averaging the previous and next active scan lines.

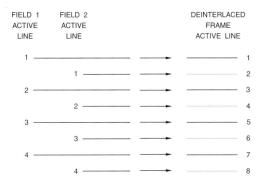

FIELD 1
ACTIVE
LINE

FIELD 2
ACTIVE
LINE

DEINTERLACED
FRAME
ACTIVE LINE

Figure 7.51. Deinterlacing Using Field Merging. Shaded scan lines are generated by using the input scan line from the next or previous field.

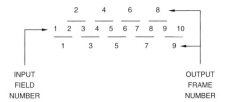

Figure 7.52. Producing Deinterlaced Frames at Field Rates.

$$
\begin{aligned}
out_n = \ &(160 * (in_{n-1} + in_{n+1}) \\
&- 48 * (in_{n-3} + in_{n+3}) \\
&+ 24 * (in_{n-5} + in_{n+5}) \\
&- 12 * (in_{n-7} + in_{n+7}) \\
&+ 6 * (in_{n-9} + in_{n+9}) \\
&- 2 * (in_{n-11} + in_{n+11})
\end{aligned}
$$

Fractional Ratio Interpolation

In many cases, there is a periodic, but non-integral, relationship between the number of input scan lines and the number of output scan lines. In this case, fractional ratio interpolation may be necessary, similar to the polyphase filtering used for scaling only performed in the vertical direction. This technique combines deinterlacing and vertical scaling into a single process.

Variable Interpolation

In a few cases, there is no periodicity in the relationship between the number of input and output scan lines. Therefore, in theory, an infi-

nite number of filter phases and coefficients are required. Since this is not feasible, the solution is to use a large, but finite, number of filter phases. The number of filter phases determines the interpolation accuracy. This technique also combines deinterlacing and vertical scaling into a single process.

Video Mode: Interfield Processing

In this method, video information from more than one field is used to generate a single progressive frame. This method can provide higher vertical resolution since it uses content from more than a single field.

Field Merging

This technique merges two consecutive fields together to produce a frame of video (Figure 7.51). At each field time, the active scan lines of that field are merged with the active scan lines of the previous field. The

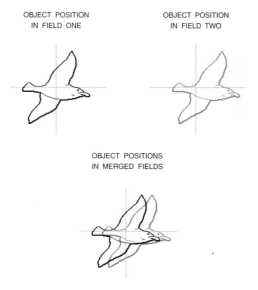

OBJECT POSITION
IN FIELD ONE

OBJECT POSITION
IN FIELD TWO

OBJECT POSITIONS
IN MERGED FIELDS

Figure 7.53. Movement Artifacts When Field Merging Is Used.

result is that for each input field time, a pair of fields combine to generate a frame (see Figure 7.52). Although simple to implement, the vertical resolution is doubled only in regions of no movement.

Moving objects will have artifacts, also called "combing," due to the time difference between two fields—a moving object is located in a different position from one field to the next. When the two fields are merged, moving objects will have a "double image" (see Figure 7.53).

It is common to soften the image slightly in the vertical direction to attempt to reduce the visibility of combing. When implemented, it causes a loss of vertical resolution and jitter on movement and pans.

The computer industry refers to this technique as "weave," but "weave" also includes the inverse telecine process to remove any 3:2 pulldown present in the source. Theoretically, this eliminates the "double image" artifacts since two identical fields are now being merged.

Motion Adaptive Deinterlacing

A good deinterlacing solution is to use field merging for still areas of the picture and scan line interpolation for areas of movement. To accomplish this, motion, on a sample-by-sample basis, must be detected over the entire picture in real time, requiring processing several fields of video.

As two fields are combined, full vertical resolution is maintained in still areas of the picture, where the eye is most sensitive to detail. The sample differences may have any value, from 0 (no movement and noise-free) to maximum (for example, a change from full intensity to black). A choice must be made when to use a sample from the previous field (which is in the wrong location due to motion) or to interpolate a new sample from adjacent scan lines in the current field. Sudden switching between methods is visible, so crossfading (also called soft switching) is used. At some magnitude of sample difference, the loss of resolution due to a double image is equal to the loss of resolution due to interpolation. That amount of motion should result in the crossfader being at the 50% point. Less motion will result in a fade towards field merging and more motion in a fade towards the interpolated values.

Rather than "per pixel" motion adaptive deinterlacing, which makes decisions for every sample, some low-cost solutions use "per field" motion adaptive deinterlacing. In this case, the algorithm is selected each field, based on the amount of motion between the fields. "Per pixel" motion adaptive deinterlacing, although difficult to implement, looks quite good when properly done. "Per field" motion adaptive deinterlacing rarely looks much better than vertical interpolation.

Motion Compensated Deinterlacing

Motion compensated (or motion vector steered) deinterlacing is several orders of magnitude more complex than motion adaptive deinterlacing, and is commonly found in pro-video format converters.

Motion compensated processing requires calculating motion vectors between fields for each sample, and interpolating along each sample's motion trajectory. Motion vectors must also be found that pass through each of any missing samples. Areas of the picture may be covered or uncovered as you move between frames. The motion vectors must also have sub-pixel accuracy, and be determined in two temporal directions between frames.

The motion vector errors used by MPEG are self-correcting since the residual difference between the predicted macroblocks is encoded. As motion compensated deinterlacing is a single-ended system, motion vector errors will produce artifacts, so different search and verification algorithms must be used.

Film Mode (using Inverse Telecine)

For sources that have 3:2 pulldown (i.e., 60 fields/second video converted from 24 frames/second film), higher deinterlacing performance may be obtained by removing duplicate fields prior to processing.

The inverse telecine process detects the 3:2 field sequence and the redundant third fields are removed. The remaining field pairs are merged (since there is no motion between them) to form progressive frames at 24 frames/second. These are then repeated in a 3:2 sequence to get to 60 frames/second.

Although this may seem to be the ideal solution, many MPEG-based sources use both 60 fields/second video (that has 3:2 pulldown) and 24 frames/second video (film-based) within a program. In addition, some programs may occasionally have both video types present simultaneously. In other cases, the 3:2 pulldown timing (cadence) doesn't stay regular, or the source was never originally from film. Thus, the deinterlacer has to detect each video type and process it differently (video mode vs. film mode). Display artifacts are common due to the delay between the video type changing and the deinterlacer detecting the change.

Frequency Response Considerations

Various two-times vertical upsampling techniques for deinterlacing may be implemented by stuffing zero values between two valid lines and filtering, as shown in Figure 7.54.

Line A shows the frequency response for line duplication, in which the lowpass filter coefficients for the filter shown are 1, 1, and 0.

Line interpolation, using lowpass filter coefficients of 0.5, 1.0, and 0.5, results in the frequency response curve of Line B. Note that line duplication results in a better high-frequency response. Vertical filters with a better frequency response than the one for line duplication are possible, at the cost of more line stores and processing.

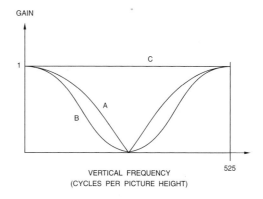

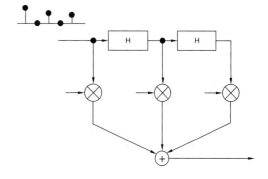

Figure 7.54. Frequency Response of Various Deinterlacing Filters. (a) Line duplication. (b) Line interpolation. (c) Field merging.

The best vertical frequency response is obtained when field merging is implemented. The spatial position of the lines is already correct and no vertical processing is required, resulting in a flat curve (Line C). Again, this applies only for stationary areas of the image.

DCT-Based Compression

The transform process of many video compression standards is based on the Discrete Cosine Transform, or DCT. The easiest way to envision it is as a filter bank with all the filters computed in parallel.

During encoding, the DCT is usually followed by several other operations, such as quantization, zig-zag scanning, run-length encoding, and variable-length encoding. During decoding, this process flow is reversed.

Many times, the terms macroblocks and blocks are used when discussing video compression. Figure 7.55 illustrates the relationship between these two terms, and shows why transform processing is usually done on 8 × 8 samples.

DCT

The 8 × 8 DCT processes an 8 × 8 block of samples to generate an 8 × 8 block of DCT coefficients, as shown in Figure 7.56. The input may be samples from an actual frame of video or motion-compensated difference (error) values, depending on the encoder mode of operation. Each DCT coefficient indicates the amount of a particular horizontal or vertical frequency within the block.

DCT coefficient (0,0) is the DC coefficient, or average sample value. Since natural images tend to vary only slightly from sample to sample, low frequency coefficients are typically larger values and high frequency coefficients are typically smaller values.

The 8 × 8 DCT is defined in Figure 7.57. f(x, y) denotes sample (x, y) of the 8 × 8 input block and F(u,v) denotes coefficient (u, v) of the DCT transformed block.

A reconstructed 8 × 8 block of samples is generated using an 8 × 8 inverse DCT (IDCT), defined in Figure 7.58. Although exact reconstruction is theoretically achievable, it is not practical due to finite-precision arithmetic, quantization and differing IDCT implementations. As a result, there are "mismatches" between different IDCT implementations.

"Mismatch control" attempts to reduce the drift between encoder and decoder IDCT results by eliminating bit patterns having the greatest contribution towards mismatches.

MPEG-1 mismatch control is known as "oddification" since it forces all quantized DCT coefficients to negative values. MPEG-2 and MPEG-4 Part 2 use an improved method called "LSB toggling" which affects only the LSB of the 63rd DCT coefficient after inverse quantization.

H.264 (also known as MPEG-4 Part 10) neatly sidesteps the issue by using an "exact-match inverse transform." Every decoder will produce exactly the same pictures, all else being equal.

Quantization

The 8 × 8 block of DCT coefficients is quantized, which reduces the overall precision of the integer coefficients and tends to eliminate high frequency coefficients, while maintaining perceptual quality. Higher frequencies are usually quantized more coarsely (fewer values allowed) than lower frequencies, due to visual perception of quantization error. The quantizer is also used for constant bit rate

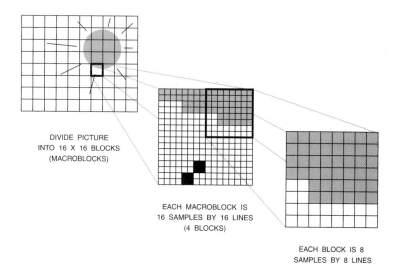

Figure 7.55. The Relationship between Macroblocks and Blocks.

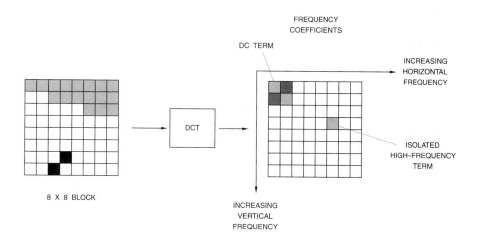

Figure 7.56. The DCT Processes the 8 × 8 Block of Samples or Error Terms to Generate an 8 × 8 Block of DCT Coefficients.

$$F(u, v) = 0.25 C(u)C(v) \sum_{x=0}^{7} \sum_{y=0}^{7} f(x, y)\cos(((2x + 1)u\pi)/16)\cos(((2y + 1)v\pi)/16)$$

u, v, x, y = 0, 1, 2, . . . 7
(x, y) are spatial coordinates in the sample domain
(u, v) are coordinates in the transform domain

Figure 7.57. 8 × 8 Two-Dimensional DCT Definition.

$$f(x, y) = 0.25 \sum_{u=0}^{7} \sum_{v=0}^{7} C(u)C(v)F(u, v)\cos(((2x + 1)u\pi)/16)\cos(((2y + 1)v\pi)/16)$$

Figure 7.58. 8 × 8 Two-Dimensional Inverse DCT (IDCT) Definition.

applications where it is varied to control the output bit rate.

Zig-Zag Scanning

The quantized DCT coefficients are re-arranged into a linear stream by scanning them in a zig-zag order. This rearrangement places the DC coefficient first, followed by frequency coefficients arranged in order of increasing frequency, as shown in Figures 7.59, 7.60, and 7.61. This produces long runs of zero coefficients.

Run Length Coding

The linear stream of quantized frequency coefficients is converted into a series of [run, amplitude] pairs. [run] indicates the number of zero coefficients, and [amplitude] the non-zero coefficient that ended the run.

Variable-Length Coding

The [run, amplitude] pairs are coded using a variable-length code, resulting in additional lossless compression. This produces shorter codes for common pairs and longer codes for less common pairs.

This coding method produces a more compact representation of the DCT coefficients, as a large number of DCT coefficients are usually quantized to zero and the re-ordering results (ideally) in the grouping of long runs of consecutive zero values.

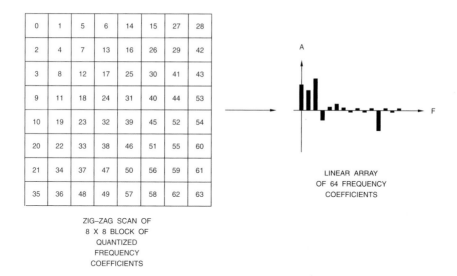

ZIG–ZAG SCAN OF
8 X 8 BLOCK OF
QUANTIZED
FREQUENCY
COEFFICIENTS

LINEAR ARRAY
OF 64 FREQUENCY
COEFFICIENTS

Figure 7.59. The 8 × 8 Block of Quantized DCT Coefficients Are Zig-Zag Scanned to Arrange in Order of Increasing Frequency. This scanning order is used for H.261, H.263, MPEG-1, MPEG-2, MPEG-4 Part 2, ITU-R BT.1618, ITU-R BT.1620, SMPTE 314M and SMPTE 370M.

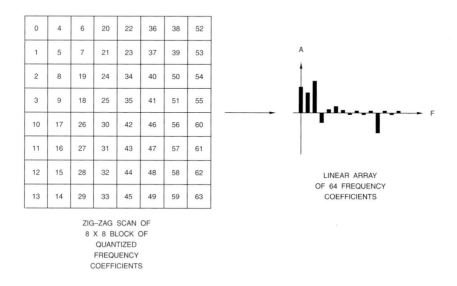

ZIG–ZAG SCAN OF
8 X 8 BLOCK OF
QUANTIZED
FREQUENCY
COEFFICIENTS

LINEAR ARRAY
OF 64 FREQUENCY
COEFFICIENTS

Figure 7.60. H.263, MPEG-2 and MPEG-4 Part 2 "Alternate-Vertical" Scanning Order.

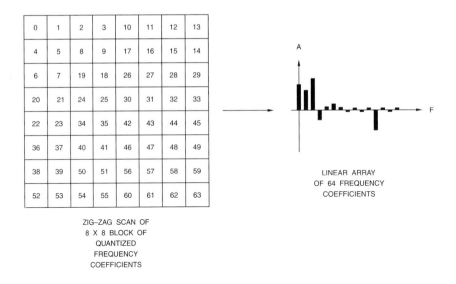

ZIG–ZAG SCAN OF
8 X 8 BLOCK OF
QUANTIZED
FREQUENCY
COEFFICIENTS

LINEAR ARRAY
OF 64 FREQUENCY
COEFFICIENTS

Figure 7.61. H.263 and MPEG-4 Part 2 "Alternate-Horizontal" Scanning Order.

Fixed Pixel Display Considerations

The unique designs and color reproduction gamuts of fixed pixel displays have resulted in new video processing technologies being developed. The result is brighter, sharper, more colorful images regardless of the video source.

Expanded Color Reproduction

Broadcast stations are usually tuned to meet the limited color reproduction characteristics of CRT-based televisions. To fit the color reproduction capabilities of PDP and LCD, manufacturers have introduced various color expansion technologies. These include using independent hue and saturation controls for each primary and complementary color, plus the flesh color.

Detail Correction

In CRT-based televisions, enhancing the image is commonly done by altering the electron beam diameter. With fixed-pixel displays, adding overshoot and undershoot to the video signals cause distortion. An acceptable implementation is to gradually change the brightness of the images before and after regions needing contour enhancement.

Non-uniform Quantization

Rather than simply increasing the number of quantization levels, the quantization steps can be changed in accordance with the intensity of the image. This is possible since people better detect small changes in brightness for dark images than for bright images. In addition, the brighter the image, the less sensitive people are to changes in brightness. This means that more quantization steps can be used for dark images than for bright ones. This technique can also be used to increase the quantization steps for shades that appear frequently.

Scaling and Deinterlacing

Fixed-pixel displays, such as LCD and plasma, usually upscale then downscale during deinterlacing to minimize moiré noise due to folded distortion. For example, a 1080i source is deinterlaced to 2160p, scaled to 1536p, then finally scaled to 768p (to drive a 1024x768 display). Alternately, some solutions deinterlace and upscale to 1500p, then scale to the display's native resolution.

Application Example

Figures 7.62 and 7.63 illustrate the typical video processing done after MPEG decoding and deinterlacing.

In addition to the primary video source (such as a MPEG decoder), additional video sources typically include an on-screen-display (OSD), second video for picture-in-picture (PIP), graphics and closed captioning or subtitles.

OSD design is unique to each product, so the OSD memory usually supports a wide variety of RGB/YCbCr formats and resolutions. Optional lookup tables add gamma correction to RGB data, convert 2-, 4- or 8-indexed color to 32-bit YCbCrA data, or translate 0–255 video levels to 16–235.

Graphics memory is usually used for JPEG images, background images, etc. It also usually supports a wide variety of RGB/YCbCr formats and resolutions. Optional lookup tables add gamma correction to RGB data, convert 2-, 4- or 8-indexed color to 32-bit YCbCrA data, or translate 0–255 video levels to 16–235.

The subtitle memory is a useful region for rendering closed captioning, DVB subtitles and DVD subpictures. Lookup tables convert 2-, 4- or 8-indexed color to 32-bit YCbCrA data.

Being able to scale each source independently offers maximum flexibility in the design of the display. In addition to being able to output any resolution regardless of the source resolutions, special effects can also be accommodated.

Chromaticity correction circuits ensure colors are accurate independent of the sources and display (SDTV vs. HDTV).

Independent brightness, contrast, saturation, hue and sharpness controls for each source and video output interface offer users the most flexibility. PIP can be adjusted without affecting the main picture, video can be adjusted without affecting still picture video quality, etc.

The optional downscaling and progressive-to-interlaced conversion block for the top NTSC/PAL encoder in Figure 7.63 enables simultaneous HD and SD, or simultaneous progressive and interlaced, outputs without affecting the HD or progressive video quality.

The second NTSC/PAL encoder shown at the bottom of Figure 7.63 is useful for recording a program without any OSD or subtitle information being accidently recorded.

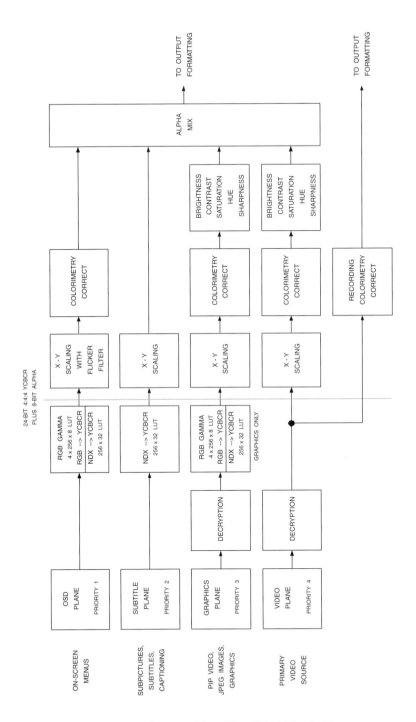

Figure 7.62. Video Composition Simplified Block Diagram.

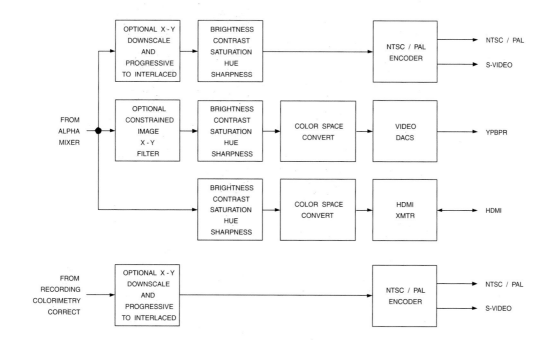

Figure 7.63. Video Output Port Processing.

References

1. Clarke, C. K. P., 1989, *Digital Video: Studio Signal Processing*, BBC Research Department Report BBC RD1989/14.
2. Devereux, V. G., 1984, *Filtering of the Colour-Difference Signals in 4:2:2 YUV Digital Video Coding Systems*, BBC Research Department Report BBC RD1984/4.
3. ITU-R BT.601–5, 1995, *Studio Encoding Parameters of Digital Television for Standard 4:3 and Widescreen 16:9 Aspect Ratios*.
4. ITU-R BT.709–5, 2002, *Parameter Values for the HDTV Standards for Production and International Programme Exchange*.
5. ITU-R BT.1358, 1998, *Studio Parameters of 625 and 525 Line Progressive Scan Television Systems*.
6. Johan G.W.M. Janssen, Jeroen H. Stessen, and Peter H.N. de With, *An Advanced Sampling Rate Conversion Technique for Video and Graphics Signals*, Philips Research Labs.
7. Sandbank, C. P., *Digital Television*, John Wiley & Sons, Ltd., New York, 1990.
8. SMPTE 274M–2003, *Television—1920 × 1080 Image Sample Structure, Digital Representation and Digital Timing Reference Sequences for Multiple Picture Rates*.
9. SMPTE 293M–2003, *Television—720 × 483 Active Line at 59.94-Hz Progressive Scan Production—Digital Representation*.
10. SMPTE 296M–2001, *Television—1280 × 720 Progressive Image Sample Structure, Analog and Digital Representation and Analog Interface*.
11. SMPTE EG36–1999, *Transformations Between Television Component Color Signals*.
12. Thomas, G. A., 1996, *A Comparison of Motion-Compensated Interlace-to-Progressive Conversion Methods*, BBC Research Department Report BBC RD1996/9.
13. Ultimatte®, Technical Bulletin No. 5, Ultimatte Corporation.
14. Watkinson, John, *The Engineer's Guide to Standards Conversion*, Snell and Wilcox Handbook Series.
15. Watkinson, John, *The Engineer's Guide to Motion Compensation*, Snell and Wilcox Handbook Series.

Chapter 8

NTSC, PAL, and SECAM Overview

To fully understand the NTSC, PAL, and SECAM encoding and decoding processes, it is helpful to review the background of these standards and how they came about.

NTSC Overview

The first color television system was developed in the United States, and on December 17, 1953, the Federal Communications Commission (FCC) approved the transmission standard, with broadcasting approved to begin January 23, 1954. Most of the work for developing a color transmission standard that was compatible with the (then current) 525-line, 60-field-per-second, 2:1 interlaced monochrome standard was done by the National Television System Committee (NTSC).

Luminance Information

The monochrome luminance (Y) signal is derived from gamma-corrected red, green, and blue (R′G′B′) signals:

$$Y = 0.299R' + 0.587G' + 0.114B'$$

Due to the sound subcarrier at 4.5 MHz, a requirement was made that the color signal fit within the same bandwidth as the monochrome video signal (0–4.2 MHz).

For economic reasons, another requirement was made that monochrome receivers must be able to display the black and white portion of a color broadcast and that color receivers must be able to display a monochrome broadcast.

Color Information

The eye is most sensitive to spatial and temporal variations in luminance; therefore, luminance information was still allowed the entire bandwidth available (0–4.2 MHz). Color information, to which the eye is less sensitive and which therefore requires less bandwidth, is represented as hue and saturation information.

The hue and saturation information is transmitted using a 3.58-MHz subcarrier, encoded so that the receiver can separate the hue, saturation, and luminance information and convert them back to RGB signals for display. Although this allows the transmission of color signals within the same bandwidth as monochrome signals, the problem still remains as to how to separate the color and luminance information cost-effectively, since they occupy the same portion of the frequency spectrum.

To transmit color information, U and V or I and Q "color difference" signals are used:

$$R' - Y = 0.701R' - 0.587G' - 0.114B'$$

$$B' - Y = -0.299R' - 0.587G' + 0.886B'$$

$$U = 0.492(B' - Y)$$

$$V = 0.877(R' - Y)$$

$$I = 0.596R' - 0.275G' - 0.321B'$$
$$= V\cos 33° - U\sin 33°$$
$$= 0.736(R' - Y) - 0.268(B' - Y)$$

$$Q = 0.212R' - 0.523G' + 0.311B'$$
$$= V\sin 33° + U\cos 33°$$
$$= 0.478(R' - Y) + 0.413(B' - Y)$$

The scaling factors to generate U and V from (B' − Y) and (R' − Y) were derived due to overmodulation considerations during transmission. If the full range of (B' − Y) and (R' − Y) were used, the modulated chrominance levels would exceed what the monochrome transmitters were capable of supporting. Experimentation determined that modulated subcarrier amplitudes of 20% of the Y signal amplitude could be permitted above white and below black. The scaling factors were then selected so that the maximum level of 75% color would be at the white level.

I and Q were initially selected since they more closely related to the variation of color acuity than U and V. The color response of the eye decreases as the size of viewed objects decreases. Small objects, occupying frequencies of 1.3–2.0 MHz, provide little color sensation. Medium objects, occupying the 0.6–1.3 MHz frequency range, are acceptable if repro-duced along the orange-cyan axis. Larger objects, occupying the 0–0.6 MHz frequency range, require full three-color reproduction.

The I and Q bandwidths were chosen accordingly, and the preferred color reproduction axis was obtained by rotating the U and V axes by 33°. The Q component, representing the green-purple color axis, was band-limited to about 0.6 MHz. The I component, representing the orange-cyan color axis, was band-limited to about 1.3 MHz.

Another advantage of limiting the I and Q bandwidths to 1.3 MHz and 0.6 MHz, respectively, is to minimize crosstalk due to asymmetrical sidebands as a result of lowpass filtering the composite video signal to about 4.2 MHz. Q is a double sideband signal; however, I is asymmetrical, bringing up the possibility of crosstalk between I and Q. The symmetry of Q avoids crosstalk into I; since Q is bandwidth limited to 0.6 MHz, I crosstalk falls outside the Q bandwidth.

U and V, both bandwidth-limited to 1.3 MHz, are now commonly used instead of I and Q. When broadcast, UV crosstalk occurs above 0.6 MHz, however, this is not usually visible due to the limited UV bandwidths used by NTSC decoders for consumer equipment.

The UV and IQ vector diagram is shown in Figure 8.1.

Color Modulation

I and Q (or U and V) are used to modulate a 3.58-MHz color subcarrier using two balanced modulators operating in phase quadrature: one modulator is driven by the subcarrier at sine phase, the other modulator is driven by the subcarrier at cosine phase. The outputs of the modulators are added together to form the modulated chrominance signal:

$$C = Q \sin(\omega t + 33°) + I \cos(\omega t + 33°)$$

$$\omega = 2\pi F_{SC}$$

$$F_{SC} = 3.579545 \text{ MHz } (\pm 10 \text{ Hz})$$

or, if U and V are used instead of I and Q:

$$C = U \sin \omega t + V \cos \omega t$$

Hue information is conveyed by the chrominance phase relative to the subcarrier. Saturation information is conveyed by chrominance amplitude. In addition, if an object has no color (such as a white, gray, or black object), the subcarrier is suppressed.

Composite Video Generation

The modulated chrominance is added to the luminance information along with appropriate horizontal and vertical sync signals, blanking information, and color burst information, to generate the composite color video waveform shown in Figure 8.2.

$$\text{composite NTSC} = Y + Q \sin(\omega t + 33°) + I \cos(\omega t + 33°) + \text{timing}$$

or, if U and V are used instead of I and Q:

$$\text{composite NTSC} = Y + U \sin \omega t + V \cos \omega t + \text{timing}$$

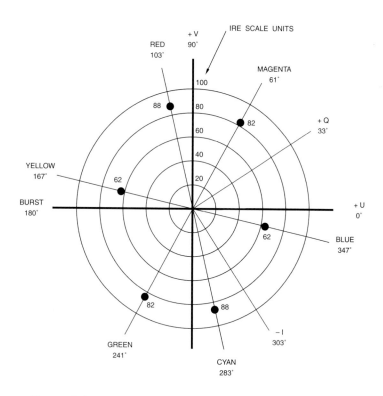

Figure 8.1. UV and IQ Vector Diagram for 75% Color Bars.

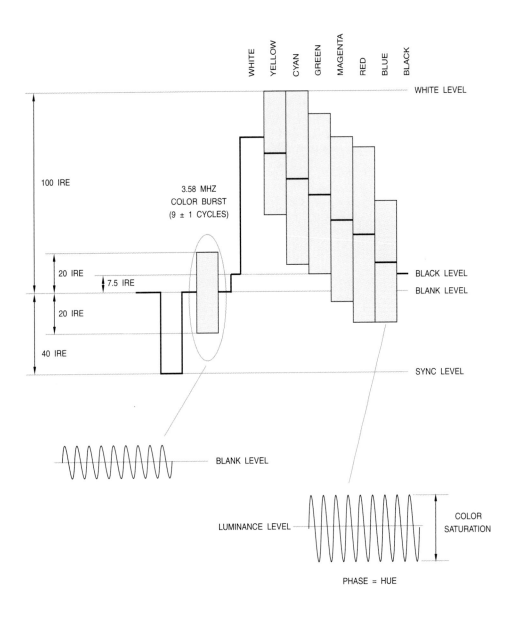

Figure 8.2. (M) NTSC Composite Video Signal for 75% Color Bars.

The bandwidth of the resulting composite video signal is shown in Figure 8.3.

The I and Q (or U and V) information can be transmitted without loss of identity as long as the proper color subcarrier phase relationship is maintained at the encoding and decoding process. A color burst signal, consisting of nine cycles of the subcarrier frequency at a specific phase, follows most horizontal sync pulses, and provides the decoder a reference signal so as to be able to recover the I and Q (or U and V) signals properly. The color burst phase is defined to be along the –U axis as shown in Figure 8.1.

Color Subcarrier Frequency

The specific choice for the color subcarrier frequency was dictated by several factors. The first was the need to provide horizontal interlace to reduce the visibility of the subcarrier, requiring that the subcarrier frequency, F_{SC}, be an odd multiple of one-half the horizontal line rate. The second factor was selection of a frequency high enough that it generated a fine interference pattern having low visibility. Third, double sidebands for I and Q (or U and V) bandwidths below 0.6 MHz had to be allowed.

The choice of the frequencies is:

$F_H = (4.5 \times 10^6/286)$ Hz = 15,734.27 Hz

$F_V = F_H/(525/2) = 59.94$ Hz

$F_{SC} = ((13 \times 7 \times 5)/2) \times F_H = (455/2) \times F_H$
$\quad = 3.579545$ MHz

The resulting F_V (field) and F_H (line) rates were slightly different from the monochrome standards, but fell well within the tolerance ranges and were therefore acceptable. Figure 8.4 illustrates the resulting spectral interleaving.

The luminance (Y) components are modulated due to the horizontal blanking process, resulting in bunches of luminance information spaced at intervals of F_H. These signals are further modulated by the vertical blanking process, resulting in luminance frequency components occurring at $NF_H \pm MF_V$. N has a maximum value of about 277 with a 4.2-MHz bandwidth-limited luminance. Thus, luminance information is limited to areas about integral harmonics of the line frequency (F_H), with additional spectral lines offset from NF_H by the 29.97-Hz vertical frame rate.

The area in the spectrum between luminance groups, occurring at odd multiples of one-half the line frequency, contains minimal spectral energy and is therefore used for the transmission of chrominance information. The harmonics of the color subcarrier are separated from each other by F_H since they are odd multiples of one-half F_H, providing a half-line offset and resulting in an interlace pattern that moves upward. Four complete fields are required to repeat a specific sample position, as shown in Figure 8.5.

NTSC Standards

Figure 8.6 shows the common designations for NTSC systems. The letter "M" refers to the monochrome standard for line and field rates (525/59.94), a video bandwidth of 4.2 MHz, an audio carrier frequency 4.5 MHz above the video carrier frequency, and a RF channel bandwidth of 6 MHz. The "NTSC" refers to the technique to add color information to the monochrome signal. Detailed timing parameters can be found in Table 8.9.

"NTSC 4.43" is commonly used for multistandard analog VCRs. The horizontal and vertical timing is the same as (M) NTSC; color

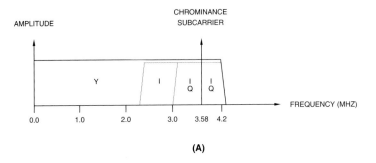

(A)

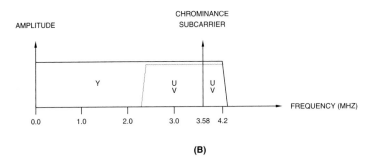

(B)

Figure 8.3. Video Bandwidths of Baseband (M) NTSC Video. (a) Using 1.3-MHz I and 0.6-MHz Q signals. (b) Using 1.3-MHz U and V signals.

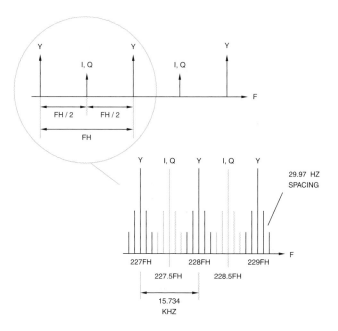

Figure 8.4. Luma and Chroma Frequency Interleave Principle.
Note that 227.5F$_H$ = F$_{SC}$.

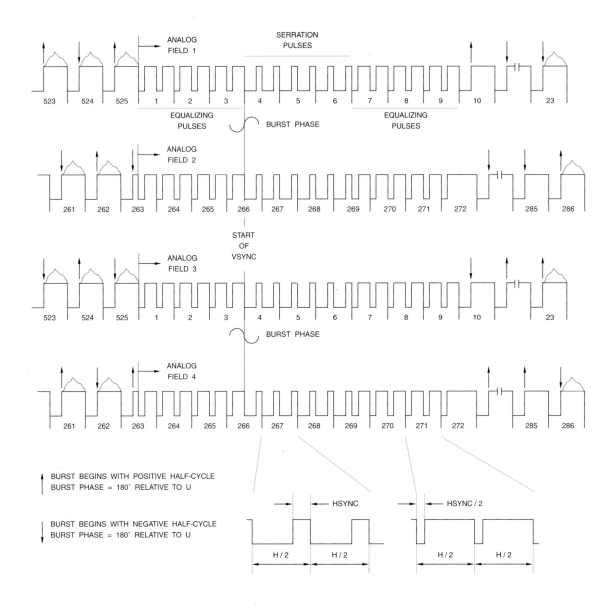

Figure 8.5. Four-field (M) NTSC Sequence and Burst Blanking.

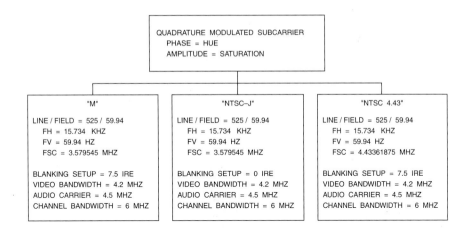

Figure 8.6. Common NTSC Systems.

encoding uses the PAL modulation format and a 4.43361875 MHz color subcarrier frequency.

"NTSC–J," used in Japan, is the same as (M) NTSC, except there is no blanking pedestal during active video. Thus, active video has a nominal amplitude of 714 mV.

"Noninterlaced NTSC" is a 262-line, 60 frames-per-second version of NTSC, as shown in Figure 8.7. This format is identical to standard (M) NTSC, except that there are 262 lines per frame.

RF Modulation

Figures 8.8, 8.9, and 8.10 illustrate the basic process of converting baseband (M) NTSC composite video to a RF (radio frequency) signal.

Figure 8.8a shows the frequency spectrum of a baseband composite video signal. It is similar to Figure 8.3. However, Figure 8.3 only shows the upper sideband for simplicity. The "video carrier" notation at 0 MHz serves only as a reference point for comparison with Figure 8.8b.

Figure 8.8b shows the audio/video signal as it resides within a 6-MHz channel (such as channel 3). The video signal has been lowpass filtered, most of the lower sideband has been removed, and audio information has been added.

Figure 8.8c details the information present on the audio subcarrier for stereo (BTSC) operation.

As shown in Figures 8.9 and 8.10, back porch clamping (see glossary) of the analog video signal ensures that the back porch level is constant, regardless of changes in the average picture level. White clipping of the video signal prevents the modulated signal from going below 10%; below 10% may result in overmodulation and "buzzing" in television receivers. The video signal is then lowpass filtered to 4.2 MHz and drives the AM (amplitude modulation) video modulator. The sync level corresponds to 100% modulation, the blanking

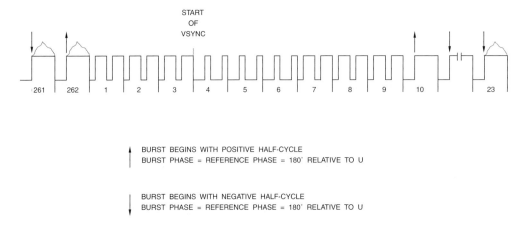

Figure 8.7. Noninterlaced NTSC Frame Sequence.

corresponds to 75%, and the white level corresponds to 10%. (M) NTSC systems use an IF (intermediate frequency) for the video of 45.75 MHz.

At this point, audio information is added on a subcarrier at 41.25 MHz. A monaural audio signal is processed as shown in Figure 8.9 and drives the FM (frequency modulation) modulator. The output of the FM modulator is added to the IF video signal.

The SAW filter, used as a vestigial sideband filter, provides filtering of the IF signal. The mixer, or up converter, mixes the IF signal with the desired broadcast frequency. Both sum and difference frequencies are generated by the mixing process, so the difference signal is extracted by using a bandpass filter.

Stereo Audio (Analog)

BTSC

This standard, defined by EIA TVSB5 and known as the BTSC system (Broadcast Television Systems Committee), is shown in Figure 8.10. Countries that use this system include

the United States, Canada, Mexico, Brazil, and Taiwan.

To enable stereo, L–R information is transmitted using a suppressed AM subcarrier. A SAP (secondary audio program) channel may also be present, used to transmit a second language or video description (descriptive audio for the visually impaired). A professional channel may also be present, allowing communication with remote equipment and people.

Zweiton M

This standard (ITU-R BS.707), also known as A2 M, is similar to that used with PAL. The L+R information is transmitted on a FM subcarrier at 4.5 MHz. The L–R information, or a second L+R audio signal, is transmitted on a second FM subcarrier at 4.724212 MHz.

If stereo or dual mono signals are present, the FM subcarrier at 4.724212 MHz is amplitude-modulated with a 55.0699 kHz subcarrier. This 55.0699 kHz subcarrier is 50% amplitude-modulated at 149.9 Hz to indicate stereo audio or 276.0 Hz to indicate dual mono audio.

This system is used in South Korea.

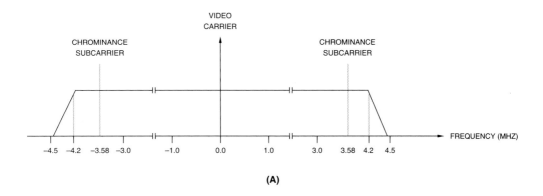

(A)

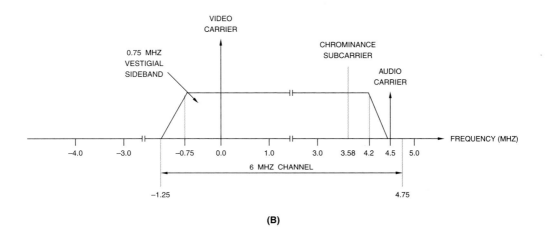

(B)

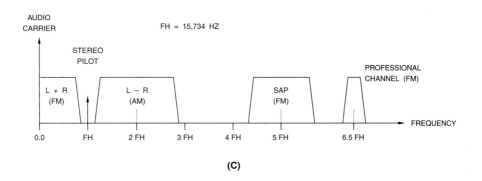

(C)

Figure 8.8. Transmission Channel for (M) NTSC. (a) Frequency spectrum of baseband composite video. (b) Frequency spectrum of typical channel including audio information. (c) Detailed frequency spectrum of BTSC stereo audio information.

EIA-J

This standard is similar to BTSC, and is used in Japan. The L+R information is transmitted on a FM subcarrier at 4.5 MHz. The L–R signal, or a second L+R signal, is transmitted on a second FM subcarrier at $+2F_H$.

If stereo or dual mono signals are present, a $+3.5F_H$ subcarrier is amplitude-modulated with either a 982.5 Hz subcarrier (stereo audio) or a 922.5 Hz subcarrier (dual mono audio).

Analog Channel Assignments

Tables 8.1 through 8.4 list the typical channel assignments for VHF, UHF and cable for various NTSC systems.

Note that cable systems routinely reassign channel numbers to alternate frequencies to minimize interference and provide multiple levels of programming (such as regular and preview premium movie channels).

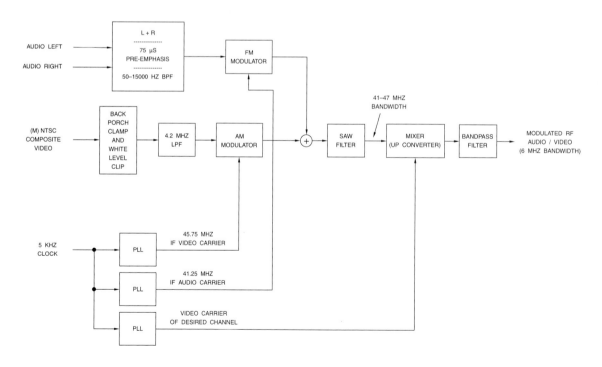

Figure 8.9. Typical RF Modulation Implementation for (M) NTSC: Mono Audio.

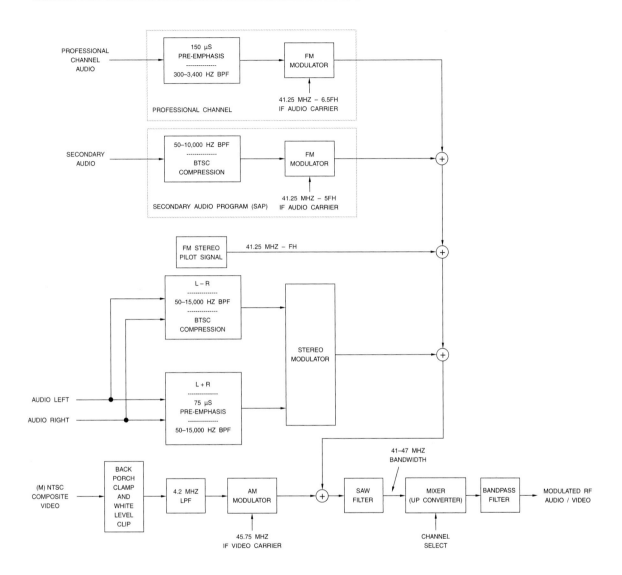

Figure 8.10. Typical RF Modulation Implementation for (M) NTSC: BTSC Stereo Audio.

Broadcast Channel	Video Carrier (MHz)	Audio Carrier (MHz)	Channel Range (MHz)	Broadcast Channel	Video Carrier (MHz)	Audio Carrier (MHz)	Channel Range (MHz)
–	–	–	–	40	627.25	631.75	626–632
–	–	–	–	41	633.25	637.75	632–638
2	55.25	59.75	54–60	42	639.25	643.75	638–644
3	61.25	65.75	60–66	43	645.25	649.75	644–650
4	67.25	71.75	66–72	44	651.25	655.75	650–656
5	77.25	81.75	76–82	45	657.25	661.75	656–662
6	83.25	87.75	82–88	46	663.25	667.75	662–668
7	175.25	179.75	174–180	47	669.25	673.75	668–674
8	181.25	185.75	180–186	48	675.25	679.75	674–680
9	187.25	191.75	186–192	49	681.25	685.75	680–686
10	193.25	197.75	192–198	50	687.25	691.75	686–692
11	199.25	203.75	198–204	51	693.25	697.75	692–698
12	205.25	209.75	204–210	52	699.25	703.75	698–704
13	211.25	215.75	210–216	53	705.25	709.75	704–710
14	471.25	475.75	470–476	54	711.25	715.75	710–716
15	477.25	481.75	476–482	55	717.25	721.75	716–722
16	483.25	487.75	482–488	56	723.25	727.75	722–728
17	489.25	493.75	488–494	57	729.25	733.75	728–734
18	495.25	499.75	494–500	58	735.25	739.75	734–740
19	501.25	505.75	500–506	59	741.25	745.75	740–746
20	507.25	511.75	506–512	60	747.25	751.75	746–752
21	513.25	517.75	512–518	61	753.25	757.75	752–758
22	519.25	523.75	518–524	62	759.25	763.75	758–764
23	525.25	529.75	524–530	63	765.25	769.75	764–770
24	531.25	535.75	530–536	64	771.25	775.75	770–776
25	537.25	541.75	536–542	65	777.25	781.75	776–782
26	543.25	547.75	542–548	66	783.25	787.75	782–788
27	549.25	553.75	548–554	67	789.25	793.75	788–794
28	555.25	559.75	554–560	68	795.25	799.75	794–800
29	561.25	565.75	560–566	69	801.25	805.75	800–806
30	567.25	571.75	566–572				
31	573.25	577.75	572–578				
32	579.25	583.75	578–584				
33	585.25	589.75	584–590				
34	591.25	595.75	590–596				
35	597.25	601.75	596–602				
36	603.25	607.75	602–608				
37	609.25	613.75	608–614				
38	615.25	619.75	614–620				
39	621.25	625.75	620–626				

Table 8.1. Analog Broadcast Nominal Frequencies for North America.

Broadcast Channel	Video Carrier (MHz)	Audio Carrier (MHz)	Channel Range (MHz)	Broadcast Channel	Video Carrier (MHz)	Audio Carrier (MHz)	Channel Range (MHz)
–	–	–	–	40	633.25	637.75	632–638
1	91.25	95.75	90–96	41	639.25	643.75	638–644
2	97.25	101.75	96–102	42	645.25	649.75	644–650
3	103.25	107.75	102–108	43	651.25	655.75	650–656
4	171.25	175.75	170–176	44	657.25	661.75	656–662
5	177.25	181.75	176–182	45	663.25	667.75	662–668
6	183.25	187.75	182–188	46	669.25	673.75	668–674
7	189.25	193.75	188–194	47	675.25	679.75	674–680
8	193.25	197.75	192–198	48	681.25	685.75	680–686
9	199.25	203.75	198–204	49	687.25	691.75	686–692
10	205.25	209.75	204–210	50	693.25	697.75	692–698
11	211.25	215.75	210–216	51	699.25	703.75	698–704
12	217.25	221.75	216–222	52	705.25	709.75	704–710
13	471.25	475.75	470–476	53	711.25	715.75	710–716
14	477.25	481.75	476–482	54	717.25	721.75	716–722
15	483.25	487.75	482–488	55	723.25	727.75	722–728
16	489.25	493.75	488–494	56	729.25	733.75	728–734
17	495.25	499.75	494–500	57	735.25	739.75	734–740
18	501.25	505.75	500–506	58	741.25	745.75	740–746
19	507.25	511.75	506–512	59	747.25	751.75	746–752
20	513.25	517.75	512–518	60	753.25	757.75	752–758
21	519.25	523.75	518–524	61	759.25	763.75	758–764
22	525.25	529.75	524–530	62	765.25	769.75	764–770
23	531.25	535.75	530–536	–	–	–	–
24	537.25	541.75	536–542	–	–	–	–
25	543.25	547.75	542–548	–	–	–	–
26	549.25	553.75	548–554	–	–	–	–
27	555.25	559.75	554–560	–	–	–	–
28	561.25	565.75	560–566	–	–	–	–
29	567.25	571.75	566–572	–	–	–	–
30	573.25	577.75	572–578				
31	579.25	583.75	578–584				
32	585.25	589.75	584–590				
33	591.25	595.75	590–596				
34	597.25	601.75	596–602				
35	603.25	607.75	602–608				
36	609.25	613.75	608–614				
37	615.25	619.75	614–620				
38	621.25	625.75	620–626				
39	627.25	631.75	626–632				

Table 8.2. Analog Broadcast Nominal Frequencies for Japan.

Cable Channel	Video Carrier (MHz)	Audio Carrier (MHz)	Channel Range (MHz)	Cable Channel	Video Carrier (MHz)	Audio Carrier (MHz)	Channel Range (MHz)
–	–	–	–	40	319.2625	323.7625	318–324
–	–	–	–	41	325.2625	329.7625	324–330
2	55.25	59.75	54–60	42	331.2750	335.7750	330–336
3	61.25	65.75	60–66	43	337.2625	341.7625	336–342
4	67.25	71.75	66–72	44	343.2625	347.7625	342–348
5	77.25	81.75	76–82	45	349.2625	353.7625	348–354
6	83.25	87.75	82–88	46	355.2625	359.7625	354–360
7	175.25	179.75	174–180	47	361.2625	365.7625	360–366
8	181.25	185.75	180–186	48	367.2625	371.7625	366–372
9	187.25	191.75	186–192	49	373.2625	377.7625	372–378
10	193.25	197.75	192–198	50	379.2625	383.7625	378–384
11	199.25	203.75	198–204	51	385.2625	389.7625	384–390
12	205.25	209.75	204–210	52	391.2625	395.7625	390–396
13	211.25	215.75	210–216	53	397.2625	401.7625	396–402
14	121.2625	125.7625	120–126	54	403.25	407.75	402–408
15	127.2625	131.7625	126–132	55	409.25	413.75	408–414
16	133.2625	137.7625	132–138	56	415.25	419.75	414–420
17	139.25	143.75	138–144	57	421.25	425.75	420–426
18	145.25	149.75	144–150	58	427.25	431.75	426–432
19	151.25	155.75	150–156	59	433.25	437.75	432–438
20	157.25	161.75	156–162	60	439.25	443.75	438–444
21	163.25	167.75	162–168	61	445.55	449.75	444–450
22	169.25	173.75	168–174	62	451.25	455.75	450–456
23	217.25	221.75	216–222	63	457.25	461.75	456–462
24	223.25	227.75	222–228	64	463.25	467.75	462–468
25	229.2625	233.7625	228–234	65	469.25	473.75	468–474
26	235.2625	239.7625	234–240	66	475.25	479.75	474–480
27	241.2625	245.7625	240–246	67	481.25	485.75	480–486
28	247.2625	251.7625	246–252	68	487.25	491.75	486–492
29	253.2625	257.7625	252–258	69	493.25	497.75	492–498
30	259.2625	263.7625	258–264	70	499.25	503.75	498–504
31	265.2625	269.7625	264–270	71	505.25	509.75	504–510
32	271.2625	275.7625	270–276	72	511.25	515.75	510–516
33	277.2625	281.7625	276–282	73	517.25	521.75	516–522
34	283.2625	287.7625	282–288	74	523.25	527.75	522–528
35	289.2625	293.7625	288–294	75	529.25	533.75	528–534
36	295.2625	299.7625	294–300	76	535.25	539.75	534–540
37	301.2625	305.7625	300–306	77	541.25	545.75	540–546
38	307.2625	311.7625	306–312	78	547.25	551.75	546–552
39	313.2625	317.7625	312–318	79	553.25	557.75	552–558

Table 8.3a. Standard Analog Cable TV Nominal Frequencies for USA.

Cable Channel	Video Carrier (MHz)	Audio Carrier (MHz)	Channel Range (MHz)	Cable Channel	Video Carrier (MHz)	Audio Carrier (MHz)	Channel Range (MHz)
80	559.25	563.75	558–564	120	769.25	773.75	768–774
81	565.25	569.75	564–570	121	775.25	779.75	774–780
82	571.25	575.75	570–576	122	781.25	785.75	780–786
83	577.25	581.75	576–582	123	787.25	791.75	786–792
84	583.25	587.75	582–588	124	793.25	797.75	792–798
85	589.25	593.75	588–594	125	799.25	803.75	798–804
86	595.25	599.75	594–600	126	805.25	809.75	804–810
87	601.25	605.75	600–606	127	811.25	815.75	810–816
88	607.25	611.75	606–612	128	817.25	821.75	816–822
89	613.25	617.75	612–618	129	823.25	827.75	822–828
90	619.25	623.75	618–624	130	829.25	833.75	828–834
91	625.25	629.75	624–630	131	835.25	839.75	834–840
92	631.25	635.75	630–636	132	841.25	845.75	840–846
93	637.25	641.75	636–642	133	847.25	851.75	846–852
94	643.25	647.75	642–648	134	853.25	857.75	852–858
95	91.25	95.75	90–96	135	859.25	863.75	858–864
96	97.25	101.75	96–102	136	865.25	869.75	864–870
97	103.25	107.75	102–108	137	871.25	875.75	870–876
98	109.2750	113.7750	108–114	138	877.25	881.75	876–882
99	115.2750	119.7750	114–120	139	883.25	887.75	882–888
100	649.25	653.75	648–654	140	889.25	893.75	888–894
101	655.25	659.75	654–660	141	895.25	899.75	894–900
102	661.25	665.75	660–666	142	901.25	905.75	900–906
103	667.25	671.75	666–672	143	907.25	911.75	906–912
104	673.25	677.75	672–678	144	913.25	917.75	912–918
105	679.25	683.75	678–684	145	919.25	923.75	918–924
106	685.25	689.75	684–690	146	925.25	929.75	924–930
107	691.25	695.75	690–696	147	931.25	935.75	930–936
108	697.25	701.75	696–702	148	937.25	941.75	936–942
109	703.25	707.75	702–708	149	943.25	947.75	942–948
110	709.25	713.75	708–714	150	949.25	953.75	948–954
111	715.25	719.75	714–720	151	955.25	959.75	954–960
112	721.25	725.75	720–726	152	961.25	965.75	960–966
113	727.25	731.75	726–732	153	967.25	971.75	966–972
114	733.25	737.75	732–738	154	973.25	977.75	972–978
115	739.25	743.75	738–744	155	979.25	983.75	978–984
116	745.25	749.75	744–750	156	985.25	989.75	984–990
117	751.25	755.75	750–756	157	991.25	995.75	990–996
118	757.25	761.75	756–762	158	997.25	1001.75	996–1002
119	763.25	767.75	762–768	–	–	–	–

Table 8.3b. Standard Analog Cable TV Nominal Frequencies for USA.

Cable Channel	Video Carrier (MHz)	Audio Carrier (MHz)	Cable Channel	Video Carrier (MHz)	Audio Carrier (MHz)
–	–	–	40	319.2625	323.7625
1	73.2625	77.7625	41	325.2625	329.7625
2	55.2625	59.7625	42	331.2750	335.7750
3	61.2625	65.7625	43	337.2625	341.7625
4	67.2625	71.7625	44	343.2625	347.7625
5	79.2625	83.7625	45	349.2625	353.7625
6	85.2625	89.7625	46	355.2625	359.7625
7	175.2625	179.7625	47	361.2625	365.7625
8	181.2625	185.7625	48	367.2625	371.7625
9	187.2625	191.7625	49	373.2625	377.7625
10	193.2625	197.7625	50	379.2625	383.7625
11	199.2625	203.7625	51	385.2625	389.7625
12	205.2625	209.7625	52	391.2625	395.7625
13	211.2625	215.7625	53	397.2625	401.7625
14	121.2625	125.7625	54	403.2625	407.7625
15	127.2625	131.7625	55	409.2625	413.7625
16	133.2625	137.7625	56	415.2625	419.7625
17	139.2625	143.7625	57	421.2625	425.7625
18	145.2625	149.7625	58	427.2625	431.7625
19	151.2625	155.7625	59	433.2625	437.7625
20	157.2625	161.7625	60	439.2625	443.7625
21	163.2625	167.7625	61	445.2625	449.7625
22	169.2625	173.7625	62	451.2625	455.7625
23	217.2625	221.7625	63	457.2625	461.7625
24	223.2625	227.7625	64	463.2625	467.7625
25	229.2625	233.7625	65	469.2625	473.7625
26	235.2625	239.7625	66	475.2625	479.7625
27	241.2625	245.7625	67	481.2625	485.7625
28	247.2625	251.7625	68	487.2625	491.7625
29	253.2625	257.7625	69	493.2625	497.7625
30	259.2625	263.7625	70	499.2625	503.7625
31	265.2625	269.7625	71	505.2625	509.7625
32	271.2625	275.7625	72	511.2625	515.7625
33	277.2625	281.7625	73	517.2625	521.7625
34	283.2625	287.7625	74	523.2625	527.7625
35	289.2625	293.7625	75	529.2625	533.7625
36	295.2625	299.7625	76	535.2625	539.7625
37	301.2625	305.7625	77	541.2625	545.7625
38	307.2625	311.7625	78	547.2625	551.7625
39	313.2625	317.7625	79	553.2625	557.7625

Table 8.3c. Analog Cable TV Nominal Frequencies for USA: Incrementally Related Carrier (IRC) Systems.

Cable Channel	Video Carrier (MHz)	Audio Carrier (MHz)	Cable Channel	Video Carrier (MHz)	Audio Carrier (MHz)
80	559.2625	563.7625	120	769.2625	773.7625
81	565.2625	569.7625	121	775.2625	779.7625
82	571.2625	575.7625	122	781.2625	785.7625
83	577.2625	581.7625	123	787.2625	791.7625
84	583.2625	587.7625	124	793.2625	797.7625
85	589.2625	593.7625	125	799.2625	803.7625
86	595.2625	599.7625	126	805.2625	809.7625
87	601.2625	605.7625	127	811.2625	815.7625
88	607.2625	611.7625	128	817.2625	821.7625
89	613.2625	617.7625	129	823.2625	827.7625
90	619.2625	623.7625	130	829.2625	833.7625
91	625.2625	629.7625	131	835.2625	839.7625
92	631.2625	635.7625	132	841.2625	845.7625
93	637.2625	641.7625	133	847.2625	851.7625
94	643.2625	647.7625	134	853.2625	857.7625
95	91.2625	95.7625	135	859.2625	863.7625
96	97.2625	101.7625	136	865.2625	869.7625
97	103.2625	107.7625	137	871.2625	875.7625
98	109.2750	113.7750	138	877.2625	881.7625
99	115.2625	119.7625	139	883.2625	887.7625
100	649.2625	653.7625	140	889.2625	893.7625
101	655.2625	659.7625	141	895.2625	899.7625
102	661.2625	665.7625	142	901.2625	905.7625
103	667.2625	671.7625	143	907.2625	911.7625
104	673.2625	677.7625	144	913.2625	917.7625
105	679.2625	683.7625	145	919.2625	923.7625
106	685.2625	689.7625	146	925.2625	929.7625
107	691.2625	695.7625	147	931.2625	935.7625
108	697.2625	701.7625	148	937.2625	941.7625
109	703.2625	707.7625	149	943.2625	947.7625
110	709.2625	713.7625	150	949.2625	953.7625
111	715.2625	719.7625	151	955.2625	959.7625
112	721.2625	725.7625	152	961.2625	965.7625
113	727.2625	731.7625	153	967.2625	971.7625
114	733.2625	737.7625	154	973.2625	977.7625
115	739.2625	743.7625	155	979.2625	983.7625
116	745.2625	749.7625	156	985.2625	989.7625
117	751.2625	755.7625	157	991.2625	995.7625
118	757.2625	761.7625	158	997.2625	1001.7625
119	763.2625	767.7625	–	–	–

Table 8.3d. Analog Cable TV Nominal Frequencies for USA: Incrementally Related Carrier (IRC) Systems.

Cable Channel	Video Carrier (MHz)	Audio Carrier (MHz)	Cable Channel	Video Carrier (MHz)	Audio Carrier (MHz)
–	–	–	40	318.0159	322.5159
1	72.0036	76.5036	41	324.0162	328.5162
2	54.0027	58.5027	42	330.0165	334.5165
3	60.0030	64.5030	43	336.0168	340.5168
4	66.0033	70.5030	44	342.0168	346.5168
5	72.0036	82.5039	45	348.0168	352.5168
6	78.0039	88.5042	46	354.0168	358.5168
7	174.0087	178.5087	47	360.0168	364.5168
8	180.0090	184.5090	48	366.0168	370.5168
9	186.0093	190.5093	49	372.0168	376.5168
10	192.0096	196.5096	50	378.0168	382.5168
11	198.0099	202.5099	51	384.0168	388.5168
12	204.0102	208.5102	52	390.0168	394.5168
13	210.0105	214.5105	53	396.0168	400.5168
14	120.0060	124.5060	54	402.0201	406.5201
15	126.0063	130.5063	55	408.0204	412.5204
16	132.0066	136.5066	56	414.0207	418.5207
17	138.0069	142.5069	57	420.0210	424.5210
18	144.0072	148.5072	58	426.0213	430.5213
19	150.0075	154.5075	59	432.0216	436.5216
20	156.0078	160.5078	60	438.0219	442.5219
21	162.0081	166.5081	61	444.0222	448.5222
22	168.0084	172.5084	62	450.0225	454.5225
23	216.0108	220.5108	63	456.0228	460.5228
24	222.0111	226.5111	64	462.0231	466.5231
25	228.0114	232.5114	65	468.0234	472.5234
26	234.0117	238.5117	66	474.0237	478.5237
27	240.0120	244.5120	67	480.0240	484.5240
28	246.0123	250.5123	68	486.0243	490.5243
29	252.0126	256.5126	69	492.0246	496.5246
30	258.0129	262.5129	70	498.0249	502.5249
31	264.0132	268.5132	71	504.0252	508.5252
32	270.0135	274.5135	72	510.0255	514.5255
33	276.0138	280.5138	73	516.0258	520.5258
34	282.0141	286.5141	74	522.0261	526.5261
35	288.0144	292.5144	75	528.0264	532.5264
36	294.0147	298.5147	76	534.0267	538.5267
37	300.0150	304.5150	77	540.0270	544.5270
38	306.0153	310.5153	78	546.0273	550.5273
39	312.0156	316.5156	79	552.0276	556.5276

Table 8.3e. Analog Cable TV Nominal Frequencies for USA: Harmonically Related Carrier (HRC) systems.

Cable Channel	Video Carrier (MHz)	Audio Carrier (MHz)	Cable Channel	Video Carrier (MHz)	Audio Carrier (MHz)
80	558.0279	562.5279	120	768.0384	772.5384
81	564.0282	568.5282	121	774.0387	778.5387
82	570.0285	574.5285	122	780.0390	784.5390
83	576.0288	580.5288	123	786.0393	790.5393
84	582.0291	586.5291	124	792.0396	796.5396
85	588.0294	592.5294	125	798.0399	802.5399
86	594.0297	598.5297	126	804.0402	808.5402
87	600.0300	604.5300	127	810.0405	814.5405
88	606.0303	610.5303	128	816.0408	820.5408
89	612.0306	616.5306	129	822.0411	826.5411
90	618.0309	622.5309	130	828.0414	832.5414
91	624.0312	628.5312	131	834.0417	838.5417
92	630.0315	634.5315	132	840.0420	844.5420
93	636.0318	640.5318	133	846.0423	850.5423
94	642.0321	646.5321	134	852.0426	856.5426
95	90.0045	94.5045	135	858.0429	862.5429
96	96.0048	100.5048	136	864.0432	868.5432
97	102.0051	106.5051	137	870.0435	874.5435
98	–	–	138	876.0438	880.5438
99	–	–	139	882.0441	888.5441
100	648.0324	652.5324	140	888.0444	892.5444
101	654.0327	658.5327	141	894.0447	898.5447
102	660.0330	664.5330	142	900.0450	904.5450
103	666.0333	670.5333	143	906.0453	910.5453
104	672.0336	676.5336	144	912.0456	916.5456
105	678.0339	682.5339	145	918.0459	922.5459
106	684.0342	688.5342	146	924.0462	928.5462
107	690.0345	694.5345	147	930.0465	934.5465
108	696.0348	700.5348	148	936.0468	940.5468
109	702.0351	706.5351	149	942.0471	946.5471
110	708.0354	712.5354	150	948.0474	952.5474
111	714.0357	718.5357	151	954.0477	958.5477
112	720.0360	724.5360	152	960.0480	964.5480
113	726.0363	730.5363	153	966.0483	970.5483
114	732.0366	736.5366	154	972.0486	976.5486
115	738.0369	742.5369	155	978.0489	982.5489
116	744.0372	748.5372	156	984.0492	988.5492
117	750.0375	754.5375	157	990.0495	994.5495
118	756.0378	760.5378	158	996.0498	1000.5498
119	762.0381	766.5381	–	–	–

Table 8.3f. Analog Cable TV Nominal Frequencies for USA: Harmonically Related Carrier (HRC) systems.

Cable Channel	Video Carrier (MHz)	Audio Carrier (MHz)	Channel Range (MHz)	Cable Channel	Video Carrier (MHz)	Audio Carrier (MHz)	Channel Range (MHz)
–	–	–	–	40	325.25	329.75	324–330
–	–	–	–	41	331.25	335.75	330–336
–	–	–	–	42	337.25	341.75	336–342
13	109.25	113.75	108–114	46	343.25	347.75	342–348
14	115.25	119.75	114–120	44	349.25	353.75	348–354
15	121.25	125.75	120–126	45	355.25	359.75	354–360
16	127.25	131.75	126–132	46	361.25	365.75	360–366
17	133.25	137.75	132–138	47	367.25	371.75	366–372
18	139.25	143.75	138–144	48	373.25	377.75	372–378
19	145.25	149.75	144–150	49	379.25	383.75	378–384
20	151.25	155.75	150–156	50	385.25	389.75	384–390
21	157.25	161.75	156–162	51	391.25	395.75	390–396
22	165.25	169.75	164–170	52	397.25	401.75	396–402
23	223.25	227.75	222–228	53	403.25	407.75	402–408
24	231.25	235.75	230–236	54	409.25	413.75	408–414
25	237.25	241.75	236–242	55	415.25	419.75	414–420
26	243.25	247.75	242–248	56	421.25	425.75	420–426
27	249.25	253.75	248–254	57	427.25	431.75	426–432
28	253.25	257.75	252–258	58	433.25	437.75	432–438
29	259.25	263.75	258–264	59	439.25	443.75	438–444
30	265.25	269.75	264–270	60	445.25	449.75	444–450
31	271.25	275.75	270–276	61	451.25	455.75	450–456
32	277.25	281.75	276–282	62	457.25	461.75	456–462
33	283.25	287.75	282–288	63	463.25	467.75	462–468
34	289.25	293.75	288–294	–	–	–	–
35	295.25	299.75	294–300	–	–	–	–
36	301.25	305.75	300–306	–	–	–	–
37	307.25	311.75	306–312	–	–	–	–
38	313.25	317.75	312–318	–	–	–	–
39	319.25	323.75	318–324	–	–	–	–

Table 8.4. Analog Cable TV Nominal Frequencies for Japan.

Luminance Equation Derivation

The equation for generating luminance from RGB is determined by the chromaticities of the three primary colors used by the receiver and what color white actually is.

The chromaticities of the RGB primaries and reference white (CIE illuminate C) were specified in the 1953 NTSC standard to be:

R: $x_r = 0.67$ $y_r = 0.33$ $z_r = 0.00$

G: $x_g = 0.21$ $y_g = 0.71$ $z_g = 0.08$

B: $x_b = 0.14$ $y_b = 0.08$ $z_b = 0.78$

white: $x_w = 0.3101$ $y_w = 0.3162$
$z_w = 0.3737$

where x and y are the specified CIE 1931 chromaticity coordinates; z is calculated by knowing that x + y + z = 1.

Luminance is calculated as a weighted sum of RGB, with the weights representing the actual contributions of each of the RGB primaries in generating the luminance of reference white. We find the linear combination of RGB that gives reference white by solving the equation:

$$\begin{bmatrix} x_r & x_g & x_b \\ y_r & y_g & y_b \\ z_r & z_g & z_b \end{bmatrix} \begin{bmatrix} K_r \\ K_g \\ K_b \end{bmatrix} = \begin{bmatrix} x_w/y_w \\ 1 \\ z_w/y_w \end{bmatrix}$$

Rearranging to solve for K_r, K_g, and K_b yields:

$$\begin{bmatrix} K_r \\ K_g \\ K_b \end{bmatrix} = \begin{bmatrix} x_w/y_w \\ 1 \\ z_w/y_w \end{bmatrix} \begin{bmatrix} x_r & x_g & x_b \\ y_r & y_g & y_b \\ z_r & z_g & z_b \end{bmatrix}^{-1}$$

Substituting the known values gives us the solution for K_r, K_g, and K_b:

$$\begin{bmatrix} K_r \\ K_g \\ K_b \end{bmatrix} = \begin{bmatrix} 0.3101/0.3162 \\ 1 \\ 0.3737/0.3162 \end{bmatrix} \begin{bmatrix} 0.67 & 0.21 & 0.14 \\ 0.33 & 0.71 & 0.08 \\ 0.00 & 0.08 & 0.78 \end{bmatrix}^{-1}$$

$$= \begin{bmatrix} 0.9807 \\ 1 \\ 1.1818 \end{bmatrix} \begin{bmatrix} 1.730 & -0.482 & -0.261 \\ -0.814 & 1.652 & -0.023 \\ 0.083 & -0.169 & 1.284 \end{bmatrix}$$

$$= \begin{bmatrix} 0.906 \\ 0.827 \\ 1.430 \end{bmatrix}$$

Y is defined to be

$$Y = (K_r y_r)R' + (K_g y_g)G' + (K_b y_b)B'$$
$$= (0.906)(0.33)R' + (0.827)(0.71)G'$$
$$+ (1.430)(0.08)B'$$

or

$$Y = 0.299R' + 0.587G' + 0.114B'$$

Modern receivers use a different set of RGB phosphors, resulting in slightly different chromaticities of the RGB primaries and reference white (CIE illuminate D_{65}):

R: $x_r = 0.630$ $y_r = 0.340$ $z_r = 0.030$

G: $x_g = 0.310$ $y_g = 0.595$ $z_g = 0.095$

B: $x_b = 0.155$ $y_b = 0.070$ $z_b = 0.775$

white: $x_w = 0.3127$ $y_w = 0.3290$
$z_w = 0.3583$

where x and y are the specified CIE 1931 chromaticity coordinates; z is calculated by knowing that x + y + z = 1. Once again, substituting the known values gives us the solution for K_r, K_g, and K_b:

$$\begin{bmatrix} K_r \\ K_g \\ K_b \end{bmatrix} = \begin{bmatrix} 0.3127/0.3290 \\ 1 \\ 0.3583/0.3290 \end{bmatrix} \begin{bmatrix} 0.630 & 0.310 & 0.155 \\ 0.340 & 0.595 & 0.070 \\ 0.030 & 0.095 & 0.775 \end{bmatrix}^{-1}$$

$$= \begin{bmatrix} 0.6243 \\ 1.1770 \\ 1.2362 \end{bmatrix}$$

Since Y is defined to be

$$Y = (K_r y_r) R' + (K_g y_g) G' + (K_b y_b) B'$$
$$= (0.6243)(0.340) R' + (1.1770)(0.595) G'$$
$$+ (1.2362)(0.070) B'$$

this results in:

$$Y = 0.212 R' + 0.700 G' + 0.086 B'$$

However, the standard Y = 0.299R' + 0.587G' + 0.114B' equation is still used. Adjustments are made in the receiver to minimize color errors.

PAL Overview

Europe delayed adopting a color television standard, evaluating various systems between 1953 and 1967 that were compatible with their 625-line, 50-field-per-second, 2:1 interlaced monochrome standard. The NTSC specification was modified to overcome the high order of phase and amplitude integrity required during broadcast to avoid color distortion. The Phase Alternation Line (PAL) system implements a line-by-line reversal of the phase of one of the color components, originally relying on the eye to average any color distortions to the correct color. Broadcasting began in 1967 in Germany and the United Kingdom, with each using a slightly different variant of the PAL system.

Luminance Information

The monochrome luminance (Y) signal is derived from R'G'B':

$$Y = 0.299 R' + 0.587 G' + 0.114 B'$$

As with NTSC, the luminance signal occupies the entire video bandwidth. PAL has several variations, depending on the video bandwidth and placement of the audio subcarrier. The composite video signal has a bandwidth of 4.2, 5.0, 5.5, or 6.0 MHz, depending on the specific PAL standard.

Color Information

To transmit color information, U and V are used:

$$U = 0.492 (B' - Y)$$
$$V = 0.877 (R' - Y)$$

U and V have a typical bandwidth of 1.3 MHz.

Color Modulation

As in the NTSC system, U and V are used to modulate the color subcarrier using two balanced modulators operating in phase quadrature: one modulator is driven by the subcarrier at sine phase, the other modulator is driven by the subcarrier at cosine phase. The outputs of the modulators are added together to form the modulated chrominance signal:

$$C = U \sin \omega t \pm V \cos \omega t$$

$$\omega = 2\pi F_{SC}$$

$$F_{SC} = 4.43361875 \text{ MHz } (\pm 5 \text{ Hz})$$
for (B, D, G, H, I, N) PAL

$$F_{SC} = 3.58205625 \text{ MHz } (\pm 5 \text{ Hz}) \text{ for } (N_C) \text{ PAL}$$

$$F_{SC} = 3.57561149 \text{ MHz } (\pm 10 \text{ Hz}) \text{ for } (M) \text{ PAL}$$

In PAL, the phase of V is reversed every other line. V was chosen for the reversal process since it has a lower gain factor than U and therefore is less susceptible to a one-half F_H switching rate imbalance. The result of alternating the V phase at the line rate is that any color subcarrier phase errors produce complementary errors, allowing line-to-line averaging at the receiver to cancel the errors and generate the correct hue with slightly reduced saturation. This technique requires the PAL receiver to be able to determine the correct V phase. This is done using a technique known as AB sync, PAL sync, PAL Switch, or "swing-ing burst," consisting of alternating the phase of the color burst by ±45° at the line rate. The UV vector diagrams are shown in Figures 8.11 and 8.12.

"Simple" PAL decoders rely on the eye to average the line-by-line hue errors. "Standard" PAL decoders use a 1-H delay line to separate U from V in an averaging process. Both implementations have the problem of Hanover bars, in which pairs of adjacent lines have a real and complementary hue error. Chrominance vertical resolution is reduced as a result of the line averaging process.

Composite Video Generation

The modulated chrominance is added to the luminance information along with appropriate horizontal and vertical sync signals, blanking signals, and color burst signals, to generate the composite color video waveform shown in Figure 8.13.

$$\text{composite PAL} = Y + U \sin \omega t$$
$$\pm V \cos \omega t + \text{timing}$$

The bandwidth of the resulting composite video signal is shown in Figure 8.14.

Like NTSC, the luminance components are spaced at F_H intervals due to horizontal blanking. Since the V component is switched symmetrically at one-half the line rate, only odd harmonics are generated, resulting in V components that are spaced at intervals of F_H. The V components are spaced at half-line intervals from the U components, which also have F_H spacing. If the subcarrier had a half-line offset like NTSC uses, the U components would be perfectly interleaved, but the V components would coincide with the Y components and thus not be interleaved, creating vertical stationary dot patterns. For this reason, PAL uses a 1/4 line offset for the subcarrier frequency:

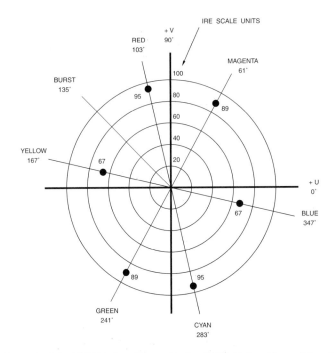

Figure 8.11. UV Vector Diagram for 75% Color Bars. Line [n], PAL Switch = zero.

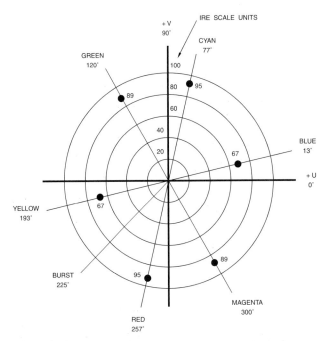

Figure 8.12. UV Vector Diagram for 75% Color Bars. Line [n + 1], PAL Switch = one.

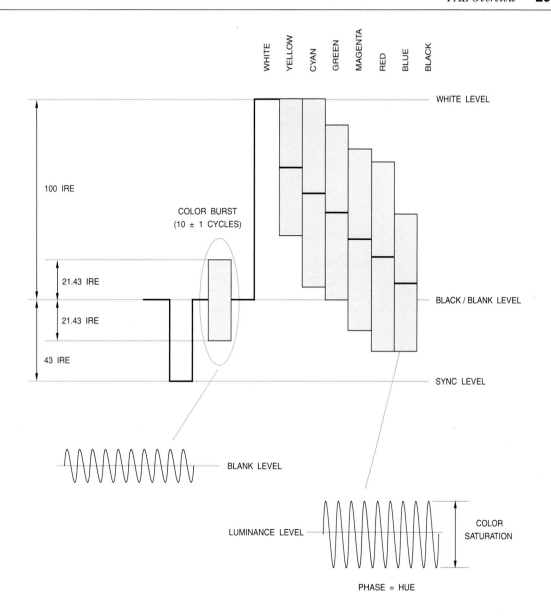

Figure 8.13. (B, D, G, H, I, N$_C$) PAL Composite Video Signal for 75% Color Bars.

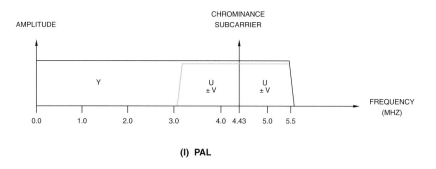

(I) PAL

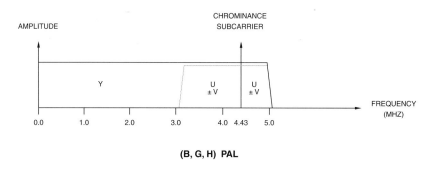

(B, G, H) PAL

Figure 8.14. Video Bandwidths of Some PAL Systems.

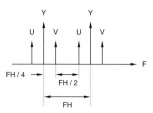

Figure 8.15. Luma and Chroma Frequency Interleave Principle.

$F_{SC} = ((1135/4) + (1/625)) \, F_H$
for (B, D, G, H, I, N) PAL

$F_{SC} = (909/4) \, F_H$ for (M) PAL

$F_{SC} = ((917/4) + (1/625)) \, F_H$ for (N_C) PAL

The additional $(1/625) \, F_H$ factor (equal to 25 Hz) provides motion to the color dot pattern, reducing its visibility. Figure 8.15 illustrates the resulting frequency interleaving. Eight complete fields are required to repeat a specific sample position, as shown in Figures 8.16 and 8.17.

PAL Standards

Figure 8.19 shows the common designations for PAL systems. The letters refer to the monochrome standard for line and field rate, video bandwidth (4.2, 5.0, 5.5, or 6.0 MHz), audio carrier relative frequency, and RF channel bandwidth (6.0, 7.0, or 8.0 MHz). The "PAL" refers to the technique to add color information to the monochrome signal. Detailed timing parameters may be found in Table 8.9.

"Noninterlaced PAL," shown in Figure 8.18, is a 312-line, 50 frames-per-second version of PAL common among video games and on-screen displays. This format is identical to standard PAL, except that there are 312 lines per frame.

RF Modulation

Figures 8.20 and 8.21 illustrate the process of converting baseband (G) PAL composite video to an RF (radio frequency) signal. The process for the other PAL standards is similar, except primarily for the different video bandwidths and subcarrier frequencies.

Figure 8.20a shows the frequency spectrum of a (G) PAL baseband composite video signal. It is similar to Figure 8.14. However, Figure 8.14 only shows the upper sideband for simplicity. The "video carrier" notation at 0 MHz serves only as a reference point for comparison with Figure 8.20b.

Figure 8.20b shows the audio/video signal as it resides within an 8 MHz channel. The video signal has been lowpass filtered, most of the lower sideband has been removed, and audio information has been added. Note that (H) and (I) PAL have a vestigial sideband of 1.25 MHz, rather than 0.75 MHz.

Figure 8.20c details the information present on the audio subcarrier for analog stereo operation.

As shown in Figure 8.21, back porch clamping of the analog video signal ensures that the back porch level is constant, regardless of changes in the average picture level. The video signal is then lowpass filtered to 5.0 MHz and drives the AM (amplitude modulation) video modulator. The sync level corresponds to 100% modulation; the blanking and white modulation levels are dependent on the specific version of PAL:

blanking level (% modulation)

B, G	75%
D, H, M, N	75%
I	76%

white level (% modulation)

B, G, H, M, N	10%
D	10%
I	20%

Note that PAL systems use a variety of video and audio IF frequencies (values in MHz):

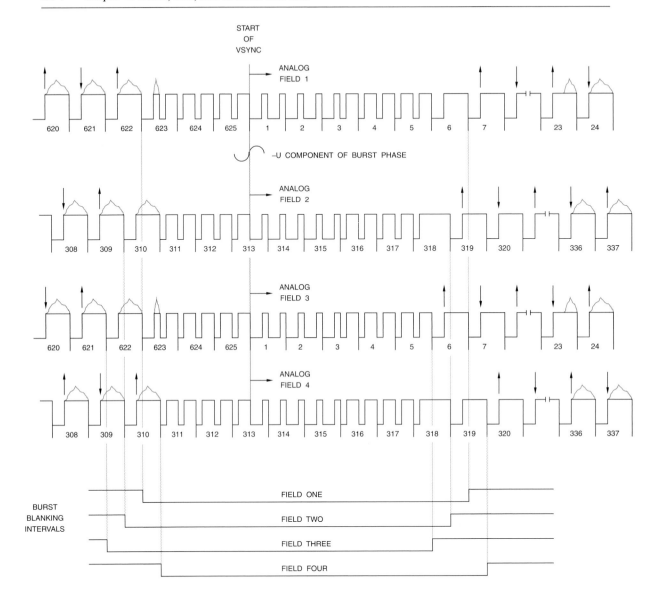

Figure 8.16a. Eight-field (B, D, G, H, I, N$_C$) PAL Sequence and Burst Blanking. See Figure 8.5 for equalization and serration pulse details.

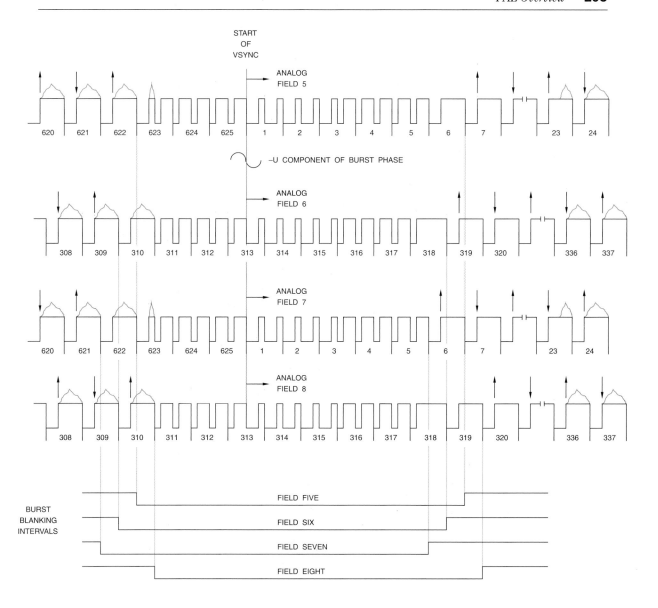

Figure 8.16b. Eight-field (B, D, G, H, I, N_C) PAL Sequence and Burst Blanking.
See Figure 8.5 for equalization and serration pulse details.

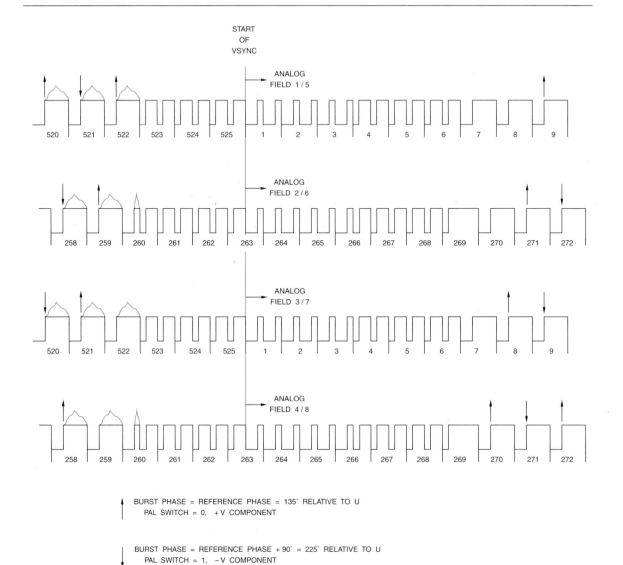

**Figure 8.17. Eight-field (M) PAL Sequence and Burst Blanking.
See Figure 8.5 for equalization and serration pulse details.**

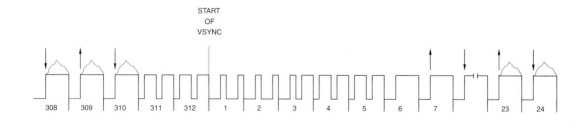

BURST PHASE = REFERENCE PHASE = 135° RELATIVE TO U
PAL SWITCH = 0, +V COMPONENT

BURST PHASE = REFERENCE PHASE + 90° = 225° RELATIVE TO U
PAL SWITCH = 1, −V COMPONENT

Figure 8.18. Noninterlaced PAL Frame Sequence.

	video	audio	
B, G	38.900	33.400	
B	36.875	31.375	Australia
D	37.000	30.500	China
D	38.900	32.400	OIRT
I	38.900	32.900	
I	39.500	33.500	U.K.
M, N	45.750	41.250	

At this point, audio information is added on the audio subcarrier. A monaural L+R audio signal is processed as shown in Figure 8.21 and drives the FM (frequency modulation) modulator. The output of the FM modulator is added to the IF video signal.

The SAW filter, used as a vestigial sideband filter, provides filtering of the IF signal. The mixer, or up converter, mixes the IF signal with the desired broadcast frequency. Both sum and difference frequencies are generated by the mixing process, so the difference signal is extracted by using a bandpass filter.

Stereo Audio (Analog)

The standard (ITU-R BS.707), also known as Zweiton or A2, is shown in Figure 8.21. The L+R information is transmitted on a FM subcarrier. The R information, or a second L+R audio signal, is transmitted on a second FM subcarrier at $+15.5F_H$.

If stereo or dual mono signals are present, the FM subcarrier at $+15.5F_H$ is amplitude-modulated with a 54.6875 kHz ($3.5F_H$) subcarrier. This 54.6875 kHz subcarrier is 50% amplitude-modulated at 117.5 Hz ($F_H / 133$) to indicate stereo audio or 274.1 Hz ($F_H / 57$) to indicate dual mono audio.

Countries that use this system include Australia, Austria, China, Germany, Italy, Malaysia, Netherlands, Slovenia, and Switzerland.

Stereo Audio (Digital)

The standard uses NICAM 728 (Near Instantaneous Companded Audio Multiplex),

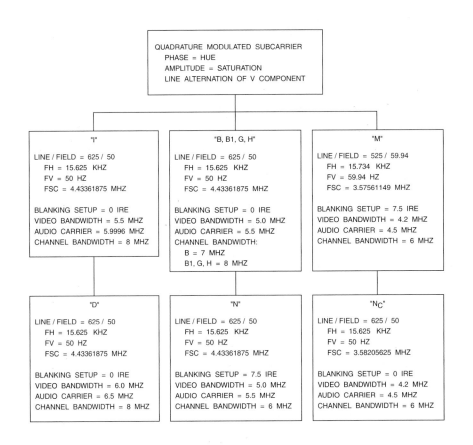

Figure 8.19. Common PAL Systems.

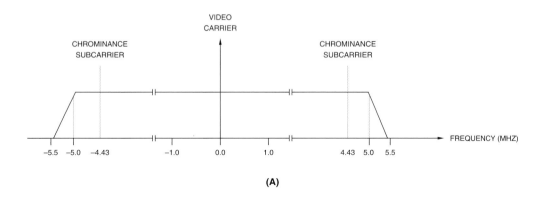

(A)

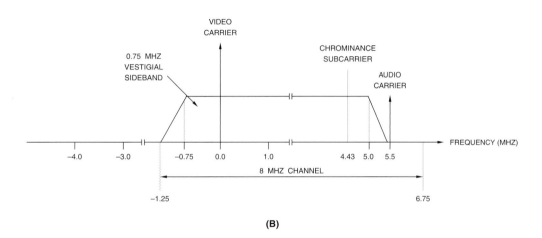

(B)

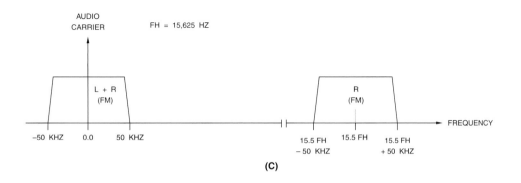

(C)

Figure 8.20. Transmission Channel for (G) PAL. (a) Frequency spectrum of baseband composite video. (b) Frequency spectrum of typical channel including audio information. (c) Detailed frequency spectrum of Zweiton analog stereo audio information.

discussed within BS.707 and ETSI EN 300 163. It was developed by the BBC and IBA to increase sound quality, provide multiple channels of digital sound or data, and be more resistant to transmission interference.

The subcarrier resides either 5.85 MHz above the video carrier for (B, D, G, H) PAL and (L) SECAM systems or 6.552 MHz above the video carrier for (I) PAL systems.

Countries that use NICAM 728 include Belgium, China, Denmark, Finland, France, Hungary, New Zealand, Norway, Singapore, South Africa, Spain, Sweden, and the United Kingdom.

NICAM 728 is a digital system that uses a 32 kHz sampling rate and 14-bit resolution. A bit rate of 728 kbps is used, giving it the name NICAM 728. Data is transmitted in frames, with each frame containing 1 ms of audio. As shown in Figure 8.22, each frame consists of:

8-bit frame alignment word (01001110)
5 control bits (C0–C4)
11 undefined bits (AD0–AD10)
704 audio data bits (A000–A703)

C0 is a "1" for eight successive frames and a "0" for the next eight frames, defining a 16-frame sequence. C1–C3 specify the format transmitted: "000" = one stereo signal with the left channel being odd-numbered samples and the right channel being even-numbered samples, "010" = two independent mono channels transmitted in alternate frames, "100" = one mono channel and one 352 kbps data channel

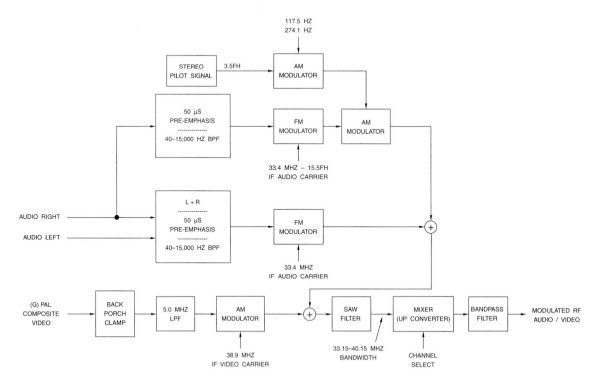

Figure 8.21. Typical RF Modulation Implementation for (G) PAL: Zweiton Stereo Audio.

transmitted in alternate frames, "110" = one 704 kbps data channel. C4 is a "1" if the analog sound is the same as the digital sound.

Stereo Audio Encoding

The thirty-two 14-bit samples (1 ms of audio, 2's complement format) per channel are pre-emphasized to the ITU-T J.17 curve.

The largest positive or negative sample of the 32 is used to determine which 10 bits of all 32 samples to transmit. Three range bits per channel ($R0_L$, $R1_L$, $R2_L$, and $R0_R$, $R1_R$, $R2_R$) are used to indicate the scaling factor. D13 is the sign bit ("0" = positive).

```
D13-D0           R2-R0    Bits Used

01xxxxxxxxxxxx   111      D13, D12-D4
001xxxxxxxxxxx   110      D13, D11-D3
0001xxxxxxxxxx   101      D13, D10-D2
00001xxxxxxxxx   011      D13, D9-D1
000001xxxxxxxx   101      D13, D8-D0
0000001xxxxxxx   010      D13, D8-D0
0000000xxxxxxx   00x      D13, D8-D0
1111111xxxxxxx   00x      D13, D8-D0
1111110xxxxxxx   010      D13, D8-D0
111110xxxxxxxx   100      D13, D8-D0
11110xxxxxxxxx   011      D13, D9-D1
1110xxxxxxxxxx   101      D13, D10-D2
110xxxxxxxxxxx   110      D13, D11-D3
10xxxxxxxxxxxx   111      D13, D12-D4
```

A parity bit for the six MSBs of each sample is added, resulting in each sample being 11 bits. The 64 samples are interleaved, generating L0, R0, L1, R1, L2, R2, ... L31, R31, and numbered 0–63.

The parity bits are used to convey to the decoder what scaling factor was used for each channel ("signalling-in-parity").

If $R2_L$ = "0," even parity for samples 0, 6, 12, 18, ... 48 is used. If $R2_L$ = "1," odd parity is used.

If $R2_R$ = "0," even parity for samples 1, 7, 13, 19, ... 49 is used. If $R2_R$ = "1," odd parity is used.

If $R1_L$ = "0," even parity for samples 2, 8, 14, 20, ... 50 is used. If $R1_L$ = "1," odd parity is used.

If $R1_R$ = "0," even parity for samples 3, 9, 15, 21, ... 51 is used. If $R1_R$ = "1," odd parity is used.

If $R0_L$ = "0," even parity for samples 4, 10, 16, 22, ... 52 is used. If $R0_L$ = "1," odd parity is used.

If $R0_R$ = "0," even parity for samples 5, 11, 17, 23, ... 53 is used. If $R0_R$ = "1," odd parity is used.

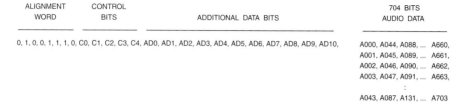

Figure 8.22. NICAM 728 Bitstream for One Frame.

The parity of samples 54–63 is normally even. However, they may be modified to transmit two additional bits of information:

> If CIB0 = "0," even parity for samples 54, 55, 56, 57, and 58 is used. If CIB0 = "1," odd parity is used.
>
> If CIB1 = "0," even parity for samples 59, 60, 61, 62, and 63 is used. If CIB1 = "1," odd parity is used.

The audio data is bit-interleaved as shown in Figure 8.22 to reduce the influence of dropouts. If the bits are numbered 0–703, they are transmitted in the order 0, 44, 88, ... 660, 1, 45, 89, ... 661, 2, 46, 90, ... 703.

The whole frame, except the frame alignment word, is exclusive-ORed with a 1-bit pseudo-random binary sequence (PRBS). The PRBS generator is reinitialized after the frame alignment word of each frame so that the first bit of the sequence processes the C0 bit. The polynomial of the PRBS is $x^9 + x^4 + 1$ with an initialization word of "111111111."

Actual transmission consists of taking bits in pairs from the 728 kbps bitstream, then generating 356k symbols per second using Differential Quadrature Phase-Shift Keying (DQPSK). If the symbol is "00," the subcarrier phase is left unchanged. If the symbol is "01," the subcarrier phase is delayed 90°. If the symbol is "11," the subcarrier phase is inverted. If the symbol is "10," the subcarrier phase is advanced 90°.

Finally, the signal is spectrum-shaped to a –30 dB bandwidth of ~700 kHz for (I) PAL or ~500 kHz for (B, G) PAL.

Stereo Audio Decoding

A PLL locks to the NICAM subcarrier frequency and recovers the phase changes that represent the encoded symbols. The symbols are decoded to generate the 728 kbps bitstream.

The frame alignment word is found and the following bits are exclusive-ORed with a locally-generated PRBS to recover the packet. The C0 bit is tested for 8 frames high, 8 frames low behavior to verify it is a NICAM 728 bitstream.

The bit-interleaving of the audio data is reversed, and the "signalling-in-parity" decoded:

> A majority vote is taken on the parity of samples 0, 6, 12, ... 48. If even, $R2_L$ = "0"; if odd, $R2_L$ = "1."
>
> A majority vote is taken on the parity of samples 1, 7, 13, ... 49. If even, $R2_R$ = "0"; if odd, $R2_R$ = "1."
>
> A majority vote is taken on the parity of samples 2, 8, 14, ... 50. If even, $R1_L$ = "0"; if odd, $R1_L$ = "1."
>
> A majority vote is taken on the parity of samples 3, 9, 15, ... 51. If even, $R1_R$ = "0"; if odd, $R1_R$ = "1."
>
> A majority vote is taken on the parity of samples 4, 10, 16, ... 52. If even, $R0_L$ = "0"; if odd, $R0_L$ = "1."
>
> A majority vote is taken on the parity of samples 5, 11, 17, ... 53. If even, $R0_R$ = "0"; if odd, $R0_R$ = "1."

A majority vote is taken on the parity of samples 54, 55, 56, 57, and 58. If even, CIB0 = "0"; if odd, CIB0 = "1."

A majority vote is taken on the parity of samples 59, 60, 61, 62, and 63. If even, CIB1 = "0"; if odd, CIB1 = "1."

Any samples whose parity disagreed with the vote are ignored and replaced with an interpolated value.

The left channel uses range bits $R2_L$, $R1_L$, and $R0_L$ to determine which bits below the sign bit were discarded during encoding. The sign bit is duplicated into those positions to generate a 14-bit sample.

The right channel is similarly processed, using range bits $R2_R$, $R1_R$, and $R0_R$. Both channels are then de-emphasized using the J.17 curve.

Dual Mono Audio Encoding

Two blocks of thirty-two 14-bit samples (2 ms of audio, 2's complement format) are pre-emphasized to the ITU-T J.17 specification. As with the stereo audio, three range bits per block ($R0_A$, $R1_A$, $R2_A$, and $R0_B$, $R1_B$, $R2_B$) are used to indicate the scaling factor. Unlike stereo audio, the samples are not interleaved.

If $R2_A$ = "0," even parity for samples 0, 3, 6, 9, ... 24 is used. If $R2_A$ = "1," odd parity is used.

If $R2_B$ = "0," even parity for samples 27, 30, 33, ... 51 is used. If $R2_B$ = "1," odd parity is used.

If $R1_A$ = "0," even parity for samples 1, 4, 7, 10, ... 25 is used. If $R1_A$ = "1," odd parity is used.

If $R1_B$ = "0," even parity for samples 28, 31, 34, ... 52 is used. If $R1_B$ = "1," odd parity is used.

If $R0_A$ = "0," even parity for samples 2, 5, 8, 11, ... 26 is used. If $R0_A$ = "1," odd parity is used.

If $R0_B$ = "0," even parity for samples 29, 32, 35, ... 53 is used. If $R0_B$ = "1," odd parity is used.

The audio data is bit-interleaved; however, odd packets contain 64 samples of audio channel 1 while even packets contain 64 samples of audio channel 2. The rest of the processing is the same as for stereo audio.

Analog Channel Assignments

Tables 8.5 through 8.7 list the channel assignments for VHF, UHF and cable for various PAL systems.

Note that cable systems routinely reassign channel numbers to alternate frequencies to minimize interference and provide multiple levels of programming (such as two versions of a premium movie channel: one for subscribers, and one for nonsubscribers during pre-view times).

Luminance Equation Derivation

The equation for generating luminance from RGB information is determined by the chromaticities of the three primary colors used by the receiver and what color white actually is.

The chromaticities of the RGB primaries and reference white (CIE illuminate D_{65}) are:

R: $x_r = 0.64$ $y_r = 0.33$ $z_r = 0.03$

G: $x_g = 0.29$ $y_g = 0.60$ $z_g = 0.11$

B: $x_b = 0.15$ $y_b = 0.06$ $z_b = 0.79$

white: $x_w = 0.3127$ $y_w = 0.3290$
$z_w = 0.3583$

Channel	Video Carrier (MHz)	Audio Carrier (MHz)	Channel Range (MHz)	Channel	Video Carrier (MHz)	Audio Carrier (MHz)	Channel Range (MHz)
(B) PAL, Australia, 7 MHz Channel				**(B) PAL, Italy, 7 MHz Channel**			
0	46.25	51.75	45–52	A	53.75	59.25	52.5–59.5
1	57.25	62.75	56–63	B	62.25	67.75	61–68
2	64.25	69.75	63–70	C	82.25	87.75	81–88
3	86.25	91.75	85–92	D	175.25	180.75	174–181
4	95.25	100.75	94–101	E	183.75	189.25	182.5–189.5
5	102.25	107.75	101–108	F	192.25	197.75	191–198
5A	138.25	143.75	137–144	G	201.25	206.75	200–207
6	175.25	180.75	174–181	H	210.25	215.75	209–216
7	182.25	187.75	181–188	H–1	217.25	222.75	216–223
8	189.25	194.75	188–195	H–2	224.25	229.75	223–230
9	196.25	201.75	195–202	–	–	–	–
10	209.25	214.75	208–215	–	–	–	–
11	216.25	221.75	215–222	–	–	–	–
12	223.25			–	–	–	–
(I) PAL, Ireland, 8 MHz Channel				**(B) PAL, New Zealand, 7 MHz Channel**			
1	45.75	51.75	44.5–52.5	1	45.25	50.75	44–51
2	53.75	59.75	52.5–60.5	2	55.25	60.75	54–61
3	61.75	67.75	60.5–68.5	3	62.25	67.75	61–68
4	175.25	181.25	174–182	4	175.25	180.75	174–181
5	183.25	189.25	182–190	5	182.25	187.75	181–188
6	191.25	197.25	190–198	6	189.25	194.75	188–195
7	199.25	205.25	198–206	7	196.25	201.75	195–202
8	207.25	213.25	206–214	8	203.25	208.75	202–209
9	215.25	221.25	214–222	9	210.25	215.75	209–216

Table 8.5. Analog Broadcast and Cable TV Nominal Frequencies for (B, I) PAL in Various Countries.

Broadcast Channel	Video Carrier (MHz)	Audio Carrier (MHz)		Channel Range (MHz)
		(G, H) PAL	(I) PAL	
2^1	45.75	51.25	51.75	44.5–52.5
3^1	53.75	59.25	59.75	52.5–60.5
4^1	61.75	67.25	67.75	60.5–68.5
5^1	175.25	180.75	181.25	174–182
6^1	183.25	188.75	189.25	182–190
7^1	191.25	196.75	197.25	190–198
8^1	199.25	204.75	205.25	198–206
9^1	207.25	212.75	213.25	206–214
10^1	215.25	220.75	221.25	214–222
2^2	48.25	53.75	–	47–54
3^2	55.25	60.75	–	54–61
4^2	62.25	67.75	–	61–68
5^2	175.25	180.75	–	174–181
6^2	182.25	187.75	–	181–188
7^2	189.25	194.75	–	188–195
8^2	196.25	201.75	–	195–202
9^2	203.25	208.75	–	202–209
10^2	210.25	215.75	–	209–216
11^2	217.25	222.75	–	216–223
12^2	224.25	229.75	–	223–230
21	471.25	476.75	477.25	470–478
22	479.25	484.75	485.25	478–486
23	487.25	492.75	493.25	486–494
24	495.25	500.75	501.25	494–502
25	503.25	508.75	509.25	502–510
26	511.25	516.75	517.25	510–518
27	519.25	524.75	525.25	518–526
28	527.25	532.75	533.25	526–534
29	535.25	540.75	541.25	534–542
30	543.25	548.75	549.25	542–550
31	551.25	556.75	557.25	550–558
32	559.25	564.75	565.25	558–566
33	567.25	572.75	573.25	566–574
34	575.25	580.75	581.25	574–582
35	583.25	588.75	589.25	582–590
36	591.25	596.75	597.25	590–598
37	599.25	604.75	605.25	598–606
38	607.25	612.75	613.25	606–614
39	615.25	620.75	621.25	614–622

Table 8.6a. Analog Broadcast Nominal Frequencies for the [1]United Kingdom, [1]Ireland, [1]South Africa, [1]Hong Kong, and [2]Western Europe.

Broadcast Channel	Video Carrier (MHz)	Audio Carrier (MHz)		Channel Range (MHz)
		(G, H) PAL	(I) PAL	
40	623.25	628.75	629.25	622–630
41	631.25	636.75	637.25	630–638
42	639.25	644.75	645.25	638–646
43	647.25	652.75	653.25	646–654
44	655.25	660.75	661.25	654–662
45	663.25	668.75	669.25	662–670
46	671.25	676.75	677.25	670–678
47	679.25	684.75	685.25	678–686
48	687.25	692.75	693.25	686–694
49	695.25	700.75	701.25	694–702
50	703.25	708.75	709.25	702–710
51	711.25	716.75	717.25	710–718
52	719.25	724.75	725.25	718–726
53	727.25	732.75	733.25	726–734
54	735.25	740.75	741.25	734–742
55	743.25	748.75	749.25	742–750
56	751.25	756.75	757.25	750–758
57	759.25	764.75	765.25	758–766
58	767.25	772.75	773.25	766–774
59	775.25	780.75	781.25	774–782
60	783.25	788.75	789.25	782–790
61	791.25	796.75	797.25	790–798
62	799.25	804.75	805.25	798–806
63	807.25	812.75	813.25	806–814
64	815.25	820.75	821.25	814–822
65	823.25	828.75	829.25	822–830
66	831.25	836.75	837.25	830–838
67	839.25	844.75	845.25	838–846
68	847.25	852.75	853.25	846–854
69	855.25	860.75	861.25	854–862

Table 8.6b. Analog Broadcast Nominal Frequencies for the United Kingdom, Ireland, South Africa, Hong Kong, and Western Europe.

Cable Channel	Video Carrier (MHz)	Audio Carrier (MHz)	Channel Range (MHz)	Cable Channel	Video Carrier (MHz)	Audio Carrier (MHz)	Channel Range (MHz)
E 2	48.25	53.75	47–54	S 11	231.25	236.75	230–237
E 3	55.25	60.75	54–61	S 12	238.25	243.75	237–244
E 4	62.25	67.75	61–68	S 13	245.25	250.75	244–251
S 01	69.25	74.75	68–75	S 14	252.25	257.75	251–258
S 02	76.25	81.75	75–82	S 15	259.25	264.75	258–265
S 03	83.25	88.75	82–89	S 16	266.25	271.75	265–272
S 1	105.25	110.75	104–111	S 17	273.25	278.75	272–279
S 2	112.25	117.75	111–118	S 18	280.25	285.75	279–286
S 3	119.25	124.75	118–125	S 19	287.25	292.75	286–293
S 4	126.25	131.75	125–132	S 20	294.25	299.75	293–300
S 5	133.25	138.75	132–139	S 21	303.25	308.75	302–310
S 6	140.75	145.75	139–146	S 22	311.25	316.75	310–318
S 7	147.75	152.75	146–153	S 23	319.25	324.75	318–326
S 8	154.75	159.75	153–160	S 24	327.25	332.75	326–334
S 9	161.25	166.75	160–167	S 25	335.25	340.75	334–342
S 10	168.25	173.75	167–174	S 26	343.25	348.75	342–350
–	–	–	–	S 27	351.25	356.75	350–358
–	–	–	–	S 28	359.25	364.75	358–366
–	–	–	–	S 29	367.25	372.75	366–374
E 5	175.25	180.75	174–181	S 30	375.25	380.75	374–382
E 6	182.25	187.75	181–188	S 31	383.25	388.75	382–390
E 7	189.25	194.75	188–195	S 32	391.25	396.75	390–398
E 8	196.25	201.75	195–202	S 33	399.25	404.75	398–406
E 9	203.25	208.75	202–209	S 34	407.25	412.75	406–414
E 10	210.25	215.75	209–216	S 35	415.25	420.75	414–422
E 11	217.25	222.75	216–223	S 36	423.25	428.75	422–430
E 12	224.25	229.75	223–230	S 37	431.25	436.75	430–438
–	–	–	–	S 38	439.25	444.75	438–446
–	–	–	–	S 39	447.25	452.75	446–454
–	–	–	–	S 40	455.25	460.75	454–462
–	–	–	–	S 41	463.25	468.75	462–470

Table 8.7. Analog Cable TV Nominal Frequencies for the United Kingdom, Ireland, South Africa, Hong Kong, and Western Europe.

where x and y are the specified CIE 1931 chromaticity coordinates; z is calculated by knowing that x + y + z = 1.

As with NTSC, substituting the known values gives us the solution for K_r, K_g, and K_b:

$$\begin{bmatrix} K_r \\ K_g \\ K_b \end{bmatrix} = \begin{bmatrix} 0.3127/0.3290 \\ 1 \\ 0.3583/0.3290 \end{bmatrix} \begin{bmatrix} 0.64 & 0.29 & 0.15 \\ 0.33 & 0.60 & 0.06 \\ 0.03 & 0.11 & 0.79 \end{bmatrix}^{-1}$$

$$= \begin{bmatrix} 0.674 \\ 1.177 \\ 1.190 \end{bmatrix}$$

Y is defined to be

$$Y = (K_r y_r)R' + (K_g y_g)G' + (K_b y_b)B'$$
$$= (0.674)(0.33)R' + (1.177)(0.60)G'$$
$$+ (1.190)(0.06)B'$$

or

$$Y = 0.222R' + 0.706G' + 0.071B'$$

However, the standard Y = 0.299R' + 0.587G' + 0.114B' equation is still used. Adjustments are made in the receiver to minimize color errors.

PALplus

PALplus (ITU-R BT.1197 and ETSI ETS 300 731) is the result of a cooperative project started in 1990, undertaken by several European broadcasters. By 1995, they wanted to provide an enhanced definition television system (EDTV), compatible with existing receivers. PALplus has been transmitted by a few broadcasters since 1994.

A PALplus picture has a 16:9 aspect ratio. On conventional TVs, it is displayed as a 16:9 letterboxed image with 430 active lines. On PALplus TVs, it is displayed as a 16:9 picture with 574 active lines, with extended vertical resolution. The full video bandwidth is available for luminance detail. Cross color artifacts are reduced by clean encoding.

Wide Screen Signalling

Line 23 contains a Widescreen Signalling (WSS) control signal, defined by ITU-R BT.1119 and ETSI EN 300 294, used by PALplus TVs. This signal indicates:

> Program Aspect Ratio:
> Full Format 4:3
> Letterbox 14:9 Center
> Letterbox 14:9 Top
> Full Format 14:9 Center
> Letterbox 16:9 Center
> Letterbox 16:9 Top
> Full Format 16:9 Anamorphic
> Letterbox > 16:9 Center
>
> Enhanced services:
> Camera Mode
> Film Mode
>
> Subtitles:
> Teletext Subtitles Present
> Open Subtitles Present

PALplus is defined as being Letterbox 16:9 center, camera mode or film mode, helper signals present using modulation, and clean encoding used. Teletext subtitles may or may not be present, and open subtitles may be present only in the active picture area.

During a PALplus transmission, any active video on lines 23 and 623 is blanked prior to encoding. In addition to WSS data, line 23 includes 48 ±1 cycles of a 300 ±9 mV subcarrier with a –U phase, starting 51 μs ±250 ns after 0_H. Line 623 contains a 10 μs ±250 ns white pulse, starting 20 μs ±250 ns after 0_H.

A PALplus TV has the option of deinterlacing a Film Mode signal and displaying it on a 50-Hz progressive-scan display or using field repeating on a 100-Hz interlaced display.

Ghost Cancellation

An optional ghost cancellation signal on line 318, defined by ITU-R BT.1124 and ETSI ETS 300 732, allows a suitably adapted TV to measure the ghost signal and cancel any ghosting during the active video. A PALplus TV may or may not support this feature.

Vertical Filtering

All PALplus sources start out as a 16:9 YCbCr anamorphic image, occupying all 576 active scan lines. Any active video on lines 23 and 623 is blanked prior to encoding (since these lines are used for WSS and reference information), resulting in 574 active lines per frame. Lines 24–310 and 336–622 are used for active video.

Before transmission, the 574 active scan lines of the 16:9 image are squeezed into 430 scan lines. To avoid aliasing problems, the vertical resolution is reduced by lowpass filtering.

For Y, vertical filtering is done using a Quadrature Mirror Filter (QMF) highpass and lowpass pair. Using the QMF process allows the highpass and lowpass information to be resampled, transmitted, and later recombined with minimal loss.

The Y QMF lowpass output is resampled into three-quarters of the original height; little information is lost to aliasing. After clean encoding, it is the letterboxed signal that conventional 4:3 TVs display.

The Y QMF highpass output contains the rest of the original vertical frequency. It is used to generate the helper signals that are transmitted using the "black" scan lines not used by the letterbox picture.

Film Mode

A film mode broadcast has both fields of a frame coming from the same image, as is usually the case with a movie scanned on a telecine.

In film mode, the maximum vertical resolution per frame is about 287 cycles per *active* picture height (cph), limited by the 574 active scan lines per frame.

The vertical resolution of Y is reduced to 215 cph so it can be transmitted using only 430 active lines. The QMF lowpass and highpass filters split the Y vertical information into DC–215 cph and 216–287 cph.

The Y lowpass information is re-scanned into 430 lines to become the letterbox image. Since the vertical frequency is limited to a maximum of 215 cph, no information is lost.

The Y highpass output is decimated so only one in four lines are transmitted. These 144 lines are used to transmit the helper signals. Because of the QMF process, no information is lost to decimation.

The 72 lines above and 72 lines below the central 430-line of the letterbox image are used to transmit the 144 lines of the helper signal. This results in a standard 574 active line picture, but with the original image in its correct aspect ratio, centered between the helper signals. The scan lines containing the 300 mV helper signals are modulated using the U subcarrier so they look black and are not visible to the viewer.

After Fixed ColorPlus processing, the 574 scan lines are PAL encoded and transmitted as a standard interlaced PAL frame.

Camera Mode

Camera (or video) mode assumes the fields of a frame are independent of each other, as would be the case when a camera scans a scene in motion. Therefore, the image may have changed between fields. Only intra-field processing is done.

In camera mode, the maximum vertical resolution per field is about 143 cycles per *active* picture height (cph), limited by the 287 active scan lines per field.

The vertical resolution of Y is reduced to 107 cph so it can be transmitted using only 215 active lines. The QMF lowpass and highpass filter pair split the Y vertical information into DC–107 cph and 108–143 cph.

The Y lowpass information is re-scanned into 215 lines to become the letterbox image. Since the vertical frequency is limited to a maximum of 107 cph, no information is lost.

The Y highpass output is decimated so only one in four lines is transmitted. These 72 lines are used to transmit the helper signals. Because of the QMF process, no information is lost to decimation.

The 36 lines above and 36 lines below the central 215-line of the letterbox image are used to transmit the 72 lines of the helper signal. This results in a 287 active line picture, but with the original image in its correct aspect ratio, centered between the helper signals. The scan lines containing the 300 mV helper signals are modulated using the U subcarrier so they look black and are not visible to the viewer.

After either Fixed or Motion Adaptive ColorPlus processing, the 287 scan lines are PAL encoded and transmitted as a PAL field.

Clean Encoding

Only the letterboxed portion of the PALplus signal is clean encoded. The helper signals are not actual PAL video. However, they are close enough to video to pass through the transmission path and remain fairly invisible on standard TVs.

ColorPlus Processing

Fixed ColorPlus

Film Mode uses a Fixed ColorPlus technique, making use of the lack of motion between the two fields of the frame.

Fixed ColorPlus depends on the subcarrier phase of the composite PAL signal being of opposite phase when 312 lines apart. If these two lines have the same luminance and chrominance information, it can be separated by adding and subtracting the composite signals from each other. Adding cancels the chrominance, leaving luminance. Subtracting cancels the luminance, leaving chrominance.

In practice, Y information above 3 MHz (Y_{HF}) is intra-frame averaged since it shares the frequency spectrum with the modulated chrominance. For line [n], Y_{HF} is calculated as follows:

$$0 \le n \le 214 \text{ for 430-line letterboxed image}$$

$$Y_{HF(60 + n)} = 0.5(Y_{HF(372 + n)} + Y_{HF(60 + n)})$$

$$Y_{HF(372 + n)} = Y_{HF(60 + n)}$$

Y_{HF} is then added to the low-frequency Y (Y_{LF}) information. The same intra-frame averaging process is also used for Cb and Cr. The 430-line letterbox image is then PAL encoded.

Thus, Y information above 3 MHz, and CbCr information, is the same on lines [n] and [n+312]. Y information below 3 MHz may be different on lines [n] and [n+312]. The full vertical resolution of 287 cph is reconstructed by the decoder with the aid of the helper signals.

Motion Adaptive ColorPlus (MACP)

Camera Mode uses either Motion Adaptive ColorPlus or Fixed ColorPlus, depending on the amount of motion between fields. This requires a motion detector in both the encoder and decoder.

To detect movement, the CbCr data on lines [n] and [n+312] are compared. If they match, no movement is assumed, and Fixed ColorPlus operation is used. If the CbCr data doesn't match, movement is assumed, and Motion Adaptive ColorPlus operation is used.

During Motion Adaptive ColorPlus operation, the amount of Y_{HF} added to Y_{LF} is dependent on the difference between $CbCr_{(n)}$ and $CbCr_{(n+312)}$. For the maximum CbCr difference, no Y_{HF} data for lines [n] and [n+312] is transmitted.

In addition, the amount of intra-frame averaged CbCr data mixed with the direct CbCr data is dependent on the difference between $CbCr_{(n)}$ and $CbCr_{(n+312)}$. For the maximum CbCr difference, only direct CbCr data is transmitted separately for lines [n] and [n+312].

SECAM Overview

SECAM (Sequentiel Couleur Avec Mémoire or Sequential Color with Memory) was developed in France, with broadcasting starting in 1967, by realizing that, if color could be bandwidth-limited horizontally, why not also vertically? The two pieces of color information (Db and Dr) added to the monochrome signal could be transmitted on alternate lines, avoiding the possibility of crosstalk.

The receiver requires memory to store one line so that it is concurrent with the next line, and also requires the addition of a line-switching identification technique.

Like PAL, SECAM is a 625-line, 50-field-per-second, 2:1 interlaced system. SECAM was adopted by other countries; however, many are changing to PAL due to the abundance of professional and consumer PAL equipment.

Luminance Information

The monochrome luminance (Y) signal is derived from R′G′B′:

$$Y = 0.299R' + 0.587G' + 0.114B'$$

As with NTSC and PAL, the luminance signal occupies the entire video bandwidth. SECAM has several variations, depending on the video bandwidth and placement of the audio subcarrier. The video signal has a bandwidth of 5.0 or 6.0 MHz, depending on the specific SECAM standard.

Color Information

SECAM transmits Db information during one line and Dr information during the next line; luminance information is transmitted each line. Db and Dr are scaled versions of B′ − Y and R′ − Y:

$$Dr = -1.902(R' - Y)$$

$$Db = 1.505(B' - Y)$$

Since there is an odd number of lines, any given line contains Db information on one field and Dr information on the next field. The decoder requires a 1-H delay, switched synchronously with the Db and Dr switching, so that Db and Dr exist simultaneously in order to convert to YCbCr or RGB.

Color Modulation

SECAM uses FM modulation to transmit the Db and Dr color difference information, with each component having its own subcarrier.

Db and Dr are lowpass filtered to 1.3 MHz and pre-emphasis is applied. The curve for the pre-emphasis is expressed by:

$$A = \frac{1 + j\left(\frac{f}{85}\right)}{1 + j\left(\frac{f}{255}\right)}$$

where f = signal frequency in kHz.

After pre-emphasis, Db and Dr frequency modulate their respective subcarriers. The frequency of each subcarrier is defined as:

F_{OB} = 272 F_H = 4.250000 MHz (± 2 kHz)

F_{OR} = 282 F_H = 4.406250 MHz (± 2 kHz)

These frequencies represent no color information. Nominal Dr deviation is ±280 kHz and the nominal Db deviation is ±230 kHz. Figure 8.23 illustrates the frequency modulation process of the color difference signals. The choice of frequency shifts reflects the idea of keeping the frequencies representing critical colors away from the upper limit of the spectrum to minimize distortion.

After modulation of Db and Dr, subcarrier pre-emphasis is applied, changing the amplitude of the subcarrier as a function of the frequency deviation. The intention is to reduce the visibility of the subcarriers in areas of low luminance and to improve the signal-to-noise ratio of highly saturated colors. This pre-emphasis is given as:

$$G = M\frac{1 + j16F}{1 + j1.26F}$$

where F = $(f/4286) - (4286/f)$, f = instantaneous subcarrier frequency in kHz, and 2M = 23 ± 2.5% of luminance amplitude.

As shown in Figure 8.24, Db and Dr information is transmitted on alternate scan lines. The initial phase of the color subcarrier is also modified as shown in Table 8.8 to further reduce subcarrier visibility. Note that subcarrier phase information in the SECAM system carries no picture information.

Composite Video Generation

The subcarrier data is added to the luminance along with appropriate horizontal and vertical sync signals, blanking signals, and burst signals to generate composite video.

As with PAL, SECAM requires some means of identifying the line-switching sequence. Modern practice has been to use a F_{OR}/F_{OB} burst after most horizontal syncs to derive the switching synchronization information, as shown in Figure 8.25.

SECAM Standards

Figure 8.26 shows the common designations for SECAM systems. The letters refer to the monochrome standard for line and field rates, video bandwidth (5.0 or 6.0 MHz), audio carrier relative frequency, and RF channel bandwidth. The SECAM refers to the technique to add color information to the monochrome signal. Detailed timing parameters may be found in Table 8.9.

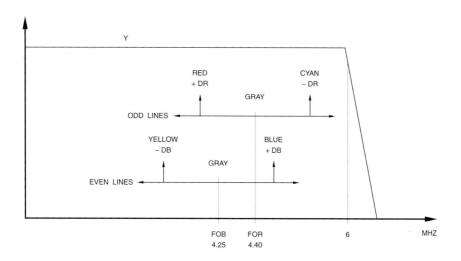

Figure 8.23. SECAM FM Color Modulation.

The initial phase subcarrier undergoes in each line a variation defined by
Frame to frame: 0°, 180°, 0°, 180° ...
Line to line: 0°, 0°, 180°, 0°, 0°, 180° ... or 0°, 0°, 0°, 180°, 180°, 180° ...

Table 8.8. SECAM Subcarrier Timing.

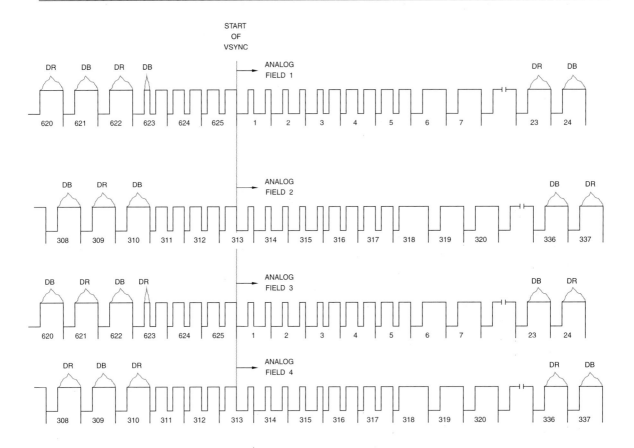

Figure 8.24. Four-field SECAM Sequence. See Figure 8.5 for equalization and serration pulse details.

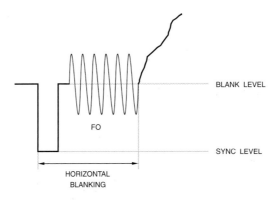

Figure 8.25. SECAM Chroma Synchronization Signals.

Luminance Equation Derivation

The equation for generating luminance from RGB information is determined by the chromaticities of the three primary colors used by the receiver and what color white actually is.

The chromaticities of the RGB primaries and reference white (CIE illuminate D_{65}) are:

R: $x_r = 0.64$ $y_r = 0.33$ $z_r = 0.03$

G: $x_g = 0.29$ $y_g = 0.60$ $z_g = 0.11$

B: $x_b = 0.15$ $y_b = 0.06$ $z_b = 0.79$

white: $x_w = 0.3127$ $y_w = 0.3290$
$z_w = 0.3583$

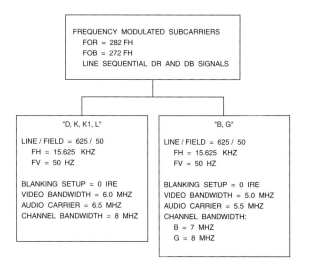

Figure 8.26. Common SECAM Systems.

where x and y are the specified CIE 1931 chromaticity coordinates; z is calculated by knowing that x + y + z = 1. Once again, substituting the known values gives us the solution for K_r, K_g, and K_b:

$$\begin{bmatrix} K_r \\ K_g \\ K_b \end{bmatrix} = \begin{bmatrix} 0.3127/0.3290 \\ 1 \\ 0.3583/0.3290 \end{bmatrix} \begin{bmatrix} 0.64 & 0.29 & 0.15 \\ 0.33 & 0.60 & 0.06 \\ 0.03 & 0.11 & 0.79 \end{bmatrix}^{-1}$$

$$= \begin{bmatrix} 0.674 \\ 1.177 \\ 1.190 \end{bmatrix}$$

Y is defined to be

$$Y = (K_r y_r)R' + (K_g y_g)G' + (K_b y_b)B'$$
$$= (0.674)(0.33)R' + (1.177)(0.60)G'$$
$$+ (1.190)(0.06)B'$$

or

$$Y = 0.222R' + 0.706G' + 0.071B'$$

However, the standard Y = 0.299R' + 0.587G' + 0.114B' equation is still used. Adjustments are made in the receiver to minimize color errors.

	M	N	B, G	H	I	D, K	K1	L
SCAN LINES PER FRAME	525	625	625					
FIELD FREQUENCY (FIELDS / SECOND)	59.94	50	50					
LINE FREQUENCY (HZ)	15,734	15,625	15,625					
PEAK WHITE LEVEL (IRE)	100	100	100					
SYNC TIP LEVEL (IRE)	−40	−40 (−43)	−43					
SETUP (IRE)	7.5 ± 2.5	7.5 ± 2.5 (0)	0					
PEAK VIDEO LEVEL (IRE)	120		133		133	115	115	125
GAMMA OF RECEIVER	2.2	2.8	2.8	2.8	2.8	2.8	2.8	2.8
VIDEO BANDWIDTH (MHZ)	4.2	5.0 (4.2)	5.0	5.0	5.5	6.0	6.0	6.0
LUMINANCE SIGNAL	Y = 0.299R´ + 0.685G´ + 0.114B´ (RGB ARE GAMMA−CORRECTED)							

[1] Values in parentheses apply to (N_C) PAL used in Argentina.

Table 8.9a. Basic Characteristics of Color Video Signals.

Characteristics	M	N	B, D, G, H, I K, K1, L, N_C
Nominal line period (µs)	63.5555	64	64
Line blanking interval (µs)	10.7 ± 0.1	10.88 ± 0.64	11.85 ± 0.15
0_H to start of active video (µs)	9.2 ± 0.1	9.6 ± 0.64	10.5
Front porch (µs)	1.5 ± 0.1	1.92 ± 0.64	1.65 ± 0.15
Line synchronizing pulse (µs)	4.7 ± 0.1	4.99 ± 0.77	4.7 ± 0.2
Rise and fall time of line blanking (10%, 90%) (ns)	140 ± 20	300 ± 100	300 ± 100
Rise and fall time of line synchronizing pulses (10%, 90%) (ns)	140 ± 20	≤ 250	250 ± 50

Notes:
1. 0_H is at 50% point of falling edge of horizontal sync.
2. In case of different standards having different specifications and tolerances, the tightest specification and tolerance is listed.
3. Timing is measured between half-amplitude points on appropriate signal edges.

Table 8.9b. Details of Line Synchronization Signals.

Characteristics	M	N	B, D, G, H, I K, K1, L, N_C
Field period (ms)	16.6833	20	20
Field blanking interval	20 lines	19–25 lines	25 lines
Rise and fall time of field blanking (10%, 90%) (ns)	140 ± 20	≤ 250	300 ± 100
Duration of equalizing and synchronizing sequences	3 H	3 H	2.5 H
Equalizing pulse width (µs)	2.3 ± 0.1	2.43 ± 0.13	2.35 ± 0.1
Serration pulse width (µs)	4.7 ± 0.1	4.7 ± 0.8	4.7 ± 0.1
Rise and fall time of synchronizing and equalizing pulses (10%, 90%) (ns)	140 ± 20	< 250	250 ± 50

Notes:
1. In case of different standards having different specifications and tolerances, the tightest specification and tolerance is listed.
2. Timing is measured between half-amplitude points on appropriate signal edges.

Table 8.9c. Details of Field Synchronization Signals.

	M / NTSC	M / PAL	B, D, G, H, I, N / PAL	B, D, G, K, K1, K / SECAM
ATTENUATION OF COLOR DIFFERENCE SIGNALS	U, V, I, Q: < 2 DB AT 1.3 MHZ > 20 DB AT 3.6 MHZ OR Q: < 2 DB AT 0.4 MHZ < 6 DB AT 0.5 MHZ > 6 DB AT 0.6 MHZ	< 2 DB AT 1.3 MHZ > 20 DB AT 3.6 MHZ	< 3 DB AT 1.3 MHZ > 20 DB AT 4 MHZ (> 20 DB AT 3.6 MHZ)	< 3 DB AT 1.3 MHZ > 30 DB AT 3.5 MHZ (BEFORE LOW FREQUENCY PRE–CORRECTION)
START OF BURST AFTER 0H (µS)	5.3 ± 0.07	5.8 ± 0.1	5.6 ± 0.1	
BURST DURATION (CYCLES)	9 ± 1	9 ± 1	10 ± 1 (9 ± 1)	
BURST PEAK AMPLITUDE	40 ± 1 IRE	42.86 ± 4 IRE	42.86 ± 4 IRE	

Note: Values in parentheses apply to (N_C) PAL used in Argentina.

Table 8.9d. Basic Characteristics of Color Video Signals.

Video Test Signals

Many industry-standard video test signals have been defined to help test the relative quality of encoding, decoding, and the transmission path, and to perform calibration. Note that some video test signals cannot properly be generated by providing RGB data to an encoder; in this case, YCbCr data may be used.

If the video standard uses a 7.5-IRE setup, typically only test signals used for visual examination use the 7.5-IRE setup. Test signals designed for measurement purposes typically use a 0-IRE setup, providing the advantage of defining a known blanking level.

Color Bars Overview

Color bars are one of the standard video test signals, and there are several variations, depending on the video standard and application. For this reason, this section reviews the most common color bar formats. Color bars have two major characteristics: amplitude and saturation.

The amplitude of a color bar signal is determined by:

$$amplitude\ (\%) \ = \frac{max\ (R,\ G,\ B)_a}{max\ (R,\ G,\ B)_b} \times 100$$

where $max(R,G,B)_a$ is the maximum value of R´G´B´ during colored bars and $max(R,G,B)_b$ is the maximum value of R´G´B´ during reference white.

The saturation of a color bar signal is less than 100% if the minimum value of any one of the R´G´B´ components is not zero. The saturation is determined by:

$$saturation\ (\%) = \left[1 - \left(\frac{min(R,\ G,\ B)}{max(R,\ G,\ B)} \right)^{\gamma} \right] \times 100$$

where min(R,G,B) and max(R,G,B) are the minimum and maximum values, respectively, of R′G′B′ during colored bars, and γ is the gamma exponent, typically [1/0.45].

NTSC Color Bars

In 1953, it was normal practice for the analog R′G′B′ signals to have a 7.5 IRE setup, and the original NTSC equations assumed this form of input to an encoder. Today, digital R′G′B′ or YCbCr signals typically do not include the 7.5 IRE setup, and the 7.5 IRE setup is added within the encoder.

The different color bar signals are described by four amplitudes, expressed in percent, separated by oblique strokes. 100% saturation is implied, so saturation is not specified. The first and second numbers are the white and black amplitudes, respectively. The third and fourth numbers are the white and black amplitudes from which the color bars are derived.

For example, 100/7.5/75/7.5 color bars would be 75% color bars with 7.5% setup in which the white bar has been set to 100% and the black bar to 7.5%. Since NTSC systems usually have the 7.5% setup, the two common color bars are 75/7.5/75/7.5 and 100/7.5/100/7.5, which are usually shortened to 75% and 100%, respectively. The 75% bars are most commonly used. Television transmitters do not pass information with an amplitude greater than about 120 IRE. Therefore, the 75% color bars are used for transmission testing. The 100% color bars may be used for testing in situations where a direct connection between equipment is possible. The 75/7.5/75/7.5 color bars are a part of the Electronic Industries Association EIA-189-A Encoded Color Bar Standard.

Figure 8.27 shows a typical vectorscope display for full-screen 75% NTSC color bars. Figure 8.28 illustrates the video waveform for 75% color bars.

Tables 8.10 and 8.11 list the luminance and chrominance levels for the two common color bar formats for NTSC.

For reference, the RGB and YCbCr values to generate the standard NTSC color bars are shown in Tables 8.12 and 8.13. RGB is assumed to have a range of 0–255; YCbCr is assumed to have a range of 16–235 for Y and 16–240 for Cb and Cr. It is assumed any 7.5 IRE setup is implemented within the encoder.

PAL Color Bars

Unlike NTSC, PAL does not support a 7.5 IRE setup; the black and blank levels are the same. The different color bar signals are usually described by four amplitudes, expressed in percent, separated by oblique strokes. The first and second numbers are the maximum and minimum percentages, respectively, of R′G′B′ values for an uncolored bar. The third and fourth numbers are the maximum and minimum percentages, respectively, of R′G′B′ values for a colored bar.

Since PAL systems have a 0% setup, the two common color bars are 100/0/75/0 and 100/0/100/0, which are usually shortened to 75% and 100%, respectively. The 75% color bars are used for transmission testing. The 100% color bars may be used for testing in situations where a direct connection between equipment is possible.

The 100/0/75/0 color bars also are referred to as EBU (European Broadcast Union) color bars. All of the color bars discussed in this section are also a part of *Specification of Television Standards for 625-line System-I Transmissions* (1971) published by the Independent Television Authority (ITA) and the British Broadcasting Corporation (BBC), and ITU-R BT.471.

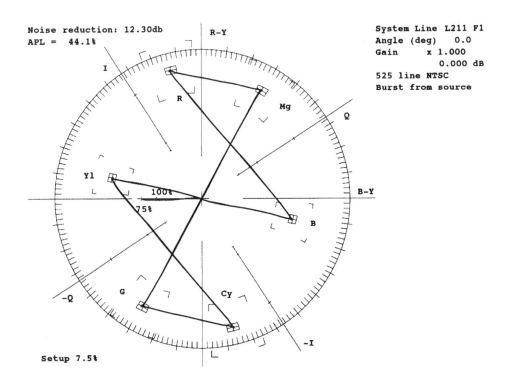

Figure 8.27. Typical Vectorscope Display for 75% NTSC Color Bars.

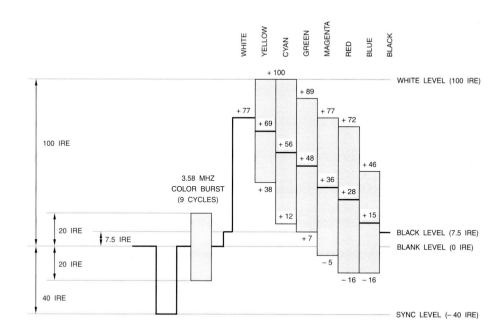

Figure 8.28. IRE Values for 75% NTSC Color Bars.

	Luminance (IRE)	Chrominance Level (IRE)	Minimum Chrominance Excursion (IRE)	Maximum Chrominance Excursion (IRE)	Chrominance Phase (degrees)
white	76.9	0	–	–	–
yellow	69.0	62.1	37.9	100.0	167.1
cyan	56.1	87.7	12.3	100.0	283.5
green	48.2	81.9	7.3	89.2	240.7
magenta	36.2	81.9	–4.8	77.1	60.7
red	28.2	87.7	–15.6	72.1	103.5
blue	15.4	62.1	–15.6	46.4	347.1
black	7.5	0	–	–	–

Table 8.10. 75/7.5/75/7.5 (75%) NTSC Color Bars.

	Luminance (IRE)	Chrominance Level (IRE)	Minimum Chrominance Excursion (IRE)	Maximum Chrominance Excursion (IRE)	Chrominance Phase (degrees)
white	100.0	0	–	–	–
yellow	89.5	82.8	48.1	130.8	167.1
cyan	72.3	117.0	13.9	130.8	283.5
green	61.8	109.2	7.2	116.4	240.7
magenta	45.7	109.2	–8.9	100.3	60.7
red	35.2	117.0	–23.3	93.6	103.5
blue	18.0	82.8	–23.3	59.4	347.1
black	7.5	0	–	–	–

Table 8.11. 100/7.5/100/7.5 (100%) NTSC Color Bars.

	White	Yellow	Cyan	Green	Magenta	Red	Blue	Black
gamma-corrected RGB (gamma = 1/0.45)								
R′	191	191	0	0	191	191	0	0
G′	191	191	191	191	0	0	0	0
B′	191	0	191	0	191	0	191	0
linear RGB								
R	135	135	0	0	135	135	0	0
G	135	135	135	135	0	0	0	0
B	135	0	135	0	135	0	135	0
YCbCr								
Y	180	162	131	112	84	65	35	16
Cb	128	44	156	72	184	100	212	128
Cr	128	142	44	58	198	212	114	128

Table 8.12. RGB and YCbCr Values for 75% NTSC Color Bars.

	White	Yellow	Cyan	Green	Magenta	Red	Blue	Black
gamma-corrected RGB (gamma = 1/0.45)								
R′	255	255	0	0	255	255	0	0
G′	255	255	255	255	0	0	0	0
B′	255	0	255	0	255	0	255	0
linear RGB								
R	255	255	0	0	255	255	0	0
G	255	255	255	255	0	0	0	0
B	255	0	255	0	255	0	255	0
YCbCr								
Y	235	210	170	145	106	81	41	16
Cb	128	16	166	54	202	90	240	128
Cr	128	146	16	34	222	240	110	128

Table 8.13. RGB and YCbCr Values for 100% NTSC Color Bars.

Figure 8.29 illustrates the video waveform for 75% color bars. Figure 8.30 shows a typical vectorscope display for full-screen 75% PAL color bars.

Tables 8.14, 8.15, and 8.16 list the luminance and chrominance levels for the three common color bar formats for PAL.

For reference, the RGB and YCbCr values to generate the standard PAL color bars are shown in Tables 8.17, 8.18, and 8.19. RGB is assumed to have a range of 0–255; YCbCr is assumed to have a range of 16–235 for Y and 16–240 for Cb and Cr.

EIA Color Bars (NTSC)

The EIA color bars (Figure 8.28 and Table 8.10) are a part of the EIA-189-A standard. The seven bars (gray, yellow, cyan, green, magenta, red and blue) are at 75% amplitude, 100% saturation. The duration of each color bar is 1/7 of the active portion of the scan line. Note that the black bar in Figure 8.28 and Table 8.10 is not part of the standard and is shown for reference only. The color bar test signal allows checking for hue and color saturation accuracy.

	Luminance (volts)	Peak-to-Peak Chrominance			Chrominance Phase (degrees)	
		U axis (volts)	V axis (volts)	Total (volts)	Line n (135° burst)	Line n + 1 (225° burst)
white	0.700	0	–	–	–	–
yellow	0.465	0.459	0.105	0.470	167	193
cyan	0.368	0.155	0.646	0.664	283.5	76.5
green	0.308	0.304	0.541	0.620	240.5	119.5
magenta	0.217	0.304	0.541	0.620	60.5	299.5
red	0.157	0.155	0.646	0.664	103.5	256.5
blue	0.060	0.459	0.105	0.470	347	13.0
black	0	0	0	0	–	–

Table 8.14. 100/0/75/0 (75%) PAL Color Bars.

	Luminance (volts)	Peak-to-Peak Chrominance			Chrominance Phase (degrees)	
		U axis (volts)	V axis (volts)	Total (volts)	Line n (135° burst)	Line n + 1 (225° burst)
white	0.700	0	–	–	–	–
yellow	0.620	0.612	0.140	0.627	167	193
cyan	0.491	0.206	0.861	0.885	283.5	76.5
green	0.411	0.405	0.721	0.827	240.5	119.5
magenta	0.289	0.405	0.721	0.827	60.5	299.5
red	0.209	0.206	0.861	0.885	103.5	256.5
blue	0.080	0.612	0.140	0.627	347	13.0
black	0	0	0	0	–	–

Table 8.15. 100/0/100/0 (100%) PAL Color Bars.

	Luminance (volts)	Peak-to-Peak Chrominance			Chrominance Phase (degrees)	
		U axis (volts)	V axis (volts)	Total (volts)	Line n (135° burst)	Line n + 1 (225° burst)
white	0.700	0	–	–	–	–
yellow	0.640	0.459	0.105	0.470	167	193
cyan	0.543	0.155	0.646	0.664	283.5	76.5
green	0.483	0.304	0.541	0.620	240.5	119.5
magenta	0.392	0.304	0.541	0.620	60.5	299.5
red	0.332	0.155	0.646	0.664	103.5	256.5
blue	0.235	0.459	0.105	0.470	347	13.0
black	0	0	0	0	–	–

Table 8.16. 100/0/100/25 (98%) PAL Color Bars.

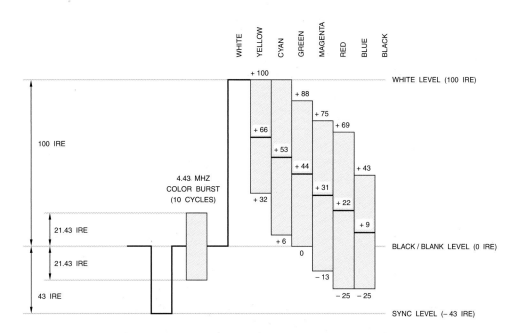

Figure 8.29. IRE Values for 75% PAL Color Bars.

VM700A Video Measurement Set

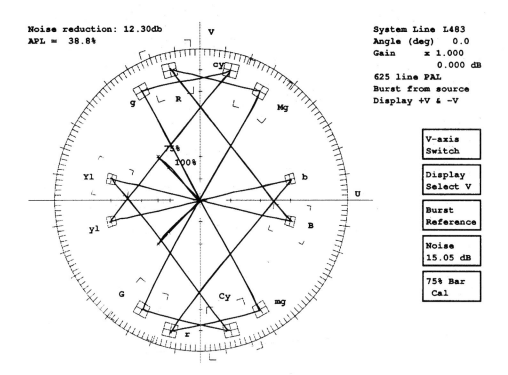

Figure 8.30. Typical Vectorscope Display for 75% PAL Color Bars.

	White	Yellow	Cyan	Green	Magenta	Red	Blue	Black
	gamma-corrected RGB (gamma = 1/0.45)							
R′	255	191	0	0	191	191	0	0
G′	255	191	191	191	0	0	0	0
B′	255	0	191	0	191	0	191	0
	linear RGB							
R	255	135	0	0	135	135	0	0
G	255	135	135	135	0	0	0	0
B	255	0	135	0	135	0	135	0
	YCbCr							
Y	235	162	131	112	84	65	35	16
Cb	128	44	156	72	184	100	212	128
Cr	128	142	44	58	198	212	114	128

Table 8.17. RGB and YCbCr Values for 75% PAL Color Bars.

	White	Yellow	Cyan	Green	Magenta	Red	Blue	Black
	gamma-corrected RGB (gamma = 1/0.45)							
R′	255	255	0	0	255	255	0	0
G′	255	255	255	255	0	0	0	0
B′	255	0	255	0	255	0	255	0
	linear RGB							
R	255	255	0	0	255	255	0	0
G	255	255	255	255	0	0	0	0
B	255	0	255	0	255	0	255	0
	YCbCr							
Y	235	210	170	145	106	81	41	16
Cb	128	16	166	54	202	90	240	128
Cr	128	146	16	34	222	240	110	128

Table 8.18. RGB and YCbCr Values for 100% PAL Color Bars.

	White	Yellow	Cyan	Green	Magenta	Red	Blue	Black
gamma-corrected RGB (gamma = 1/0.45)								
R'	255	255	44	44	255	255	44	44
G'	255	255	255	255	44	44	44	44
B'	255	44	255	44	255	44	255	44
linear RGB								
R	255	255	5	5	255	255	5	5
G	255	255	255	255	5	5	5	5
B	255	5	255	5	255	5	255	5
YCbCr								
Y	235	216	186	167	139	120	90	16
Cb	128	44	156	72	184	100	212	128
Cr	128	142	44	58	198	212	114	128

Table 8.19. RGB and YCbCr Values for 98% PAL Color Bars.

EBU Color Bars (PAL)

The EBU color bars are similar to the EIA color bars, except a 100 IRE white level is used (see Figure 8.29 and Table 8.14). The six colored bars (yellow, cyan, green, magenta, red, and blue) are at 75% amplitude, 100% saturation, while the white bar is at 100% amplitude. The duration of each color bar is 1/7 of the active portion of the scan line. Note that the black bar in Figure 8.29 and Table 8.14 is not part of the standard and is shown for reference only. The color bar test signal allows checking for hue and color saturation accuracy.

SMPTE Bars (NTSC)

This split-field test signal is composed of the EIA color bars for the first 2/3 of the field, the Reverse Blue bars for the next 1/12 of the field, and the PLUGE test signal for the remainder of the field.

Reverse Blue Bars

The Reverse Blue bars are composed of the blue, magenta, and cyan colors bars from the EIA/EBU color bars, but are arranged in a different order—blue, black, magenta, black, cyan, black, and white. The duration of each color bar is 1/7 of the active portion of the scan line. Typically, Reverse Blue bars are used with the EIA or EBU color bar signal in a split-field arrangement, with the EIA/EBU color bars comprising the first 3/4 of the field and the Reverse Blue bars comprising the remainder of the field. This split-field arrangement eases adjustment of chrominance and hue on a color monitor.

PLUGE

PLUGE (Picture Line-Up Generating Equipment) is a visual black reference, with one area blacker-than-black, one area at black, and one area lighter-than-black. The brightness of the monitor is adjusted so that the black and blacker-than-black areas are indistinguishable from each other and the lighter-than-black area is slightly lighter (the contrast should be at the normal setting). Additional test signals, such as a white pulse and modulated IQ signals, are usually added to facilitate testing and monitor alignment.

The NTSC PLUGE test signal (shown in Figure 8.31) is composed of a 7.5 IRE (black level) pedestal with a 40 IRE "–I" phase modulation, a 100 IRE white pulse, a 40 IRE "+Q" phase modulation, and 3.5 IRE, 7.5 IRE, and 11.5 IRE pedestals. Typically, PLUGE is used as part of the SMPTE bars.

For PAL, each country has its own slightly different PLUGE configuration, with most differences being in the black pedestal level used, and work is being done on a standard test signal. Figure 8.32 illustrates a typical PAL PLUGE test signal. Usually used as a full-screen test signal, it is composed of a 0 IRE

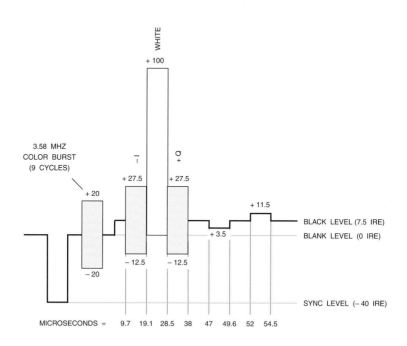

Figure 8.31. PLUGE Test Signal for NTSC. IRE values are indicated.

pedestal with PLUGE (–2 IRE, 0 IRE, and 2 IRE pedestals) and a white pulse. The white pulse may have five levels of brightness (0, 25, 50, 75, and 100 IRE), depending on the scan line number, as shown in Figure 8.32. The PLUGE is displayed on scan lines that have non-zero IRE white pulses. ITU-R BT.1221 discusses considerations for various PAL systems.

Y Bars

The Y bars consist of the luminance-only levels of the EIA/EBU color bars; however, the black level (7.5 IRE for NTSC and 0 IRE for PAL) is included and the color burst is still present. The duration of each luminance bar is therefore 1/8 of the active portion of the scan line. Y bars are useful for color monitor adjustment and measuring luminance nonlinearity. Typically, the Y bars signal is used with the EIA or EBU color bar signal in a split-field arrangement, with the EIA/EBU color bars comprising the first 3/4 of the field and the Y bars signal comprising the remainder of the field.

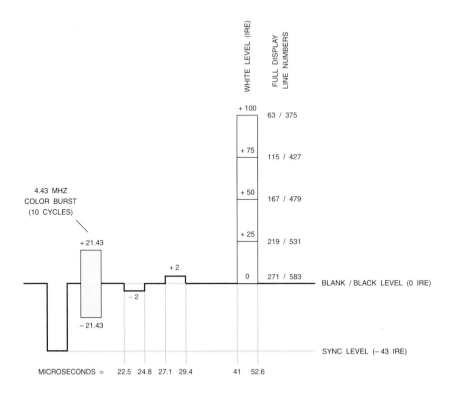

Figure 8.32. PLUGE Test Signal for PAL. IRE values are indicated.

Red Field

The Red Field signal consists of a 75% amplitude, 100% saturation red chrominance signal. This is useful as the human eye is sensitive to static noise intermixed in a red field. Distortions that cause small errors in picture quality can be examined visually for the effect on the picture. Typically, the Red Field signal is used with the EIA/EBU color bars signal in a split-field arrangement, with the EIA/EBU color bars comprising the first 3/4 of the field, and the Red Field signal comprising the remainder of the field.

10-Step Staircase

This test signal is composed of ten unmodulated luminance steps of 10 IRE each, ranging from 0 IRE to 100 IRE, shown in Figure 8.33. This test signal may be used to measure luminance nonlinearity.

Modulated Ramp

The modulated ramp test signal, shown in Figure 8.34, is composed of a luminance ramp from 0 IRE to either 80 or 100 IRE, superimposed with modulated chrominance that has a phase of 0° ±1° relative to the burst. The 80 IRE ramp provides testing of the normal operating range of the system; a 100 IRE ramp may optionally be used to test the entire operating range. The peak-to-peak modulated chrominance is 40 ±0.5 IRE for (M) NTSC and 42.86 ±0.5 IRE for (B, D, G, H, I) PAL. Note a 0 IRE setup is used. The rise and fall times at the start and end of the modulated ramp envelope are 400 ±25 ns (NTSC systems) or approximately 1 μs (PAL systems). This test signal may be used to measure differential gain. The modulated ramp signal is preferred over a 5-step or 10-step modulated staircase signal when testing digital systems.

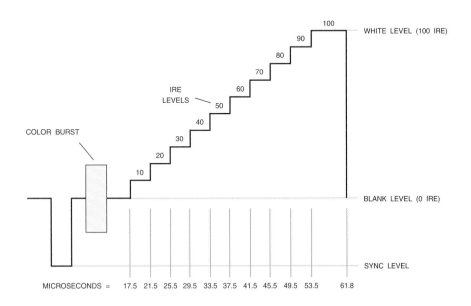

Figure 8.33. Ten-Step Staircase Test Signal for NTSC and PAL.

Modulated Staircase

The 5-step modulated staircase signal (a 10-step version is also used), shown in Figure 8.35, consists of 5 luminance steps, superimposed with modulated chrominance that has a phase of 0° ±1° relative to the burst. The peak-to-peak modulated chrominance amplitude is 40 ±0.5 IRE for (M) NTSC and 42.86 ±0.5 IRE for (B, D, G, H, I) PAL. Note that a 0 IRE setup is used. The rise and fall times of each modulation packet envelope are 400 ±25 ns (NTSC systems) or approximately 1 µs (PAL systems). The luminance IRE levels for the 5-step modulated staircase signal are shown in Figure 8.35. This test signal may be used to measure differential gain. The modulated ramp signal is preferred over a 5-step or 10-step modulated staircase signal when testing digital systems.

Modulated Pedestal

The modulated pedestal test signal (also called a three-level chrominance bar), shown in Figure 8.36, is composed of a 50 IRE luminance pedestal, superimposed with three amplitudes of modulated chrominance that has a phase relative to the burst of –90° ±1°. The peak-to-peak amplitudes of the modulated chrominance are 20 ±0.5, 40 ±0.5, and 80 ±0.5 IRE for (M) NTSC and 20 ±0.5, 60 ±0.5, and 100 ±0.5 IRE for (B, D, G, H, I) PAL. Note a 0 IRE setup is used. The rise and fall times of each modulation packet envelope are 400 ±25 ns (NTSC systems) or approximately 1 µs (PAL systems). This test signal may be used to measure chrominance-to-luminance intermodulation and chrominance nonlinear gain.

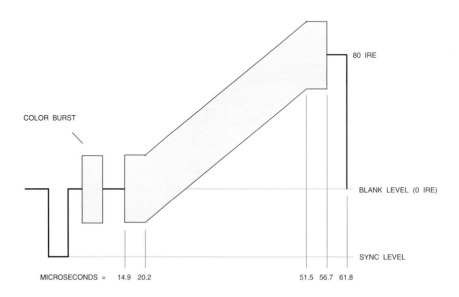

Figure 8.34. 80 IRE Modulated Ramp Test Signal for NTSC and PAL.

Figure 8.35. Five-Step Modulated Staircase Test Signal for NTSC and PAL.

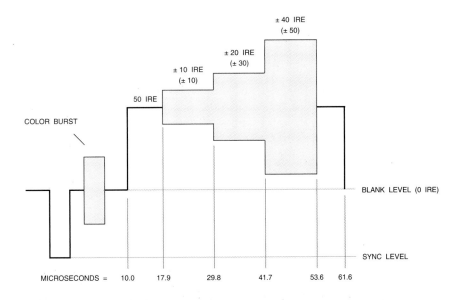

Figure 8.36. Modulated Pedestal Test Signal for NTSC and PAL.
PAL IRE values are shown in parentheses.

Multiburst

The multiburst test signal for (M) NTSC, shown in Figure 8.37, consists of a white flag with a peak amplitude of 100 ±1 IRE and six frequency packets, each a specific frequency. The packets have a 40 ±1 IRE pedestal with peak-to-peak amplitudes of 60 ±0.5 IRE. Note a 0 IRE setup is used and the starting and ending point of each packet is at zero phase.

The ITU multiburst test signal for (B, D, G, H, I) PAL, shown in Figure 8.38, consists of a 4 µs white flag with a peak amplitude of 80 ±1 IRE and six frequency packets, each a specific frequency. The packets have a 50 ±1 IRE pedestal with peak-to-peak amplitudes of 60 ±0.5 IRE. Note the starting and ending points of each packet are at zero phase. The gaps between packets are 0.4–2.0 µs. The ITU multiburst test signal may be present on line 18.

The multiburst signals are used to test the frequency response of the system by measuring the peak-to-peak amplitudes of the packets.

Line Bar

The line bar is a single 100 ±0.5 IRE (reference white) pulse of 10 µs (PAL), 18 µs (NTSC), or 25 µs (PAL) that occurs anywhere within the active scan line time (rise and fall times are ≤ 1 µs). Note that the color burst is not present, and a 0 IRE setup is used. This test signal is used to measure line time distortion (line tilt or H tilt). A digital encoder or decoder does not generate line time distortion; the distortion is generated primarily by the analog filters and transmission channel.

Multipulse

The (M) NTSC multipulse contains a 2T pulse and 25T and 12.5T pulses with various high-frequency components, as shown in Figure 8.39. The (B, D, G, H, I) PAL multipulse is similar, except 20T and 10T pulses are used, and there is no 7.5 IRE setup. This test signal is typically used to measure the frequency response of the transmission channel.

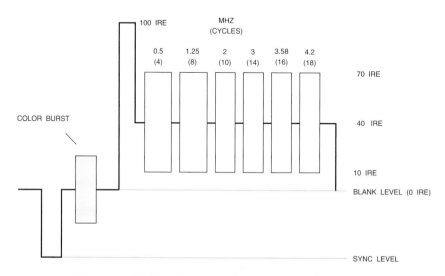

Figure 8.37. Multiburst Test Signal for NTSC.

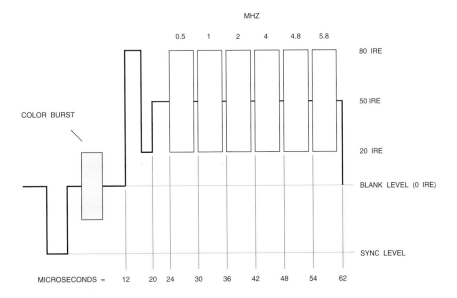

Figure 8.38. ITU Multiburst Test Signal for PAL.

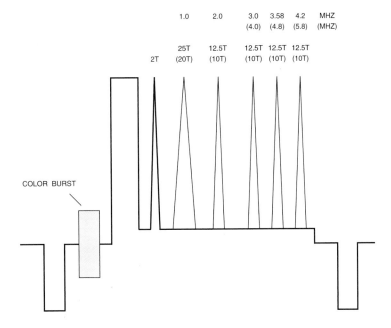

Figure 8.39. Multipulse Test Signal for NTSC and PAL. PAL values are shown in parentheses.

Field Square Wave

The field square wave contains 100 ±0.5 IRE pulses for the entire active line time for Field 1 and blanked scan lines for Field 2. Note that the color burst is not present and a 0 IRE setup is used. This test signal is used to measure field time distortion (field tilt or V tilt). A digital encoder or decoder does not generate field time distortion; the distortion is generated primarily by the analog filters and transmission channel.

Composite Test Signal

NTC-7 Version for NTSC

The NTC (U. S. Network Transmission Committee) has developed a composite test signal that may be used to test several video parameters, rather than using multiple test signals. The NTC-7 composite test signal for NTSC systems (shown in Figure 8.40) consists of a 100 IRE line bar, a 2T pulse, a 12.5T chrominance pulse, and a 5-step modulated staircase signal.

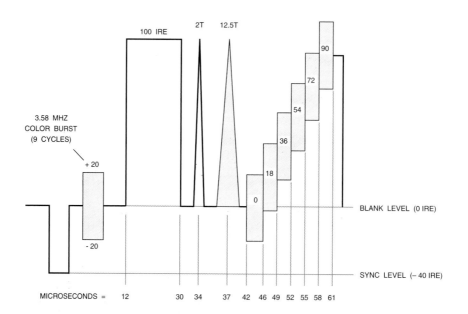

Figure 8.40. NTC-7 Composite Test Signal for NTSC, With Corresponding IRE Values.

The line bar has a peak amplitude of 100 ±0.5 IRE, and 10–90% rise and fall times of 125 ±5 ns with an integrated sine-squared shape. It has a width at the 60 IRE level of 18 μs.

The 2T pulse has a peak amplitude of 100 ±0.5 IRE, with a half-amplitude width of 250 ±10 ns.

The 12.5T chrominance pulse has a peak amplitude of 100 ±0.5 IRE, with a half-amplitude width of 1562.5 ±50 ns.

The 5-step modulated staircase signal consists of 5 luminance steps superimposed with a 40 ±0.5 IRE subcarrier that has a phase of 0° ±1° relative to the burst. The rise and fall times of each modulation packet envelope are 400 ±25 ns.

The NTC-7 composite test signal may be present on line 17.

ITU Version for PAL

The ITU (BT.628 and BT.473) has developed a composite test signal that may be used to test several video parameters, rather than using multiple test signals. The ITU composite test signal for PAL systems (shown in Figure 8.41) consists of a white flag, a 2T pulse, and a 5-step modulated staircase signal.

The white flag has a peak amplitude of 100 ±1 IRE and a width of 10 μs.

The 2T pulse has a peak amplitude of 100 ±0.5 IRE, with a half-amplitude width of 200 ±10 ns.

The 5-step modulated staircase signal consists of 5 luminance steps (whose IRE values are shown in Figure 8.41) superimposed with a 42.86 ±0.5 IRE subcarrier that has a phase of 60° ±1° relative to the U axis. The rise and fall times of each modulation packet envelope are approximately 1 μs.

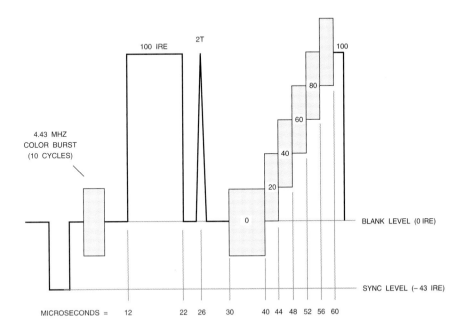

Figure 8.41. ITU Composite Test Signal for PAL, With Corresponding IRE Values.

The ITU composite test signal may be present on line 330.

U.K. Version

The United Kingdom allows the use of a slightly different test signal since the 10T pulse is more sensitive to delay errors than the 20T pulse (at the expense of occupying less chrominance bandwidth). Selection of an appropriate pulse width is a trade-off between occupying the PAL chrominance bandwidth as fully as possible and obtaining a pulse with sufficient sensitivity to delay errors. Thus, the national test signal (developed by the British Broadcasting Corporation and the Independent Television Authority) in Figure 8.42 may be present on lines 19 and 332 for (I) PAL systems in the United Kingdom.

The white flag has a peak amplitude of 100 ±1 IRE and a width of 10 μs.

The 2T pulse has a peak amplitude of 100 ±0.5 IRE, with a half-amplitude width of 200 ±10 ns.

The 10T chrominance pulse has a peak amplitude of 100 ±0.5 IRE.

The 5-step modulated staircase signal consists of 5 luminance steps (whose IRE values are shown in Figure 8.42) superimposed with a 21.43 ±0.5 IRE subcarrier that has a phase of 60° ±1° relative to the U axis. The rise and fall times of each modulation packet envelope is approximately 1 μs.

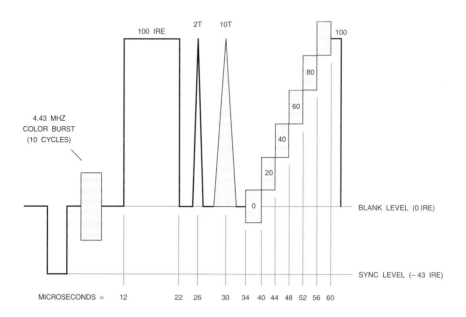

Figure 8.42. United Kingdom (I) PAL National Test Signal #1, With Corresponding IRE Values.

Combination Test Signal

NTC-7 Version for NTSC

The NTC (U. S. Network Transmission Committee) has also developed a combination test signal that may be used to test several video parameters, rather than using multiple test signals. The NTC-7 combination test signal for NTSC systems (shown in Figure 8.43) consists of a white flag, a multiburst, and a modulated pedestal signal.

The white flag has a peak amplitude of 100 ±1 IRE and a width of 4 µs.

The multiburst has a 50 ±1 IRE pedestal with peak-to-peak amplitudes of 50 ±0.5 IRE. The starting point of each frequency packet is at zero phase. The width of the 0.5 MHz packet is 5 µs; the width of the remaining packets is 3 µs.

The 3-step modulated pedestal is composed of a 50 IRE luminance pedestal, superimposed with three amplitudes of modulated chrominance (20 ±0.5, 40 ±0.5, and 80 ±0.5 IRE peak-to-peak) that have a phase of –90° ±1° relative to the burst. The rise and fall times of each modulation packet envelope are 400 ±25 ns.

The NTC-7 combination test signal may be present on line 280.

ITU Version for PAL

The ITU (BT.473) has developed a combination test signal that may be used to test several video parameters, rather than using multiple test signals. The ITU combination test signal for PAL systems (shown in Figure 8.44) consists of a white flag, a 2T pulse, a 20T modulated chrominance pulse, and a 5-step luminance staircase signal.

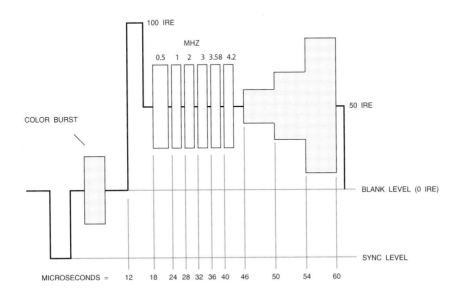

Figure 8.43. NTC-7 Combination Test Signal for NTSC.

The line bar has a peak amplitude of 100 ±1 IRE and a width of 10 µs.

The 2T pulse has a peak amplitude of 100 ±0.5 IRE, with a half-amplitude width of 200 ±10 ns.

The 20T chrominance pulse has a peak amplitude of 100 ±0.5 IRE, with a half-amplitude width of 2.0 ±0.06 µs.

The 5-step luminance staircase signal consists of 5 luminance steps, at 20, 40, 60, 80 and 100 ±0.5 IRE.

The ITU combination test signal may be present on line 17.

ITU ITS Version for PAL

The ITU (BT.473) has developed a combination ITS (insertion test signal) that may be used to test several PAL video parameters, rather than using multiple test signals. The ITU combination ITS for PAL systems (shown in Figure 8.45) consists of a 3-step modulated pedestal with peak-to-peak amplitudes of 20,

60, and 100 ±1 IRE, and an extended subcarrier packet with a peak-to-peak amplitude of 60 ±1 IRE. The rise and fall times of each subcarrier packet envelope are approximately 1 µs. The phase of each subcarrier packet is 60° ±1° relative to the U axis. The tolerance on the 50 IRE level is ±1 IRE.

The ITU composite ITS may be present on line 331.

U. K. Version

The United Kingdom allows the use of a slightly different test signal, as shown in Figure 8.46. It may be present on lines 20 and 333 for (I) PAL systems in the United Kingdom.

The test signal consists of a 50 IRE luminance bar, part of which has a 100 IRE subcarrier superimposed that has a phase of 60° ±1° relative to the U axis, and an extended burst of subcarrier on the second half of the scan line.

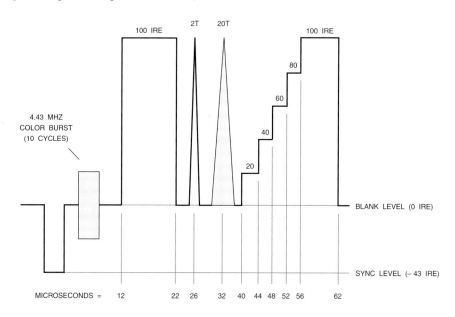

Figure 8.44. ITU Combination Test Signal for PAL.

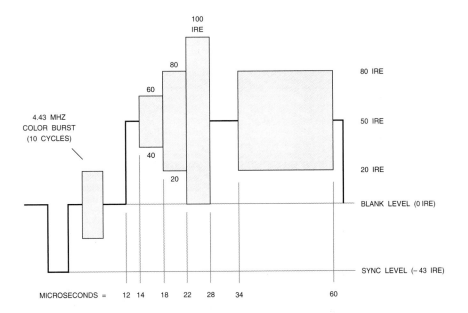

Figure 8.45. ITU Combination ITS Test Signal for PAL.

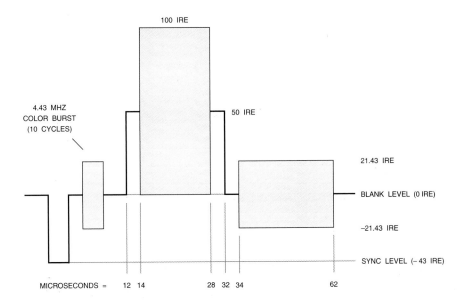

Figure 8.46. United Kingdom (I) PAL National Test Signal #2.

T Pulse

Square waves with fast rise times cannot be used for testing video systems, since attenuation and phase shift of out-of-band components cause ringing in the output signal, obscuring the in-band distortions being measured. T, or $\sin^2$, pulses are bandwidth-limited, so are used for testing video systems.

The 2T pulse is shown in Figure 8.47 and, like the T pulse, is obtained mathematically by squaring a half-cycle of a sine wave. T pulses are specified in terms of half amplitude duration (HAD), which is the pulse width measured at 50% of the pulse amplitude. Pulses with HADs that are multiples of the time interval T are used to test video systems. As seen in Figures 8.39 through 8.44, T, 2T, 12.5T and 25T pulses are common when testing NTSC video systems, whereas T, 2T, 10T, and 20T pulses are common for PAL video systems.

T is the Nyquist interval or

$$1/2F_C$$

where F_C is the cutoff frequency of the video system. For NTSC, F_C is 4 MHz, whereas F_C for PAL systems is 5 MHz. Therefore, T for NTSC systems is 125 ns and for PAL systems it is 100 ns. For a T pulse with a HAD of 125 ns, a 2T pulse has a HAD of 250 ns, and so on. The frequency spectra for the 2T pulse is shown in Figure 8.47 and is representative of the energy content in a typical character generator waveform.

To generate smooth rising and falling edges of most video signals, a T step (generated by integrating a T pulse) is typically used.

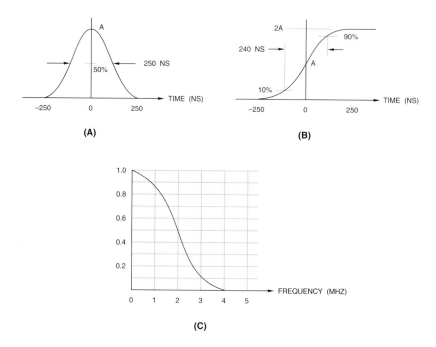

Figure 8.47. The T Pulse. (a) 2T pulse. (b) 2T step. (c) Frequency spectra of the 2T pulse.

T steps have 10–90% rise/fall times of 0.964T and a well-defined bandwidth. The 2T step generated from a 2T pulse is shown in Figure 8.47.

The 12.5T chrominance pulse, illustrated in Figure 8.48, is a good test signal to measure any chrominance-to-luminance timing error since its energy spectral distribution is bunched in two relatively narrow bands. Using this signal detects differences in the luminance and chrominance phase distortion, but not between other frequency groups.

VBI Data

VBI (vertical blanking interval) data may be inserted up to about five scan lines into the active picture region to ensure it won't be deleted by equipment replacing the VBI, by DSS MPEG which deletes the VBI, or by cable systems inserting their own VBI data. This is common practice by Neilsen and others to ensure their programming and commercial tracking data gets through the distribution systems to the receivers. In most cases, this will be unseen since it is masked by the TV's overscan.

Timecode

Two types of time coding are commonly used, as defined by ANSI/SMPTE 12M and IEC 461: longitudinal timecode (LTC) and vertical interval timecode (VITC).

The LTC is recorded on a separate audio track; as a result, the analog VCR must use

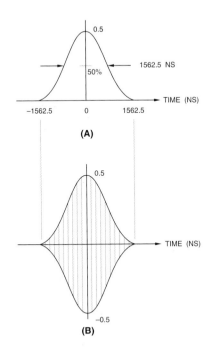

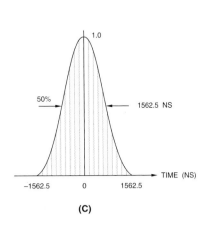

Figure 8.48. The 12.5T Chrominance Pulse. (a) Luma component. (b) Chroma component. (c) Addition of (a) and (b).

high-bandwidth amplifiers and audio heads. This is due to the time code frequency increasing as tape speed increases, until the point that the frequency response of the system results in a distorted time code signal that may not be read reliably. At slower tape speeds, the time code frequency decreases, until at very low tape speeds or still pictures, the time code information is no longer recoverable.

The VITC is recorded as part of the video signal; as a result, the time code information is always available, regardless of the tape speed. However, the LTC allows the time code signal to be written without writing a video signal; the VITC requires the video signal to be changed if a change in time code information is required. The LTC therefore is useful for synchronizing multiple audio or audio/video sources.

Frame Dropping

If the field rate is 60/1.001 fields per second, straight counting at 60 fields per second yields an error of about 108 frames for each hour of running time. This may be handled in one of three ways:

> Nondrop frame: During a continuous recording, each time count increases by 1 frame. In this mode, the drop frame flag will be a "0."

> Drop frame: To minimize the timing error, the first two frame numbers (00 and 01) at the start of each minute, except for minutes 00, 10, 20, 30, 40, and 50, are omitted from the count. In this mode, the drop frame flag will be a "1."

> Drop frame for (M) PAL: To minimize the timing error, the first four frame numbers (00 to 03) at the start of every second minute (even minute numbers) are omitted from the count, except for minutes 00, 20, and 40. In this mode, the drop frame flag will be a "1."

Even with drop framing, there is a long-term error of about 2.26 frames per 24 hours. This error accumulation is the reason time-code generators must be periodically reset if they are to maintain any correlation to the correct time-of-day. Typically, this "reset-to-real-time" is referred to as a "jam sync" procedure. Some jam sync implementations reset the timecode to 00:00:00.00 and, therefore, must occur at midnight; others allow a true re-sync to the correct time-of-day.

One inherent problem with jam sync correction is the interruption of the timecode. Although this discontinuity may be brief, it may cause timecode readers to "hiccup" due to the interruption.

Longitudinal Timecode (LTC)

The LTC information is transferred using a separate serial interface, using the same electrical interface as the AES/EBU digital audio interface standard, and is recorded on a separate track. The basic structure of the time data is based on the BCD system. Tables 8.20 and 8.21 list the LTC bit assignments and arrangement. Note that the 24-hour clock system is used.

LTC Timing

The modulation technique is such that a transition occurs at the beginning of every bit period. "1" is represented by a second transition one-half a bit period from the start of the bit. "0" is represented when there is no transition within the bit period (see Figure 8.49). The signal has a peak-to-peak amplitude of 0.5–4.5V, with rise and fall times of 40 ±10 µs (10% to 90% amplitude points).

Because the entire frame time is used to generate the 80-bit LTC information, the bit rate (in bits per second) is determined by:

Bit(s)	Function	Note	Bit(s)	Function	Note
0–3	units of frames		58	flag 5	note 5
4–7	user group 1		59	flag 6	note 6
8–9	tens of frames		60–63	user group 8	
10	flag 1	note 1	64	sync bit	fixed"0"
11	flag 2	note 2	65	sync bit	fixed"0"
12–15	user group 2		66	sync bit	fixed"1"
16–19	units of seconds		67	sync bit	fixed"1"
20–23	user group 3		68	sync bit	fixed"1"
24–26	tens of seconds		69	sync bit	fixed"1"
27	flag 3	note 3	70	sync bit	fixed"1"
28–31	user group 4		71	sync bit	fixed"1"
32–35	units of minutes		72	sync bit	fixed"1"
36–39	user group 5		73	sync bit	fixed"1"
40–42	tens of minutes		74	sync bit	fixed"1"
43	flag 4	note 4	75	sync bit	fixed"1"
44–47	user group 6		76	sync bit	fixed"1"
48–51	units of hours		77	sync bit	fixed"1"
52–55	user group 7		78	sync bit	fixed"0"
56–57	tens of hours		79	sync bit	fixed"1"

Notes:
1. Drop frame flag. 525-line and 1125-line systems: "1" if frame numbers are being dropped, "0" if no frame dropping is done. 625-line systems: "0."
2. Color frame flag. 525-line systems: "1" if even units of frame numbers identify fields 1 and 2 and odd units of field numbers identify fields 3 and 4. 625-line systems: "1" if timecode is locked to the video signal in accordance with 8-field sequence and the video signal has the "preferred subcarrier-to-line-sync phase." 1125-line systems: "0."
3. 525-line and 1125-line systems: Phase correction. This bit shall be put in a state so that every 80-bit word contains an even number of "0"s. 625-line systems: Binary group flag 0.
4. 525-line and 1125-line systems: Binary group flag 0. 625-line systems: Binary group flag 2.
5. Binary group flag 1.
6. 525-line and 1125-line systems: Binary group flag 2. 625-line systems: Phase correction. This bit shall be put in a state so that every 80-bit word contains an even number of "0"s.

Table 8.20. LTC Bit Assignments.

Frames (count 0–29 for 525-line and 1125-line systems, 0–24 for 625-line systems)	
units of frames (bits 0–3)	4-bit BCD (count 0–9); bit 0 is LSB
tens of frames (bits 8–9)	2-bit BCD (count 0–2); bit 8 is LSB

Seconds	
units of seconds (bits 16–19)	4-bit BCD (count 0–9); bit 16 is LSB
tens of seconds (bits 24–26)	3-bit BCD (count 0–5); bit 24 is LSB

Minutes	
units of minutes (bits 32–35)	4-bit BCD (count 0–9); bit 32 is LSB
tens of minutes (bits 40–42)	3-bit BCD (count 0–5); bit 40 is LSB

Hours	
units of hours (bits 48–51)	4-bit BCD (count 0–9); bit 48 is LSB
tens of hours (bits 56–57)	2-bit BCD (count 0–2); bit 56 is LSB

Table 8.21. LTC Bit Arrangement.

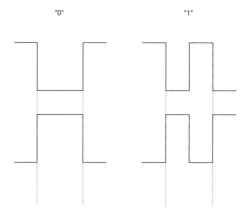

Figure 8.49. LTC Data Bit Transition Format.

$$F_C = 80 \, F_V$$

where F_V is the vertical frame rate in frames per second. The 80 bits of time code information are output serially, with bit 0 being first. The LTC word occupies the entire frame time, and the data must be evenly spaced throughout this time. The start of the LTC word occurs at the beginning of line 5 ±1.5 lines for 525-line systems, at the beginning of line 2 ±1.5 lines for 625-line systems, and at the vertical sync timing reference of the frame ±1 line for 1125-line systems.

Vertical Interval Time Code (VITC)

The VITC is recorded during the vertical blanking interval of the video signal in both fields. Since it is recorded with the video, it can be read in still mode. However, it cannot be re-recorded (or restriped). Restriping requires dubbing down a generation, deleting and inserting a new time code. For YPbPr and S-video interfaces, VITC is present on the Y signal. For analog RGB interfaces, VITC is present on all three signals.

As with the LTC, the basic structure of the time data is based on the BCD system. Tables 8.22 and 8.23 list the VITC bit assignments and arrangement. Note that the 24-hour clock system is used.

VITC Cyclic Redundancy Check

Eight bits (82–89) are reserved for the code word for error detection by means of cyclic redundancy checking. The generating polynomial, $x^8 + 1$, applies to all bits from 0 to 81, inclusive. Figure 8.50 illustrates implementing the polynomial using a shift register. During passage of timecode data, the multiplexer is in position 0 and the data is output while the CRC calculation is done simultaneously by the shift register. After all the timecode data has been output, the shift register contains the CRC value, and switching the multiplexer to position 1 enables the CRC value to be output. Repeating the process on decoding, the shift register contains all zeros if no errors exist.

VITC Timing

The modulation technique is such that each state corresponds to a binary state, and a transition occurs only when there is a change in the data between adjacent bits from a "1" to "0" or "0" to "1." No transitions occur when adjacent bits contain the same data. This is commonly referred to as "non-return to zero" (NRZ). Synchronization bit pairs are inserted throughout the VITC data to assist the receiver in maintaining the correct frequency lock.

The bit rate (F_C) is defined to be:

$$F_C = 115 \, F_H \pm 2\%$$

where F_H is the horizontal line frequency. The 90 bits of time code information are output serially, with bit 0 being first. For 625i (576i) systems, lines 19 and 332 (or 21 and 334) are commonly used for the VITC. For 525i (480i) systems, lines 14 and 277 are commonly used. For 1125i (1080i) systems, lines 9 and 571 are commonly used. To protect the VITC against drop-outs, it may also be present two scan lines later, although any two nonconsecutive scan lines per field may be used.

Figure 8.51 illustrates the timing of the VITC data on the scan line. The data must be evenly spaced throughout the VITC word. The 10% to 90% rise and fall times of the VITC bit data should be 200 ±50 ns (525-line and 625-line systems) or 100 ±25 ns (1125-line systems) before adding it to the video signal to avoid possible distortion of the VITC signal by downstream chrominance circuits. In most circumstances, the analog lowpass filters after the video D/A converters should suffice for the filtering.

Bit(s)	Function	Note	Bit(s)	Function	Note
0	sync bit	fixed "1"	42–45	units of minutes	
1	sync bit	fixed "0"	46–49	user group 5	
2–5	units of frames		50	sync bit	fixed "1"
6–9	user group 1		51	sync bit	fixed "0"
10	sync bit	fixed "1"	52–54	tens of minutes	
11	sync bit	fixed "0"	55	flag 4	note 4
12–13	tens of frames		56–59	user group 6	
14	flag 1	note 1	60	sync bit	fixed "1"
15	flag 2	note 2	61	sync bit	fixed"0"
16–19	user group 2		62–65	units of hours	
20	sync bit	fixed "1"	66–69	user group 7	
21	sync bit	fixed "0"	70	sync bit	fixed"1"
22–25	units of seconds		71	sync bit	fixed"0"
26–29	user group 3		72–73	tens of hours	
30	sync bit	fixed "1"	74	flag 5	note 5
31	sync bit	fixed "0"	75	flag 6	note 6
32–34	tens of seconds		76–79	user group 8	
35	flag 3	note 3	80	sync bit	fixed"1"
36–39	user group 4		81	sync bit	fixed"0"
40	sync bit	fixed "1"	82–89	CRC group	
41	sync bit	fixed "0"			

Notes:
1. Drop frame flag. 525-line and 1125-line systems: "1" if frame numbers are being dropped, "0" if no frame dropping is done. 625-line systems: "0."
2. Color frame flag. 525-line systems: "1" if even units of frame numbers identify fields 1 and 2 and odd units of field numbers identify fields 3 and 4. 625-line systems: "1" if timecode is locked to the video signal in accordance with 8-field sequence and the video signal has the "preferred subcarrier-to-line-sync phase." 1125-line systems: "0."
3. 525-line systems: Field flag. "0" during fields 1 and 3, "1" during fields 2 and 4. 625-line systems: Binary group flag 0. 1125-line systems: Field flag. "0" during field 1, "1" during field 2.
4. 525-line and 1125-line systems: Binary group flag 0. 625-line systems: Binary group flag 2.
5. Binary group flag 1.
6. 525-line and 1125-line systems: Binary group flag 2. 625-line systems: Field flag. "0" during fields 1, 3, 5, and 7, "1" during fields 2, 4, 6, and 8.

Table 8.22. VITC Bit Assignments.

Frames (count 0–29 for 525-line and 1125-line systems, 0–24 for 625-line systems)	
units of frames (bits 2–5)	4-bit BCD (count 0–9); bit 2 is LSB
tens of frames (bits 12–13)	2-bit BCD (count 0–2); bit 12 is LSB

Seconds	
units of seconds (bits 22–25)	4-bit BCD (count 0–9); bit 22 is LSB
tens of seconds (bits 32–34)	3-bit BCD (count 0–5); bit 32 is LSB

Minutes	
units of minutes (bits 42–45)	4-bit BCD (count 0–9); bit 42 is LSB
tens of minutes (bits 52–54)	3-bit BCD (count 0–5); bit 52 is LSB

Hours	
units of hours (bits 62–65)	4-bit BCD (count 0–9); bit 62 is LSB
tens of hours (bits 72–73)	2-bit BCD (count 0–2); bit 72 is LSB

Table 8.23. VITC Bit Arrangement.

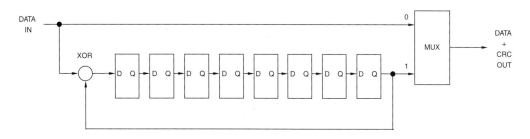

Figure 8.50. VITC CRC Generation.

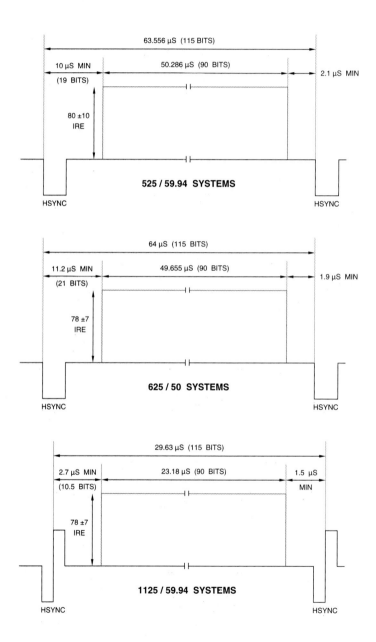

Figure 8.51. VITC Position and Timing.

User Bits Content	Timecode Referenced to External Clock	BGF2	BGF1	BGF0
user defined	no	0	0	0
8-bit character set[1]	no	0	0	1
user defined	yes	0	1	0
reserved	unassigned	0	1	1
date and time zone[3]	no	1	0	0
page / line[2]	no	1	0	1
date and time zone[3]	yes	1	1	0
page / line[2]	yes	1	1	1

Notes:
1. Conforming to ISO/IEC 646 or 2022.
2. Described in SMPTE 262M.
3. Described in SMPTE 309M. See Tables 8.25 through 8.27.

Table 8.24. LTC and VITC Binary Group Flag (BGF) Bit Definitions.

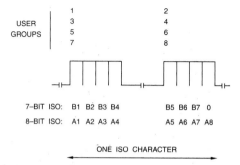

**Figure 8.52. Use of Binary Groups to Describe
ISO Characters Coded With 7 or 8 Bits.**

User Group 8				User Group 7			
Bit 3	Bit 2	Bit 1	Bit 0	Bit 3	Bit 2	Bit 1	Bit 0
MJD Flag	0			time zone offset code 00_H–$3F_H$			

Notes:
1. MJD flag: "0" = YYMMDD format, "1" = MJD format.

Table 8.25. Date and Time Zone Format Coding.

User Group	Assignment	Value	Description
1	D	0–9	day units
2	D	0–3	day units
3	M	0–9	month units
4	M	0, 1	month units
5	Y	0–9	year units
6	Y	0–9	year units

Table 8.26. YYMMDD Date Format.

User Bits

The binary group flag (BGF) bits shown in Table 8.24 specify the content of the 32 user bits. The 32 user bits are organized as eight groups of four bits each.

The user bits are intended for storage of data by users. The 32 bits may be assigned in any manner without restriction, if indicated as user-defined by the binary group flags.

If an 8-bit character set conforming to ISO/IEC 646 or 2022 is indicated by the binary group flags, the characters are to be inserted as shown in Figure 8.52. Note that some user bits will be decoded before the binary group flags are decoded; therefore, the decoder must store the early user data before any processing is done.

When the user groups are used to transfer time zone and date information, user groups 7 and 8 specify the time zone and the format of the date in the remaining six user groups, as shown in Tables 8.25 and 8.27. The date may be either a six-digit YYMMDD format (Table 8.26) or a six-digit modified Julian date (MJD), as indicated by the MJD flag.

EIA–608 Closed Captioning

This section reviews EIA–608 closed captioning for the hearing impaired in the United States. Closed captioning and text are transmitted during the blanked active line-time portion of lines 21 and 284. However, due to video editing they may occasionally reside on any line between 21–25 and 284–289.

Code	Hours	Code	Hours	Code	Hours
00	UTC	16	UTC + 10.00	2C	UTC + 09.30
01	UTC − 01.00	17	UTC + 09.00	2D	UTC + 08.30
02	UTC − 02.00	18	UTC + 08.00	2E	UTC + 07.30
03	UTC − 03.00	19	UTC + 07.00	2F	UTC + 06.30
04	UTC − 04.00	1A	UTC − 06.30	30	TP–1
05	UTC − 05.00	1B	UTC − 07.30	31	TP–0
06	UTC − 06.00	1C	UTC − 08.30	32	UTC + 12.45
07	UTC − 07.00	1D	UTC − 09.30	33	reserved
08	UTC − 08.00	1E	UTC − 10.30	34	reserved
09	UTC − 09.00	1F	UTC − 11.30	35	reserved
0A	UTC − 00.30	20	UTC + 06.00	36	reserved
0B	UTC − 01.30	21	UTC + 05.00	37	reserved
0C	UTC − 02.30	22	UTC + 04.00	38	user defined
0D	UTC − 03.30	23	UTC + 03.00	39	unknown
0E	UTC − 04.30	24	UTC + 02.00	3A	UTC + 05.30
0F	UTC − 05.30	25	UTC + 01.00	3B	UTC + 04.30
10	UTC − 10.00	26	reserved	3C	UTC + 03.30
11	UTC − 11.00	27	reserved	3D	UTC + 02.30
12	UTC − 12.00	28	TP–3	3E	UTC + 01.30
13	UTC + 13.00	29	TP–2	3F	UTC + 00.30
14	UTC + 12.00	2A	UTC + 11.30		
15	UTC + 11.00	2B	UTC + 10.30		

Table 8.27. Time Zone Offset Codes.

Extended data services (XDS) also may be transmitted during the blanked active line-time portion of line 284. XDS may indicate the program name, time into the show, time remaining to the end, and so on.

Note that due to editing before transmission, it may be possible that the caption information is occasionally moved down a scan line or two. Therefore, caption decoders should monitor more than just lines 21 and 284 for caption information.

Waveform

The data format for both lines consists of a clock run-in signal, a start bit, and two 7-bit plus parity words of ASCII data (per X3.4-1967). For YPbPr and S-video interfaces, captioning is present on the Y signal. For analog RGB interfaces, captioning is present on all three signals.

Figure 8.53 illustrates the waveform and timing for transmitting the closed captioning and XDS information and conforms to the Television Synchronizing Waveform for Color Transmission in Subpart E, Part 73 of the FCC Rules and Regulations and EIA-608. The clock run-in is a 7-cycle sinusoidal burst that is frequency-locked and phase-locked to the caption data and is used to provide synchronization for the decoder. The nominal data rate is $32\times F_H$. However, decoders should not rely on this timing relationship due to possible horizontal timing variations introduced by video processing circuitry and VCRs. After the clock run-in signal, the blanking level is maintained for a two data bit duration, followed by a "1" start bit.

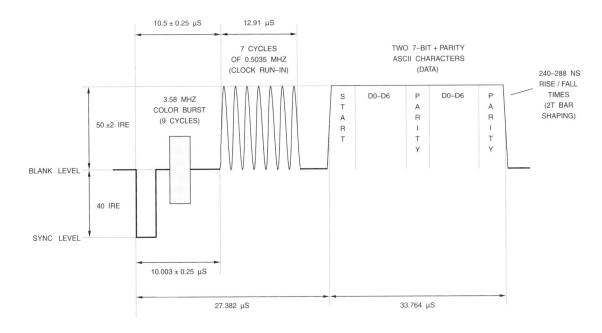

Figure 8.53. 525-Line Lines 21 and 284 Closed Captioning Timing.

The start bit is followed by 16 bits of data, composed of two 7-bit + odd parity ASCII characters. Caption data is transmitted using a non–return-to-zero (NRZ) code; a "1" corresponds to the 50 ± 2 IRE level and a "0" corresponds to the blanking level (0–2 IRE). The negative-going crossings of the clock are coherent with the data bit transitions.

Typical decoders specify the time between the 50% points of sync and clock run-in to be 10.5 ± 0.5 µs, with a $\pm 3\%$ tolerance on F_H, 50 ± 12 IRE for a "1" bit, and –2 to +12 IRE for a "0" bit. Decoders must also handle bit rise/fall times of 240–480 ns.

NUL characters (00_H) should be sent when no display or control characters are being transmitted. This, in combination with the clock run-in, enables the decoder to determine whether captioning or text transmission is being implemented.

If using only line 21, the clock run-in and data do not need to be present on line 284. However, if using only line 284, the clock run-in and data should be present on both lines 21 and 284; data for line 21 would consist of NUL characters.

At the decoder, as shown in Figure 8.54, the display area of a 525-line 4:3 interlaced display is typically 15 rows high and 34 columns wide. The vertical display area begins on lines 43 and 306 and ends on lines 237 and 500. The horizontal display area begins 13 µs and ends 58 µs, after the leading edge of horizontal sync.

In text mode, all rows are used to display text; each row contains a maximum of 32 characters, with at least a one-column wide space on the left and right of the text. The only transparent area is around the outside of the text area.

In caption mode, text usually appears only on rows 1–4 or 12–15; the remaining rows are usually transparent. Each row contains a maximum of 32 characters, with at least a one-column wide space on the left and right of the text.

Some caption decoders support up to 48 columns per row, and up to 16 rows, allowing some customization for the display of caption data.

Basic Services

There are two types of display formats: text and captioning. In understanding the operation of the decoder, it is easier to visualize an invisible cursor that marks the position where the next character will be displayed. Note that if you are designing a decoder, you should obtain the latest FCC Rules and Regulations and EIA-608 to ensure correct operation, as this section is only a summary.

Text Mode

Text mode uses 7–15 rows of the display and is enabled upon receipt of the Resume Text Display or Text Restart code. When text mode has been selected, and the text memory is empty, the cursor starts at the top-most row, character 1 position. Once all the rows of text are displayed, scrolling is enabled.

With each carriage return received, the top-most row of text is erased, the text is rolled up one row (over a maximum time of 0.433 seconds), the bottom row is erased, and the cursor is moved to the bottom row, character 1 position. If new text is received while scrolling, it is seen scrolling up from the bottom of the display area. If a carriage return is received while scrolling, the rows are immediately moved up one row to their final position.

Once the cursor moves to the character 32 position on any row, any text received before a carriage return, preamble address code, or backspace will be displayed at the character 32 position, replacing any previous character at that position. The Text Restart command erases all characters on the display and moves the cursor to the top row, character 1 position.

Captioning Mode

Captioning has several modes available, including roll-up, pop-on, and paint-on.

Roll-up captioning is enabled by receiving one of the miscellaneous control codes to select the number of rows displayed. "Roll-up captions, 2 rows" enables rows 14 and 15; "roll-up captions, 3 rows" enables rows 13–15, "roll-up captions, 4 rows" enables rows 12–15. Regardless of the number of rows enabled, the cursor remains on row 15. Once row 15 is full, the rows are scrolled up one row (at the rate of one dot per frame), and the cursor is moved back to row 15, character 1.

Pop-on captioning may use rows 1–4 or 12–15, and is initiated by the Resume Caption Loading command. The display memory is essentially double-buffered. While memory buffer 1 is displayed, memory buffer 2 is being loaded with caption data. At the receipt of a End of Caption code, memory buffer 2 is displayed while memory buffer 1 is being loaded with new caption data.

Paint-on captioning, enabled by the Resume Direct Captioning command, is similar to Pop-on captioning, but no double-buffering is used; caption data is loaded directly into display memory.

Three types of control codes (preamble address codes, midrow codes, and miscellaneous control codes) are used to specify the format, location, and attributes of the charac-

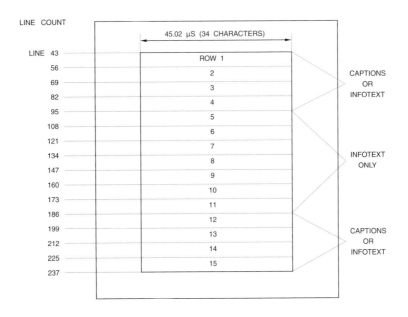

Figure 8.54. Closed Captioning Display Format.

ters. Each control code consists of two bytes, transmitted together on line 21 or line 284. On line 21, they are normally transmitted twice in succession to help ensure correct reception. They are not transmitted twice on line 284 to minimize bandwidth used for captioning.

The first byte is a nondisplay control byte with a range of 10_H to $1F_H$; the second byte is a display control byte in the range of 20_H to $7F_H$. At the beginning of each row, a control code is sent to initialize the row. Caption roll-up and text modes allow either a preamble address code or midrow control code at the start of a row; the other caption modes use a preamble address code to initialize a row. The preamble address codes are illustrated in Figure 8.55 and Table 8.28.

The midrow codes are typically used within a row to change the color, italics, underline, and flashing attributes and should occur only between words. Color, italics, and underline are controlled by the preamble address and midrow codes; flash on is controlled by a miscellaneous control code. An attribute remains in effect until another control code is received or the end of row is reached. Each row starts with a control code to set the color and underline attributes (white nonunderlined is the default if no control code is received before the first character on an empty row). The color attribute can be changed only by the midrow code of another color; the italics attribute does not change the color attribute. However, a color attribute turns off the italics attribute. The flash on command does not alter the status of the color, italics, or underline attributes. However, a color or italics midrow control code turns off the flash. Note that the underline color is the same color as the character being underlined; the underline resides on dot row 11 and covers the entire width of the character column.

Table 8.29, Figure 8.56, and Table 8.30 illustrate the midrow and miscellaneous control code operation. For example, if it were the end of a caption, the control code could be End

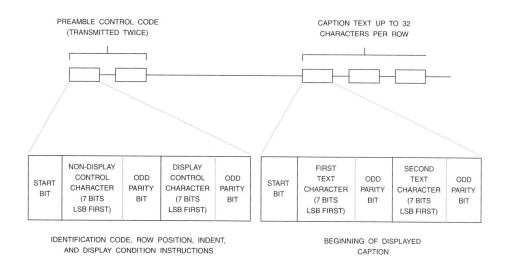

Figure 8.55. Closed Captioning Preamble Address Code Format.

Non-display Control Byte							Display Control Byte							Row Position
D6	D5	D4	D3	D2	D1	D0	D6	D5	D4	D3	D2	D1	D0	
0	0	1	CH	0	0	1	1	0	A	B	C	D	U	1
							1	1						2
				0	1	0	1	0						3
							1	1						4
				1	0	1	1	0						5
							1	1						6
				1	1	0	1	0						7
							1	1						8
				1	1	1	1	0						9
							1	1						10
				0	0	0	1	0						11
				0	1	1	1	0						12
							1	1						13
				1	0	0	1	0						14
							1	1						15

Notes:
1. U: "0" = no underline, "1" = underline.
2. CH: "0" = data channel 1, "1" = data channel 2.

A	B	C	D	Attribute
0	0	0	0	white
0	0	0	1	green
0	0	1	0	blue
0	0	1	1	cyan
0	1	0	0	red
0	1	0	1	yellow
0	1	1	0	magenta
0	1	1	1	italics
1	0	0	0	indent 0, white
1	0	0	1	indent 4, white
1	0	1	0	indent 8, white
1	0	1	1	indent 12, white
1	1	0	0	indent 16, white
1	1	0	1	indent 20, white
1	1	1	0	indent 24, white
1	1	1	1	indent 28, white

Table 8.28. Closed Captioning Preamble Address Codes. In text mode, the indent codes may be used to perform indentation; in this instance, the row information is ignored.

of Caption (transmitted twice). It could be followed by a preamble address code (transmitted twice) to start another line of captioning.

Characters are displayed using a dot matrix format. Each character cell is typically 16 samples wide and 26 samples high (16 × 26), as shown in Figure 8.57. Dot rows 2–19 are usually used for actual character outlines. Dot rows 0, 1, 20, 21, 24, and 25 are usually blanked to provide vertical spacing between characters, and underlining is typically done on dot rows 22 and 23. Dot columns 0, 1, 14 and 15 are blanked to provide horizontal spacing between characters, except on dot rows 22 and 23 when the underline is displayed. This results in 12 × 18 characters stored in character ROM. Table 8.31 shows the basic character set.

Some caption decoders support multiple character sizes within the 16 × 26 region, including 13 × 16, 13 × 24, 12 × 20, and 12 × 26.

Not all combinations generate a sensible result due to the limited display area available.

Optional Captioning Features

Three sets of optional features are available for advanced captioning decoders.

Optional Attributes

Additional color choices are available for advanced captioning decoders, as shown in Table 8.32.

If a decoder doesn't support semitransparent colors, the opaque colors may be used instead. If a specific background color isn't supported by a decoder, it should default to the black background color. However, if the black foreground color is supported in a decoder, all the background colors should be implemented.

A background attribute appears as a standard space on the display, and the attribute

Non-display Control Byte							Display Control Byte							Attribute
D6	D5	D4	D3	D2	D1	D0	D6	D5	D4	D3	D2	D1	D0	
0	0	1	CH	0	0	1	0	1	0	0	0	0	U	white
										0	0	1		green
										0	1	0		blue
										0	1	1		cyan
										1	0	0		red
										1	0	1		yellow
										1	1	0		magenta
										1	1	1		italics

Notes:
1. U: "0" = no underline, "1" = underline.
2. CH: "0" = data channel 1, "1" = data channel 2.
3. Italics is implemented as a two-dot slant to the right over the vertical range of the character. Some decoders implement a one dot slant for every four scan lines. Underline resides on dot rows 22 and 23, and covers the entire column width.

Table 8.29. Closed Captioning Midrow Codes.

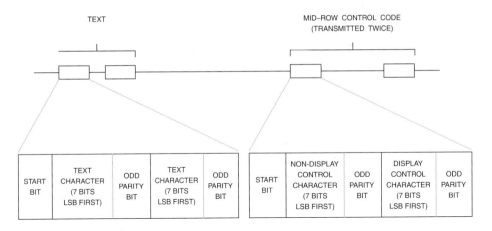

Figure 8.56. Closed Captioning Midrow Code Format. Miscellaneous control codes may also be transmitted in place of the midrow control code.

Non-display Control Byte							Display Control Byte							Command
D6	D5	D4	D3	D2	D1	D0	D6	D5	D4	D3	D2	D1	D0	
										0	0	0	0	resume caption loading
										0	0	0	1	backspace
										0	0	1	0	reserved
										0	0	1	1	reserved
										0	1	0	0	delete to end of row
										0	1	0	1	roll-up captions, 2 rows
										0	1	1	0	roll-up captions, 3 rows
										0	1	1	1	roll-up captions, 4 rows
0	0	1	CH	1	0	F	0	1	0	1	0	0	0	flash on
										1	0	0	1	resume direct captioning
										1	0	1	0	text restart
										1	0	1	1	resume text display
										1	1	0	0	erase displayed memory
										1	1	0	1	carriage return
										1	1	1	0	erase nondisplayed memory
										1	1	1	1	end of caption (flip memories)
										0	0	0	1	tab offset (1 column)
0	0	1	CH	1	1	1	0	1	0	0	0	1	0	tab offset (2 columns)
										0	0	1	1	tab offset (3 columns)

Notes:
1. F: "0" = line 21, "1" = line 284. CH: "0" = data channel 1, "1" = data channel 2.
2. "Flash on" blanks associated characters for 0.25 seconds once per second.

Table 8.30. Closed Captioning Miscellaneous Control Codes.

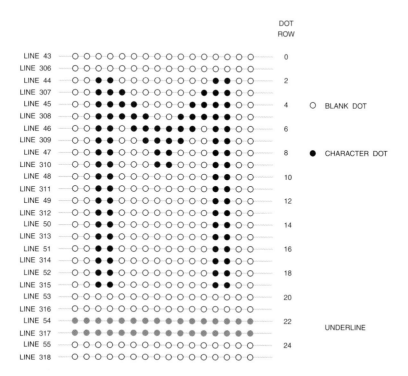

Figure 8.57. Typical 16×26 Closed Captioning Character Cell Format for Row 1.

Nondisplay Control Byte							Display Control Byte							Special Characters
D6	D5	D4	D3	D2	D1	D0	D6	D5	D4	D3	D2	D1	D0	
0	0	1	CH	0	0	1	0	1	1	0	0	0	0	®
										0	0	0	1	°
										0	0	1	0	1/2
										0	0	1	1	¿
										0	1	0	0	™
										0	1	0	1	¢
										0	1	1	0	£
										0	1	1	1	music note
										1	0	0	0	à
										1	0	0	1	transparent space
										1	0	1	0	è
										1	0	1	1	â
										1	1	0	0	ê
										1	1	0	1	î
										1	1	1	0	ô
										1	1	1	1	û

D6 D5 D4 D3 → D2 D1 D0 ↓	0100	0101	0110	0111	1000	1001	1010	1011	1100	1101	1110	1111
000		(	0	8	@	H	P	X	ú	h	p	x
001	!	)	1	9	A	I	Q	Y	a	i	q	y
010	"	á	2	:	B	J	R	Z	b	j	r	z
011	#	+	3	;	C	K	S	[	c	k	s	ç
100	$	,	4	<	D	L	T	é	d	l	t	÷
101	%	–	5	=	E	M	U	]	e	m	u	Ñ
110	&	.	6	>	F	N	V	í	f	n	v	ñ
111	'	/	7	?	G	O	W	ó	g	o	w	

Table 8.31. Closed Captioning Basic Character Set.

Non-display Control Byte							Display Control Byte							Background Attribute
D6	D5	D4	D3	D2	D1	D0	D6	D5	D4	D3	D2	D1	D0	
0	0	1	CH	0	0	0	0	1	0	0	0	0	T	white
										0	0	1		green
										0	1	0		blue
										0	1	1		cyan
										1	0	0		red
										1	0	1		yellow
										1	1	0		magenta
										1	1	1		black
0	0	1	CH	1	1	1	0	1	0	1	1	0	1	transparent
D6	**D5**	**D4**	**D3**	**D2**	**D1**	**D0**	**D6**	**D5**	**D4**	**D3**	**D2**	**D1**	**D0**	**Foreground Attribute**
0	0	1	CH	1	1	1	0	1	0	1	1	1	0	black
													1	black underline

Notes:
1. F: "0" = opaque, "1" = semi-transparent.
2. CH: "0" = data channel 1, "1" = data channel 2.
3. Underline resides on dot rows 22 and 23, and covers the entire column width.

Table 8.32. Closed Captioning Optional Attribute Codes.

remains in effect until the end of the row or until another background attribute is received.

The foreground attributes provide an eighth color (black) as a character color. As with midrow codes, a foreground attribute code turns off italics and blinking, and the least significant bit controls underlining.

Background and foreground attribute codes have an automatic backspace for backward compatibility with current decoders. Thus, an attribute must be preceded by a standard space character. Standard decoders display the space and ignore the attribute. Extended decoders display the space, and on receiving the attribute, backspace, then display a space that changes the color and opacity. Thus, text formatting remains the same regardless of the type of decoder.

Optional Closed Group Extensions

To support custom features and characters not defined by the standards, the EIA/CEG maintains a set of code assignments requested by various caption providers and decoder manufacturers. These code assignments (currently used to select various character sets) are not compatible with caption decoders in the United States and videos using them should not be distributed in the U. S. market.

Closed group extensions require two bytes. Table 8.33 lists the currently assigned closed group extensions to support captioning in the Asian languages.

Optional Extended Characters

An additional 64 accented characters (eight character sets of eight characters each) may be supported by decoders, permitting the display of other languages such as Spanish, French, Portuguese, German, Danish, Italian, Finnish, and Swedish. If supported, these accented characters are available in all caption and text modes.

Each of the extended characters incorporates an automatic backspace for backward compatibility with current decoders. Thus, an extended character must be preceded by the standard ASCII version of the character. Standard decoders display the ASCII character and ignore the accented character. Extended decoders display the ASCII character, and on receiving the accented character, backspace, then display the accented character. Thus, text formatting remains the same regardless of the type of decoder.

Extended characters require two bytes. The first byte is 12_H or 13_H for data channel one ($1A_H$ or $1B_H$ for data channel two), followed by a value of 20_H–$3F_H$.

Extended Data Services

Line 284 may contain extended data service information, interleaved with the caption and text information, as bandwidth is available. In this case, control codes are not transmitted twice, as they may be for the caption and text services.

Information is transmitted as packets and operates as a separate unique data channel. Data for each packet may or may not be contig-

Non-display Control Byte							Display Control Byte							Background Attribute
D6	D5	D4	D3	D2	D1	D0	D6	D5	D4	D3	D2	D1	D0	
										0	1	0	0	standard character set (normal size)
										0	1	0	1	standard character set (double size)
										0	1	1	0	first private character set
0	0	1	CH	1	1	1	0	1	0	0	1	1	1	second private character set
										1	0	0	0	People's Republic of China character set (GB 2312)
										1	0	0	1	Korean Standard character set (KSC 5601-1987)
										1	0	1	0	first registered character set

Notes:
1. CH: "0" = data channel 1, "1" = data channel 2.

Table 8.33. Closed Captioning Optional Closed Group Extensions.

uous and may be separated into subpackets that can be inserted anywhere space is available in the line 284 information stream.

There are four types of extended data characters:

> *Control:* Control characters are used as a mode switch to enable the extended data mode. They are the first character of two and have a value of 01_F to $0F_H$.
>
> *Type:* Type characters follow the control character (thus, they are the second character of two) and identify the packet type. They have a value of 01_F to $0F_H$.
>
> *Checksum*: Checksum characters always follow the "end of packet" control character. Thus, they are the second character of two and have a value of 00_F to $7F_H$.
>
> *Informational*: These characters may be ASCII or non-ASCII data. They are transmitted in pairs up to and including 32 characters. A NUL character (00_H) is used to ensure pairs of characters are always sent.

Control Characters

Table 8.34 lists the control codes. The *current class* describes a program currently being transmitted. The *future class* describes a program to be transmitted later. It contains the same information and formats as the current class. The *channel class* describes non-program-specific information about the channel. The *miscellaneous class* describes miscellaneous information. The *public class* transmits data or messages of a public service nature. The *undefined class* is used in proprietary systems for whatever that system wishes.

Type Characters (Current, Future Class)
Program Identification Number (01_H)

This packet uses four characters to specify the program start time and date relative to Coordinated Universal Time (UTC). The format is shown in Table 8.35.

Minutes have a range of 0–59. Hours have a range of 0–23. Dates have a range of 1–31. Months have a range of 1–12. "T" indicates if a program is routinely tape delayed for the Mountain and Pacific time zones. The "D," "L," and "Z" bits are ignored by the decoder.

Program Length (02_H)

This packet has 2, 4, or 6 characters and indicates the scheduled length of the program and elapsed time for the program. The format is shown in Table 8.36.

Minutes and seconds have a range of 0–59. Hours have a range of 0–63.

Program Name (03_H)

This packet contains 2–32 ASCII characters that specify the title of the program.

Program Type (04_H)

This packet contains 2–32 characters that specify the type of program. Each character is assigned a keyword, as shown in Table 8.37.

Program Rating (05_H)

This packet, commonly referred to regarding the "V-chip", contains the information shown in Table 8.38 to indicate the program rating.

V indicates if violence is present. S indicates if sexual situations are present. L indicates if adult language is present. D indicates if sexually suggestive dialog is present.

Control Code	Function	Class
01_H 02_H	start continue	current
03_H 04_H	start continue	future
05_H 06_H	start continue	channel
07_H 08_H	start continue	miscellaneous
09_H $0A_H$	start continue	public service
$0B_H$ $0C_H$	start continue	reserved
$0D_H$ $0E_H$	start continue	undefined
$0F_H$	end	all

Table 8.34. EIA-608 Control Codes.

D6	D5	D4	D3	D2	D1	D0	Character
1	m5	m4	m3	m2	m1	m0	minute
1	D	h4	h3	h2	h1	h0	hour
1	L	d4	d3	d2	d1	d0	date
1	Z	T	m3	m2	m1	m0	month

Table 8.35. EIA-608 Program Identification Number Format.

D6	D5	D4	D3	D2	D1	D0	Character
1	m5	m4	m3	m2	m1	m0	length, minute
1	h5	h4	h3	h2	h1	h0	length, hour
1	m5	m4	m3	m2	m1	m0	elapsed time, minute
1	h5	h4	h3	h2	h1	h0	elapsed time, hour
1	s5	s4	s3	s2	s1	s0	elapsed time, second
0	0	0	0	0	0	0	null character

Table 8.36. EIA-608 Program Length Format.

Code (hex)	Keyword	Code (hex)	Keyword	Code (hex)	Keyword
20	education	30	business	40	fantasy
21	entertainment	31	classical	41	farm
22	movie	32	college	42	fashion
23	news	33	combat	43	fiction
24	religious	34	comedy	44	food
25	sports	35	commentary	45	football
26	other	36	concert	46	foreign
27	action	37	consumer	47	fund raiser
28	advertisement	38	contemporary	48	game/quiz
29	animated	39	crime	49	garden
2A	anthology	3A	dance	4A	golf
2B	automobile	3B	documentary	4B	government
2C	awards	3C	drama	4C	health
2D	baseball	3D	elementary	4D	high school
2E	basketball	3E	erotica	4E	history
2F	bulletin	3F	exercise	4F	hobby
50	hockey	60	music	70	romance
51	home	61	mystery	71	science
52	horror	62	national	72	series
53	information	63	nature	73	service
54	instruction	64	police	74	shopping
55	international	65	politics	75	soap opera
56	interview	66	premiere	76	special
57	language	67	prerecorded	77	suspense
58	legal	68	product	78	talk
59	live	69	professional	79	technical
5A	local	6A	public	7A	tennis
5B	math	6B	racing	7B	travel
5C	medical	6C	reading	7C	variety
5D	meeting	6D	repair	7D	video
5E	military	6E	repeat	7E	weather
5F	miniseries	6F	review	7F	western

Table 8.37. EIA-608 Program Types.

Program Audio Services (06$_H$)

This packet contains two characters as shown in Table 8.39 to indicate the program audio services available.

Program Caption Services (07$_H$)

This packet contains 2–8 characters as shown in Table 8.40 to indicate the program caption services available. L2–L0 are coded as shown in Table 8.39.

Copy Generation Management System (08$_H$)

This CGMS-A (Copy Generation Management System—Analog) packet contains 2 characters as shown in Table 8.41.

In the case where either B3 or B4 is a "0," there is no Analog Protection System (B1 and B2 are "0"). B0 is the analog source bit.

Program Aspect Ratio (09$_H$)

This packet contains two or four characters as shown in Table 8.42 to indicate the aspect ratio of the program.

S0–S5 specify the first line containing active picture information. The value of S0–S5 is calculated by subtracting 22 from the first line containing active picture information. The valid range for the first line containing active picture information is 22–85.

E0–E5 specify the last line containing active picture information. The last line containing active video is calculated by subtracting the value of E0–E5 from 262. The valid range for the last line containing active picture information is 199–262.

When this packet contains all zeros for both characters, or the packet is not detected, an aspect ratio of 4:3 is assumed.

The Q0 bit specifies whether the video is squeezed ("1") or normal ("0"). Squeezed video (anamorphic) is the result of compressing a 16:9 aspect ratio picture into a 4:3 aspect ratio picture without cropping side panels.

The aspect ratio is calculated as follows:

$$320 / (E - S) : 1$$

Program Description (10$_H$–17$_H$)

This packet contains 1–8 packet rows, with each packet row containing 0–32 ASCII characters. A packet row corresponds to a line of text on the display.

Each packet is used in numerical sequence, and if a packet contains no ASCII characters, a blank line will be displayed.

Type Characters (Channel Class)

Network Name (01$_H$)

This packet uses 2–32 ASCII characters to specify the network name.

Network Call Letters (02$_H$)

This packet uses four or six ASCII characters to specify the call letters of the channel. When six characters are used, they reflect the over-the-air channel number (2–69) assigned by the FCC. Single-digit channel numbers are preceded by a zero or a null character.

Channel Tape Delay (03$_H$)

This packet uses two characters to specify the number of hours and minutes the local station typically delays network programs. The format of this packet is shown in Table 8.43.

Minutes have a range of 0–59. Hours have a range of 0–23. This delay applies to all programs on the channel that have the "T" bit set in their Program ID packet (Table 8.35).

D6	D5	D4	D3	D2	D1	D0	Character
1	D / a2	a1	a0	r2	r1	r0	MPAA movie rating
1	V	S	L / a3	g2	g1	g0	TV rating

r2–r0: Movie Rating
- 000   not applicable
- 001   G
- 010   PG
- 011   PG-13
- 100   R
- 101   NC-17
- 110   X
- 111   not rated

g2–g0: USA TV Rating
- 000   not rated
- 001   TV-Y
- 010   TV-Y7
- 011   TV-G
- 100   TV-PG
- 101   TV-14
- 110   TV-MA
- 111   not rated

a3–a0:
- xxx0   MPAA movie rating
- LD01   USA TV rating
- 0011   Canadian English TV rating
- 0111   Canadian French TV rating
- 1011   reserved
- 1111   reserved

g2–g0: Canadian English TV Rating
- 000   exempt
- 001   C
- 010   C8 +
- 011   G
- 100   PG
- 101   14 +
- 110   18 +
- 111   reserved

g2–g0: Canadian French TV Rating
- 000   exempt
- 001   G
- 010   8 ans +
- 011   13 ans +
- 100   16 ans +
- 101   18 ans +
- 110   reserved
- 111   reserved

Table 8.38. EIA-608 Program Rating Format.

D6	D5	D4	D3	D2	D1	D0	Character
1	L2	L1	L0	T2	T1	T0	main audio program
1	L2	L1	L0	S2	S1	S0	second audio program (SAP)

L2–L0:
- 000   unknown
- 001   english
- 010   spanish
- 011   french
- 100   german
- 101   italian
- 110   other
- 111   none

T2–T0:
- 000   unknown
- 001   mono
- 010   simulated stereo
- 011   true stereo
- 100   stereo surround
- 101   data service
- 110   other
- 111   none

S2–S0:
- 000   unknown
- 001   mono
- 010   video descriptions
- 011   non-program audio
- 100   special effects
- 101   data service
- 110   other
- 111   none

Table 8.39. EIA-608 Program Audio Services Format.

D6	D5	D4	D3	D2	D1	D0	Character
1	L2	L1	L0	F	C	T	service code

FCT: 000   line 21, data channel 1 captioning
      001   line 21, data channel 1 text
      010   line 21, data channel 2 captioning
      011   line 21, data channel 2 text
      100   line 284, data channel 1 captioning
      101   line 284, data channel 1 text
      110   line 284, data channel 2 captioning
      111   line 284, data channel 2 text

Table 8.40. EIA-608 Program Caption Services Format.

D6	D5	D4	D3	D2	D1	D0	Character
1	0	B4	B3	B2	B1	B0	CGMS
0	0	0	0	0	0	0	null

B4–B3 CGMS–A Services:

    00   copying permitted without restriction
    01   condition not to be used
    10   one generation copy allowed
    11   no copying permitted

B2–B1 Analog Protection Services: (APS)

    00   no APS
    01   pseudo-sync pulse on; color striping off
    10   pseudo-sync pulse on; 2-line color striping on
    11   pseudo-sync pulse on; 4-line color striping on

Table 8.41. EIA-608 CGMS–A Format.

D6	D5	D4	D3	D2	D1	D0	Character
1	S5	S4	S3	S2	S1	S0	start
1	E5	E4	E3	E2	E1	E0	end
1	–	–	–	–	–	Q0	other
0	0	0	0	0	0	0	null

Table 8.42. EIA-608 Program Aspect Ratio Format.

D6	D5	D4	D3	D2	D1	D0	Character
1	m5	m4	m3	m2	m1	m0	minute
1	–	h4	h3	h2	h1	h0	hour

Table 8.43. EIA-608 Channel Tape Delay Format.

D6	D5	D4	D3	D2	D1	D0	Character
1	m5	m4	m3	m2	m1	m0	minute
1	D	h4	h3	h2	h1	h0	hour
1	L	d4	d3	d2	d1	d0	date
1	Z	T	m3	m2	m1	m0	month
1	–	–	–	D2	D1	D0	day
1	Y5	Y4	Y3	Y2	Y1	Y0	year

Table 8.44. EIA-608 Time of Day Format.

Type Characters (Miscellaneous Class)

Time of Day (01_H)

This packet uses six characters to specify the current time of day, month, and date relative to Coordinated Universal Time (UTC). The format is shown in Table 8.44.

Minutes have a range of 0–59. Hours have a range of 0–23. Dates have a range of 1–31. Months have a range of 1–12. Days have a range of 1 (Sunday) to 7 (Saturday). Years have a range of 0–63 (added to 1990).

"T" indicates if a program is routinely tape delayed for the Mountain and Pacific time zones. "D" indicates whether daylight savings time currently is being observed. "L" indicates whether the local day is February 28th or 29th when it is March 1st UTC. "Z" indicates whether the seconds should be set to zero (to allow calibration without having to transmit the full 6 bits of seconds data).

Impulse Capture ID (02_H)

This packet carries the program start time and length, and can be used to tell a VCR to record this program. The format is shown in Table 8.45.

Start and length minutes have a range of 0–59. Start hours have a range of 0–23; length hours have a range of 0–63. Dates have a range of 1–31. Months have a range of 1–12. "T" indicates if a program is routinely tape delayed for the Mountain and Pacific time zones. The "D," "L." and "Z" bits are ignored by the decoder.

Supplemental Data Location (03_H)

This packet uses 2–32 characters to specify other lines where additional VBI data may be found. Table 8.46 shows the format.

"F" indicates field one ("0") or field two ("1"). N may have a value of 7–31, and indicates a specific line number.

Local Time Zone (04_H)

This packet uses two characters to specify the viewer time zone and whether the locality observes daylight savings time. The format is shown in Table 8.47.

Hours have a range of 0–23. This is the nominal time zone offset, in hours, relative to UTC. "D" is a "1" when the area is using daylight savings time.

Out-of-Band Channel Number (40_H)

This packet uses two characters to specify a channel number to which all subsequent out-of-band packets refer. This is the CATV channel number to which any following out-of-band packets belong to. The format is shown in Table 8.48.

Caption (CC) and Text (T) Channels

CC1, CC2, T1 and T2 are on line 21. CC3, CC4, T3 and T4 are on line 284. A fifth channel on line 284 carries the Extended Data Services. T1-T4 are similar to CC1-CC4, but take over all or half of the screen to display scrolling text information.

CC1 is usually the main caption channel. CC2 or CC3 is occasionally used for supporting a second language version.

Closed Captioning for PAL

For (M) PAL, caption data may be present on lines 18 and 281, however, it may occasionally reside on any line between 18–22 and 281–285 due to editing.

For (B, D, G, H, I, N, NC) PAL video tapes, caption data may be present on lines 22 and 335, however, it may occasionally reside on any line between 22–26 and 335–339 due to editing. The data format, amplitudes, and rise and fall times match those used in the United States. The timing, as shown in Figure 8.58, is slightly different due to the 625-line horizontal timing.

D6	D5	D4	D3	D2	D1	D0	Character
1	m5	m4	m3	m2	m1	m0	start, minute
1	D	h4	h3	h2	h1	h0	start, hour
1	L	d4	d3	d2	d1	d0	start, date
1	Z	T	m3	m2	m1	m0	start, month
1	m5	m4	m3	m2	m1	m0	length, minute
1	h5	h4	h3	h2	h1	h0	length, hour

Table 8.45. EIA-608 Impulse Capture ID Format.

D6	D5	D4	D3	D2	D1	D0	Character
1	F	N4	N3	N2	N1	N0	location

Table 8.46. EIA-608 Supplemental Data Format.

D6	D5	D4	D3	D2	D1	D0	Character
1	D	h4	h3	h2	h1	h0	hour
0	0	0	0	0	0	0	null

Table 8.47. EIA-608 Local Time Zone Format.

D6	D5	D4	D3	D2	D1	D0	Character
1	c5	c4	c3	c2	c1	c0	channel low
1	c11	c10	c9	c8	c7	c6	channel high

Table 8.48. EIA-608 Out-of-Band Channel Number Format.

Widescreen Signalling

To facilitate the handling of various aspect ratios of program material received by TVs, a widescreen signalling (WSS) system has been developed. This standard allows a WSS-enhanced 16:9 TV to display programs in their correct aspect ratio.

625i Systems

625i (576i) systems are based on ITU-R BT.1119 and ETSI EN 300 294. For YPbPr and S-video interfaces, WSS is present on the Y signal. For analog RGB interfaces, WSS is present on all three signals.

The Analog Copy Generation Management System (CGMS-A) is also supported by the WSS signal.

Data Timing

For (B, D, G, H, I, N, NC) PAL, WSS data is normally on line 23, as shown in Figure 8.59. However, due to video editing, WSS data may reside on any line between 23–27.

The clock frequency is 5 MHz (±100 Hz). The signal waveform should be a sine-squared pulse, with a half-amplitude duration of 200 ±10 ns. The signal amplitude is 500 mV ±5%.

The NRZ data bits are processed by a bi-phase code modulator, such that one data period equals 6 elements at 5 MHz.

Data Content

The WSS consists of a run-in code, a start code, and 14 bits of data, as shown in Table 8.49.

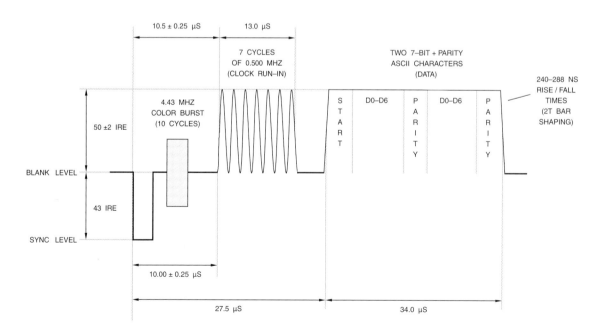

Figure 8.58. 625-Line Lines 22 and 335 Closed Captioning Timing.

Run-In

The run-in consists of 29 elements at 5 MHz of a specific sequence, shown in Table 8.49.

Start Code

The start code consists of 24 elements at 5 MHz of a specific sequence, shown in Table 8.49.

Group A Data

The group A data consists of 4 data bits that specify the aspect ratio. Each data bit generates 6 elements at 5 MHz. b0 is the LSB.

Table 8.50 lists the data bit assignments and usage. The number of active lines listed in Table 8.50 are for the exact aspect ratio (a = 1.33, 1.56, or 1.78).

The aspect ratio label indicates a range of possible aspect ratios (a) and number of active lines:

4:3	$a \leq 1.46$	527–576
14:9	$1.46 < a \leq 1.66$	463–526
16:9	$1.66 < a \leq 1.90$	405–462
>16:9	$a > 1.90$	< 405

To allow automatic selection of the display mode, a 16:9 receiver should support the following minimum requirements:

Case 1: The 4:3 aspect ratio picture should be centered on the display, with black bars on the left and right sides.

Case 2: The 14:9 aspect ratio picture should be centered on the display, with black bars on the left and right sides. Alternately, the picture may be displayed using the full display width by using a small (typically 8%) horizontal geometrical error.

Case 3: The 16:9 aspect ratio picture should be displayed using the full width of the display.

Case 4: The >16:9 aspect ratio picture should be displayed as in Case 3 or use the full height of the display by zooming in.

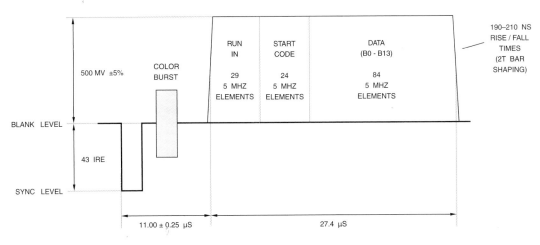

Figure 8.59. 625-Line Line 23 WSS Timing.

run-in	29 elements at 5 MHz	1 1111 0001 1100 0111 0001 1100 0111 (1F1C 71C7$_H$)
start code	24 elements at 5 MHz	0001 1110 0011 1100 0001 1111 (1E 3C1F$_H$)
group A (aspect ratio)	24 elements at 5 MHz "0" = 000 111 "1" = 111 000	b0, b1, b2, b3
group B (enhanced services)	24 elements at 5 MHz "0" = 000 111 "1" = 111 000	b4, b5, b6, b7 (b7 = "0" since reserved)
group C (subtitles)	18 elements at 5 MHz "0" = 000 111 "1" = 111 000	b8, b9, b10
group D (reserved)	18 elements at 5 MHz "0" = 000 111 "1" = 111 000	b11, b12, b13

Table 8.49. 625-Line WSS Information.

b3, b2, b1, b0	Aspect Ratio Label	Format	Position On 4:3 Display	Active Lines	Minimum Requirements
1000	4:3	full format	–	576	case 1
0001	14:9	letterbox	center	504	case 2
0010	14:9	letterbox	top	504	case 2
1011	16:9	letterbox	center	430	case 3
0100	16:9	letterbox	top	430	case 3
1101	> 16:9	letterbox	center	–	case 4
1110	14:9	full format	center	576	–
0111	16:9	full format (anamorphic)	–	576	–

Table 8.50.   625-Line WSS Group A (Aspect Ratio) Data Bit Assignments and Usage.

Group B Data

The group B data consists of four data bits that specify enhanced services. Each data bit generates six elements at 5 MHz. Data bit b4 is the LSB. Bits b5 and b6 are used for PALplus.

b4: mode
- 0   camera mode
- 1   film mode

b5: color encoding
- 0   normal PAL
- 1   Motion Adaptive ColorPlus

b6: helper signals
- 0   not present
- 1   present

Group C Data

The group C data consists of three data bits that specify subtitles. Each data bit generates six elements at 5 MHz. Data bit b8 is the LSB.

b8: teletext subtitles
- 0   no
- 1   yes

b10, b9: open subtitles
- 00   no
- 01   inside active picture
- 10   outside active picture
- 11   reserved

Group D Data

The group D data consists of three data bits that specify surround sound and copy protection. Each data bit generates six elements at 5 MHz. Data bit b11 is the LSB.

b11: surround sound
- 0   no
- 1   yes

b12: copyright
- 0   no copyright asserted or unknown
- 1   copyright asserted

b13: copy protection
- 0   copying not restricted
- 1   copying restricted

525i Systems

EIA-J CPR–1204 and IEC 61880 define a widescreen signalling standard for 525i (480i) systems. For YPbPr and S-video interfaces, WSS is present on the Y signal. For analog RGB interfaces, WSS is present on all three signals.

Data Timing

Lines 20 and 283 are used to transmit the WSS information, as shown in Figure 8.60. However, due to video editing, it may reside on any line between 20–24 and 283–287.

The clock frequency is $F_{SC}/8$ or about 447.443 kHz; F_{SC} is the color subcarrier frequency of 3.579545 MHz. The signal waveform should be a sine-squared pulse, with a half-amplitude duration of 2.235 µs ±50 ns. The signal amplitude is 70 ±10 IRE for a "1," and 0 ±5 IRE for a "0."

Data Content

The WSS consists of 2 bits of start code, 14 bits of data, and 6 bits of CRC, as shown in Table 8.51. The CRC used is $X^6 + X + 1$, all preset to "1."

Start Code

The start code consists of a "1" data bit followed by a "0" data bit, as shown in Table 8.51.

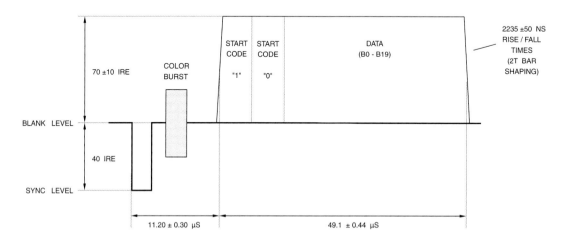

Figure 8.60. 525-Line Lines 20 and 283 WSS Timing.

start code	"1"
start code	"0"
word 0	b0, b1
word 1	b2, b3, b4, b5
word 2	b6, b7, b8, b9, b10, b11, b12, b13
CRC	b14, b15, b16, b17, b18, b19

Table 8.51. 525-Line WSS Data Bit Assignments and Usage.

Word 0 Data

Word 0 data consists of 2 data bits:

b1, b0:
00	4:3 aspect ratio	normal
01	16:9 aspect ratio	anamorphic
10	4:3 aspect ratio	letterbox
11	reserved	

Word 1 Data

Word 1 data consists of 4 data bits:

b5, b4, b3, b2:
0000	copy control information
1111	default

Copy control information is transmitted in Word 2 data when Word 1 data is "0000." When copy control information is not to be transferred, Word 1 data must be set to the default value "1111."

Word 2 Data

Word 2 data consists of 14 data bits. When Word 1 data is "0000," Word 2 data consists of copy control information. Word 2 copy control data must be transferred at the rate of two or more frames per two seconds.

Bits b6 and b7 specify the copy generation management system in an analog signal (CGMS-A). CGMS-A consists of two bits of digital information:

b7, b6:
00	copying permitted
01	one copy permitted
10	reserved
11	no copying permitted

This CGMS-A information must also usually be conveyed via the line 284 Extended Data Services *Copy Generation Management System* packet discussed in the closed captioning section.

Bits b8 and b9 specify the operation of the the Macrovision copy protection signals added to the analog NTSC video signal:

b9, b8:
00	PSP off
01	PSP on, 2-line split burst on
10	PSP on, split burst off
11	PSP on, 4-line split burst on

PSP is the Macrovision pseudo-sync pulse operation that, if on, will be present on the composite, s-video and Y (of YPbPr) analog video outputs. Split burst operation inverts the normal phase of the first half of the color burst signal on certain scan lines on the composite and s-video analog video outputs.

This Analog Protection System (APS) information must also usually be conveyed via the line 284 Extended Data Services *Copy Generation Management System* packet discussed in the closed captioning section.

Bit b10 specifies whether the source originated from an analog pre-recorded medium.

b10:
0	not analog pre-recorded medium
1	analog pre-recorded medium

Bits b11, b12, and b13 are reserved and are "000."

Teletext

Teletext allows the transmission of text, graphics, and data. Data may be transmitted on any line, although the VBI interval is most commonly used. The teletext standards are specified by ETSI EN 300 706, ITU-R BT.653 and EIA–516.

For YPbPr and S-video interfaces, teletext is present on the Y signal. For analog RGB interfaces, teletext is present on all three signals.

There are many systems that use the teletext physical layer to transmit proprietary information. The advantage is that teletext has already been approved in many countries for broadcast, so certification for a new transmission technique is not required.

The data rate for teletext is much higher than that used for closed captioning, approaching up to 7 Mbps in some cases. Therefore, ghost cancellation is needed to recover the transmitted data reliably.

There are seven teletext systems defined, as shown in Table 8.52. System B (also known as "World System Teletext" or "WST") has become the defacto standard and most widely adopted solution.

EIA–516, also referred to as NABTS (North American Broadcast Teletext Specification), was used a little in the United States, and was an expansion of the BT.653 525-line system C standard.

Figure 8.61 illustrates the teletext data on a scan line. If a line normally contains a color burst signal, it will still be present if teletext data is present. The 16 bits of clock run-in (or clock sync) consists of alternating "1's" and "0's."

Figures 8.62 and 8.63 illustrates the structure of teletext systems B and C, respectively.

System B Teletext Overview

Since teletext System B is the defacto teletext standard, a basic overview is presented here.

A teletext service typically consists of pages, with each page corresponding to a screen of information. The pages are transmitted one at time, and after all pages have been transmitted, the cycle repeats, with a typical cycle time of about 30 seconds. However, the broadcaster may transmit some pages more frequently than others, if desired.

The teletext service is usually based on up to eight magazines (allowing up to eight independent teletext services), with each magazine containing up to 100 pages. Magazine 1 uses page numbers 100–199, magazine 2 uses page numbers 200–299, etc. Each page may also have sub-pages, used to extend the number of pages within a magazine.

Parameter	System A	System B	System C	System D
625-Line Video Systems				
bit rate (Mbps)	6.203125	6.9375	5.734375	5.6427875
data amplitude	67 IRE	66 IRE	70 IRE	70 IRE
data per line	40 bytes	45 bytes	36 bytes	37 bytes
525-Line Video Systems				
bit rate (Mbps)	–	5.727272	5.727272	5.727272
data amplitude	–	70 IRE	70 IRE	70 IRE
data per line	–	37 bytes	36 bytes	37 bytes

Table 8.52. Summary of Teletext Systems and Parameters.

Each page contains 24 rows, with up to 40 characters per row. A character may be a letter, number, symbol, or simple graphic. There are also control codes to select colors and other attributes such as blinking and double height.

In addition to teletext information, the teletext protocol may be used to transmit other information, such as subtitling, program delivery control (PDC), and private data.

Subtitling

Subtitling is similar to the closed captioning used in the United States. "Open" subtitles are the insertion of text directly into the picture prior to transmission. "Closed" subtitles are transmitted separately from the picture. The transmission of closed subtitles in the UK uses teletext page 888. In the case where multiple languages are transmitted using teletext, separate pages are used for each language.

Program Delivery Control (PDC)

Program Delivery Control (defined by ETSI EN 300 231 and ITU-R BT.809) is a system that controls VCR recording using teletext information. The VCR can be programmed to look for and record various types of programs or a specific program. Programs are recorded even if the transmission time changes for any reason.

There are two methods of transmitting PDC information via teletext: methods A and B.

Method A places the data on a viewable teletext page, and is usually transmitted on scan line 16. This method is also known as the Video Programming System (VPS).

Method B places the data on a hidden packet (packet 26) in the teletext signal. This packet 26 data contains the data on each program, including channel, program data, and start time.

Data Broadcasting

Data broadcasting may be used to transmit information to private receivers. Typical applications include real-time financial information, airport flight schedules for hotels and travel agents, passenger information for railroads, software upgrades, etc.

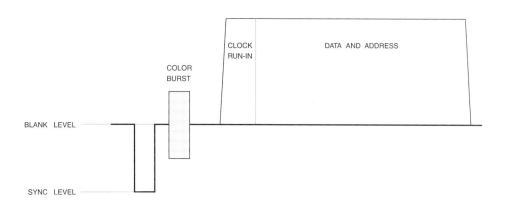

Figure 8.61. Teletext Line Format.

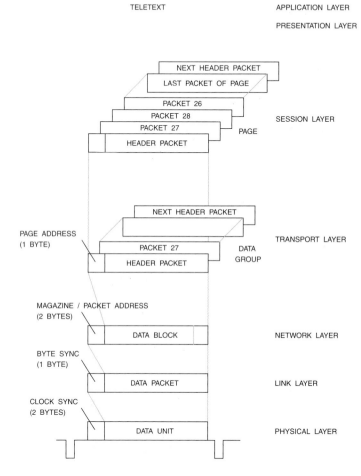

FIgure 8.62. Teletext System B Structure.

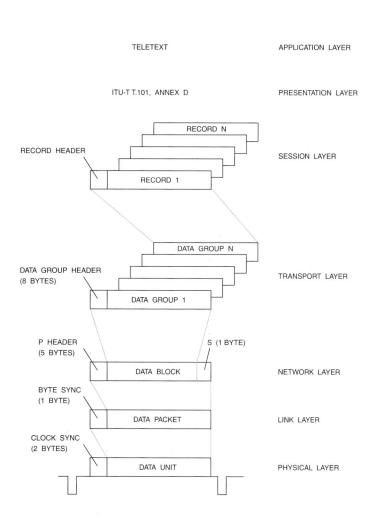

Figure 8.63. Teletext System C Structure.

Packets 0–23

A typical teletext page uses 24 packets, numbered 0–23, that correspond to the 24 rows on a displayed page. Packet 24 can add a status row at the bottom for user prompting. For each packet, three bits specify the magazine address (1–8), and five bits specify the row address (0–23). The magazine and row address bits are Hamming error protected to permit single-bit errors to be corrected.

To save bandwidth, the whole address isn't sent with all packets. Only packet 0 (also called the header packet) has all the address information such as row, page, and magazine address data. Packets 1–28 contain information that is part of the page identified by the most recent packet 0 of the same magazine.

The transmission of a page starts with a header packet. Subsequent packets with the same magazine address provide additional data for that page. These packets may be transmitted in any order, and interleaved with packets from other magazines. A page is considered complete when the next header packet for that magazine is received.

The general format for packet 0 is:

clock run-in	2 bytes
framing code	1 byte
magazine and row address	2 bytes
page number	2 bytes
subcode	4 bytes
control codes	2 bytes
display data	32 bytes

The general format for packets 1–23 is:

clock run-in	2 bytes
framing code	1 byte
magazine and row address	2 bytes
display data	40 bytes

Packet 24

This packet defines an additional row for user prompting. Teletext decoders may use the data in packet 27 to react to prompts in the packet 24 display row.

Packet 25

This packet defines a replacement header line. If present, the 40 bytes of data are displayed instead of the channel, page, time, and date from packet 8.30.

Packet 26

Packet 26 consists of:

clock run-in	2 bytes
framing code	1 byte
magazine and row address	2 bytes
designation code	1 byte
13 3-byte data groups, each consisting of	
7 data bits	
6 address bits	
5 mode bits	
6 Hamming bits	

There are 15 variations of packet 26, defined by the designation code. Each of the 13 data groups specify a specific display location and data relating to that location.

This packet is also used to extend the addressable range of the basic character set in order to support other languages, such as Arabic, Spanish, Hungarian, Chinese, etc.

For PDC, packet 26 contains data for each program, identifying the channel, program date, start time, and the cursor position of the program information on the page. When the user selects a program, the cursor position is linked to the appropriate packet 26 preselection data. This data is then used to program the VCR. When the program is transmitted, the program information is transmitted using packet 8.30 format 2. A match between the pre-selection data and the packet 8.30 data turns the VCR record mode on.

Packet 27

Packet 27 tells the teletext decoder how to respond to user selections for packet 24. There may be up to four packet 27s (packets 27/0 through 27/3), allowing up to 24 links. Packet 27 consists of:

clock run-in	2 bytes
framing code	1 byte
magazine and row address	2 bytes
designation code	1 byte
link 1 (red)	6 bytes
link 2 (green)	6 bytes
link 3 (yellow)	6 bytes
link 4 (cyan)	6 bytes
link 5 (next page)	6 bytes
link 6 (index)	6 bytes
link control data	1 byte
page check digit	2 bytes

Each link consists of:
7 data bits
6 address bits
5 mode bits
6 hamming bits

This packet contains information linking the current page to six page numbers (links). The four colored links correspond to the four colored Fastext page request keys on the remote. Typically, these four keys correspond to four colored menu selections at the bottom of the display using packet 24. Selection of one of the colored page request keys results in the selection of the corresponding linked page.

The fifth link is used for specifying a page the user might want to see after the current page, such as the next page in a sequence.

The sixth link corresponds to the Fastext index key on the remote, and specifies the page address to go to when the index is selected.

Packets 28 and 29

These are used to define level 2 and level 3 pages to support higher resolution graphics, additional colors, alternate character sets, etc. They are similar in structure to packet 26.

Packet 8.30 Format 1

Packet 8.30 (magazine 8, packet 30) isn't associated with any page, but is sent once per second. This packet is also known as the Television Service Data Packet, or TSDP. It contains data that notifies the teletext decoder about the transmission in general and the time.

clock run-in	2 bytes
framing code	1 byte
magazine and row address	2 bytes
designation code	1 byte
initial teletext page	6 bytes
network ID	2 bytes
time offset from UTC	1 byte
date (Modified Julian Day)	3 bytes
UTC time	3 bytes
TV program label	4 bytes
status display	20 bytes

The *Designation Code* indicates whether the transmission is during the VBI or full-field.

Initial Teletext Page tells the decoder which page should be captured and stored on power-up. This is usually an index or menu page.

The *Network Identification* code identifies the transmitting network.

The *TV Program Label* indicates the program label for the current program.

Status Display is used to display a transmission status message.

Packet 8.30 Format 2

This format is used for PDC recorder control, and is transmitted once per second per stream. It contains a program label indicating the start of each program, usually transmitted about 30 seconds before the start of the program to allow the VCR to detect it and get ready to record.

clock run-in	2 bytes
framing code	1 byte
magazine and row address	2 bytes
designation code	1 byte
initial teletext page	6 bytes
label channel ID	1 byte
program control status	1 byte
country and network ID	2 bytes
program ID label	5 bytes
country and network ID	2 bytes
program type	2 bytes
status display	20 bytes

The content is the same as for Format 1, except for the 13 bytes of information before the status display information.

Label channel ID (LCI) identifies each of up to four PDC streams that may be transmitted simultaneously.

The *Program Control Status* (PCS) indicates real-time status information, such as the type of analog sound transmission.

The *Country and Network ID* (CNI) is split into two groups. The first part specifies the country and the second part specifies the network.

Program ID Label (PIL) specifies the month, day, and local time of the start of the program.

Program Type (PTY) is a code that indicates an intended audience or a particular series. Examples are "adult," "children," "music," "drama," etc.

Packet 31

Packet 31 is used for the transmission of data to private receivers. It consists of:

clock run-in	2 bytes
framing code	1 byte
data channel group	1 byte
message bits	1 byte
format type	1 byte
address length	1 byte
address	0–6 bytes
repeat indicator	0–1 byte
continuity indicator	0–1 byte
data length	0–1 byte
user data	28–36 bytes
CRC	2 bytes

"Raw" VBI Data

"Raw," or oversampled, VBI data is simply digitized VBI data. It is typically oversampled using a 2× video sample clock, such as 27 MHz for 480i and 54 MHz for 480p video. Use of the 2× video sample clock enables transferring the "raw" VBI data over a standard 8-bit BT.656 interface. VBI data may be present on any scan line, except during the serration and equalization intervals.

The "raw" VBI data is then converted to binary (or "sliced") data and processed and/or passed through to the composite, s-video and YPbPr analog video outputs so it may be decoded by the TV.

In the conversion from "raw" to "sliced" VBI data, the VBI decoders must compensate for varying DC offsets, amplitude variations, ghosting, and timing variations.

Hysteresis must also be used to prevent the VBI decoders from turning on and off rapidly due to noise and transmission errors. Once the desired VBI signal is found for (typically) 32 consecutive frames, VBI decoding should commence. When the desired VBI signal is not found on the appropriate scan lines for (typically) 32 consecutive frames, VBI decoding should stop.

"Sliced" VBI Data

"Sliced," or binary, VBI data is commonly available from NTSC/PAL video decoders. This has the advantage of lower data rates since binary, rather than oversampled, data is present. The primary disadvantage is the variety of techniques NTSC/PAL video decoder chip manufacturers use to transfer the "sliced" VBI data over the video interface.

NTSC/PAL Decoder Considerations

Closed Captioning

In addition to caption and text commands that clear the display, five other events typically force the display to be cleared:

(1) A change in the caption display mode, such as switching from CC1 to T1.

(2) A loss of video lock, such as on a channel change, forces the display to be cleared. The currently active display mode does not change. For example, if CC1 was selected before loss of video lock, it remains selected.

(3) Activation of autoblanking. If the caption signal has not been detected for (typically) 32 consecutive frames, or no new data for the selected channel has been received for (typically) 512 frames, the display memory is cleared. Once the caption signal has been detected for (typically) 32 consecutive frames, or new data has been received, it is displayed.

(4) A "clear" command (from the remote control for example) forces the display to be cleared.

(5) Disabling caption decoding also forces the display to be cleared.

Widescreen Signaling

The decoder must be able to handle a variety of WSS inputs including:

(1) PAL or NTSC WSS signal on composite, s-video or Y (of YPbPr).

(2) SCART analog inputs (DC offset indicator)

(3) S-Video analog inputs (DC offset indicator)

In addition to automatically processing the video signal to fit a 4:3 or 16:9 display based on the WSS data, the decoder should also support manual overrides in case the user wishes a specific mode of operation due to personal preferences. Software uses this aspect ratio information, user preferences, and display format to assist in properly processing the program for display.

Ghost Cancellation

Ghost cancellation (the removal of undesired reflections present in the signal) is required due to the high data rate of some services, such as teletext. Ghosting greater than 100 ns and –12 dB corrupts teletext data. Ghosting greater than –3 dB is difficult to remove cost-effectively in hardware or software, while ghosting less than –12 dB need not be removed. Ghost cancellation for VBI data is not as complex as ghost cancellation for active video.

Unfortunately, the GCR (ghost cancellation reference) signal is not usually present. Thus, a ghost cancellation algorithm must determine the amount of ghosting using other available signals, such as the serration and equalization pulses.

The NTSC GCR signal is specified in ATSC A/49 and ITU-R BT.1124. If present, it occupies lines 19 and 282. The GCR permits the detection of ghosting from –3 to +45 µs, and follows an 8-field sequence.

The PAL GCR signal is specified in BT.1124 and ETSI ETS 300 732. If present, it occupies line 318. The GCR permits the detection of ghosting from –3 to +45 µs, and follows a 4-frame sequence.

Enhanced Television Programming

The enhanced television programming standard (SMPTE 363M) is used for creating and delivering enhanced and interactive programs. The enhanced content can be delivered over a variety of mediums—including analog and digital television broadcasts—using terrestrial, cable and satellite networks. In defining how to create enhanced content, the specification defines the minimum receiver functionality. To minimize the creation of new specifications, it leverages Internet technologies such as HTML and Java-script. The benefits of doing this are that there are already millions of pages of potential content, and the ability to use existing web-authoring tools.

The specification mandates that receivers support, as a minimum, HTML 4.0, Javascript 1.1, and Cascading Style Sheets. Supporting additional capabilities, such as Java and VRML, is optional. This ensures content is available to the maximum number of viewers.

For increased capability, a new "tv:" attribute is added to the HTML. This attribute enables the insertion of the television program into the content, and may be used in a HTML document anywhere that a regular image may be placed. Creating an enhanced content page that displays the current television channel anywhere on the display is as easy as inserting an image in a HTML document.

The specification also defines how the receivers obtain the content and how they are informed that enhancements are available. The latter task is accomplished with triggers.

Triggers

Triggers alert receivers to content enhancements, and contain information about the enhancements. Among other things, triggers contain a Universal Resource Locator (URL) that defines the location of the enhanced content. Content may reside locally—such as when delivered over the network and cached to a local hard drive—or it may reside on the Internet or another network.

Triggers may also contain a human-readable description of the content. For example, it may contain the description "Press ORDER to order this product," which can be displayed for the viewer. Triggers also may contain expiration information, indicating how long the enhancement should be offered to the viewer.

Lastly, triggers may contain scripts that trigger the execution of Javascript within the associated HTML page, to support synchronization of the enhanced content with the video signal and updating of dynamic screen data.

The processing of triggers is defined in SMPTE 363M and is independent of the method used to carry them.

Transports

Besides defining how content is displayed and how the receiver is notified of new content, the specification also defines how content is delivered. Because a receiver may not have an Internet connection, the specification describes two models for delivering content. These two models are called transports, and the two transports are referred to as Transport Type A and Transport Type B.

If the receiver has a back-channel (or return path) to the Internet, Transport Type A will broadcast the trigger and the content will be pulled over the Internet.

If the receiver does not have an Internet connection, Transport Type B provides for delivery of both triggers and content via the broadcast medium. Announcements are sent over the network to associate triggers with content streams. An announcement describes the content, and may include information regarding bandwidth, storage requirements, and language.

Delivery Protocols

For traditional bi-directional Internet communication, the Hypertext Transfer Protocol (HTTP) defines how data is transferred at the application level. For uni-directional broadcasts where a two-way connection is not available, SMPTE 364M defines a uni-directional application-level protocol for data delivery: Uni-directional Hypertext Transfer Protocol (UHTTP).

Like HTTP, UHTTP uses traditional URL naming schemes to reference content. Content can reference enhancement pages using the standard "http:" and "ftp:" naming schemes. A "lid:," or local identifier, URL is also available to allow reference to content that exists locally (such as on the receiver's hard drive) as opposed to on the Internet or other network.

Bindings

How data is delivered over a specific network is called "binding." Bindings have been defined for NTSC and PAL.

NTSC Bindings

Transport Type A triggers are broadcast on data channel 2 of the EIA-608 captioning signal.

Transport Type B binding also includes a mechanism for delivering IP multicast packets over the vertical blanking interval (VBI), otherwise known as IP over VBI (IP/VBI). At the lowest level, the television signal transports NABTS (North American Basic Teletext Standard) packets during the VBI. These NABTS packets are recovered to form a sequential data stream (encapsulated in a SLIP-like protocol) that is unframed to produce IP packets.

PAL Bindings

Both transport types are based on carriage of IP multicast packets in VBI lines of a PAL system by means of teletext packets 30 or 31.

Transport Type A triggers are carried in UDP/IP multicast packets, delivered to address 224.0.23.13 and port 2670.

Transport Type B (described in SMPTE 357M) carries a single trigger in a single UDP/IP multicast packet, delivered on the address and port defined in the SDP announcement for the enhanced television program. The trigger protocol is very lightweight in order to provide quick synchronization.

References

1. Advanced Television Enhancement Forum, *Enhanced Content Specification,* 1999.
2. ATSC A/49, 13 May 1993, *Ghost Cancelling Reference Signal for NTSC.*
3. *BBC Technical Requirements for Digital Television Services,* Version 1.0, February 3, 1999, BBC Broadcast.
4. EIA–189–A, July 1976, *Encoded Color Bar Signal.*
5. EIA–516, May 1988, *North American Basic Teletext Specification (NABTS).*
6. EIA–608, September 1994, *Recommended Practice for Line 21 Data Service.*
7. EIA-J CPR–1204, *Transfer Method of Video ID Information using Vertical Blanking Interval (525-line System)*, March 1997.
8. ETSI EN 300 163, *Television Systems: NICAM 728: Transmission of Two Channel Digital Sound with Terrestrial Television Systems B, G, H, I, K1, and L,* March 1998.
9. ETSI EN 300 231, *Television Systems: Specification of the Domestic Video Programme Delivery Control System (PDC)*, December 2002.
10. ETSI EN 300 294, *Television Systems: 625-line Television Widescreen Signalling (WSS)*, December 2002.
11. ETSI EN 300 706, *Enhanced Teletext Specification*, December 2002.
12. ETSI EN 300 708, *Television Systems: Data Transmission within Teletext*, December 2002.
13. ETSI ETS 300 731, *Television Systems: Enhanced 625-Line Phased Alternate Line (PAL) Television: PALplus*, March 1997.
14. ETSI ETS 300 732, *Television Systems: Enhanced 625-Line PAL/SECAM Television; Ghost Cancellation Reference (GCR) Signals*, January 1997.
15. Faroudja, Yves Charles, *NTSC and Beyond, IEEE Transactions on Consumer Electronics*, Vol. 34, No. 1, February 1988.
16. IEC 61880, 1998–1, *Video Systems (525/60)—Video and Accompanied Data Using the Vertical Blanking Interval—Analog Interface.*
17. ITU-R BS.707–3, 1998, *Transmission of Multisound in Terrestrial Television Systems PAL B, G, H, and I and SECAM D, K, K1, and L.*
18. ITU-R BT.470–6, 1998, *Conventional Television Systems.*
19. ITU-R BT.471–1, 1986, *Nomenclature and Description of Colour Bar Signals.*
20. ITU-R BT.472–3, 1990, *Video Frequency Characteristics of a Television System to Be Used for the International Exchange of Programmes Between Countries that Have Adopted 625-Line Colour or Monochrome Systems.*
21. ITU-R BT.473–5, 1990, *Insertion of Test Signals in the Field-Blanking Interval of Monochrome and Colour Television Signals.*
22. ITU-R BT.569–2, 1986, *Definition of Parameters for Simplified Automatic Measurement of Television Insertion Test Signals.*
23. ITU-R BT.653–3, 1998, *Teletext Systems.*
24. ITU-R BT.809, 1992, *Programme Delivery Control (PDC) System for Video Recording.*

25. ITU-R BT.1118, 1994, *Enhanced Compatible Widescreen Television Based on Conventional Television Systems.*

26. ITU-R BT.1119–2, 1998, *Wide-Screen Signalling for Broadcasting.*

27. ITU-R BT.1124, 1994, *Reference Signals for Ghost Cancelling in Analogue Television Systems.*

28. ITU-R BT.1197–1, 1998, *Enhanced Wide-Screen PAL TV Transmission System (the PALplus System).*

29. ITU-R BT.1298, 1997, *Enhanced Wide-Screen NTSC TV Transmission System.*

30. *Multichannel TV Sound System BTSC System Recommended Practices*, EIA Television Systems Bulletin No. 5 (TVSB5).

31. *NTSC Video Measurements*, Tektronix, Inc., 1997.

32. SMPTE 12M–1999, *Television, Audio and Film—Time and Control Code.*

33. SMPTE 170M–1999, *Television—Composite Analog Video Signal—NTSC for Studio Applications.*

34. SMPTE 262M–1995, *Television, Audio and Film—Binary Groups of Time and Control Codes—Storage and Transmission of Data.*

35. SMPTE 309M–1999, *Television—Transmission of Date and Time Zone Information in Binary Groups of Time and Control Code.*

36. SMPTE 357M–2002, *Television—Declarative Data Essence – Internet Protocol Multicast Encapsulation.*

37. SMPTE 361M–2002, *Television—NTSC IP and Trigger Binding to VBI.*

38. SMPTE 363M–2002, *Television—Declarative Data Essence – Content Level 1.*

39. SMPTE 364M–2001, *Declarative Data Essence – Unidirectional Hypertext Transport Protocol.*

40. SMPTE RP-164–1996, *Location of Vertical Interval Time Code.*

41. SMPTE RP-186–1995, *Video Index Information Coding for 525- and 625-Line Television Systems.*

42. SMPTE RP-201–1999, *Encoding Film Transfer Information Using Vertical Interval Time Code.*

43. *Specification of Television Standards for 625-Line System-I Transmissions*, 1971, Independent Television Authority (ITA) and British Broadcasting Corporation (BBC).

44. *Television Measurements, NTSC Systems*, Tektronix, Inc., 1998.

45. *Television Measurements, PAL Systems*, Tektronix, Inc., 1990.

Chapter 9

NTSC and PAL Digital Encoding and Decoding

Although not exactly "digital" video, the NTSC and PAL composite color video formats are currently the most common formats for video. Although the video signals themselves are analog, they can be encoded and decoded almost entirely digitally.

Analog NTSC and PAL encoders and decoders have been available for some time. However, they have been difficult to use, required adjustment, and offered limited video quality. Using digital techniques to implement NTSC and PAL encoding and decoding offers many advantages such as ease of use, minimum analog adjustments, and excellent video quality.

In addition to composite video, S-video is supported by consumer and pro-video equipment, and should also be implemented. S-video uses separate luminance (Y) and chrominance (C) analog video signals so higher quality may be maintained by eliminating the Y/C separation process.

This chapter discusses the design of a digital encoder (Figure 9.1) and decoder (Figure 9.21) that support composite and S-video (M) NTSC and (B, D, G, H, I, N_C) PAL video signals. (M) and (N) PAL are easily accommodated with some slight modifications.

NTSC encoders and decoders are usually based on the YCbCr, YUV, or YIQ color space. PAL encoders and decoders are usually based on the YCbCr or YUV color space.

Video Standard	Sample Clock Rate	Applications	Active Resolution	Total Resolution	Field Rate (per second)
(M) NTSC, (M) PAL	9 MHz	SVCD	$480 \times 480i$	$572 \times 525i$	59.94 interlaced
	13.5 MHz	BT.601	$720^1 \times 480i$	$858 \times 525i$	
		MPEG-2	$704 \times 480i$		
		DV	$720 \times 480i$		
	12.27 MHz	square pixels	$640 \times 480i$	$780 \times 525i$	
(B, D, G, H, I, N, N_C) PAL	9 MHz	SVCD	$480 \times 576i$	$576 \times 625i$	50 interlaced
	14.75 MHz	square pixels	$768 \times 576i$	$944 \times 625i$	
	13.5 MHz	BT.601	$720^2 \times 576i$	$864 \times 625i$	
		MPEG-2	$704 \times 576i$		
		DV	$720 \times 576i$		

Table 9.1. Common NTSC/PAL Sample Rates and Resolutions. [1]Typically 716 true active samples between 10% blanking points. [2]Typically 702 true active samples between 50% blanking points.

NTSC and PAL Encoding

YCbCr input data has a nominal range of 16–235 for Y and 16–240 for Cb and Cr. RGB input data has a range of 0–255; pro-video applications may use a nominal range of 16–235.

As YCbCr values outside these ranges result in overflowing the standard YIQ or YUV ranges for some color combinations, one of three things may be done, in order of preference: (a) allow the video signal to be generated using the extended YIQ or YUV ranges; (b) limit the color saturation to ensure a legal video signal is generated; or (c) clip the YIQ or YUV levels to the valid ranges.

4:1:1, 4:2:0, or 4:2:2 YCbCr data must be converted to 4:4:4 YCbCr data before being converted to YIQ or YUV data. The chrominance lowpass filters will not perform the interpolation properly.

Table 9.1 lists some of the common sample rates and resolutions.

2× Oversampling

2× oversampling generates 8:8:8 YCbCr or RGB data, simplifying the analog output filters. The oversampler is also a convenient place to convert from 8-bit to 10-bit data, providing an increase in video quality.

Color Space Conversion

Choosing the 10-bit video levels to be white = 800 and sync = 16, and knowing that the sync-to-white amplitude is 1V, the full-scale output of the D/A converters (DACs) is therefore set to 1.305V.

(M) NTSC, (M, N) PAL

Since (M) NTSC and (M, N) PAL have a 7.5 IRE blanking pedestal and a 40 IRE sync amplitude, the color space conversion equations are derived so as to generate 0.660V of active video.

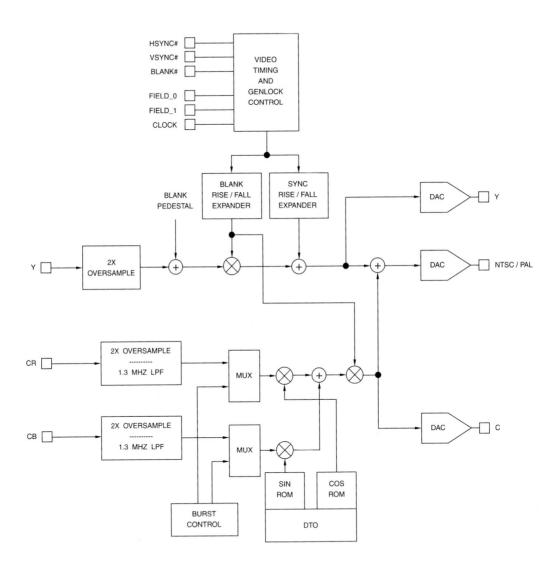

Figure 9.1. Typical NTSC/PAL Digital Encoder Implementation.

YUV Color Space Processing

Modern encoder designs are now based on the YUV color space. For these encoders, the YCbCr to YUV equations are:

$$Y = 0.591(Y_{601} - 64)$$

$$U = 0.504(Cb - 512)$$

$$V = 0.711(Cr - 512)$$

The R′G′B′ to YUV equations are:

$$Y = 0.151R' + 0.297G' + 0.058B'$$

$$U = -0.074R' - 0.147G' + 0.221B'$$

$$V = 0.312R' - 0.261G' - 0.051B'$$

For pro-video applications using a 10-bit nominal range of 64–940 for RGB, the R′G′B′ to YUV equations are:

$$Y = 0.177(R' - 64) + 0.347(G' - 64) + 0.067(B' - 64)$$

$$U = -0.087(R' - 64) - 0.171(G' - 64) + 0.258(B' - 64)$$

$$V = 0.364(R' - 64) - 0.305(G' - 64) - 0.059(B' - 64)$$

Y has a nominal range of 0 to 518, U a nominal range of 0 to ±226, and V a nominal range of 0 to ±319. Negative values of Y should be supported to allow test signals, keying information, and real-world video to be passed through the encoder with minimum corruption.

YIQ Color Space Processing

For older NTSC encoder designs based on the YIQ color space, the YCbCr to YIQ equations are:

$$Y = 0.591(Y_{601} - 64)$$

$$I = 0.596(Cr - 512) - 0.274(Cb - 512)$$

$$Q = 0.387(Cr - 512) + 0.423(Cb - 512)$$

The R′G′B′ to YIQ equations are:

$$Y = 0.151R' + 0.297G' + 0.058B'$$

$$I = 0.302R' - 0.139G' - 0.163B'$$

$$Q = 0.107R' - 0.265G' + 0.158B'$$

For pro-video applications using a 10-bit nominal range of 64–940 for R′G′B′, the R′G′B′ to YIQ equations are:

$$Y = 0.177(R' - 64) + 0.347(G' - 64) + 0.067(B' - 64)$$

$$I = 0.352(R' - 64) - 0.162(G' - 64) - 0.190(B' - 64)$$

$$Q = 0.125(R' - 64) - 0.309(G' - 64) + 0.184(B' - 64)$$

Y has a nominal range of 0 to 518, I a nominal range of 0 to ±309, and Q a nominal range of 0 to ±271. Negative values of Y should be supported to allow test signals, keying information, and real-world video to be passed through the encoder with minimum corruption.

YCbCr Color Space Processing

If the design is based on the YUV color space, the Cb and Cr conversion to U and V may be avoided by scaling the sin and cos values during the modulation process or scaling the color difference lowpass filter coefficients. This has the advantage of reducing data path processing.

NTSC–J

Since the version of (M) NTSC used in Japan has a 0 IRE blanking pedestal, the color space conversion equations are derived so as to generate 0.714V of active video.

YUV Color Space Processing

The YCbCr to YUV equations are:

$Y = 0.639(Y_{601} - 64)$

$U = 0.545(Cb - 512)$

$V = 0.769(Cr - 512)$

The R'G'B' to YUV equations are:

$Y = 0.164R' + 0.321G' + 0.062B'$

$U = -0.080R' - 0.159G' + 0.239B'$

$V = 0.337R' - 0.282G' - 0.055B'$

For pro-video applications using a 10-bit nominal range of 64–940 for R'G'B', the R'G'B' to YUV equations are:

$Y = 0.191(R' - 64) + 0.375(G' - 64) + 0.073(B' - 64)$

$U = -0.094(R' - 64) - 0.185(G' - 64) + 0.279(B' - 64)$

$V = 0.393(R' - 64) - 0.329(G' - 64) - 0.064(B' - 64)$

Y has a nominal range of 0 to 560, U a nominal range of 0 to ±244, and V a nominal range of 0 to ±344. Negative values of Y should be supported to allow test signals, keying information, and real-world video to be passed through the encoder with minimum corruption.

YIQ Color Space Processing

For older encoder designs based on the YIQ color space, the YCbCr to YIQ equations are:

$Y = 0.639(Y_{601} - 64)$

$I = 0.645(Cr - 512) - 0.297(Cb - 512)$

$Q = 0.419(Cr - 512) + 0.457(Cb - 512)$

The R'G'B' to YIQ equations are:

$Y = 0.164R' + 0.321G' + 0.062B'$

$I = 0.326R' - 0.150G' - 0.176B'$

$Q = 0.116R' - 0.286G' + 0.170B'$

For pro-video applications using a 10-bit nominal range of 64–940 for R'G'B', the R'G'B' to YIQ equations are:

$Y = 0.191(R' - 64) + 0.375(G' - 64) + 0.073(B' - 64)$

$I = 0.381(R' - 64) - 0.176(G' - 64) - 0.205(B' - 64)$

$Q = 0.135(R' - 64) - 0.334(G' - 64) + 0.199(B' - 64)$

Y has a nominal range of 0 to 560, I a nominal range of 0 to ±334, and Q a nominal range of 0 to ±293. Negative values of Y should be supported to allow test signals, keying information, and real-world video to be passed through the encoder with minimum corruption.

YCbCr Color Space Processing

If the design is based on the YUV color space, the Cb and Cr conversion to U and V may be avoided by scaling the sin and cos values during the modulation process or scaling the color difference lowpass filter coefficients. This has the advantage of reducing data path processing.

(B, D, G, H, I, N$_C$) PAL

Since these PAL standards have a 0 IRE blanking pedestal and a 43 IRE sync amplitude, the color space conversion equations are derived so as to generate 0.7V of active video.

YUV Color Space Processing

The YCbCr to YUV equations are:

$$Y = 0.625(Y_{601} - 64)$$

$$U = 0.533(Cb - 512)$$

$$V = 0.752(Cr - 512)$$

The R′G′B′ to YUV equations are:

$$Y = 0.160R' + 0.314G' + 0.061B'$$

$$U = -0.079R' - 0.155G' + 0.234B'$$

$$V = 0.329R' - 0.275G' - 0.054B'$$

For pro-video applications using a 10-bit nominal range of 64–940 for R′G′B′, the R′G′B′ to YUV equations are:

$$Y = 0.187(R' - 64) + 0.367(G' - 64) + 0.071(B' - 64)$$

$$U = -0.092(R' - 64) - 0.181(G' - 64) + 0.273(B' - 64)$$

$$V = 0.385(R' - 64) - 0.322(G' - 64) - 0.063(B' - 64)$$

Y has a nominal range of 0 to 548, U a nominal range of 0 to ±239, and V a nominal range of 0 to ±337. Negative values of Y should be supported to allow test signals, keying information, and real-world video to be passed through the encoder with minimum corruption.

YCbCr Color Space Processing

If the design is based on the YUV color space, the Cb and Cr conversion to U and V may be avoided by scaling the sin and cos values during the modulation process or scaling the color difference lowpass filter coefficients. This has the advantage of reducing data path processing.

Luminance (Y) Processing

Lowpass filtering to about 6 MHz must be done to remove high-frequency components generated as a result of the 2x oversampling process.

An optional notch filter may also be used to remove the color subcarrier frequency from the luminance information. This improves decoded video quality for decoders that use simple Y/C separation. The notch filter should be disabled when generating S-video, RGB, or YPbPr video signals.

Next, any blanking pedestal is added during active video, and the blanking and sync information are added.

(M) NTSC, (M, N) PAL

As (M) NTSC and (M, N) PAL have a 7.5 IRE blanking pedestal, a value of 42 is added to the luminance data during active video. 0 is added during the blank time.

After the blanking pedestal is added, the luminance data is clamped by a blanking signal that has a raised cosine distribution to slow the slew rate of the start and end of the video signal. Typical blank rise and fall times are 140 ±20 ns for NTSC and 300 ±100 ns for PAL.

Digital composite sync information is added to the luminance data after the blank processing has been performed. Values of 16 (sync present) or 240 (no sync) are assigned. The sync rise and fall times should be processed to generate a raised cosine distribution (between 16 and 240) to slow the slew rate of the sync signal. Typical sync rise and fall times are 140 ±20 ns for NTSC and 250 ±50 ns for

PAL, although the encoder should generate sync edges of about 130 or 240 ns to compensate for the analog output filters slowing the sync edges.

At this point, we have digital luminance with sync and blanking information, as shown in Table 9.2.

NTSC–J

When generating NTSC–J video, there is a 0 IRE blanking pedestal. Thus, no blanking pedestal is added to the luminance data during active video. Otherwise, the processing is the same as for (M) NTSC.

(B, D, G, H, I, N_C) PAL

When generating (B, D, G, H, I, N_C) PAL video, there is a 0 IRE blanking pedestal. Thus, no blanking pedestal is added to the luminance data during active video.

Blanking information is inserted using the same technique as used for (M) NTSC. However, typical blank rise and fall times are 300 ±100 ns.

Composite sync information is added using the same technique as used for (M) NTSC, except values of 16 (sync present) or 252 (no sync) are used. Typical sync rise and fall times are 250 ±50 ns, although the encoder should generate sync edges of about 240 ns to compensate for the analog output filters slowing the sync edges.

At this point, we have digital luminance with sync and blanking information, as shown in Table 9.2.

Analog Luminance (Y) Generation

The digital luminance data may drive a 10-bit DAC that generates a 0–1.305V output to generate the Y video signal of a S-video (Y/C) interface.

Figures 9.2 and 9.3 show the luminance video waveforms for 75% color bars. The numbers on the luminance levels indicate the data value for a 10-bit DAC with a full-scale output value of 1.305V. The video signal at the connector should have a source impedance of 75Ω.

As the sample-and-hold action of the DAC introduces a $(\sin x)/x$ characteristic, the video data may be digitally filtered by a $[(\sin x)/x]^{-1}$ filter to compensate. Alternately, as an analog lowpass filter is usually present after the DAC, the correction may take place in the analog filter.

As an option, the ability to delay the digital Y information a programmable number of clock cycles before driving the DAC may be useful. If the analog luminance video is lowpass filtered after the DAC, and the analog chrominance video is bandpass filtered after its

Video Level	(M) NTSC	NTSC–J	(B, D, G, H, I, N_C) PAL	(M, N) PAL
white	800	800	800	800
black	282	240	252	282
blank	240	240	252	240
sync	16	16	16	16

Table 9.2. 10-Bit Digital Luminance Values.

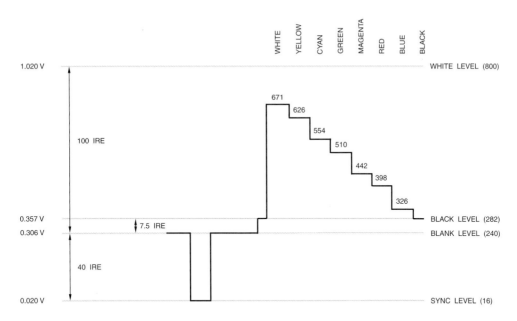

Figure 9.2. (M) NTSC Luminance (Y) Video Signal for 75% Color Bars. Indicated luminance levels are 10-bit values.

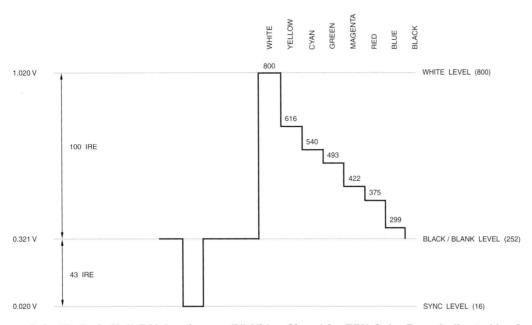

Figure 9.3. (B, D, G, H, I) PAL Luminance (Y) Video Signal for 75% Color Bars. Indicated luminance levels are 10-bit values.

DAC, the chrominance video path may have a longer delay (typically up to about 400 ns) than the luminance video path. By adjusting the delay of the Y data, the analog luminance and chrominance video will be aligned more closely after filtering, simplifying the analog design.

Color Difference Processing

Lowpass Filtering

The color difference signals (CbCr, UV, or IQ) should be lowpass filtered using a Gaussian filter. This filter type minimizes ringing and overshoot, avoiding the generation of visual artifacts on sharp edges.

If the encoder is used in a video editing application, the filters should have a maximum ripple of ±0.1 dB in the passband. This minimizes the cumulation of gain and loss artifacts due to the filters, especially when multiple passes through the encoding and decoding processes are done. At the final encoding point, Gaussian filters may be used.

YCbCr and YUV Color Space

Cb and Cr, or U and V, are lowpass filtered to about 1.3 MHz. Typical filter characteristics are <2 dB attenuation at 1.3 MHz and >20 dB attenuation at 3.6 MHz. The filter characteristics are shown in Figure 9.4.

YIQ Color Space

Q is lowpass filtered to about 0.6 MHz. Typical filter characteristics are <2 dB attenuation at 0.4 MHz, <6 dB attenuation at 0.5 MHz, and >6 dB attenuation at 0.6 MHz. The filter characteristics are shown in Figure 9.5.

Typical filter characteristics for I are the same as for U and V.

Filter Considerations

The modulation process is shown in spectral terms in Figures 9.6 through 9.9. The frequency spectra of the modulation process are the same as those if the modulation process were analog, but are repeated at harmonics of the sample rate.

Using wide-band (1.3 MHz) filters, the modulated chrominance spectra overlap near the zero frequency regions, resulting in aliasing. Also, there may be considerable aliasing just above the subcarrier frequency. For these reasons, the use of narrower-band lowpass filters (0.6 MHz) may be more appropriate.

Wide-band Gaussian filters ensure optimum compatibility with monochrome displays by minimizing the artifacts at the edges of colored objects. A narrower, sharper-cut lowpass filter would emphasize the subcarrier signal at these edges, resulting in ringing. If monochrome compatibility can be ignored, a beneficial effect of narrower filters would be to reduce the spread of the chrominance into the low-frequency luminance (resulting in low-frequency cross-luminance), which is difficult to suppress in a decoder.

Also, although the encoder may maintain a wide chrominance bandwidth, the bandwidth of the color difference signals in a decoder is usually much narrower. In the decoder, loss of the chrominance upper sidebands (due to lowpass filtering the video signal to 4.2–5.5 MHz) contributes to ringing and color difference crosstalk on color transitions. Any increase in the decoder chrominance bandwidth causes a proportionate increase in cross-color.

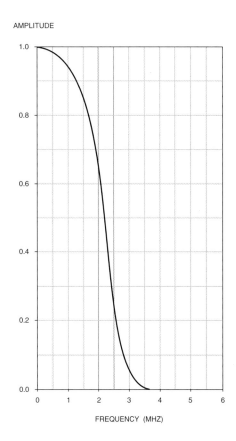

Figure 9.4. Typical 1.3-MHz Lowpass Digital Filter Characteristics.

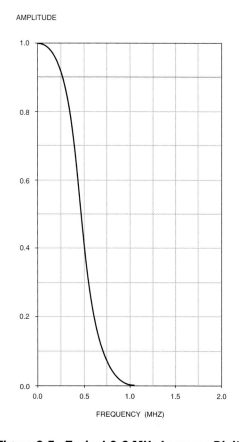

Figure 9.5. Typical 0.6-MHz Lowpass Digital Filter Characteristics.

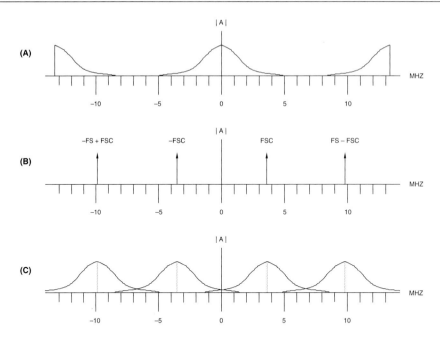

Figure 9.6. Frequency Spectra for NTSC Digital Chrominance Modulation (F_S = 13.5 MHz, F_{SC} = 3.58 MHz). (a) Lowpass filtered U and V signals. (b) Color subcarrier. (c) Modulated chrominance spectrum produced by convolving (a) and (b).

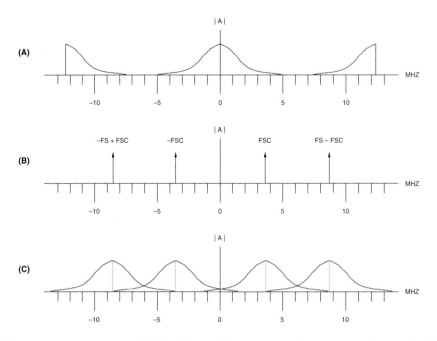

Figure 9.7. Frequency Spectra for NTSC Digital Chrominance Modulation (F_S = 12.27 MHz, F_{SC} = 3.58 MHz). (a) Lowpass filtered U and V signals. (b) Color subcarrier. (c) Modulated chrominance spectrum produced by convolving (a) and (b).

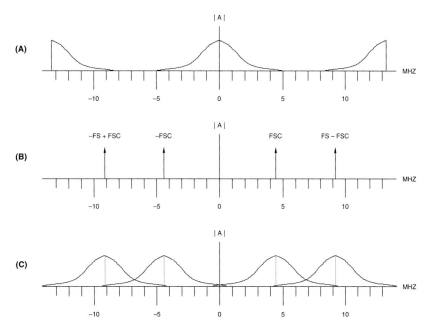

Figure 9.8. Frequency Spectra for PAL Digital Chrominance Modulation (F_S = 13.5 MHz, F_{SC} = 4.43 MHz). (a) Lowpass filtered U and V signals. (b) Color subcarrier. (c) Modulated chrominance spectrum produced by convolving (a) and (b).

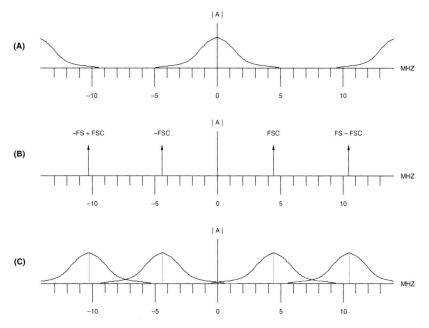

Figure 9.9. Frequency Spectra for PAL Digital Chrominance Modulation (F_S = 14.75 MHz, F_{SC} = 4.43 MHz). (a) Lowpass filtered U and V signals. (b) Color subcarrier. (c) Modulated chrominance spectrum produced by convolving (a) and (b).

Chrominance (C) Modulation

(M) NTSC, NTSC–J

During active video, the CbCr, UV, or IQ data modulate sin and cos subcarriers, as shown in Figure 9.1, resulting in digital chrominance (C) data. For this design, the 11-bit reference subcarrier phase (see Figure 9.17) and the burst phase are the same (180°).

For YUV and YCbCr processing, 180° must be added to the 11-bit reference subcarrier phase during active video time so the output of the sin and cos ROMs have the proper subcarrier phases (0° and 90°, respectively).

For YIQ processing, 213° must be added to the 11-bit reference subcarrier phase during active video time so the output of the sin and cos ROMs have the proper subcarrier phases (33° and 123°, respectively).

For the following equations,

$$\omega = 2\pi F_{SC}$$

$$F_{SC} = 3.579545 \text{ MHz } (\pm 10 \text{ Hz})$$

YUV Color Space

As discussed in Chapter 8, the chrominance signal may be represented by:

$$(U \sin \omega t) + (V \cos \omega t)$$

Chrominance amplitudes are $\pm \text{sqrt}(U^2 + V^2)$

YCbCr Color Space

If the encoder is based on the YCbCr color space, the chrominance signal may be represented by:

$$(Cb - 512)(0.504)(\sin \omega t) + \\ (Cr - 512)(0.711)(\cos \omega t)$$

For NTSC–J systems, the equations are:

$$(Cb - 512)(0.545)(\sin \omega t) + \\ (Cr - 512)(0.769)(\cos \omega t)$$

In these cases, the values in the sin and cos ROMs are scaled by the indicated values to allow the modulator multipliers to accept Cb and Cr data directly, instead of U and V data.

YIQ Color Space

As discussed in Chapter 8, the chrominance signal may also be represented by:

$$(Q \sin (\omega t + 33°)) + (I \cos (\omega t + 33°))$$

Chrominance amplitudes are $\pm \text{sqrt}(I^2 + Q^2)$

(B, D, G, H, I, M, N, N$_C$) PAL

During active video, the CbCr or UV data modulate sin and cos subcarriers, as shown in Figure 9.1, resulting in digital chrominance (C) data. For this design, the 11-bit reference subcarrier phase (see Figure 9.17) is 135°.

For the following equations,

$$\omega = 2\pi F_{SC}$$

$$F_{SC} = 4.43361875 \text{ MHz } (\pm 5 \text{ Hz}) \\ \text{for (B, D, G, H, I, N) PAL}$$

$$F_{SC} = 3.58205625 \text{ MHz } (\pm 5 \text{ Hz}) \text{ for (N}_C\text{) PAL}$$

$$F_{SC} = 3.57561149 \text{ MHz } (\pm 5 \text{ Hz}) \text{ for (M) PAL}$$

PAL Switch

In theory, since the [sin ωt] and [cos ωt] subcarriers are orthogonal, the U and V signals can be perfectly separated from each other in the decoder. However, if the video signal is subjected to distortion, such as asymmetrical attenuation of the sidebands due to lowpass filtering, the orthogonality is degraded, resulting in crosstalk between the U and V signals.

PAL uses alternate line switching of the V signal to provide a frequency offset between the U and V subcarriers, in addition to the 90° subcarrier phase offset. When decoded, crosstalk components appear modulated onto the alternate line carrier frequency, in solid color areas producing a moving pattern known as Hanover bars. This pattern may be suppressed in the decoder by a comb filter that averages equal contributions from switched and unswitched lines.

When PAL Switch = zero, the 11-bit reference subcarrier phase (see Figure 9.17) and the burst phase are the same (135°). Thus, 225° must be added to the 11-bit reference subcarrier phase during active video so the output of the sin and cos ROMs have the proper subcarrier phases (0° and 90°, respectively).

When PAL Switch = one, 90° is added to the 11-bit reference subcarrier phase, resulting in a 225° burst phase. Thus, an additional 135° must be added to the 11-bit reference subcarrier phase during active video so the output of the sin and cos ROMs have the proper phases (0° and 90°, respectively).

Note that in Figure 9.17, while PAL Switch = one, the –V subcarrier is generated, implementing the –V component.

YUV Color Space

As discussed in Chapter 8, the chrominance signal is represented by:

$$(U \sin \omega t) \pm (V \cos \omega t)$$

with the sign of V alternating from one line to the next (known as the PAL Switch).

Chrominance amplitudes are $\pm \text{sqrt}(U^2 + V^2)$.

YCbCr Color Space

If the encoder is based on the YCbCr color space, the chrominance signal for (B, D, G, H, I, N_C) PAL may be represented by:

$$(Cb - 512)(0.533)(\sin \omega t) \pm \\ (Cr - 512)(0.752)(\cos \omega t)$$

The chrominance signal for (M, N) PAL may be represented by:

$$(Cb - 512)(0.504)(\sin \omega t) \pm \\ (Cr - 512)(0.711)(\cos \omega t)$$

In these cases, the values in the sin and cos ROMs are scaled by the indicated values to allow the modulator multipliers to accept Cb and Cr data directly, instead of U and V data.

General Processing

The subcarrier sin and cos values should have a minimum of nine bits plus sign of accuracy. The modulation multipliers must have saturation logic on the outputs to ensure overflow and underflow conditions are saturated to the maximum and minimum values, respectively.

After the modulated color difference signals are added together, the result is rounded to nine bits plus sign. At this point, the digital modulated chrominance has the ranges shown in Table 9.3. The resulting digital chrominance data is clamped by a blanking signal that has the same raised cosine values and timing as the signal used to blank the luminance data.

Burst Generation

As shown in Figure 9.1, the lowpass filtered color difference data are multiplexed with the color burst envelope information. During the color burst time, the color difference data should be ignored and the burst envelope signal inserted on the Cb, U, or Q channel (the Cr, V, or I channel is forced to zero).

The burst envelope rise and fall times should generate a raised cosine distribution to slow the slew rate of the burst envelope. Typical burst envelope rise and fall times are 300 ±100 ns.

The burst envelope should be wide enough to generate nine or ten cycles of burst information with an amplitude of 50% or greater. When the burst envelope signal is multiplied by the output of the sin ROM, the color burst is generated and will have the range shown in Table 9.3.

For pro-video applications, the phase of the color burst should be programmable over a 0° to 360° range to provide optional system phase matching with external video signals. This can be done by adding a programmable value to the 11-bit subcarrier reference phase during the burst time (see Figure 9.17).

Analog Chrominance (C) Generation

The digital chrominance data may drive a 10-bit DAC that generates a 0–1.305V output to generate the C video signal of an S-video (Y/C) interface. The video signal at the connector should have a source impedance of 75Ω.

Figures 9.10 and 9.11 show the modulated chrominance video waveforms for 75% color bars. The numbers in parentheses indicate the data value for a 10-bit DAC with a full-scale output value of 1.305V. If the DAC can't handle the generation of bipolar video signals, an offset must be added to the chrominance data (and the sign information dropped) before driving the DAC. In this instance, an offset of +512 was used, positioning the blanking level at the midpoint of the 10-bit DAC output level.

As the sample-and-hold action of the DAC introduces a $(\sin x)/x$ characteristic, the video data may be digitally filtered by a $[(\sin x)/x]^{-1}$ filter to compensate. Alternately, as an analog lowpass filter is usually present after the DAC, the correction may take place in the analog filter.

Video Level	(M) NTSC	NTSC–J	(B, D, G, H, I, N_C) PAL	(M, N) PAL
peak chroma	328	354	347	328
peak burst	112	112	117	117
blank	0	0	0	0
peak burst	–112	–112	–117	–117
peak chroma	–328	–354	–347	–328

Table 9.3. 10-Bit Digital Chrominance Values.

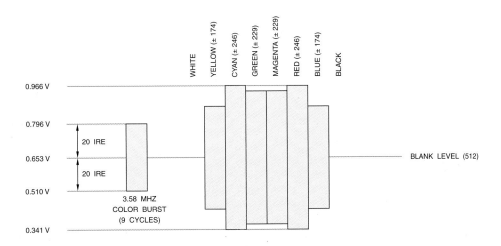

Figure 9.10. (M) NTSC Chrominance (C) Video Signal for 75% Color Bars. Indicated video levels are 10-bit values.

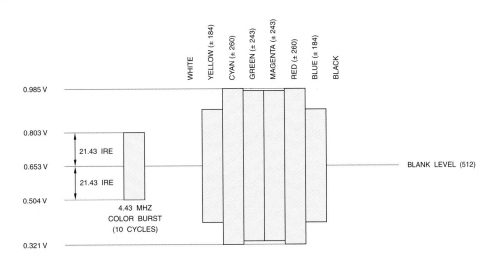

Figure 9.11. (B, D, G, H, I) PAL Chrominance (C) Video Signal for 75% Color Bars. Indicated video levels are 10-bit values.

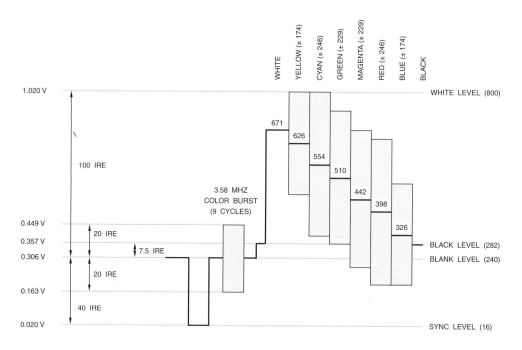

Figure 9.12. (M) NTSC Composite Video Signal for 75% Color Bars. Indicated video levels are 10-bit values.

Analog Composite Video

The digital luminance (Y) data and the digital chrominance (C) data are added together, generating digital composite color video with the levels shown in Table 9.4.

The result may drive a 10-bit DAC that generates a 0–1.305V output to generate the composite video signal. The video signal at the connector should have a source impedance of 75Ω.

Figures 9.12 and 9.13 show the video waveforms for 75% color bars. The numbers in parentheses indicate the data value for a 10-bit DAC with a full-scale output value of 1.305V.

As the sample-and-hold action of the DAC introduces a $(\sin x)/x$ characteristic, the video data may be digitally filtered by a $[(\sin x)/x]^{-1}$ filter to compensate. Alternately, as an analog lowpass filter is usually present after the DAC, the correction may take place in the analog filter.

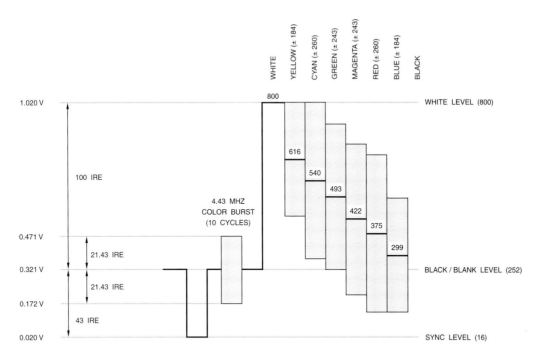

Figure 9.13. (B, D, G, H, I) PAL Composite Video Signal for 75% Color Bars. Indicated video levels are 10-bit values.

Video Level	(M) NTSC	NTSC–J	(B, D, G, H, I, N$_C$) PAL	(M, N) PAL
peak chroma	973	987	983	973
white	800	800	800	800
peak burst	352	352	369	357
black	282	240	252	282
blank	240	240	252	240
peak burst	128	128	135	123
peak chroma	109	53	69	109
sync	16	16	16	16

Table 9.4. 10-Bit Digital Composite Video Levels.

Black Burst Video Signal

As an option, the encoder can generate a black burst (or house sync) video signal that can be used to synchronize multiple video sources. Figures 9.14 and 9.15 illustrate the black burst video signals. Note that these are the same as analog composite, but do not contain any active video information. The numbers in parentheses indicate the data value for a 10-bit DAC with a full-scale output value of 1.305V.

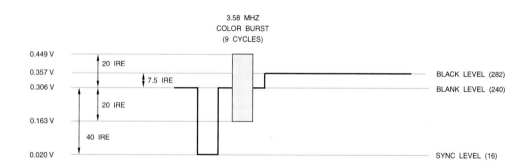

Figure 9.14. (M) NTSC Black Burst Video Signal. Indicated video levels are 10-bit values.

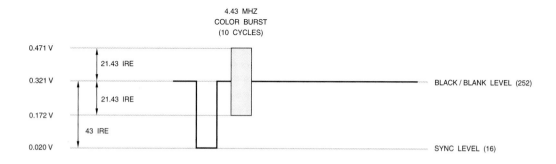

Figure 9.15. (B, D, G, H, I) PAL Black Burst Video Signal. Indicated video levels are 10-bit values.

Color Subcarrier Generation

The color subcarrier can be generated from the sample clock using a discrete time oscillator (DTO).

When generating video that may be used for editing, it is important to maintain the phase relationship between the color subcarrier and sync information. Unless the subcarrier phase relative to the sync phase is properly maintained, an edit may result in a momentary color shift. PAL also requires the addition of a PAL Switch, which is used to invert the polarity of the V data every other scan line. Note that the polarity of the PAL Switch should be maintained through the encoding and decoding process.

Since in this design the color subcarrier is derived from the sample clock, any jitter in the sample clock will result in a corresponding subcarrier frequency jitter. In some PCs, the sample clock is generated using a phase-lock loop (PLL), which may not have the necessary clock stability to keep the subcarrier phase jitter below $2°$–$3°$.

Frequency Relationships

(M) NTSC, NTSC–J

As shown in Chapter 8, there is a defined relationship between the subcarrier frequency (F_{SC}) and the line frequency (F_H):

$$F_{SC}/F_H = 910/4$$

Assuming (for example only) a 13.5-MHz sample clock rate (F_S):

$$F_S = 858\ F_H$$

Combining these equations produces the relationship between F_{SC} and F_S:

$$F_{SC}/F_S = 35/132$$

which may also be expressed in terms of the sample clock period (T_S) and the subcarrier period (T_{SC}):

$$T_S/T_{SC} = 35/132$$

The color subcarrier phase must be advanced by this fraction of a subcarrier cycle each sample clock.

(B, D, G, H, I, N) PAL

As shown in Chapter 8, there is a defined relationship between the subcarrier frequency (F_{SC}) and the line frequency (F_H):

$$F_{SC}/F_H = (1135/4) + (1/625)$$

Assuming (for example only) a 13.5-MHz sample clock rate (F_S):

$$F_S = 864\ F_H$$

Combining these equations produces the relationship between F_{SC} and F_S:

$$F_{SC}/F_S = 709379/2160000$$

which may also be expressed in terms of the sample clock period (T_S) and the subcarrier period (T_{SC}):

$$T_S/T_{SC} = 709379/2160000$$

The color subcarrier phase must be advanced by this fraction of a subcarrier cycle each sample clock.

(N_C) PAL

In the (N_C) PAL video standard used in Argentina, there is a different relationship between the subcarrier frequency (F_{SC}) and the line frequency (F_H):

$$F_{SC}/F_H = (917/4) + (1/625)$$

Assuming (for example only) a 13.5-MHz sample clock rate (F_S):

$$F_S = 864 \, F_H$$

Combining these equations produces the relationship between F_{SC} and F_S:

$$F_{SC}/F_S = 573129/2160000$$

which may also be expressed in terms of the sample clock period (T_S) and the subcarrier period (T_{SC}):

$$T_S/T_{SC} = 573129/2160000$$

The color subcarrier phase must be advanced by this fraction of a subcarrier cycle each sample clock.

Quadrature Subcarrier Generation

A DTO consists of an accumulator in which a smaller number [p] is added modulo to another number [q]. The counter consists of an adder and a register as shown in Figure 9.16. The contents of the register are constrained so that if they exceed or equal [q], [q] is subtracted from the contents. The output signal (X_N) of the adder is:

$$X_N = (X_{N-1} + p) \text{ modulo } q$$

With each clock cycle, [p] is added to produce a linearly increasing series of digital values. It is important that [q] not be an integer multiple of [p] so that the generated values are continuously different and the remainder changes from one cycle to the next.

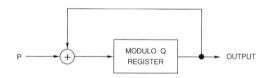

Figure 9.16. Single Stage DTO.

The DTO is used to reduce the sample clock frequency, F_S, to the color subcarrier frequency, F_{SC}:

$$F_{SC} = (p/q) \, F_S$$

Since [p] is of finite word length, the DTO output frequency can be varied only in steps. With a [p] word length of [w], the lowest [p] step is 0.5w and the lowest DTO frequency step is:

$$F_{SC} = F_S/2^w$$

Note that the output frequency cannot be greater than half the input frequency. This means that the output frequency F_{SC} can only be varied by the increment [p] and within the range:

$$0 < F_{SC} < F_S/2$$

In this application, an overflow corresponds to the completion of a full cycle of the subcarrier.

Since only the remainder (which represents the subcarrier phase) is required, the number of whole cycles completed is of no interest. During each clock cycle, the output of the [q] register shows the relative phase of a subcarrier frequency in qths of a subcarrier period. By using the [q] register contents to address a ROM containing a sine wave characteristic, a numerical representation of the sampled subcarrier sine wave can be generated.

Single Stage DTO

A single 24-bit or 32-bit modulo [q] register may be used, with the 11 most significant bits providing the subcarrier reference phase. An example of this architecture is shown in Figure 9.16.

Multi-Stage DTO

More long-term accuracy may be achieved if the ratio is partitioned into two or three fractions, the more significant of which provides the subcarrier reference phase, as shown in Figure 9.17.

To use the full capacity of the ROM and make the overflow automatic, the denominator of the most significant fraction is made a power of two. The 4× HCOUNT denominator of the least significant fraction is used to simplify hardware calculations.

Subdividing the subcarrier period into 2048 phase steps, and using the total number of samples per scan line (HCOUNT), the ratio may be partitioned as follows:

$$\frac{FSC}{FS} = \frac{P1 + \dfrac{(P2)}{(4)(HCOUNT)}}{2048}$$

P1 and P2 are programmed to generate the desired color subcarrier frequency (F_{SC}). The modulo 4× HCOUNT and modulo 2048 counters should be reset at the beginning of each vertical sync of field one to ensure the generation of the correct subcarrier reference (as shown in Figures 8.5 and 8.16).

The less significant stage produces a sequence of carry bits which correct the approximate ratio of the upper stage by altering the counting step by one: from P1 to P1 + 1. The upper stage produces an 11-bit subcarrier phase used to address the sine and cosine ROMs.

Although the upper stage adder automatically overflows to provide modulo 2048 operation, the lower stage requires additional circuitry because 4× HCOUNT may not be (and usually isn't) an integer power of two. In this case, the 16-bit register has a maximum capacity of 65535 and the adder generates a carry for any value greater than this. To produce the correct carry sequence, it is necessary, each time the adder overflows, to adjust the next number added to make up the difference between 65535 and 4× HCOUNT. This requires:

$$P3 = 65536 - (4)(HCOUNT) + P2$$

Although this changes the contents of the lower stage register, the sequence of carry bits is unchanged, ensuring that the correct phase values are generated.

The P1 and P2 values are determined for (M) NTSC operation using the following equation:

$$\frac{FSC}{FS} = \frac{P1 + \dfrac{(P2)}{(4)(HCOUNT)}}{2048}$$

$$= \left(\frac{910}{4}\right)\left(\frac{1}{HCOUNT}\right)$$

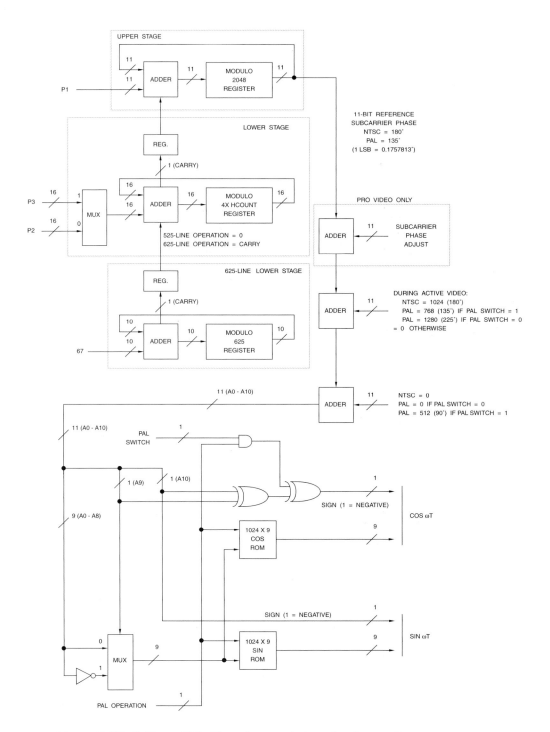

Figure 9.17. 3-Stage DTO Chrominance Subcarrier Generation.

The P1 and P2 values are determined for (B, D, G, H, I, N) PAL operation using the following equation:

$$\frac{FSC}{FS} = \frac{P1 + \dfrac{P2}{(4)(HCOUNT)}}{2048}$$

$$= \left(\frac{1135}{4} + \frac{1}{625}\right)\left(\frac{1}{HCOUNT}\right)$$

The P1 and P2 values are determined for the version of (N_C) PAL used in Argentina using the following equation:

$$\frac{FSC}{FS} = \frac{P1 + \dfrac{P2}{(4)(HCOUNT)}}{2048}$$

$$= \left(\frac{917}{4} + \frac{1}{625}\right)\left(\frac{1}{HCOUNT}\right)$$

The modulo 625 counter, with a [p] value of 67, is used during 625-line operation to more accurately adjust subcarrier generation due to the 0.1072 remainder after calculating the P1 and P2 values. During 525-line operation, the carry signal should always be forced to be zero. Table 9.5 lists some of the common horizontal resolutions, sample clock rates, and their corresponding HCOUNT, P1, and P2 values.

Sine and Cosine Generation

Regardless of the type of DTO used, each value of the 11-bit subcarrier phase corresponds to one of 2048 waveform values taken at a particular point in the subcarrier cycle period and stored in ROM. The sample points are taken at odd multiples of one 4096th of the total period to avoid end-effects when the sample values are read out in reverse order.

Note that only one quadrant of the subcarrier wave shape is stored in ROM, as shown in Figure 9.18. The values for the other quadrants are produced using the symmetrical properties of the sinusoidal waveform. The maximum phase error using this technique is ±0.09° (half of 360/2048), which corresponds to a maximum amplitude error of ±0.08%, relative to the peak-to-peak amplitude, at the steepest part of the sine wave signal.

Figure 9.17 also shows a technique for generating quadrature subcarriers from an 11-bit subcarrier phase signal. It uses two ROMs to store quadrants of sine and cosine waveforms. XOR gates invert the addresses for generating time-reversed portions of the waveforms and to invert the output polarity to make negative portions of the waveforms. An additional gate is provided in the sign bit for the V subcarrier to allow injection of a PAL Switch square wave to implement phase inversion of the V signal on alternate scan lines.

Horizontal and Vertical Timing

Vertical and horizontal counters are used to control the video timing.

Timing Control

To control the horizontal and vertical counters, separate horizontal sync (HSYNC#) and vertical sync (VSYNC#) signals are commonly used. A BLANK# control signal is usually used to indicate when to generate active video.

If HSYNC#, VSYNC#, and BLANK# are inputs, controlling the horizontal and vertical counters, this is referred to as "slave" timing. HSYNC#, VSYNC#, and BLANK# are generated by another device in the system, and used by the encoder to generate the video.

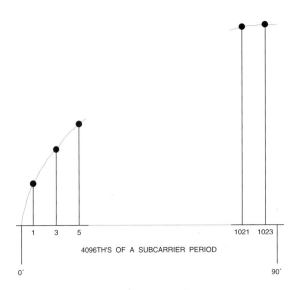

Figure 9.18. Positions of the 512 Stored Sample Values in the sin and cos ROMs for One Quadrant of a Subcarrier Cycle. Samples for other quadrants are generated by inverting the addresses and/or sign values.

Typical Application	Total Samples per Scan Line (HCOUNT)	4× HCOUNT	P1	P2
13.5 MHz (M) NTSC	858	3432	543	104
13.5 MHz (B, D, G, H, I) PAL	864	3456	672	2061
12.27 MHz (M) NTSC	780	3120	597	1040
14.75 MHz (B, D, G, H, I) PAL	944	3776	615	2253

Table 9.5. Typical HCOUNT, P1, and P2 Values for the 3-Stage DTO in Figure 9.17.

The horizontal and vertical counters may also be used to generate the basic video timing. In this case, referred to as "master" timing, HSYNC#, VSYNC#, and BLANK# are outputs from the encoder, and used elsewhere in the system.

For a BT.656 video interface, horizontal blanking (H), vertical blanking (V), and field (F) information are used. In this application, the encoder would use the H, V, and F timing bits directly, rather than depending on HSYNC#, VSYNC#, and BLANK# control signals.

Table 9.6 lists the typical horizontal blank timing for common sample clock rates. A blanking control signal (BLANK#) is used to specify when to generate active video.

Horizontal Timing

An 11-bit horizontal counter is incremented on each rising edge of the sample clock, and reset by HSYNC#. The counter value is monitored to determine when to assert and negate various control signals each scan line, such as the start of burst envelope, end of burst envelope, etc.

During "slave" timing operation, if there is no HSYNC# pulse at the end of a line, the counter can either continue incrementing (recommended) or automatically reset (not recommended).

Vertical Timing

A 10-bit vertical counter is incremented on each leading edge of HSYNC#, and reset when coincident leading edges of VSYNC# and HSYNC# occur. Rather than exactly coincident falling edges, a "coincident window" of about ±64 clock cycles should be used to ease interfacing to some video timing controllers. If both the HSYNC# and VSYNC# leading edges are detected within 64 clock cycles of each other, it is assumed to be the beginning of Field 1. The counter value is monitored to determine which scan line is being generated.

For interlaced (M) NTSC, color burst information should be disabled on scan lines 1–9 and 264–272, inclusive. On the remaining scan lines, color burst information should be enabled and disabled at the appropriate horizontal count values.

Typical Application	Sync + Back Porch Blanking (Samples)	Front Porch Blanking (Samples)
13.5 MHz (M) NTSC	122	16
13.5 MHz (B, D, G, H, I) PAL	132	12
12.27 MHz (M) NTSC	126	14
14.75 MHz (B, D, G, H, I) PAL	163	13

Table 9.6. Typical BLANK# Horizontal Timing.

For noninterlaced (M) NTSC, color burst information should be disabled on scan lines 1–9, inclusive. A 29.97 Hz (30/1.001) offset may be added to the color subcarrier frequency so the subcarrier phase will be inverted from field to field. On the remaining scan lines, color burst information should be enabled and disabled at the appropriate horizontal count values.

For interlaced (B, D, G, H, I, N, N_C) PAL, during fields 1, 2, 5, and 6, color burst information should be disabled on scan lines 1–6, 310–318, and 623–625, inclusive. During fields 3, 4, 7, and 8, color burst information should be disabled on scan lines 1–5, 311–319, and 622–625, inclusive. On the remaining scan lines, color burst information should be enabled and disabled at the appropriate horizontal count values.

For noninterlaced (B, D, G, H, I, N, N_C) PAL, color burst information should be disabled on scan lines 1–6 and 310–312, inclusive. On the remaining scan lines, color burst information should be enabled and disabled at the appropriate horizontal count values.

For interlaced (M) PAL, during fields 1, 2, 5, and 6, color burst information should be disabled on scan lines 1–8, 260–270, and 523–525, inclusive. During fields 3, 4, 7, and 8, color burst information should be disabled on scan lines 1–7, 259–269, and 522–525, inclusive. On the remaining scan lines, color burst information should be enabled and disabled at the appropriate horizontal count values.

For noninterlaced (M) PAL, color burst information should be disabled on scan lines 1–8 and 260–262, inclusive. On the remaining scan lines, color burst information should be enabled and disabled at the appropriate horizontal count values.

Early PAL receivers produced colored "twitter" at the top of the picture due to the swinging burst. To fix this, Bruch blanking was implemented to ensure that the phase of the first burst is the same following each vertical sync pulse. Analog encoders used a "meander gate" to control the burst reinsertion time by shifting one line at the vertical field rate. A digital encoder simply keeps track of the scan line and field number. Modern receivers do not require Bruch blanking, but it is useful for determining which field is being processed.

During "slave" timing operation, if there is no VSYNC# pulse at the end of a frame, the counter can either continue incrementing (recommended) or automatically reset (not recommended).

During "master" timing operation, for pro-video applications, it may be desirable to generate 2.5 scan line VSYNC# pulses during 625-line operation. However, this may cause Field 1 vs. Field 2 detection problems in some commercially available video chips.

Field ID Signals

Although the timing relationship between HSYNC# and VSYNC#, or the BT.656 F bit, is used to specify Field 1 or Field 2, additional signals may be used to specify which one of four or eight fields to generate, as shown in Table 9.7.

FIELD_0 should change state coincident with the leading edge of VSYNC# during fields 1, 3, 5, and 7. FIELD_1 should change state coincident with the leading edge of VSYNC# during fields 1 and 5.

For BT.656 video interface, FIELD_0 and FIELD_1 may be transmitted using ancillary data.

Clean Encoding

Typically, the only filters present in a conventional encoder are the color difference low-pass filters. This results in considerable spectral overlap between the luminance and chrominance components, making it impossible to separate the signals completely at the decoder.

However, additional processing at the encoder can be used to reduce cross-color (luminance-to-chrominance crosstalk) and cross-luminance (chrominance-to-luminance crosstalk) decoder artifacts. Cross-color appears as a coarse rainbow pattern or random colors in regions of fine detail. Cross-luminance appears as a fine pattern on chrominance edges.

Cross-color in a decoder may be reduced by removing some of the high-frequency luminance data in the encoder, using a notch filter at F_{SC}. However, while reducing the cross-color, luminance detail is lost.

A better method is to pre-comb filter the luminance and chrominance information in the encoder (see Figure 9.19). High-frequency luminance information is precombed to minimize interference with chrominance frequencies in that spectrum. Chrominance information also is pre-combed by averaging over a number of lines, reducing cross-luminance or the "hanging dot" pattern.

This technique allows fine, moving luminance (which tends to generate cross-color at the decoder) to be removed while retaining full resolution for static luminance. However, there is a small loss of diagonal luminance resolution due to it being averaged over multiple lines. This is offset by an improvement in the chrominance signal-to-noise ratio (SNR).

FIELD_1 Signal	FIELD_0 Signal	HSYNC# and VSYNC# Timing Relationship or BT.656 F Bit	NTSC Field Number		PAL Field Number	
0	0	field 1	1	odd field	1	even field
0	0	field 2	2	even field	2	odd field
0	1	field 1	3	odd field	3	even field
0	1	field 2	4	even field	4	odd field
1	0	field 1	–	–	5	even field
1	0	field 2	–	–	6	odd field
1	1	field 1	–	–	7	even field
1	1	field 2	–	–	8	odd field

Table 9.7. Field Numbering.

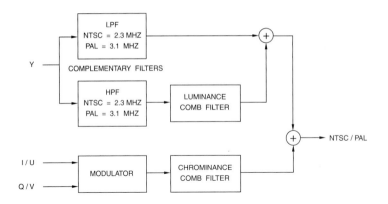

Figure 9.19. Clean Encoding Example.

Bandwidth-Limited Edge Generation

Smooth sync and blank edges may be generated by integrating a T, or raised cosine, pulse to generate a T step (Figure 9.20). NTSC systems use a T pulse with T = 125 ns; therefore, the 2T step has little signal energy beyond 4 MHz. PAL systems use a T pulse with T = 100 ns; in this instance, the 2T step has little signal energy beyond 5 MHz.

The T step provides a fast risetime, without ringing, within a well-defined bandwidth. The risetime of the edge between the 10% and 90% points is 0.964T. By choosing appropriate sample values for the sync edges, blanking edges, and burst envelope, these values can be stored in a small ROM, which is triggered at the appropriate horizontal count. By reading the contents of the ROM forward and backward, both rising and falling edges may be generated.

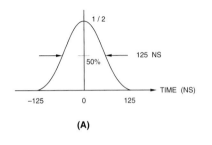

(A)

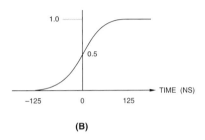

(B)

Figure 9.20. Bandwidth-limited Edge Generation. (a) NTSC T pulse. (b) The T step, the result of integrating the T pulse.

Level Limiting

Certain highly saturated colors produce composite video levels that may cause problems in downstream equipment.

Invalid video levels greater than 100 IRE or less than –20 IRE (relative to the blank level) may be transmitted, but may cause distortion in VCRs or demodulators and cause sync separation problems.

Illegal video levels greater than 120 IRE (NTSC) or 133 IRE (PAL), or below the sync tip level, may not be transmitted.

Although usually not a problem in a conventional video application, computer systems commonly use highly saturated colors, which may generate invalid or illegal video levels. It may be desirable to optionally limit these signal levels to around 110 IRE, compromising between limiting the available colors and generating legal video levels.

One method of correction is to adjust the luminance or saturation of invalid and illegal pixels until the desired peak limits are attained. Alternately, the frame buffer contents may be scanned, and pixels flagged that would generate an invalid or illegal video level (using a separate overlay plane or color change). The user then may change the color to a more suitable one.

In a professional editing application, the option of transmitting all the video information (including invalid and illegal levels) between equipment is required to minimize editing and processing artifacts.

Encoder Video Parameters

Many industry-standard video parameters have been defined to specify the relative quality of NTSC and PAL encoders. To measure these parameters, the output of the encoder (while generating various video test signals such as those described in Chapter 8) is monitored using video test equipment. Along with a description of several of these parameters, typical AC parameter values for both consumer and studio-quality encoders are shown in Table 9.8.

Several AC parameters, such as group delay and K factors, are dependent on the quality of the output filters and are not discussed here. In addition to the AC parameters discussed in this section, there are several others that should be included in an encoder specification, such as burst frequency and tolerance, horizontal frequency, horizontal blanking time, sync rise and fall times, burst envelope rise and fall times, video blanking rise and fall times, and the bandwidths of the YIQ or YUV components.

There are also several DC parameters (such as white level and tolerance, blanking level and tolerance, sync height and tolerance, peak-to-peak burst amplitude and tolerance) that should be specified, as shown in Table 9.9.

Differential Phase

Differential phase distortion, commonly referred to as differential phase, specifies how much the chrominance phase is affected by the luminance level—in other words, how much hue shift occurs when the luminance level changes. Both positive and negative phase errors may be present, so differential phase is expressed as a peak-to-peak measurement, expressed in degrees of subcarrier phase.

This parameter is measured using chroma of uniform phase and amplitude superimposed on different luminance levels, such as the modulated ramp test signal or the modulated 5-step portion of the composite test signal. The differential phase parameter for a studio-quality encoder may approach 0.2° or less.

Parameter	Consumer Quality		Studio Quality		Units
	NTSC	**PAL**	**NTSC**	**PAL**	
differential phase	4		≤ 1		degrees
differential gain	4		≤ 1		%
luminance nonlinearity	2		≤ 1		%
hue accuracy	3		≤ 1		degrees
color saturation accuracy	3		≤ 1		%
residual subcarrier	0.5		0.1		IRE
SNR (per EIA-250-C)	48		> 60		dB
SCH phase	0 ±40	0 ±20	0 ±2		degrees
analog Y/C output skew	5		≤ 2		ns
H tilt	< 1		< 1		%
V tilt	< 1		< 1		%
subcarrier tolerance	10	5	10	5	Hz

Table 9.8. Typical AC Video Parameters for (M) NTSC and (B, D, G, H, I) PAL Encoders.

Parameter	Consumer Quality		Studio Quality		Units
	NTSC	**PAL**	**NTSC**	**PAL**	
white relative to blank	714 ±70	700 ±70	714 ±7	700 ±7	mV
black relative to blank	54 ±5	0	54 ±0.5	0	mV
sync relative to blank	−286 ±30	−300 ±30	−286 ±3	−300 ±3	mV
burst amplitude	286 ±30	300 ±30	286 ±3	300 ±3	mV

Table 9.9. Typical DC Video Parameters for (M) NTSC and (B, D, G, H, I) PAL Encoders.

Differential Gain

Differential gain distortion, commonly referred to as differential gain, specifies how much the chrominance gain is affected by the luminance level—in other words, how much color saturation shift occurs when the luminance level changes. Both attenuation and amplification may occur, so differential gain is expressed as the largest amplitude change between any two levels, expressed as a percentage of the largest chrominance amplitude.

This parameter is measured using chroma of uniform phase and amplitude superimposed on different luminance levels, such as the modulated ramp test signal or the modulated 5-step portion of the composite test signal. The differential gain parameter for a studio-quality encoder may approach 0.2% or less.

Luminance Nonlinearity

Luminance nonlinearity, also referred to as differential luminance and luminance nonlinear distortion, specifies how much the luminance gain is affected by the luminance level—in other words, a nonlinear relationship between the generated and ideal luminance levels.

Using an unmodulated 5-step or 10-step staircase test signal, the difference between the largest and smallest steps, expressed as a percentage of the largest step, is used to specify the luminance nonlinearity. Although this parameter is included within the differential gain and phase parameters, it is traditionally specified independently.

Chrominance Nonlinear Phase Distortion

Chrominance nonlinear phase distortion specifies how much the chrominance phase (hue) is affected by the chrominance amplitude (saturation)—in other words, how much hue shift occurs when the saturation changes.

Using a modulated pedestal test signal, or the modulated pedestal portion of the combination test signal, the phase differences between each chrominance packet and the burst are measured. The difference between the largest and the smallest measurements is the peak-to-peak value, expressed in degrees of subcarrier phase. This parameter is usually not independently specified, but is included within the differential gain and phase parameters.

Chrominance Nonlinear Gain Distortion

Chrominance nonlinear gain distortion specifies how much the chrominance gain is affected by the chrominance amplitude (saturation)—in other words, a nonlinear relationship between the generated and ideal chrominance amplitude levels, usually seen as an attenuation of highly saturated chrominance signals.

Using a modulated pedestal test signal, or the modulated pedestal portion of the combination test signal, the test equipment is adjusted so that the middle chrominance packet is 40 IRE. The largest difference between the measured and nominal values of the amplitudes of the other two chrominance packets specifies the chrominance nonlinear gain distortion, expressed in IRE or as a percentage of the nominal amplitude of the worst-case packet. This parameter is usually not independently specified, but is included within the differential gain and phase parameters.

Chrominance-to-Luminance Intermodulation

Chrominance-to-luminance intermodulation, commonly referred to as cross-modulation, specifies how much the luminance level is affected by the chrominance. This may be the result of clipping highly saturated chrominance levels or quadrature distortion and may show up as irregular brightness variations due to changes in color saturation.

Using a modulated pedestal test signal, or the modulated pedestal portion of the combination test signal, the largest difference between the ideal 50 IRE pedestal level and the measured luminance levels (after removal of chrominance information) specifies the chrominance-to-luminance intermodulation, expressed in IRE or as a percentage. This parameter is usually not independently specified, but is included within the differential gain and phase parameters.

Hue Accuracy

Hue accuracy specifies how closely the generated hue is to the ideal hue value. Both positive and negative phase errors may be present, so hue accuracy is the difference between the worst-case positive and worst-case negative measurements from nominal, expressed in degrees of subcarrier phase. This parameter is measured using EIA or EBU 75% color bars as a test signal.

Color Saturation Accuracy

Color saturation accuracy specifies how closely the generated saturation is to the ideal saturation value, using EIA or EBU 75% color bars as a test signal. Both gain and attenuation may be present, so color saturation accuracy is the difference between the worst-case gain and worst-case attenuation measurements from nominal, expressed as a percentage of nominal.

Residual Subcarrier

The residual subcarrier parameter specifies how much subcarrier information is present during white or gray (note that, ideally,

none should be present). Excessive residual subcarrier is visible as noise during white or gray portions of the picture.

Using an unmodulated 5-step or 10-step staircase test signal, the maximum peak-to-peak measurement of the subcarrier (expressed in IRE) during active video is used to specify the residual subcarrier relative to the burst amplitude.

SCH Phase

SCH (Subcarrier to Horizontal) phase refers to the phase relationship between the leading edge of horizontal sync (at the 50% amplitude point) and the zero crossings of color burst (by extrapolating the color burst to the leading edge of sync). The error is referred to as SCH phase and is expressed in degrees of subcarrier phase.

For PAL, the definition of SCH phase is slightly different due to the more complicated relationship between the sync and subcarrier frequencies—the SCH phase relationship for a given line repeats only once every eight fields. Therefore, PAL SCH phase is defined, per EBU Technical Statement D 23-1984 (E), as "the phase of the +U component of the color burst extrapolated to the half-amplitude point of the leading edge of the synchronizing pulse of line 1 of field 1."

SCH phase is important when merging two or more video signals. To avoid color shifts or "picture jumps," the video signals must have the same horizontal, vertical, and subcarrier timing and the phases must be closely matched. To achieve these timing constraints, the video signals must have the same SCH phase relationship since the horizontal sync and subcarrier are continuous signals with a defined relationship. It is common for an encoder to allow adjustment of the SCH phase to simplify merging two or more video signals. Maintaining proper SCH phase is also important since NTSC and PAL decoders may monitor the SCH phase to determine which color field is being decoded.

Analog Y/C Video Output Skew

The output skew between the analog luminance (Y) and chrominance (C) video signals should be minimized to avoid phase shift errors between the luminance and chrominance information. Excessive output skew is visible as artifacts along sharp vertical edges when viewed on a monitor.

H Tilt

H tilt, also known as line tilt and line time distortion, causes a tilt in line-rate signals, predominantly white bars. This type of distortion causes variations in brightness between the left and right edges of an image. For a digital encoder, such as that described in this chapter, H tilt is primarily an artifact of the analog output filters and the transmission medium.

H tilt is measured using a line bar (such as the one in the NTC-7 NTSC composite test signal) and measuring the peak-to-peak deviation of the tilt (in IRE or percent of white bar amplitude), ignoring the first and last microsecond of the white bar.

V Tilt

V tilt, also known as field tilt and field time distortion, causes a tilt in field-rate signals, predominantly white bars. This type of distortion causes variations in brightness between the top and bottom edges of an image. For a digital encoder, such as that described in this chapter, V tilt is primarily an artifact of the analog output filters and the transmission medium.

V tilt is measured using a 18 µs, 100 IRE white bar in the center of 130 lines in the center of the field or using a field square wave. The peak-to-peak deviation of the tilt is measured (in IRE or percent of white bar amplitude), ignoring the first three and last three lines.

Genlocking Support

In many instances, it is desirable to be able to genlock the output (align the timing signals) of an encoder to another composite analog video signal to facilitate downstream video processing. This requires locking the horizontal, vertical, and color subcarrier frequencies and phases together, as discussed in the NTSC/PAL decoder section of this chapter. In addition, the luminance and chrominance amplitudes must be matched. A major problem in genlocking is that the regenerated sample clock may have excessive jitter, resulting in color artifacts.

One genlocking variation is to send an advance house sync (also known as black burst or advance sync) to the encoder. The advancement compensates for the delay from the house sync generator to the encoder output being used in the downstream processor, such as a mixer. Each video source has its own advanced house sync signal, so each video source is time-aligned at the mixing or processing point.

Another genlocking option allows adjustment of the subcarrier phase so it can be matched with other video sources at the mixing or processing point. The subcarrier phase must be able to be adjusted from 0° to 360°. Either zero SCH phase is always maintained or another adjustment is allowed to independently position the sync and luminance information in about 10 ns steps.

The output delay variation between products should be within about ±0.8 ns to allow video signals from different genlocked devices to be mixed properly. Mixers usually assume the two video signals are perfectly genlocked, and excessive time skew between the two video signals results in poor mixing performance.

Alpha Channel Support

An encoder designed for pro-video editing applications may support an alpha channel. Eight or ten bits of digital alpha data are input, pipelined to match the pipeline of the encoding process, and converted to an analog alpha signal (discussed in Chapter 7). Alpha is usually linear, with the data generating an analog alpha signal (also called a key) with a range of 0–100 IRE. There is no blanking pedestal or sync information present.

In computer systems that support 32-bit pixels, 8 bits are typically available for alpha information.

NTSC and PAL Digital Decoding

Although the luminance and chrominance components in a NTSC/PAL encoder are usually combined by simply adding the two signals together, separating them in a decoder is much more difficult. Analog NTSC and PAL decoders have been around for some time. However, they have been difficult to use, required adjustment, and offered limited video quality.

Using digital techniques to implement NTSC and PAL decoding offers many advantages, such as ease of use, minimum analog adjustments, and excellent video quality. The use of digital circuitry also enables the design of much more robust and sophisticated Y/C separator and genlock implementations.

A general block diagram of a NTSC/PAL digital decoder is shown in Figure 9.21.

Digitizing the Analog Video

The first step in digital decoding of composite video signals is to digitize the entire composite video signal using an A/D converter (ADC). For our example, 10-bit ADCs are used; therefore, indicated values are 10-bit values.

The composite and S-video signals are illustrated in Figures 9.2, 9.3. 9.10, 9.11, 9.12, and 9.13.

Video inputs are usually AC-coupled and have a 75-Ω AC and DC input impedance. As a result, the video signal must be DC restored every scan line during horizontal sync to position the sync tips at a known voltage level.

The video signal must also be lowpass filtered (typically to about 6 MHz) to remove any high-frequency components that may result in aliasing. Although the video bandwidth for broadcast is rigidly defined, there is no standard for consumer equipment. The video source generates as much bandwidth as it can; the receiving equipment accepts as much bandwidth as it can process.

Video signals with amplitudes of 0.25× to 2× ideal are common in the consumer market. The active video and/or sync signal may change amplitude, especially in editing situations where the video signal may be composed of several different video sources merged together.

In addition, the decoder should be able to handle 100% colors. Although only 75% colors may be broadcast, there is no such limitation for baseband video. With the frequent use of computer-generated text and graphics, highly-saturated colors are becoming more common.

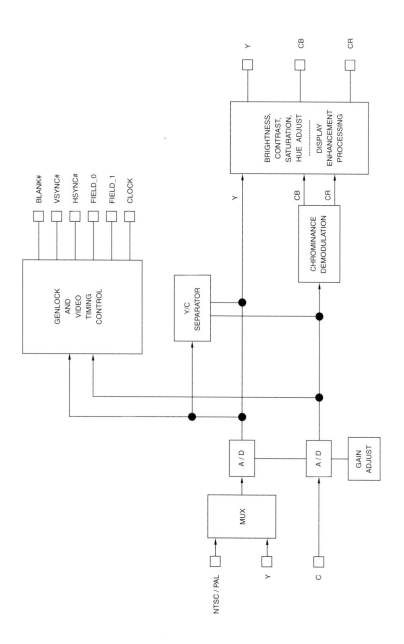

Figure 9.21. Typical NTSC/PAL Digital Decoder Implementation.

DC Restoration

To remove any DC offset that may be present in the video signal, and position it at a known level, DC restoration (also called clamping) is done.

For composite or luminance (Y) video signals, the analog video signal is DC restored to the REF– voltage of the ADC during each horizontal sync time. Thus, the ADC generates a code of 0 during the sync level.

For chrominance (C) video signals, the analog video signal is DC restored to the midpoint of the ADC during the horizontal sync time. Thus, the ADC generates a code of 512 during the blanking level.

Automatic Gain Control

An automatic gain control (AGC) is used to ensure that a constant value for the blanking level is generated by the ADC. If the blanking level is low or high, the video signal is amplified or attenuated until the blanking level is correct.

In S-video applications, the same amount of gain that is applied to the luminance video signal should also be applied to the chrominance video signal.

After DC restoration and AGC processing, an offset of 16 is added to the digitized composite and luminance signals to match the levels used by the encoder.

Tables 9.2, 9.3, and 9.4 show the ideal ADC values for composite and s-video sources after DC restoration and automatic gain control has been done.

Blank Level Determination

The most common method of determining the blanking level is to digitally lowpass filter the video signal to about 0.5 MHz to remove subcarrier information and noise. The back porch is then sampled multiple times to determine an average blank level value.

To limit line-to-line variations and clamp streaking (the result of quantizing errors), the result should be averaged over 3–32 consecutive scan lines. Alternately, the back porch level may be determined during the vertical blanking interval and the result used for the entire field.

Video Gain Options

The difference from the ideal blanking level is processed and used in one of several ways to generate the correct blanking level:

(a) controlling a voltage-controlled amplifier

(b) adjusting the REF+ voltage of the ADC

(c) multiplying the outputs of the ADC

In (a) and (b), an analog signal for controlling the gain may be generated by either a DAC or a charge pump. If a DAC is used, it should have twice the resolution of the ADC to avoid quantizing noise. For this reason, a charge pump implementation may be more suitable.

Option (b) is dependent on the ADC being able to operate over a wide range of reference voltages, and is therefore rarely implemented.

Option (c) is rarely used due to the resulting quantization errors from processing in the digital domain.

Sync Amplitude AGC

This is the most common mode of AGC, and is used where the characteristics of the video signal are not known. The difference between the measured and the ideal blanking level is used to determine how much to increase or decrease the gain of the entire video signal.

Burst Amplitude AGC

Another method of AGC is based on the color burst amplitude. This is commonly used in pro-video applications when the sync amplitude may not be related to the active video amplitude.

First, the blanking level is adjusted to the ideal value, regardless of the sync tip position. This may be done by adding or subtracting a DC offset to the video signal.

Next, the burst amplitude is determined. To limit line-to-line variations, the burst amplitude may be averaged over 3–32 consecutive scan lines.

The difference between the measured and the ideal burst amplitude is used to determine how much to increase or decrease the gain of the entire video signal. During the gain adjustment, the blanking value should not change.

AGC Options

For some pro-video applications, such as if the video signal levels are known to be correct, if all the video levels except the sync height are correct, or if there is excessive noise in the video signal, it may be desirable to disable the automatic gain control.

The AGC value to use may be specified by the user, or the AGC value frozen once determined.

Y/C Separation

When decoding composite video, the luminance (Y) and chrominance (C) must be separated. The many techniques for doing this are discussed in detail later in the chapter.

After Y/C separation, Y has the nominal values shown in Table 9.2. Note that the luminance still contains sync and blanking information. Modulated chrominance has the nominal values shown in Table 9.3.

The quality of Y/C separation is a major factor in the overall video quality generated by the decoder.

Color Difference Processing

Chrominance (C) Demodulation

The chrominance demodulator (Figure 9.22) accepts modulated chroma data from either the Y/C separator or the chroma ADC. It generates CbCr, UV, or IQ color difference data.

(M) NTSC, NTSC–J

During active video, the chrominance data is demodulated using sin and cos subcarrier data, as shown in Figure 9.22, resulting in CbCr, UV, or IQ data. For this design, the 11-bit reference subcarrier phase (see Figure 9.32) and the burst phase are the same (180°).

For YUV or YCbCr processing, 180° must be added to the 11-bit reference subcarrier phase during active video time so the output of the sin and cos ROMs have the proper subcarrier phases (0° and 90°, respectively).

For YIQ processing, 213° must be added to the 11-bit reference subcarrier phase during active video time so the output of the sin and cos ROMs have the proper subcarrier phases (33° and 123°, respectively).

For all the equations,

$$\omega = 2\pi F_{SC}$$

$$F_{SC} = 3.579545 \text{ MHz}$$

YUV Color Space Processing

As shown in Chapter 8, the chrominance signal processed by the demodulator may be represented by:

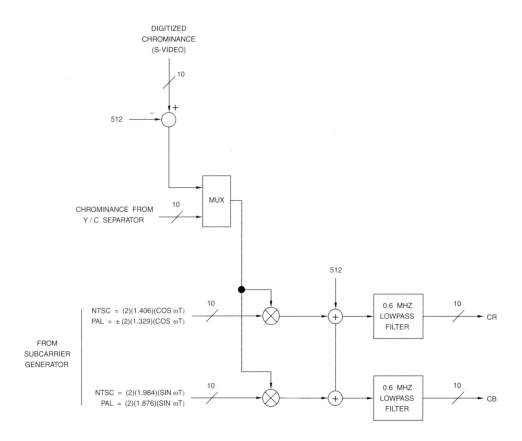

Figure 9.22. Chrominance Demodulation Example That Generates CbCr Directly.

(U sin ωt) + (V cos ωt)

U is obtained by multiplying the chrominance data by [2 sin ωt], and V is obtained by multiplying by [2 cos ωt]:

((U sin ωt) + (V cos ωt)) (2 sin ωt)
= U − (U cos 2ωt) + (V sin 2ωt)

((U sin ωt) + (V cos ωt)) (2 cos ωt)
= V + (V cos 2ωt) + (U sin 2ωt)

The 2ωt components are removed by low-pass filtering, resulting in the U and V signals

being recovered. The demodulator multipliers should ensure overflow and underflow conditions are saturated to the maximum and minimum values, respectively. The UV signals are then rounded to nine bits plus sign and lowpass filtered.

For (M) NTSC, U has a nominal range of 0 to ±226, and V has a nominal range of 0 to ±319.

For NTSC–J used in Japan, U has a nominal range of 0 to ±244, and V has a nominal range of 0 to ±344.

YIQ Color Space Processing

As shown in Chapter 8, for older decoders, the chrominance signal processed by the demodulator may be represented by:

$$(Q \sin (\omega t + 33°)) + (I \cos (\omega t + 33°))$$

The subcarrier generator of the decoder provides a 33° phase offset during active video, cancelling the 33° phase terms in the equation.

Q is obtained by multiplying the chrominance data by [2 sin ωt], and I is obtained by multiplying by [2 cos ωt]:

$$((Q \sin \omega t) + (I \cos \omega t)) \, (2 \sin \omega t)$$
$$= Q - (Q \cos 2\omega t) + (I \sin 2\omega t)$$

$$((Q \sin \omega t) + (I \cos \omega t)) \, (2 \cos \omega t)$$
$$= I + (I \cos 2\omega t) + (Q \sin 2\omega t)$$

The 2ωt components are removed by low-pass filtering, resulting in the I and Q signals being recovered. The demodulator multipliers should ensure overflow and underflow conditions are saturated to the maximum and minimum values, respectively. The IQ signals are then rounded to nine bits plus sign and low-pass filtered.

For (M) NTSC, I has a nominal range of 0 to ±309, and Q has a nominal range of 0 to ±271.

For NTSC–J used in Japan, I has a nominal range of 0 to ±334, and Q has a nominal range of 0 to ±293.

YCbCr Color Space Processing

If the decoder is based on the YCbCr color space, the chrominance signal may be represented by:

$$(Cb - 512) (0.504) (\sin \omega t) +$$
$$(Cr - 512) (0.711) (\cos \omega t)$$

For NTSC–J systems, the equations are:

$$(Cb - 512) (0.545) (\sin \omega t) +$$
$$(Cr - 512) (0.769) (\cos \omega t)$$

In these cases, the values in the sin and cos ROMs are scaled by the reciprocal of the indicated values to allow the demodulator to generate Cb and Cr data directly, instead of U and V data.

(B, D, G, H, I, M, N, N$_C$) PAL

During active video, the digital chrominance (C) data is demodulated using sin and cos subcarrier data, as shown in Figure 9.22, resulting in CbCr or UV data. For this design, the 11-bit reference subcarrier phase (see Figure 9.32) and the burst phase are the same (135°).

For all the equations,

$$\omega = 2\pi F_{SC}$$

$$F_{SC} = 4.43361875 \text{ MHz}$$
$$\text{for (B, D, G, H, I, N) PAL}$$

$$F_{SC} = 3.58205625 \text{ MHz for (N}_C) \text{ PAL}$$

$$F_{SC} = 3.57561149 \text{ MHz for (M) PAL}$$

Using a switched subcarrier waveform in the Cr or V channel also removes the PAL Switch modulation. Thus, [+2 cos ωt] is used while the PAL Switch is a logical zero (burst phase = +135°) and [–2 cos ωt] is used while the PAL Switch is a logical one (burst phase = 225°).

YUV Color Space

As shown in Chapter 8, the chrominance signal is represented by:

$$(U \sin \omega t) \pm (V \cos \omega t)$$

U is obtained by multiplying the chrominance data by [2 sin ωt] and V is obtained by multiplying by [±2 cos ωt]:

$$((U \sin \omega t) \pm (V \cos \omega t)) \ (2 \sin \omega t)$$
$$= U - (U \cos 2\omega t) \pm (V \sin 2\omega t)$$

$$((U \sin \omega t) \pm (V \cos \omega t)) \ (\pm 2 \cos \omega t)$$
$$= V \pm (U \sin 2\omega t) + (V \cos 2\omega t)$$

The 2ωt components are removed by low-pass filtering, resulting in the U and V signals being recovered. The demodulation multipliers should ensure overflow and underflow conditions are saturated to the maximum and minimum values, respectively. The UV signals are then rounded to nine bits plus sign and lowpass filtered.

For (B, D, G, H, I, N_C) PAL, U has a nominal range of 0 to ±239, and V has a nominal range of 0 to ±337.

For (M, N) PAL, U has a nominal range of 0 to ±226, and V has a nominal range of 0 to ±319.

YCbCr Color Space

If the decoder is based on the YCbCr color space, the chrominance signal for (B, D, G, H, I, N_C) PAL may be represented by:

$$(Cb - 512)(0.533)(\sin \omega t) \pm$$
$$(Cr - 512)(0.752)(\cos \omega t)$$

The chrominance signal for (M, N) PAL may be represented by:

$$(Cb - 512)(0.504)\sin \omega t \pm$$
$$(Cr - 512)(0.711)\cos \omega t$$

In these cases, the values in the sin and cos ROMs are scaled by the reciprocal of the indicated values to allow the demodulator to generate Cb and Cr data directly, instead of U and V data.

Hanover Bars

If the locally generated subcarrier phase is incorrect, a line-to-line pattern known as Hanover bars results in which pairs of adjacent lines have a real and complementary hue error. As shown in Figure 9.23 with an ideal color of green, two adjacent lines of the display have a hue error (towards yellow), the next two have the complementary hue error (towards cyan), and so on.

This can be shown by introducing a phase error (θ) in the locally generated subcarrier:

$$((U \sin \omega t) \pm (V \cos \omega t)) \ (2 \sin (\omega t - \theta))$$
$$= (U \cos \theta) -/+ (V \sin \theta)$$

$$((U \sin \omega t) \pm (V \cos \omega t)) \ (\pm 2 \cos (\omega t - \theta))$$
$$= (V \cos \theta) +/- (U \sin \theta)$$

In areas of constant color, averaging equal contributions from even and odd lines (either visually or using a delay line), cancels the alternating crosstalk component, leaving only a desaturation of the true component by [cos θ].

Lowpass Filtering

The decoder requires sharper roll-off filters than the encoder to ensure adequate suppression of the sampling alias components. Note that with a 13.5-MHz sampling frequency, they start to become significant above 3 MHz.

The demodulation process for (M) NTSC is shown spectrally in Figures 9.24 and 9.25; the process is similar for PAL. In both figures, (a) represents the spectrum of the video signal and (b) represents the spectrum of the subcarrier used for demodulation. Convolution of (a) and (b), equivalent to multiplication in the time domain, produces the spectrum shown in (c), in which the baseband spectrum has been shifted to be centered about F_{SC} and $-F_{SC}$. The chrominance is now a baseband signal, which

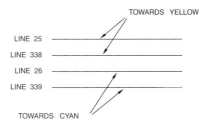

Figure 9.23. Example Display of Hanover Bars. Green is the ideal color.

may be separated from the low-frequency luminance, centered at F_{SC}, by a lowpass filter.

The lowpass filters after the demodulator are a compromise between several factors. Simply using a 1.3-MHz filter, such as the one shown in Figure 9.26, increases the amount of cross-color since a greater number of luminance frequencies are included. When using lowpass filters with a passband greater than about 0.6 MHz for NTSC (4.2 – 3.58) or 1.07 MHz for PAL (5.5 – 4.43), the loss of the upper sidebands of chrominance also introduces ringing and color difference crosstalk. If a 1.3-MHz lowpass filter is used, it may include some gain for frequencies between 0.6 MHz and 1.3 MHz to compensate for the loss of part of the upper sideband.

Filters with a sharp cutoff accentuate chrominance edge ringing; for these reasons slow roll-off 0.6-MHz filters, such as the one shown in Figure 9.27, are usually used. These result in poorer color resolution but minimize cross-color, ringing, and color difference crosstalk on edges.

If the decoder is to be used in a pro-video editing environment, the filters should have a maximum ripple of ±0.1 dB in the passband. This is needed to minimize the cumulation of gain and loss artifacts due to the filters, especially when multiple passes through the encoding and decoding processes are required.

Luminance (Y) Processing

To remove the sync and blanking information, Y data from either the Y/C separator or the luma ADC has the black level subtracted from it. At this point, negative Y values should be supported to allow test signals, keying information, and real-world video to pass through without corruption.

A notch filter, with a center frequency of F_{SC}, is usually optional. It may be used to remove any remaining chroma information from the Y data. The notch filter is especially useful to help "clean up" the Y data when comb filtering Y/C separation is used for PAL, due to the closeness of the PAL frequency packets.

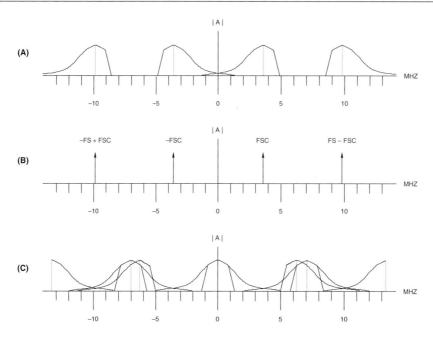

Figure 9.24. Frequency Spectra for NTSC Digital Chrominance Demodulation (F_S = 13.5 MHz, F_{SC} = 3.58 MHz). (a) Modulated chrominance. (b) Color subcarrier. (c) U and V spectrum produced by convolving (a) and (b).

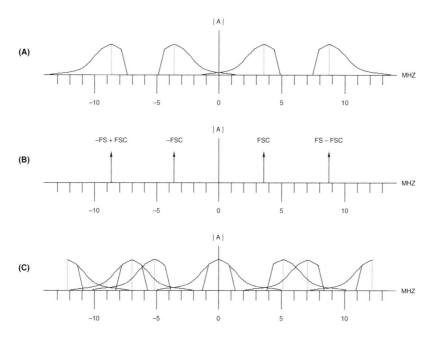

Figure 9.25. Frequency Spectra for NTSC Digital Chrominance Demodulation (F_S = 12.27 MHz, F_{SC} = 3.58 MHz). (a) Modulated chrominance. (b) Color subcarrier. (c) U and V spectrum produced by convolving (a) and (b).

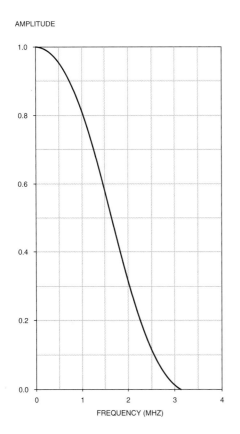

Figure 9.26. Typical 1.3-MHz Lowpass Digital Filter Characteristics.

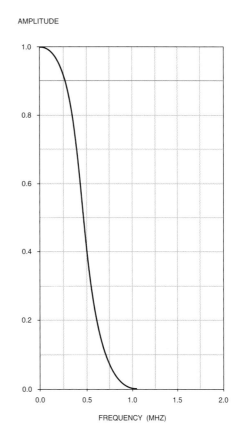

Figure 9.27. Typical 0.6-MHz Lowpass Digital Filter Characteristics.

User Adjustments

Contrast, Brightness, and Sharpness

Programmable contrast, brightness, and sharpness adjustments may be implemented, as discussed in Chapter 7. In addition, color transient improvement may be used to improve the image quality.

Hue

A programmable hue adjustment may be implemented, as discussed in Chapter 7.

Alternately, to reduce circuitry in the data path, the hue adjustment is usually implemented as a subcarrier phase offset that is added to the 11-bit reference subcarrier phase during the active video time (see Figure 9.32). The result is to shift the phase of the sin and cos subcarriers by a constant amount. An 11-bit hue adjustment allows adjustments in hue from 0° to 360°, in increments of 0.176°.

Due to the alternating sign of the V component in PAL decoders, the sign of the phase offset (θ) is set to be the opposite of the V component. A negative sign of the phase offset (θ) is equivalent to adding 180° to the desired phase shift. PAL decoders do not usually have a hue adjustment feature.

Saturation

A programmable saturation adjustment may be implemented, as discussed in Chapter 7.

Alternately, to reduce circuitry in the data path, the saturation adjustment may be done on the sin and cos values in the demodulator.

In either case, a "burst level error" signal and the user-programmable saturation value are multiplied together, and the result is used to adjust the gain or attenuation of the color difference signals. The intent here is to minimize the amount of circuitry in the color difference signal path. The "burst level error" signal is used in the event the burst (and thus the modulated chrominance information) is not at the correct amplitude and adjusts the saturation of the color difference signals appropriately.

For more information on the "burst level error" signal, please see the Color Killer section.

Automatic Flesh Tone Correction

Flesh tone correction may be used in NTSC decoders since the eye is very sensitive to flesh tones, and the actual colors may become slightly corrupted during the broadcast process. If the grass is not quite the proper color of green, it is not noticeable; however, a flesh tone that has a green or orange tint is unacceptable. Since the flesh tones are located close to the +I axis, a typical flesh tone corrector looks for colors in a specific area (Figure 9.28), and any colors within that area are made a color that is closer to the flesh tone.

A simple flesh tone corrector may halve the Q value for all colors that have a corresponding +I value. However, this implementation also changes nonflesh tone colors. A more sophisticated implementation is if the color has a value between 25% and 75% of full-scale, and is within ±30° of the +I axis, then Q is halved. This moves any colors within the flesh tone region closer to "ideal" flesh tone.

It should be noted that the phase angle for flesh tone varies between companies. Phase angles from 116° to 126° are used; however, using 123° (the +I axis) simplifies the processing.

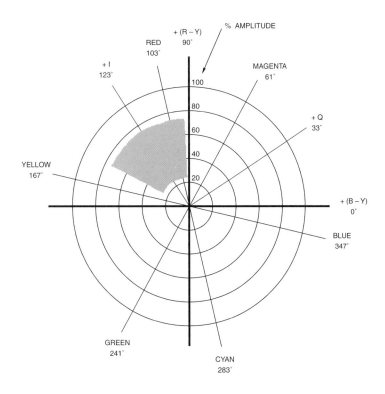

Figure 9.28. Typical Flesh Tone Color Range.

Color Killer

If a color burst of 12.5% or less of ideal amplitude is detected for 128 consecutive scan lines, the color difference signals should be forced to zero. Once a color burst of 25% or more of ideal amplitude is detected for 128 consecutive scan lines, the color difference signals may again be enabled. This hysteresis prevents wandering back and forth between enabling and disabling the color information in the event the burst amplitude is borderline.

The burst level may be determined by forcing all burst samples positive and sampling the result multiple times to determine an average value. This should be averaged over three scan lines to limit line-to-line variations.

The "burst level error" is the ideal amplitude divided by the average result. If no burst is detected, this should be used to force the color difference signals to zero and to disable any filtering in the luminance path, allowing maximum resolution luminance to be output.

Providing the ability to force the color decoding on or off optionally is useful in some applications, such as video editing.

Color Space Conversion

YUV or YIQ data is usually converted to YCbCr or R′G′B′ data before being output from the decoder. If converting to R′G′B′ data, the R′G′B′ data must be clipped at the 0 and 1023 values to prevent wrap-around errors.

(M) NTSC, (M, N) PAL

YUV Color Space Processing

Modern decoder designs are now based on the YUV color space, For these decoders, the YUV to YCbCr equations are:

$Y_{601} = 1.691Y + 64$

$Cb = [1.984U \cos \theta_B] + [1.984V \sin \theta_B] + 512$

$Cr = [1.406U \cos \theta_R] + [1.406V \sin \theta_R] + 512$

To generate R′G′B′ data with a range of 0–1023, the YUV to R′G′B′ equations are:

$R′ = 1.975Y + [2.251U \cos \theta_R] + [2.251V \sin \theta_R]$

$G′ = 1.975Y - 0.779U - 1.146V$

$B′ = 1.975Y + [4.013U \cos \theta_B] + [4.013V \sin \theta_B]$

To generate R′G′B′ data with a nominal range of 64–940 for pro-video applications, the YUV to R′G′B′ equations are:

$R′ = 1.691Y + 1.928V + 64$

$G′ = 1.691Y - 0.667U - 0.982V + 64$

$B′ = 1.691Y + 3.436U + 64$

The ideal values for θ_R and θ_B are 90° and 0°, respectively. However, for consumer televisions sold in the United States, θ_R and θ_B usually have values of 110° and 0°, respectively, or 100° and –10°, respectively, to reduce the visibility of differential phase errors, at the cost of color accuracy.

YIQ Color Space Processing

For older NTSC decoder designs based on the YIQ color space, the YIQ to YCbCr equations are:

$Y_{601} = 1.692Y + 64$

$Cb = -1.081I + 1.664Q + 512$

$Cr = 1.181I + 0.765Q + 512$

To generate R′G′B′ data with a range of 0–1023, the YIQ to R′G′B′ equations are:

$R′ = 1.975Y + 1.887I + 1.224Q$

$G′ = 1.975Y - 0.536I - 1.278Q$

$B′ = 1.975Y - 2.189I + 3.367Q$

To generate R′G′B′ data with a nominal range of 64–940 for pro-video applications, the YIQ to R′G′B′ equations are:

$R′ = 1.691Y + 1.616I + 1.048Q + 64$

$G′ = 1.691Y - 0.459I - 1.094Q + 64$

$B′ = 1.691Y - 1.874I + 2.883Q + 64$

YCbCr Color Space Processing

If the design is based on the YUV color space, the UV to CbCr conversion may be avoided by scaling the sin and cos values during the demodulation process, or scaling the color difference lowpass filter coefficients.

NTSC–J

Since the version of (M) NTSC used in Japan has a 0 IRE blanking pedestal, the color space conversion equations are slightly different than those for standard (M) NTSC.

YUV Color Space Processing

Modern decoder designs are now based on the YUV color space, For these decoders, the YUV to YCbCr equations are:

$$Y_{601} = 1.564Y + 64$$

$$Cb = 1.835U + 512$$

$$Cr = [1.301U \cos \theta_R] + [1.301V \sin \theta_R] + 512$$

To generate R′G′B′ data with a range of 0–1023, the YUV to R′G′B′ equations are:

$$R' = 1.827Y + [2.082U \cos \theta_R] + [2.082V \sin \theta_R]$$

$$G' = 1.827Y - 0.721U - 1.060V$$

$$B' = 1.827Y + 3.712U$$

To generate R′G′B′ data with a nominal range of 64–940 for pro-video applications, the YUV to R′G′B′ equations are:

$$R' = 1.564Y + 1.783V + 64$$

$$G' = 1.564Y - 0.617U - 0.908V + 64$$

$$B' = 1.564Y + 3.179U + 64$$

The ideal value for θ_R is 90°. However, for televisions sold in the Japan, θ_R usually has a value of 95° to reduce the visibility of differential phase errors, at the cost of color accuracy.

YIQ Color Space Processing

For older NTSC decoder designs based on the YIQ color space, the YIQ to YCbCr equations are:

$$Y_{601} = 1.565Y + 64$$

$$Cb = -1.000I + 1.539Q + 512$$

$$Cr = 1.090I + 0.708Q + 512$$

To generate R′G′B′ data with a range of 0–1023, the YIQ to R′G′B′ equations are:

$$R' = 1.827Y + 1.746I + 1.132Q$$

$$G' = 1.827Y - 0.496I - 1.182Q$$

$$B' = 1.827Y - 2.024I + 3.115Q$$

To generate R′G′B′ data with a nominal range of 64–940 for pro-video applications, the YIQ to R′G′B′ equations are:

$$R' = 1.564Y + 1.495I + 0.970Q + 64$$

$$G' = 1.564Y - 0.425I - 1.012Q + 64$$

$$B' = 1.564Y - 1.734I + 2.667Q + 64$$

YCbCr Color Space Processing

If the design is based on the YUV color space, the UV to CbCr conversion may be avoided by scaling the sin and cos values during the demodulation process, or scaling the color difference lowpass filter coefficients.

(B, D, G, H, I, N$_C$) PAL

YUV Color Space Processing

The YUV to YCbCr equations are:

$$Y_{601} = 1.599Y + 64$$

$$Cb = 1.875U + 512$$

$$Cr = 1.329V + 512$$

To generate R′G′B′ data with a range of 0–1023, the YUV to R′G′B′ equations are:

$$R' = 1.867Y + 2.128V$$

$$G' = 1.867Y - 0.737U - 1.084V$$

$$B' = 1.867Y + 3.793U$$

To generate R′G′B′ data with a nominal range of 64–940 for pro-video applications, the YUV to R′G′B′ equations are:

$$R' = 1.599Y + 1.822V + 64$$

$$G' = 1.599Y - 0.631U - 0.928V + 64$$

$$B' = 1.599Y + 3.248U + 64$$

YCbCr Color Space Processing

The UV to CbCr conversion may be avoided by scaling the sin and cos values during the demodulation process, or scaling the color difference lowpass filter coefficients.

Genlocking

The purpose of the genlock circuitry is to recover a sample clock and the timing control signals (such as horizontal sync, vertical sync, and the color subcarrier) from the video signal. Since the original sample clock is not available, it is usually generated by multiplying the horizontal line frequency, F_H, by the desired number of samples per line, using a phase-lock loop (PLL). Also, the color subcarrier must be regenerated and locked to the color subcarrier of the video signal being decoded.

There are, however, several problems. Video signals may contain noise, making the determination of sync edges unreliable. The amount of time between horizontal sync edges may vary slightly each line, particularly in analog video tape recorders (VCRs) due to mechanical limitations. For analog VCRs, instantaneous line-to-line variations are up to ±100 ns; line variations between the beginning and end of a field are up to ±5 μs. When analog VCRs are in a "special feature" mode, such as fast-forwarding or still-picture, the amount of time between horizontal sync signals may vary up to ±20% from nominal.

Vertical sync, as well as horizontal sync, information must be recovered. Unfortunately, analog VCRs, in addition to destroying the SCH phase relationship, perform head switching at field boundaries, usually somewhere between the end of active video and the start of vertical sync. When head switching occurs, one video signal (field n) is replaced by another video signal (field n + 1) which has an unknown time offset from the first video signal. There may be up to a ±1/2 line variation in vertical timing each field. As a result, longer-than-normal horizontal or vertical syncs may be generated.

By monitoring the horizontal line timing, it is possible to determine automatically whether the video source is in the "normal" or "special feature" mode. During "normal" operation, the horizontal line time typically varies by no more than ±5 μs over an entire field. Line timing outside this ±5 μs window may be used to enable "special feature" mode timing. Hysteresis should be used in the detection algorithm to prevent wandering back and forth between the "normal" and "special feature" operations in the event the video timing is borderline between the two modes. A typical circuit for performing the horizontal and vertical sync detection is shown in Figure 9.29.

In the absence of a video signal, the decoder should be designed to optionally free-run, continually generating the video timing to the system, without missing a beat. During the loss of an input signal, any automatic gain circuits should be disabled and the decoder should provide the option either to be transparent (so the input source can be monitored), to auto-freeze the output data (to compensate for short duration dropouts), or to autoblack the

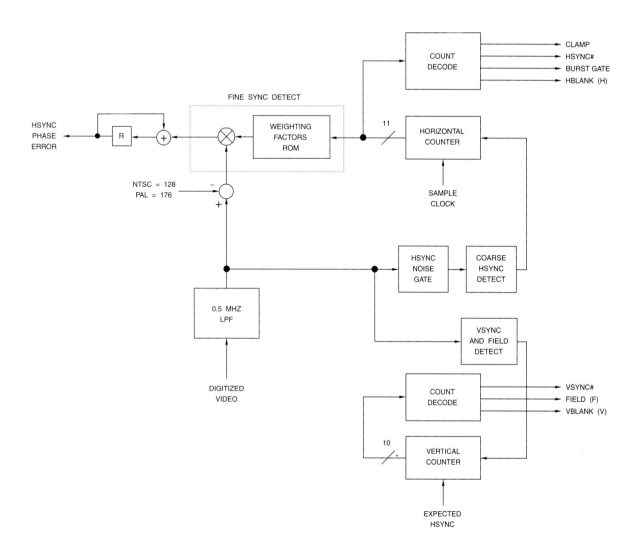

Figure 9.29. Sync Detection and Phase Comparator Circuitry.

output data (to avoid potential problems driving a mixer or VCR).

Horizontal Sync Detection

Early decoders typically used analog sync slicing techniques to determine the midpoint of the leading edge of the sync pulse and used a PLL to multiply the horizontal frequency rate up to the sample clock rate. However, the lack of accuracy of the analog sync slicer, combined with the limited stability of the PLL, resulted in sample clock jitter and noise amplification. When using comb filters for Y/C separation, the long delay between writing and reading the video data means that even a small sample clock frequency error results in a delay that is a significant percentage of the subcarrier period, negating the effectiveness of the comb filter.

Coarse Horizontal Sync Locking

The coarse sync locking enables a faster lock-up time to be achieved. Digitized video is lowpass filtered to about 0.5 MHz to remove high-frequency information, such as noise and color subcarrier information. Performing the sync detection on lowpass filtered data also provides edge shaping in the event that fast sync edges (rise and fall times less than one clock cycle) are present.

An 11-bit horizontal counter is incremented each sample clock cycle, resetting to 001_H after counting up to the HCOUNT value, where HCOUNT specifies the total number of samples per line. A value of 001_H indicates that the beginning of a horizontal sync is expected. When the horizontal counter value is (HCOUNT – 64), a sync gate is enabled, allowing recovered sync information to be detected.

Up to five consecutive missing sync pulses should be detected before any correction to the clock frequency or other adjustments are done. Once sync information has been detected, the sync gate is disabled until the next time the horizontal counter value is (HCOUNT – 64). This helps filter out noise, serration and equalization pulses. If the leading edge of recovered horizontal sync is not within ±64 clock cycles (approximately ±5 μs) of where it is expected to be, the horizontal counter is reset to 001_H to realign the edges more closely.

Additional circuitry may be included to monitor the width of the recovered horizontal sync pulse. If the horizontal sync pulse is not approximately the correct pulse width, ignore it and treat it as a missing sync pulse.

If the leading edge of recovered horizontal sync is within ±64 sample clock cycles (approximately ±5 μs) of where it is expected to be, the fine horizontal sync locking circuitry is used to fine-tune the timing.

Fine Horizontal Sync Locking

One-half the sync amplitude is subtracted from the 0.5-MHz lowpass-filtered video data so the sync timing reference point (50% sync amplitude) is at zero.

The leading horizontal sync edge may be determined by summing a series of weighted samples from the region of the sync edge. To perform the filtering, the weighting factors are read from a ROM by a counter triggered by the horizontal counter. When the central weighting factor (A0) is coincident with the 50% amplitude point of the leading edge of sync, the result integrates to zero. Typical weighting factors are:

A0 = 102/4096
A1 = 90/4096
A2 = 63/4096
A3 = 34/4096
A4 = 14/4096
A5 = 5/4096
A6 = 2/4096

This arrangement uses more of the timing information from the sync edge and suppresses noise. Note that circuitry should be included to avoid processing the trailing edge of horizontal sync.

Figure 9.30 shows the operation of the fine sync phase comparator. Figure 9.30a shows the leading sync edge for NTSC. Figure 9.30b shows the weighting factors being generated, and when multiplied by the sync information, produces the waveform shown in Figure 9.30c. When the A0 coefficient is coincident with the 50% amplitude point of sync, the waveform integrates to zero. Distortion of sync edges, resulting in the locking point being slightly shifted, is minimized by the lowpass filtering, effectively shaping the sync edges prior to processing.

Sample Clock Generation

The horizontal sync phase error signal from Figure 9.29 is used to adjust the frequency of a line-locked PLL, as shown in Figure 9.31. A line-locked PLL always generates a constant number of clock cycles per line, regardless of any line time variations. The free-running frequency of the PLL should be the nominal sample clock frequency required (for example, 13.5 MHz).

Using a VCO-based PLL has the advantage of a wider range of sample clock frequency adjustments, useful for handling video timing variations outside the normal video specifications. A disadvantage is that, due to jitter in the sample clock, there may be visible hue artifacts and poor Y/C separation.

A VCXO-based PLL has the advantage of minimal sample clock jitter. However, the sample clock frequency range may be adjusted only a small amount, limiting the ability of the decoder to handle nonstandard video timing.

Ideally, with either design, the rising edge of the sample clock is aligned with the half-amplitude point of the leading edge of horizontal sync, and a fixed number of sample clock cycles per line (HCOUNT) are always generated.

An alternate method is to asynchronously sample the video signal with a fixed-frequency clock (for example, 13.5 MHz). Since in this case the sample clock is not aligned with horizontal sync, there is a phase difference between the actual sample position and the ideal sample position. As with the conventional genlock solution, this phase difference is determined by the difference between the recovered and expected horizontal syncs.

The ideal sample position is defined to be aligned with a sample clock generated by a line-locked PLL. Rather than controlling the sample clock frequency, the horizontal sync phase error signal is used to control interpolation between two samples of data to generate the ideal sample value. If using comb filtering for Y/C separation, the digitized composite video may be interpolated to generate the ideal sample points, providing better Y/C separation by aligning the samples more precisely.

Vertical Sync Detection

Digitized video is lowpass filtered to about 0.5 MHz to remove high-frequency information, such as noise and color subcarrier information. The 10-bit vertical counter is incremented by each expected horizontal sync,

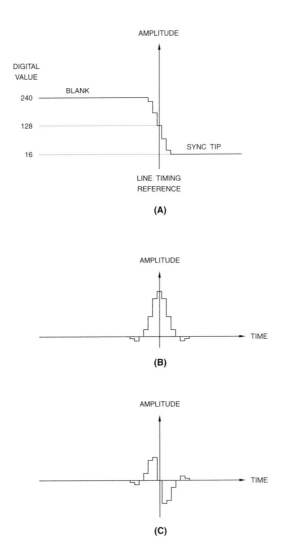

Figure 9.30. Fine Lock Phase Comparator Waveforms. (a) The NTSC sync leading edge. (b) The series of weighting factors. (c) The weighted leading edge samples.

Figure 9.31. Typical Line-Locked Sample Clock Generation.

resetting to 001_H after counting up to 525 or 625. A value of 001_H indicates that the beginning of a vertical sync for Field 1 is expected.

The end of vertical sync intervals is detected and used to set the value of the vertical counter according to the mode of operation. By monitoring the relationship of recovered vertical and horizontal syncs, Field 1 vs. Field 2 information is detected. If a recovered horizontal sync occurs more than 64, but less than (HCOUNT/2), clock cycles after expected horizontal sync, the vertical counter is not adjusted to avoid double incrementing the vertical counter. If a recovered horizontal sync occurs (HCOUNT/2) or more clock cycles after the vertical counter has been incremented, the vertical counter is again incremented.

During "special feature" operation, there is no longer any correlation between the vertical and horizontal timing information, so Field 1 vs. Field 2 detection cannot be done. Thus, every other detection of the end of vertical sync should set the vertical counter accordingly in order to synthesize Field 1 and Field 2 timing.

Subcarrier Generation

As with the encoder, the color subcarrier is generated from the sample clock using a DTO (Figure 9.32), and the same frequency relationships apply as those discussed in the encoder section.

Unlike the encoder, the phase of the generated subcarrier must be continuously adjusted to match that of the video signal being decoded.

The subcarrier locking circuitry phase compares the generated subcarrier and the incoming subcarrier, resulting in an F_{SC} error signal indicating the amount of phase error. This F_{SC} error signal is added to the [p] value to continually adjust the step size of the DTO, adjusting the phase of the generated subcarrier to match that of the video signal being decoded.

As a 22-bit single-stage DTO is used to divide down the sample clock to generate the subcarrier in Figure 9.32, the [p] value is determined as follows:

$$F_{SC}/F_S = (P/4194303) = (P/(2^{22} - 1))$$

where F_{SC} = the desired subcarrier frequency and F_S = the sample clock rate. Some values of [p] for popular sample clock rates are shown in Table 9.10.

Subcarrier Locking

The purpose of the subcarrier locking circuitry (Figure 9.33) is to phase lock the generated color subcarrier to the color subcarrier of the video signal being decoded.

Digital composite video (or digital chrominance video) has the blanking level subtracted from it. It is also gated with a "burst gate" to ensure that the data has a value of zero outside the burst time. The burst gate signal should be

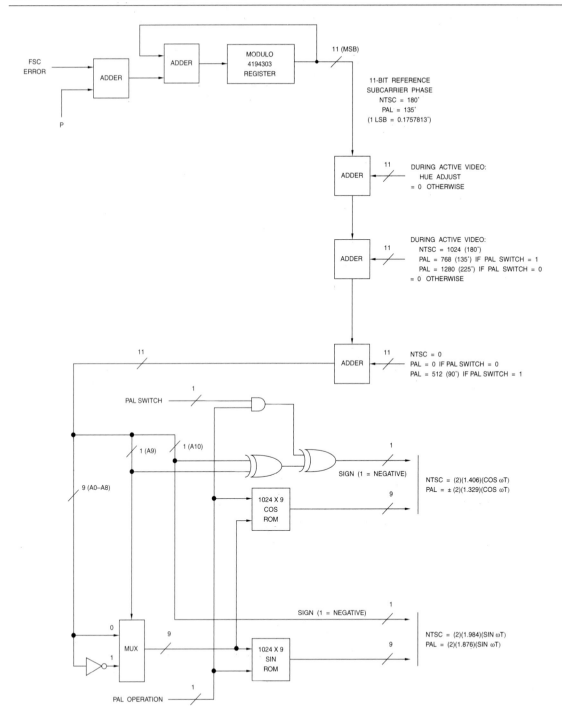

Figure 9.32. Chrominance Subcarrier Generator.

Typical Application	Total Samples per Scan Line (HCOUNT)	P
13.5 MHz (M) NTSC	858	1,112,126
13.5 MHz (B, D, G, H, I) PAL	864	1,377,477
12.27 MHz (M) NTSC	780	1,223,338
14.75 MHz (B, D, G, H, I) PAL	944	1,260,742

Table 9.10. Typical HCOUNT and P Values for the 1-Stage 22-Bit DTO in Figure 9.32.

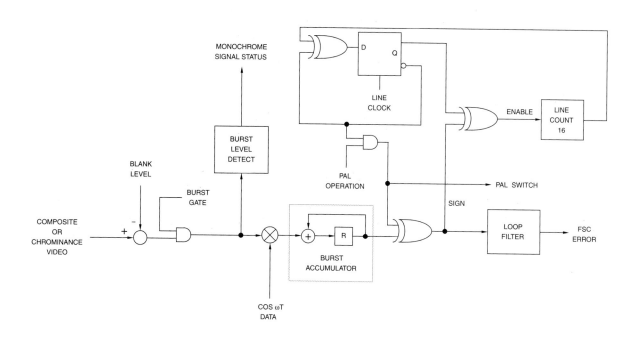

Figure 9.33. Subcarrier Phase Comparator Circuitry.

timed to eliminate the edges of the burst, which may have transient distortions that will reduce the accuracy of the phase measurement.

The color burst data is phase compared to the locally generated burst. Note that the sign information must also be compared so lock will not occur on 180° out-of-phase signals. The burst accumulator averages the sixteen samples, and the accumulated values from two adjacent lines are averaged to produce the error signal. When the local subcarrier is correctly phased, the accumulated values from alternate lines cancel, and the phase error signal is zero. The error signal is sampled at the line rate and processed by the loop filter, which should be designed to achieve a lock-up time of about ten lines (50 or more lines may be required for noisy video signals). It is desirable to avoid updating the error signal during vertical intervals due to the lack of burst. The resulting F_{SC} error signal is used to adjust the DTO that generates the local subcarrier (Figure 9.32).

During PAL operation, the phase detector also recovers the PAL Switch information used in generating the switched V subcarrier. The PAL Switch D flip-flop is synchronized to the incoming signal by comparing the local switch sense with the sign of the accumulated burst values. If the sense is consistently incorrect for sixteen lines, then the flip-flop is reset.

Note the subcarrier locking circuit should be able to handle short-term frequency variations (over a few frames) of ±200 Hz, long-term frequency variations of ±500 Hz, and color burst amplitudes of 25–200% of normal with short-term amplitude variations (over a few frames) of up to 5%. The lock-up time of ten lines is desirable to accommodate video signals that may have been incorrectly edited (i.e., not careful about the SCH phase relation-

ship) or nonstandard video signals due to freeze-framing, special effects, and so on. The ten lines enable the subcarrier to be locked before the active video time, ensuring correct color representation at the beginning of the picture.

Video Timing Generation

HSYNC# (Horizontal Sync) Generation

An 11-bit horizontal counter is incremented on each rising edge of the sample clock. The count is monitored to determine when to generate the burst gate, HSYNC# output, horizontal blanking, etc. Typically, each time the counter is reset to 001_H, the HSYNC# output is asserted. The exact timing of HSYNC# is dependent on the video interface used, as discussed in Chapter 6.

H (Horizontal Blanking) Generation

A horizontal blanking signal, H, may be implemented to specify when the horizontal blanking interval occurs. The timing of H is dependent on the video interface used, as discussed in Chapter 6.

The horizontal blank timing may be user programmable by incorporating start and stop blank registers. The values of these registers are compared to the horizontal counter value, and used to assert and negate the H control signal.

VSYNC# (Vertical Sync) Generation

A 10-bit vertical counter is incremented on each rising edge of HSYNC#. Typically, each time the counter is reset to 001_H, the VSYNC# output is asserted. The exact timing of VSYNC# is dependent on the video interface used, as discussed in Chapter 6.

F (FIELD) Generation

A field signal, F, may be implemented to specify whether Field 1 or Field 2 is being decoded. The exact timing of F is dependent on the video interface used, as discussed in Chapter 6.

In instances where the output of an analog VCR is being decoded, and the VCR is in a special effects mode (such as still or fast-forward), there is no longer enough timing information to determine Field 1 vs. Field 2 timing. Thus, the Field 1 and Field 2 timing as specified by the VSYNC#/HSYNC# relationship (or the F signal) should be synthesized and may not reflect the true field timing of the video signal being decoded.

V (Vertical Blanking) Generation

A vertical blanking signal, V, may be implemented to specify when the vertical blanking interval occurs. The exact timing of V is dependent on the video interface used, as discussed in Chapter 6.

The vertical blank timing may be user programmable by incorporating start and stop blank registers. The values of these registers are compared to the vertical counter value, and used to assert and negate the V control signal.

BLANK# Generation

The composite blanking signal, BLANK#, is the logical NOR of the H and V signals.

While BLANK# is asserted, RGB data may be forced to be a value of 0. YCbCr data may be forced to an 8-bit value of 16 for Y and 128 for Cb and Cr. Alternately, the RGB or YCbCr data outputs may not be blanked, allowing vertical blanking interval (VBI) data, such as closed captioning, teletext, widescreen signalling and other information to be output.

Field Identification

Although the timing relationship between the horizontal sync (HSYNC#) and vertical sync (VSYNC#) signals, or the F signal, may be used to specify whether a Field 1 vs. Field 2 is being decoded, one or two additional signals may be used to specify which one of four or eight fields is being decoded, as shown in Table 9.7. We refer to these additional control signals as FIELD_0 and FIELD_1.

FIELD_0 should change state at the beginning of VSYNC#, or coincident with F, during fields 1, 3, 5, and 7. FIELD_1 should change state at the beginning of VSYNC#, or coincident with F, during fields 1 and 5.

NTSC Field Identification

The beginning of fields 1 and 3 may be determined by monitoring the relationship of the subcarrier phase relative to sync. As shown in Figure 8.5, at the beginning of field 1, the subcarrier phase is ideally 0° relative to sync; at the beginning of field 3, the subcarrier phase is ideally 180° relative to sync.

In the real world, there is a tolerance in the SCH phase relationship. For example, although the ideal SCH phase relationship may be perfect at the source, transmitting the video signal over a coaxial cable may result in a shift of the SCH phase relationship due to cable characteristics. Thus, the ideal phase plus or minus a tolerance should be used. Although ±40° (NTSC) or ±20° (PAL) is specified as an acceptable tolerance by the video standards, many decoder designs use a tolerance of up to ±80°.

In the event that a SCH phase relationship not within the proper tolerance is detected, the decoder should proceed as if nothing were wrong. If the condition persists for several frames, indicating that the video source may no longer be a "stable" video source, operation should change to that for an "unstable" video source.

For "unstable" video sources that do not maintain the proper SCH relationship (such as analog VCRs), synthesized FIELD_0 and FIELD_1 outputs should be generated (for example, by dividing the F output signal by two and four) in the event the signal is required for memory addressing or downstream processing.

PAL Field Identification

The beginning of fields 1 and 5 may be determined by monitoring the relationship of the –U component of the extrapolated burst relative to sync. As shown in Figure 8.16, at the beginning of field 1, the phase is ideally 0° relative to sync; at the beginning of field 5, the phase is ideally 180° relative to sync. Either the burst blanking sequence or the subcarrier phase may be used to differentiate between fields 1 and 3, fields 2 and 4, fields 5 and 7, and fields 6 and 8. All of the considerations discussed for NTSC in the previous section also apply for PAL.

Auto-Detection of Video Signal Type

If the decoder can automatically detect the type of video signal being decoded, and configure itself automatically, the user will not have to guess at the type of video signal being processed. This information can be passed via status information to the rest of the system.

If the decoder detects less than 575 lines per frame for at least 16 consecutive frames, the decoder can assume the video signal is (M) NTSC or (M) PAL. First, assume the video signal is (M) NTSC as that is much more popular. If the vertical and horizontal timing remain locked, but the decoder is unable to

maintain subcarrier locking, the video signal may be (M) PAL. In that case, try (M) PAL operation and verify the burst timing.

If the decoder detects more than 575 lines per frame for at least 16 consecutive frames, it can assume the video signal is (B, D, G, H, I, N, N_C) PAL or a version of SECAM.

First, assume the video signal is (B, D, G, H, I, N) PAL. If the vertical and horizontal timing remain locked, but the decoder is unable to maintain a subcarrier lock, it may mean the video signal is (N_C) PAL or SECAM. In that case, try SECAM operation (as that is much more popular), and if that doesn't subcarrier lock, try (N_C) PAL operation.

If the decoder detects a video signal format to which it cannot lock, this should be indicated so the user can be notified.

Note that auto-detection cannot be performed during "special feature" modes of analog VCRs, such as fast-forwarding. If the decoder detects a "special feature" mode of operation, it should disable the auto-detection circuitry. Auto-detection should only be done when a video signal has been detected after the loss of an input video signal.

Y/C Separation Techniques

The encoder typically combines the luminance and chrominance signals by simply adding them together; the result is that chrominance and high-frequency luminance signals occupy the same portion of the frequency spectrum. As a result, separating them in the decoder is difficult. When the signals are decoded, some luminance information is decoded as color information (referred to as cross-color), and some chrominance information remains in the luminance signal (referred to as cross-luminance). Due to the stable performance of digital decoders, much more complex separation techniques can be used than is possible with analog decoders.

The presence of crosstalk is bad news in editing situations; crosstalk components from the first decoding are encoded, possibly causing new or additional artifacts when decoded the next time. In addition, when a still frame is captured from a decoded signal, the frozen residual subcarrier on edges may beat with the subcarrier of any following encoding process, resulting in edge flicker in colored areas. Although the crosstalk problem cannot be solved entirely at the decoder, more elaborate Y/C separation minimizes the problem.

If the decoder is used in an editing environment, the suppression of cross-luminance and cross-chrominance is more important than the appearance of the decoded picture. When a picture is decoded, processed, encoded, and again decoded, cross-effects can introduce substantial artifacts. It may be better to limit the luminance bandwidth (to reduce cross-luminance), producing "softer" pictures. Also, limiting the chrominance bandwidth to less than 1 MHz reduces cross-color, at the expense of losing chrominance definition.

Complementary Y/C separation preserves all of the input signal. If the separated chrominance and luminance signals are added together again, the original composite video signal is generated.

Noncomplementary Y/C separation introduces some irretrievable loss, resulting in gaps in the frequency spectrum if the separated chrominance and luminance signals are again added together to generate a composite video signal. The loss is due to the use of narrower filters to reduce cross-color and cross-luminance. Therefore, noncomplementary filtering is usually unsuitable when multiple encoding and decoding operations must be performed, as the frequency spectrum gaps continually

increase as the number of decoding operations increase. It does, however, enable the "tweaking" of luminance and chrominance response for optimum viewing.

Simple Y/C Separation

With all of these implementations, there is no loss of vertical chrominance resolution, but there is also no suppression of cross-color. For PAL, line-to-line errors due to differential phase distortion are not suppressed, resulting in the vertical pattern known as Hanover bars.

Noticeable artifacts of simple Y/C separators are color artifacts on vertical edges. These include color ringing, color smearing, and the display of color rainbows in place of high-frequency gray-scale information.

Lowpass and Highpass Filtering

The most basic Y/C separator assumes frequencies below a certain point are luminance and above this point are chrominance. An example of this simple Y/C separator is shown in Figure 9.34.

Frequencies below 3.0 MHz (NTSC) or 3.8 MHz (PAL) are assumed to be luminance. Frequencies above these are assumed to be chrominance. Not only is high-frequency luminance information lost, but it is assumed to be chrominance information, resulting in cross-color.

Notch Filtering

Although broadcast NTSC and PAL systems are strictly bandwidth-limited, this may not be true of other video sources. Luminance information may be present all the way out to 6 or 7 MHz or even higher. For this reason, the designs in Figure 6.35 are usually more appropriate, as they allow high-frequency luminance to pass, resulting in a sharper picture.

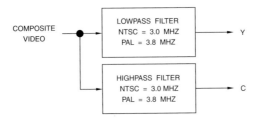

Figure 9.34. Typical Simple Y/C Separator.

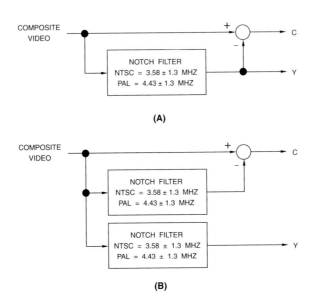

Figure 9.35. Typical Simple Y/C Separator. (a) Complementary filtering. (b) Noncomplementary filtering.

Many designs based on the notch filter also incorporate comb filters in the Y and color difference data paths to reduce cross-color and cross-luma artifacts. However, the notch filter still limits the overall Y/C separation quality.

PAL Considerations

As mentioned before, PAL uses "normal" and "inverted" scan lines, referring to whether the V component is normal or inverted, to help correct color shifting effects due to differential phase distortions.

For example, differential phase distortion may cause the green vector angle on "normal" scan lines to lag by 45° from the ideal 241° shown in Figure 8.11. This results in a vector at 196°, effectively shifting the resulting color towards yellow. On "inverted" scan lines, the vector angle also will lag by 45° from the ideal 120° shown in Figure 8.12. This results in a vector at 75°, effectively shifting the resulting color towards cyan.

PAL Delay Line

Figure 9.36, made by flipping Figure 8.12 180° about the U axis and overlaying the result onto Figure 8.11, illustrates the cancellation of the phase errors. The average phase of the two errors, 196° on "normal" scan lines and 286° on "inverted" scan lines, is 241°, which is the correct phase for green. For this reason, simple PAL decoders usually use a delay line (or line store) to facilitate averaging between two scan lines.

Using delay lines in PAL Y/C separators has unique problems. The subcarrier reference changes by –90° (or 270°) over one line period, and the V subcarrier is inverted on alternate lines. Thus, there is a 270° phase difference between the input and output of a line delay. If we want to do a simple addition or subtraction between the input and output of the delay line to recover chrominance information,

the phase difference must be 0° or 180°. And there is still that switching V floating around. Thus, we would like to find a way to align the subcarrier phases between lines and compensate for the switching V.

Simple circuits, such as the noncomplementary Y/C separator shown in Figure 9.37, use a delay line that is not a whole line (283.75 subcarrier periods), but rather 284 subcarrier periods. This small difference acts as a 90° phase shift at the subcarrier frequency.

Since there are an integral number of subcarrier periods in the delay, the U subcarriers at the input and output of the 284 T_{SC} delay line are in phase, and they can simply be added together to recover the U subcarrier. The V subcarriers are 180° out of phase at the input and output of the 284 T_{SC} delay line, due to the switching V, so the adder cancels them out. Any remaining high-frequency vertical V components are rejected by the U demodulator.

Due to the switching V, subtracting the input and output of the 284 T_{SC} delay line recovers the V subcarrier while cancelling the U subcarrier. Any remaining high-frequency vertical U components are rejected by the V demodulator.

Since the phase shift through the 284 T_{SC} delay line is a function of frequency, the subcarrier sidebands are not phase shifted exactly 90°, resulting in hue errors on vertical chrominance transitions. Also, the chrominance and luminance are not vertically aligned since the chrominance is shifted down by one half line.

PAL Modifier

Although the performance of the circuit in Figure 9.37 usually is adequate, the 284 T_{SC} delay line may be replaced by a line delay followed by a PAL modifier, as shown in Figure 9.38. The PAL modifier provides a 90° phase shift and inversion of the V subcarrier. Chrominance from the PAL modifier is now in phase

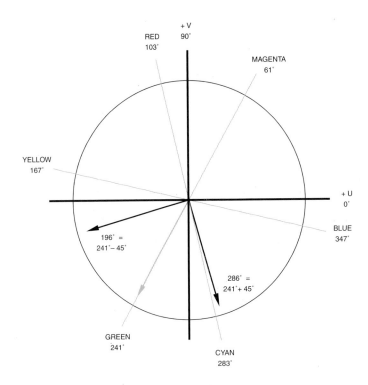

Figure 9.36. Phase Error "Correction" for PAL.

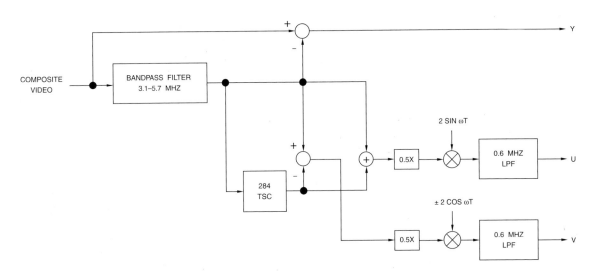

Figure 9.37. Single Delay Line PAL Y/C Separator.

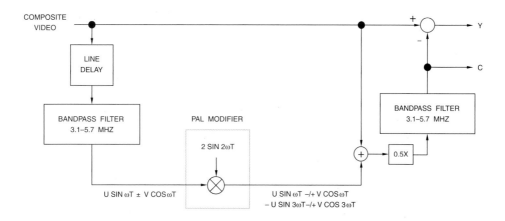

Figure 9.38. Single Line Delay PAL Y/C Separator Using a PAL Modifier.

with the line delay input, allowing the two to be combined using a single adder and share a common path to the demodulators. The averaging sacrifices some vertical resolution, however Hanover bars are suppressed.

Since the chrominance at the demodulator input is in phase with the composite video, it can be used to cancel the chrominance in the composite signal to leave luminance. However, the chrominance and luminance are still not vertically aligned since the chrominance is shifted down by one half line.

The PAL modifier produces a luminance alias centered at twice the subcarrier frequency. Without the bandpass filter before the PAL modifier and the averaging between lines, mixing the original and aliased luminance components would result in a 12.5-Hz beat frequency, noticeable in high-contrast areas of the picture.

2D Comb Filtering

In the previous Y/C separators, high-frequency luminance information is treated as chrominance information; no attempt is made to differentiate between the two. As a result, the luminance information is interpreted as chrominance information (cross-color) and passed on to the chroma demodulator to recover color information. The demodulator cannot differentiate between chrominance and high-frequency luminance, so it generates color where color should not exist. Thus, occasional display artifacts are generated.

2D (or intra-field) comb filtering attempts to improve the separation of chrominance and luminance at the expense of reduced vertical resolution. Comb filters get their name by having luminance and chrominance frequency responses that look like a comb. Ideally, these frequency responses would match the "comb-like" frequency responses of the interleaved luminance and chrominance signals shown in Figures 8.4 and 8.15.

Modern 3-line comb filters typically use two line delays for storing the last two lines of video information (there is a one-line delay in decoding using this method). Using more than two line delays usually results in excessive vertical filtering, reducing vertical resolution.

Two Line Delay Comb Filters

The BBC has done research (Reference 4) on various PAL comb filtering implementations (Figures 9.39 through 9.42). Each was evaluated for artifacts and frequency response. The vertical frequency response for each comb filter is shown in Figure 9.43.

In the comb filter design of Figure 9.39, the chrominance phase is inverted over two lines of delay. A subtracter cancels most of the luminance, leaving double-amplitude, vertically filtered chrominance. A PAL modifier provides a 90° phase shift and removal of the PAL switch inversion to phase align the chrominance with the one line-delayed composite video signal. Subtracting the chrominance from the composite signal leaves luminance. This design has the advantage of vertical alignment of the chrominance and luminance. However, there is a loss of vertical resolution and no suppression of Hanover bars. In addition, it is possible under some circumstances to generate double-amplitude luminance due to the aliased luminance components produced by the PAL modifier.

The comb filter design of Figure 9.40 is similar to the one in Figure 9.39. However, the chrominance after the PAL modifier and one line-delayed composite video signal are added to generate double-amplitude chrominance (since the subcarriers are in phase). Again, subtracting the chrominance from the composite signal leaves luminance. In this design, luminance over-ranging is avoided since both the true and aliased luminance signals are halved. There is less loss of vertical resolution

and Hanover bars are suppressed, at the expense of increased cross-color.

The comb filter design in Figure 9.41 has the advantage of not using a PAL modifier. Since the chrominance phase is inverted over two lines of delay, adding them together cancels most of the chrominance, leaving double-amplitude luminance. This is subtracted from the one line-delayed composite video signal to generate chrominance. Chrominance is then subtracted from the one line-delayed composite video signal to generate luminance (this is to maintain vertical luminance resolution). UV crosstalk is present as a 12.5-Hz flicker on horizontal chrominance edges, due to the chrominance signals not cancelling in the adder since the line-to-line subcarrier phases are not aligned. Since there is no PAL modifier, there is no luminance aliasing or luminance over-ranging.

The comb filter design in Figure 9.42 is a combination of Figures 9.39 and 9.41. The chrominance phase is inverted over two lines of delay. An adder cancels most of the chrominance, leaving double-amplitude luminance. This is subtracted from the one line-delayed composite video signal to generate chrominance signal (A). In a parallel path, a subtracter cancels most of the luminance, leaving double-amplitude, vertically filtered chrominance. A PAL modifier provides a 90° phase shift and removal of the PAL switch inversion to phase align to the (A) chrominance signal. These are added together, generating double-amplitude chrominance. Chrominance then is subtracted from the one line-delayed composite signal to generate luminance. The chrominance and luminance vertical frequency responses are the average of those for Figures 9.39 and 9.41. UV crosstalk is similar to that for Figure 9.41, but has half the amplitude. The luminance alias is also half that of Figure 9.39, and Hanover bars are suppressed.

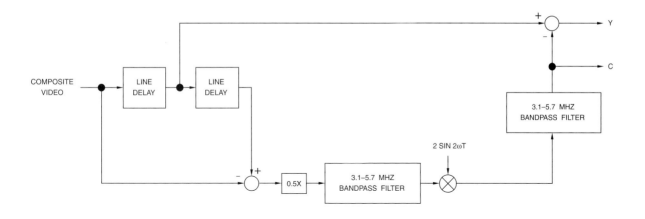

Figure 9.39. Two Line Roe PAL Y/C Separator.

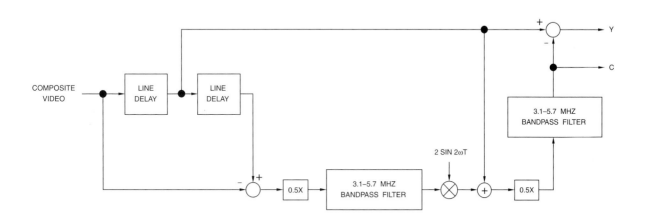

Figure 9.40. Two Line –6 dB Roe PAL Y/C Separator.

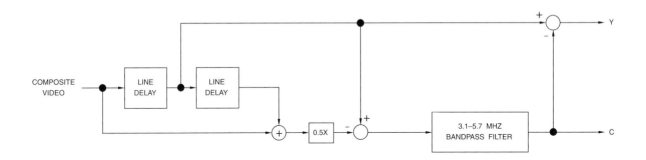

Figure 9.41. Two Line Cosine PAL Y/C Separator.

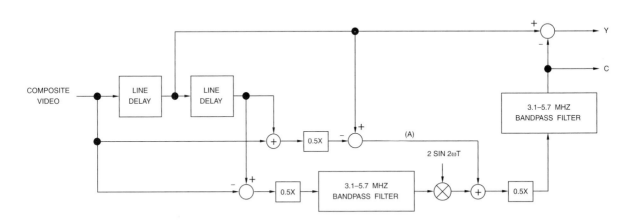

Figure 9.42. Two Line Weston PAL Y/C Separator.

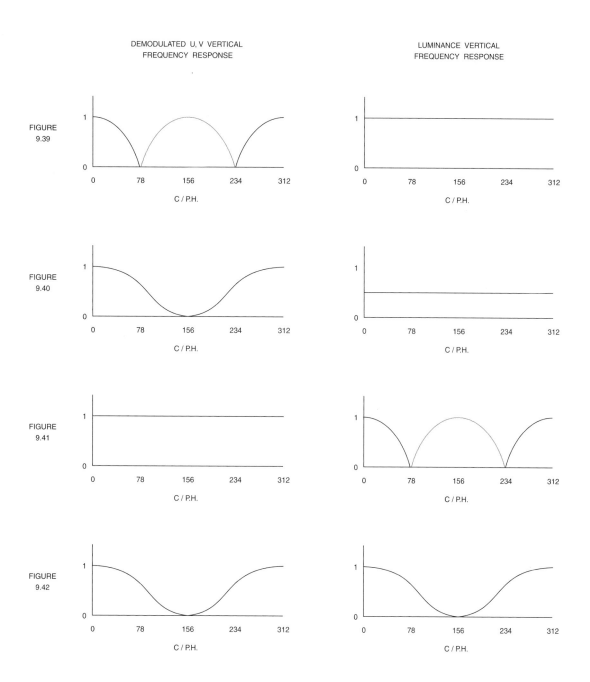

Figure 9.43. Vertical Frequency Characteristics of the Comb Filters in Figures 9.39 Through 9.42.

From these comb filter designs, the BBC has derived designs optimized for general viewing (Figure 9.44) and standards conversion (Figure 9.45).

For PAL applications, the best luminance processing (Figure 9.41) was combined with the optimum chrominance processing (Figure 9.40). The difference between the two designs is the chrominance recovery. For standards conversion (Figure 9.45), the chrominance signal is just the full-bandwidth composite video signal. Standards conversion uses vertical interpolation which tends to reduce moving and high vertical frequency components, including cross-luminance and cross-color. Thus, vertical chrominance resolution after processing usually will be better than that obtained from the circuits for general viewing. The circuit for general viewing (Figure 9.44) recovers chrominance with a goal of reducing cross-effects, at the expense of chrominance vertical resolution.

For NTSC applications, the design of comb filters is easier. There are no switched subcar-

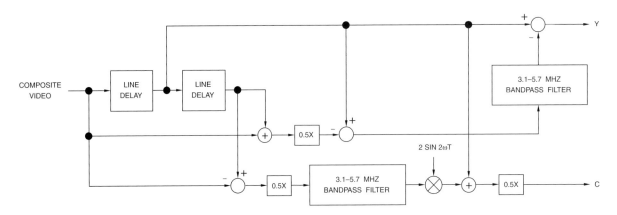

Figure 9.44. Two Line Delay PAL Y/C Separator Optimized for General Viewing.

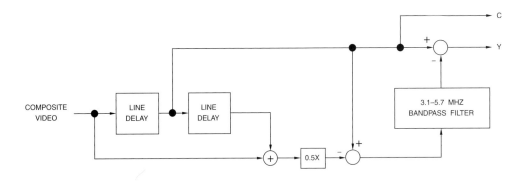

Figure 9.45. Two Line Delay PAL Y/C Separator Optimized for Standards Conversion and Video Processing.

riers to worry about, and the chrominance phases are 180° per line, rather than 270°. In addition, there is greater separation between the luminance and chrominance frequency bands than in PAL, simplifying the separation requirements.

In Figures 9.46 and 9.47, the adder generates a double-amplitude composite video signal since the subcarriers are in phase. There is a 180° subcarrier phase difference between the output of the adder and the one line-delayed composite video signal, so subtracting the two cancels most of the luminance, leaving double amplitude chrominance.

The main disadvantage of the design in Figure 9.46 is the unsuppressed cross-luminance on vertical color transitions. However, this is offset by the increased luminance resolution over simple lowpass filtering. The reasons for processing chrominance in Figure 9.47 are the same as for PAL in Figure 9.45.

Adaptive Comb Filtering

Conventional comb filters still have problems with diagonal lines and vertical color changes since only vertically-aligned samples are used for processing.

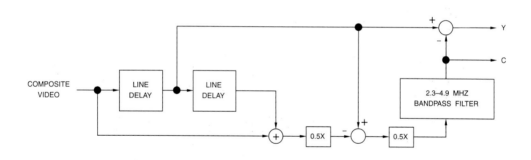

Figure 9.46. Two Line Delay NTSC Y/C Separator for General Viewing.

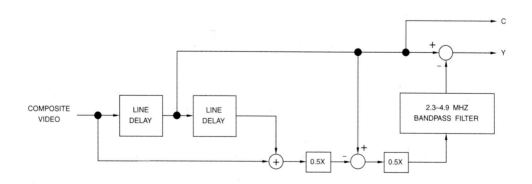

Figure 9.47. Two Line Delay NTSC Y/C Separator for Standards Conversion and Video Processing.

With diagonal lines, after standard comb filtering, the chrominance information also includes the difference between adjacent luminance values, which may also be interpreted as chrominance information. This shows up as cross-color artifacts, such as a rainbow appearance along the edge of the line.

Sharp vertical color transitions generate the "hanging dot" pattern commonly seen on the scan line between the two color changes. After standard comb filtering, the luminance information contains the color subcarrier. The amplitude of the color subcarrier is determined by the difference between the two colors. Thus, different colors modulate the luminance intensity differently, creating a "dot" pattern on the scan line between two colors. To eliminate these "hanging dots," a chroma trap filter is sometimes used after the comb filter.

The adaptive comb filter attempts to solve these problems by processing a 3×3, 5×5, or larger block of samples. The values of the samples are used to determine which Y/C separation algorithm to use for the center sample. As many as 32 or more algorithms may be available. By looking for sharp vertical transitions of luminance, or sharp color subcarrier phase changes, the operation of the comb filter is changed to avoid generating artifacts.

Due to the cost of integrated line stores, the consumer market commonly uses 3-line adaptive comb filtering, with the next level of improvement being 3D motion adaptive comb filtering.

3D Comb Filtering

This method (also called inter-field Y/C separation) uses composite video data from the current field and from two fields (NTSC) or four fields (PAL) earlier. Adding the two cancels the chrominance (since it is 180° out of phase), leaving luminance. Subtracting the two

cancels the luminance, leaving chrominance. For PAL, an adequate design may be obtained by replacing the line delays in Figure 9.42 with frame delays.

This technique provides nearly-perfect Y/C separation for stationary pictures. However, if there is any change between fields, the resulting Y/C separation is erroneous. For this reason, inter-field Y/C separators usually are not used, unless as part of a 3D motion adaptive comb filter.

3D Motion Adaptive Comb Filter

A typical implementation that uses 3D (inter-field) comb filtering for still areas, and 2D (intra-field) comb filtering for areas of the picture that contain motion, is shown in Figure 9.48. The motion detector generates a value (K) of 0–1, allowing the luminance and chrominance signals from the two comb filters to be proportionally mixed. Hard switching between algorithms is usually visible.

Figure 9.49 illustrates a simple motion detector block diagram. The concept is to compare frame-to-frame changes in the low frequency luminance signal. Its performance determines, to a large degree, the quality of the image. The motion signal (K) is usually rectified, smoothed by averaging horizontally and vertically over a few samples, multiplied by a gain factor, and clipped before being used. The only error the motion detector should make is to use the 2D comb filter on stationary areas of the image.

Alpha Channel Support

By incorporating an additional ADC within the NTSC/PAL decoder, an analog alpha signal (also called a key) may be digitized, and pipelined with the video data to maintain synchronization. This allows the designer to change decoders (which may have different

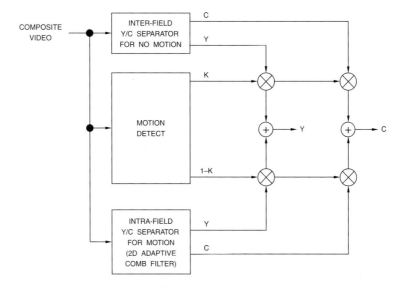

Figure 9.48. 3D Motion Adaptive Y/C Separator.

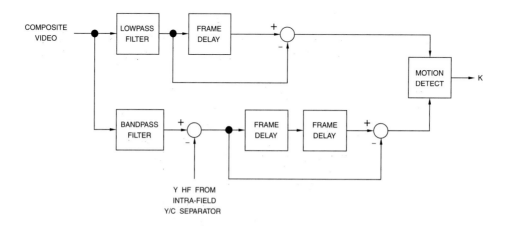

Figure 9.49. Simple Motion Detector Block Diagram for NTSC.

pipeline delays) to fit specific applications without worrying about the alpha channel pipeline delay. Alpha is usually linear, with an analog range of 0–100 IRE. There is no blanking pedestal or sync information present.

Decoder Video Parameters

Many industry-standard video parameters have been defined to specify the relative quality of NTSC/PAL decoders. To measure these parameters, the output of the NTSC/PAL decoder (while decoding various video test signals such as those described in Chapter 8) is monitored using video test equipment. Along with a description of several of these parameters, typical AC parameter values for both consumer and studio-quality decoders are shown in Table 9.11.

Several AC parameters, such as short-time waveform distortion, group delay, and K factors, are dependent on the quality of the analog video filters and are not discussed here. In addition to the AC parameters discussed in this section, there are several others that should be included in a decoder specification, such as burst capture and lock frequency range, and the bandwidths of the decoded YIQ or YUV video signals.

There are also several DC parameters that should be specified, as shown in Table 9.12. Although genlock capabilities are not usually specified, except for "clock jitter," we have attempted to generate a list of genlock parameters, shown in Table 9.13.

Differential Phase

Differential phase distortion, commonly referred to as differential phase, specifies how much the chrominance phase is affected by the luminance level—in other words, how much hue shift occurs when the luminance level changes. Both positive and negative phase errors may be present, so differential phase is expressed as a peak-to-peak measurement, expressed in degrees of subcarrier phase.

This parameter is measured using a test signal of uniform phase and amplitude chrominance superimposed on different luminance levels, such as the modulated ramp test signal, or the modulated five-step portion of the composite test signal. The differential phase parameter for a studio-quality decoder may approach 1° or less.

Differential Gain

Differential gain distortion, commonly referred to as differential gain, specifies how much the chrominance gain is affected by the luminance level—in other words, how much color saturation shift occurs when the luminance level changes. Both attenuation and amplification may occur, so differential gain is expressed as the largest amplitude change between any two levels, expressed as a percentage of the largest chrominance amplitude.

This parameter is measured using a test signal of uniform phase and amplitude chrominance superimposed on different luminance levels, such as the modulated ramp test signal, or the modulated five-step portion of the composite test signal. The differential gain parameter for a studio-quality decoder may approach 1% or less.

Luminance Nonlinearity

Luminance nonlinearity, also referred to as differential luminance and luminance nonlinear distortion, specifies how much the luminance gain is affected by the luminance level. In other words, there is a nonlinear relation-

Parameter	Consumer Quality	Studio Quality	Units
differential phase	4	≤ 1	degrees
differential gain	4	≤ 1	%
luminance nonlinearity	2	≤ 1	%
hue accuracy	3	≤ 1	degrees
color saturation accuracy	3	≤ 1	%
SNR (per EIA/TIA RS-250-C)	48	> 60	dB
chrominance-to-luminance crosstalk	< –40	< –50	dB
luminance-to-chrominance crosstalk	< –40	< –50	dB
H tilt	< 1	< 1	%
V tilt	< 1	< 1	%
Y/C sampling skew	< 5	< 2	ns
demodulation quadrature	90 ±2	90 ±0.5	degrees

Table 9.11. Typical AC Video Parameters for NTSC and PAL Decoders.

Parameter	(M) NTSC	(B, D, G, H, I) PAL	Units
sync input amplitude	40 ±20	43 ±22	IRE
burst input amplitude	40 ±20	42.86 ±22	IRE
video input amplitude (1v nominal)	0.5 to 2.0	0.5 to 2.0	volts

Table 9.12. Typical DC Video Parameters for NTSC and PAL Decoders.

Parameter	Min	Max	Units
sync locking time[1]		2	fields
sync recovery time[2]		2	fields
short-term sync lock range[3]	±100		ns
long-term sync lock range[4]	±5		μs
number of consecutive missing horizontal sync pulses before any correction	5		sync pulses
vertical correlation[5]		±5	ns
short-term subcarrier locking range[6]	±200		Hz
long-term subcarrier locking range[7]	±500		Hz
subcarrier locking time[8]		10	lines
subcarrier accuracy		±2	degrees

Notes:
1. Time from start of genlock process to vertical correlation specification is achieved.
2. Time from loss of genlock to vertical correlation specification is achieved.
3. Range over which vertical correlation specification is maintained. Short-term range assumes line time changes by amount indicated slowly between two consecutive lines.
4. Range over which vertical correlation specification is maintained. Long-term range assumes line time changes by amount indicated slowly over one field.
5. Indicates vertical sample accuracy. For a genlock system that uses a VCO or VCXO, this specification is the same as sample clock jitter.
6. Range over which subcarrier locking time and accuracy specifications are maintained. Short-term time assumes subcarrier frequency changes by amount indicated slowly over 2 frames.
7. Range over which subcarrier locking time and accuracy specifications are maintained. Long-term time assumes subcarrier frequency changes by amount indicated slowly over 24 hours.
8. After instantaneous 180° phase shift of subcarrier, time to lock to within ±2°. Subcarrier frequency is nominal ±500 Hz.

Table 9.13. Typical Genlock Parameters for NTSC and PAL Decoders. Parameters assume a video signal with ≥ 30 dB SNR and over the range of DC parameters in Table 9.12.

ship between the decoded luminance level and the ideal luminance level.

Using an unmodulated five-step or ten-step staircase test signal, or the modulated five-step portion of the composite test signal, the difference between the largest and smallest steps, expressed as a percentage of the largest step, is used to specify the luminance nonlinearity. Although this parameter is included within the differential gain and phase parameters, it is traditionally specified independently.

Chrominance Nonlinear Phase Distortion

Chrominance nonlinear phase distortion specifies how much the chrominance phase (hue) is affected by the chrominance amplitude (saturation)—in other words, how much hue shift occurs when the saturation changes.

Using a modulated pedestal test signal, or the modulated pedestal portion of the combination test signal, the decoder output for each chrominance packet is measured. The difference between the largest and the smallest hue measurements is the peak-to-peak value. This parameter is usually not specified independently, but is included within the differential gain and phase parameters.

Chrominance Nonlinear Gain Distortion

Chrominance nonlinear gain distortion specifies how much the chrominance gain is affected by the chrominance amplitude (saturation). In other words, there is a nonlinear relationship between the decoded chrominance amplitude levels and the ideal chrominance amplitude levels—this is usually seen as an attenuation of highly saturated chrominance signals.

Using a modulated pedestal test signal, or the modulated pedestal portion of the combination test signal, the decoder is adjusted so

that the middle chrominance packet (40 IRE) is decoded properly. The largest difference between the measured and nominal values of the amplitudes of the other two decoded chrominance packets specifies the chrominance nonlinear gain distortion, expressed in IRE or as a percentage of the nominal amplitude of the worst-case packet. This parameter is usually not specified independently, but is included within the differential gain and phase parameters.

Chrominance-to-Luminance Intermodulation

Chrominance-to-luminance intermodulation, commonly referred to as cross-modulation, specifies how much the luminance level is affected by the chrominance. This may be the result of clipping highly saturated chrominance levels or quadrature distortion and may show up as irregular brightness variations due to changes in color saturation.

Using a modulated pedestal test signal, or the modulated pedestal portion of the combination test signal, the largest difference between the decoded 50 IRE luminance level and the decoded luminance levels specifies the chrominance-to-luminance intermodulation, expressed in IRE or as a percentage. This parameter is usually not specified independently, but is included within the differential gain and phase parameters.

Hue Accuracy

Hue accuracy specifies how closely the decoded hue is to the ideal hue value. Both positive and negative phase errors may be present, so hue accuracy is the difference between the worst-case positive and worst-case negative measurements from nominal, expressed in degrees of subcarrier phase. This parameter is measured using EIA or EBU 75% color bars as a test signal.

Color Saturation Accuracy

Color saturation accuracy specifies how close the decoded saturation is to the ideal saturation value, using EIA or EBU 75% color bars as a test signal. Both gain and attenuation may be present, so color saturation accuracy is the difference between the worst-case gain and worst-case attenuation measurements from nominal, expressed as a percentage of nominal.

H Tilt

H tilt, also known as line tilt and line time distortion, causes a tilt in line-rate signals, predominantly white bars. This type of distortion causes variations in brightness between the left and right edges of an image. For a digital decoder, H tilt is primarily an artifact of the analog input filters and the transmission medium. H tilt is measured using a line bar (such as the one in the NTC-7 NTSC composite test signal) and measuring the peak-to-peak deviation of the tilt (in IRE or percentage of white bar amplitude), ignoring the first and last microsecond of the white bar.

V Tilt

V tilt, also known as field tilt and field time distortion, causes a tilt in field-rate signals, predominantly white bars. This type of distortion causes variations in brightness between the top and bottom edges of an image. For a digital decoder, V tilt is primarily an artifact of the analog input filters and the transmission medium. V tilt is measured using an 18-μs, 100-IRE white bar in the center of 130 lines in the center of the field or using a field square wave. The peak-to-peak deviation of the tilt is measured (in IRE or percentage of white bar amplitude), ignoring the first and last three lines.

References

1. Benson, K. Blair, 1986, *Television Engineering Handbook*, McGraw-Hill, Inc.
2. Clarke, C.K.P., 1986, *Colour encoding and decoding techniques for line-locked sampled PAL and NTSC television signals*, BBC Research Department Report BBC RD1986/2.
3. Clarke, C.K.P., 1982, *Digital Standards Conversion: comparison of colour decoding methods*, BBC Research Department Report BBC RD1982/6.
4. Clarke, C.K.P., 1982, *High quality decoding for PAL inputs to digital YUV studios*, BBC Research Department Report BBC RD1982/12.
5. Clarke, C.K.P., 1988, *PAL Decoding: Multidimensional filter design for chrominance-luminance separation*, BBC Research Department Report BBC RD1988/11.
6. Drewery, J.O., 1996, *Advanced PAL Decoding: Exploration of Some Adaptive Techniques*, BBC Research Department Report BBC RD1996/1.
7. ITU-R BT.470–6, 1998, *Conventional Television Systems*.
8. *NTSC Video Measurements*, Tektronix, Inc., 1997.
9. Perlman, Stuart S. et. al., *An Adaptive Luma-Chroma Separator Circuit for PAL and NTSC TV Signals*, International Conference on Consumer Electronics, Digest of Technical Papers, June 6–8, 1990.
10. Sandbank, C. P., *Digital Television*, John Wiley & Sons, Ltd., 1990.
11. SMPTE 170M–1999, *Television—Composite Analog Video Signal—NTSC for Studio Applications*.
12. *Television Measurements, NTSC Systems*, Tektronix, Inc., 1998.
13. *Television Measurements, PAL Systems*, Tektronix, Inc., 1990.

Chapter 10

H.261 and H.263

There are several standards for video conferencing, as shown in Table 10.1. Figures 10.1 through 10.3 illustrate the block diagrams of several common video conferencing systems.

H.261

ITU-T H.261 was the first video compression and decompression standard developed for video conferencing. Originally designed for bit rates of p × 64 kbps, where p is in the range 1–30, H.261 is now the minimum requirement of all video conferencing standards, as shown in Table 10.1.

A typical H.261 encoder block diagram is shown in Figure 10.4. The video encoder provides a self-contained digital video bitstream which is multiplexed with other signals, such as control and audio. The video decoder performs the reverse process.

H.261 video data uses the 4:2:0 YCbCr format shown in Figure 3.7, with the primary specifications listed in Table 10.2. The maximum picture rate may be restricted by having 0, 1, 2, or 3 non-transmitted pictures between transmitted ones.

Two picture (or frame) types are supported:

Intra or *I Frame*: A frame having no reference frame for prediction.

Inter or *P Frame*: A frame based on a previous frame.

Video Coding Layer

As shown in Figure 10.4, the basic functions are prediction, block transformation, and quantization.

The prediction error (inter mode) or the input picture (intra mode) is subdivided into 8 sample × 8 line blocks that are segmented as transmitted or non-transmitted. Four luminance blocks and the two spatially corresponding color difference blocks are combined to form a 16 sample × 16 line macroblock as shown in Figure 10.5.

The criteria for choice of mode and transmitting a block are not recommended and may be varied dynamically as part of the coding strategy. Transmitted blocks are transformed and the resulting coefficients quantized and variable-length coded.

	H.310	H.320	H.321	H.322	H.323	H.324	H.324/C
network	Broadband ISDN ATM LAN	Narrowband Switched Digital ISDN	Broadband ISDN ATM LAN	Guaranteed Bandwidth Packet Switched Networks	Non-guaranteed Bandwidth Packet Switched Networks (Ethernet)	PSTN or POTS	Mobile
video codec	MPEG-2 H.261	H.261 H.263	H.261 H.263	H.261 H.263	H.261 H.263	H.261 H.263	H.261 H.263
audio codec	MPEG-2 G.711 G.722 G.728	G.711 G.722 G.728	G.711 G.722 G.728	G.711 G.722 G.728	G.711 G.722 G.723 G.728 G.729	G.723	G.723
multiplexing	H.222.0 H.222.1	H.221	H.221	H.221	H.225.0	H.223	H.223A
control	H.245	H.230 H.242	H.242	H.230 H.242	H.245	H.245	H.245
multipoint		H.231	H.231	H.231	H.323		
data	T.120	T.120	T.120	T.120	T.120	T.120	T.120
communications interface	AAL I.363 AJM I.361 PHY I.432	I.400	AAL I.363 AJM I.361 PHY I.400	I.400 and TCP/IP	TCP/IP	V.34 modem	Mobile Radio

Table 10.1. Video Conferencing Family of Standards.

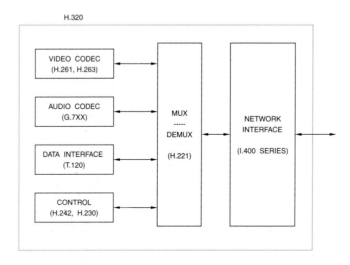

Figure 10.1. Typical H.320 System.

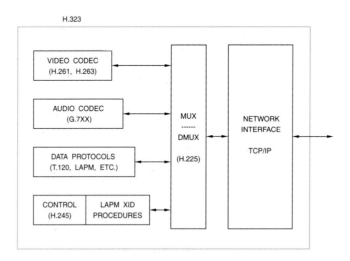

Figure 10.2. Typical H.323 System.

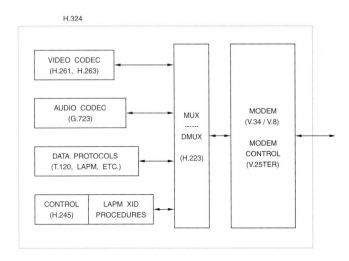

Figure 10.3. Typical H.324 System.

Prediction

The prediction is inter-picture and may include motion compensation and a spatial filter. The coding mode using prediction is called *inter*; the coding mode using no prediction is called *intra*.

Motion Compensation

Motion compensation is optional in the encoder. The decoder must support the acceptance of one motion vector per macroblock. Motion vectors are restricted—all samples referenced by them must be within the coded picture area.

The horizontal and vertical components of motion vectors have integer values not exceeding ±15. The motion vector is used for all four Y blocks in the macroblock. The motion vector for both the Cb and Cr blocks is derived by halving the values of the macroblock vector.

A positive value of the horizontal or vertical component of the motion vector indicates that the prediction is formed from samples in the previous picture that are spatially to the right or below the samples being predicted.

Loop Filter

The prediction process may use a 2D spatial filter that operates on samples within a predicted 8 × 8 block.

The filter is separated into horizontal and vertical functions. Both are non-recursive with coefficients of 0.25, 0.5, 0.25 except at block edges where one of the taps fall outside the block. In such cases, the filter coefficients are changed to 0, 1, 0.

The filter is switched on or off for all six blocks in a macroblock according to the macroblock type.

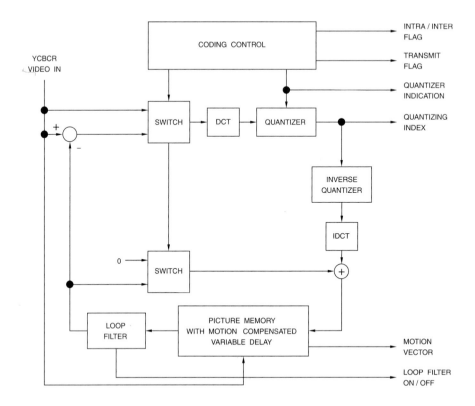

Figure 10.4. Typical H.261 Encoder.

Parameters	CIF	QCIF
active resolution (Y)	352×288	176×144
frame refresh rate	29.97 Hz	
YCbCr sampling structure	4:2:0	
form of YCbCr coding	Uniformly quantized PCM, 8 bits per sample.	

Table 10.2. H.261 YCbCr Parameters.

DCT, IDCT

Transmitted blocks are first processed by an 8×8 DCT (discrete cosine transform). The output from the IDCT (inverse DCT) ranges from -256 to $+255$ after clipping, represented using nine bits.

The procedures for computing the transforms are not defined, but the inverse transform must meet the specified error tolerance.

Quantization

Within a macroblock, the same quantizer is used for all coefficients, except the one for intra DC. The intra DC coefficient is usually linearly quantized with a step size of 8 and no dead zone. The other coefficients use one of 31 possible linear quantizers, but with a central dead zone about zero and a step size of an even value in the range of 2–62.

Clipping of Reconstructed Picture

Clipping functions are used to prevent quantization distortion of transform coefficient amplitudes, possibly causing arithmetic overflows in the encoder and decoder loops. The clipping function is applied to the reconstructed picture, formed by summing the prediction and the prediction error. Clippers force sample values less than 0 to be 0 and values greater than 255 to be 255.

Coding Control

Although not included as part of H.261, several parameters may be varied to control the rate of coded video data. These include processing prior to coding, the quantizer, block significance criterion, and temporal subsampling. Temporal subsampling is performed by discarding complete pictures.

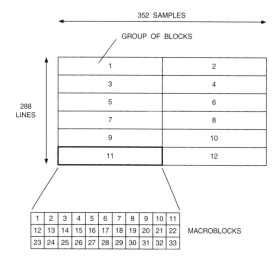
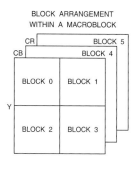

Figure 10.5. H.261 Arrangement of Group of Blocks, Macroblocks, and Blocks.

Forced Updating

This is achieved by forcing the use of the intra mode of the coding algorithm. To control the accumulation of inverse transform mismatch errors, a macroblock should be forcibly updated at least once every 132 times it is transmitted.

Video Bitstream

Unless specified otherwise, the most significant bits are transmitted first. This is bit 1 and is the leftmost bit in the code tables. Unless specified otherwise, all unused or spare bits are set to "1."

The video bitstream is a hierarchical structure with four layers. From top to bottom the layers are:

> Picture
> Group of Blocks (GOB)
> Macroblock (MB)
> Block

Picture Layer

Data for each picture consists of a picture header followed by data for group of blocks (GOBs). The structure is shown in Figure 10.6. Picture headers for dropped pictures are not transmitted.

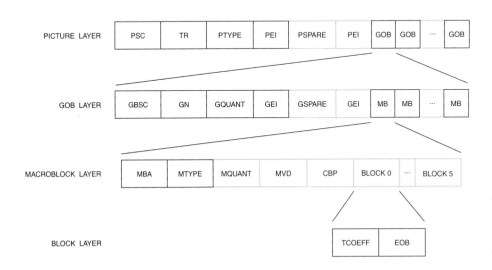

Figure 10.6. H.261 Video Bitstream Layer Structures.

Picture Start Code (PSC)

PSC is a 20-bit word with a value of 0000 0000 0000 0001 0000.

Temporal Reference (TR)

TR is a 5-bit binary number representing 32 possible values. It is generated by incrementing the value in the previous picture header by one plus the number of non-transmitted pictures (at 29.97 Hz). The arithmetic is performed with only the five LSBs.

Type Information (PTYPE)

Six bits of information about the picture are:

Bit 1	Split screen indicator "0" = off, "1" = on
Bit 2	Document camera indicator "0" = off, "1" = on
Bit 3	Freeze picture release "0" = off, "1" = on
Bit 4	Source format "0" = QCIF, "1" = CIF
Bit 5	Optional still image mode "0" = on, "1" = off
Bit 6	Spare

Extra Insertion Information (PEI)

PEI is a bit which when set to "1" indicates the presence of the following optional data field.

Spare Information (PSPARE)

If PEI is set to "1," then these nine bits follow consisting of eight bits of data (PSPARE) and another PEI bit to indicate if a further nine bits follow, and so on.

Group of Blocks (GOB) Layer

Each picture is divided into groups of blocks (GOB). A GOB comprises one-twelfth of the CIF picture area or one-third of the QCIF picture area (see Figure 10.5). A GOB relates to 176 samples × 48 lines of Y and the corresponding 88 × 24 array of Cb and Cr data.

Data for each GOB consists of a GOB header followed by macroblock data, as shown in Figure 10.6. Each GOB header is transmitted once between picture start codes in the CIF or QCIF sequence numbered in Figure 10.5, even if no macroblock data is present in that GOB.

Group of Blocks Start Code (GBSC)

GBSC is a 16-bit word with a value of 0000 0000 0000 0001.

Group Number (GN)

GN is a 4-bit binary value indicating the position of the group of blocks. The bits are the binary representation of the number in Figure 10.5. Numbers 13, 14, and 15 are reserved for future use.

Quantizer Information (GQUANT)

GQUANT is a 5-bit binary value that indicates the quantizer used for the group of blocks until overridden by any subsequent MQUANT. Values of 1–31 are allowed.

Extra Insertion Information (GEI)

GEI is a bit which, when set to "1," indicates the presence of the following optional data field.

Spare Information (GSPARE)

If GEI is set to "1," then these nine bits follow consisting of eight bits of data (GSPARE) and then another GEI bit to indicate if a further nine bits follow, and so on.

Macroblock (MB) Layer

Each GOB is divided into 33 macroblocks as shown in Figure 10.5. A macroblock relates to 16 samples × 16 lines of Y and the corresponding 8 × 8 array of Cb and Cr data.

Data for a macroblock consists of a macroblock header followed by data for blocks (see Figure 10.6).

Macroblock Address (MBA)

MBA is a variable-length codeword indicating the position of a macroblock within a group of blocks. The transmission order is shown in Figure 10.5. For the first macroblock in a GOB, MBA is the absolute address in Figure 10.5. For subsequent macroblocks, MBA is the difference between the absolute addresses of the macroblock and the last transmitted macroblock. The code table for MBA is given in Table 10.3.

A codeword is available for bit stuffing immediately after a GOB header or a coded macroblock (called MBA stuffing). This codeword is discarded by decoders.

The codeword for the start code is also shown in Table 10.3. MBA is always included in transmitted macroblocks. Macroblocks are not transmitted when they contain no information for that part of the picture.

Type Information (MTYPE)

MTYPE is a variable-length codeword containing information about the macroblock and data elements that are present. Macroblock types, included elements, and variable-length codewords are listed in Table 10.4. MTYPE is always included in transmitted macroblocks.

Quantizer (MQUANT)

MQUANT is present only if indicated by MTYPE. It is a 5-bit codeword indicating the quantizer to use for this and any following blocks in the group of blocks, until overridden by any subsequent MQUANT. Codewords for MQUANT are the same as for GQUANT.

Motion Vector Data (MVD)

Motion vector data is included for all motion-compensated (MC) macroblocks, as indicated by MTYPE. MVD is obtained from the macroblock vector by subtracting the vector of the preceding macroblock. The vector of the previous macroblock is regarded as zero for the following situations:

(a) Evaluating MVD for macroblocks 1, 12, and 23.

(b) Evaluating MVD for macroblocks where MBA does not represent a difference of 1.

(c) MTYPE of the previous macroblock was not motion-compensated.

Motion vector data consists of a variable-length codeword for the horizontal component, followed by a variable-length codeword for the vertical component. The variable-length codes are listed in Table 10.5.

Coded Block Pattern (CBP)

The variable-length CBP is present if indicated by MTYPE. It indicates which blocks in the macroblock have at least one transform coefficient transmitted. The pattern number is represented as:

$$P_0P_1P_2P_3P_4P_5$$

where P_n = "1" for any coefficient present for block [n], else P_n = "0." Block numbering (decimal format) is given in Figure 10.5.

The code words for the CBP number are given in Table 10.6.

MBA	Code			MBA	Code			
1	1			17	0000	0101	10	
2	011			18	0000	0101	01	
3	010			19	0000	0101	00	
4	0011			20	0000	0100	11	
5	0010			21	0000	0100	10	
6	0001	1		22	0000	0100	011	
7	0001	0		23	0000	0100	010	
8	0000	111		24	0000	0100	001	
9	0000	110		25	0000	0100	000	
10	0000	1011		26	0000	0011	111	
11	0000	1010		27	0000	0011	110	
12	0000	1001		28	0000	0011	101	
13	0000	1000		29	0000	0011	100	
14	0000	0111		30	0000	0011	011	
15	0000	0110		31	0000	0011	010	
16	0000	0101	11	32	0000	0011	001	
				33	0000	0011	000	
				MBA stuffing	0000	0001	111	
				start code	0000	0000	0000	0001

Table 10.3. H.261 Variable-Length Code Table for MBA.

Prediction	MQUANT	MVD	CBP	TCOEFF	Code		
intra				×	0001		
intra	×			×	0000	001	
inter			×	×	1		
inter	×		×	×	0000	1	
inter + MC		×			0000	0000	1
inter + MC		×	×	×	0000	0001	
inter + MC	×	×	×	×	0000	0000	01
inter + MC + FIL		×			001		
inter + MC + FIL		×	×	×	01		
inter + MC + FIL	×	×	×	×	0000	01	

Table 10.4. H.261 Variable-Length Code Table for MTYPE.

Block Layer

A macroblock is made up of four Y blocks, a Cb block, and a Cr block (see Figure 10.5).

Data for an 8 sample × 8 line block consists of codewords for the transform coefficients followed by an end of block (EOB) marker as shown in Figure 10.6. The order of block transmission is shown in Figure 10.5.

Transform Coefficients (TCOEFF)

When MTYPE indicates intra, transform coefficient data is present for all six blocks in a macroblock. Otherwise, MTYPE and CBP signal which blocks have coefficient data transmitted for them. The quantized DCT coefficients are transmitted in the order shown in Figure 7.59.

Vector Difference	Code			Vector Difference	Code		
–16 & 16	0000	0011	001	1	010		
–15 & 17	0000	0011	011	2 & –30	0010		
–14 & 18	0000	0011	101	3 & –29	0001	0	
–13 & 19	0000	0011	111	4 & –28	0000	110	
–12 & 20	0000	0100	001	5 & –27	0000	1010	
–11 & 21	0000	0100	011	6 & –26	0000	1000	
–10 & 22	0000	0100	11	7 & –25	0000	0110	
–9 & 23	0000	0101	01	8 & –24	0000	0101	10
–8 & 24	0000	0101	11	9 & –23	0000	0101	00
–7 & 25	0000	0111		10 & –22	0000	0100	10
–6 & 26	0000	1001		11 & –21	0000	0100	010
–5 & 27	0000	1011		12 & –20	0000	0100	000
–4 & 28	0000	111		13 & –19	0000	0011	110
–3 & 29	0001	1		14 & –18	0000	0011	100
–2 & 30	0011			15 & –17	0000	0011	010
–1	011						
0	1						

Table 10.5. H.261 Variable-Length Code Table for MVD.

CBP	Code		CBP	Code	
60	111		62	0100	0
4	1101		24	0011	11
8	1100		36	0011	10
16	1011		3	0011	01
32	1010		63	0011	00
12	1001	1	5	0010	111
48	1001	0	9	0010	110
20	1000	1	17	0010	101
40	1000	0	33	0010	100
28	0111	1	6	0010	011
44	0111	0	10	0010	010
52	0110	1	18	0010	001
56	0110	0	34	0010	000
1	0101	1	7	0001	1111
61	0101	0	11	0001	1110
2	0100	1	19	0001	1101

Table 10.6a. H.261 Variable-Length Code Table for CBP.

CBP	Code		CBP	Code	
35	0001	1100	38	0000	1100	
13	0001	1011	29	0000	1011	
49	0001	1010	45	0000	1010	
21	0001	1001	53	0000	1001	
41	0001	1000	57	0000	1000	
14	0001	0111	30	0000	0111	
50	0001	0110	46	0000	0110	
22	0001	0101	54	0000	0101	
42	0001	0100	58	0000	0100	
15	0001	0011	31	0000	0011	1
51	0001	0010	47	0000	0011	0
23	0001	0001	55	0000	0010	1
43	0001	0000	59	0000	0010	0
25	0000	1111	27	0000	0001	1
37	0000	1110	39	0000	0001	0
26	0000	1101				

Table 10.6b. H.261 Variable-Length Code Table for CBP.

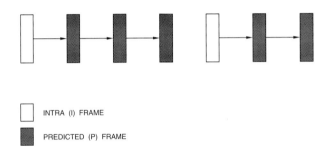

INTRA (I) FRAME

PREDICTED (P) FRAME

Figure 10.7. Typical H.261 Decoded Sequence.

The most common combinations of successive zeros (RUN) and the following value (LEVEL) are encoded using variable-length codes, listed in Table 10.7. Since CBP indicates blocks with no coefficient data, EOB cannot occur as the first coefficient. The last bit "s" denotes the sign of the level: "0" = positive, "1" = negative.

Other combinations of (RUN, LEVEL) are encoded using a 20-bit word: six bits of escape (ESC), six bits of RUN, and eight bits of LEVEL, as shown in Table 10.8.

Two code tables are used for the variable-length coding: one is used for the first transmitted LEVEL in inter, inter + MC, and inter + MC + FIL blocks; another is used for all other LEVELs, except for the first one in intra blocks, which is fixed-length coded with eight bits.

All coefficients, except for intra DC, have reconstruction levels (REC) in the range –2048 to 2047. Reconstruction levels are recovered by the following equations, and the results are clipped. QUANT ranges from 1 to 31 and is transmitted by either GQUANT or MQUANT.

QUANT = odd:
  for LEVEL > 0
    REC = QUANT × (2 × LEVEL + 1)
  for LEVEL < 0
    REC = QUANT × (2 × LEVEL – 1)

QUANT = even:
  for LEVEL > 0
    REC = (QUANT × (2 × LEVEL + 1)) – 1
  for LEVEL < 0
    REC = (QUANT × (2 × LEVEL – 1)) + 1
  for LEVEL = 0
    REC = 0

For intra DC blocks, the first coefficient is typically the transform value quantized with a step size of 8 and no dead zone, resulting in an 8-bit coded value, *n*. Black has a coded value of 0001 0000 (16), and white has a coded value of 1110 1011 (235). A transform value of 1024 is coded as 1111 1111. Coded values of 0000 0000 and 1000 0000 are not used. The decoded value is 8*n*, except that an *n* value of 255 results in a reconstructed transform value of 1024.

Run	Level	Code			
EOB		10			
0	**1**	1s	if first coefficient in block*		
0	**1**	11s	if not first coefficient in block		
0	**2**	0100	s		
0	**3**	0010	1s		
0	**4**	0000	110s		
0	**5**	0010	0110	s	
0	**6**	0010	0001	s	
0	**7**	0000	0010	10s	
0	**8**	0000	0001	1101	s
0	**9**	0000	0001	1000	s
0	**10**	0000	0001	0011	s
0	**11**	0000	0001	0000	s
0	**12**	0000	0000	1101	0s
0	**13**	0000	0000	1100	1s
0	**14**	0000	0000	1100	0s
0	**15**	0000	0000	1011	1s
1	**1**	011s			
1	**2**	0001	10s		
1	**3**	0010	0101	s	
1	**4**	0000	0011	00s	
1	**5**	0000	0001	1011	s
1	**6**	0000	0000	1011	0s
1	**7**	0000	0000	1010	1s
2	**1**	0101	s		
2	**2**	0000	100s		
2	**3**	0000	0010	11s	
2	**4**	0000	0001	0100	s
2	**5**	0000	0000	1010	0s
3	**1**	0011	1s		
3	**2**	0010	0100	s	
3	**3**	0000	0001	1100	s
3	**4**	0000	0000	1001	1s

Table 10.7a. H.261 Variable-Length Code Table for TCOEFF.
***Never used in intra macroblocks.**

Run	Level	Code			
4	1	0011	0s		
4	2	0000	0011	11s	
4	3	0000	0001	0010	s
5	1	0001	11s		
5	2	0000	0010	01s	
5	3	0000	0000	1001	0s
6	1	0001	01s		
6	2	0000	0001	1110	s
7	1	0001	00s		
7	2	0000	0001	0101	s
8	1	0000	111s		
8	2	0000	0001	0001	s
9	1	0000	101s		
9	2	0000	0000	1000	1s
10	1	0010	0111	s	
10	2	0000	0000	1000	0s
11	1	0010	0011	s	
12	1	0010	0010	s	
13	1	0010	0000	s	
14	1	0000	0011	10s	
15	1	0000	0011	01s	
16	1	0000	0010	00s	
17	1	0000	0001	1111	s
18	1	0000	0001	1010	s
19	1	0000	0001	1001	s
20	1	0000	0001	0111	s
21	1	0000	0001	0110	s
22	1	0000	0000	1111	1s
23	1	0000	0000	1111	0s
24	1	0000	0000	1110	1s
25	1	0000	0000	1110	0s
26	1	0000	0000	1101	1s
ESC		0000	01		

Table 10.7b. H.261 Variable-Length Code Table for TCOEFF.

Run	Code	Level	Code
0	0000 00	−128	forbidden
1	0000 01	−127	1000 0001
:	:	:	:
63	1111 11	−2	1111 1110
		−1	1111 1111
		0	forbidden
		1	0000 0001
		2	0000 0010
		:	:
		127	0111 1111

Table 10.8. H.261 Run, Level Codes.

Still Image Transmission

H.261 allows the transmission of a still image of four times the resolution of the currently selected video format. If the video format is QCIF, a still image of CIF resolution may be transmitted; if the video format is CIF, a still image of 704×576 resolution may be transmitted.

H.263

ITU-T H.263 improves on H.261 by providing improved video quality at lower bit rates.

The video encoder provides a self-contained digital bitstream which is combined with other signals (such as H.223). The video decoder performs the reverse process. The primary specifications of H.263 regarding YCbCr video data are listed in Table 10.9. It is also possible to negotiate a custom picture size. The 4:2:0 YCbCr sampling is shown in Figure 3.7.

With H.263 version 2 (formally known as H.263+), seven frame (or picture) types are now supported, with the first two being mandatory (baseline H.263):

Intra or *I Frame*: A frame having no reference frame for prediction.

Inter or *P Frame*: A frame based on a previous frame.

PB Frame and *Improved PB Frame*: A frame representing two frames and based on a previous frame.

B Frame: A frame based two reference frames, one previous and one afterwards.

EI Frame: A frame having a temporally simultaneous frame which has either the same or smaller frame size.

EP Frame: A frame having a two reference frames, one previous and one simultaneous.

Video Coding Layer

A typical encoder block diagram is shown in Figure 10.8. The basic functions are prediction, block transformation, and quantization.

The prediction error or the input picture are subdivided into 8 × 8 blocks which are segmented as transmitted or non-transmitted. Four luminance blocks and the two spatially corresponding color difference blocks are combined to form a macroblock as shown in Figure 10.9.

The criteria for choice of mode and transmitting a block are not recommended and may be varied dynamically as part of the coding strategy. Transmitted blocks are transformed and the resulting coefficients are quantized and variable-length coded.

Prediction

The prediction is interpicture and may include motion compensation. The coding mode using prediction is called *inter*; the coding mode using no prediction is called *intra*.

Intra coding is signalled at the picture level (I frame for intra or P frame for inter) or at the macroblock level in P frames. In the optional *PB frame* mode, B frames always use the inter mode.

Motion Compensation

Motion compensation is optional in the encoder. The decoder must support accepting one motion vector per macroblock (one or four motion vectors per macroblock in the optional *advanced prediction* or *deblocking filter* modes).

In the optional *PB frame* mode, each macroblock may have an additional vector. In the optional *improved PB frame* mode, each macroblock can include an additional forward motion vector. In the optional *B frame* mode, macroblocks can be transmitted with both a forward and backward motion vector.

For baseline H.263, motion vectors are restricted such that all samples referenced by them are within the coded picture area. Many of the optional modes remove this restriction. The horizontal and vertical components of motion vectors have integer or half-integer values not exceeding −16 to +15.5. Several of the optional modes increase the range to [−31.5, +31.5] or [−31.5, +30.5].

A positive value of the horizontal or vertical component of the motion vector typically indicates that the prediction is formed from samples in the previous frame which are spatially to the right or below the samples being predicted. However, for backward motion vectors in B frames, a positive value of the horizontal or vertical component of the motion vector indicates that the prediction is formed from samples in the next frame which are spatially to the left or above the samples being predicted.

Quantization

The number of quantizers is 1 for the first intra coefficient and 31 for all other coefficients. Within a macroblock, the same quantizer is used for all coefficients except the first one of intra blocks. The first intra coefficient is usually the transform DC value linearly quantized with a step size of 8 and no dead zone. Each of the other 31 quantizers are also linear, but with a central dead zone around zero and a step size of an even value in the range of 2–62.

Coding Control

Although not a part of H.263, several parameters may be varied to control the rate of coded video data. These include processing prior to coding, the quantizer, block significance criterion, and temporal subsampling.

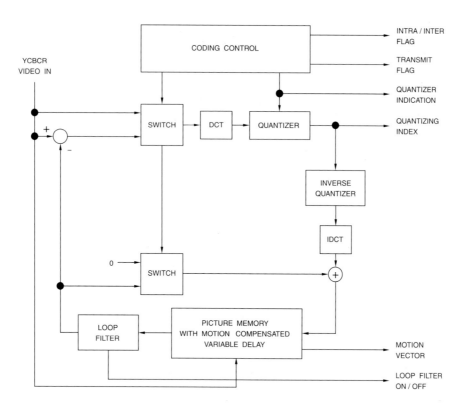

Figure 10.8. Typical Baseline H.263 Encoder.

Parameters	16CIF	4CIF	CIF	QCIF	SQCIF
active resolution (Y)	1408 × 1152	704 × 576	352 × 288	176 × 144	128 × 96
frame refresh rate	29.97 Hz				
YCbCr sampling structure	4:2:0				
form of YCbCr coding	Uniformly quantized PCM, 8 bits per sample.				

Table 10.9. Baseline H.263 YCbCr Parameters.

Forced Updating

This is achieved by forcing the use of the intra mode. To control the accumulation of inverse transform mismatch errors, a macroblock should be forcibly updated at least once every 132 times it is transmitted.

Video Bitstream

Unless specified otherwise, the most significant bits are transmitted first. Bit 1, the leftmost bit in the code tables, is the most significant. Unless specified otherwise, all unused or spare bits are set to "1."

The video multiplexer is arranged in a hierarchical structure with four layers. From top to bottom the layers are:

> Picture
> Group of Blocks (GOB) or Slice
> Macroblock (MB)
> Block

Picture Layer

Data for each picture consists of a picture header followed by data for a group of blocks (GOBs), followed by an end-of-sequence (EOS) and stuffing bits (PSTUF). The baseline structure is shown in Figure 10.10. Picture headers for dropped pictures are not transmitted.

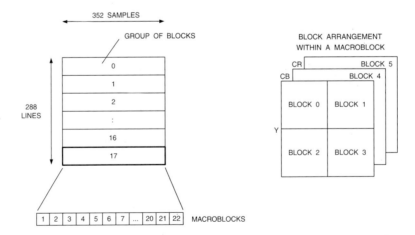

Figure 10.9. H.263 Arrangement of Group of Blocks, Macroblocks, and Blocks.

Picture Start Code (PSC)

PSC is a 22-bit word with a value of 0000 0000 0000 0000 1 00000. It must be byte-aligned; therefore, 0–7 zero bits are added before the start code to ensure the first bit of the start code is the first, and most significant, bit of a byte.

Temporal Reference (TR)

TR is an 8-bit binary number representing 256 possible values. It is generated by incrementing its value in the previously transmitted picture header by one and adding the number of non-transmitted 29.97 Hz pictures since the last transmitted one. The arithmetic is performed with only the eight LSBs.

If a custom picture clock frequency (PCF) is indicated, Extended TR (ETR) and TR form a 10-bit number where TR stores the eight LSBs and ETR stores the two MSBs. The arithmetic in this case is performed with the ten LSBs.

In the *PB frame* and *improved PB frame* mode, TR only addresses P frames.

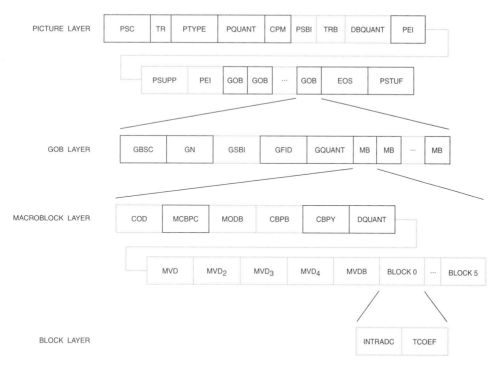

Figure 10.10. Baseline H.263 Video Bitstream Layer Structures (Without Optional PLUSPTYPE Related Fields in the Picture Layer).

Type Information (PTYPE)

PTYPE contains 13 bits of information about the picture:

Bit 1   "1"

Bit 2   "0"

Bit 3   Split screen indicator
         "0" = off, "1" = on

Bit 4   Document camera indicator
         "0" = off, "1" = on

Bit 5   Freeze picture release
         "0" = off, "1" = on

Bit 6–8 Source format
         "000" = reserved
         "001" = SQCIF
         "010" = QCIF
         "011" = CIF
         "100" = 4CIF
         "101" = 16CIF
         "110" = reserved
         "111" = extended PTYPE

If bits 6-8 are not "111," the following five bits are present in PTYPE:

Bit 9   Picture coding type
         "0" = intra, "1" = inter

Bit 10   Optional unrestricted motion
         vector mode
         "0" = off, "1" = on

Bit 11   Optional syntax-based arithmetic
         coding mode
         "0" = off, "1" = on

Bit 12   Optional advanced prediction
         mode
         "0" = off, "1" = on

Bit 13   Optional PB frames mode
         "0" = normal picture
         "1" = PB frame

If bit 9 is set to "0," bit 13 must be set to a "0." Bits 10–13 are optional modes that are negotiated between the encoder and decoder.

Quantizer Information (PQUANT)

PQUANT is a 5-bit binary number (value of 1–31) representing the quantizer to be used until updated by a subsequent GQUANT or DQUANT.

Continuous Presence Multipoint (CPM)

CPM is a 1-bit value that signals the use of the optional *continuous presence multipoint* and *video multiplex* mode; "0" = off, "1" = on. CPM immediately follows PQUANT if PLUSPTYPE is not present, and is immediately after PLUSPTYPE if PLUSPTYPE is present.

Picture Sub-Bitstream Indicator (PSBI)

PSBI is an optional 2-bit binary number that is only present if the optional *continuous presence multipoint* and *video multiplex* mode is indicated by CPM.

Temporal Reference of B Frames in PB Frames (TRB)

TRB is present if PTYPE or PLUSTYPE indicate a *PB frame* or *improved PB frame*. TRB is a 3-bit or 5-bit binary number of the [number + 1] of nontransmitted pictures (at 29.97 Hz or the custom picture clock frequency indicated in CPCFC) since the last I or P frame or the P-part of a *PB frame* or *improved PB frame* and before the B-part of the *PB frame* or *improved PB frame*. The value of TRB is extended to 5 bits when a custom picture clock frequency is in use.

The maximum number of non-transmitted pictures is six for 29.97 Hz, or thirty when a custom picture clock frequency is used.

Quantizer Information for B Frames in PB Frames (DBQUANT)

DBQUANT is present if PTYPE or PLUSTYPE indicate a *PB frame* or *improved PB frame*. DBQUANT is a 2-bit codeword indicating the relationship between QUANT and BQUANT as shown in Table 10.10. The division is done using truncation. BQUANT has a range of 1–31. If the result is less than 1 or greater than 31, BQUANT is clipped to 1 and 31, respectively.

DBQUANT	BQUANT
00	(5 * QUANT) / 4
01	(6 * QUANT) / 4
10	(7 * QUANT) / 4
11	(8 * QUANT) / 4

Table 10.10. Baseline H.263 DBQUANT Codes and QUANT/BQUANT Relationship.

Extra Insertion Information (PEI)

PEI is a bit which when set to "1" signals the presence of the PSUPP data field.

Supplemental Enhancement Information (PSUPP)

If PEI is set to "1," then nine bits follow consisting of eight bits of data (PSUPP) and another PEI bit to indicate if a further nine bits follow, and so on.

End of Sequence (EOS)

EOS is a 22-bit word with a value of 0000 0000 0000 0000 1 11111. EOS must be byte aligned by inserting 0–7 zero bits before the code so that the first bit of the EOS code is the first, and most significant, bit of a byte.

Stuffing (PSTUF)

PSTUF is a variable-length word of zero bits. The last bit of PSTUF must be the last, and least significant, bit of a byte.

Group of Blocks (GOB) Layer

As shown in Figure 10.9, each picture is divided into groups of blocks (GOBs). A GOB comprises 16 lines for the SQCIF, QCIF, and CIF resolutions, 32 lines for the 4CIF resolution, and 64 lines for the 16CIF resolution. Thus, a SQCIF picture contains six GOBs (96/16) each with one row of macroblock data. QCIF pictures have nine GOBs (144/16) each with one row of macroblock data. A CIF picture contains eighteen GOBs (288/16) each with one row of macroblock data. 4CIF pictures have eighteen GOBs (576/32) each with two rows of macroblock data. A 16CIF picture has eighteen GOBs (1152/64) each with four rows of macroblock data. GOB numbering starts with 0 at the top of picture, and increases going down vertically.

Data for each GOB consists of a GOB header followed by macroblock data, as shown in Figure 10.10. Macroblock data is transmitted in increasing macroblock number order. For GOB number 0 in each picture, no GOB header is transmitted. A decoder can signal an encoder to transmit only non-empty GOB headers.

Group of Blocks Start Code (GBSC)

GBSC is a 17-bit word with a value of 0000 0000 0000 0000 1. It must be byte-aligned; therefore, 0–7 zero bits are added before the start code to ensure the first bit of the start code is the first, and most significant, bit of a byte.

Group Number (GN)

GN is a 5-bit binary number indicating the number of the GOB. Group numbers 1–17 are used with the standard picture formats. Group numbers 1–24 are used with custom picture formats. Group numbers 16–29 are emulated in the slice header. Group number 30 is used in the end of sub-bitstream indicators (EOSBS) code and group number 31 is used in the end of sequence (EOS) code.

GOB Sub-Bitstream Indicator (GSBI)

GSBI is a 2-bit binary number representing the sub-bitstream number until the next picture or GOB start code. GSBI is present only if *continuous presence multipoint* and *video multiplex* (CPM) mode is enabled.

GOB Frame ID (GFID)

GFID is a 2-bit value indicating the frame ID. It must have the same value in every GOB (or slice) header of a given frame. In general, if PTYPE is the same as for the previous picture header, the GFID value must be the same as the previous frame. If PTYPE has changed

from the previous picture header, GFID must have a different value from the previous frame.

Quantizer Information (GQUANT)

GQUANT is a 5-bit binary number that indicates the quantizer to be used in the group of blocks until overridden by any subsequent GQUANT or DQUANT. The codewords are the binary representations of the values 1–31.

Macroblock (MB) Layer

Each GOB is divided into macroblocks, as shown in Figure 10.9. A macroblock relates to 16 samples × 16 lines of Y and the corresponding 8 samples × 8 lines of Cb and Cr. Macroblock numbering increases left-to-right and top-to-bottom. Macroblock data is transmitted in increasing macroblock numbering order.

Data for a macroblock consists of a MB header followed by block data (Figure 10.10).

Coded Macroblock Indication (COD)

COD is a single bit that indicates whether or not the block is coded. "0" indicates coded; "1" indicates not coded, and the rest of the macroblock layer is empty. COD is present only in pictures that are not intra.

If not coded, the decoder processes the macroblock as an inter block with motion vectors equal to zero for the whole block and no coefficient data.

Macroblock Type and Coded Block Pattern for Chrominance (MCBPC)

MCBPC is a variable-length codeword indicating the macroblock type and the coded block pattern for Cb and Cr.

Codewords for MCBPC are listed in Tables 10.11 and 10.12. A codeword is available for bit stuffing, and should be discarded by decoders. In some cases, bit stuffing must not occur before the first macroblock of the pic-

MB Type	CBPC (Cb, Cr)	Code		
3	0, 0	1		
3	0, 1	001		
3	1, 0	010		
3	1, 1	011		
4	0, 0	0001		
4	0, 1	0000	01	
4	1, 0	0000	10	
4	1, 1	0000	11	
stuffing		0000	0000	1

Table 10.11. Baseline H.263 Variable-Length Code Table for MCBPC for I Frames.

ture to avoid start code emulation. The macroblock types (MB Type) are listed in Tables 10.13 and 10.14.

The coded block pattern for chrominance (CBPC) signifies when a non-intra DC transform coefficient is transmitted for Cb or Cr. A "1" indicates a non-intra DC coefficient is present that block.

Macroblock Mode for B Blocks (MODB)

MODB is present for macroblock types 0–4 if PTYPE indicates *PB frame*. It is a variable-length codeword indicating whether B coefficients and/or motion vectors are transmitted for this macroblock. Table 10.15 lists the codewords for MODB. MODB is coded differently for *improved PB frames*.

Coded Block Pattern for B Blocks (CBPB)

The 6-bit CBPB is present if indicated by MODB. It indicates which blocks in the macroblock have at least one transform coefficient transmitted. The pattern number is represented as:

$$P_0 P_1 P_2 P_3 P_4 P_5$$

where P_n = "1" for any coefficient present for block [n], else P_n = "0." Block numbering (decimal format) is given in Figure 10.9.

Coded Block Pattern for Luminance (CBPY)

CBPY is a variable-length codeword specifying the Y blocks in the macroblock for which at least one non-intra DC transform coefficient is transmitted. However, in the *advanced intra coding* mode, intra DC is indicated in the same manner as the other coefficients.

Table 10.16 lists the codes for CBPY. Y_N is a "1" if any non-intra DC coefficient is present for that Y block. Y block numbering (decimal format) is as shown in Figure 10.9.

Quantizer Information (DQUANT)

DQUANT is a 2-bit codeword signifying the change in QUANT. Table 10.17 lists the differential values for the codewords.

QUANT has a range of 1–31. If the value of QUANT as a result of the indicated change is less than 1 or greater than 31, it is made 1 and 31, respectively.

MB Type	CBPC (Cb, Cr)	Code			
0	0, 0	1			
0	0, 1	0011			
0	1, 0	0010			
0	1, 1	0001	01		
1	0, 0	011			
1	0, 1	0000	111		
1	1, 0	0000	110		
1	1, 1	0000	0010	1	
2	0, 0	010			
2	0, 1	0000	101		
2	1, 0	0000	100		
2	1, 1	0000	0101		
3	0, 0	0001	1		
3	0, 1	0000	0100		
3	1, 0	0000	0011		
3	1, 1	0000	011		
4	0, 0	0001	00		
4	0, 1	0000	0010	0	
4	1, 0	0000	0001	1	
4	1, 1	0000	0001	0	
stuffing		0000	0000	1	
5	0, 0	0000	0000	010	
5	0, 1	0000	0000	0110	0
5	1, 0	0000	0000	0111	0
5	1, 1	0000	0000	0111	1

Table 10.12. Baseline H.263 Variable-Length Code Table for MCBPC for P Frames.

Frame Type	MB Type	Name	COD	MCBPC	CBPY	DQUANT	MVD	MVD$_{2-4}$
inter	not coded	–	×					
inter	0	inter	×	×	×		×	
inter	1	inter + q	×	×	×	×	×	
inter	2	inter4v	×	×	×		×	×
inter	3	intra	×	×	×			
inter	4	intra + q	×	×	×	×		
inter	5	inter4v + q	×	×	×	×	×	×
inter	stuffing	–	×	×				
intra	3	intra		×	×			
intra	4	intra + q		×	×	×		
intra	stuffing	–		×				

Table 10.13. Baseline H.263 Macroblock Types and Included Data for Normal Frames.

Frame Type	MB Type	Name	COD	MCBPC	MODB	CBPY
inter	not coded	–	×			
inter	0	inter	×	×	×	×
inter	1	inter + q	×	×	×	×
inter	2	inter4v	×	×	×	×
inter	3	intra	×	×	×	×
inter	4	intra + q	×	×	×	×
inter	5	inter4v + q	×	×	×	×
inter	stuffing	–	×	×		

Table 10.14a. Baseline H.263 Macroblock Types and Included Data for PB Frames.

Frame Type	MB Type	Name	CBPB	DQUANT	MVD	MVDB	MVD$_{2-4}$
inter	not coded	–					
inter	0	inter	×		×	×	
inter	1	inter + q	×	×	×	×	
inter	2	inter4v	×		×	×	×
inter	3	intra	×		×	×	
inter	4	intra + q	×	×	×	×	
inter	5	inter4v + q	×	×	×	×	×
inter	stuffing	–					

Table 10.14b. Baseline H.263 Macroblock Types and Included Data for PB Frames.

CBPB	MVDB	Code
		0
	×	10
×	×	11

Table 10.15. Baseline H.263 Variable-Length Code Table for MODB.

CBPY (Y0, Y1, Y2, Y3)		Code	
Intra	Inter		
0, 0, 0, 0	1, 1, 1, 1	0011	
0, 0, 0, 1	1, 1, 1, 0	0010	1
0, 0, 1, 0	1, 1, 0, 1	0010	0
0, 0, 1, 1	1, 1, 0, 0	1001	
0, 1, 0, 0	1, 0, 1, 1	0001	1
0, 1, 0, 1	1, 0, 1, 0	0111	
0, 1, 1, 0	1, 0, 0, 1	0000	10
0, 1, 1, 1	1, 0, 0, 0	1011	
1, 0, 0, 0	0, 1, 1, 1	0001	0
1, 0, 0, 1	0, 1, 1, 0	0000	11
1, 0, 1, 0	0, 1, 0, 1	0101	
1, 0, 1, 1	0, 1, 0, 0	1010	
1, 1, 0, 0	0, 0, 1, 1	0100	
1, 1, 0, 1	0, 0, 1, 0	1000	
1, 1, 1, 0	0, 0, 0, 1	0110	
1, 1, 1, 1	0, 0, 0, 0	11	

Table 10.16. Baseline H.263 Variable-Length Code Table for CBPY.

Differential Value of QUANT	DQUANT
−1	00
−2	01
1	10
2	11

Table 10.17. Baseline H.263 DQUANT Codes for QUANT Differential Values.

Motion Vector Data (MVD)

Motion vector data is included for all inter macroblocks and intra blocks when in *PB frame* mode.

Motion vector data consists of a variable-length codeword for the horizontal component, followed by a variable-length codeword for the vertical component. The variable-length codes are listed in Table 10.18. For the *unrestricted motion vector* mode, other motion vector coding may be used.

Motion Vector Data (MVD2–4)

The three codewords MVD_2, MVD_3, and MVD_4 are present if indicated by PTYPE and MCBPC during the *advanced prediction* or *deblocking filter* modes. Each consists of a variable-length codeword for the horizontal component followed by a variable-length codeword for the vertical component. The variable-length codes are listed in Table 10.18.

Motion Vector Data for B Macroblock (MVDB)

MVDB is present if indicated by MODB during the *PB frame* and *improved PB frame* modes. It consists of a variable-length codeword for the horizontal component followed by a variable-length codeword for the vertical component of each vector. The variable-length codes are listed in Table 10.18.

Block Layer

If not in *PB frames* mode, a macroblock is made up of four Y blocks, a Cb block, and a Cr block (see Figure 10.9). Data for an 8 sample × 8 line block consists of codewords for the intra DC coefficient and transform coefficients as shown in Figure 10.10. The order of block transmission is shown in Figure 10.9.

In *PB frames* mode, a macroblock is made up of four Y blocks, a Cb block, a Cr block, and data for six B blocks.

The quantized DCT coefficients are transmitted in the order shown in Figure 7.59. In the *modified quantization* mode, quantized DCT coefficients are transmitted in the order shown in Figure 7.60.

DC Coefficient for Intra Blocks (Intra DC)

Intra DC is an 8-bit codeword. The values and their corresponding reconstruction levels are listed in Table 10.19.

If not in *PB frames* mode, the intra DC coefficient is present for every block of the macroblock if MCBPC indicates macroblock type 3 or 4. In *PB frames* mode, the intra DC coefficient is present for every P block if MCBPC indicates macroblock type 3 or 4 (the intra DC coefficient is not present for B blocks).

Transform Coefficient (TCOEF)

If not in *PB frames* mode, TCOEF is present if indicated by MCBPC or CBPY. In *PB frames* mode, TCOEF is present for B blocks if indicated by CBPB.

An event is a combination of a last non-zero coefficient indication (LAST = "0" if there are more non-zero coefficients in the block; LAST = "1" if there are no more non-zero coefficient in the block), the number of successive zeros preceding the coefficient (RUN), and the non-zero coefficient (LEVEL).

The most common events are coded using a variable-length code, shown in Table 10.20. The "s" bit indicates the sign of the level; "0" for positive, and "1" for negative.

Vector Difference		Code			
−16	16	0000	0000	0010	1
−15.5	16.5	0000	0000	0011	1
−15	17	0000	0000	0101	
−14.5	17.5	0000	0000	0111	
−14	18	0000	0000	1001	
−13.5	18.5	0000	0000	1011	
−13	19	0000	0000	1101	
−12.5	19.5	0000	0000	1111	
−12	20	0000	0001	001	
−11.5	20.5	0000	0001	011	
−11	21	0000	0001	101	
−10.5	21.5	0000	0001	111	
−10	22	0000	0010	001	
−9.5	22.5	0000	0010	011	
−9	23	0000	0010	101	
−8.5	23.5	0000	0010	111	
−8	24	0000	0011	001	
−7.5	24.5	0000	0011	011	
−7	25	0000	0011	101	
−6.5	25.5	0000	0011	111	
−6	26	0000	0100	001	
−5.5	26.5	0000	0100	011	
−5	27	0000	0100	11	
−4.5	27.5	0000	0101	01	
−4	28	0000	0101	11	
−3.5	28.5	0000	0111		
−3	29	0000	1001		
−2.5	29.5	0000	1011		
−2	30	0000	111		
−1.5	30.5	0001	1		
−1	31	0011			
−0.5	31.5	011			
0		1			

Table 10.18a. Baseline H.263 Variable-Length Code Table for MVD, MVD$_{2-4}$, and MVDB.

Vector Difference		Code			
0.5	−31.5	010			
1	−31	0010			
1.5	−30.5	0001	0		
2	−30	0000	110		
2.5	−29.5	0000	1010		
3	−29	0000	1000		
3.5	−28.5	0000	0110		
4	−28	0000	0101	10	
4.5	−27.5	0000	0101	00	
5	−27	0000	0100	10	
5.5	−26.5	0000	0100	010	
6	−26	0000	0100	000	
6.5	−25.5	0000	0011	110	
7	−25	0000	0011	100	
7.5	−24.5	0000	0011	010	
8	−24	0000	0011	000	
8.5	−23.5	0000	0010	110	
9	−23	0000	0010	100	
9.5	−22.5	0000	0010	010	
10	−22	0000	0010	000	
10.5	−21.5	0000	0001	110	
11	−21	0000	0001	100	
11.5	−20.5	0000	0001	010	
12	−20	0000	0001	000	
12.5	−19.5	0000	0000	1110	
13	−19	0000	0000	1100	
13.5	−18.5	0000	0000	1010	
14	−18	0000	0000	1000	
14.5	−17.5	0000	0000	0110	
15	−17	0000	0000	0100	
15.5	−16.5	0000	0000	0011	0

Table 10.18b. Baseline H.263 Variable-Length Code Table for MVD, MVD$_{2-4}$, and MVDB.

Intra DC Value	Reconstruction Level
0000 0000	not used
0000 0001	8
0000 0010	16
0000 0011	24
⋮	⋮
0111 1111	1016
1111 1111	1024
1000 0001	1032
⋮	⋮
1111 1101	2024
1111 1110	2032

Table 10.19. Baseline H.263 Reconstruction Levels for Intra DC.

Other combinations of (LAST, RUN, LEVEL) are encoded using a 22-bit word: 7 bits of escape (ESC), 1 bit of LAST, 6 bits of RUN, and 8 bits of LEVEL. The codes for RUN and LEVEL are shown in Table 10.21. Code 1000 0000 is forbidden unless in the *modified quantization* mode.

All coefficients, except for intra DC, have reconstruction levels (REC) in the range −2048 to 2047. Reconstruction levels are recovered by the following equations, and the results are clipped.

if LEVEL = 0, REC = 0

if QUANT = odd:
$$|REC| = QUANT \times (2 \times |LEVEL| + 1)$$

if QUANT = even:
$$|REC| = QUANT \times (2 \times |LEVEL| + 1) - 1$$

After calculation of |REC|, the sign is added to obtain REC. Sign (LEVEL) is specified by the "s" bit in the TCOEF code in Table 10.20.

$$REC = \text{sign}(LEVEL) \times |REC|$$

For intra DC blocks, the reconstruction level is:

$$REC = 8 \times LEVEL$$

Last	Run	|Level|	Code			
0	0	1	10s			
0	0	2	1111	s		
0	0	3	0101	01s		
0	0	4	0010	111s		
0	0	5	0001	1111	s	
0	0	6	0001	0010	1s	
0	0	7	0001	0010	0s	
0	0	8	0000	1000	01s	
0	0	9	0000	1000	00s	
0	0	10	0000	0000	111s	
0	0	11	0000	0000	110s	
0	0	12	0000	0100	000s	
0	1	1	110s			
0	1	2	0101	00s		
0	1	3	0001	1110	s	
0	1	4	0000	0011	11s	
0	1	5	0000	0100	001s	
0	1	6	0000	0101	0000	s
0	2	1	1110	s		
0	2	2	0001	1101	s	
0	2	3	0000	0011	10s	
0	2	4	0000	0101	0001	s
0	3	1	0110	1s		
0	3	2	0001	0001	1s	
0	3	3	0000	0011	01s	
0	4	1	0110	0s		
0	4	2	0001	0001	0s	
0	4	3	0000	0101	0010	s
0	5	1	0101	1s		
0	5	2	0000	0011	00s	
0	5	3	0000	0101	0011	s
0	6	1	0100	11s		
0	6	2	0000	0010	11s	
0	6	3	0000	0101	0100	s
0	7	1	0100	10s		

Table 10.20a. Baseline H.263 Variable-Length Code Table for TCOEF.

Last	Run	|Level|	Code			
0	7	2	0000	0010	10s	
0	8	1	0100	01s		
0	8	2	0000	0010	01s	
0	9	1	0100	00s		
0	9	2	0000	0010	00s	
0	10	1	0010	110s		
0	10	2	0000	0101	0101	s
0	11	1	0010	101s		
0	12	1	0010	100s		
0	13	1	0001	1100	s	
0	14	1	0001	1011	s	
0	15	1	0001	0000	1s	
0	16	1	0001	0000	0s	
0	17	1	0000	1111	1s	
0	18	1	0000	1111	0s	
0	19	1	0000	1110	1s	
0	20	1	0000	1110	0s	
0	21	1	0000	1101	1s	
0	22	1	0000	1101	0s	
0	23	1	0000	0100	010s	
0	24	1	0000	0100	011s	
0	25	1	0000	0101	0110	s
0	26	1	0000	0101	0111	s
1	0	1	0111	s		
1	0	2	0000	1100	1s	
1	0	3	0000	0000	101s	
1	1	1	0011	11s		
1	1	2	0000	0000	100s	
1	2	1	0011	10s		
1	3	1	0011	01s		
1	4	1	0011	00s		
1	5	1	0010	011s		
1	6	1	0010	010s		
1	7	1	0010	001s		

Table 10.20b. Baseline H.263 Variable-Length Code Table for TCOEF.

Last	Run	|Level|	Code			
1	8	1	0010	000s		
1	9	1	0001	1010	s	
1	10	1	0001	1001	s	
1	11	1	0001	1000	s	
1	12	1	0001	0111	s	
1	13	1	0001	0110	s	
1	14	1	0001	0101	s	
1	15	1	0001	0100	s	
1	16	1	0001	0011	s	
1	17	1	0000	1100	0s	
1	18	1	0000	1011	1s	
1	19	1	0000	1011	0s	
1	20	1	0000	1010	1s	
1	21	1	0000	1010	0s	
1	22	1	0000	1001	1s	
1	23	1	0000	1001	0s	
1	24	1	0000	1000	1s	
1	25	1	0000	0001	11s	
1	26	1	0000	0001	10s	
1	27	1	0000	0001	01s	
1	28	1	0000	0001	00s	
1	29	1	0000	0100	100s	
1	30	1	0000	0100	101s	
1	31	1	0000	0100	110s	
1	32	1	0000	0100	111s	
1	33	1	0000	0101	1000	s
1	34	1	0000	0101	1001	s
1	35	1	0000	0101	1010	s
1	36	1	0000	0101	1011	s
1	37	1	0000	0101	1100	s
1	38	1	0000	0101	1101	s
1	39	1	0000	0101	1110	s
1	40	1	0000	0101	1111	s
	ESC		0000	011		

Table 10.20c. Baseline H.263 Variable-Length Code Table for TCOEF.

Run	Code	Level	Code
0	0000 00	–128	forbidden
1	0000 01	–127	1000 0001
:	:	:	:
63	1111 11	–2	1111 1110
		–1	1111 1111
		0	forbidden
		1	0000 0001
		2	0000 0010
		:	:
		127	0111 1111

Table 10.21. Baseline H.263 Run, Level Codes.

PLUSPTYPE Picture Layer Option

PLUSTYPE is present when indicated by bits 6–8 of PTYPE, and is used to enable the H.263 version 2 options. When present, the PLUSTYPE and related fields immediately follow PTYPE, preceding PQUANT.

If PLUSPTYPE is present, then CPM immediately follows PLUSTYPE. If PLUSP-TYPE is not present, then CPM immediately follows PQUANT. PSBI always immediately follows CPM (if CPM = "1").

PLUSTYPE is a 12- or 30-bit codeword, comprised of up to three subfields: UFEP, OPPTYPE, and MPPTYPE. The PLUSTYPE and related fields are illustrated in Figure 10.11.

Update Full Extended PTYPE (UFEP)

UFEP is a 3-bit codeword present if "extended PTYPE" is indicated by PTYPE.

A value of "000" indicates that only MPP-TYPE is included in the picture header.

A value "001" indicates that both OPP-TYPE and MPPTYPE are included in the picture header. If the picture type is intra or EI, this field must be set to "001."

In addition, if PLUSPTYPE is present in each of a continuing sequence of pictures, this field shall be set to "001" every five seconds or every five frames, whichever is larger. UFEP should be set to "001" more often in error-prone environments.

Values other than "000" and "001" are reserved.

Optional Part of PLUSPTYPE (OPPTYPE)

This field contains features that are not likely to be changed from one frame to another. If UFEP is "001," the following bits are present in OPPTYPE:

Bit 1–3   Source format
"000" = reserved
"001" = SQCIF
"010" = QCIF
"011" = CIF
"100" = 4CIF
"101" = 16CIF
"110" = custom source format
"111" = reserved

Bit 4   Custom picture clock frequency
"0" = standard, "1" = custom

Bit 5   Unrestricted motion vector
(UMV) mode
"0" = off, "1" = on

Bit 6   Syntax-based arithmetic coding
(SAC) mode
"0" = off, "1" = on

Bit 7   Advanced prediction (AP) mode
"0" = off, "1" = on

Bit 8   Advanced intra coding
(AIC) mode
"0" = off, "1" = on

Bit 9   Deblocking filter (DF) mode
"0" = off, "1" = on

Bit 10   Slice-structured (SS) mode
mode
"0" = off, "1" = on

Bit 11   Reference picture selection
(RPS) mode
"0" = off, "1" = on

Bit 12   Independent segment decoding
(ISD) mode
"0" = off, "1" = on

Bit 13   Alternative Inter VLC
(AIV) mode
"0" = off, "1" = on

Bit 14   Modified quantization (MQ) mode
"0" = off, "1" = on

Bit 15   "1"

Bit 16   "0"

Bit 17   "0"

Bit 18   "0"

Figure 10.11.   H.263 PLUSPTYPE and Related Fields.

Mandatory Part of PLUSPTYPE (MPPTYPE)

Regardless of the value of UFEP, the following 9 bits are also present in MPPTYPE:

Bit 1–3   Picture code type
"000" = I frame (intra)
"001" = P frame (inter)
"010" = Improved PB frame
"011" = B frame
"100" = EI frame
"101" = EP frame
"110" = reserved
"111" = reserved

Bit 4   Reference picture resampling (RPR) mode
"0" = off, "1" = on

Bit 5   Reduced resolution update (RRU) mode
"0" = off, "1" = on

Bit 6   Rounding type (RTYPE) mode
"0" = off, "1" = on

Bit 7   "0"

Bit 8   "0"

Bit 9   "1"

Custom Picture Format (CPFMT)

CPFMT is a 23-bit value that is present if the use of a custom picture format is specified by PLUSPTYPE and UFEP is "001."

Bit 1–4   Pixel aspect ratio code
"0000" = reserved
"0001" = 1:1
"0010" = 12:11
"0011" = 10:11
"0100" = 16:11
"0101" = 40:33
"0110" – "1110" = reserved
"1111" = extended PAR

Bit 5–13   Picture width indication (PWI) number of samples per line = (PWI + 1) × 4

Bit 14   "1"

Bit 15–23   Picture height indication (PHI) number of lines per frame = (PHI + 1) × 4

Extended Pixel Aspect Ratio (EPAR)

EPAR is a 16-bit value present if CPFMT is present and "extended PAR" is indicated by CPFMT.

Bit 1–8   PAR width

Bit 9–16   PAR height

Custom Picture Clock Frequency Code (CPCFC)

CPCFC is an 8-bit value present only if PLUSPTYPE is present, UFEP is "001," and PLUSPTYPE indicates a custom picture clock frequency. The custom picture clock frequency (in Hz) is:

1,800,000 / (clock divisor × clock conversion factor)

Bit 1   Clock conversion factor code
"0" = 1000, "1" = 1001

Bit 2–8   Clock divisor

Extended Temporal Reference (ETR)

ETR is a 2-bit value present if a custom picture clock frequency is in use. It is the two MSBs of the 10-bit TR value.

Unlimited Unrestricted Motion Vectors Indicator (UUI)

UUI is a 1- or 2-bit variable-length value indicating the effective range limit of motion vectors. It is present if the optional *unrestricted motion vector* mode is indicated in PLUSP-TYPE and UFEP is "001."

A value of "1" indicates the motion vector range is limited according to Tables 10.22 and 10.23. A value of "01" indicates the motion vector range is not limited except by the picture size.

Picture Width	Horizontal Motion Vector Range
4–352	–32, +31.5
356–704	–64, +63.5
708–1408	–128, +127.5
1412–2048	–256, +255.5

Table 10.22. Optional Horizontal Motion Range.

Picture Height	Vertical Motion Vector Range
4–288	–32, +31.5
292–576	–64, +63.5
580–1152	–128, +127.5

Table 10.23. Optional Vertical Motion Range.

Slice Structured Submode Bits (SSS)

SSS is a 2-bit value present only if the optional *slice structured* mode is indicated in PLUSPTYPE and UFEP is "001." If the *slice structured* mode is in use but UFEP is not "001," the last SSS value remains in effect.

Bit 1	Rectangular slices "0" = no, "1" = yes
Bit 2	Arbitrary slice ordering "0" = sequential, "1" = arbitrary

Enhancement Layer Number (ELNUM)

ELNUM is a 4-bit value present only during the *temporal, SNR*, and *spatial scalability* mode. It identifies a specific enhancement layer. The first enhancement layer above the base layer is designated as enhancement layer number 2, and the base layer is number 1.

Reference Layer Number (RLNUM)

RLNUM is a 4-bit value present only during the *temporal, SNR*, and *spatial scalability* mode UFEP is "001." The layer number for the frames used as reference anchors is identified by the RLNUM.

Reference Picture Selection Mode Flags (RPSMF)

RPSMF is a 3-bit codeword present only during the *reference picture selection* mode and UFEP is "001." When present, it indicates which back-channel messages are needed by the encoder. If the *reference picture selection* mode is in use but RPSMF is not present, the last value of RPSMF that was sent remains in effect.

"000" – "011" = reserved
"100" = neither ACK nor NACK needed
"101" = need ACK
"110" = need NACK
"111" = need both ACK and NACK

Temporal Reference for Prediction Indication (TRPI)

TRPI is a 1-bit value present only during the *reference picture selection* mode. When present, it indicates the presence of the following TRP field. "0" = TRP field not present; "1" = TRP field present. TRPI is "0" whenever the picture header indicates an I frame or EI frame.

Temporal Reference for Prediction (TRP)

TRP is a 10-bit value indicating the temporal reference used for encoding prediction, except in the case of B frames. For B frames, the frame having the temporal reference specified by TRP is used for the prediction in the forward direction.

If the custom picture clock frequency is not being used, the two MSBs of TRP are zero and the LSBs contain the 8-bit TR value in the picture header of the reference picture. If a custom picture clock frequency is being used, TRP is a 10-bit number consisting of the concatenation of ETR and TR from the reference picture header.

If TRP is not present, the previous anchor picture is used for prediction, as when not in the *reference picture selection* mode. TRP is valid until the next PSC, GSC, or SSC.

Back-Channel message Indication (BCI)

BCI is a 1- or 2-bit variable-length codeword present only during the optional *reference picture selection* mode. "1" indicates the presence of the optional back-channel message (BCM) field. "01" indicates the absence or the end of the back-channel message field. BCM and BCI may be repeated when present.

Back-Channel Message (BCM)

The variable-length back-channel message is present if the preceding BCI field is set to "1."

Reference Picture Resampling Parameters (RPRP)

A variable-length field present only during the optional *reference picture resampling* mode. This field carries the parameters of the *reference picture resampling* mode.

Optional H.263 Modes

Unrestricted Motion Vector Mode

In this optional mode, motion vectors are allowed to point outside the picture. The edge samples are used as prediction for the "nonexisting" samples. The edge sample is found by limiting the motion vector to the last full sample position within the picture area. Motion vector limiting is done separately for the horizontal and vertical components.

Additionally, this mode includes an extension of the motion vector range so that larger motion vectors can be used (Tables 10.22 and 10.23). These longer motion vectors improve the coding efficiency for the larger picture formats, such 4CIF or 16CIF. A significant gain is also achieved for the other picture formats if there is movement along the picture edges, camera movement, or background movement.

When this mode is employed within H.263 version 2, new reversible variable-length codes (RVLCs) are used for encoding the motion vectors, as shown in Table 10.24. These codes are single-valued, as opposed to the baseline double-valued VLCs. The double-valued codes were not popular due to limitations in their extensibility and their high cost of implementation. The RVLCs are also easier to implement.

Each row in Table 10.24 represents a motion vector difference in half-pixel units. "...x_1x_0" denotes all bits following the leading "1" in the binary representation of the absolute value of the motion vector difference. The "s" bit denotes the sign of the motion vector differ-

Absolute Value of Motion Vector Difference in Half-Pixel Units	Code
0	1
1	0s0
"x_0" + 2 (2–3)	$0x_01s0$
"x_1x_0" + 4 (4–7)	$0x_11x_01s0$
"$x_2x_1x_0$" + 8 (8–15)	$0x_21x_11x_01s0$
"$x_3x_2x_1x_0$" + 16 (16–31)	$0x_31x_21x_11x_01s0$
"$x_4x_3x_2x_1x_0$" + 32 (32–63)	$0x_41x_31x_21x_11x_01s0$
"$x_5x_4x_3x_2x_1x_0$" + 64 (64–127)	$0x_51x_41x_31x_21x_11x_01s0$
"$x_6x_5x_4x_3x_2x_1x_0$" + 128 (128–255)	$0x_61x_51x_41x_31x_21x_11x_01s0$
"$x_7x_6x_5x_4x_3x_2x_1x_0$" + 256 (256–511)	$0x_71x_61x_51x_41x_31x_21x_11x_01s0$
"$x_8x_7x_6x_5x_4x_3x_2x_1x_0$" + 512 (512–1023)	$0x_81x_71x_61x_51x_41x_31x_21x_11x_01s0$
"$x_9x_8x_7x_6x_5x_4x_3x_2x_1x_0$" + 1024 (1024–2047)	$0x_91x_81x_71x_61x_51x_41x_31x_21x_11x_01s0$
"$x_{10}x_9x_8x_7x_6x_5x_4x_3x_2x_1x_0$" + 2048 (2048–4095)	$0x_{10}1x_91x_81x_71x_61x_51x_41x_31x_21x_11x_01s0$

Table 10.24. H.263 Reversible Variable-Length Codes for Motion Vectors.

ence: "0" for positive and "1" for negative. The binary representation of the motion vector difference is interleaved with bits that indicate if the code continues or ends. The "0" in the last position indicates the end of the code.

RVLCs can also be used to increase resilience to channel errors. Decoding can be performed by processing the motion vectors in the forward and reverse directions. If an error is detected while decoding in one direction, the decoder can proceed in the reverse direction, improving the error resilience of the bit stream. In addition, the motion vector range is extended up to [–256, +255.5], depending on the picture size.

Syntax-based Arithmetic Coding Mode

In this optional mode, the variable-length coding is replaced with arithmetic coding. The SNR and reconstructed pictures will be the same, but the bit rate can be reduced by about 5% since the requirement of a fixed number of bits for information is removed.

The syntax of the picture, group of blocks, and macroblock layers remains exactly the same. The syntax of the block layer changes slightly in that any number of TCOEF entries may be present.

It is worth noting that use of this mode is not widespread.

Advanced Prediction Mode

In this optional mode, four motion vectors per macroblock (one for each Y block) are used instead of one. In addition, *overlapped block motion compensation* (OBMC) is used for the Y blocks of P frames.

If one motion vector is used for a macroblock, it is defined as four motion vectors with the same value. If four motion vectors are used for a macroblock, the first motion vector is the MVD codeword and applies to Y_1 in Figure 10.9. The second motion vector is the MVD_2 codeword that applies to Y_2, the third motion vector is the MVD_3 codeword that applies to Y_3, and the fourth motion vector is the MVD_4 codeword that applies to Y_4. The motion vector for Cb and Cr of the macroblock is derived from the four Y motion vectors.

The encoder has to decide which type of vectors to use. Four motion vectors use more bits, but provide improved prediction. This mode improves inter-picture prediction and yields a significant improvement in picture quality for the same bit rate by reducing blocking artifacts.

PB Frames Mode

Like MPEG, H.263 optionally supports PB frames. A PB frame consists of one P frame (predicted from the previous P frame) and one B frame (bi-directionally predicted from the previous and current P frame), as shown in Figure 10.12.

With this coding option, the picture rate can be increased without substantially increasing the bit rate. However, an *improved PB frames* mode is supported in Annex M. This original *PB frames* mode is retained only for purposes of compatibility with systems made prior to the adoption of Annex M.

Continuous Presence Multipoint and Video Multiplex Mode

In this optional mode, up to four independent H.263 bitstreams can be multiplexed into a single bitstream. The sub-bitstream with the lowest identifier number (sent via the SBI field) is considered to have the highest priority unless a different priority convention is established by external means.

This feature is designed for use in continuous presence multipoint application or other situations in which separate logical channels are not available, but the use of multiple video bitstreams is desired. It is not to be used with H.324.

Forward Error Correction Mode

This optional mode provides forward error correction (code and framing) for transmission of H.263 video data. It is not to be used with H.324.

Both the framing and the forward error correction code are the same as in H.261.

Advanced Intra Coding Mode

This optional mode improves compression for intra macroblocks. It uses intra-block prediction from neighboring intra blocks, a modified inverse quantization of intra DCT coefficients, and a separate VLC table for intra coefficients. This mode significantly improves the compression performance over the intra coding of baseline H.263.

An additional 1- or 2-bit variable-length codeword, INTRA_MODE, is added to the macroblock layer immediately following the MCBPC field to indicate the prediction mode:

"0" = DC only
"10" = Vertical DC and AC
"11" = Horizontal DC and AC

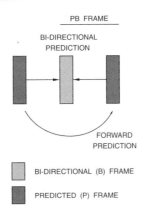

Figure 10.12. Baseline H.263 PB Frames.

For intra-coded blocks, if the prediction mode is DC only, the zig-zag scan order in Figure 7.59 is used. If the prediction mode is vertical DC and AC, the "alternate-vertical" scanning order in Figure 7.60 is used. If the prediction mode is horizontal DC and AC, the "alternate-horizontal" scanning order in Figure 7.61 is used.

For non-intra blocks, the zig-zag scan order in Figure 7.59 is used.

Deblocking Filter Mode

This optional mode introduces a deblocking filter inside the coding loop. The filter is applied to the edge boundaries of 8×8 blocks to reduce blocking artifacts.

The filter coefficients depend on the macroblock's quantizer step size, with larger coefficients used for a coarser quantizer. This mode also allows the use of four motion vectors per macroblock, as specified in the *advanced prediction* mode, and also allows motion vectors to point outside the picture, as in the *unrestricted motion vector* mode. The computationally expensive overlapping motion compensation operation of the *advanced pre-*

diction mode is not used so as to keep the complexity of this mode minimal.

The result is better prediction and a reduction in blocking artifacts.

Slice Structured Mode

In this optional mode, a slice layer is substituted for the GOB layer. This mode provides error resilience, makes the bitstream easier to use with a packet transport delivery scheme, and minimizes video delay.

The slice layer consists of a slice header followed by consecutive complete macroblocks. Two additional modes can be signaled to reflect the order of transmission (sequential or arbitrary) and the shape of the slices (rectangular or not). These add flexibility to the slice structure so that it can be designed for different applications.

Supplemental Enhancement Information

With this optional mode, additional supplemental information may be included in the bitstream to signal enhanced display capability.

Typical enhancement information can signal full- or partial-picture freezes, picture freeze releases, or chroma keying for video compositing.

The supplemental information may be present in the bitstream even though the decoder may not be capable of using it. The decoder simply discards the supplemental information, unless a requirement to support the capability has been negotiated by external means.

Improved PB Frames Mode

This optional mode represents an improvement compared to the baseline H.263 *PB frames* option. This mode permits forward, backward, and bi-directional prediction for B frames in a PB frame. The operation of the MODB field changes are shown in Table 10.25

Bi-directional prediction methods are the same in both PB frames modes except that, in the *improved PB frames* mode, no delta vector is transmitted.

In forward prediction, the B macroblock is predicted from the previous P macroblock, and a separate motion vector is then transmitted.

In backwards prediction, the predicted macroblock is equal to the future P macroblock, and therefore no motion vector is transmitted.

Improved PB frames are less susceptible to changes that may occur between frames, such as when there is a scene cut between the previous P frame and the PB frame.

Reference Picture Selection Mode

In baseline H.263, a frame may be predicted from the previous frame. If a portion of the reference frame is lost due to errors or packet loss, the quality of future frames is degraded. Using this optional mode, it is possible to select which reference frame to use for prediction, minimizing error propagation.

Four back-channel messaging signals (NEITHER, ACK, NACK and ACK+NACK) are used by the encoder and decoder to specify which picture segment will be used for prediction. For example, a NACK sent to the encoder from the decoder indicates that a given frame has been degraded by errors. Thus, the encoder may choose not to use this frame for future prediction, and instead use a different, unaffected, reference frame. This reduces error propagation, maintaining improved picture quality in error-prone environments.

Temporal, SNR and Spatial Scalability Mode

In this optional mode, there is support for temporal, SNR, and spatial scalability. Scalability allows for the decoding of a sequence at more than one quality level. This is done by using a hierarchy of pictures and enhancement pictures partitioned into one or more layers. The lowest layer is called the base layer.

The base layer is a separately decodable bitstream. The enhancement layers can be decoded in conjunction with the base layer to increase the picture rate, increase the picture quality, or increase the picture size.

Temporal scalability is achieved using bi-directionally predicted pictures, or B frames. They allow prediction from either or both a previous and subsequent picture in the base layer. This results in improved compression as compared to that of P frames. These B frames differ from the B-picture part of a *PB frame* or *improved PB frame* in that they are separate entities in the bitstream.

SNR scalability refers to enhancement information that increases the picture quality without increasing resolution. Since compression introduces artifacts, the difference between a decoded picture and the original is the coding error. Normally, the coding error is

CBPB	MVDB	Code	Coding Mode
		0	bi-directional prediction
×		10	bi-directional prediction
	×	110	forward prediction
×	×	1110	forward prediction
		11110	backward prediction
×		111111	backward prediction

Table 10.25. H.263 Variable-Length Code Table for MODB for Improved PB Frames Mode.

lost at the encoder and never recovered. With SNR scalability, the coding errors are sent to the decoder, enabling an enhancement to the decoded picture. The extra data serves to increase the signal-to-noise ratio (SNR) of the picture, hence the term "SNR scalability."

Spatial scalability is closely related to SNR scalability. The only difference is that before the picture in the reference layer is used to predict the picture in the spatial enhancement layer, it is interpolated by a factor of two either horizontally or vertically (1D spatial scalability), or both horizontally and vertically (2D spatial scalability). Other than the upsampling process, the processing and syntax for a spatial scalability picture is the same as for a SNR scalability picture.

Since there is very little syntactical distinction between frames using SNR scalability and frames using spatial scalability, the frames used for either purpose are called EI frames and EP frames.

The frame in the base layer which is used for upward prediction in an EI or EP frame may be an I frame, a P frame, the P-part of a PB frame, or the P-part of an improved PB frame (but not a B frame, the B-part of a PB frame, or the B-part an improved PB frame).

This mode can be useful for networks having varying bandwidth capacity.

Reference Picture Resampling Mode

In this optional mode, the reference frame is resampled to a different size prior to using it for prediction.

This allows having a different source reference format than the frame being predicted. It can also be used for global motion estimation, or estimation of rotating motion, by warping the shape, size and location of the reference frame.

Reduced Resolution Update Mode

An optional mode is provided which allows the encoder to send update information for a frame encoded at a lower resolution, while still maintaining a higher resolution for the reference frame, to create a final frame at the higher resolution.

This mode is best used when encoding a highly active scene, allowing an encoder to increase the frame rate for moving parts of a scene, while maintaining a higher resolution in more static areas of the scene.

The syntax is the same as baseline H.263, but interpretation of the semantics is different. The dimensions of the macroblocks are doubled, so the macroblock data size is one-quarter of what it would have been without this mode enabled. Therefore, motion vectors must be doubled in both dimensions. To produce the final picture, the macroblock is upsampled to the intended resolution. After upsampling, the full resolution frame is added to the motion-compensated frame to create the full resolution frame for future reference.

Independent Segment Decoding Mode

In this optional mode, picture segment boundaries are treated as picture boundaries—no data dependencies across segment boundaries are allowed.

Use of this mode prevents the propagation of errors, providing error resilience and recovery. This mode is best used with slice layers, where, for example, the slices can be sized to match a specific packet size.

Alternative Inter VLC Mode

The intra VLC table used in the *advanced intra coding* mode can also be used for inter block coding when this optional mode is enabled.

Large quantized coefficients and small runs of zeros, typically present in intra blocks, become more frequent in inter blocks when small quantizer step sizes are used. When bit savings are obtained, and the use of the intra quantized DCT coefficient table can be detected at the decoder, the encoder will use the intra table. The decoder will first try to decode the quantized coefficients using the inter table. If this results in addressing coefficients beyond the 64 coefficients of the 8×8 block, the decoder will use the intra table.

Modified Quantization Mode

This optional mode improves the bit rate control for encoding, reduces CbCr quantization error, expands the range of DCT coefficients, and places certain restrictions on coefficient values.

In baseline H.263, the quantizer value may be modified at the macroblock level. However, only a small adjustment (± 1 or ± 2) in the value of the most recent quantizer is permitted. The *modified quantization* mode allows the modification of the quantizer to any value.

In baseline H.263, the Y and CbCr quantizers are the same. The *modified quantization* mode also increases CbCr picture quality by using a smaller quantizer step size for the Cb and Cr blocks relative to the Y blocks.

In baseline H.263, when a quantizer smaller than eight is employed, quantized coefficients exceeding the range of [−127, +127] are clipped. The *modified quantization* mode also allows coefficients that are outside the range of [−127, +127] to be represented. Therefore, when a very fine quantizer step size is selected, an increase in Y quality is obtained.

Enhanced Reference Picture Selection Mode

An optional *Enhanced Reference Picture Selection* (ERPS) *mode* offers enhanced coding efficiency and error resilience. It manages a multi-picture buffer of stored pictures.

Data-Partitioned Slice Mode

An optional *Data-Partitioned Slice* (DPS) *mode* offers enhanced error resilience. It separates the header and motion vector data from the DCT coefficient data and protects the motion vector data by using a reversible representation.

Additional Supplemental Enhancement Information Specification

An optional *Additional Supplemental Enhancement Information Specification* provides backward-compatible enhancements, such as:

(a) Indication of using a specific fixed-point IDCT

(b) Picture Messages, including message types:

- Arbitrary binary data

- Text (arbitrary, copyright, caption, video description or Uniform Resource Identifier)

- Picture header repetition (current, previous, next with reliable temporal reference or next with unreliable temporal reference)

- Interlaced field indications (top or bottom)

- Spare reference picture identification

References

1. *Efficient Motion Vector Estimation and Coding for H.263-Based Very Low Bit Rate Video Compression*, by Guy Cote, Michael Gallant and Faouzi Kossentini, Department of Electrical and Computer Engineering, University of British Columbia.

2. *H.263+: Video Coding at Low Bit Rates*, by Guy Cote, Berna Erol, Michael Gallant and Faouzi Kossentini, Department of Electrical and Computer Engineering, University of British Columbia.

3. ITU-T H.261, *Video Codec for Audiovisual Services at $p \times 64$ kbits*, 3/93.

4. ITU-T H.263, *Video Coding for Low Bit Rate Communication*, 2/98.

5. ITU-T H.263 Annex U, *Enhanced Reference Picture Selection Mode*, 11/00.

6. ITU-T H.263 Annex V, *Data-partitioned Slice Mode*, 11/00.

7. ITU-T H.263 Annex W, *Additional Supplemental Enhancement Information Specification*, 11/00.

Chapter 11

Consumer DV

The DV (digital video) format is used by tape-based digital camcorders, and is based on IEC 61834 (25 Mbps bit rate) and the newer SMPTE 314M and 370M specifications (25, 50 or 100 Mbps bit rate). The compression algorithm used is neither motion-JPEG nor MPEG, although it shares much in common with MPEG I frames. A proprietary compression algorithm is used that can be edited since it is an intra-frame technique.

The digitized video is stored in memory before compression is done. The correlation between the two fields stored in the buffer is measured. If the correlation is low, indicating inter-field motion, the two fields are individually compressed. Normally, the entire frame is compressed. In either case, DCT-based compression is used.

To achieve a constant 25, 50 or 100 Mbps bit rate, DV uses adaptive quantization, which uses the appropriate DCT quantization table for each frame.

Figure 11.1 illustrates the contents of one track as written on tape. The *ITI sector* (insert and track information) contains information on track status and serves in place as a conventional control track during video editing.

The *audio sector*, shown in Figure 11.2, contains both audio data and auxiliary audio data (AAUX).

The *video sector*, shown in Figure 11.3, contains video data and auxiliary video data (VAUX). VAUX data includes recording date and time, lens aperture, shutter speed, color balance, and other camera settings.

The *subcode sector* stores a variety of information, including timecode, teletext, closed captioning in multiple languages, subtitles and karaoke lyrics in multiple languages, titles, table of contents, chapters, etc. The subcode sector, AAUX data, and VAUX data use 5-byte blocks of data called packs.

Chapter 6 reviews the transmission of DV over IEEE 1394 (IEC 61883), which is a common interface for digital camcorders. This chapter reviews the transmission of DV over SDTI (SMPTE 321M and 322M), which is aimed at the pro-video environment.

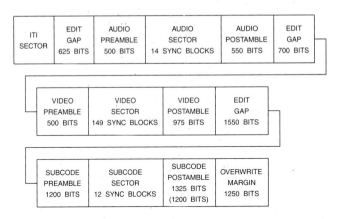

Figure 11.1. Sector Arrangement for One Track for a 480i System. The total bits per track, excluding the overwrite margin, is 134,975 (134,850). There are 10 (12) of these tracks per video frame. 576i system parameters (if different) are shown in parentheses.

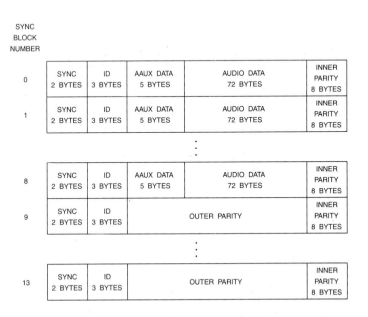

Figure 11.2. Structure of Sync Blocks in an Audio Sector.

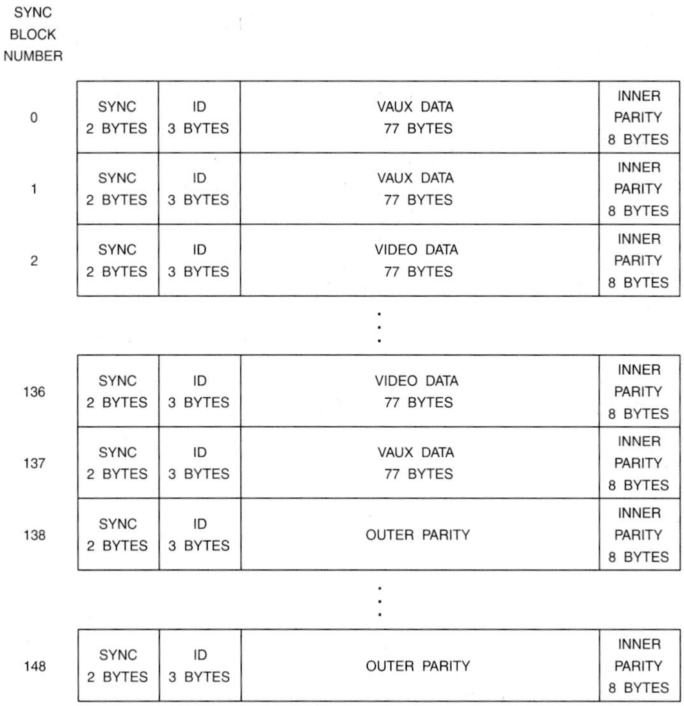

SYNC
BLOCK
NUMBER

	SYNC 2 BYTES	ID 3 BYTES	VAUX DATA 77 BYTES	INNER PARITY 8 BYTES
0	SYNC 2 BYTES	ID 3 BYTES	VAUX DATA 77 BYTES	INNER PARITY 8 BYTES
1	SYNC 2 BYTES	ID 3 BYTES	VAUX DATA 77 BYTES	INNER PARITY 8 BYTES
2	SYNC 2 BYTES	ID 3 BYTES	VIDEO DATA 77 BYTES	INNER PARITY 8 BYTES

⋮

136	SYNC 2 BYTES	ID 3 BYTES	VIDEO DATA 77 BYTES	INNER PARITY 8 BYTES
137	SYNC 2 BYTES	ID 3 BYTES	VAUX DATA 77 BYTES	INNER PARITY 8 BYTES
138	SYNC 2 BYTES	ID 3 BYTES	OUTER PARITY	INNER PARITY 8 BYTES

⋮

148	SYNC 2 BYTES	ID 3 BYTES	OUTER PARITY	INNER PARITY 8 BYTES

Figure 11.3. Structure of Sync Blocks in a Video Sector.

Audio

An audio frame starts with an audio sample within –50 samples of the beginning of line 1 (480i systems) or the middle of line 623 (576i systems).

Each track contains nine audio sync blocks, with each audio sync block containing five bytes of audio auxiliary data (AAUX) and 72 bytes of audio data, as illustrated in Figure 11.2. Audio samples are shuffled over tracks and data-sync blocks within a frame. The remaining five audio sync blocks are used for error correction.

Two 44.1-kHz, 16-bit channels require a data rate of about 1.64 Mbps. Four 32-kHz, 12-bit channels require a data rate of about 1.536 Mbps. Two 48-kHz, 16-bit channels require a data rate of about 1.536 Mbps.

IEC 61834

IEC 61834 supports a variety of audio sampling rates:

48 kHz (16 bits, 2 channels)

44.1 kHz (16 bits, 2 channels)

32 kHz (16 bits, 2 channels)

32 kHz (12 bits, 4 channels)

Audio sampling may be either locked or unlocked to the video frame frequency.

Audio data is processed in frames. At a locked 48 kHz sample rate, each frame contains either 1600 or 1602 audio samples (480i system) or 1920 audio samples (576i system). For the 480i system, the number of audio samples per frame follows a five-frame sequence:

> 1600, 1602, 1602, 1602, 1602, 1600, ...

With a locked 32 kHz sample rate, each frame contains either 1066 or 1068 audio samples (480i system) or 1280 audio samples (576i system). For the 480i system, the number of audio samples per frame follows a fifteen-frame sequence:

> 1066, 1068, 1068, 1068, 1068, 1068, 1068, 1066, 1068, 1068, 1068, 1068, 1068, 1068, ...

For unlocked audio sampling, there is no exact number of audio samples per frame, although minimum and maximum values are specified.

SMPTE 314M

SMPTE 314M supports a more limited option, with audio sampling locked to the video frame frequency:

> 48 kHz (16 bits, 2 channels) for 25 Mbps
>
> 48 kHz (16 bits, 4 channels) for 50 Mbps

Audio data is processed in frames. At a locked 48 kHz sample rate, each frame contains either 1600 or 1602 audio samples (480i system) or 1920 audio samples (576i system). For the 480i system, the number of audio samples per frame follows a five-frame sequence:

> 1600, 1602, 1602, 1602, 1602, 1600, ...

The audio capacity is capable of 1620 samples per frame for the 480i system or 1944 samples per frame for the 576i system. The unused space at the end of each frame is filled with arbitrary data.

Audio Auxiliary Data (AAUX)

AAUX information is added to the shuffled audio data as shown in Figure 11.2. The AAUX pack includes a 1-byte pack header and four bytes of data (payload), resulting in a 5-byte AAUX pack. Since there are nine of them per video frame, they are numbered from 0 to 8. An AAUX source (AS) pack and an AAUX source control (ASC) pack must be included in the compressed stream. Only the AS and ASC packs are currently supported by SMPTE 314M, although IEC 61834 supports many other pack formats.

AAUX Source (AS) Pack

The format for this pack is shown in Table 11.1.

LF	Locked audio sample rate "0" = locked to video "1" = unlocked to video
AF	Audio frame size. Specifies the number of audio samples per frame.
SM	Stereo mode "0" = multi-stereo audio "1" = lumped audio

IEC 61834	D7	D6	D5	D4	D3	D2	D1	D0
PC0	0	1	0	1	0	0	0	0
PC1	LF	1	AF					
PC2	SM	CHN		PA	AM			
PC3	1	ML	50/60	ST				
PC4	EF	TC	SMP			QU		

SMPTE 314M	D7	D6	D5	D4	D3	D2	D1	D0
PC0	0	1	0	1	0	0	0	0
PC1	LF	1	AF					
PC2	0	CHN		1	AM			
PC3	1	1	50/60	ST				
PC4	1	1	SMP			QU		

Table 11.1. AAUX Source (AS) Pack.

PA   Specifies if the audio signals recorded in CH1 (CH3) are related to the audio signals recorded in CH2 (CH4)
"0" = one of pair channels
"1" = independent channels

CHN   Number of audio channels within an audio block
"00" = one channel per block
"01" = two channels per block
"10" = reserved
"11" = reserved

AM   Specifies the content of the audio signal on each channel

ML   Multi-language flag
"0" = recorded in multi-language
"1" = not recorded in multi-language

50/60   50- or 59.94-Hz video system
"0" = 59.94-Hz field system
"1" = 50-Hz field system

ST   For SMPTE 314M, this specifies the number of audio blocks per frame.
"00000" = 2 audio blocks
"00001" = reserved
"00010" = 4 audio blocks
"00011" to "11111" = reserved

For IEC 61834, this specifies the video system.
"00000" = standard definition
"00010" = reserved
"00010" = high definition
"00011" to "11111" = reserved

EF   Audio emphasis flag
"0" = on
"1" = off

TC   Emphasis time constant
"1" = 50/15 µs
"0" = reserved

SMP   Audio sampling frequency
"000" = 48 kHz
"001" = 44.1 kHz
"010" = 32 kHz
"011" to "111" = reserved

QU   Audio quantization
"000" = 16 bits linear
"001" = 12 bits nonlinear
"010" = 20 bits linear
"011" to "111" = reserved

AAUX Source Control (ASC) Pack

The format for this pack is shown in Table 11.2.

CGMS   Copy generation management system
"00" = copying permitted without restriction
"01" = reserved
"10" = one copy permitted
"11" = no copy permitted

ISR   Previous input source
"00" = analog input
"01" = digital input
"10" = reserved
"11" = no information

CMP   Number of times of compression
"00" = once
"01" = twice
"10" = three or more
"11" = no information

SS   Source and recorded situation
"00" = scrambled source with audience restrictions and recorded without descrambling
"01" = scrambled source without audience restrictions and recorded without descrambling
"10" = source with audience restrictions or descrambled source with audience restrictions
"11" = no information

EFC   Audio emphasis flags
"00" = emphasis off
"01" = emphasis on
"10" = reserved
"11" = reserved

REC S   Recording start point
"0" = at recording start point
"1" = not at recording start point

REC E   Recording end point
"0" = at recording end point
"1" = not at recording end point

REC M   Recording mode
"001" = original
"011" = one CH insert
"100" = four CHs insert
"101" = two CHs insert
"111" = invalid recording

FADE S   Fading of recording start point
"0" = fading off
"1" = fading on

IEC 61834	D7	D6	D5	D4	D3	D2	D1	D0
PC0	0	1	0	1	0	0	0	1
PC1	CGMS		ISR		CMP		SS	
PC2	REC S	REC E	REC M				ICH	
PC3	DRF	SPD						
PC4	1	GEN						

SMPTE 314M	D7	D6	D5	D4	D3	D2	D1	D0
PC0	0	1	0	1	0	0	0	1
PC1	CGMS		1	1	1	1	EFC	
PC2	REC S	REC E	FADE S	FADE E	1	1	1	1
PC3	DRF	SPD						
PC4	1	1	1	1	1	1	1	1

Table 11.2. AAUX Source Control (ASC) Pack.

FADE E   Fading of recording end point
    "0" = fading off
    "1" = fading on

ICH   Insert audio channel
    "000" = CH1
    "001" = CH2
    "010" = CH3
    "011" = CH4
    "100" = CH1, CH2
    "101" = CH3, CH4
    "110" = CH1, CH2, CH3, CH4
    "111" = no information

DRF   Direction flag
    "0" = reverse direction
    "1" = forward direction

SP   Playback speed. This is defined by a 3-bit coarse value plus a 4-bit fine value. For normal recording, it is set to "0100000."

GEN   Indicates the category of the audio source

Video

As shown in Table 11.3, IEC 61834 uses 4:1:1 YCbCr for 720 × 480i video (Figure 3.5) and 4:2:0 YCbCr for 720 × 576i video (Figure 3.11).

SMPTE 314M uses 4:1:1 YCbCr (Figure 3.5) for both video standards for the 25 Mbps implementation. 4:2:2 YCbCr (Figure 3.3) is used for both video standards for the 50 and 100 Mbps implementations.

DCT Blocks

The Y, Cb, and Cr samples for one frame are divided into 8 × 8 blocks, called DCT blocks. Each DCT block, with the exception of the right-most DCT blocks for Cb and Cr during 4:1:1 mode, transform 8 samples × 8 lines of video data. Rows 1, 3, 5, and 7 of the DCT block process field 1, while rows 0, 2, 4, and 6 process field 2.

For 480i systems, there are either 10,800 (4:2:2) or 8,100 (4:1:1) DCT blocks per video frame.

For 576i systems, there are either 12,960 (4:2:2) or 9,720 (4:1:1, 4:2:0) DCT blocks per video frame.

Macroblocks

As shown in Figure 11.4, each macroblock in the 4:2:2 mode consists of four DCT blocks. As shown in Figures 11.5 and 11.6, each macroblock in the 4:1:1 and 4:2:0 modes consists of six DCT blocks.

For 480i systems, the macroblock arrangement for one frame of 4:1:1 and 4:2:2 YCbCr data is shown in Figures 11.7 and 11.8, respectively. There are either 2,700 (4:2:2) or 1,350 (4:1:1) macroblocks per video frame.

For 576i systems, the macroblock arrangement for one frame of 4:2:0, 4:1:1, and 4:2:2 YCbCr data is shown in Figures 11.9, 11.10, and 11.11, respectively. There are either 3,240 (4:2:2) or 1,620 (4:1:1, 4:2:0) macroblocks per video frame.

Super Blocks

Each super block consists of 27 macroblocks.

For 480i systems, the super block arrangement for one frame of 4:1:1 and 4:2:2 YCbCr data is shown in Figures 11.7 and 11.8, respectively. There are either 100 (4:2:2) or 50 (4:1:1) super blocks per video frame.

For 576i systems, the super block arrangement for one frame of 4:2:0, 4:1:1, and 4:2:2 YCbCr data is shown in Figures 11.9, 11.10, and 11.11, respectively. There are either 120 (4:2:2) or 60 (4:1:1, 4:2:0) super blocks per video frame.

Compression

Like MPEG and H.263, DV uses DCT-based video compression. However, in this case, DCT blocks are comprised from two fields, with each field providing samples from four scan lines and eight horizontal samples.

Two DCT modes, called 8-8-DCT and 2-4-8-DCT, are available for the transform process, depending upon the degree of content variation between the two fields of a video frame. The 8-8-DCT is your normal 8 × 8 DCT, and is used when there a high degree of correlation (little motion) between the two fields. The 2-4-8-DCT uses two 4 × 8 DCTs (one for each field), and is used when there is a low degree of correlation (lots of motion) between the two fields. Which DCT is used is stored in the DC coefficient area using a single bit.

The DCT coefficients are quantized to nine bits, then divided by a quantization number so as to limit the amount of data in one video segment to five compressed macroblocks.

Each DCT block is classified into one of four classes based on quantization noise and maximum absolute values of the AC coefficients. The 2-bit class number is stored in the DC coefficient area.

An area number is used for the selection of the quantization step. The area number, of which there are four, is based on the horizontal and vertical frequencies.

The quantization step is decided by the class number, area number, and quantization number (QNO). Quantization information is passed in the DIF header of video blocks.

Variable-length coding converts the quantized AC coefficients to variable-length codes.

Figures 11.12 and 11.13 illustrate the arrangement of compressed macroblocks.

Video Auxiliary Data (VAUX)

VAUX information is added to the shuffled video data as shown in Figure 11.3. The VAUX pack includes a 1-byte pack header and four bytes of data (payload), resulting in a 5-byte VAUX pack. Since there are 45 of them per video frame, they are numbered from 0 to 44. A VAUX source (VS) pack and an VAUX source control (VSC) pack must be included in the compressed stream. Only the VS and VSC packs are currently supported by SMPTE 314M, although IEC 61834 supports many other pack formats.

VAUX Source (VS) Pack

The format for this pack is shown in Table 11.4.

TVCH   The number of the television channel, from 0–999. A value of EEE_H is reserved for pre-recorded tape or a line input. A value of FFF_H is reserved for "no information."

B/W   Black and white flag
        "0" = black and white video
        "1" = color video

Parameters	480i System	576i System
active resolution (Y)	720 × 480i	720 × 576i
frame refresh rate	29.97 Hz	25 Hz

YCbCr sampling structure
   IEC 61834
   SMPTE 314M
	4:1:1
	4:1:1, 4:2:2
		4:2:0
		4:1:1, 4:2:2

form of YCbCr coding	Uniformly quantized PCM, 8 bits per sample.	
active line numbers	23–262, 285–524	23–310, 335–622

Table 11.3. IEC 61834 and SMPTE 314M YCbCr Parameters. Note that the active line numbers are slightly different than those used by MPEG.

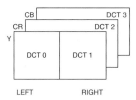

Figure 11.4. 4:2:2 Macroblock Arrangement.

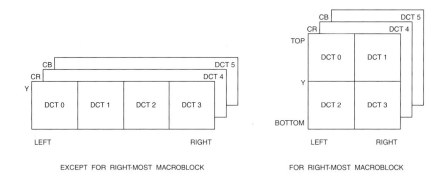

Figure 11.5. 4:1:1 Macroblock Arrangement.

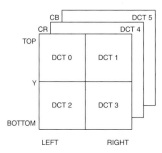

Figure 11.6. 4:2:0 Macroblock Arrangement.

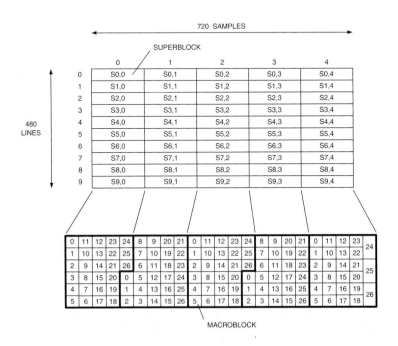

Figure 11.7. Relationship Between Super Blocks and Macroblocks (4:1:1 YCbCr, 720 × 480i).

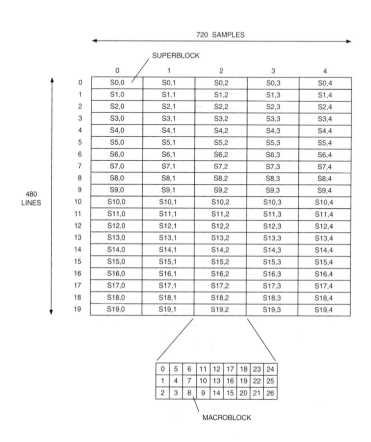

Figure 11.8. Relationship Between Super Blocks and Macroblocks (4:2:2 YCbCr, 720 × 480i).

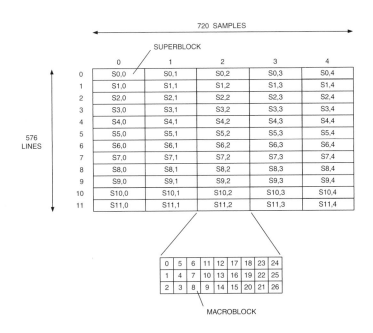

Figure 11.9. Relationship Between Super Blocks and Macroblocks (4:2:0 YCbCr, 720 × 576i).

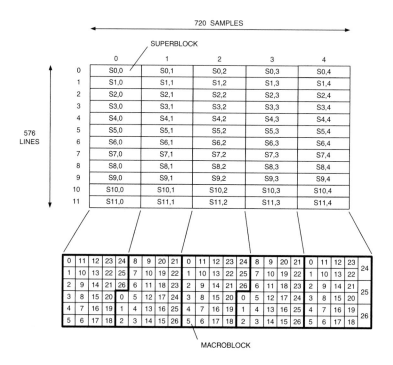

Figure 11.10. Relationship Between Super Blocks and Macroblocks (4:1:1 YCbCr, 720 × 576i).

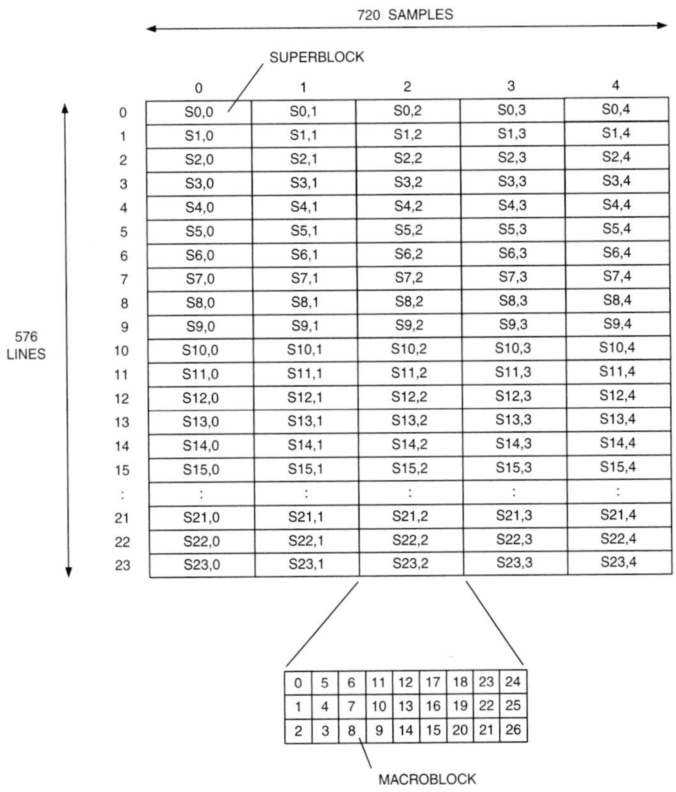

Figure 11.11. Relationship Between Super Blocks and Macroblocks (4:2:2 YCbCr, 720 × 576i).

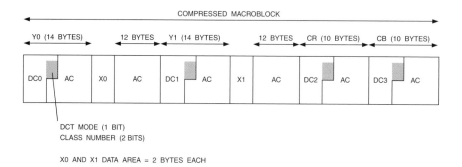

Figure 11.12. 4:2:2 Compressed Macroblock Arrangement.

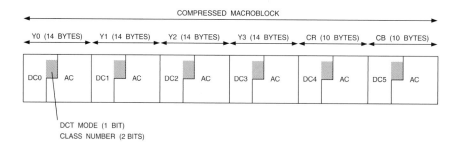

Figure 11.13. 4:2:0 and 4:1:1 Compressed Macroblock Arrangement.

IEC 61834	D7	D6	D5	D4	D3	D2	D1	D0
PC0	0	1	1	0	0	0	0	0
PC1	TVCH (tens of units, 0–9)				TVCH (units, 0–9)			
PC2	B/W	EN	CLF		TVCH (hundreds of units, 0–9)			
PC3	SRC		50/60	ST				
PC4	TUN							

SMPTE 314M	D7	D6	D5	D4	D3	D2	D1	D0
PC0	0	1	1	0	0	0	0	0
PC1	1	1	1	1	1	1	1	1
PC2	B/W	EN	CLF		1	1	1	1
PC3	1	1	50/60	ST				
PC4	VISC							

Table 11.4. VAUX Source (VS) Pack.

EN   CLF valid flag
"0" = CLF is valid
"1" = CLF is invalid

CLF   Color frames identification code

For 480i systems:
"00" = color frame A
"01" = color frame B
"10" = reserved
"11" = reserved

For 576i systems:
"00" = 1st, 2nd field
"01" = 3rd, 4th field
"10" = 5th, 6th field
"11" = 7th, 8th field

SRC   Defines the input source of the video signal

50/60   Same as for AAUX

ST   Same as for AAUX

TUN   Tuner Category consists of 3-bit area number and a 5-bit satellite number. "11111111" indicates no information is available.

VISC

    "10001000" = –180

            ⋮

    "00000000" = 0

            ⋮

    "01111000" = 180
    "01111111" = no information
    other values = reserved

VAUX Source Control (VSC) Pack

The format for this pack is shown in Table 11.5.

CGMS   Same as for AAUX

ISR   Same as for AAUX

CMP   Same as for AAUX

SS   Same as for AAUX

REC S   Same as for AAUX

REC M   Same as for AAUX

BCS   Broadcast system. Indicates the type information of display format with DISP.
    "00" = type 0 (IEC 61880, EIA-608)
    "01" = type 1 (ETS 300 294)
    "10" = reserved
    "11" = reserved

DISP   Aspect ratio information

FF   Frame/Field flag. Indicates whether both fields are output in order or only one of them is output twice during one frame period.
    "0" = one field output twice
    "1" = both fields output in order

FS   First/Second flag. Indicates which field which should be output during field 1 period.
    "0" = field 2
    "1" = field 1

FC   Frame change flag. Indicates if the picture of the current frame is the same picture of the immediate previous frame.
    "0" = same picture
    "1" = different picture

IL   Interlace flag. Indicates if the data of two fields which construct one frame are interlaced or non-interlaced.
    "0" = noninterlaced
    "1" = interlaced or unrecognized

SF   Still-field picture flag. Indicates the time difference between the two fields within a frame.
    "0" = 0 seconds
    "1" = 1,001/60 or 1/50 second

SC   Still camera picture flag
    "0" = still camera picture
    "1" = not still camera picture

GEN   Indicates the category of the video source

Digital Interfaces

IEC 61834 and SMPTE 314M both specify the data format for a generic digital interface. This data format may be sent via IEEE 1394 or SDTI, for example. Figure 11.14 illustrates the frame data structure.

IEC 61834	D7	D6	D5	D4	D3	D2	D1	D0
PC0	0	1	1	0	0	0	0	1
PC1	CGMS		ISR		CMP		SS	
PC2	REC S	1	REC M		1	DISP		
PC3	FF	FS	FC	IL	SF	SC	BCS	
PC4	1	GEN						

SMPTE 314M	D7	D6	D5	D4	D3	D2	D1	D0
PC0	0	1	1	0	0	0	0	1
PC1	CGMS		1	1	1	1	1	1
PC2	1	1	0	0	1	DISP		
PC3	FF	FS	FC	IL	1	1	0	0
PC4	1	1	1	1	1	1	1	1

Table 11.5. VAUX Source Control (VSC) Pack.

Each of the 720 × 480i 4:1:1 YCbCr frames are compressed to 103,950 bytes. Including overhead and audio increases the amount of data to 120,000 bytes.

The compressed 720 × 480i frame is divided into ten DIF (data in frame) sequences. Each DIF sequence contains 150 DIF blocks of 80 bytes each, used as follows:

135 DIF blocks for video

9 DIF blocks for audio

6 DIF blocks used for Header, Subcode, and Video Auxiliary (VAUX) information

Figure 11.14 illustrates the DIF sequence structure in detail. Each video DIF block contains 80 bytes of compressed macroblock data:

3 bytes for DIF block ID information

1 byte for the header that includes the quantization number (QNO) and block status (STA)

14 bytes each for Y0, Y1, Y2, and Y3

10 bytes each for Cb and Cr

720 × 576i frames may use either the 4:2:0 YCbCr format (IEC 61834) or the 4:1:1 YCbCr format (SMPTE 314M), and require 12 DIF sequences. Each 720 × 576i frame is com-

pressed to 124,740 bytes. Including overhead and audio increases the amount of data to 144,000 bytes, requiring 300 packets to transfer.

Note that the organization of data transferred over the interface differs from the actual DV recording format since error correction is not required for digital transmission. In addition, although the video blocks are numbered in sequence in Figure 11.15, the sequence does not correspond to the left-to-right, top-to-bottom transmission of blocks of video data. Compressed macroblocks are shuffled to minimize the effect of errors and aid in error concealment. Audio data is also shuffled. Data is transmitted in the same shuffled order as recorded.

To illustrate the video data shuffling, DV video frames are organized as super blocks, with each super block being composed of 27 compressed macroblocks, as shown in Figures 11.7 through 11.11. A group of five super blocks (one from each super block column) make up one DIF sequence. Tables 11.6 and 11.7 illustrate the transmission order of the DIF blocks.

For the 50 Mbps SMPTE 314M format, each compressed 720 × 480i or 720 × 576i frame is divided into two channels. Each channel uses either ten (480i systems) or twelve DIF sequences (576i systems).

IEEE 1394

Using the IEEE 1394 interface for transferring DV information is discussed in Chapter 6.

SDTI

The general concept of SDTI is discussed in Chapter 6.

SMPTE 314M Data

SMPTE 221M details how to transfer SMPTE 314M DV data over SDTI. Figure 11.16 illustrates the basic implementation.

IEC 61834 Data

SMPTE 222M details how to transfer IEC 61834 DV data over SDTI.

100 Mbps DV Differences

The 100 Mbps SMPTE 370M format supports 1920 × 1080i and 1280 × 720p sources.

1920 × 1080i30 and 1920 × 1080i25 sources are scaled to 1280 × 1080i30 and 1440 × 1080i25, respectively. 1280 × 720p60 sources are scaled to 960 × 720p60. 4:2:2 YCbCr sampling is used.

Each compressed frame is divided into four channels. Each channel uses either ten (1080i30 or 720p60 systems) or twelve DIF sequences (1080i25 systems).

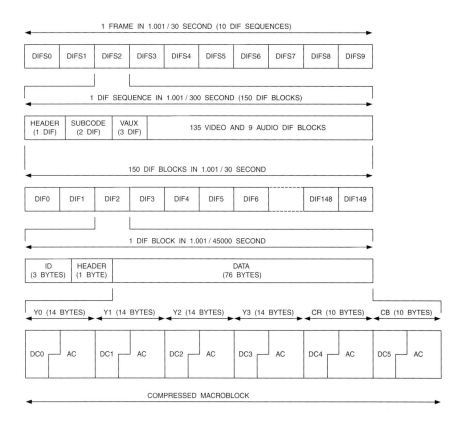

Figure 11.14. Packet Formatting for 25 Mbps 4:1:1 YCbCr 720 × 480i Systems.

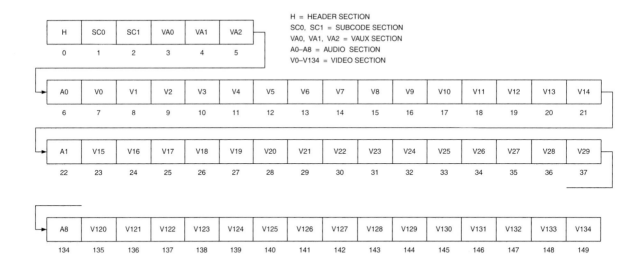

H = HEADER SECTION
SC0, SC1 = SUBCODE SECTION
VA0, VA1, VA2 = VAUX SECTION
A0–A8 = AUDIO SECTION
V0–V134 = VIDEO SECTION

Figure 11.15. DIF Sequence Detail.

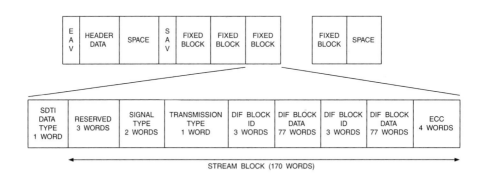

Figure 11.16. Transferring DV Data Using SDTI.

DIF Sequence Number	Video DIF Block Number	Compressed Macroblock		DIF Sequence Number	Video DIF Block Number	Compressed Macroblock	
		Superblock Number	Macroblock Number			Superblock Number	Macroblock Number
0	0	2, 2	0			:	
	1	6, 1	0	n–1	0	1, 2	0
	2	8, 3	0		1	5, 1	0
	3	0, 0	0		2	7, 3	0
	4	4, 4	0		3	n–1, 0	0
	:				4	3, 4	0
	133	0, 0	26		:		
	134	4, 4	26		133	n–1, 0	26
1	0	3, 2	0		134	3, 4	26
	1	7, 1	0				
	2	9, 3	0				
	3	1, 0	0				
	4	5, 4	0				
	:						
	133	1, 0	26				
	134	5, 4	26				

Notes:
1. n = 10 for 480i systems, n = 12 for 576i systems.

Table 11.6. Video DIF Blocks and Compressed Macroblocks for 25 Mbps (4:1:1 or 4:2:0 YCbCr).

References

1. IEC 61834–1, *Recording—Helical-scan digital video cassette recording system using 6.35mm magnetic tape for consumer use (525-60, 625-50, 1125-60 and 1250-50 systems)–Part 1: General specifications.*

2. IEC 61834–2, *Recording—Helical-scan digital video cassette recording system using 6.35mm magnetic tape for consumer use (525-60, 625-50, 1125-60 and 1250-50 systems)–Part 2: SD format for 525-60 and 625-50 systems.*

3. IEC 61834–4, *Recording—Helical-scan digital video cassette recording system using 6.35mm magnetic tape for consumer use (525-60, 625-50, 1125-60 and 1250-50 systems)–Part 4: Pack header table and contents.*

DIF Sequence Number	Video DIF Block Number	Compressed Macroblock		DIF Sequence Number	Video DIF Block Number	Compressed Macroblock	
		Superblock Number	Macroblock Number			Superblock Number	Macroblock Number
0	0, 0	4, 2	0	n–1	:		
	0, 1	5, 2	0		0, 0	2, 2	0
	1, 0	12, 1	0		0, 1	3, 2	0
	1, 1	13, 1	0		1, 0	10, 1	0
	2, 0	16, 3	0		1, 1	11, 1	0
	:				2, 0	14, 3	0
	134, 0	8, 4	26		:		
	134, 1	9, 4	26		134, 0	6, 4	26
1	0, 0	6, 2	0		134, 1	7, 4	26
	0, 1	7, 2	0				
	1, 0	14, 1	0				
	1, 1	15, 1	0				
	2, 0	18, 3	0				
	:						
	134, 0	10, 4	26				
	134, 1	11, 4	26				

Notes:
1. n = 10 for 480i systems, n = 12 for 576i systems.

Table 11.7. Video DIF Blocks and Compressed Macroblocks for 50 Mbps (4:2:2 YCbCr).

4. SMPTE 314M–1999, *Television—Data Structure for DV-Based Audio, Data and Compressed Video–25 and 50 Mbps.*

5. SMPTE 370M–2002, *Television—Data Structure for DV-Based Audio, Data and Compressed Video at 100 Mb/s 1080/60i, 1080/50i, 720-60p.*

6. SMPTE 321M–2002, *Television—Data Stream Format for the Exchange of DV-Based Audio, Data and Compressed Video Over a Serial Data Transport Interface.*

7. SMPTE 322M–1999, *Television—Format for Transmission of DV Compressed Video, Audio and Data Over a Serial Data Transport Interface.*

Chapter 12

MPEG-1

MPEG-1 audio and video compression was developed for storing and distributing digital audio and video. Features include random access, fast forward, and reverse playback. MPEG-1 is used as the basis for the original video CDs (VCD).

The channel bandwidth and image resolution were set by the available media at the time (CDs). The goal was playback of digital audio and video using a standard compact disc with a bit rate of 1.416 Mbps (1.15 Mbps of this is for video).

MPEG-1 is an ISO standard (ISO/IEC 11172), and consists of six parts:

system	ISO/IEC 11172–1
video	ISO/IEC 11172–2
audio	ISO/IEC 11172–3
low bit rate audio	ISO/IEC 13818–3
conformance testing	ISO/IEC 11172–4
simulation software	ISO/IEC 11172–5

The bitstreams implicitly define the decompression algorithms. The compression algorithms are up to the individual manufacturers, allowing a proprietary advantage to be obtained within the scope of an international standard.

MPEG vs. JPEG

JPEG (ISO/IEC 10918) was designed for still continuous-tone grayscale and color images. It doesn't handle bi-level (black and white) images efficiently, and pseudo-color images have to be expanded into the unmapped color representation prior to processing. JPEG images may be of any resolution and color space, with both lossy and lossless algorithms available.

Since JPEG is such a general purpose standard, it has many features and capabilities. By adjusting the various parameters, compressed image size can be traded against reconstructed image quality over a wide range. Image quality ranges from "browsing" (100:1 compression ratio) to "indistinguishable from the source" (about 3:1 compression ratio). Typically, the threshold of visible difference between the source and reconstructed images is somewhere between a 10:1 to 20:1 compression ratio.

JPEG does not use a single algorithm, but rather a family of four, each designed for a certain application. The most familiar lossy algorithm is *sequential DCT*. Either Huffman encoding (baseline JPEG) or arithmetic encoding may be used. When the image is decoded, it is decoded left-to-right, top-to-bottom.

Progressive DCT is another lossy algorithm, requiring multiple scans of the image. When the image is decoded, a coarse approximation of the full image is available right away, with the quality progressively improving until complete. This makes it ideal for applications such as image database browsing. Either spectral selection, successive approximation, or both may be used. The spectral selection option encodes the lower-frequency DCT coefficients first (to obtain an image quickly), followed by the higher-frequency ones (to add more detail). The successive approximation option encodes the more significant bits of the DCT coefficients first, followed by the less significant bits.

The *hierarchical* mode represents an image at multiple resolutions. For example, there could be 512 × 512, 1024 × 1024, and 2048 × 2048 versions of the image. Higher-resolution images are coded as differences from the next smaller image, requiring fewer bits than they would if stored independently. Of course, the *total* number of bits is greater than that needed to store just the highest-resolution image. Note that the individual images in a hierarchical sequence may be coded progressively if desired.

Also supported is a *lossless* spatial algorithm that operates in the pixel domain as opposed to the transform domain. A prediction is made of a sample value using up to three neighboring samples. This prediction then is subtracted from the actual value and the difference is losslessly coded using either Huffman or arithmetic coding. Lossless operation achieves about a 2:1 compression ratio.

Since video is just a series of still images, and baseline JPEG encoders and decoders were readily available, people used baseline JPEG to compress real-time video (also called motion JPEG or MJPEG). However, this technique does not take advantage of the frame-to-frame redundancies to improve compression, as does MPEG.

Perhaps most important, JPEG is symmetrical, meaning the cost of encoding and decoding is roughly the same. MPEG, on the other hand, *was designed primarily for mastering a video once and playing it back many times on many platforms*. To minimize the cost of MPEG hardware decoders, MPEG was designed to be asymmetrical, with the encoding process requiring about 100× the computing power of the decoding process.

Since MPEG is targeted for specific applications, the hardware usually supports only a few specific resolutions. Also, only one color space (YCbCr) is supported using 8-bit samples. MPEG is also optimized for a limited range of compression ratios.

If capturing video for editing, you can use either baseline JPEG or I-frame-only (intra frame) MPEG to compress to disc in real-time. Using JPEG requires that the system be able to transfer data and access the hard disk at bit rates of about 4 Mbps for SIF (Standard Input Format) resolution. Once the editing is done, the result can be converted into MPEG for maximum compression.

Quality Issues

At bit rates of about 3–4 Mbps, "broadcast quality" is achievable with MPEG-1. However, sequences with complex spatial-temporal activity (such as sports) may require up to 5–6 Mbps due to the frame-based processing of MPEG-1. MPEG-2 allows similar "broadcast quality" at bit rates of about 4–6 Mbps by supporting field-based processing.

Several factors affect the quality of MPEG-compressed video:

- the resolution of the original video source
- the bit rate (channel bandwidth) allowed after compression
- motion estimator effectiveness

One limitation of the quality of the compressed video is determined by the resolution of the original video source. If the original resolution was too low, there will be a general lack of detail.

Motion estimator effectiveness determines motion artifacts, such as a reduction in video quality when movement starts or when the amount of movement is above a certain threshold. Poor motion estimation will contribute to a general degradation of video quality.

Most importantly, the higher the bit rate (channel bandwidth), the more information that can be transmitted, allowing fewer motion artifacts to be present or a higher resolution image to be displayed. Generally speaking, decreasing the bit rate does not result in a "graceful degradation" of the decoded video quality. The video quality rapidly degrades, with the 8×8 blocks becoming clearly visible once the bit rate drops below a given threshold.

Audio Overview

MPEG-1 uses a family of three audio coding schemes, called Layer I, Layer II, and Layer III, with increasing complexity and sound quality. The three layers are hierarchical: a Layer III decoder handles Layers I, II, and III; a Layer II decoder handles only Layers I and II; a Layer I decoder handles only Layer I. All layers support 16-bit digitized audio using 16, 22.05, 24, 32, 44.1 or 48 kHz sampling rates.

For each layer, the bitstream format and the decoder are specified. The encoder is not specified, to allow for future improvements. All layers work with similar bit rates:

Layer I: 32–448 kbps
Layer II: 8–384 kbps
Layer III: 8–320 kbps

Two audio channels are supported with four modes of operation:

normal stereo
joint (intensity and/or ms) stereo
dual channel mono
single channel mono

For normal stereo, one channel carries the left audio signal and one channel carries the right audio signal. For intensity stereo (supported by all layers), high frequencies (above 2 kHz) are combined. The stereo image is preserved but only the temporal envelope is transmitted. For ms stereo (supported by Layer III only), one channel carries the sum signal (L+R) and the other the difference (L–R) signal. In addition, pre-emphasis, copyright marks, and original/copy indication are supported.

Sound Quality

To determine which layer should be used for a specific application, look at the available bit rate, as each layer was designed to support certain bit rates with a minimum degradation of sound quality.

Layer I, a simplified version of Layer 2, has a target bit rate 192 kbps per channel or higher.

Layer II is identical to MUSICAM, and has a target bit rate 128 kbps per channel. It was designed as a trade-off between sound quality and encoder complexity. It is most useful for bit rates around 96–128 kbps per channel.

Layer III (also known as "mp3") merges the best ideas of MUSICAM and ASPEC and has a target bit rate of about 64 kbps per channel. The Layer III format specifies a set of advanced features that all address a single goal: to preserve as much sound quality as possible, even at relatively low bit rates.

Background Theory

All layers use a coding scheme based on psychoacoustic principles—in particular, "masking" effects where, for example, a loud tone at one frequency prevents another, quieter, tone at a nearby frequency from being heard.

Suppose you have a strong tone with a frequency of 1000 Hz, and a second tone at 1100 Hz that is 18 dB lower in intensity. The 1100 Hz tone will not be heard; it is masked by the 1000 Hz tone. However, a tone at 2000 Hz 18 dB below the 1000 Hz tone will be heard. In order to have the 1000 Hz tone mask it, the 2000 Hz tone will have to be about 45 dB down. Any relatively weak frequency near a strong frequency is masked; the further you get from a frequency, the smaller the masking effect.

Curves have been developed that plot the relative energy versus frequency that is masked (concurrent masking). Masking effects also occur before (premasking) and after (postmasking) a strong frequency if there is a significant (30–40 dB) shift in level. The reason is believed to be that the brain needs processing time. Premasking time is about 2–5 ms; postmasking can last up to 100 ms.

Adjusting the noise floor reduces the amount of needed data, enabling further compression. CDs use 16 bits of resolution to achieve a signal-to-noise ratio (SNR) of about 96 dB, which just happens to match the dynamic range of hearing pretty well (meaning most people will not hear noise during silence). If 8-bit resolution were used, there would be a noticeable noise during silent moments in the music or between words. However, noise isn't noticed during loud passages. due to the masking effect, which means that around a strong sound you can raise the noise floor since the noise will be masked anyway.

For a stereo signal, there usually is redundancy between channels. All layers may exploit these stereo effects by using a "joint stereo" mode, with the most flexible approach being used by Layer III.

Video Coding Layer

MPEG-1 permits resolutions up to 4095 × 4095 at 60 frames per second (progressive scan). What many people think of as MPEG-1 is a subset known as Constrained Parameters Bitstream (CPB). The CPB is a limited set of sampling and bit rate parameters designed to standardize buffer sizes and memory bandwidths, allowing a nominal guarantee of interoperability for decoders and encoders, while still addressing the widest possible range of applications. Devices not capable of handling these are not considered to be true MPEG-1. Table 12.1 lists some of the constrained parameters.

The CPB limits video to 396 macroblocks (101,376 pixels). Therefore, MPEG-1 video is typically coded at SIF resolutions of 352 × 240p or 352 × 288p. During encoding, the original BT.601 resolution of 704 × 480i or 704 × 576i is scaled down to SIF resolution. This is usually done by ignoring field 2 and scaling down field 1 horizontally. During decoding, the SIF resolution is scaled up to the 704 × 480i or 704 × 576i resolution. Note that some entire active scan lines and samples on a scan line are ignored to ensure the number of Y samples can be evenly divided by 16. Table 12.2 lists some of the more common MPEG-1 resolutions.

The coded video rate is limited to 1.856 Mbps. However, the bit rate is the most-often waived parameter, with some applications using up to 6 Mbps or higher.

MPEG-1 video data uses the 4:2:0 YCbCr format shown in Figure 3.7.

Interlaced Video

MPEG-1 was designed to handle progressive (also referred to as noninterlaced) video. Early on, in an effort to improve video quality, several schemes were devised to enable the use of both fields of an interlaced picture.

For example, both fields can be combined into a single frame of $704 \times 480p$ or $704 \times 576p$ resolution and encoded. During decoding, the fields are separated. This, however, results in motion artifacts due to a moving object being in slightly different places in the two fields. Coding the two fields separately avoids motion artifacts, but reduces the compression ratio since the redundancy between fields isn't used.

There were many other schemes for handling interlaced video, so MPEG-2 defined a standard way of handling it (covered in Chapter 13).

Encode Preprocessing

Better images can be obtained by preprocessing the video stream prior to MPEG encoding.

To avoid serious artifacts during encoding of a particular picture, prefiltering can be applied over the entire picture or just in specific problem areas. Prefiltering before compression processing is analogous to anti-alias filtering prior to A/D conversion. Prefiltering may take into account texture patterns, motion, and edges, and may be applied at the picture, slice, macroblock, or block level.

MPEG encoding works best on scenes with little fast or random movement and good lighting. For best results, foreground lighting should be clear and background lighting diffused. Foreground contrast and detail should be normal, but low contrast backgrounds containing soft edges are preferred. Editing tools typically allow you to preprocess potential problem areas.

The MPEG-1 specification has example filters for scaling down from BT.601 to SIF resolution. In this instance, field 2 is ignored, throwing away half the vertical resolution, and a decimation filter is used to reduce the horizontal resolution of the remaining scan lines by a factor of two. Appropriate decimation of the Cb and Cr components must still be carried out.

Better video quality may be obtained by deinterlacing prior to scaling down to SIF resolution. When working on macroblocks (defined later), if the difference between macroblocks between two fields is small, average both to generate a new macroblock. Otherwise, use the macroblock area from the field of the same parity to avoid motion artifacts.

Coded Frame Types

There are four types of coded frames. I (intra) frames (~1 bit/pixel) are frames coded as a stand-alone still image. They allow random access points within the video stream. As such, I frames should occur about two times a second. I frames should also be used where scene cuts occur.

horizontal resolution	≤ 768 samples
vertical resolution	≤ 576 scan lines
picture area	≤ 396 macroblocks
pel rate	≤ 396 × 25 macroblocks per second
picture rate	≤ 30 frames per second
bit rate	≤ 1.856 Mbps

Table 12.1. Some of the Constrained Parameters for MPEG-1.

Resolution	Frames per Second
352 × 240p	29.97
352 × 240p	23.976
352 × 288p	25
320 × 240p[1]	29.97
384 × 288p[1]	25

Notes:
1. Square pixel format.

Table 12.2. Common MPEG-1 Resolutions.

P (predicted) frames (~0.1 bit/pixel) are coded relative to the nearest previous I or P frame, resulting in forward prediction processing, as shown in Figure 12.1. P frames provide more compression than I frames, through the use of motion compensation, and are also a reference for B frames and future P frames.

B (bi-directional) frames (~0.015 bit/pixel) use the closest past and future I or P frame as a reference, resulting in bi-directional prediction, as shown in Figure 12.1. B frames provide the most compression and decrease noise by averaging two frames. Typically, there are two B frames separating I or P frames.

D (DC) frames are frames coded as a stand-alone still image, using only the DC component of the DCTs. D frames may not be in a sequence containing any other frame types and are rarely used.

A group of pictures (GOP) is a series of one or more coded frames intended to assist in random accessing and editing. The GOP value is configurable during the encoding process. The smaller the GOP value, the better the response to movement (since the I frames are closer together), but the lower the compression.

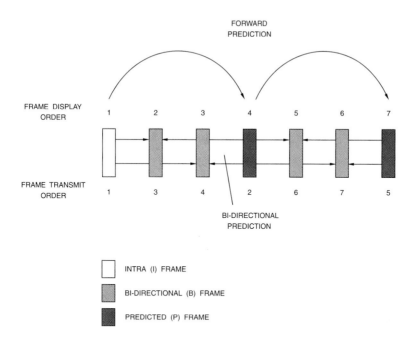

Figure 12.1. MPEG-1 I, P, and B Frames. Some frames are transmitted out of display sequence, complicating the interpolation process, and requiring frame reordering by the MPEG decoder. Arrows show inter-frame dependencies.

In the coded bitstream, a GOP must start with an I frame and may be followed by any number of I, P, or B frames in any order. In display order, a GOP must start with an I or B frame and end with an I or P frame. Thus, the smallest GOP size is a single I frame, with the largest size unlimited.

Originally, each GOP was to be coded and displayed independently of any other GOP. However, this is not possible unless no B frames precede I frames, or if they do, they use only backward motion compensation. This results in both open and closed GOP formats. A *closed GOP* is a GOP that can be decoded without using frames of the previous GOP for motion compensation. An *open GOP* requires that they be available.

Motion Compensation

Motion compensation improves compression of P and B frames by removing temporal redundancies between frames. It works at the macroblock level (defined later).

The technique relies on the fact that within a short sequence of the same general image, most objects remain in the same location, while others move only a short distance. The motion is described as a two-dimensional motion vector that specifies where to retrieve a macroblock from a previously decoded frame to predict the sample values of the current macroblock.

After a macroblock has been compressed using motion compensation, it contains both the spatial difference (motion vectors) and content difference (error terms) between the reference macroblock and macroblock being coded.

Note that there are cases where information in a scene cannot be predicted from the previous scene, such as when a door opens. The previous scene doesn't contain the details of the area behind the door. In cases such as this, when a macroblock in a P frame cannot be represented by motion compensation, it is coded the same way as a macroblock in an I frame (using intra-picture coding).

Macroblocks in B frames are coded using either the closest previous or future I or P frames as a reference, resulting in four possible codings:

- intra coding
    no motion compensation

- forward prediction
    closest previous I or P frame is the reference

- backward prediction
    closest future I or P frame is the reference

- bi-directional prediction
    two frames are used as the reference:
    the closest previous I or P frame and
    the closest future I or P frame

Backward prediction is used to predict "uncovered" areas that appear in previous frames.

I Frames

Image blocks and prediction error blocks have a high spatial redundancy. Several steps are used to remove this redundancy within a frame to improve the compression. The inverse of these steps is used by the decoder to recover the data.

Macroblock

A macroblock (shown in Figure 7.55) consists of a 16-sample × 16-line set of Y components and the corresponding two 8-sample × 8-line Cb and Cr components.

A block is an 8-sample × 8-line set of Y, Cb, or Cr values. Note that a Y block refers to one-fourth the image size as the corresponding Cb or Cr blocks. Thus, a macroblock contains four Y blocks, one Cb block, and one Cr block, as seen in Figure 12.2.

There are two types of macroblocks in I frames, both using intra coding, as shown in Table 12.9. One (called *intra-d*) uses the current quantizer scale; the other (called *intra-q*) defines a new value for the quantizer scale

If the macroblock type is intra-q, the macroblock header specifies a 5-bit quantizer scale factor. The decoder uses this to calculate the DCT coefficients from the transmitted quantized coefficients. Quantizer scale factors may range from 1–31, with zero not allowed.

If the macroblock type is intra-d, no quantizer scale is sent, and the decoder uses the current one.

DCT

Each 8 × 8 block (of input samples or prediction error terms) is processed by an 8 × 8 DCT (discrete cosine transform), resulting in an 8 × 8 block of horizontal and vertical frequency coefficients, as shown in Figure 7.56.

Input sample values are 0–255, resulting in a range of 0–2,040 for the DC coefficient and a range of about –1,000 to +1,000 for the AC coefficients.

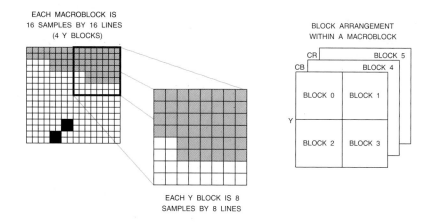

Figure 12.2. MPEG-1 Macroblocks and Blocks.

Quantizing

The 8 × 8 block of frequency coefficients are uniformly quantized, limiting the number of allowed values. The quantizer step scale is derived from the quantization matrix and quantizer scale and may be different for different coefficients and may change between macroblocks.

The quantizer step size of the DC coefficients is fixed at eight. The DC quantized coefficient is determined by dividing the DC coefficient by eight and rounding to the nearest integer. AC coefficients are quantized using the intra-quantization matrix.

Zig-Zag Scan

Zig-zag scanning, starting with the DC component, generates a linear stream of quantized frequency coefficients arranged in order of increasing frequency, as shown in Figure 7.59. This produces long runs of zero coefficients.

Coding of Quantized DC Coefficients

After the DC coefficients have been quantized, they are losslessly coded.

Coding of Y blocks within a macroblock follows the order shown in Figure 12.2. The DC value of block 4 is the DC predictor for block 1 of the next macroblock. At the beginning of each slice, the DC predictor is set to 1,024.

The DC values of each Cb and Cr block are coded using the DC value of the corresponding block of the previous macroblock as a predictor. At the beginning of each slice, both DC predictors are set to 1,024.

The DCT DC differential values are organized by their absolute value as shown in Table 12.16. [size], which specifies the number of additional bits to define the level uniquely, is transmitted by a variable-length code, and is different for Y and CbCr since the statistics are different. For example, a size of four is followed by four additional bits.

The decoder reverses the procedure to recover the quantized DC coefficients.

Coding of Quantized AC Coefficients

After the AC coefficients have been quantized, they are scanned in the zig-zag order shown in Figure 7.59 and coded using run-length and level. The scan starts in position 1, as shown in Figure 7.59, as the DC coefficient in position 0 is coded separately.

The run-lengths and levels are coded as shown in Table 12.18. The "s" bit denotes the sign of the level; "0" is positive and "1" is negative.

For run-level combinations not shown in Table 12.18, an escape sequence is used, consisting of the escape code (ESC), followed by the run-length and level codes from Table 12.19.

After the last DCT coefficient has been coded, an EOB code is added to tell the decoder that there are no more quantized coefficients in this 8×8 block.

P Frames

Macroblocks

There are eight types of macroblocks in P frames, as shown in Table 12.10, due to the additional complexity of motion compensation.

Skipped macroblocks are predicted macroblocks with a zero motion vector. Thus, no correction is available; the decoder copies skipped macroblocks from the previous frame into the current frame. The advantage of skipped macroblocks is that they require very few bits to transmit. They have no code; they are coded by having the macroblock address increment code skip over them.

If the [macroblock quant] column in Table 12.10 has a "1," the quantizer scale is transmitted. For the remaining macroblock types, the DCT correction is coded using the previous value for quantizer scale.

If the [motion forward] column in Table 12.10 has a "1," horizontal and vertical forward motion vectors are successively transmitted.

If the [coded pattern] column in Table 12.10 has a "1," the 6-bit coded block pattern is transmitted as a variable-length code. This tells the decoder which of the six blocks in the macroblock are coded ("1") and which are not coded ("0"). Table 12.14 lists the codewords assigned to the 63 possible combinations. There is no code for when none of the blocks are coded; it is indicated by the macroblock type. For macroblocks in I frames and for intra-coded macroblocks in P and B frames, the coded block pattern is not transmitted, but is assumed to be a value of 63 (all blocks are coded).

To determine which type of macroblock to use, the encoder typically makes a series of decisions, as shown in Figure 12.3.

DCT

Intra block AC coefficients are transformed in the same manner as they are for I frames. Intra block DC coefficients are transformed differently; the predicted values are set to 1,024, unless the previous block was intra-coded.

Non-intra block coefficients represent differences between sample values rather than actual sample values. They are obtained by subtracting the motion compensated values of the previous frame from the values in the current macroblock. There is no prediction of the DC value.

Input sample values are –255 to +255, resulting in a range of about –2,000 to +2,000 for the AC coefficients.

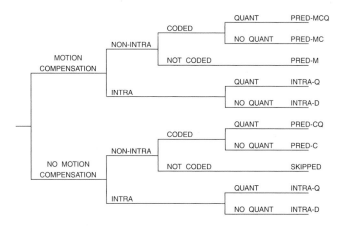

Figure 12.3. MPEG-1 P Frame Macroblock Type Selection.

Quantizing

Intra blocks are quantized in the same manner as they are for I frames.

Non-intra blocks are quantized using the quantizer scale and the non-intra quantization matrix. The AC and DC coefficients are quantized in the same manner.

Coding of Intra Blocks

Intra blocks are coded the same way as I frame intra blocks. There is a difference in the handling of the DC coefficients in that the predicted value is 128, unless the previous block was intra coded.

Coding of Non-Intra Blocks

The coded block pattern (CBP) is used to specify which blocks have coefficient data. These are coded similarly to the coding of intra blocks, except the DC coefficient is coded in the same manner as the AC coefficients.

B Frames

Macroblocks

There are twelve types of macroblocks in B frames, as shown in Table 12.11, due to the additional complexity of backward motion compensation.

Skipped macroblocks are macroblocks having the same motion vector and macroblock type as the previous macroblock, which cannot be intra coded. The advantage of skipped macroblocks is that they require very few bits to transmit. They have no code; they are coded by having the macroblock address increment code skip over them.

If the [macroblock quant] column in Table 12.11 has a "1," the quantizer scale is transmitted. For the rest of the macroblock types, the DCT correction is coded using the previous value for the quantizer scale.

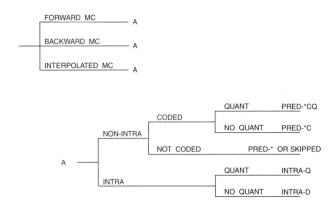

Figure 12.4. MPEG-1 B Frame Macroblock Type Selection.

If the [motion forward] column in Table 12.11 has a "1," horizontal and vertical forward motion vectors are successively transmitted. If the [motion backward] column in Table 12.11 has a "1," horizontal and vertical backward motion vectors are successively transmitted. If both forward and backward motion types are present, the vectors are transmitted in this order:

    horizontal forward
    vertical forward
    horizontal backward
    vertical backward

If the [coded pattern] column in Table 12.11 has a "1," the 6-bit coded block pattern is transmitted as a variable-length code. This tells the decoder which of the six blocks in the macroblock are coded ("1") and which are not coded ("0"). Table 12.14 lists the codewords assigned to the 63 possible combinations. There is no code for when none of the blocks

are coded; this is indicated by the macroblock type. For macroblocks in I frames and for intra-coded macroblocks in P and B frames, the coded block pattern is not transmitted, but is assumed to be a value of 63 (all blocks are coded).

To determine which type of macroblock to use, the encoder typically makes a series of decisions, shown in Figure 12.4.

Coding

DCT coefficients of blocks are transformed into quantized coefficients and coded in the same way they are for P frames.

D Frames

D frames contain only DC-frequency data and are intended to be used for fast visible search applications. The data contained in a D frame should be just sufficient for the user to locate the desired video.

Video Bitstream

Figure 12.5 illustrates the video bitstream, a hierarchical structure with seven layers. From top to bottom the layers are:

>   Video Sequence
>   Sequence Header
>   Group of Pictures (GOP)
>   Picture
>   Slice
>   Macroblock (MB)
>   Block

Note that start codes ($000001xx_H$) must be byte aligned by inserting 0–7 "0" bits before the start code.

Video Sequence

Sequence_end_code

This 32-bit field has a value of $000001B7_H$ and terminates a video sequence.

Sequence Header

Data for each sequence consists of a sequence header followed by data for group of pictures (GOPs). The structure is shown in Figure 12.5.

Sequence_header_code

This 32-bit field has a value of $000001B3_H$ and indicates the beginning of a sequence header.

Horizontal_size

This 12-bit binary value specifies the width of the viewable portion of the Y component. The width in macroblocks is defined as (*horizontal_size* + 15)/16.

Vertical_size

This 12-bit binary value specifies the height of the viewable portion of the Y component. The height in macroblocks is defined as (*vertical_size* + 15)/16.

Pel_aspect_ratio

This 4-bit codeword indicates the pixel aspect ratio, as shown in Table 12.3.

Picture_rate

This 4-bit codeword indicates the frame rate, as shown in Table 12.4.

Bit_rate

An 18-bit binary value specifying the bitstream bit rate, measured in units of 400 bps rounded upwards. A zero value is not allowed; a value of $3FFFF_H$ specifies variable bit rate operation. If *constrained_parameters_flag* is a "1," the bit rate must be ≤1.856 Mbps.

Marker_bit

Always a "1."

Vbv_buffer_size

This 10-bit binary number specifies the minimum size of the video buffering verifier needed by the decoder to properly decode the sequence. It is defined as:

$$B = 16 \times 1024 \times vbv\_buffer\_size$$

If the *constrained_parameters_flag* bit is a "1," the *vbv_buffer_size* must be ≤40 kB.

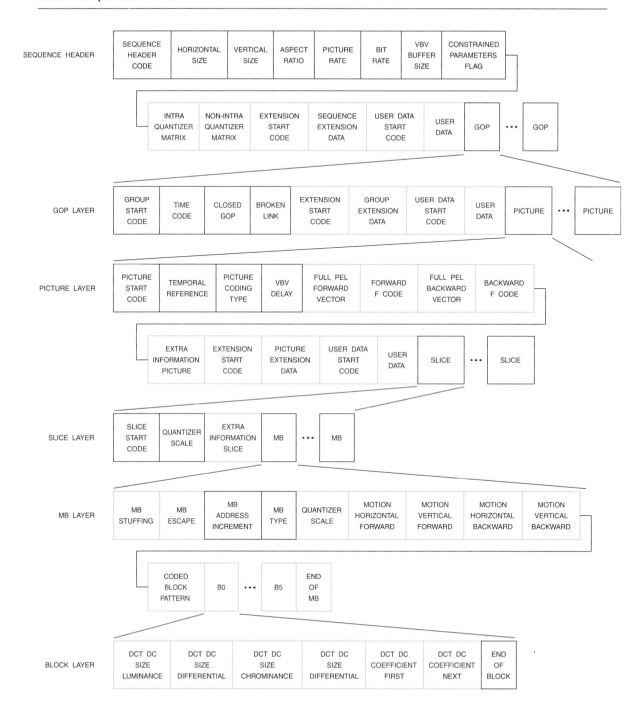

Figure 12.5. MPEG-1 Video Bitstream Layer Structures. Marker and reserved bits not shown.

Height / Width	Example	Aspect Ratio Code
forbidden		0000
1.0000	square pixel	0001
0.6735		0010
0.7031	576-line 16:9	0011
0.7615		0100
0.8055		0101
0.8437	480-line 16:9	0110
0.8935		0111
0.9157	576-line 4:3	1000
0.9815		1001
1.0255		1010
1.0695		1011
1.0950	480-line 4:3	1100
1.1575		1101
1.2015		1110
reserved		1111

Table 12.3. MPEG-1 *pel_aspect_ratio* Codewords.

Frames Per Second	Picture Rate Code
forbidden	0000
24/1.001	0001
24	0010
25	0011
30/1.001	0100
30	0101
50	0110
60/1.001	0111
60	1000
reserved	1001
reserved	1010
reserved	1011
reserved	1100
reserved	1101
reserved	1110
reserved	1111

Table 12.4. MPEG-1 *picture_rate* Codewords.

Constrained_parameters_flag

This bit is set to a "1" if the following constraints are met:

horizontal_size $\leq$ 768 samples
vertical_size $\leq$ 576 lines
$((horizontal\_size + 15)/16) \times ((vertical\_size + 15)/16) \leq 396$
$((horizontal\_size + 15)/16) \times ((vertical\_size + 15)/16) \times picture\_rate \leq 396*25$
picture_rate $\leq$ 30 frames per second
forward_f_code $\leq$ 4
backward_f_code $\leq$ 4

Load_intra_quantizer_matrix

This bit is set to a "1" if *intra_quantizer_matrix* follows. If set to a "0," the default values below are used until the next occurrence of a sequence header.

8	16	19	22	26	27	29	34
16	16	22	24	27	29	34	37
19	22	26	27	29	34	34	38
22	22	26	27	29	34	37	40
22	26	27	29	32	35	40	48
26	27	29	32	35	40	48	58
26	27	29	34	38	46	56	69
27	29	35	38	46	56	69	83

Intra_quantizer_matrix

An optional list of sixty-four 8-bit values that replace the current intra quantizer values. A value of zero is not allowed. The value for intra_quant [0, 0] is always eight. These values take effect until the next occurrence of a sequence header.

Load_non_intra_quantizer_matrix

This bit is set to a "1" if *non_intra_quantizer_matrix* follows. If set to a "0," the default values below are used until the next occurrence of a sequence header.

16	16	16	16	16	16	16	16
16	16	16	16	16	16	16	16
16	16	16	16	16	16	16	16
16	16	16	16	16	16	16	16
16	16	16	16	16	16	16	16
16	16	16	16	16	16	16	16
16	16	16	16	16	16	16	16
16	16	16	16	16	16	16	16

Non_intra_quantizer_matrix

An optional list of sixty-four 8-bit values that replace the current non-intra quantizer values. A value of zero is not allowed. These values take effect until the next occurrence of a sequence header.

Extension_start_code

This optional 32-bit string of $000001B5_H$ indicates the beginning of *sequence_extension_data*. *sequence_extension_data* continues until the detection of another start code.

Sequence_extension_data

These $n \times 8$ bits are present only if *extension_start_code* is present.

User_data_start_code

This optional 32-bit string of $000001B2_H$ indicates the beginning of *user_data*. *user_data* continues until the detection of another start code.

User_data

These $n \times 8$ bits are present only if *user_data_start_code* is present. *user_data* must not contain a string of 23 or more consecutive zero bits.

Time Code	Range of Value	Number of Bits
drop_frame_flag		1
time_code_hours	0–23	5
time_code_minutes	0–59	6
marker_bit	1	1
time_code_seconds	0–59	6
time_code_pictures	0–59	6

Table 12.5. MPEG-1 *time_code* Field.

Group of Pictures (GOP) Layer

Data for each group of pictures consists of a GOP header followed by picture data. The structure is shown in Figure 12.5.

Group_start_code

This 32-bit value of $000001B8_H$ indicates the beginning of a group of pictures.

Time_code

These 25 bits indicate timecode information, as shown in Table 12.5. [drop_frame_flag] may be set to "1" only if the picture rate is 30/1.001 (29.97) Hz.

Closed_gop

This 1-bit flag is set to "1" if the group of pictures has been encoded without motion vectors referencing the previous group of pictures. This bit allows support of editing the compressed bitstream.

Broken_link

This 1-bit flag is set to a "0" during encoding. It is set to a "1" during editing when the B frames following the first I frame of a group of pictures cannot be correctly decoded.

Extension_start_code

This optional 32-bit string of $000001B5_H$ indicates the beginning of *group_extension_data*. *group_extension_data* continues until the detection of another start code.

Group_extension_data

These n × 8 bits are present only if *extension_start_code* is present.

User_data_start_code

This optional 32-bit string of $000001B2_H$ indicates the beginning of *user_data*. *user_data* continues until the detection of another start code.

User_data

These n × 8 bits are present only if *user_data_start_code* is present. *user_data* must not contain a string of 23 or more consecutive zero bits.

Picture Layer

Data for each picture layer consists of a picture header followed by slice data. The structure is shown in Figure 12.5.

Picture_start_code

This has a 32-bit value of 00000100_H.

Temporal_reference

For the first frame in display order of each group of pictures, the *temporal_reference* value is zero. This 10-bit binary value then increments by one, modulo 1024 for each frame in display order.

Picture_coding_type

This 3-bit codeword indicates the frame type (I frame, P frame, B frame, or D frame), as shown in Table 12.6. D frames are not to be used in the same video sequence as other frames.

Coding Type	Code
forbidden	000
I frame	001
P frame	010
B frame	011
D frame	100
reserved	101
reserved	110
reserved	111

Table 12.6. MPEG-1 *picture_coding_type* **Code.**

Vbv_delay

For constant bit rates, the 16-bit *vbv_delay* binary value sets the initial occupancy of the decoding buffer at the start of decoding a picture so that it doesn't overflow or underflow. For variable bit rates, *vbv_delay* has a value of $FFFF_H$.

Full_pel_forward_vector

This 1-bit flag is present if *picture_coding_type* is "010" (P frames) or "011" (B frames). If a "1," the forward motion vectors are based on integer samples, rather than half-samples.

Forward_f_code

This 3-bit binary number is present if *picture_coding_type* is "010" (P frames) or "011" (B frames). Values of "001" to "111" are used; a value of "000" is forbidden.

Two parameters used by the decoder to decode the forward motion vectors are derived from this field: *forward_r_size* and *forward_f*. *forward_r_size* is one less than *forward_f_code*. *forward_f* is defined in Table 12.7.

Forward F Code	Forward F Value
001	1
010	2
011	4
100	8
101	16
110	32
111	64

Table 12.7. MPEG-1 *forward_f_code* **Values.**

Full_pel_backward_vector

This 1-bit flag is present if *picture_coding_type* is "011" (B frames). If a "1," the backward motion vectors are based on integer samples, rather than half-samples.

Backward_f_code

This 3-bit binary number is present if *picture_coding_type* is "011" (B frames). Values of "001" to 111" are used; a value of "000" is forbidden.

Two parameters used by the decoder to decode the backward motion vectors are derived from this field: *backward_r_size* and *backward_f*. *backward_r_size* is one less than *backward_f_code*. *backward_f* is defined the same as *forward_f*.

Extra_bit_picture

A bit which, when set to "1," indicates that *extra_information_picture* follows.

Extra_information_picture

If *extra_bit_picture* = "1," then these nine bits follow consisting of eight bits of data (*extra_information_picture*) and then another *extra_bit_picture* to indicate if a further nine bits follow, and so on.

Extension_start_code

This optional 32-bit string of $000001B5_H$ indicates the beginning of *picture_extension_data*. *picture_extension_data* continues until the detection of another start code.

Picture_extension_data

These n × 8 bits are present only if *extension_start_code* is present.

User_data_start_code

This optional 32-bit string of $000001B2_H$ indicates the beginning of *user_data*. *user_data* continues until the detection of another start code.

User_data

These n × 8 bits are present only if *user_data_start_code* is present. User data must not contain a string of 23 or more consecutive zero bits.

Slice Layer

Data for each slice layer consists of a slice header followed by macroblock data. The structure is shown in Figure 12.5.

Slice_start_code

The first 24 bits of this 32-bit field have a value of 000001_H. The last eight bits are the *slice_vertical_position*, and have a value of 01_H–AF_H.

slice_vertical_position specifies the vertical position in macroblock units of the first macroblock in the slice. The value of the first row of macroblocks is one.

Quantizer_scale

This 5-bit binary number has a value of 1–31 (a value of 0 is forbidden). It specifies the scale factor of the reconstruction level of the DCT coefficients. The decoder uses this value until another *quantizer_scale* is received at the either the slice or macroblock layer.

Extra_bit_slice

A bit which, when set to "1," indicates that *extra_information_slice* follows.

Extra_information_slice

If *extra_bit_slice* = "1," then these nine bits follow consisting of eight bits of data (*extra_information_slice*) and then another *extra_bit_slice* to indicate if a further nine bits follow, and so on.

Macroblock (MB) Layer

Data for each macroblock layer consists of a macroblock header followed by motion vectors and block data. The structure is shown in Figure 12.5.

Macroblock_stuffing

This optional 11-bit field is a fixed bit string of "0000 0001 111" and may be used to increase the bit rate to match the storage or transmission requirements. Any number of consecutive *macroblock_stuffing* fields may be used.

Macroblock_escape

This optional 11-bit field is a fixed bit string of "0000 0001 000" and is used when the difference between the current macroblock address and the previous macroblock address is greater than 33. It forces the value of *macroblock_address_increment* to be increased by 33. Any number of consecutive *macroblock_escape* fields may be used.

Macroblock_address_increment

This is a variable-length codeword that specifies the difference between the current macroblock address and the previous macroblock address. It has a maximum value of 33. Values greater than 33 are encoded using the *macroblock_escape* field. The variable-length codes are listed in Table 12.8.

Macroblock_type

This is a variable-length codeword that specifies the coding method and macroblock content. The variable-length codes are listed in Tables 12.9 through 12.12.

Quantizer_scale

This optional 5-bit binary number has a value of 1–31 (a value of 0 is forbidden). It specifies the scale factor of the reconstruction level of the received DCT coefficients. The decoder uses this value until another *quantizer_scale* is received at the either the slice or macroblock layer. The *quantizer_scale* field is present only when [macroblock quant] = "1" in Tables 12.9 through 12.12.

Motion_horizontal_forward_code

This optional variable-length codeword contains forward motion vector information as defined in Table 12.13. It is present only when [motion forward] = "1" in Tables 12.9 through 12.12.

Motion_horizontal_forward_r

This optional binary number (of *forward_r_size* bits) is used to help decode the forward motion vectors. It is present only when [motion forward] = "1" in Tables 12.9 through 12.12, *forward_f_code* ≠ "001," and *motion_horizontal_forward_code* ≠ "0."

Motion_vertical_forward_code

This optional variable-length codeword contains forward motion vector information as defined in Table 12.13. It is present only when [motion forward] = "1" in Tables 12.9 through 12.12.

Increment Value	Code	Increment Value	Code
1	1	17	0000 0101 10
2	011	18	0000 0101 01
3	010	19	0000 0101 00
4	0011	20	0000 0100 11
5	0010	21	0000 0100 10
6	0001 1	22	0000 0100 011
7	0001 0	23	0000 0100 010
8	0000 111	24	0000 0100 001
9	0000 110	25	0000 0100 000
10	0000 1011	26	0000 0011 111
11	0000 1010	27	0000 0011 110
12	0000 1001	28	0000 0011 101
13	0000 1000	29	0000 0011 100
14	0000 0111	30	0000 0011 011
15	0000 0110	31	0000 0011 010
16	0000 0101 11	32	0000 0011 001
		33	0000 0011 000

Table 12.8. MPEG-1 Variable-Length Code Table for *macroblock_address_increment*.

Macroblock Type	Macroblock Quant	Motion Forward	Motion Backward	Coded Pattern	Intra Macroblock	Code
intra-d	0	0	0	0	1	1
intra-q	1	0	0	0	1	01

Table 12.9. MPEG-1 Variable-Length Code Table for *macroblock_type* for I Frames.

Macroblock Type	Macroblock Quant	Motion Forward	Motion Backward	Coded Pattern	Intra Macroblock	Code
pred-mc	0	1	0	1	0	1
pred-c	0	0	0	1	0	01
pred-m	0	1	0	0	0	001
intra-d	0	0	0	0	1	0001 1
pred-mcq	1	1	0	1	0	0001 0
pred-cq	1	0	0	1	0	0000 1
intra-q	1	0	0	0	1	0000 01
skipped						

Table 12.10. MPEG-1 Variable-Length Code Table for *macroblock_type* for P Frames.

Macroblock Type	Macroblock Quant	Motion Forward	Motion Backward	Coded Pattern	Intra Macroblock	Code
pred-i	0	1	1	0	0	10
pred-ic	0	1	1	1	0	11
pred-b	0	0	1	0	0	010
intra-bc	0	0	1	1	0	011
pred-f	0	1	0	0	0	0010
pred-fc	0	1	0	1	0	0011
intra-d	0	0	0	0	1	0001 1
pred-icq	1	1	1	1	0	0001 0
pred-fcq	1	1	0	1	0	0000 11
pred-bcq	1	0	1	1	0	0000 10
intra-q	1	0	0	0	1	0000 01
skipped						

Table 12.11. MPEG-1 Variable-Length Code Table for *macroblock_type* for B Frames.

Macroblock Quant	Motion Forward	Motion Backward	Coded Pattern	Intra Macroblock	Code
0	0	0	0	1	1

Table 12.12. MPEG-1 Variable-Length Code Table for *macroblock_type* for D Frames.

Motion Vector Difference	Code	Motion Vector Difference	Code
−16	0000 0011 001	1	010
−15	0000 0011 011	2	0010
−14	0000 0011 101	3	0001 0
−13	0000 0011 111	4	0000 110
−12	0000 0100 001	5	0000 1010
−11	0000 0100 011	6	0000 1000
−10	0000 0100 11	7	0000 0110
−9	0000 0101 01	8	0000 0101 10
−8	0000 0101 11	9	0000 0101 00
−7	0000 0111	10	0000 0100 10
−6	0000 1001	11	0000 0100 010
−5	0000 1011	12	0000 0100 000
−4	0000 111	13	0000 0011 110
−3	0001 1	14	0000 0011 100
−2	0011	15	0000 0011 010
−1	011	16	0000 0011 000
0	1		

Table 12.13. MPEG-1 Variable-Length Code Table for *motion_horizontal_forward_code, motion_vertical_forward_code, motion_horizontal_backward_code,* **and** *motion_vertical_backward_code.*

Motion_vertical_forward_r

This optional binary number (of *forward_r_size* bits) is used to help decode the forward motion vectors. It is present only when [motion forward] = "1" in Tables 12.9 through 12.12, *forward_f_code* ≠ "001," and *motion_vertical_forward_code* ≠ "0."

Motion_horizontal_backward_code

This optional variable-length codeword contains backward motion vector information as defined in Table 12.13. It is present only when [motion backward] = "1" in Tables 12.9 through 12.12.

Motion_horizontal_backward_r

This optional binary number (of *backward_r_size* bits) is used to help decode the backward motion vectors. It is present only when [motion backward] = "1" in Tables 12.9 through 12.12, *backward_f_code* ≠ "001," and *motion_horizontal_backward_code* ≠ "0."

Motion_vertical_backward_code

This optional variable-length codeword contains backward motion vector information as defined in Table 12.13. The decoded value helps decide if *motion_vertical_backward_r* appears in the bitstream. This parameter is present only when [motion backward] = "1" in Tables 12.9 through 12.12.

Motion_vertical_backward_r

This optional binary number (of *backward_r_size* bits) is used to help decode the backward motion vectors. It is present only when [motion backward] = "1" in Tables 12.9 through 12.12, *backward_f_code* ≠ "001," and *motion_vertical_backward_code* ≠ "0."

Coded_block_pattern

This optional variable-length codeword is used to derive the coded block pattern (CBP) as shown in Table 12.14. It is present only if [coded pattern] = "1" in Tables 12.9 through 12.12, and indicates which blocks in the macroblock have at least one transform coefficient transmitted. The coded block pattern binary number is represented as:

$$P_0 P_1 P_2 P_3 P_4 P_5$$

where P_n = "1" for any coefficient present for block [n], else P_n = "0." Block numbering (decimal format) is given in Figure 12.2.

End_of_macroblock

This optional 1-bit field has a value of "1." It is present only for D frames.

Block Layer

Data for each block layer consists of coefficient data. The structure is shown in Figure 12.5.

Dct_dc_size_luminance

This optional variable-length codeword is used with intra-coded Y blocks. It specifies the number of bits used for *dct_dc_differential*. The variable-length codewords are shown in Table 12.15.

Coded Block Pattern	Code	Coded Block Pattern	Code	Coded Block Pattern	Code
60	111	9	0010 110	43	0001 0000
4	1101	17	0010 101	25	0000 1111
8	1100	33	0010 100	37	0000 1110
16	1011	6	0010 011	26	0000 1101
32	1010	10	0010 010	38	0000 1100
12	1001 1	18	0010 001	29	0000 1011
48	1001 0	34	0010 000	45	0000 1010
20	1000 1	7	0001 1111	53	0000 1001
40	1000 0	11	0001 1110	57	0000 1000
28	0111 1	19	0001 1101	30	0000 0111
44	0111 0	35	0001 1100	46	0000 0110
52	0110 1	13	0001 1011	54	0000 0101
56	0110 0	49	0001 1010	58	0000 0100
1	0101 1	21	0001 1001	31	0000 0011 1
61	0101 0	41	0001 1000	47	0000 0011 0
2	0100 1	14	0001 0111	55	0000 0010 1
62	0100 0	50	0001 0110	59	0000 0010 0
24	0011 11	22	0001 0101	27	0000 0001 1
36	0011 10	42	0001 0100	39	0000 0001 0
3	0011 01	15	0001 0011		
63	0011 00	51	0001 0010		
5	0010 111	23	0001 0001		

Table 12.14. MPEG-1 Variable-Length Code Table for *coded_block_pattern*.

DCT DC Size Luminance	Code	DCT DC Size Luminance	Code
0	100	5	1110
1	00	6	1111 0
2	01	7	1111 10
3	101	8	1111 110
4	110		

Table 12.15. MPEG-1 Variable-Length Code Table for *dct_dc_size_luminance*.

Dct_dc_differential

This optional variable-length codeword is present after *dct_dc_size_luminance* if *dct_dc_size_luminance* ≠ "0." The values are shown in Table 12.16.

Dct_dc_size_chrominance

This optional variable-length codeword is used with intra-coded Cb and Cr blocks. It specifies the number of bits used for *dct_dc_differential*. The variable-length codewords are shown in Table 12.17.

Dct_dc_differential

This optional variable-length codeword is present after *dct_dc_size_chrominance* if *dct_dc_size_chrominance* ≠ "0." The values are shown in Table 12.16.

Dct_coefficient_first

This optional variable-length codeword is used for the first DCT coefficient in non-intra-coded blocks, and is defined in Tables 12.18 and 12.19.

Dct_coefficient_next

Up to 63 optional variable-length codewords present only for I, P, and B frames. They are the DCT coefficients after the first one, and are defined in Tables 12.18 and 12.19.

End_of_block

This 2-bit value (present only for I, P, and B frames) is used to indicate that no additional non-zero coefficients are present. The value of this parameter is "10."

DCT DC Differential	Size	Code (Y)	Code (CbCr)	Additional Code
–255 to –128	8	1111110	11111110	00000000 to 01111111
–127 to –64	7	111110	1111110	0000000 to 0111111
–63 to –32	6	11110	111110	000000 to 011111
–31 to –16	5	1110	11110	00000 to 01111
–15 to –8	4	110	1110	0000 to 0111
–7 to –4	3	101	110	000 to 011
–3 to –2	2	01	10	00 to 01
–1	1	00	01	0
0	0	100	00	
1	1	00	01	1
2 to 3	2	01	10	10 to 11
4 to 7	3	101	110	100 to 111
8 to 15	4	110	1110	1000 to 1111
16 to 31	5	1110	11110	10000 to 11111
32 to 63	6	11110	111110	100000 to 111111
64 to 127	7	111110	1111110	1000000 to 1111111
128 to 255	8	1111110	11111110	10000000 to 11111111

Table 12.16. MPEG-1 Variable-Length Code Table for *dct_dc_differential*.

DCT DC Size Chrominance	Code	DCT DC Size Chrominance	Code
0	00	5	1111 0
1	01	6	1111 10
2	10	7	1111 110
3	110	8	1111 1110
4	1110		

Table 12.17. MPEG-1 Variable-Length Code Table for *dct_dc_size_chrominance*.

Run	Level	Code	Run	Level	Code
end_of_block		10	escape		0000 01
0 (note 2)	1	1 s	0	5	0010 0110 s
0 (note 3)	1	11 s	0	6	0010 0001 s
1	1	011 s	1	3	0010 0101 s
0	2	0100 s	3	2	0010 0100 s
2	1	0101 s	10	1	0010 0111 s
0	3	0010 1 s	11	1	0010 0011 s
3	1	0011 1 s	12	1	0010 0010 s
4	1	0011 0 s	13	1	0010 0000 s
1	2	0001 10 s	0	7	0000 0010 10 s
5	1	0001 11 s	1	4	0000 0011 00 s
6	1	0001 01 s	2	3	0000 0010 11 s
7	1	0001 00 s	4	2	0000 0011 11 s
0	4	0000 110 s	5	2	0000 0010 01 s
2	2	0000 100 s	14	1	0000 0011 10 s
8	1	0000 111 s	15	1	0000 0011 01 s
9	1	0000 101 s	16	1	0000 0010 00 s

Notes:

1. s = sign of level; "0" for positive; s = "1" for negative.
2: Used for *dct_coefficient_first*
3: Used for *dct_coefficient_next*.

Table 12.18a. MPEG-1 Variable-Length Code Table for *dct_coefficient_first* and *dct_coefficient_next*.

Run	Level	Code	Run	Level	Code
0	8	0000 0001 1101 s	0	12	0000 0000 1101 0 s
0	9	0000 0001 1000 s	0	13	0000 0000 1100 1 s
0	10	0000 0001 0011 s	0	14	0000 0000 1100 0 s
0	11	0000 0001 0000 s	0	15	0000 0000 1011 1 s
1	5	0000 0001 1011 s	1	6	0000 0000 1011 0 s
2	4	0000 0001 0100 s	1	7	0000 0000 1010 1 s
3	3	0000 0001 1100 s	2	5	0000 0000 1010 0 s
4	3	0000 0001 0010 s	3	4	0000 0000 1001 1 s
6	2	0000 0001 1110 s	5	3	0000 0000 1001 0 s
7	2	0000 0001 0101 s	9	2	0000 0000 1000 1 s
8	2	0000 0001 0001 s	10	2	0000 0000 1000 0 s
17	1	0000 0001 1111 s	22	1	0000 0000 1111 1 s
18	1	0000 0001 1010 s	23	1	0000 0000 1111 0 s
19	1	0000 0001 1001 s	24	1	0000 0000 1110 1 s
20	1	0000 0001 0111 s	25	1	0000 0000 1110 0 s
21	1	0000 0001 0110 s	26	1	0000 0000 1101 1 s

Notes:
1. s = sign of level; "0" for positive; s = "1" for negative.

Table 12.18b. MPEG-1 Variable-Length Code Table for *dct_coefficient_first* and *dct_coefficient_next*.

Run	Level	Code	Run	Level	Code
0	16	0000 0000 0111 11 s	0	40	0000 0000 0010 000 s
0	17	0000 0000 0111 10 s	1	8	0000 0000 0011 111 s
0	18	0000 0000 0111 01 s	1	9	0000 0000 0011 110 s
0	19	0000 0000 0111 00 s	1	10	0000 0000 0011 101 s
0	20	0000 0000 0110 11 s	1	11	0000 0000 0011 100 s
0	21	0000 0000 0110 10 s	1	12	0000 0000 0011 011 s
0	22	0000 0000 0110 01 s	1	13	0000 0000 0011 010 s
0	23	0000 0000 0110 00 s	1	14	0000 0000 0011 001 s
0	24	0000 0000 0101 11 s	1	15	0000 0000 0001 0011 s
0	25	0000 0000 0101 10 s	1	16	0000 0000 0001 0010 s
0	26	0000 0000 0101 01 s	1	17	0000 0000 0001 0001 s
0	27	0000 0000 0101 00 s	1	18	0000 0000 0001 0000 s
0	28	0000 0000 0100 11 s	6	3	0000 0000 0001 0100 s
0	29	0000 0000 0100 10 s	11	2	0000 0000 0001 1010 s
0	30	0000 0000 0100 01 s	12	2	0000 0000 0001 1001 s
0	31	0000 0000 0100 00 s	13	2	0000 0000 0001 1000 s
0	32	0000 0000 0011 000 s	14	2	0000 0000 0001 0111 s
0	33	0000 0000 0010 111 s	15	2	0000 0000 0001 0110 s
0	34	0000 0000 0010 110 s	16	2	0000 0000 0001 0101 s
0	35	0000 0000 0010 101 s	27	1	0000 0000 0001 1111 s
0	36	0000 0000 0010 100 s	28	1	0000 0000 0001 1110 s
0	37	0000 0000 0010 011 s	29	1	0000 0000 0001 1101 s
0	38	0000 0000 0010 010 s	30	1	0000 0000 0001 1100 s
0	39	0000 0000 0010 001 s	31	1	0000 0000 0001 1011 s

Notes:
1. s = sign of level; "0" for positive; s = "1" for negative.

Table 12.18c. MPEG-1 Variable-Length Code Table for *dct_coefficient_first* and *dct_coefficient_next*.

Run	Level	Fixed Length Code
0		0000 00
1		0000 01
2		0000 10
:		:
63		1111 11
	−256	forbidden
	−255	1000 0000 0000 0001
	−254	1000 0000 0000 0010
	:	:
	−129	1000 0000 0111 1111
	−128	1000 0000 1000 0000
	−127	1000 0001
	−126	1000 0010
	:	:
	−2	1111 1110
	−1	1111 1111
	0	forbidden
	1	0000 0001
	:	:
	127	0111 1111
	128	0000 0000 1000 0000
	129	0000 0000 1000 0001
	:	:
	255	0000 0000 1111 1111

Table 12.19. Run, Level Encoding Following An Escape Code for *dct_coefficient_first* **and** *dct_coefficient_next.*

System Bitstream

The system bitstream multiplexes the audio and video bitstreams into a single bitstream, and formats it with control information into a specific protocol as defined by MPEG-1.

Packet data may contain either audio or video information. Up to 32 audio and 16 video streams may be multiplexed together. Two types of private data streams are also supported. One type is completely private; the other is used to support synchronization and buffer management.

Maximum packet sizes usually are about 2,048 bytes, although much larger sizes are supported. When stored on CDROM, the length of the packs coincides with the sectors. Typically, there is one audio packet for every six or seven video packets.

Figure 12.6 illustrates the system bitstream, a hierarchical structure with three layers. From top to bottom the layers are:

> ISO/IEC 11172 Layer
> Pack
> Packet

Note that start codes $(000001xx_H)$ must be byte aligned by inserting 0–7 "0" bits before the start code.

ISO/IEC 11172 Layer

ISO_11172_end_code
This 32-bit field has a value of $000001B9_H$ and terminates a system bitstream.

Pack Layer

Data for each pack consists of a pack header followed by a system header (optional) and packet data. The structure is shown in Figure 12.6.

Pack_start_code
This 32-bit field has a value of $000001BA_H$ and identifies the start of a pack.

Fixed_bits
These four bits always have a value of "0010."

System_clock_reference_32–30
The *system_clock_reference* (SCR) is a 33-bit number coded using three fields separated by marker bits.

system_clock_reference indicates the intended time of arrival of the last byte of the *system_clock_reference* field at the input of the decoder. The value of *system_clock_reference* is the number of 90 kHz clock periods.

Marker_bit
This bit always has a value of "1."

System_clock_reference_29–15

Marker_bit
This bit always has a value of "1."

System_clock_reference_14–0

Marker_bit
This bit always has a value of "1."

Marker_bit
This bit always has a value of "1."

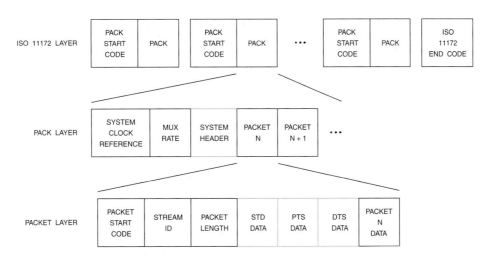

Figure 12.6. MPEG-1 System Bitstream Layer Structures. Marker and reserved bits not shown.

Mux_rate

This 22-bit binary number specifies the rate at which the decoder receives the bitstream. It specifies units of 50 bytes per second, rounded upwards. A value of zero is not allowed.

Marker_bit

This bit always has a value of "1."

System Header

System_header_start_code

This 32-bit field has a value of $000001BB_H$ and identifies the start of a system header.

Header_length

This 16-bit binary number specifies the number of bytes in the system header following *header_length*.

Marker_bit

This bit always has a value of "1."

Rate_bound

This 22-bit binary number specifies an integer value greater than or equal to the maximum value of *mux_rate*. It may be used by the decoder to determine if it is capable of decoding the entire bitstream.

Marker_bit

This bit always has a value of "1."

Audio_bound

This 6-bit binary number, with a range of 0–32, specifies an integer value greater than or equal to the maximum number of simultaneously active audio streams.

Fixed_flag

This bit specifies fixed bit rate ("1") or variable bit rate ("0") operation.

CSPS_flag

This bit specifies whether the bitstream is a constrained system parameter stream ("1") or not ("0").

System_audio_lock_flag

This bit has a value of "1" if there is a constant relationship between the audio sampling rate and the decoder's system clock frequency.

System_video_lock_flag

This bit has a value of "1" if there is a constant relationship between the video picture rate and the decoder's system clock frequency.

Marker_bit

This bit always has a value of "1."

Video_bound

This 5-bit binary number, with a range of 0–16, specifies an integer value greater than or equal to the maximum number of simultaneously active video streams.

Reserved_byte

These eight bits always have a value of "1111 1111."

Stream_ID

This optional 8-bit field, as defined in Table 12.20, indicates the type and stream number to which the following *STD_buffer_bound_scale* and *STD_buffer_size_bound* fields refer to. Each audio and video stream present in the system bitstream must be specified only once in each system header.

Fixed_bits

This optional 2-bit field has a value of "11." It is present only if *stream_ID* is present.

Stream Type	Stream ID
all audio streams	1011 1000
all video streams	1011 1001
reserved stream	1011 1100
private stream 1	1011 1101
padding stream	1011 1110
private stream 2	1011 1111
audio stream number xxxxx	110x xxxx
video stream number xxxx	1110 xxxx
reserved data stream number xxxx	1111 xxxx

Table 12.20. MPEG-1 *stream_ID* Code.

STD_buffer_bound_scale

This optional 1-bit field specifies the scaling factor used to interpret *STD_buffer_size_bound*. For an audio stream, it has a value of "0." For a video stream, it has a value of "1." For other stream types, it can be either a "0" or a "1." It is present only if *stream_ID* is present.

STD_buffer_size_bound

This optional 13-bit binary number specifies a value greater than or equal to the maximum decoder input buffer size. If *STD_buffer_bound_scale* = "0," then *STD_buffer_size_bound* measures the size in units of 128 bytes. If *STD_buffer_bound_scale* = "1," then *STD_buffer_size_bound* measures the size in units of 1024 bytes. It is present only if *stream_ID* is present.

Packet Layer

Packet_start_code_prefix

This 24-bit field has a value of 000001_H. Together with the *stream_ID* that follows, it indicates the start of a packet.

Stream_ID

This 8-bit binary number specifies the type and number of the bitstream present, as defined in Table 12.20.

Packet_length

This 16-bit binary number specifies the number of bytes in the packet after the *packet_length* field.

Stuffing_byte

This optional parameter has a value of "1111 1111." Up to 16 consecutive *stuffing_bytes* may used to meet the requirements of the storage medium. It is present only if *stream_ID* ≠ private stream 2.

STD_bits

These optional two bits have a value of "01" and indicate the *STD_buffer_scale* and *STD_buffer_size* fields follow. This field may be present only if *stream_ID* ≠ private stream 2.

STD_buffer_scale

This optional 1-bit field specifies the scaling factor used to interpret *STD_buffer_size*. For an audio stream, it has a value of "0." For a video stream, it has a value of "1." For other stream types, it can be either a "0" or a "1." This field is present only if *STD_bits* is present and *stream_ID* ≠ private stream 2.

STD_buffer_size

This optional 13-bit binary number specifies the size of the decoder input buffer. If *STD_buffer_scale* = "0," then *STD_buffer_size* measures the size in units of 128 bytes. If *STD_buffer_scale* = "1," then *STD_buffer_size* measures the size in units of 1024 bytes. This field is present only if *STD_bits* is present and *stream_ID* ≠ private stream 2.

PTS_bits

These optional four bits have a value of "0010" and indicate the following presentation time stamps are present. This field may be present only if *stream_ID* ≠ private stream 2.

Presentation_time_stamp_32–30

The optional *presentation_time_stamp* (PTS) is a 33-bit number coded using three fields, separated by marker bits. PTS indicates the intended time of display by the decoder. The value of PTS is the number of periods of a 90 kHz system clock. This field is present only if *PTS_bits* is present and *stream_ID* ≠ private stream 2.

Marker_bit

This optional bit always has a value of "1." It is present only if *PTS_bits* is present and *stream_ID* ≠ private stream 2.

Presentation_time_stamp_29–15

This optional field is present only if *PTS_bits* is present and *stream_ID* ≠ private stream 2.

Marker_bit

This optional bit always has a value of "1." It is present only if *PTS_bits* is present and *stream_ID* ≠ private stream 2.

Presentation_time_stamp_14–0

This optional field is present only if *PTS_bits* is present and *stream_ID* ≠ private stream 2.

Marker_bit

This optional 1-bit field always has a value of "1." It is present only if *PTS_bits* is present and *stream_ID* ≠ private stream 2.

DTS_bits

These optional four bits have a value of "0011" and indicate the following presentation and decoding time stamps are present. This field may be present only if *stream_ID* ≠ private stream 2.

Presentation_time_stamp_32–30

The optional *presentation_time_stamp* (PTS) is a 33-bit number coded using three fields, separated by marker bits. PTS indicates the intended time of display by the decoder. The value of PTS is the number of periods of a 90 kHz system clock. This field is present only if *DTS_bits* is present and *stream_ID* ≠ private stream 2.

Marker_bit

This optional 1-bit field always has a value of "1." It is present only if *DTS_bits* is present and *stream_ID* ≠ private stream 2.

Presentation_time_stamp_29–15

This optional field is present only if *DTS_bits* is present and *stream_ID* ≠ private stream 2.

Marker_bit

This optional 1-bit field always has a value of "1." It is present only if *DTS_bits* is present and *stream_ID* ≠ private stream 2.

Presentation_time_stamp_14–0

This optional field is present only if *DTS_bits* is present and *stream_ID* ≠ private stream 2.

Marker_bit

This optional 1-bit field always has a value of "1." It is present only if *DTS_bits* is present and *stream_ID* ≠ private stream 2.

Fixed_bits

This optional 4-bit field has a value of "0001." It is present only if *DTS_bits* is present and *stream_ID* ≠ private stream 2.

Decoding_time_stamp_32–30

The optional *decoding_time_stamp* (DTS) is a 33-bit number coded using three fields, separated by marker bits. DTS indicates the intended time of decoding by the decoder of the first access unit that commences in the packet. The value of DTS is the number of periods of a 90 kHz system clock. It is present only if *DTS_bits* is present and *stream_ID* ≠ private stream 2.

Marker_bit

This optional 1-bit field always has a value of "1." It is present only if *DTS_bits* is present and *stream_ID* ≠ private stream 2.

Decoding_time_stamp_29–15

This optional field is present only if *DTS_bits* is present and *stream_ID* ≠ private stream 2.

Marker_bit

This optional 1-bit field always has a value of "1." It is present only if *DTS_bits* is present and *stream_ID* ≠ private stream 2.

Decoding_time_stamp_14–0

This optional field is present only if *DTS_bits* is present and *stream_ID* ≠ private stream 2.

Marker_bit

This optional 1-bit field always has a value of "1." It is present only if *DTS_bits* is present and *stream_ID* ≠ private stream 2.

NonPTS_nonDTS_bits

These optional eight bits have a value of "0000 1111" and are present if the *STD_bits* field, *PTS_bits* field or, *DTS_bits* field (and their corresponding following fields) are not present.

Packet_data_byte

This is [n] bytes of data from the bitstream specified by the packet layer *stream_ID*. The number of data bytes may be determined from the *packet_length* parameter.

Video Decoding

A system demultiplexer parses the system bitstream, demultiplexing the audio and video bitstreams.

The video decoder essentially performs the inverse of the encoder. From the coded video bitstream, it reconstructs the I frames. Using I frames, additional coded data, and motion vectors, the P and B frames are generated. Finally, the frames are output in the proper order.

Fast Playback Considerations

Fast forward operation can be implemented by using D frames or the decoding only of I frames. However, decoding only I frames at the faster rate places a major burden on the transmission medium and the decoder.

Alternately, the source may be able to sort out the desired I frames and transmit just those frames, allowing the bit rate to remain constant.

Pause Mode Considerations

This requires the decoder to be able to control the incoming bitstream. If it doesn't, when playback resumes there may be a delay and skipped frames.

Reverse Playback Considerations

This requires the decoder to be able to decode each group of pictures in the forward direction, store them, and display them in reverse order. To minimize the storage requirements of the decoder, groups of pictures should be small or the frames may be reordered. Reordering can be done by transmitting frames in another order or by reordering the coded pictures in the decoder buffer.

Decode Postprocessing

The SIF data usually is converted to 720 × 480i or 720 × 576i. Suggested upsampling filters are discussed in the MPEG-1 specification. The original decoded lines correspond to field 1. Field 2 uses interpolated lines.

Real-World Issues

System Bitstream Termination

A common error is the improper placement of *sequence_end_code* in the system bitstream. When this happens, some decoders may not know that the end of the video occurred, and output garbage.

Another problem occurs when a system bitstream is shortened just by eliminating trailing frames, removing *sequence_end_code* altogether. In this case, the decoder may be unsure when to stop.

Timecodes

Since some decoders rely on the timecode information, it should be implemented. To minimize problems, the video bitstream should start with a timecode of zero and increment by one each frame.

Variable Bit Rates

Although variable bit rates are supported, a constant bit rate should be used if possible. Since *vbv_delay* doesn't make sense for a variable bit rate, the MPEG-1 standard specifies that it be set to the maximum value.

However, some decoders use *vbv_delay* with variable bit rates. This could result in a 2–3 second delay before starting video, causing the first 60–90 frames to be skipped.

Constrained Bitstreams

Most MPEG-1 decoders can handle only the constrained parameters subset of MPEG-1. To ensure maximum compatibility, only the constrained parameters subset should be used.

Source Sample Clock

Good compression with few artifacts requires a video source that generates or uses a very stable sample clock. This ensures that the vertical alignment of samples over the entire picture is maintained. With poorly designed sample clock generation, the artifacts usually get worse towards the right side of the picture.

References

1. Digital Video Magazine, "*Not All MPEGs Are Created Equal,*" by John Toebes, Doug Walker, and Paul Kaiser, August 1995.
2. Digital Video Magazine, "*Squeeze the Most From MPEG,*" by Mark Magel, August 1995.
3. ISO/IEC 11172–1, Coding of moving pictures and associated audio for digital storage media at up to about 1.5 Mbit/s, Part 1: Systems.
4. ISO/IEC 11172–2, Coding of moving pictures and associated audio for digital storage media at up to about 1.5 Mbit/s, Part 2: Video.
5. ISO/IEC 11172–3, Coding of moving pictures and associated audio for digital storage media at up to about 1.5 Mbit/s, Part 3: Audio.
6. ISO/IEC 11172–4, Coding of moving pictures and associated audio for digital storage media at up to about 1.5 Mbit/s, Part 4: Compliance testing.
7. ISO/IEC 11172–5, Coding of moving pictures and associated audio for digital storage media at up to about 1.5 Mbit/s, Part 5: Software simulation.
8. Watkinson, John, *The Engineer's Guide to Compression*, Snell and Wilcox Handbook Series.

Chapter 13

MPEG-2

MPEG-2 extends MPEG-1 to cover a wider range of applications. The MPEG-1 chapter should be reviewed to become familiar with the basics of MPEG before reading this chapter.

The primary application targeted during the definition process was all-digital transmission of broadcast-quality video at bit rates of 4–9 Mbps. However, MPEG-2 is useful for many other applications, such as HDTV, and now supports bit rates of 1.5–60 Mbps.

MPEG-2 is an ISO standard (ISO/IEC 13818), and consists of eleven parts:

systems	ISO/IEC 13818–1
video	ISO/IEC 13818–2
audio	ISO/IEC 13818–3
conformance testing	ISO/IEC 13818–4
software simulation	ISO/IEC 13818–5
DSM-CC extensions	ISO/IEC 13818–6
advanced audio coding	ISO/IEC 13818–7
RTI extension	ISO/IEC 13818–9
DSM-CC conformance	ISO/IEC 13818–10
IPMP	ISO/IEC 13818–11

As with MPEG-1, the compressed bit-streams implicitly define the decompression algorithms. The compression algorithms are up to the individual manufacturers, within the scope of an international standard.

The Digital Storage Media Command and Control (DSM-CC) extension (ISO/IEC 13818–6) is a toolkit for developing control channels associated with MPEG-2 streams. In addition to providing VCR-type features such as fast-forward, rewind, pause, etc., it may be used for a wide variety of other purposes, such as packet data transport. DSM-CC works in conjunction with next-generation packet networks, working alongside Internet protocols as RSVP, RTSP, RTP and SCP.

The Real Time Interface (RTI) extension (ISO/IEC 13818-9) defines a common interface point to which terminal equipment manufacturers and network operators can design. RTI specifies a delivery model for the bytes of a MPEG-2 System stream at the input of a real decoder, whereas MPEG-2 System defines an idealized byte delivery schedule.

IPMP (Intellectual Property Management and Protection) is a digital rights management (DRM) standard, adapted from the MPEG-4 IPMP extension specification. Rather than a complete system, a variety of functions are provided within a framework.

Audio Overview

In addition to the non-backwards-compatible audio extension (ISO/IEC 13818–7), MPEG-2 supports up to five full-bandwidth channels compatible with MPEG-1 audio coding. It also extends the coding of MPEG-1 audio to half sampling rates (16 kHz, 22.05 kHz, and 24 kHz) for improved quality for bit rates at or below 64 kbps per channel.

MPEG-2.5 is an unofficial, yet common, extension to the audio capabilities of MPEG-2. It adds sampling rates of 8 kHz, 11.025 kHz, and 12 kHz.

Video Overview

With MPEG-2, *profiles* specify the syntax (i.e., algorithms) and *levels* specify various parameters (resolution, frame rate, bit rate, etc.). Main Profile@Main Level is targeted for SDTV applications, while Main Profile@High Level is targeted for HDTV applications.

Levels

MPEG-2 supports four levels, which specify resolution, frame rate, coded bit rate, and so on for a given profile.

Low Level (LL)

MPEG-1 Constrained Parameters Bitstream (CPB), supporting up to 352×288 at up to 30 frames per second. Maximum bit rate is 4 Mbps.

Main Level (ML)

MPEG-2 Constrained Parameters Bitstream (CPB) supports up to 720×576 at up to 30 frames per second and is intended for SDTV applications. Maximum bit rate is 15–20 Mbps.

High 1440 Level

This Level supports up to 1440×1088 at up to 60 frames per second and is intended for HDTV applications. Maximum bit rate is 60–80 Mbps.

High Level (HL)

High Level supports up to 1920×1088 at up to 60 frames per second and is intended for HDTV applications. Maximum bit rate is 80–100 Mbps.

Profiles

MPEG-2 supports six profiles, which specify which coding syntax (algorithms) is used. Tables 13.1 through 13.8 illustrate the various combinations of levels and profiles allowed.

Simple Profile (SP)

Main profile without the B frames, intended for software applications and perhaps digital cable TV.

Main Profile (MP)

Supported by most MPEG-2 decoder chips, it should satisfy 90% of the consumer SDTV and HDTV applications. Typical resolutions are shown in Table 13.6.

Multiview Profile (MVP)

By using existing MPEG-2 tools, it is possible to encode video from two cameras shooting the same scene with a small angle difference.

4:2:2 Profile (422P)

Previously known as "studio profile," this profile uses 4:2:2 YCbCr instead of 4:2:0, and with main level, increases the maximum bit rate up to 50 Mbps (300 Mbps with high level). It was added to support pro-video SDTV and HDTV requirements.

Level	Profile						
	Nonscalable				Scalable		
	Simple	Main	Multiview	4:2:2	SNR	Spatial	High
High	–	yes	–	yes	–	–	yes
High 1440	–	yes	–	–	–	yes	yes
Main	yes	yes	yes	yes	yes	–	yes
Low	–	yes	–	–	yes	–	–

Table 13.1. MPEG-2 Acceptable Combinations of Levels and Profiles.

Constraint	Profile						
	Nonscalable				Scalable		
	Simple	Main	Multiview	4:2:2	SNR	Spatial	High
chroma format	4:2:0	4:2:0	4:2:0	4:2:0 or 4:2:2	4:2:0	4:2:0	4:2:0 or 4:2:2
picture types	I, P	I, P, B	I, P, B	I, P, B	I, P, B	I, P, B	I, P, B
scalable modes	–	–	Temporal	–	SNR	SNR or Spatial	SNR or Spatial
intra dc precision (bits)	8, 9, 10	8, 9, 10	8, 9, 10	8, 9, 10, 11	8, 9, 10	8, 9, 10	8, 9, 10, 11
sequence scalable extension	no	no	yes	no	yes	yes	yes
picture spatial scalable extension	no	no	no	no	no	yes	yes
picture temporal scalable extension	no	no	yes	no	no	no	no
repeat first field	constrained		unconstrained	constrained	unconstrained		

Table 13.2. Some MPEG-2 Profile Constraints.

Level	Maximum Number of Layers	SNR	Spatial	High	Multiview
High	All layers (base + enhancement)	–	–	3	2
	Spatial enhancement layers			1	0
	SNR enhancement layers			1	0
	Temporal auxiliary layers			0	1
High 1440	All layers (base + enhancement)	–	3	3	2
	Spatial enhancement layers		1	1	0
	SNR enhancement layers		1	1	0
	Temporal auxiliary layers		0	0	1
Main	All layers (base + enhancement)	2	–	3	2
	Spatial enhancement layers	0		1	0
	SNR enhancement layers	1		1	0
	Temporal auxiliary layers	0		0	1
Low	All layers (base + enhancement)	2	–	–	2
	Spatial enhancement layers	0			0
	SNR enhancement layers	1			0
	Temporal auxiliary layers	0			1

Table 13.3. MPEG-2 Number of Permissible Layers for Scalable Profiles.

Profile	Base Layer	Enhancement Layer 1	Enhancement Layer 2	Profile at Level for Base Decoder
SNR	4:2:0	SNR, 4:2:0	–	MP@ same level
Spatial	4:2:0	SNR, 4:2:0	–	MP@same level
	4:2:0	Spatial, 4:2:0	–	MP@ (level–1)
	4:2:0	SNR, 4:2:0	Spatial, 4:2:0	
	4:2:0	Spatial, 4:2:0	SNR, 4:2:0	
High	4:2:0 or 4:2:2	–	–	HP@same level
	4:2:0	SNR, 4:2:0	–	
	4:2:0 or 4:2:2	SNR, 4:2:2	–	
	4:2:0	Spatial, 4:2:0	–	HP@ (level–1)
	4:2:0 or 4:2:2	Spatial, 4:2:2	–	
	4:2:0	SNR, 4:2:0	Spatial, 4:2:0 or 4:2:2	
	4:2:0 or 4:2:2	SNR, 4:2:2	Spatial, 4:2:2	
	4:2:0	Spatial, 4:2:0	SNR, 4:2:0 or 4:2:2	
	4:2:0	Spatial, 4:2:2	SNR, 4:2:2	
	4:2:2	Spatial, 4:2:2	SNR, 4:2:2	
Multiview	4:2:0	Temporal, 4:2:0	–	MP@same level

Table 13.4. Some MPEG-2 Video Decoder Requirements for Various Profiles.

Level	Spatial Resolution Layer	Parameter	Simple	Main	Multiview	4:2:2	SNR / Spatial	High
High	Enhancement	Samples per line Lines per frame Frames per second	–	1920 1088 60	1920 1088 60	1920 1088 60	–	1920 1088 60
High	Lower	Samples per line Lines per frame Frames per second	–	–	1920 1088 60	.	–	960 576 30
High 1440	Enhancement	Samples per line Lines per frame Frames per second	–	1440 1088 60	1440 1088 60	–	1440 1088 60	1440 1088 60
High 1440	Lower	Samples per line Lines per frame Frames per second	–	–	1440 1088 60	–	720 576 30	720 576 30
Main	Enhancement	Samples per line Lines per frame Frames per second	720 576 30	720 576 30	720 576 30	720 608 30	720 576 30	720 576 30
Main	Lower	Samples per line Lines per frame Frames per second	–	–	720 576 30	–	–	352 288 30
Low	Enhancement	Samples per line Lines per frame Frames per second	–	352 288 30	352 288 30	–	352 288 30	–
Low	Lower	Samples per line Lines per frame Frames per second	–	–	352 288 30	–	–	–

Notes:

1. The above levels and profiles that originally specified 1152 maximum lines per frame were changed to 1088 lines per frame.

Table 13.5. MPEG-2 Upper Limits of Resolution and Temporal Parameters. In the case of single layer or SNR scalability coding, the "Enhancement Layer" parameters apply.

Level	Maximum Bit Rate (Mbps)	Typical Active Resolutions	Frame Rate (Hz)[2]										
			23.976p	24p	25p	29.97p	30p	50p	59.94p	60p	25i	29.97i	30i
High	80 (100 for High Profile) (300 for 4:2:2 Profile)	1920×1080^1	×	×	×	×	×				×	×	×
High 1440	60 (80 for High Profile)	1280×720	×	×	×	×	×	×	×	×			
		1440×1080^1	×	×	×	×	×				×	×	×
Main	15 (20 for High Profile) (50 for 4:2:2 Profile)	352×480	×	×		×	×		×	×		×	×
		352×576		×	×			×			×		
		544×480	×	×		×	×		×	×		×	×
		544×576		×	×			×			×		
		640×480	×	×		×	×		×	×		×	×
		$704 \times 480, 720 \times 480$	×	×		×	×		×	×		×	×
		$704 \times 576, 720 \times 576$		×	×			×			×		
Low	4	320×240	×	×		×	×		×	×		×	×
		352×240	×	×		×	×		×	×		×	×
		352×288		×	×			×			×		

Notes:

1. The video coding system requires that the number of active scan lines be a multiple of 32 for interlaced pictures, and a multiple of 16 for progressive pictures. Thus, for the 1080-line interlaced format, the video encoder and decoder must actually use 1088 lines. The extra eight lines are "dummy" lines having no content, and designers choose dummy data that simplifies the implementation. The extra eight lines are always the last eight lines of the encoded image. These dummy lines do not carry useful information, but add little to the data required for transmission.

2. p = progressive; i = interlaced.

Table 13.6. Example Levels and Resolutions for MPEG-2 Main Profile.

Level	Spatial Resolution Layer	Profile					
		Simple	**Main**	**Multiview**	**SNR / Spatial**	**High**	**4:2:2**
High	Enhancement	–	62.668800	62.668800	–	62.668800 (4:2:2) 83.558400 (4:2:0)	62.668800
	Lower	–	–	62.668800	–	14.745600 (4:2:2) 19.660800 (4:2:0)	–
High 1440	Enhancement	–	47.001600	47.001600	47.001600	47.001600 (4:2:2) 62.668800 (4:2:0)	–
	Lower	–	–	47.001600	10.368000	11.059200 (4:2:2) 14.745600 (4:2:0)	–
Main	Enhancement	10.368000	10.368000	10.368000	10.368000	11.059200 (4:2:2) 14.745600 (4:2:0)	11.059200
	Lower	–	–	10.368000	–	3.041280 (4:2:0)	–
Low	Enhancement	–	3.041280	3.041280	3.041280	–	–
	Lower	–	–	3.041280	–	–	–

Table 13.7. MPEG-2 Upper Limits for Y Sample Rate (Msamples/second). In the case of single layer or SNR scalability coding, the "Enhancement Layer" parameters apply.

Level	Profile					
	Nonscalable				**Scalable**	
	Simple	**Main**	**Multiview**	**4:2:2**	**SNR/Spatial**	**High**
High	–	80	130 (both layers) 80 (base layer)	300	–	100 (all layers) 80 (middle + base layers) 25 (base layer)
High 1440	–	60	100 (both layers) 60 (base layer)	–	60 (all layers) 40 (middle + base layers) 15 (base layer)	80 (all layers) 60 (middle + base layers) 20 (base layer)
Main	15	15	25 (both layers) 5 (base layer)	50	15 (both layers) 10 (base layer)	20 (all layers) 15 (middle + base layers) 4 (base layer)
Low	–	4	8 (both layers) 4 (base layer)	–	4 (both layers) 3 (base layer)	–

Table 13.8.   MPEG-2 Upper Limits for Bit Rates (Mbps).

SNR and Spatial Profiles

Adds support for SNR scalability and/or spatial scalability.

High Profile (HP)

Targeted for pro-video HDTV applications.

Scalability

The MPEG-2 SNR, Spatial, and High profiles support four scalable modes of operation. These modes break MPEG-2 video into layers for the purpose of prioritizing video data. Scalability is not commonly used since efficiency decreases by about 2 dB (or about 30% more bits are required).

SNR Scalability

This mode is targeted for applications that desire multiple quality levels. All layers have the same spatial resolution. The base layer provides the basic video quality. The enhancement layer increases the video quality by providing refinement data for the DCT coefficients of the base layer.

Spatial Scalability

Useful for simulcasting, each layer has a different spatial resolution. The base layer provides the basic spatial resolution and temporal rate. The enhancement layer uses the spatially interpolated base layer to increase the spatial resolution. For example, the base layer may implement 352×240 resolution video, with the enhancement layers used to generate 704×480 resolution video.

Temporal Scalability

This mode allows migration from low temporal rate to higher temporal rate systems. The base layer provides the basic temporal rate. The enhancement layer uses temporal prediction relative to the base layer. The base and enhancement layers can be combined to produce a full temporal rate output. All layers have the same spatial resolution and chroma formats. In case of errors in the enhancement layers, the base layer can be used for concealment.

Data Partitioning

This mode is targeted for cell loss resilience in ATM networks. It breaks the 64 quantized transform coefficients into two bitstreams. The higher priority bitstream contains critical lower-frequency DCT coefficients and side information such as headers and motion vectors. A lower-priority bitstream carries higher-frequency DCT coefficients that add detail.

Transport and Program Streams

The MPEG-2 Systems Standard specifies two methods for multiplexing the audio, video, and other data into a format suitable for transmission and storage.

The *program stream* is designed for applications where errors are unlikely. It contains audio, video, and data bitstreams (also called *elementary* bitstreams) all merged into a single bitstream. The program stream, as well as each of the elementary bitstreams, may be a fixed or variable bit rate. DVDs and SVCDs use program streams, carrying the DVD- and SVCD-specific data in private data streams interleaved with the video and audio streams.

The *transport stream*, using fixed-size packets of 188 bytes, is designed for applications where data loss is likely. Also containing audio, video, and data bitstreams all merged into a single bitstream, multiple programs can be carried. The ARIB, ATSC, DVB and Open-Cable™ standards use transport streams.

Both the *transport stream* and *program stream* are based on a common packet structure, facilitating common decoder implementations and conversions. Both streams are designed to support a large number of known and anticipated applications, while retaining flexibility.

Video Coding Layer

YCbCr Color Space

MPEG-2 uses the YCbCr color space, supporting 4:2:0, 4:2:2 and 4:4:4 sampling. The 4:2:2 and 4:4:4 sampling options increase the chroma resolution over 4:2:0, resulting in better picture quality.

The 4:2:0 sampling structure for MPEG-2 is shown in Figures 3.8 through 3.10. The 4:2:2 and 4:4:4 sampling structures are shown in Figures 3.2 and 3.3.

Coded Picture Types

There are three types of coded pictures. I (intra) pictures are fields or frames coded as a stand-alone still image. They allow random access points within the video stream. As such, I pictures should occur about two times a second. I pictures also should be used where scene cuts occur.

P (predicted) pictures are fields or frames coded relative to the nearest previous I or P picture, resulting in forward prediction processing, as shown in Figure 13.1. P pictures provide more compression than I pictures, through the use of motion compensation, and are also a reference for B pictures and future P pictures.

B (bidirectional) pictures are fields or frames that use the closest past and future I or P picture as a reference, resulting in bidirectional prediction, as shown in Figure 13.1. B pictures provide the most compression. and decrease noise by averaging two pictures. Typically, there are two B pictures separating I or P pictures.

D (DC) pictures are not supported in MPEG-2, except for decoding to support backwards compatibility with MPEG-1.

A group of pictures (GOP) is a series of one or more coded pictures intended to assist in random accessing and editing. The GOP value is configurable during the encoding process. The smaller the GOP value, the better the response to movement (since the I pictures are closer together), but the lower the compression.

In the coded bitstream, a GOP must start with an I picture and may be followed by any number of I, P, or B pictures in any order. In display order, a GOP must start with an I or B picture and end with an I or P picture. Thus, the smallest GOP size is a single I picture, with the largest size unlimited.

Each GOP should be coded independently of any other GOP. However, this is not true unless no B pictures precede the first I picture, or if they do, they use only backward motion compensation. This results in both open and closed GOP formats. A *closed GOP* is a GOP that can be decoded without using pictures of the previous GOP for motion compensation. An *open GOP*, identified by the *broken_link* flag, indicates that the first B pictures (if any) immediately following the first I picture after the

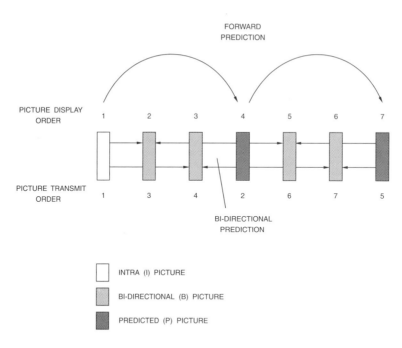

FORWARD
PREDICTION

PICTURE DISPLAY
ORDER

1   2   3   4   5   6   7

PICTURE TRANSMIT
ORDER

1   3   4   2   6   7   5

BI-DIRECTIONAL
PREDICTION

INTRA (I) PICTURE

BI-DIRECTIONAL (B) PICTURE

PREDICTED (P) PICTURE

Figure 13.1. MPEG-2 I, P, and B Pictures. Some pictures can be transmitted out of sequence, complicating the interpolation process and requiring picture reordering by the MPEG decoder. Arrows show inter-frame dependencies.

GOP header may not be decoded correctly (and thus not be displayed) since the reference picture used for prediction is not available due to editing.

Motion Compensation

Motion compensation for MPEG-2 is more complex due to the introduction of fields. After a macroblock has been compressed using motion compensation, it contains both the spatial difference (motion vectors) and content difference (error terms) between the reference macroblock and macroblock being coded.

The two major classifications of prediction are *field* and *frame*. Within field pictures, only field predictions are used. Within frame pic-

tures, either field or frame predictions can be used (selectable at the macroblock level).

Motion vectors for MPEG-2 are always coded in half-pixel units. MPEG-1 supports either half-pixel or full-pixel units.

16 × 8 Motion Compensation Option

Two motion vectors (four for B pictures) per macroblock are used, one for the upper 16 × 8 region of a macroblock and one for the lower 16 × 8 region of a macroblock. It is only used with field pictures.

Dual-Prime Motion Compensation Option

This is only used with P pictures that have no B pictures between the predicted and reference fields of frames. One motion vector is used, together with a small differential motion vector. All of the necessary predictions are derived from these.

Macroblocks

Three types of macroblocks are available in MPEG-2.

The 4:2:0 macroblock (Figure 13.2) consists of four Y blocks, one Cb block, and one Cr block. The block ordering is shown in the figure.

The 4:2:2 macroblock (Figure 13.3) consists of four Y blocks, two Cb blocks, and two Cr blocks. The block ordering is shown in the figure.

The 4:4:4 macroblock (Figure 13.4) consists of four Y blocks, four Cb blocks, and four Cr blocks. The block ordering is shown in the figure.

Macroblocks in P pictures are coded using the closest previous I or P picture as a reference, resulting in two possible codings:

- intra coding
    no motion compensation
- forward prediction
    closest previous I or P picture is the reference

Macroblocks in B pictures are coded using the closest previous and/or future I or P picture as a reference, resulting in four possible codings:

- intra coding
    no motion compensation
- forward prediction
    closest previous I or P picture is the reference
- backward prediction
    closest future I or P picture is the reference

- bi-directional prediction
    two pictures used as the reference:
    the closest previous I or P picture and
    the closest future I or P picture

I Pictures

Macroblocks

There are ten types of macroblocks in I pictures, as shown in Table 13.28.

If the [macroblock quant] column in Table 13.28 has a "1," the quantizer scale is transmitted. For the remaining macroblock types, the DCT correction is coded using the previous value for quantizer scale.

If the [coded pattern] column in Table 13.28 has a "1," the 6-bit coded block pattern is transmitted as a variable-length code. This tells the decoder which of the six blocks in the 4:2:0 macroblock are coded ("1") and which are not coded ("0"). Table 13.33 lists the codewords assigned to the 63 possible combinations. There is no code for when none of the blocks are coded; it is indicated by the macroblock type. For 4:2:2 and 4:4:4 macroblocks, an additional two or six bits, respectively, are used to extend the coded block pattern.

DCT

Each 8×8 block (of input samples or prediction error terms) is processed by an 8×8 DCT (discrete cosine transform), resulting in an 8×8 block of horizontal and vertical frequency coefficients, as shown in Figure 7.56.

Input sample values are 0–255, resulting in a range of 0–2,040 for the DC coefficient and a range of about –2,048 to 2,047 for the AC coefficients.

Due to spatial and SNR scalability, non-intra blocks (blocks within a non-intra macroblock) are also possible. Non-intra block coefficients represent differences between sample

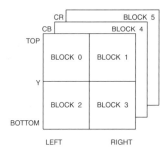

Figure 13.2. MPEG-2 4:2:0 Macroblock Structure.

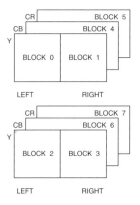

Figure 13.3. MPEG-2 4:2:2 Macroblock Structure.

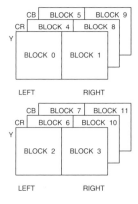

Figure 13.4. MPEG-2 4:4:4 Macroblock Structure.

values rather than actual sample values. They are obtained by subtracting the motion-compensated values from the previous picture from the values in the current macroblock.

Quantizing

The 8 × 8 block of frequency coefficients are uniformly quantized, limiting the number of allowed values. The quantizer step scale is derived from the quantization matrix and quantizer scale and may be different for different coefficients and may change between macroblocks.

Since the eye is sensitive to large luma areas, the quantizer step size of the DC coefficient is selectable to 8, 9, 10, or 11 bits of precision. The quantized DC coefficient is determined by dividing the DC coefficient by 8, 4, 2, or 1 and rounding to the nearest integer.

AC coefficients are quantized using two quantization matrices: one for intra macroblocks and one for non-intra macroblocks. When using 4:2:2 or 4:4:4 data, different matrices may be used for Y and CbCr data. Each quantization matrix has a default set of values that may be overwritten.

If the [macroblock quant] column in Table 13.28 has a "1," the quantizer scale is transmitted. For the remaining macroblock types, the DCT correction is coded using the previous value for quantizer scale.

Zig-Zag Scan

Zig-zag scanning, starting with the DC component, generates a linear stream of quantized frequency coefficients arranged in order of increasing frequency, as shown in Figures 7.59 and 7.60. This produces long runs of zero coefficients.

Coding of Quantized DC Coefficients

After the DC coefficients have been quantized, they are losslessly coded.

Coding of Y blocks within a macroblock follows the order shown in Figures 13.2 through 13.4. The DC value of block 4 is the DC predictor for block 1 of the next macroblock. At the beginning of each slice, whenever a macroblock is skipped, or whenever a non-intra macroblock is decoded, the DC predictor is set to 128 (if 8 bits of DC precision), 256 (if 9 bits of DC precision), 512 (if 10 bits of DC precision), or 1,024 (if 11 bits of DC precision).

The DC values of each Cb and Cr block are coded using the DC value of the corresponding block of the previous macroblock as a predictor. At the beginning of each slice, whenever a macroblock is skipped, or whenever a non-intra block is decoded, the DC predictors are set to 128 (8 bits of DC precision), 256 (9 bits of DC precision), 512 (10 bits of DC precision), or 1,024 (11 bits of DC precision).

However, a common implementation is to reset the DC predictors to zero and center the intra-block DC terms about zero instead of the 50% grey level. Decoders then only have to handle the different intra DC precisions in the quantizer (which already has a multiplier that can be used to reconstruct the right value) instead of the parser (which generally doesn't touch that data and has no multiplier).

Coding of Quantized AC Coefficients

After the AC coefficients have been quantized, they are scanned in the order shown in Figure 7.59 or 7.60 and coded using run-length and level. The scan starts in position 1, as shown in Figures 7.59 and 7.60, as the DC coefficient in position 0 is coded separately.

The run-lengths and levels are coded as shown in Tables 13.37 and 13.38. The "s" bit denotes the sign of the level; "0" is positive and "1" is negative. For intra blocks, either Table 13.37 or Table 13.38 may be used, as specified by *intra_vlc_format* in the bitstream. For non-intra blocks, only Table 13.37 is used.

For run-level combinations not shown in Tables 13.37 and 13.38, an escape sequence is used, consisting of the escape code (ESC), followed by the run-length and level codes from Tables 13.39 and 13.40.

After the last DCT coefficient has been coded, an EOB code is added to tell the decoder that there are no more quantized coefficients in this 8×8 block.

P Pictures

Macroblocks

There are 26 types of macroblocks in P pictures, as shown in Table 13.29, due to the additional complexity of motion compensation.

Skipped macroblocks are present when the *macroblock_address_increment* parameter in the bitstream is greater than 1. For P field pictures, the decoder predicts from the field of the same parity as the field being predicted, motion vector predictors are set to 0, and the motion vector is set to 0. For P frame pictures, the decoder sets the motion vector predictors to 0, and the motion vector is set to 0.

If the [macroblock quant] column in Table 13.29 has a "1," the quantizer scale is transmitted. For the remaining macroblock types, the DCT correction is coded using the previous value for quantizer scale.

If the [motion forward] column in Table 13.29 has a "1," horizontal and vertical forward motion vectors are successively transmitted.

If the [coded pattern] column in Table 13.29 has a "1," the 6-bit coded block pattern is transmitted as a variable-length code. This tells the decoder which of the six blocks in the macroblock are coded ("1") and which are not coded ("0"). Table 13.33 lists the codewords assigned to the 63 possible combinations. There is no code for when none of the blocks are coded; it is indicated by the macroblock type. For intra-coded macroblocks in P and B

pictures, the coded block pattern is not transmitted, but is assumed to be a value of 63 (all blocks are coded). For 4:2:2 and 4:4:4 macroblocks, an additional two or six bits, respectively, are used to extend the coded block pattern.

DCT

Intra block AC coefficients are transformed in the same manner as they are for I pictures. Intra block DC coefficients are transformed differently; the predicted values are set to 1,024, unless the previous block was intra-coded.

Non-intra block coefficients represent differences between sample values rather than actual sample values. They are obtained by subtracting the motion compensated values of the previous picture from the values in the current macroblock. There is no prediction of the DC value.

Input sample values are –255 to +255, resulting in a range of about –2,000 to +2,000 for the AC coefficients.

Quantizing

Intra blocks are quantized in the same manner as they are for I pictures.

Non-intra blocks are quantized using the quantizer scale and the non-intra quantization matrix. The AC and DC coefficients are quantized in the same manner.

Coding of Intra Blocks

Intra blocks are coded the same way as I picture intra blocks. There is a difference in the handling of the DC coefficients in that the predicted value is 128, unless the previous block was intra coded.

Coding of Non-Intra Blocks

The coded block pattern (CBP) is used to specify which blocks have coefficient data.

These are coded similarly to the coding of intra blocks, except the DC coefficient is coded in the same manner as the AC coefficients.

B Pictures

Macroblocks

There are 34 types of macroblocks in B pictures, as shown in Table 13.30, due to the additional complexity of backward motion compensation.

For B field pictures, the decoder predicts from the field of the same parity as the field being predicted. The direction of prediction (forward, backward, or bidirectional) is the same as the previous macroblock, motion vector predictors are unaffected, and the motion vectors are taken from the appropriate motion vector predictors. For B frame pictures, the direction of prediction (forward, backward, or bidirectional) is the same as the previous macroblock, motion vector predictors are unaffected, and the motion vectors are taken from the appropriate motion vector predictors.

If the [macroblock quant] column in Table 13.30 has a "1," the quantizer scale is transmitted. For the rest of the macroblock types, the DCT correction is coded using the previous value for the quantizer scale.

If the [motion forward] column in Table 13.30 has a "1," horizontal and vertical forward motion vectors are successively transmitted. If the [motion backward] column in Table 13.30 has a "1," horizontal and vertical backward motion vectors are successively transmitted. If both forward and backward motion types are present, the vectors are transmitted in this order:

horizontal forward
vertical forward
horizontal backward
vertical backward

If the [coded pattern] column in Table 13.30 has a "1," the 6-bit coded block pattern is transmitted as a variable-length code. This tells the decoder which of the six blocks in the macroblock are coded ("1") and which are not coded ("0"). Table 13.33 lists the codewords assigned to the 63 possible combinations. There is no code for when none of the blocks are coded; this is indicated by the macroblock type. For intra-coded macroblocks in P and B pictures, the coded block pattern is not transmitted, but is assumed to be a value of 63 (all blocks are coded). For 4:2:2 and 4:4:4 macroblocks, an additional two or six bits respectively, are used to extend the coded block pattern.

Coding

DCT coefficients of blocks are transformed into quantized coefficients and coded in the same way they are for P pictures.

Video Bitstream

Figure 13.5 illustrates the video bitstream, a hierarchical structure with seven layers. From top to bottom the layers are:

Video Sequence
Sequence Header
Group of Pictures (GOP)
Picture
Slice
Macroblock (MB)
Block

Several extensions may be used to support various levels of capability. These extensions are:

Sequence Extension
Sequence Display Extension
Sequence Scalable Extension

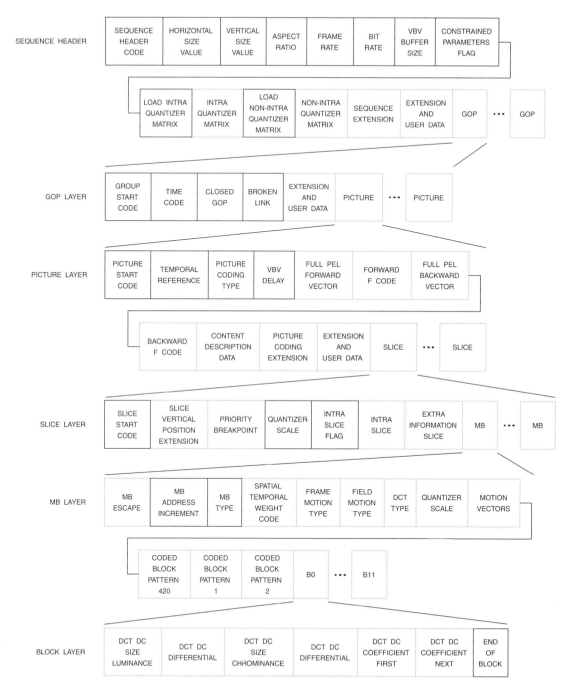

Figure 13.5. MPEG-2 Video Bitstream Layer Structures. Marker and reserved bits not shown.

Picture Coding Extension
Quant Matrix Extension
Picture Display Extension
Picture Temporal Scalable Extension
Picture Spatial Scalable Extension

If the first sequence header of a video sequence is not followed by an extension start code ($000001B5_H$), then the video bitstream must conform to the MPEG-1 video bitstream.

For MPEG-2 video bitstreams, an extension start code ($000001B5_H$) and a sequence extension must follow each sequence header.

Note that start codes ($000001xx_H$) must be byte aligned by inserting 0–7 "0" bits before the start code.

Video Sequence

Sequence_end_code

This 32-bit field has a value of $000001B7_H$ and terminates a video sequence.

Sequence Header

A sequence header should occur about every one-half second. The structure is shown in Figure 13.5. If not followed by a sequence extension, the bitstream conforms to MPEG-1.

Sequence_header_code

This 32-bit string has a value of $000001B3_H$ and indicates the beginning of a sequence header.

Horizontal_size_value

This is the twelve least significant bits of the width (in samples) of the viewable portion of the Y component. The two most significant bits of the 14-bit value are specified in the *horizontal_size_extension*. A value of zero is not allowed.

Vertical_size_value

This is the twelve least significant bits of the height (in scan lines) of the viewable portion of the Y component. The two most significant bits of the 14-bit value are specified in the *vertical_size_extension*. A value of zero is not allowed.

Aspect_ratio_information

This 4-bit codeword indicates either the sample aspect ratio (SAR) or display aspect ratio (DAR) as shown in Table 13.9.

If *sequence_display_extension* is not present, the SAR is determined as follows:

$$SAR = DAR \times (horizontal\_size / vertical\_size)$$

If *sequence_display_extension* is present, the SAR is determined as follows:

$$SAR = DAR \times (display\_horizontal\_size / display\_vertical\_size)$$

Frame_rate_code

This 4-bit codeword indicates the frame rate, as shown in Table 13.10.

The actual frame rate is determined as follows:

$$frame\_rate = frame\_rate\_value \times (frame\_rate\_extension\_n + 1) / (frame\_rate\_extension\_d + 1)$$

When an entry is specified in Table 13.10, both *frame_rate_extension_n* and *frame_rate_extension_d* are "00." If *progressive_sequence* is "1," the time between two frames at the output of the decoder is the

SAR	DAR	Code
forbidden	forbidden	0000
1.0000	–	0001
–	3/4	0010
–	9/16	0011
–	1/2.21	0100
–		0101
–		0110
–		0111
–		1000
–		1001
–	reserved	1010
–		1011
–		1100
–		1101
–		1110
–		1111

Table 13.9. MPEG-2 *aspect_ratio_information* Codewords.

Frames Per Second	Code
forbidden	0000
24/1.001	0001
24	0010
25	0011
30/1.001	0100
30	0101
50	0110
60/1.001	0111
60	1000
reserved	1001
reserved	1010
reserved	1011
reserved	1100
reserved	1101
reserved	1110
reserved	1111

Table 13.10. MPEG-2 *frame_rate_code* Codewords.

reciprocal of the *frame_rate*. If *progressive_sequence* is "0," the time between two frames at the output of the decoder is one-half of the reciprocal of the *frame_rate*.

Bit_rate_value

The eighteen least significant bits of a 30-bit binary number. The twelve most significant bits are in the *bit_rate_extension*. This specifies the bitstream bit rate, measured in units of 400 bps, rounded upwards. A zero value is not allowed. For the ATSC standard, the value must be $\leq 48500_D$ ($\leq 97000_D$ for high data rate mode). For the OpenCable™ standard, the value must be $\leq 67500_D$ for 64QAM systems ($\leq 97000_D$ for 256QAM systems).

Marker_bit

Always a "1."

Vbv_buffer_size_value

The ten least significant bits of a 18-bit binary number. The eight most significant bits are in the *vbv_buffer_size_extension*. Defines the size of the Video Buffering Verifier needed to decode the sequence. It is defined as:

$$B = 16 \times 1024 \times vbv\_buffer\_size$$

For the ATSC and OpenCable™ standards, the value must be $\leq 488_D$.

Constrained_parameters_flag

This bit is set to a "0" since it has no meaning for MPEG-2.

Load_intra_quantizer_matrix

This bit is set to a "1" if an *intra_quantizer_matrix* follows. If set to a "0," the default values below are used for intra blocks (both Y and CbCr) until the next occurrence of a sequence header or *quant_matrix_extension*.

8	16	19	22	26	27	29	34
16	16	22	24	27	29	34	37
19	22	26	27	29	34	34	38
22	22	26	27	29	34	37	40
22	26	27	29	32	35	40	48
26	27	29	32	35	40	48	58
26	27	29	34	38	46	56	69
27	29	35	38	46	56	69	83

Intra_quantizer_matrix

An optional list of sixty-four 8-bit values that replace the current values. A value of zero is not allowed. The value for intra_quant [0, 0] is always 8. These values take effect until the next occurrence of a sequence header or *quant_matrix_extension*. For 4:2:2 and 4:4:4 data formats, the new values are used for both the Y and CbCr intra matrix, unless a different CbCr intra matrix is loaded.

Load_non_intra_quantizer_matrix

This bit is set to a "1" if a *non_intra_quantizer_matrix* follows. If set to a "0," the default values below are used for non-intra blocks (both Y and CbCr) until the next occurrence of a sequence header or *quant_matrix_extension*.

16	16	16	16	16	16	16	16
16	16	16	16	16	16	16	16
16	16	16	16	16	16	16	16
16	16	16	16	16	16	16	16
16	16	16	16	16	16	16	16
16	16	16	16	16	16	16	16
16	16	16	16	16	16	16	16
16	16	16	16	16	16	16	16

Non_intra_quantizer_matrix

An optional list of sixty-four 8-bit values that replace the current values. A value of zero is not allowed. These values take effect until the next occurrence of a sequence header or *quant_matrix_extension*. For 4:2:2 and 4:4:4

data formats, the new values are used for both Y and CbCr non-intra matrix, unless a different CbCr non-intra matrix is loaded.

User Data

User_data_start_code
This optional 32-bit string of $000001B2_H$ indicates the beginning of *user_data*. *user_data* continues until the detection of another start code.

User_data
These n × 8 bits are present only if *user_data_start_code* is present. *user_data* must not contain a string of 23 or more consecutive zero bits.

Sequence Extension

A sequence extension may only occur after a sequence header.

Extension_start_code
This 32-bit string of $000001B5_H$ indicates the beginning of extension data beyond MPEG-1.

Extension_start_code_ID
This 4-bit field has a value of "0001" and indicates the beginning of a sequence extension. For MPEG-2 video bitstreams, a sequence extension must follow each sequence header.

Profile_and_level_indication
This 8-bit field specifies the profile and level, as shown in Table 13.11.

    Bit 7: escape bit
    Bits 6–4: profile ID
    Bits 3–0: level ID

Progressive_sequence
A "1" for this bit indicates only progressive pictures are present. A "0" indicates both frame and field pictures may be present, and frame pictures may be progressive or interlaced. For the SVCD standard, this value must be "0."

Chroma_format
This 2-bit codeword indicates the CbCr format, as shown in Table 13.12. For the ATSC and OpenCable™ standards, the value must be "01."

Horizontal_size_extension
The two most significant bits of *horizontal_size*. For the ATSC and OpenCable™ standards, the value must be "00."

Vertical_size_extension
The two most significant bits of *vertical_size*. For the ATSC and OpenCable™ standards, the value must be "00."

Bit_rate_extension
The twelve most significant bits of *bit_rate*. For the ATSC and OpenCable™ standards, the value must be "0000 0000 0000."

Marker_bit
Always a "1."

vbv_buffer_size_extension
The eight most significant bits of *vbv_buffer_size*. For the ATSC and OpenCable™ standards, the value must be "0000 0000."

Low_delay
A "1" for this bit indicates that no B pictures are present, so no frame reordering delay. For the SVCD standard, this value must be "0."

Profile	Profile ID Code	Level	Level ID Code
reserved	000	reserved	0000
high	001	reserved	0001
spatial scalable	010	reserved	0010
SNR scalable	011	reserved	0011
main	100	high	0100
simple	101	reserved	0101
reserved	110	high 1440	0110
reserved	111	reserved	0111
		main	1000
		reserved	1001
		low	1010
		reserved	1011
		reserved	1100
		reserved	1101
		reserved	1110
		reserved	1111

Table 13.11. MPEG-2 *profile_and_level_indication* Codewords.

Chroma Format	Code
reserved	00
4:2:0	01
4:2:2	10
4:4:4	11

Table 13.12. MPEG-2 *chroma_format* Codewords.

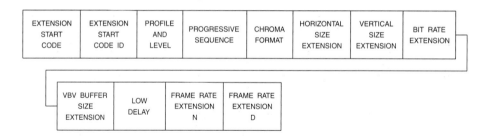

Figure 13.6. MPEG-2 Sequence Extension Structure. Marker bits not shown.

Frame_rate_extension_n

See *frame_rate_code* regarding this 2-bit binary value. For the ATSC and OpenCable™ standards, the value must be "00."

Frame_rate_extension_d

See *frame_rate_code* regarding this 5-bit binary value. For the ATSC and OpenCable™ standards, the value must be "00000."

Sequence Display Extension

This optional extension may only occur after a sequence extension.

Extension_start_code

This 32-bit string of 000001B5$_H$ indicates the beginning of a new set of extension data.

Extension_start_code_ID

This 4-bit field has a value of "0010" and indicates the beginning of a sequence display extension. Information provided by this extension does not affect the decoding process and may be ignored. It allows the display of the decoded pictures to be as accurate as possible.

Video_format

This 3-bit codeword indicates the source of the pictures prior to MPEG encoding, as shown in Table 13.13. For the ATSC and Open-Cable™ standards, the value must be "000."

Color_description

A "1" for this bit indicates that *color_primaries*, *transfer_characteristics*, and *matrix_coefficients* are present in the bit-stream.

Color_primaries

This optional 8-bit codeword describes the chromaticity coordinates of the source primaries, as shown in Table 13.14. If *sequence_display_extension* is not present, or *color_description* = "0," the indicated default value must be used.

This information may be used to adjust the color processing after MPEG-2 decoding to compensate for the color primaries of the display.

Video Format	Code
component	000
PAL	001
NTSC	010
SECAM	011
MAC	100
unspecified	101
reserved	110
reserved	111

Table 13.13. MPEG-2 *video_format* Codewords.

Color Primaries	Code	Application Default
forbidden	0000 0000	
BT.709, SMPTE 274M	0000 0001	MPEG-2, ATSC, DVB 25Hz HDTV, DVB 30Hz HDTV
unspecified	0000 0010	
reserved	0000 0011	
BT.470 system M	0000 0100	DVD-Video 30 Hz
BT.470 system B, G, I	0000 0101	DVD-Video 25 Hz, DVB 25Hz SDTV
SMPTE 170M	0000 0110	DVD-Video 30 Hz, DVB 30Hz SDTV
SMPTE 240M	0000 0111	
reserved	0000 1000	
:	:	:
reserved	1111 1111	

Table 13.14. MPEG-2 *color_primaries* Codewords.

Opto-Electronic Transfer Characteristics	Code	Application Default
forbidden	0000 0000	
BT.709, SMPTE 274M	0000 0001	MPEG-2, ATSC, DVB 25Hz HDTV, DVB 30Hz HDTV
unspecified	0000 0010	
reserved	0000 0011	
BT.470 system M	0000 0100	DVD-Video 30 Hz
BT.470 system B, G, I	0000 0101	DVD-Video 25 Hz, DVB 25Hz SDTV
SMPTE 170M	0000 0110	DVD-Video 30 Hz, DVB 30Hz SDTV
SMPTE 240M	0000 0111	
linear	0000 1000	
reserved	0000 1001	
:	:	:
reserved	1111 1111	

Table 13.15. MPEG-2 *transfer_characteristics* Codewords.

Matrix Coefficients	Code	Application Default
forbidden	0000 0000	
BT.709, SMPTE 274M	0000 0001	MPEG-2, ATSC, DVB 25Hz HDTV, DVB 30Hz HDTV
unspecified	0000 0010	
reserved	0000 0011	
FCC	0000 0100	
BT.470 system B, G, I	0000 0101	DVD-Video 25 Hz, DVB 25Hz SDTV
SMPTE 170M	0000 0110	DVD-Video 30 Hz, DVB 30Hz SDTV
SMPTE 240M	0000 0111	
reserved	0000 1000	
:	:	:
reserved	1111 1111	

Table 13.16. MPEG-2 *matrix_coefficients* Codewords.

EXTENSION START CODE	EXTENSION START CODE ID	VIDEO FORMAT	COLOR DESCRIPTION	COLOR PRIMARIES	TRANSFER CHARACTERISTICS	MATRIX COEFFICIENTS	DISPLAY HORIZONTAL SIZE	DISPLAY VERTICAL SIZE

Figure 13.7. MPEG-2 Sequence Display Extension Structure. Marker bits not shown.

Transfer_characteristics

This optional 8-bit codeword describes the optoelectronic transfer characteristic of the source picture, as shown in Table 13.15. If *sequence_display_extension* is not present, or *color_description* = "0," the indicated default value must be used.

This information may be used to adjust the processing after MPEG-2 decoding to compensate for the gamma of the display.

Matrix_coefficients

This optional 8-bit codeword describes the coefficients used in deriving YCbCr from R′G′B′, as shown in Table 13.16. If *sequence_display_extension* is not present, or *color_description* = "0," the indicated default value must be used.

This information is used to select the proper YCbCr-to-RGB matrix, if needed, after MPEG-2 decoding.

Display_horizontal_size

See *display_vertical_size* regarding this 14-bit binary number.

Marker_bit

Always a "1."

Display_vertical_size

This 14-bit binary number, in conjunction with *display_horizontal_size*, defines the active region of the display. If the display region is smaller than the encoded picture size, only a portion of the picture will be displayed. If the display region is larger than the picture size, the picture will be displayed on a portion of the display.

Sequence Scalable Extension

This optional extension may only occur after a sequence extension.

Extension_start_code

This 32-bit string of $000001B5_H$ indicates the beginning of a new set of extension data.

Extension_start_code_ID

This 4-bit field has a value of "0101" and indicates the beginning of a sequence scalable extension. This extension specifies the scalability modes implemented for the video bitstream. If *sequence_scalable_extension* is not present in the bitstream, no scalability is used. The base layer of a scalable hierarchy does not have a *sequence_scalable_extension*, except in the case of data partitioning.

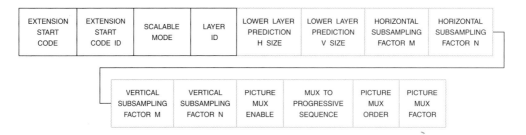

Figure 13.8.   MPEG-2 Sequence Scalable Extension Structure. Marker bits not shown.

Scalable_mode

This 2-bit codeword indicates the scalability type of the video sequence as shown in Table 13.17.

Scalable Mode	Code
data partitioning	00
spatial scalability	01
SNR scalability	10
temporal scalability	11

Table 13.17. MPEG-2 *scalable_mode* Codewords.

Layer_ID

This 4-bit binary number identifies the layers in a scalable hierarchy. The base layer has an ID of "0000." During data partitioning, *layer_ID* "0000" is assigned to partition layer zero and *layer_ID* "0001" is assigned to partition layer one.

Lower_layer_prediction_horizontal_size

This optional 14-bit binary number is present only if *scalable_mode* = "01." It indicates the horizontal size of the lower layer frame used for prediction. It contains the value of *horizontal_size* in the lower layer bitstream.

Marker_bit

Always a "1." It is present only if *scalable_mode* = "01."

Lower_layer_prediction_vertical_size

This optional 14-bit binary number is present only if *scalable_mode* = "01." It indicates the vertical size of the lower layer frame used for prediction. It contains the value of *vertical_size* in the lower layer bitstream.

Horizontal_subsampling_factor_m

This optional 5-bit binary number is present only if *scalable_mode* = "01," and affects the spatial upsampling process. A value of "00000" is not allowed.

Horizontal_subsampling_factor_n

This optional 5-bit binary number is present only if *scalable_mode* = "01," and affects the spatial upsampling process. A value of "00000" is not allowed.

Vertical_subsampling_factor_m

This optional 5-bit binary number is present only if *scalable_mode* = "01," and affects the spatial upsampling process. A value of "00000" is not allowed.

Vertical_subsampling_factor_n

This optional 5-bit binary number is present only if *scalable_mode* = "01," and affects the spatial upsampling process. A value of "00000" is not allowed.

Picture_mux_enable

This optional 1-bit field is present only if *scalable_mode* = "11." If set to a "1," the *picture_mux_order* and *picture_mux_factor* parameters are used for remultiplexing prior to display.

Mux_to_progressive_sequence

This optional 1-bit field is present only if *scalable_mode* = "11" and *picture_mux_enable* = "1." If set to a "1," it indicates the decoded pictures are to be temporally multiplexed to generate a progressive sequence for display. When temporal multiplexing is to generate an interlaced sequence, this flag is a "0."

Picture_mux_order

This optional 3-bit binary number is present only if *scalable_mode* = "11." It specifies the number of enhancement layer pictures prior to the first base layer picture. It is used to assist the decoder in properly remultiplexing pictures prior to display.

Picture_mux_factor

This optional 3-bit binary number is present only if *scalable_mode* = "11." It denotes the number of enhancement layer pictures between consecutive base layer pictures, and is used to assist the decoder in properly remultiplexing pictures prior to display.

Group of Pictures (GOP) Layer

A GOP header should occur about every two seconds. Data for each group of pictures consists of a GOP header followed by picture data. The structure is shown in Figure 13.5. The DVD standard uses user data extensions at this layer for closed captioning data.

Group_start_code

This 32-bit string has a value of $000001B8_H$ and indicates the beginning of a group of pictures.

Time_code

These 25 bits indicate timecode information, as shown in Table 13.18. *Drop_frame_flag* may be set to "1" only if the frame rate is 30/1.001 (29.97) Hz.

Closed_gop

This 1-bit flag is set to "1" if the group of pictures has been encoded without motion vectors referencing the previous group of pictures. This bit allows support of editing the compressed bitstream.

Broken_link

This 1-bit flag is set to a "0" during encoding. It is set to a "1" during editing when the B frames following the first I frame of a group of pictures cannot be correctly decoded.

Picture Layer

Data for each picture consists of a picture header followed by slice data. The structure is shown in Figure 13.5. If a sequence extension is present, each picture header is followed by a picture coding extension.

Some implementations enable frame-accurate switching of aspect ratio information via user data extensions at this layer. The ATSC standard also uses user data extensions at this layer for EIA–708 closed captioning data.

Picture_start_code

This 32-bit string has a value of 00000100_H.

Temporal_reference

For the first frame in a GOP, the 10-bit binary number *temporal_reference* is zero. It then increments by one, modulo 1024, for each frame in the display order. When a frame is coded as two fields, the temporal reference of both fields is the same.

Picture_coding_type

This 3-bit codeword indicates the picture type (I picture, P picture, or B picture) as shown in Table 13.19.

Picture Type	Code
forbidden	000
I picture	001
P picture	010
B picture	011
forbidden	100
reserved	101
reserved	110
reserved	111

Table 13.19. MPEG-2 *picture_coding_type* Codewords.

Time Code	Range of Value	Number of Bits
drop_frame_flag		1
time_code_hours	0–23	5
time_code_minutes	0–59	6
marker_bit	1	1
time_code_seconds	0–59	6
time_code_pictures	0–59	6

Table 13.18. MPEG-2 *time_code* Field.

Vbv_delay

For constant bit rates, this 16-bit binary number sets the initial occupancy of the decoding buffer at the start of decoding a picture so that it doesn't overflow or underflow. For the ATSC and OpenCable™ standards, unless *vbv_delay* has the value $FFFF_H$, the value must be $\leq 45000_D$.

Full_pel_forward_vector

This optional 1-bit field is not used for MPEG-2, so has a value of "0." It is present only if *picture_coding_type* = "010" or "011."

Forward_f_code

This optional 3-bit field is not used for MPEG-2, so has a value of "111." It is present only if *picture_coding_type* = "010" or "011."

Full_pel_backward_vector

This optional 1-bit field is not used for MPEG-2, so has a value of "0." It is present only if *picture_coding_type* = "011."

Backward_f_code

This optional 3-bit field is not used for MPEG-2, so has a value of "111." It is present only if *picture_coding_type* = "011."

Extra_bit_picture

A bit which, when set to "1," indicates that *content_description_data* follows. A "0" indicates that no *content_description_data* follows.

Content_description_data

If *extra_bit_picture* = "1," then this optional variable-length field is present, with every 9th bit having the value of "1."

Extra_bit_picture

This optional bit has a value of "0" and is present only if *content_description_data* is present.

Content Description Data

This optional data is only present when indicated by *extra_bit_picture* in the picture header.

Data_type_upper

This 8-bit field contains the eight most significant bits of the 16-bit binary *data_type* that defines the type of content description data, as shown in Table 13.20.

Marker_bit

Always a "1."

Data_type_lower

This 8-bit field contains the eight least significant bits of the 16-bit binary *data_type* that defines the type of content description data, as shown in Table 13.20.

Data Type	Code
reserved	0000 0000 0000 0000
padding bytes	0000 0000 0000 0001
capture timecode	0000 0000 0000 0010
pan-scan parameters	0000 0000 0000 0011
active region window	0000 0000 0000 0100
coded picture length	0000 0000 0000 0101
reserved	0000 0000 0000 0110
:	:
reserved	1111 1111 1111 1111

Table 13.20. MPEG-2 *data_type* Codewords.

Marker_bit
Always a "1."

Data_length
This 8-bit binary number specifies the remaining amount of data that follows, expressed in units of nine bits.

Note: The following fields are present when "padding bytes" is indicated by data_type. The two fields are repeated for the number of times indicated by the data_length field.

Marker_bit
Always a "1."

Padding_byte
This 8-bit field has the value of "0000 0000." All other values are forbidden.

Note: The following fields are present when "capture timecode" is indicated by data_type. It contains timestamps that indicate the source capture or creation time of the fields or frames. It does not take precedence over any timecode present at the system level.

Marker_bit
Always a "1."

Timecode_type
This 2-bit codeword indicates the number of timecodes associated with the picture, as shown in Table 13.21.

Timecode Type	Code
one timecode for the frame	00
one timecode for the first or only field	01
one timecode for the second field	10
two timecodes, one for each of two fields	11

Table 13.21. MPEG-2 *timecode_type* Codewords.

Counting_type
This optional 3-bit codeword specifies the method used for compensating the nframes counting parameter to reduce drift accumulation.

Reserved_bit
Always a "0."

Reserved_bit
Always a "0."

Reserved_bit
Always a "0."

Marker_bit
This optional bit is always a "1." This field is present only when *counting_type* ≠ "000."

Nframes_conversion_code
This optional bit specifies the conversion factor (1000 + *nframes_conversion_code*) in determining the amount of time indicated by the nframes parameter. This field is present only when *counting_type* ≠ "000."

Clock_divisor

This optional 7-bit binary number specifies the number of divisions of the 27 MHz system clock to be applied for generating the equivalent timestamp. This field is present only when *counting_type* ≠ "000."

Marker_bit

This optional bit is always a "1." This field is present only when *counting_type* ≠ "000."

Nframes_multiplier_upper

This optional 8-bit value is the eight most significant bits of the 16-bit *nframes_multiplier* value. This field is present only when *counting_type* ≠ "000."

Marker_bit

This optional bit is always a "1." This field is present only when *counting_type* ≠ "000."

Nframes_multiplier_lower

This optional 8-bit value is the eight least significant bits of the 16-bit *nframes_multiplier* value. This field is present only when *counting_type* ≠ "000."

"Field or frame capture timestamp" information follows.

Marker_bit

This optional bit is always a "1." This field is present only when *counting_type* ≠ "000."

Nframes

This optional 8-bit binary number specifies the number of frame time increments to add in deriving the equivalent timestamp. This field is present only when *counting_type* ≠ "000."

Marker_bit

Always a "1."

Time_discontinuity

A "1" for this 1-bit flag indicates that a discontinuity in the timecode sequence has occurred.

Prior_count_dropped

This 1-bit flag indicates if the counting of one or more values of *nframes* was dropped.

Time_offset_part_a

A 6-bit value containing the six most significant bits of *time_offset*. *Time_offset* is a 30-bit signed value that specifies the number of clock cycles offset from the time specified by other timestamp parameters to specify the equivalent timestamp for when the current field or frame was captured.

Marker_bit

Always a "1."

Time_offset_part_b

An 8-bit value containing the eight second most significant bits of *time_offset*.

Marker_bit

Always a "1."

Time_offset_part_c

An 8-bit value containing the eight third most significant bits of *time_offset*.

Marker_bit

Always a "1."

Time_offset_part_d

An 8-bit value containing the eight least significant bits of *time_offset*.

Marker_bit
Always a "1."

Units_of_seconds
A 4-bit binary number that indicates the seconds timestamp value. It may have a value of "0000" to "1001."

Tens_of_seconds
A 4-bit binary number that indicates the tens of seconds timestamp value. It may have a value of "0000" to "0101."

Marker_bit
Always a "1."

Units_of_minutes
A 4-bit binary number that indicates the seconds timestamp value. It may have a value of "0000" to "1001."

Tens_of_minutes
A 4-bit binary number that indicates the tens of minutes timestamp value. It may have a value of "0000" to "0101."

Marker_bit
Always a "1."

Units_of_hours
A 4-bit binary number that indicates the hours timestamp value. It may have a value of "0000" to "1001."

Tens_of_hours
A 4-bit binary number that indicates the tens of hours timestamp value. It may have a value of "0000" to "0010."

When timecode_type = "11", the "field or frame capture timestamp" fields are again present to convey the information for the second field.

Note: The following fields are present when "pan-scan parameters" is indicated by data_type. This allows the transmission of additional pan-scan information for a display that has a different aspect ratio.

Marker_bit
Always a "1."

Aspect_ratio_information
This 4-bit codeword is the same as used by the sequence header.

Reserved_bit
Always a "0."

Reserved_bit
Always a "0."

Reserved_bit
Always a "0."

Display_size_present
A 1-bit flag that indicates whether or not the *display_horizontal_size* and *display_vertical_size* fields follow.

Marker_bit
Always a "1." This optional field is present only if *display_size_present* = "1."

Reserved_bit
Always a "0." This optional field is present only if *display_size_present* = "1."

Reserved_bit
Always a "0." This optional field is present only if *display_size_present* = "1."

Display_horizontal_size_upper
These are the six most significant bits of *display_horizontal_size*. This optional field is present only if *display_size_present* = "1."

Marker_bit
Always a "1." This optional field is present only if *display_size_present* = "1."

Display_horizontal_size_lower
These are the eight least significant bits of *display_horizontal_size*. This optional field is present only if *display_size_present* = "1."

Marker_bit
Always a "1." This optional field is present only if *display_size_present* = "1."

Reserved_bit
Always a "0." This optional field is present only if *display_size_present* = "1."

Reserved_bit
Always a "0." This optional field is present only if *display_size_present* = "1."

Display_vertical_size_upper
These are the six most significant bits of *display_vertical_size*. This optional field is present only if *display_size_present* = "1."

Marker_bit
Always a "1." This optional field is present only if *display_size_present* = "1."

Display_vertical_size_lower
These are the eight least significant bits of *display_vertical_size*. This optional field is present only if *display_size_present* = "1."

Note: The following fields are present for each of the frame center offsets present.

Marker_bit
Always a "1."

Frame_center_horizontal_offset_upper
These are the eight most significant bits of *frame_center_horizontal_offset*. The definition of *frame_center_horizonal_offset* is specified in the picture display extension.

Marker_bit
Always a "1."

Frame_center_horizontal_offset_lower
These are the eight least significant bits of *frame_center_horizontal_offset*.

Marker_bit
Always a "1."

Frame_center_vertical_offset_upper
These are the eight most significant bits of *frame_center_vertical_offset*. The definition of *frame_center_horizonal_offset* is specified in the picture display extension.

Marker_bit
Always a "1."

Frame_center_vertical_offset_lower
These are the eight least significant bits of *frame_center_vertical_offset*.

Note: The following fields are present when "active region window" is indicated by data_type. The active_region_window defines the rectangle the decoded picture is intended to be displayed. It must not be larger than the region defined by horizontal_size and vertical_size.

Marker_bit
Always a "1."

Top_left_x_upper
This 8-bit field is the eight most significant bits of the 16-bit *top_left_x*. *Top_left_x* defines the Y sample number that, together with *top_left_y*, defines the upper left corner of the *active_region_window* rectangle.

Marker_bit
Always a "1."

Top_left_x_lower
This 8-bit field is the eight least significant bits of the 16-bit *top_left_x*.

Marker_bit
Always a "1."

Top_left_y_upper
This 8-bit field is the eight most significant bits of the 16-bit *top_left_y*. *Top_left_y* defines the Y line number that, together with *top_left_x*, defines the upper left corner of the *active_region_window* rectangle.

Marker_bit
Always a "1."

Top_left_y_lower
This 8-bit field is the eight least significant bits of the 16-bit *top_left_y*.

Marker_bit
Always a "1."

Active_region_horizontal_size_upper
This 8-bit field is the eight most significant bits of the 16-bit *active_region_horizontal_size*. *Active_region_horizontal_size*, along with *active_region_vertical_size*, defines the bottom right corner of the *active_region_window* rectangle. A value of 0000_H for *active_region_horizontal_size* indicates the size is unknown.

Marker_bit
Always a "1."

Active_region_horizontal_size_lower
This 8-bit field is the eight least significant bits of the 16-bit *active_region_horizontal_size*.

Marker_bit
Always a "1."

Active_region_vertical_size_upper
This 8-bit field is the eight most significant bits of the 16-bit *active_region_vertical_size*. *Active_region_vertical_size*, along with *active_region_horizontal_size*, defines the bottom right corner of the *active_region_window* rectangle. A value of 0000_H for *active_region_vertical_size* indicates the size is unknown.

Marker_bit
Always a "1."

Active_region_vertical_size_lower
This 8-bit field is the eight least significant bits of the 16-bit *active_region_vertical_size*.

Note: The following fields are present when "coded picture length" is indicated by data_type.

Marker_bit
Always a "1."

Picture_byte_count_part_a
This 8-bit field is the eight most significant bits of the 32-bit *picture_byte_count*. *Picture_byte_count* indicates the number of bytes starting with the first byte of the first *slice_start_code* of the current picture and ending with the byte preceding the start code prefix immediately following the last macroblock of the picture. A value of 0000_H indicates the length is unknown.

Marker_bit
Always a "1."

Picture_byte_count_part_b
This 8-bit field is the eight second most significant bits of the 32-bit *picture_byte_count*.

Marker_bit
Always a "1."

Picture_byte_count_part_c
This 8-bit field is the eight third most significant bits of the 32-bit *picture_byte_count*.

Marker_bit
Always a "1."

Picture_byte_count_part_d
This 8-bit field is the eight least significant bits of the 32-bit *picture_byte_count*.

Note: The following two fields are present when no other data is present as indicated by data_type. The two fields are repeated for the number of times indicated by the data_length field.

Marker_bit
Always a "1."

Reserved_content_description_data
This 8-bit field is reserved.

Picture Coding Extension

A picture coding extension may only occur following a picture header.

Extension_start_code
This 32-bit string of $000001B5_H$ indicates the beginning of a new set of extension data.

Extension_start_code_ID
This 4-bit field has a value of "1000" and indicates the beginning of a picture coding extension.

f_code [0,0]
A 4-bit binary number, having a range of "0001" to "1001," that is used for the decoding of forward horizontal motion vectors. A value of "0000" is not allowed; a value of "1111" indicates this field is ignored.

f_code [0,1]
A 4-bit binary number, having a range of "0001" to "1001," that is used for the decoding of forward vertical motion vectors. A value of "0000" is not allowed; a value of "1111" indicates this parameter is ignored.

f_code [1,0]

A 4-bit binary number, having a range of "0001" to "1001," that is used for the decoding of backward horizontal motion vectors. A value of "0000" is not allowed; a value of "1111" indicates this field is ignored.

f_code [1,1]

A 4-bit code, having a range of "0001" to "1001," that is used for the decoding of backward vertical motion vectors. A value of "0000" is not allowed; a value of "1111" indicates this field is ignored.

Intra_dc_precision

This 2-bit codeword specifies the intra DC precision as shown in Table 13.22.

Picture_structure

This 2-bit codeword specifies the picture structure as shown in Table 13.23.

Intra DC Precision (Bits)	Code
8	00
9	01
10	10
11	11

Table 13.22. MPEG-2 *intra_dc_precision* Codewords.

Picture Structure	Code
reserved	00
top field	01
bottom field	10
frame picture	11

Table 13.23. MPEG-2 *picture_structure* Codewords.

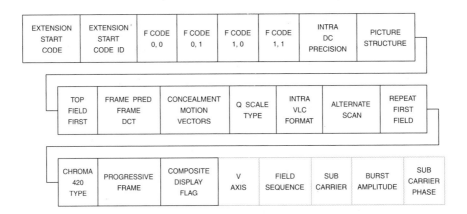

Figure 13.9.   MPEG-2 Picture Coding Extension Structure. Marker bits not shown.

Top_field_first

If *progressive_sequence* = "0," this bit indicates what field is output first by the decoder. In a field, this bit has a value of "0." In a frame, a "1" indicates the first field of the decoded frame is the top field. A value of "0" indicates the first field is the bottom field.

If *progressive_sequence* = "1" and *repeat_first_field* = "0," this bit is a "0" and the decoder generates a progressive frame.

If *progressive_sequence* = "1," *repeat_first_field* = "1," and this bit is a "0," the decoder generates two identical progressive frames.

If *progressive_sequence* = "1," *repeat_first_field* = "1," and this bit is a "1," the decoder generates three identical progressive frames.

Frame_pred_frame_dct

If this bit is a "1," only frame-DCT and frame prediction are used. For field pictures, it is always a "0." This parameter is a "1" if *progressive_frame* is "1."

Concealment_motion_vectors

If this bit is a "1," it indicates that the motion vectors are coded for intra macroblocks.

Q_scale_type

This bit indicates which of two mappings between *quantizer_scale_code* and *quantizer_scale* are used by the decoder.

Intra_vlc_format

This bit indicates which table is to be used for DCT coefficients for intra blocks. Table 13.37 is used when *intra_vlc_format* = "0." Table 13.38 is used when *intra_vlc_format* = "1." For non-intra blocks, Table 13.37 is used regardless of the value of *intra_vlc_format*.

Alternate_scan

This bit indicates which scanning pattern is to be used by the decoder for transform coefficient data. "0" = Figure 7.59; "1" = Figure 7.60.

Repeat_first_field

See *top_field_first* for the use of this bit. For field pictures, it has a value of "0."

Chroma_420_type

If *chroma_format* is 4:2:0, this bit is the same as *progressive_frame*. Otherwise, it is a "0."

Progressive_frame

If a "0," this bit indicates the two fields of the frame are interlaced fields, with a time interval between them. If a "1," the two fields of the frame are from the same instant in time.

Composite_display_flag

This bit indicates whether or not *v_axis*, *field_sequence*, *sub_carrier*, *burst_amplitude*, and *sub_carrier_phase* are present in the bitstream.

V_axis

This bit is present only when *composite_display_flag* = "1." It is used when the original source was a PAL video signal. *v_axis* = "1" on a positive V sign, "0" otherwise.

This information can be obtained from a PAL decoder that is driving the MPEG-2 encoder. It can be used to enable a MPEG-2 decoder to set the V switching of a PAL encoder to the same as the original.

Field_sequence

This 3-bit codeword is present only when *composite_display_flag* = "1." It specifies the number of the field in the original four- or eight-field sequence as shown in Table 13.24.

This information can be obtained from a NTSC/PAL decoder that is driving the MPEG-2 encoder. It can be used to enable a MPEG-2 decoder to set the field sequence of a NTSC/PAL encoder to the same as the original.

Frame Sequence	Field Sequence	Code
1	1	000
1	2	001
2	3	010
2	4	011
3	5	100
3	6	101
4	7	110
4	8	111

Table 13.24. MPEG-2 *field_sequence* Codewords.

Sub_carrier

This bit is present only when *composite_display_flag* = "1." A "0" indicates that the original subcarrier-to-line frequency relationship was correct.

This information can be obtained from the NTSC/PAL decoder that is driving the MPEG-2 encoder.

Burst_amplitude

This 7-bit binary number is present only when *composite_display_flag* = "1." It specifies the original PAL or NTSC burst amplitude when quantized per BT.601 (ignoring the MSB).

This information can be obtained from a NTSC/PAL decoder that is driving the MPEG-2 encoder. It can be used to enable a MPEG-2 decoder to set the color burst amplitude of a NTSC/PAL encoder to the same as the original.

Sub_carrier_phase

This 8-bit binary number is present only when *composite_display_flag* = "1." It specifies the original PAL or NTSC subcarrier phase as defined in BT.470. The value is defined as: $(360° / 256) \times sub\_carrier\_phase$.

This information can be obtained from an NTSC/PAL decoder that is driving the MPEG-2 encoder. It can be used to enable a MPEG-2 decoder to set the color subcarrier phase of a NTSC/PAL encoder to the same as the original.

Quant Matrix Extension

Each quantization matrix has default values. When a sequence header is decoded, all matrices reset to their default values. User-defined matrices may be downloaded during a sequence header or using this extension. This optional extension may only occur after a picture coding extension.

Extension_start_code

This 32-bit string of $000001B5_H$ indicates the beginning of a new set of extension data.

Extension_start_code_ID

This 4-bit string has a value of "0011" and indicates the beginning of a *quant_matrix_extension*. This extension also allows quantizer matrices to be transmitted for the 4:2:2 and 4:4:4 chroma formats.

Load_intra_quantizer_matrix

This bit is set to a "1" if an *intra_quantizer_matrix* follows. If set to a "0," the default values below are used for intra blocks until the next occurrence of a sequence header or *quant_matrix_extension*.

8	16	19	22	26	27	29	34
16	16	22	24	27	29	34	37
19	22	26	27	29	34	34	38
22	22	26	27	29	34	37	40
22	26	27	29	32	35	40	48
26	27	29	32	35	40	48	58
26	27	29	34	38	46	56	69
27	29	35	38	46	56	69	83

Intra_quantizer_matrix

An optional list of sixty-four 8-bit values that replace the default values shown above. A value of zero is not allowed. The value for *intra_quant* [0, 0] is always 8. These values take effect until the next occurrence of a sequence header or *quant_matrix_extension*. The order follows that shown in Figure 7.59.

For 4:2:2 and 4:4:4 data formats, the new values are used for both the Y and CbCr intra matrix, unless a different CbCr intra matrix is loaded.

Load_non_intra_quantizer_matrix

This bit is set to a "1" if a *non_intra_quantizer_matrix* follows. If set to a "0," the default values below are used for non-intra blocks until the next occurrence of a sequence header or *quant_matrix_extension*.

16	16	16	16	16	16	16	16
16	16	16	16	16	16	16	16
16	16	16	16	16	16	16	16
16	16	16	16	16	16	16	16
16	16	16	16	16	16	16	16
16	16	16	16	16	16	16	16
16	16	16	16	16	16	16	16
16	16	16	16	16	16	16	16

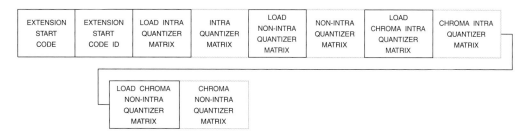

Figure 13.10.   MPEG-2 Quant Matrix Extension Structure. Marker bits not shown.

Non-intra_quantizer_matrix

An optional list of sixty-four 8-bit values that replace the default values shown above. A value of zero is not allowed. These values take effect until the next occurrence of a sequence header or *quant_matrix_extension*. The order follows that shown in Figure 7.59.

For 4:2:2 and 4:4:4 data formats, the new values are used for both the Y and CbCr non-intra matrix, unless a new CbCr non-intra matrix is loaded.

Load_chroma_intra_quantizer_matrix

This bit is set to a "1" if a *chroma_intra_quantizer_matrix* follows. If set to a "0," there is no change in the values used. If *chroma_format* is 4:2:0, this bit is a "0."

Chroma_intra_quantizer_matrix

An optional list of sixty-four 8-bit values that replace the previous or default values used for CbCr data. A value of zero is not allowed. The value for *chroma_intra_quant* [0,0] is always 8. These values take effect until the next occurrence of a sequence header or *quant_matrix_extension*. The order follows that shown in Figure 7.59.

Load_chroma_non_intra_quantizer_matrix

This bit is set to a "1" if a *chroma_non_intra_quantizer_matrix* follows. If set to a "0," there is no change in the values used. If *chroma_format* is 4:2:0, this bit is a "0."

Chroma_non_intra_quantizer_matrix

An optional list of sixty-four 8-bit values that replace the previous or default values used for CbCr data. A value of zero is not allowed. These values take effect until the next occurrence of a sequence header or *quant_matrix_extension*. The order follows that shown in Figure 7.59.

Picture Display Extension

This extension allows the position of the display rectangle to be moved on a picture-by-picture basis. A typical application would be implementing pan-and-scan. This optional extension may only occur after a picture coding extension.

Extension_start_code

This 32-bit string of 000001B5$_\mathrm{H}$ indicates the beginning of a new set of extension data.

EXTENSION START CODE	EXTENSION START CODE ID	FRAME CENTER HORIZONTAL OFFSET	FRAME CENTER VERTICAL OFFSET

Figure 13.11. MPEG-2 Picture Display Extension Structure. Marker bits not shown.

EXTENSION START CODE	EXTENSION START CODE ID	REFERENCE SELECT CODE	FORWARD TEMPORAL REFERENCE	BACKWARD TEMPORAL REFERENCE

Figure 13.12. MPEG-2 Picture Temporal Scalable Extension Structure. Marker bits not shown.

Extension_start_code_ID

This 4-bit field has a value of "0111" and indicates the beginning of a picture display extension.

In the case of an interlaced sequence, a picture may relate to one, two, or three decoded fields. Thus, there may be up to three sets of the following four fields present in the bitstream.

Frame_center_horizontal_offset

This 16-bit 2's complement number specifies the horizontal offset in units of 1/16th of a sample. A positive value positions the center of the decoded picture to the right of the center of the display region.

Marker_bit

Always a "1."

Frame_center_vertical_offset

This 16-bit 2's complement number specifies the vertical offset in units of 1/16th of a scan line. A positive value positions the center of the decoded picture below the center of the display region.

Marker_bit

Always a "1."

Picture Temporal Scalable Extension

This optional extension may only occur after a picture coding extension.

Extension_start_code

This 32-bit string of $000001B5_H$ indicates the beginning of a new set of extension data.

Extension_start_code_ID

This 4-bit value of "1010" indicates the beginning of a picture temporal scalable extension.

Reference_select_code

This 2-bit codeword identifies reference frames or fields for prediction.

Forward_temporal_reference

This 10-bit binary number indicates the temporal reference of the lower layer to be used to provide the forward prediction. If more than 10 bits are required to specify the temporal reference, only the 10 LSBs are used.

Marker_bit

Always a "1."

Lower Layer Deinterlaced Field Select	Lower Layer Progressive Frame	Progressive Frame	Apply Deinterlace Process	Use For Prediction
0	0	1	yes	top field
1	0	1	yes	bottom field
1	1	1	no	frame
1	1	0	no	frame
1	0	0	yes	both fields

Table 13.25. MPEG-2 Picture Spatial Scalable Extension Upsampling Process.

Backward_temporal_reference

This 10-bit binary number indicates the temporal reference of the lower layer to be used to provide the backward prediction. If more than 10 bits are required to specify the temporal reference, only the 10 LSBs are used.

Picture Spatial Scalable Extension

This optional extension may only occur after a picture coding extension.

Extension_start_code

This 32-bit string of $000001B5_H$ indicates the beginning of a new set of extension data.

Extension_start_code_ID

This 4-bit value of "1001" indicates the beginning of a picture spatial scalable extension.

Lower_layer_temporal_reference

This 10-bit binary number indicates the temporal reference of the lower layer to be used to provide the prediction. If more than 10 bits are required to specify the temporal reference, only the 10 LSBs are used.

Marker_bit

Always a "1."

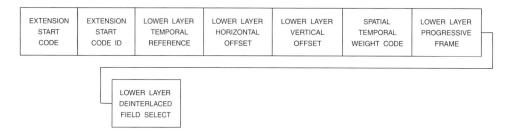

Figure 13.13. MPEG-2 Picture Spatial Scalable Extension Structure. Marker bits not shown.

Lower_layer_horizontal_offset

This 15-bit 2's complement number indicates the horizontal offset of the top-left corner of the upsampled lower layer picture relative to the enhancement layer picture. This parameter must be an even number for the 4:2:0 and 4:2:2 formats.

Marker_bit

Always a "1."

Lower_layer_vertical_offset

This 15-bit 2's complement number indicates the vertical offset of the top-left corner of the upsampled lower layer picture relative to the enhancement layer picture. This parameter must be an even number for the 4:2:0 format.

Spatial_temporal_weight_code_table_index

This 2-bit codeword indicates which spatial temporal weight codes are to be used.

Lower_layer_progressive_frame

This bit is "1" if the lower layer picture is progressive.

Lower_layer_deinterlaced_field_select

This bit is used in conjunction with other parameters to assist the decoder. See Table 13.25.

Copyright Extension

This optional extension may only occur after a picture coding extension.

Extension_start_code

This 32-bit string of $000001B5_H$ indicates the beginning of a new set of extension data.

Extension_start_code_ID

This 4-bit value of "0100" indicates the beginning of a copyright extension.

Copyright_flag

A "1" for this bit specifies the following video content, up to the next copyright extension, is copyrighted. A "0" does not indicate whether the following video content is copyrighted or not.

Copyright_identifier

This 8-bit binary number indicates the copyright holder. A value of "0000 0000" indicates the information is not available. When *copyright_flag* = "0," *copyright_identifier* must be "0000 0000."

Original_or_copy

A "1" for this bit indicates original material; a "0" indicates that it is a copy.

Reserved

These seven bits are always a "000 0000."

Marker_bit

Always a "1."

Copyright_number_1

These 20 bits represent bits 44-63 of the copyright number.

Marker_bit

Always a "1."

Copyright_number_2

These 22 bits represent bits 22-43 of the copyright number.

Marker_bit

Always a "1."

Copyright_number_3

These 22 bits represent bits 0-21 of the copyright number. The 64-bit copyright_number uniquely identifies the copyrighted content. When *copyright_identifier* = "0000 0000," or *copyright_flag* = "0," the *copyright_number* must be zero.

Camera Parameters Extension

This optional extension may only occur after a picture coding extension.

After the 32-bit *extension_start_code* of $000001B5_H$, and 4-bit extension_start_code_ID of "1011", there are several fields that specify the location and characteristics of the camera used.

ITU-T ext. D Extension

This optional extension may only occur after a picture coding extension.

After the 32-bit *extension_start_code* of $000001B5_H$, and 4-bit extension_start_code_ID of "1100", there one bit of data. The use of this extension is defined in ITU-T H.320 Annex A.

Active Format Description (AFD)

AFD (a part of ATSC A/53 and ETSI TR 101 154) describes the "area of interest" of the picture, enabling receivers to present the picture in an optimum fashion. The functionality of AFD is similar to that of Widescreen Signaling (WSS), described in Chapter 8.

AFD is optionally carried in the user data of video elementary bitstreams, after the sequence extension, GOP header, and/or picture coding extension. Support for the AFD in decoders is optional.

User_data_start_code

This 32-bit string of $000001B2_H$ indicates the beginning of *user_data*.

AFD_identifier

This 32-bit string of 44544731_H indicates that the syntax of the user data is for AFD.

Reserved_bit

Always a "1."

Active_format_flag

If bit is set to a "1," an active format is described in this data structure.

Reserved_bits

Always "00 0001."

Reserved_bits

These optional bits are always "1111." They are present if *active_format_flag* = "1."

Active_format

These optional bits specify the area of interest as shown in Table 13.26. They are present only if *active_format_flag* = "1."

Slice Layer

Data for each slice layer consists of a slice header followed by macroblock data. The structure is shown in Figure 13.5.

Slice_start_code

The first 24 bits have a value of 000001_H. The last eight bits are *slice_vertical_position*, and have a value of 01_H–AF_H.

The *slice_vertical_position* specifies the vertical position in macroblock units of the first macroblock in the slice. The *slice_vertical_position* of the first row of macroblocks is one.

AFD Active_Format Value	WSS Bits b3, b2, b1, b0	Area of Interest Aspect Ratio	Position On 4:3 Display	Active Lines	
				480-Line System	576-Line System
0000 - 0001	reserved				
0010	0100	16:9	letterbox, top	360	430
0011	0010	14:9	letterbox, top	420	504
0100	1101	> 16:9	letterbox, center	–	–
0101 - 0111	reserved				
1000	active format same as coded frame				
1001	1000	4:3	–	480	576
1010	1011	16:9	letterbox, center	360	430
1011	0001	14:9	letterbox, center	420	504
1100	reserved				
1101	1110	4:3	-	480	576
1110	–	16:9	letterbox, center	480	576
1111	–	16:9	letterbox, center	480	576

Table 13.26.   AFD *active_format* Values.

Slice_vertical_position_extension

This optional 3-bit binary number represents the three MSBs of an 11-bit *slice_vertical_position* value if the vertical size of the frame is >2800 lines. If the vertical size of the frame is ≤2800 lines, this field is not present.

Priority_breakpoint

This optional 7-bit binary number is present only when *sequence_scalable_extension* is present in the bitstream and *scalable_mode* = data partitioning. It specifies where in the bitstream to partition.

Quantizer_scale_code

This 5-bit binary value has a value of 1 to 31 (a value of zero is forbidden). It specifies the scale factor of the reconstruction level of the received DCT coefficients. The decoder uses this value until another *quantizer_scale_code* is received at either the slice or macroblock layer.

Slice_extension_flag

If this optional bit is set to a "1," *intra_slice*, *slice_picture_ID_enable* and *slice_picture_ID* fields follow.

Intra_slice

This optional bit is present only if *slice_extension_flag* = "1." It must be set to a "0" if any macroblocks in the slice are non-intra macroblocks.

Slice_picture_ID_enable

This optional bit is present only if *slice_extension_flag* = "1." A value of "1" indicates that *slice_picture_ID* is used.

Slice_picture_ID

These optional six bits are intended to aid in the recovery on severe error bursts for errors. *Slice_picture_ID* must have the same value for all slices of a picture. This field is present only if *slice_extension_flag* = "1." If *slice_picture_ID_enable* = "0", these bits must be "00 0000."

Extra_bit_slice

A bit which, when set to "1," indicates that *extra_information_slice* follows. A value of "0" indicates no data is after this field.

Extra_information_slice

If *extra_bit_slice* = "1," these nine bits follow consisting of eight bits of data (*extra_information_slice*) and then another *extra_bit_slice* to indicate if a further nine bits follow, and so on.

Macroblock Layer

Data for each macroblock layer consists of a macroblock header followed by motion vector and block data, as shown in Figure 13.5.

Macroblock_escape

This optional 11-bit field is a fixed bit string of "0000 0001 000" and is used when the difference between the current macroblock address and the previous macroblock address is greater than 33. It forces the value of *macroblock_address_increment* to be increased by 33. Any number of consecutive *macroblock_escape* fields may be used.

Macroblock_address_increment

This is a variable-length codeword that specifies the difference between the current macroblock address and the previous macroblock address. It has a maximum value of 33. Values greater than 33 are encoded using the *macroblock_escape* field. The variable-length codes are listed in Table 13.27.

Macroblock_type

This variable-length codeword indicates the method of coding and macroblock content according to Tables 13.28, 13.29, and 13.30.

Spatial_temporal_weight_code

This optional 2-bit codeword indicates, in the case of spatial scalability, how the spatial and temporal predictions are combined to do the prediction for the macroblock. This field is present only if the [spatial temporal weight class] = 1 in Tables 13.28, 13.29, and 13.30, and *spatial_temporal_weight_code_table_index* ≠ "00."

Frame_motion_type

This optional 2-bit codeword indicates the macroblock motion prediction, as shown in Table 13.31. It is present only if *picture_structure* = frame, *frame_pred_frame_dct* = "0" and [motion forward] or [motion backward] = "1" in Tables 13.28, 13.29, and 13.30.

Field_motion_type

This optional 2-bit codeword indicates the macroblock motion prediction, as shown in Table 13.32. It is present only if [motion forward] or [motion backward] = "1" in Tables 13.28, 13.29, and 13.30 and the *frame_motion_type* field is not present.

Dct_type

This optional bit indicates whether the macroblock is frame or field DCT coded. "1" = field, "0" = frame. It is present only if *picture_structure* = "11," *frame_pred_frame_dct* = "0," and [macroblock intra] or [coded pattern] = "1" in Tables 13.28, 13.29, and 13.30.

Quantizer_scale_code

This optional 5-bit binary number has a value of 1–31 (a value of zero is forbidden). It specifies the scale factor of the reconstruction level of the received DCT coefficients. The decoder uses this value until another *quantizer_scale_code* is received. This field is present only when [macroblock quant] = "1" in Tables 13.28, 13.29, and 13.30.

Optional Motion Vectors

Marker_bit

This always has a value of "1." It is present only if *concealment_motion_vectors* = "1" and [macroblock intra] = "1" in Tables 13.28, 13.29, and 13.30.

Coded_block_pattern_420

This optional variable-length codeword is used to derive the 4:2:0 coded block pattern (CBP) as shown in Table 13.33. It is present only if [coded pattern] = "1" in Tables 13.28, 13.29, and 13.30, and indicates which blocks in the macroblock have at least one transform coefficient transmitted. The coded block pattern number is represented as:

$$P_1P_2P_3P_4P_5P_6$$

where P_n = "1" for any coefficient present for block [n], else P_n = "0." Block numbering (decimal format) is given in Figure 13.2.

Increment Value	Code	Increment Value	Code
1	1	17	0000 0101 10
2	011	18	0000 0101 01
3	010	19	0000 0101 00
4	0011	20	0000 0100 11
5	0010	21	0000 0100 10
6	0001 1	22	0000 0100 011
7	0001 0	23	0000 0100 010
8	0000 111	24	0000 0100 001
9	0000 110	25	0000 0100 000
10	0000 1011	26	0000 0011 111
11	0000 1010	27	0000 0011 110
12	0000 1001	28	0000 0011 101
13	0000 1000	29	0000 0011 100
14	0000 0111	30	0000 0011 011
15	0000 0110	31	0000 0011 010
16	0000 0101 11	32	0000 0011 001
		33	0000 0011 000
macroblock_escape			0000 0001 000

Table 13.27. MPEG-2 Variable-Length Code Table for *macroblock_address_increment*.

Type	Macroblock Quant	Motion Forward	Motion Backward	Coded Pattern	Intra Macroblock	Spatial Temporal Weight Code Flag	Permitted Spatial Temporal Weight Class	Code
intra	0	0	0	0	1	0	0	1
intra, quant	1	0	0	0	1	0	0	01
I Pictures with Spatial Scalability								
coded, compatible	0	0	0	1	0	0	4	1
coded, compatible, quant	1	0	0	1	0	0	4	01
intra	0	0	0	0	1	0	0	0011
intra, quant	1	0	0	0	1	0	0	0010
not coded, compatible	0	0	0	0	0	0	4	0001
I Pictures with SNR Scalability								
coded	0	0	0	1	0	0	0	1
coded, quant	1	0	0	1	0	0	0	01
not coded	0	0	0	0	0	0	0	001

Table 13.28. MPEG-2 Variable-Length Code Table for I Picture *macroblock_type*.

Type	Macroblock Quant	Motion Forward	Motion Backward	Coded Pattern	Intra Macroblock	Spatial Temporal Weight Code Flag	Permitted Spatial Temporal Weight Class	Code
mc, coded	0	1	0	1	0	0	0	1
no mc, coded	0	0	0	1	0	0	0	01
mc, not coded	0	1	0	0	0	0	0	001
intra	0	0	0	0	1	0	0	0001 1
mc, coded, quant	1	1	0	1	0	0	0	0001 0
no mc, coded, quant	1	0	0	1	0	0	0	0000 1
intra, quant	1	0	0	0	1	0	0	0000 01
P Pictures with SNR Scalability								
coded	0	0	0	1	0	0	0	1
coded, quant	1	0	0	1	0	0	0	01
not coded	0	0	0	0	0	0	0	001

Table 13.29a. MPEG-2 Variable-Length Code Table for P Picture *macroblock_type*.

Type	Macroblock Quant	Motion Forward	Motion Backward	Coded Pattern	Intra Macroblock	Spatial Temporal Weight Code Flag	Permitted Spatial Temporal Weight Class	Code
P Pictures with Spatial Scalability								
mc, coded	0	1	0	1	0	0	0	10
mc, coded, compatible	0	1	0	1	0	1	1, 2, 3	011
no mc, coded	0	0	0	1	0	0	0	0000 100
no mc, coded, compatible	0	0	0	1	0	1	1, 2, 3	0001 11
mc, not coded	0	1	0	0	0	0	0	0010
intra	0	0	0	0	1	0	0	0000 111
mc, not coded, compatible	0	1	0	0	0	1	1, 2, 3	0011
mc, coded, quant	1	1	0	1	0	0	0	010
no mc, coded, quant	1	0	0	1	0	0	0	0001 00
intra, quant	1	0	0	0	1	0	0	0000 110
mc, coded, compatible, quant	1	1	0	1	0	1	1, 2, 3	11
no mc, coded, compatible, quant	1	0	0	1	0	1	1, 2, 3	0001 01
no mc, not coded, compatible	0	0	0	0	0	1	1, 2, 3	0001 10
coded, compatible	0	0	0	1	0	0	4	0000 101
coded, compatible, quant	1	0	0	1	0	0	4	0000 010
not coded, compatible	0	0	0	0	0	0	4	0000 0011

Table 13.29b. MPEG-2 Variable-Length Code Table for P Picture *macroblock_type*.

Type	Macroblock Quant	Motion Forward	Motion Backward	Coded Pattern	Intra Macroblock	Spatial Temporal Weight Code Flag	Permitted Spatial Temporal Weight Class	Code
interp, not coded	0	1	1	0	0	0	0	10
interp, coded	0	1	1	1	0	0	0	11
bwd, not coded	0	0	1	0	0	0	0	010
bwd, coded	0	0	1	1	0	0	0	011
fwd, not coded	0	1	0	0	0	0	0	0010
fwd, coded	0	1	0	1	0	0	0	0011
intra	0	0	0	0	1	0	0	0001 1
interp, coded, quant	1	1	1	1	0	0	0	0001 0
fwd, coded, quant	1	1	0	1	0	0	0	0000 11
bwd, coded, quant	1	0	1	1	0	0	0	0000 10
intra, quant	1	0	0	0	1	0	0	0000 01
B Pictures with Spatial Scalability								
interp, not coded	0	1	1	0	0	0	0	10
interp, coded	0	1	1	1	0	0	0	11
bwd, not coded	0	0	1	0	0	0	0	010
bwd, coded	0	0	1	1	0	0	0	011
fwd, not coded	0	1	0	0	0	0	0	0010
fwd, coded	0	1	0	1	0	0	0	0011
bwd, not coded, compatible	0	0	1	0	0	1	1, 2, 3	0001 10
bwd, coded, compatible	0	0	1	1	0	1	1, 2, 3	0001 11
fwd, not coded, compatible	0	1	0	0	0	1	1, 2, 3	0001 00
fwd, coded, compatible	0	1	0	1	0	1	1, 2, 3	0001 01
intra	0	0	0	0	1	0	0	0000 110

Table 13.30a. MPEG-2 Variable-Length Code Table for B Picture *macroblock_type*.

Type	Macroblock Quant	Motion Forward	Motion Backward	Coded Pattern	Intra Macroblock	Spatial Temporal Weight Code Flag	Permitted Spatial Temporal Weight Class	Code
B Pictures with Spatial Scalability (continued)								
interp, coded, quant	1	1	1	1	0	0	0	0000 111
fwd, coded, quant	1	1	0	1	0	0	0	0000 100
bwd, coded, quant	1	0	1	1	0	0	0	0000 101
intra, quant	1	0	0	0	1	0	0	0000 0100
fwd, coded, compatible, quant	1	1	0	1	0	1	1, 2, 3	0000 0101
bwd, coded, compatible, quant	1	0	1	1	0	1	1, 2, 3	0000 0110 0
not coded, compatible	0	0	0	0	0	0	4	0000 0111 0
coded, quant, compatible	1	0	0	1	0	0	4	0000 0110 1
coded, compatible	0	0	0	1	0	0	4	0000 0111 1
B Pictures with SNR Scalability								
coded	0	0	0	1	0	0	0	1
coded, quant	1	0	0	1	0	0	0	01
not coded	0	0	0	0	0	0	0	001

Table 13.30b. MPEG-2 Variable-Length Code Table for B Picture *macroblock_type*.

Spatial Temporal Weight Class	Prediction Type	Motion Vector Count	Motion Vector Format	Code
	reserved			00
0, 1	field	2	field	01
2, 3	field	1	field	01
0, 1, 2, 3	frame	1	frame	10
0, 2, 3	dual prime	1	field	11

Table 13.31. MPEG-2 *frame_motion_type* Codewords.

Spatial Temporal Weight Class	Prediction Type	Motion Vector Count	Motion Vector Format	Code
	reserved			00
0, 1	field	1	field	01
0, 1	16 × 8 mc	2	field	10
0	dual prime	1	field	11

Table 13.32. MPEG-2 *field_motion_type* Codewords.

Coded_block_pattern_1

Present only if *chroma_format* = 4:2:2 and [coded pattern] = "1" in Tables 13.28, 13.29, and 13.30. This optional 2-bit field is used to extend the coded block pattern by two bits.

Coded_block_pattern_2

Present only if *chroma_format* = 4:4:4 and [coded pattern] = "1" in Tables 13.28, 13.29, and 13.30. This optional 6-bit field is used to extend the coded block pattern by six bits.

Block Layer

Data for each block layer consists of coefficient data. The structure is shown in Figure 13.5.

Dct_dc_size_luminance

This optional variable-length code is present only for Y intra-coded blocks, and specifies the number of bits in the following *dct_dc_differential*. The values are shown in Table 13.34.

Dct_dc_differential

If *dct_dc_size_luminance* ≠ 0, this optional variable-length code is present. The values are shown in Table 13.36.

Dct_dc_size_chrominance

This optional variable-length code is present only for CbCr intra-coded blocks, and specifies the number of bits in the following *dct_dc_differential*. The values are shown in Table 13.35.

Dct_dc_differential

If *dct_dc_size_chrominance* ≠ "0," this optional variable-length code is present. The values are shown in Table 13.36.

Dct_coefficient_first

This optional variable-length codeword is used for the first DCT coefficient in non-intra-coded blocks, and is defined in Tables 13.37, 13.38, 13.39 and 13.40.

Dct_coefficient_next

Up to 63 optional variable-length codewords, present only for I, P, and B frames. They are the DCT coefficients after the first one, and are defined in Tables 13.37, 13.38, 13.39 and 13.40.

End_of_block

This 2-bit or 4-bit value is used to indicate that no additional non-zero coefficients are present. The value of this parameter is "10" or "0110."

Motion Compensation

Figure 13.14 illustrates the basic motion compensation process. Motion compensation forms predictions from previously decoded pictures, which are in turn combined with the coefficient data (error terms) from the IDCT.

Field Prediction

Prediction for P pictures is made from the two most recently decoded reference fields. The simplest case is shown in Figure 13.15, used when predicting the first picture of a frame or when using field prediction within a frame.

Predicting the second field of a frame also requires the two most recently decoded reference fields. This is shown in Figure 13.16 where the second picture is the bottom field and in Figure 13.17 where the second picture is the top field.

Field prediction for B pictures is made from the two fields of the two most recent reference frames, as shown in Figure 13.18.

Frame Prediction

Prediction for P pictures is made from the most recently decoded picture, as shown in Figure 13.19. The reference picture may have been coded as either two fields or a single frame.

Frame prediction for B pictures is made from the two most recent reference frames, as shown in Figure 13.20. Each reference frame may have been coded as either two fields or a single frame.

CBP	Code			CBP	Code		
60	111			62	0100	0	
4	1101			24	0011	11	
8	1100			36	0011	10	
16	1011			3	0011	01	
32	1010			63	0011	00	
12	1001	1		5	0010	111	
48	1001	0		9	0010	110	
20	1000	1		17	0010	101	
40	1000	0		33	0010	100	
28	0111	1		6	0010	011	
44	0111	0		10	0010	010	
52	0110	1		18	0010	001	
56	0110	0		34	0010	000	
1	0101	1		7	0001	1111	
61	0101	0		11	0001	1110	
2	0100	1		19	0001	1101	
35	0001	1100		38	0000	1100	
13	0001	1011		29	0000	1011	
49	0001	1010		45	0000	1010	
21	0001	1001		53	0000	1001	
41	0001	1000		57	0000	1000	
14	0001	0111		30	0000	0111	
50	0001	0110		46	0000	0110	
22	0001	0101		54	0000	0101	
42	0001	0100		58	0000	0100	
15	0001	0011		31	0000	0011	1
51	0001	0010		47	0000	0011	0
23	0001	0001		55	0000	0010	1
43	0001	0000		59	0000	0010	0
25	0000	1111		27	0000	0001	1
37	0000	1110		39	0000	0001	0
26	0000	1101		0*	0000	0000	1

*Not to be used with 4:2:0 chroma structure.

Table 13.33. MPEG-2 Variable-Length Code Table for *coded_block_pattern_420*.

DCT DC Size Luminance	Code	DCT DC Size Luminance	Code
0	100	6	1111 0
1	00	7	1111 10
2	01	8	1111 110
3	101	9	1111 1110
4	110	10	1111 1111 0
5	1110	11	1111 1111 1

Table 13.34.   MPEG-2 Variable-Length Code Table for *dct_dc_size_luminance*.

DCT DC Size Chrominance	Code	DCT DC Size Chrominance	Code
0	00	6	1111 10
1	01	7	1111 110
2	10	8	1111 1110
3	110	9	1111 1111 0
4	1110	10	1111 1111 10
5	1111 0	11	1111 1111 11

Table 13.35.   MPEG-2 Variable-Length Code Table for *dct_dc_size_chrominance*.

DCT DC Differential	DCT DC Size	Code (Y)	Code (CbCr)	Additional Code
–2048 to –1024	11	111111111	1111111111	00000000000 to 01111111111
–1023 to –512	10	111111110	1111111110	0000000000 to 0111111111
–511 to –256	9	11111110	111111110	000000000 to 011111111
–255 to –128	8	1111110	11111110	00000000 to 01111111
–127 to –64	7	111110	1111110	0000000 to 0111111
–63 to –32	6	11110	111110	000000 to 011111
–31 to –16	5	1110	11110	00000 to 01111
–15 to –8	4	110	1110	0000 to 0111
–7 to –4	3	101	110	000 to 011
–3 to –2	2	01	10	00 to 01
–1	1	00	01	0
0	0	100	00	
1	1	00	01	1
2 to 3	2	01	10	10 to 11
4 to 7	3	101	110	100 to 111
8 to 15	4	110	1110	1000 to 1111
16 to 31	5	1110	11110	10000 to 11111
32 to 63	6	11110	111110	100000 to 111111
64 to 127	7	111110	1111110	1000000 to 1111111
128 to 255	8	1111110	11111110	10000000 to 11111111
256 to 511	9	11111110	111111110	100000000 to 111111111
512 to 1023	10	111111110	1111111110	1000000000 to 1111111111
1024 to 2047	11	111111111	1111111111	10000000000 to 11111111111

Table 13.36. MPEG-2 Variable-Length Code Table for *dct_dc_differential*.

Run	Level	Code			
EOB		10			
0	1	1s	if first coefficient		
0	1	11s	not first coefficient		
0	2	0100	s		
0	3	0010	1s		
0	4	0000	110s		
0	5	0010	0110	s	
0	6	0010	0001	s	
0	7	0000	0010	10s	
0	8	0000	0001	1101	s
0	9	0000	0001	1000	s
0	10	0000	0001	0011	s
0	11	0000	0001	0000	s
0	12	0000	0000	1101	0s
0	13	0000	0000	1100	1s
0	14	0000	0000	1100	0s
0	15	0000	0000	1011	1s
0	16	0000	0000	0111	11s
0	17	0000	0000	0111	10s
0	18	0000	0000	0111	01s
0	19	0000	0000	0111	00s
0	20	0000	0000	0110	11s
0	21	0000	0000	0110	10s
0	22	0000	0000	0110	01s
0	23	0000	0000	0110	00s
0	24	0000	0000	0101	11s
0	25	0000	0000	0101	10s
0	26	0000	0000	0101	01s
0	27	0000	0000	0101	00s
0	28	0000	0000	0100	11s
0	29	0000	0000	0100	10s
0	30	0000	0000	0100	01s

Notes:
1. s = sign of level; "0" for positive; s = "1" for negative.

Table 13.37a. MPEG-2 Variable-Length Code Table Zero for *dct_coefficient_first* and *dct_coefficient_next*.

Run	Level	Code			
0	31	0000	0000	0100	00s
0	32	0000	0000	0011	000s
0	33	0000	0000	0010	111s
0	34	0000	0000	0010	110s
0	35	0000	0000	0010	101s
0	36	0000	0000	0010	100s
0	37	0000	0000	0010	011s
0	38	0000	0000	0010	010s
0	39	0000	0000	0010	001s
0	40	0000	0000	0010	000s
1	1	011s			
1	2	0001	10s		
1	3	0010	0101	s	
1	4	0000	0011	00s	
1	5	0000	0001	1011	s
1	6	0000	0000	1011	0s
1	7	0000	0000	1010	1s
1	8	0000	0000	0011	111s
1	9	0000	0000	0011	110s
1	10	0000	0000	0011	101s
1	11	0000	0000	0011	100s
1	12	0000	0000	0011	011s
1	13	0000	0000	0011	010s
1	14	0000	0000	0011	001s
1	15	0000	0000	0001	0011s
1	16	0000	0000	0001	0010s
1	17	0000	0000	0001	0001s
1	18	0000	0000	0001	0000s
2	1	0101	s		
2	2	0000	100s		
2	3	0000	0010	11s	
2	4	0000	0001	0100	s

Notes:
1. s = sign of level; "0" for positive; s = "1" for negative.

Table 13.37b. MPEG-2 Variable-Length Code Table Zero for *dct_coefficient_first* and *dct_coefficient_next*.

Run	Level	Code			
2	5	0000	0000	1010	0s
3	1	0011	1s		
3	2	0010	0100	s	
3	3	0000	0001	1100	s
3	4	0000	0000	1001	1s
4	1	0011	0s		
4	2	0000	0011	11s	
4	3	0000	0001	0010	s
5	1	0001	11s		
5	2	0000	0010	01s	
5	3	0000	0000	1001	0s
6	1	0001	01s		
6	2	0000	0001	1110	s
6	3	0000	0000	0001	0100s
7	1	0001	00s		
7	2	0000	0001	0101	s
8	1	0000	111s		
8	2	0000	0001	0001	s
9	1	0000	101s		
9	2	0000	0000	1000	1s
10	1	0010	0111	s	
10	2	0000	0000	1000	0s
11	1	0010	0011	s	
11	2	0000	0000	0001	1010s
12	1	0010	0010	s	
12	2	0000	0000	0001	1001s
13	1	0010	0000	s	
13	2	0000	0000	0001	1000s
14	1	0000	0011	10s	
14	2	0000	0000	0001	0111s
15	1	0000	0011	01s	
15	2	0000	0000	0001	0110s

Notes:
1. s = sign of level; "0" for positive; s = "1" for negative.

Table 13.37c. MPEG-2 Variable-Length Code Table Zero for *dct_coefficient_first* and *dct_coefficient_next*.

Run	Level	Code			
16	1	0000	0010	00s	
16	2	0000	0000	0001	0101s
17	1	0000	0001	1111	s
18	1	0000	0001	1010	s
19	1	0000	0001	1001	s
20	1	0000	0001	0111	s
21	1	0000	0001	0110	s
22	1	0000	0000	1111	1s
23	1	0000	0000	1111	0s
24	1	0000	0000	1110	1s
25	1	0000	0000	1110	0s
26	1	0000	0000	1101	1s
27	1	0000	0000	0001	1111s
28	1	0000	0000	0001	1110s
29	1	0000	0000	0001	1101s
30	1	0000	0000	0001	1100s
31	1	0000	0000	0001	1011s
ESC		0000	01		

Notes:

1. s = sign of level; "0" for positive; s = "1" for negative.

Table 13.37d. MPEG-2 Variable-Length Code Table Zero for *dct_coefficient_first* and *dct_coefficient_next*.

Run	Level	Code			
EOB		0110			
0	1	10s			
0	2	110s			
0	3	0111	s		
0	4	1110	0s		
0	5	1110	1s		
0	6	0001	01s		
0	7	0001	00s		
0	8	1111	011s		
0	9	1111	100s		
0	10	0010	0011	s	
0	11	0010	0010	s	
0	12	1111	1010	s	
0	13	1111	1011	s	
0	14	1111	1110	s	

Notes:

1. s = sign of level; "0" for positive; s = "1" for negative.

Table 13.38a. MPEG-2 Variable-Length Code Table One for *dct_coefficient_first* and *dct_coefficient_next*.

Run	Level	Code			
0	15	1111	1111	s	
0	16	0000	0000	0111	11s
0	17	0000	0000	0111	10s
0	18	0000	0000	0111	01s
0	19	0000	0000	0111	00s
0	20	0000	0000	0110	11s
0	21	0000	0000	0110	10s
0	22	0000	0000	0110	01s
0	23	0000	0000	0110	00s
0	24	0000	0000	0101	11s
0	25	0000	0000	0101	10s
0	26	0000	0000	0101	01s
0	27	0000	0000	0101	00s
0	28	0000	0000	0100	11s
0	29	0000	0000	0100	10s
0	30	0000	0000	0100	01s
0	31	0000	0000	0100	00s
0	32	0000	0000	0011	000s
0	33	0000	0000	0010	111s
0	34	0000	0000	0010	110s
0	35	0000	0000	0010	101s
0	36	0000	0000	0010	100s
0	37	0000	0000	0010	011s
0	38	0000	0000	0010	010s
0	39	0000	0000	0010	001s
0	40	0000	0000	0010	000s
1	1	010s			
1	2	0011	0s		
1	3	1111	001s		
1	4	0010	0111	s	
1	5	0010	0000	s	
1	6	0000	0000	1011	0s

Notes:
1. s = sign of level; "0" for positive; s = "1" for negative.

Table 13.38b. MPEG-2 Variable-Length Code Table One for *dct_coefficient_first* **and** *dct_coefficient_next.*

Run	Level	Code			
1	7	0000	0000	1010	1s
1	8	0000	0000	0011	111s
1	9	0000	0000	0011	110s
1	10	0000	0000	0011	101s
1	11	0000	0000	0011	100s
1	12	0000	0000	0011	011s
1	13	0000	0000	0011	010s
1	14	0000	0000	0011	001s
1	15	0000	0000	0001	0011s
1	16	0000	0000	0001	0010s
1	17	0000	0000	0001	0001s
1	18	0000	0000	0001	0000s
2	1	0010	1s		
2	2	0000	111s		
2	3	1111	1100	s	
2	4	0000	0011	00s	
2	5	0000	0000	1010	0s
3	1	0011	1s		
3	2	0010	0110	s	
3	3	0000	0001	1100	s
3	4	0000	0000	1001	1s
4	1	0001	10s		
4	2	1111	1101	s	
4	3	0000	0001	0010	s
5	1	0001	11s		
5	2	0000	0010	0s	
5	3	0000	0000	1001	0s
6	1	0000	110s		
6	2	0000	0001	1110	s
6	3	0000	0000	0001	0100s
7	1	0000	100s		
7	2	0000	0001	0101	s

Notes:
1. s = sign of level; "0" for positive; s = "1" for negative.

Table 13.38c. MPEG-2 Variable-Length Code Table One for *dct_coefficient_first* and *dct_coefficient_next*.

Run	Level	Code			
8	1	0000	101s		
8	2	0000	0001	0001	s
9	1	1111	000s		
9	2	0000	0000	1000	1s
10	1	1111	010s		
10	2	0000	0000	1000	0s
11	1	0010	0001	s	
11	2	0000	0000	0001	1010s
12	1	0010	0101	s	
12	2	0000	0000	0001	1001s
13	1	0010	0100	s	
13	2	0000	0000	0001	1000s
14	1	0000	0010	1s	
14	2	0000	0000	0001	0111s
15	1	0000	0011	1s	
15	2	0000	0000	0001	0110s
16	1	0000	0011	01s	
16	2	0000	0000	0001	0101s
17	1	0000	0001	1111	s
18	1	0000	0001	1010	s
19	1	0000	0001	1001	s
20	1	0000	0001	0111	s
21	1	0000	0001	0110	s
22	1	0000	0000	1111	1s
23	1	0000	0000	1111	0s
24	1	0000	0000	1110	1s
25	1	0000	0000	1110	0s
26	1	0000	0000	1101	1s
27	1	0000	0000	0001	1111s
28	1	0000	0000	0001	1110s
29	1	0000	0000	0001	1101s
30	1	0000	0000	0001	1100s
31	1	0000	0000	0001	1011s
ESC		0000	01		

Notes:
1. s = sign of level; "0" for positive; s = "1" for negative.

Table 13.38d. MPEG-2 Variable-Length Code Table One for
dct_coefficient_first **and** ***dct_coefficient_next*.**

Run Length	Code	
0	0000	00
1	0000	01
2	0000	10
:	:	:
62	1111	10
63	1111	11

**Table 13.39. Run Encoding Following An Escape Code
for *dct_coefficient_first* and *dct_coefficient_next*.**

Level	Code		
–2047	1000	0000	0001
–2046	1000	0000	0010
:	:		
–1	1111	1111	1111
forbidden	0000	0000	0000
1	0000	0000	0001
:	:		
2047	0111	1111	1111

**Table 13.40. Level Encoding Following An Escape Code
for *dct_coefficient_first* and *dct_coefficient_next*.**

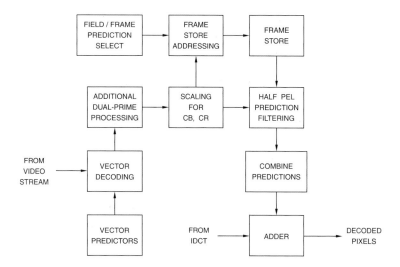

Figure 13.14. Simplified Motion Compensation Process.

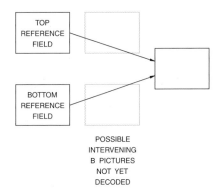

Figure 13.15. P Picture Prediction of First Field or Field Prediction in a Frame Picture.

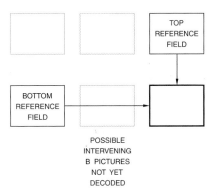

Figure 13.16. P Picture Prediction of Second Field Picture (Bottom Field).

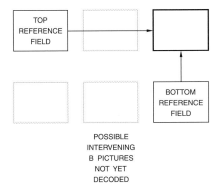

Figure 13.17. P Picture Prediction of Second Field Picture (Top Field).

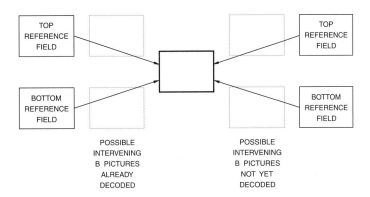

Figure 13.18. Field Prediction of B Field or Frame Pictures.

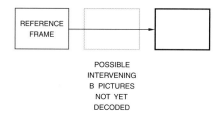

Figure 13.19. Frame Prediction For P Pictures.

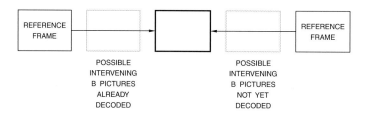

Figure 13.20. Frame Prediction For B Pictures.

PES Packet

A packetized elementary stream (PES), consists of a single elementary stream (ES) which has been made into packets, each starting with an added packet header. A PES contains only one type of data (audio, video, etc.) from one source.

The general format of the PES packet is shown in Figure 13.21. Note that start codes ($000001xx_H$) must be byte aligned by inserting 0–7 "0" bits before the start code.

Packet_start_code_prefix

This 24-bit field has a value of 000001_H and in conjunction with *stream_ID*, indicates the beginning of a packet.

Stream_ID

This 8-bit code specifies the type and number of elementary streams, as shown in Table 13.41. For the ATSC and OpenCable™ standards, the value for audio streams must be "1011 1101" to indicate Dolby® Digital.

PES_packet_length

This 16-bit binary number specifies the number of bytes in the PES packet following this field. A value of zero indicates it is neither specified nor bounded, and is used only in transport streams. For the ATSC standard, the value must be 0000_H for video streams.

Note: The following fields (until the next note) are not present if stream_ID = program stream map, padding stream, private stream 2, ECM stream, EMM stream, DSM-CC stream, H.222.1 type E, or program stream directory.

Marker_bits

These optional two bits have a value of "10."

PES_scrambling_control

This optional 2-bit code specifies the scrambling mode. "00" = not scrambled, "01" = reserved, "10" = scrambled with even key, "11" = scrambled with odd key. For the SVCD, ATSC and OpenCable™ standards, the value must be "00."

PES_priority

This optional bit specifies the priority of the payload of the PES packet. A "1" has a higher priority than a "0." For the DVB standard, this field is optional, and may be ignored by the decoder if present. For the SVCD standard, this value must be "0."

Data_alignment_indicator

A "1" for this optional bit indicates that the PES packet header is immediately followed by the video start code or audio syncword specified by the *Data Stream Alignment Descriptor* (if present). For the ATSC standard, the value must be "0" for SMPTE VC-9 video streams; "1" otherwise. For the SVCD standard, this value must be "0."

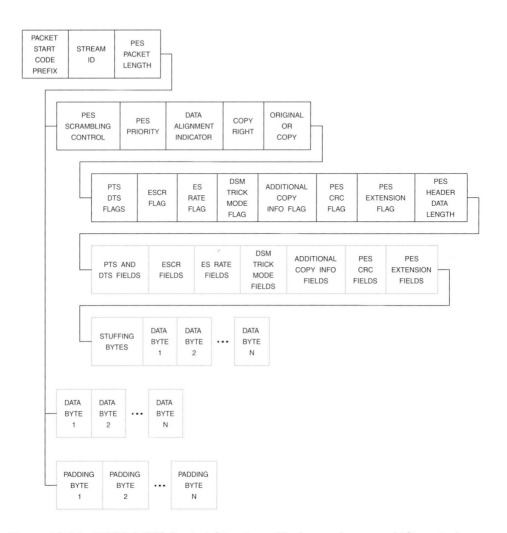

Figure 13.21. MPEG-2 PES Packet Structure. Marker and reserved bits not shown.

Stream	Code
all audio streams	1011 1000
all video streams	1011 1001
program stream map	1011 1100
private stream 1 (includes Dolby and DTS audio)	1011 1101
padding stream	1011 1110
private stream 2	1011 1111
MPEG-1, -2, -4 Part 3 MPEG-2 Part 7 audio stream	110x xxxx
MPEG-1, -2, -4 Part 2 MPEG-4 Part 10 video stream	1110 xxxx
ECM stream	1111 0000
EMM stream	1111 0001
DSM-CC stream	1111 0010
ISO/IEC 13552 stream	1111 0011
ITU-T H.222.1 type A	1111 0100
ITU-T H.222.1 type B	1111 0101
ITU-T H.222.1 type C	1111 0110
ITU-T H.222.1 type D	1111 0111
ITU-T H.222.1 type E	1111 1000
ancillary_stream	1111 1001
MPEG-4 SL-packetized stream	1111 1010
MPEG-4 FlexMux stream	1111 1011
metadata stream	1111 1100
extended_stream_ID	1111 1101
reserved	1111 1110
program stream directory	1111 1111

Table 13.41. MPEG-2 *stream_ID* **Codewords.**

Stream	Code
IPMP control information stream	000 0000
IPMP stream	000 0001
reserved_data_stream	000 0010
:	
reserved_data_stream	011 1111
private_stream	100 0000
:	
private_stream	111 1111

Table 13.42. MPEG-2 *stream_ID_extension* **Codewords. PES packets for IPMP streams have unique syntax described in MPEG-2 Part 11.**

Copyright

A "1" for this bit indicates that the material is copyrighted. For the SVCD standard, this value must be "0."

Original_or_copy

A "1" for this bit indicates that the material is original. A "0" indicates it is a copy. For the SVCD standard, this value must be "1."

PTS_DTS_flags

A value of "10" for these two bits indicates a PTS (presentation time stamp) field is present in the PES packet header. A value of "11" indicates both a PTS and DTS (decoding time stamp) fields are present. A value of "00" indicates neither PTS or DTS fields are present. For the SVCD standard, this value must be "00", "10" or "11" ("11" only for video).

ESCR_flag

A "1" for this bit indicates the ESCR (elementary stream clock reference) base and extension fields are present in the PES packet header. For the SVCD, ATSC and OpenCable™ standards, the value must be "0." For the DVB standard, the ESCR fields are optional, and may be ignored by the decoder if present.

ES_rate_flag

A "1" for this bit indicates *ES_rate* (elementary stream rate) is present in the PES packet header. For the SVCD, ATSC and Open-Cable™ standards, the value must be "0." For the DVB standard, the ES fields are optional, and may be ignored by the decoder if present.

DSM_trick_mode_flag

A "1" for this bit indicates that the *trick_mode_control* field is present. For the SVCD standard, this value must be "0."

Additional_copy_info_flag

A "1" for this bit indicates that the *additional_copy_info* field is present. For the SVCD standard, this value must be "0."

PES_CRC_flag

A "1" for this bit indicates that the *previous_PES_packet_CRC* field is present. For the SVCD, ATSC and OpenCable™ standards, the value must be "0."

PES_extension_flag

A "1" for this bit indicates that an extension field is present in this PES packet header.

PES_header_data_length

This 8-bit binary number specifies the number of bytes for optional fields and stuffing in this PES packet header.

Marker_bits

These optional four bits have a value of "0010." This field is present only if *PTS_DTS_flags* = "10."

PTS [32–30]

This optional field is present only if *PTS_DTS_flags* = "10."

Marker_bit

This optional bit has a value of "1." This field is present only if *PTS_DTS_flags* = "10."

PTS [29–15]

This optional field is present only if *PTS_DTS_flags* = "10."

Marker_bit

This optional bit has a value of "1." This field is present only if *PTS_DTS_flags* = "10."

PTS [14–0]

The optional 33-bit presentation time stamp (PTS) indicates the intended time of display by the decoder. It is specified in periods of the 27-MHz clock divided by 300. This field is present only if *PTS_DTS_flags* = "10."

Marker_bit

This optional bit has a value of "1." This field is present only if *PTS_DTS_flags* = "10."

Marker_bits

These optional four bits have a value of "0011." This field is present only if *PTS_DTS_flags* = "11."

PTS [32–30]

This optional field is present only if *PTS_DTS_flags* = "11."

Marker_bit

This optional bit has a value of "1." This field is present only if *PTS_DTS_flags* = "11."

PTS [29–15]

This optional field is present only if *PTS_DTS_flags* = "11."

Marker_bit

This optional bit has a value of "1." This field is present only if *PTS_DTS_flags* = "11."

PTS [14–0]

This optional field is present only if *PTS_DTS_flags* = "11."

Marker_bit

This optional bit has a value of "1." This field is present only if *PTS_DTS_flags* = "11."

Marker_bits

These optional four bits have a value of "0001." This field is present only if *PTS_DTS_flags* = "11."

DTS [32–30]

This optional field is present only if *PTS_DTS_flags* = "11."

Marker_bit

This optional bit has a value of "1." This field is present only if *PTS_DTS_flags* = "11."

DTS [29–15]

This optional field is present only if *PTS_DTS_flags* = "11."

Marker_bit

This optional bit has a value of "1." This field is present only if *PTS_DTS_flags* = "11."

DTS [14–0]

The optional 33-bit decoding time stamp (DTS) indicates the intended time of decoding. It is specified in periods of the 27-MHz clock divided by 300. This field is present only if *PTS_DTS_flags* = "11."

Marker_bit

This optional bit has a value of "1." This field is present only if *PTS_DTS_flags* = "11."

Reserved_bits

These optional two bits have a value of "11." This field is present only if *ESCR_flag* = "1."

ESCR_base [32–30]

This optional field is present only if *ESCR_flag* = "1."

Marker_bit

This optional bit has a value of "1." This field is present only if *ESCR_flag* = "1."

ESCR_base [29–15]

This optional field is present only if *ESCR_flag* = "1."

Marker_bit

This optional bit has a value of "1." This field is present only if *ESCR_flag* = "1."

ESCR_base [14–0]

This optional field is present only if *ESCR_flag* = "1."

Marker_bit

This optional bit has a value of "1." This field is present only if *ESCR_flag* = "1."

ESCR_extension

The optional 9-bit elementary stream clock reference (ESCR) extension and the 33-bit ESCR base are combined into a 42-bit value. It indicates the intended time of arrival of the byte containing the last bit of *ESCR_base*. The value of *ESCR_base* specifies the number of 90-kHz clock periods.

The value of *ESCR_extension* specifies the number of 27-MHz clock periods after the 90 kHz period starts.

This field is present only if *ESCR_flag* = "1."

Marker_bit

This optional bit has a value of "1." This field is present only if *ESCR_flag* = "1."

Marker_bit

This optional bit has a value of "1." This field is present only if *ES_rate_flag* = "1."

ES_rate

This optional 22-bit elementary stream rate (*ES_rate*) indicates the rate the decoder receives bytes of the PES packet. It is specified in units of 50 bytes per second. This field is present only if *ES_rate_flag* = "1."

Marker_bit

This optional bit has a value of "1." This field is present only if *ES_rate_flag* = "1."

Trick_mode_control

This optional 3-bit codeword indicates which trick mode is applied to the video stream, as shown in Table 13.43. This field is present only if *DSM_trick_mode_flag* = "1."

Trick Mode	Code
fast forward	000
slow forward	001
freeze frame	010
fast reverse	011
slow reverse	100
reserved	101
reserved	110
reserved	111

Table 13.43. MPEG-2 *trick_mode_control* Codewords.

Field_ID

This optional 2-bit codeword indicates which fields are to be displayed, as shown in Table 13.44. This field is present only if *DSM_trick_mode_flag* = "1" and *trick_mode_control* = "000" or "011."

Field ID	Code
top field only	00
bottom field only	01
complete frame	10
reserved	11

Table 13.44. MPEG-2 *field_ID* Codewords.

Intra_slice_refresh

A "1" for this optional bit indicates that there may be missing macroblocks between slices. This field is present only if *DSM_trick_mode_flag* = "1" and *trick_mode_control* = "000" or "011."

Frequency_truncation

This optional 2-bit codeword indicates that a restricted set of coefficients may have been used in coding the data, as shown in Table 13.45. This field is present only if *DSM_trick_mode_flag* = "1" and *trick_mode_control* = "000" or "011."

Description	Code
only DC coefficients are non-zero	00
first 3 coefficients are non-zero	01
first 6 coefficients are non-zero	10
all coefficients may be non-zero	11

Table 13.45. MPEG-2 *frequency_truncation* Codewords.

Rep_cntrl

This optional 5-bit binary number indicates the number of times each interlaced field or progressive frame should be displayed. A value of "00000" is not allowed. This field is present only if *DSM_trick_mode_flag* = "1" and *trick_mode_control* = "001" or "100."

Field_ID

This optional 2-bit codeword shown in Table 13.43 indicates which fields are to be displayed. This field is present only if *DSM_trick_mode_flag* = "1" and *trick_mode_control* = "010."

Reserved_bits

These three optional bits have a value of "111." This field is present only if *DSM_trick_mode_flag* = "1" and *trick_mode_control* = "010."

Reserved_bits

These five bits have a value of "1 1111." This field is present only if *DSM_trick_mode_flag* = "1" and *trick_mode_control* = "101," "110," or "111."

Marker_bit

This optional bit is always a "1." This field is present only if *additional_copy_info_flag* = "1."

Additional_copy_info

This optional 7-bit field contains private data regarding copyright information. This field is present only if *additional_copy_info_flag* = "1."

Previous_PES_packet_CRC

These optional 16 bits are present only if *PES_CRC_flag* = "1." For the DVB standard, this field is optional, and may be ignored by the decoder if present.

PES_private_data_flag

A "1" for this optional bit indicates that private data is present. This field is present only if *PES_extension_flag* = "1." For the SVCD, ATSC and OpenCable™ standards, the value must be "0."

Pack_header_field_flag

A "1" for this bit indicates that an MPEG-1 pack header or program stream pack header is in this PES packet header. This field is present only if *PES_extension_flag* = "1." For the SVCD, ATSC and OpenCable™ standards, the value must be "0."

Program_packet_sequence_counter_flag

A "1" for this optional bit indicates that the *program_packet_sequence_counter*, *MPEG1_MPEG2_identifier*, and *original_stuff_length* fields are present. This field is present only if *PES_extension_flag* = "1." For the SVCD, ATSC and OpenCable™ standards, the value must be "0."

P-STD_buffer_flag

A "1" for this optional bit indicates that *P-STD_buffer_scale* and *P-STD_buffer_size* are present. This field is present only if *PES_extension_flag* = "1." For the ATSC and OpenCable™ standards, the value must be "0." For the SVCD standard, this value must be "1."

Reserved_bits

These optional three bits are always "111." This field is present only if *PES_extension_flag* = "1."

PES_extension_flag_2

A "1" for this optional bit indicates that *PES_extension_field_length* and associated fields are present. This field is present only if *PES_extension_flag* = "1." For the SVCD standard, this value must be "0."

PES_private_data

These optional 128 bits of private data, combined with the fields before and after, must not emulate the *packet_start_code_prefix*. This field is present only if *PES_extension_flag* = "1" and *PES_private_data_flag* = "1." For the DVB standard, this field is optional, and may be ignored by the decoder if present.

Pack_field_length

This optional 8-bit binary number indicates the length, in bytes, of an immediately following pack header. This field, and the immediately following pack header, are present only if *PES_extension_flag* = "1" and *pack_header_field_flag* = "1."

Marker_bit

This optional bit is always a "1." This field is present only if *PES_extension_flag* = "1" and *program_packet_sequence_counter_flag* = "1."

Program_packet_sequence_counter

This optional 7-bit binary number increments with each successive PES packet in a program stream or MPEG-1 system stream. It wraps around to zero after reaching its maximum value. No two consecutive PES packets can have the same values. This field is present only if *PES_extension_flag* = "1" and *program_packet_sequence_counter_flag* = "1." For the DVB standard, this field is optional, and may be ignored by the decoder if present.

Marker_bit

This optional bit is always a "1." This field is present only if *PES_extension_flag* = "1" and *program_packet_sequence_counter_flag* = "1."

MPEG1_MPEG2_identifier

A "1" for this optional bit indicates the PES packet has information from a MPEG-1 system stream. A "0" indicates the PES packet has information from a program stream. This field is present only if *PES_extension_flag* = "1" and *program_packet_sequence_counter_flag* = "1." For the DVB standard, this field is optional, and may be ignored by the decoder if present.

Original_stuff_length

This optional 6-bit binary number specifies the number of stuffing bytes used in the original PES or MPEG-1 packet header. This field is present only if *PES_extension_flag* = "1" and *program_packet_sequence_counter_flag* = "1." For the DVB standard, this field is optional, and may be ignored by the decoder if present.

Marker_bits

These optional two bits are always "01." This field is present only if *PES_extension_flag* and *P-STD_buffer_flag* = "1."

P-STD_buffer_scale

This optional bit indicates the scaling factor for the following *P-STD_buffer_size* parameter. For audio streams, a value of "0" is present. For video streams, a value of "1" is present. For all other types of streams, a value of "0" or "1" may be used. This field is present only if *PES_extension_flag* and *P-STD_buffer_flag* = "1." For the DVB standard, this field is optional, and may be ignored by the decoder if present

P-STD_buffer_size

This optional 13-bit binary number specifies the size of the decoder input buffer. If *P-STD_buffer_scale* is a "0," the unit is 128 bytes. If *P-STD_buffer_scale* is a "1," the unit is 1024 bytes. This field is present only if *PES_extension_flag* and *P-STD_buffer_flag* = "1." For the DVB standard, this field is optional, and may be ignored by the decoder if present.

Marker_bit

This optional bit is always a "1." This field is present only if *PES_extension_flag* and *PES_extension_flag_2* = "1."

PES_extension_field_length

An optional 7-bit binary number that indicates the total number of bytes for the next three fields. This field is present only if *PES_extension_flag* and *PES_extension_flag_2* = "1."

Stream_ID_extension_flag

A "0" for this optional 1-bit flag indicates that a *stream_ID_extension* field follows. A "1" for this flag is reserved. This field is present only if *PES_extension_flag* and *PES_extension_flag_2* = "1."

Stream_ID_extension

This 7-bit codeword is used as an extension to *stream_ID* to specify the elementary stream type as defined in Table 13.42. This field is not used unless *stream_ID* = "1111 1101." This optional field is present only if *PES_extension_flag* and *PES_extension_flag_2* = "1" and *stream_ID_extension_flag* = "0."

Reserved_byte

[n] bytes of reserved data with a value of "1111 1111." This optional field is present only if *PES_extension_flag* and *PES_extension_flag_2* = "1" and *stream_ID_extension_flag* = "0."

Stuffing_byte

[n] optional bytes with a value of "1111 1111." Up to 32 stuffing bytes may be used. They are ignored by the decoder.

PES_packet_data_byte

[n] bytes of data from the audio stream, video stream, private stream 1, ancillary stream, H.222.1 types A–D stream, or ISO/IEC 13552 stream. The number of bytes is derived from the *PES_packet_length* field.

Note: The following field is present if stream_ID = program stream map, private stream 2, ECM stream, EMM stream, DSM-CC stream, H.222.1 type E, or program stream directory.

PES_packet_data_byte

[n] bytes of data from program stream map, private stream 2, ECM stream, EMM stream, DSM-CC stream, H.222.1 type E stream, or program stream directory descriptors. The number of bytes is derived from the *PES_packet_length* field.

Note: The following field is present if stream_ID = padding stream.

Padding_byte

[n] bytes that have a value of "1111 1111." The number of bytes is specified by *PES_packet_length*. It is ignored by the decoder.

Program Stream

The program stream, used by the DVD and SVCD standards, is designed for use in relatively error-free environments. It consists of one or more PES packets multiplexed together and coded with data that allows them to be decoded in synchronization. Program stream packets may be of variable and relatively great length.

The general format of the program stream is shown in Figure 13.22. Note that start codes

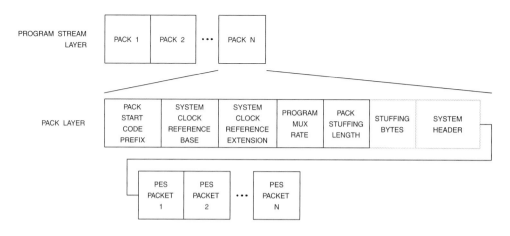

Figure 13.22. MPEG-2 Program Stream Structure. Marker and reserved bits not shown.

(000001xx$_H$) must be byte aligned by inserting 0–7 "0" bits before the start code.

Pack Layer

Data for each pack consists of a pack header followed by an optional system header and one or more PES packets.

Pack_start_code
This 32-bit string has a value of 000001BA$_H$ and indicates the beginning of a pack.

Marker_bits
These 2 bits have a value of "01."

System_clock_reference_base [32–30]

Marker_bit
This bit has a value of "1."

System_clock_reference_base [29–15]

Marker_bit
This bit has a value of "1."

System_clock_reference_base [14–0]

Marker_bit
This bit has a value of "1."

System_clock_reference_extension
This 9-bit field, along with the *system_clock_reference_base* field, comprises the system clock reference (SCR). The *system_clock_reference_base* field is specified in units of 1/300 multiplied by 90 kHz. The *system_clock_reference_extension* is specified in units of 27 MHz. SCR indicates the intended time of arrival of the byte containing the last bit of the *system_clock_reference_base*.

Marker_bit
This bit has a value of "1."

Program_mux_rate
This 22-bit binary number specifies a value measured in 50 bytes per second, with a value of zero forbidden. It indicates the rate at which the decoder receives the program stream. For the SVCD standard, this value must be $\leq 6972_D$.

Marker_bit
This bit has a value of "1."

Marker_bit
This bit has a value of "1."

Reserved_bits
These five bits have a value of "1 1111."

Pack_stuffing_length
This 3-bit binary number specifies the number of *stuffing_byte* fields following this field.

Stuffing_byte
0–7 stuffing bytes may be present. They are ignored by the decoder. Each byte has a value of "1111 1111."

System Header

This field contains a summary of the bit-stream parameters. There must be one following the first pack header, and then it may be optionally repeated in future pack headers.

System_header_start_code
This 32-bit string has a value of 000001BB$_H$ and indicates the beginning of a system header.

Header_length

This 16-bit binary number specifies the number of bytes of the system header following this field.

Marker_bit

This bit has a value of "1."

Rate_bound

This 22-bit binary number specifies a value greater than or equal to the maximum value of *program_mux_rate*. For the SVCD standard, this value must be 6972_D.

Marker_bit

This bit has a value of "1."

Audio_bound

This 6-bit binary number specifies a value (0 to 32) that is greater than or equal to the maximum number of active audio streams. For the SVCD standard, this value must be 0, 1 or 2.

Fixed_flag

A "1" for this bit indicates fixed bit rate operation. A "0" indicates variable bit rate operation. For the SVCD standard, this value must be "0."

CSPS_flag

This bit is a "1" if the program stream is a "constrained system parameters stream" (CSPS).

System_audio_lock_flag

This bit is a "1" if there is a specified, constant relationship between the audio sampling rate and the decoder system clock frequency. For the SVCD standard, this value must be "1."

System_video_lock_flag

This bit is a "1" if there is a specified, constant relationship between the video picture rate and the decoder system clock frequency. For the SVCD standard, this value must be "1."

Marker_bit

This bit has a value of "1."

Video_bound

This 5-bit binary number specifies a value (0 to 16) that is greater than or equal to the maximum number of active video streams. For the SVCD standard, this value must be 0 or 1.

Reserved_bits

These seven bits have a value of "111 1111."

Stream_ID

This optional 8-bit code shown in Table 13.41 indicates the stream to which the *P-STD_buffer_bound_scale* and *P-STD_buffer_size_bound* fields apply.

Marker_bits

These optional two bits have a value of "11." They are present only if *stream_ID* is present.

P-STD_buffer_bound_scale

This optional bit is present only when *stream_ID* is present and indicates the scaling factor used for *P_STD_buffer_size_bound*. A "0" indicates the *stream_ID* specifies an audio stream. A "1" indicates the *stream_ID* specifies a video stream. For other types of stream IDs, the value may be either a "0" or a "1."

P-STD_buffer_size_bound

This optional 13-bit binary number specifies a value greater than or equal to the maximum decoder input buffer size. It is present only when *stream_ID* is present. If *P_STD_buffer_bound_scale* is a "0," the unit is 128 bytes. If *P_STD_buffer_bound_scale* is a "1," the unit is 1024 bytes.

Program Stream Map (PSM)

The program stream map (PSM) provides a description of the bitstreams in the program stream, and their relationship to one another. It is present as PES packet data if *stream_ID* = program stream map.

Packet_start_code_prefix

This 24-bit string has a value of 000001_H and indicates the beginning of a stream map.

Map_stream_ID

This 8-bit string has a value of "1011 1100."

Program_stream_map_length

This 16-bit binary number indicates the number of bytes following this field. It has a maximum value of 1018_D.

Current_next_indicator

A "1" for this bit indicates that the program stream map is currently applicable. A "0" indicates the program stream map is not applicable yet and will be the next one valid.

Reserved_bits

These two bits have a value of "11."

Program_stream_map_version

This 5-bit binary number specifies the version number of the program stream map. It must be incremented by one when the program stream map changes, wrapping around to zero after reaching a value of 31.

Reserved_bits

These 7 bits have a value of "111 1111."

Marker_bit

This bit always has a value of "1."

Program_stream_info_length

This 16-bit binary number specifies the total length in bytes of the descriptors immediately following this field.

Descriptor_loop

Various descriptors may be present in this *descriptor_loop*.

Elementary_stream_map_length

This 16-bit binary number indicates the number of bytes of all elementary stream information in this program stream map.

Note: The following four fields are present for each stream that has a unique stream_type value.

Stream_type

This 8-bit codeword specifies the type of stream as shown in Table 13.46.

Elementary_stream_ID

This 8-bit field specifies the value of *stream_ID*, as listed in Table 13.41, in the PES packet header of PES packets containing this bitstream.

Stream Type	Code	Stream Type	Code
reserved	0000 0000	MPEG-4 Part 2 visual	0001 0000
MPEG-1 video	0000 0001	MPEG-4 audio	0001 0001
MPEG-2 video	0000 0010	MPEG-4 SL-packetized stream or FlexMux stream carried in PES packets	0001 0010
MPEG-1 audio	0000 0011	MPEG-4 SL-packetized stream or FlexMux stream carried in MPEG-4 sections	0001 0011
MPEG-2 Part 3 audio	0000 0100	MPEG-2 Synchronized Download Protocol	0001 0100
MPEG-2 private sections	0000 0101	Metadata carried in PES packets	0001 0101
MPEG-2 PES packets containing private data	0000 0110	Metadata carried in metadata_sections	0001 0110
ISO/IEC 13522 MHEG	0000 0111	Metadata carried in MPEG-2 Data Carousel	0001 0111
MPEG-2 DSM CC	0000 1000	Metadata carried in MPEG-2 Object Carousel	0001 1000
ITU-T H.222.1	0000 1001	Metadata carried in MPEG-2 Synchronized Download Protocol	0001 1001
MPEG-2 DSM-CC Multi-protocol Encapsulation	0000 1010	IPMP stream	0001 1010
MPEG-2 DSM-CC U-N Messages	0000 1011	MPEG-4 Part 10 video	0001 1011
MPEG-2 DSM-CC Stream Descriptors	0000 1100	reserved	0001 1100 – 0111 1111
MPEG-2 DSM-CC Sections	0000 1101	user private	1000 0000 – 1111 1111
MPEG-2 auxiliary	0000 1110		
MPEG-2 Part 7 audio	0000 1111		
user private details			
Dolby Digital	1000 0001	DTS audio	1001 0001
OpenCable™ subtitle	1000 0010	ATSC Data Service Table and Network Resources Table	1001 0101
OpenCable™ isochronous data	1000 0011	ATSC synchronous data stream	1100 0010
ATSC program identifier	1000 0101	OpenCable™ asynchronous data	1100 0011
Dolby Digital Plus	1000 0111		
SMPE VC-9 video	1000 1000		

Table 13.46. *stream_type* **Codewords. The Code Point Registry at www.atsc.org provides a complete listing of** *stream_type* **codes.**

Elementary_stream_info_length

This 16-bit binary number specifies the total length in bytes of the descriptors immediately following this field.

Descriptor_loop

Various descriptors may be present in this *descriptor_loop*.

CRC_32

This 32-bit CRC is for the entire program stream map.

Program Stream Directory

The program stream directory provides a description of the bitstreams in the program stream, and their relationship to one another. It is present as PES packet data if *stream_ID* = program stream map.

Transport Stream

Designed for use in environments where errors are likely, such as transmission over long distances or noisy environments, transport streams are used by the ARIB, ATSC, DVB, digital cable and OpenCable™ standards.

A transport stream combines one or more programs, with one or more independent time bases, into a single stream. Each program in a transport stream may have its own time base. The time bases of different programs within a transport stream may be different.

The transport stream consists of one or more 188-byte packets. The data for each packet is from PES packets, PSI (Program Specific Information) sections, stuffing bytes, or private data. In addition to MPEG-2 data,

MPEG-4, H.264, Microsoft® Windows Media® 9 and other data may also be sent using MPEG-2 transport streams.

At the start of each packet is a Packet IDentifier (PID) that enables the decoder to determine what to do with the packet. If the MPEG data is sent using "multiple channels per carrier", the decoder uses the PIDs to determine which packets are part of the current channel being watched or recorded and therefore process them, discarding the rest. System Information (SI), such program guides, channel frequencies, etc. are also assigned a unique PID values.

The general format of the transport stream is shown in Figure 13.23. Note that start codes ($000001xx_H$) must be byte aligned by inserting 0–7 "0" bits before the start code.

Packet Layer

Data for each packet consists of a packet header followed by an optional adaptation field and/or one or more data packets.

Sync_byte

This 8-bit string has a value of "0100 0111."

Transport_error_indicator

A "1" for this bit indicates that at least one uncorrectable bit error is present in the packet.

Payload_unit_start_indicator

The meaning of this bit is dependent on the payload.

For PES packet data, a "1" indicates the data block in this packet starts with the first byte of a PES packet. A "0" indicates no PES packet starts in the data block of this packet.

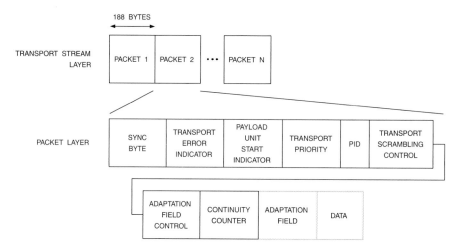

Figure 13.23. MPEG-2 Transport Stream Structure. Marker and reserved bits not shown.

For PSI data, a "1" indicates the data block of this packet contains the first byte of a PSI section.

Transport_priority

A "1" for this bit indicates that this packet is of higher priority than other packets having the same PID. For the DVB standard, this field is optional, and may be ignored by the decoder if present.

PID

This 13-bit codeword indicates the type of data in the data block, as shown in Table 13.47. The Code Point Registry at www.atsc.org provides a complete listing of *PID* codes.

Transport_scrambling_control

This 2-bit code indicates the scrambling mode of the payload. "00" = not scrambled, "01" = not scrambled (private use), "10" = scrambled with even key, "11" = scrambled with odd key. A value other than "00" requires a CA descriptor be present in the elementary stream.

Description	Code
program association table	0 0000 0000 0000
conditional access table	0 0000 0000 0001
transport stream description table	0 0000 0000 0010
IPMP control information table	0 0000 0000 0011
reserved	0 0000 0000 0100 through 0 0000 0000 1111
used by DVB	0 0000 0001 0000 through 0 0000 0001 1111
used by ARIB	0 0000 0010 0000 through 0 0000 0010 1111
may be assigned as network_PID, program_map_PID, elementary_PID, or for other purposes	0 0000 0011 0000 through 1 1111 1110 1111
CEA OSD PES packet	1 1010 1011 1100
ATSC, OpenCable™ PSIP base	1 1111 1111 1011
DOCSIS	1 1111 1111 1110
null packet	1 1111 1111 1111

Table 13.47. MPEG-2 *PID* Codewords.

Adaptation_field_control

This 2-bit code indicates whether this packet header is followed by an adaptation field or data block, as shown in Table 13.48.

Description	Code
reserved	00
data only	01
adaptation field only	10
adaptation field followed by data	11

Table 13.48. MPEG-2 *adaptation_field_control* Codewords.

Continuity_counter

This 4-bit binary number increments with each packet with the same PID. After reaching the maximum value, it wraps around. It does not increment when no data block is present in the packet.

Adaptation_field

See the Adaptation Field section.

Data_byte

These [n] data bytes are contiguous bytes of data from PES packets, PSI sections, stuffing bytes, or private data. [n] = 184 minus the number of data bytes in the Adaptation Field (if present). This field is present if *adaptation_field_control* = "01" or "11."

Adaptation Field

This field contains the 42-bit program clock references (PCRs), random access indicator and splice countdown, as well as other MPEG and private information. The PCRs are used to recreate the same 27 MHz time base clock at the decoder that was used at the encoder. This is the clock on which the presentation time stamps (PTS) are based. PCRs typically occur every 0.1 seconds in the stream. This field is present if *adaptation_field_control* = "10" or "11."

Adaptation_field_length

This 8-bit binary number specifies the number of bytes immediately following this field. The value "0000 0000" is for inserting a single stuffing byte in a transport stream packet. When *adaptation_field_control* = "11," the value is $0-182_D$. When *adaptation_field_control* = "10," the value is 183_D.

Note: None of the following fields are present if adaptation_field_length = "0000 0000."

Discontinuity_indicator

If this 1-bit flag is a "1," it indicates that there is a discontinuity state for the current transport stream packet.

Random_access_indicator

This 1-bit flag that indicates if the current transport stream packet, and possibly subsequent transport stream packets with the same PID, contain some information to aid random access.

Elementary_stream_priority_indicator

This 1-bit flag indicates, among packets with the same PID, the priority of the elementary stream data carried within this transport stream packet. A "1" indicates that the payload has a higher priority than the payloads of other transport stream packets. A "0" indicates that the payload has the same priority as all other packets which do not have this bit set to "1."

PCR_flag

A value of "1" for this 1-bit flag indicates that the adaptation field contains a PCR field. A value of "0" indicates that the adaptation field does not contain any PCR field.

OPCR_flag

A value of "1" for this optional 1-bit flag indicates that the adaptation_field contains an OPCR field. A value of "0" indicates that the adaptation field does not contain any OPCR field.

Splicing_point_flag

A value of "1" for this 1-bit flag indicates that a *splice_countdown* field is present in the associated adaptation field, specifying the occurrence of a splicing point.

Transport_private_data_flag

A value of "1" for this 1-bit flag indicates that the adaptation field contains one or more private_data bytes.

Adaptation_field_extension_flag

A value of "1" for this 1-bit flag indicates the presence of an adaptation field extension.

Program_clock_reference_base

The 33 LSBs of the optional 42-bit *program_clock_reference* field. This field is present if *PCR_flag* = "1."

Reserved_bits

These 6 optional bits have a value of "11 1111." This field is present if *PCR_flag* = "1."

Program_clock_reference_extension

This 9-bit optional field, along with the *program_clock_reference_base* field, comprises the 42-bit program clock reference (PCR). PCR indicates the intended time of arrival of the byte containing the last bit of the *program_clock_reference_base* field at the input of the decoder. This field is present if *PCR_flag* = "1."

Original_program_clock_reference_base

The 33 LSBs of the optional 42-bit *original_program_clock_reference* field. This field is present if *OPCR_flag* = "1."

Reserved_bits

These six optional bits have a value of "11 1111." This field is present if *OPCR_flag* = "1."

Original_program_clock_reference_extension

This 9-bit optional field, along with the *original_program_clock_reference_base* field, comprises the 42-bit original program clock reference (OPCR). OPCR is present only in transport stream packets that have a PCR. OPCR assists in the reconstruction of a single program transport stream from another transport stream. This field is present if *OPCR_flag* = "1."

Splice_countdown

This 8-bit 2's complement number represents a value which may be positive or negative. A positive value specifies the remaining number of transport stream packets, of the same PID, following the associated transport stream packet until a splicing point is reached. A negative number indicates that the associated transport stream packet is the nth packet following the splicing point. This field is present if *splicing_point_flag* = "1."

Transport_private_data_length

This 8-bit optional binary number [n] specifies the number of bytes immediately following this field. This field is present if *transport_private_data_flag* = "1."

Private_data_byte

These optional [n] data bytes are not specified by the MPEG-2 standard. This field is present if *transport_private_data_flag* = "1."

Adaptation_field_extension_length

This optional 8-bit binary number indicates the number of bytes of the extended adaptation field data immediately following this field, including reserved bytes if present. This field is present if *adaptation_field_extension_flag* = "1."

Ltw_flag

A "1" for this optional 1-bit flag (legal time window flag) indicates the presence of the *ltw_offset* field. This field is present if *adaptation_field_extension_flag* = "1."

Piecewise_rate_flag

A "1" for this optional 1-bit flag indicates the presence of the *piecewise_rate* field. This field is present if *adaptation_field_extension_flag* = "1."

Seamless_splice_flag

A "1" for this optional 1-bit flag indicates that the *splice_type* and *DTS_next_AU* fields are present. This field is present if *adaptation_field_extension_flag* = "1."

Reserved_bits

These five optional bits have a value of "1 1111." This field is present if *adaptation_field_extension_flag* = "1."

Ltw_valid_flag

A "1" for this optional 1-bit flag indicates that the value of *ltw_offset* is valid. This field is present if *adaptation_field_extension_flag* = "1" and *ltw_flag* = "1."

Ltw_offset

This optional 15-bit binary number specifies the legal time window offset in units of $(300/f_s)$ seconds, where f_s is the system clock frequency of the program that this PID belongs to. This field is present if *adaptation_field_extension_flag* = "1" and *ltw_flag* = "1."

Reserved_bits

These two optional bits have a value of "11." This field is present if *adaptation_field_extension_flag* = "1" and *piecewise_rate_flag* = "1."

Piecewise_rate

This optional 22-bit binary number specifies a hypothetical bitrate used to define the end times of the Legal Time Windows of transport stream packets of the same PID that follow this packet but do not include the legal_time_window_offset field. This field is present if *adaptation_field_extension_flag* = "1" and *piecewise_rate_flag* = "1."

Splice_type

This optional 4-bit binary number has the same value in all the subsequent transport stream packets of the same PID in which it is present, until the packet in which the *splice_countdown* reaches zero (including this packet). If the elementary stream carried in that PID is an audio stream, this field shall have the value "0000." If the elementary stream carried in that PID is a video stream, this field indicates the conditions that shall be respected by this elementary stream for splicing purposes. This field is present if *adaptation_field_extension_flag* = "1" and *seamless_splice_flag* = "1."

DTS_next_AU[32..30]

This optional 33-bit field indicates the decoding time of the first access unit following the splicing point. This field is present if *adaptation_field_extension_flag* = "1" and *seamless_splice_flag* = "1."

Marker_bit

This optional bit always has a value of "1." This field is present if *adaptation_field_extension_flag* = "1" and *seamless_splice_flag* = "1."

DTS_next_AU[29..15]

This optional field is present if *adaptation_field_extension_flag* = "1" and *seamless_splice_flag* = "1."

Marker_bit

This optional bit always has a value of "1." This field is present if *adaptation_field_extension_flag* = "1" and *seamless_splice_flag* = "1."

DTS_next_AU[14..0]

This optional field is present if *adaptation_field_extension_flag* = "1" and *seamless_splice_flag* = "1."

Marker_bit

This optional bit always has a value of "1." This field is present if *adaptation_field_extension_flag* = "1" and *seamless_splice_flag* = "1."

Reserved_bits

These optional [n] data bytes have a value of "1111 1111." They are present if *ltw_flag, piecewise_rate_flag* and *seamless_splice_flag* = "0," and *adaptation_field_extension_flag* = "1."

Stuffing_byte

These optional [n] data bytes have a value of "1111 1111." They are present if *OPCR_flag, adaptation_field_extension_flag, PCR_flag, transport_private_data_flag* and *splicing_point_flag* = "0."

Program Specific Information (PSI)

Program Specific Information (PSI) is additional data that enables decoders to find desired content more efficiently in a single transport stream and assemble a user-friendly electronic program guide (EPG).

Programs are composed of one or more packetized audio, video, data, etc. elementary streams (PES), each assigned a 13-bit Packet IDentification number (PID). In addition, transport stream packets carrying the same PES are assigned the same, but unique, PID.

The MPEG-2 decoder can process the correct packets only if it knows what the correct PIDs are. This is the function of the PSI. PSI is

Stream Type	Abbreviation	PID	Description
Program Association Table	PAT	0000_H	Associates program number and Program Map Table PID
Conditional Access Table	CAT	0001_H	Associates one or more (private) EMM streams each with a unique PID value
Transport Stream Description Table	TSDT	0002_H	Associates one or more descriptors to an entire transport stream
IPMP Control Information Table	ICIT	0003_H	Contains IPMP tool list, rights container, tool container defined in MPEG-2 Part 11
Program Map Table	PMT	Assigned by PAT	Specifies PID values for components of one or more programs
Network Information Table	NIT	Assigned by PAT[1]	Physical network parameters such as FDM frequencies, transponder numbers, etc.

Table 13.49. Program Specific Information Tables. [1]PID = 0010_H for ARIB and many DVB systems.

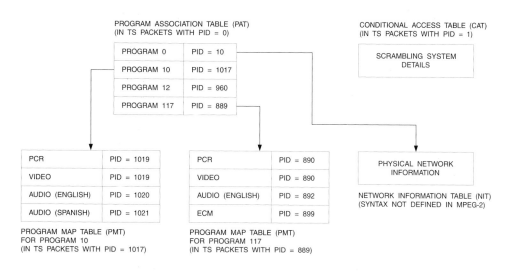

Figure 13.24. MPEG-2 PAT and PMT Example.

carried in transport stream packets having unique PIDs so the MPEG-2 decoder can easily find it.

PSI has six tables, as shown in Table 13.49. The program association table (PAT), conditional access table (CAT), transport stream description table (TSDT), IPMP control information table (ICIT) and null packet are the only fixed PIDs. The MPEG-2 decoder determines the remaining PIDs by accessing the appropriate table.

Upon first receiving a transport stream, the MPEG-2 decoder looks for the PAT, CAT, TSDT and ICIT. As shown in Figure 13.24, from the PAT, the PIDs of the network information table (NIT) and each PMT are read. From the PMTs, the PIDs of each elementary stream are read. If a program is encrypted, access to the CAT is also required.

Program Association Table (PAT)

Each transport stream contains one or more transport stream packets with a PID value of 0000_H. Together, all these packets make up the complete program association table (PAT).

The PAT provides a complete list of all programs within the transport stream, as illustrated in Figure 13.24. Included for each program is the PID value of transport stream packets containing its corresponding program map table (PMT).

As demultiplexing is impossible without a PAT, the locking speed to a new program depends on how often the PATs are sent. MPEG-2 specifies a maximum of 0.5 seconds between PATs and any PMTs that are referred to in the PATs.

The PAT may be segmented into one or sections before insertion into transport stream packets, using the following syntax:

Table_ID

This 8-bit codeword identifies the type of content and has a value of 00_H as shown in Table 13.50. The Code Point Registry at www.atsc.org provides a complete listing of *table_ID* codes.

Description	Code
PAT section	00_H
CAT section	01_H
PMT section	02_H
TSDT section	03_H
MPEG-4_scene_description_section	04_H
MPEG-4_object_descriptor_section	05_H
Metadata section	06_H
ICIT section	07_H
MPEG-2 reserved	08_H–37_H
MPEG-2 Part 6 DSM-CC sections	38_H–39_H
MPEG-2 Part 6 DSM-CC sections containing multi-protocol	$3A_H$
MPEG-2 Part 6 DSM-CC sections containing U-N messages	$3B_H$
MPEG-2 Part 6 DSM-CC sections containing download data	$3C_H$
MPEG-2 Part 6 DSM-CC sections containing stream descriptors	$3D_H$
MPEG-2 Part 6 DSM-CC sections containing private data	$3E_H$
MPEG-2 Part 6 DSM-CC addressable sections	$3F_H$
user private	40_H–FE_H
forbidden	FF_H

Table 13.50. *Table_ID* **Codewords.**

Section_syntax_indicator

The 1-bit flag is always "1."

Reserved_bits

These 2 bits have a value of "01."

Section_length

This is 12-bit binary number specifies the number of bytes of this section, starting immediately following this field and including the CRC. The value in this field may not exceed 1021_D or $3FD_H$.

Transport_stream_ID

This 16-bit binary number serves as a label to identify this transport stream from any other multiplex within a network.

Reserved_bits

These two bits have a value of "11."

Version_number

This 5-bit binary number is the version number of the whole PAT. The version number is incremented by 1 (modulo 32) whenever the definition of the PAT changes.

When *current_next_indicator* is "1," *version_number* shall be that of the currently applicable PAT. When *current_next_indicator* is "0," *version_number* is the next applicable PAT.

Current_next_indicator

When this 1-bit flag is "1," the PAT sent is currently applicable. When it is "0," the PAT sent is not yet applicable and shall be the next PAT to become valid.

Section_number

This 8-bit binary number specifies the number of this section. The *section_number* of the first section in the PAT must be 00_H. It is incremented by 1 with each additional section in the PAT.

Last_section_number

This 8-bit binary number specifies the number of the last section (that is, the section with the highest *section_number*) of the complete PAT.

Note: The following four fields are repeated for each program.

Program_number

This 16-bit binary number specifies the program that the *program_map_PID* is assigned.

Reserved_bits

These three bits have a value of "111."

Network_PID

This 13-bit binary number specifies the PID of transport stream packets that contain the network information table (NIT). It may have a value of 0010_H–$1FFE_H$. This field is present only if *program_number* = 0000_H.

Program_map_PID

This 13-bit binary number specifies the PID of transport stream packets that contain the PMT section applicable for the program as specified by *program_number*. It may have a value of 0010_H–$1FFE_H$. This field is present only if *program_number* ≠ 0000_H.

CRC_32

32-bit CRC value.

Program Map Table (PMT)

As illustrated in Figure 13.24, the program map table (PMT) provides the mappings between the program number and the program elements (video, audio, etc.). This is done by indicating the PID values of the audio, video and other streams that belong to a given program.

Note that for the ATSC, DVB and OpenCable™ standards, some PMTs require specific PIDs. Therefore, MPEG-2 and DVB/ATSC/OpenCable™ bitstreams are not fully interchangeable.

The PMT has only one section identified by the *program_number* field, using the following syntax:

Table_ID

This 8-bit codeword identifies the type of content and has a value of 02_H as shown in Table 13.50. The Code Point Registry at www.atsc.org provides a complete listing of *table_ID* codes.

Section_syntax_indicator

This 1-bit flag is always "1."

Reserved_bits

These three bits have a value of "011."

Section_length

This 12-bit binary number specifies the number of bytes of this section, starting immediately following this field and including the CRC. The value in this field may not exceed 1021_D or $3FD_H$.

Program_number

This 16-bit binary number specifies the program to which the *program_map_PID* is applicable. One program definition is carried within one program map section. A program definition is never longer than 1016_D ($3F8_H$).

Reserved_bits

These two bits have a value of "11."

Version_number

This 5-bit binary number is the version number of the program map section. The version number is incremented by 1 (modulo 32) when a change in the information carried within the section occurs.

When *current_next_indicator* is "1," *version_number* is the currently applicable program map section. When *current_next_indicator* is "0," *version_number* is the next applicable program map section.

Current_next_indicator

When this 1-bit flag is "1," the program map section sent is currently applicable. When it is "0," the program map section sent is not yet applicable and is the next program map section to become valid.

Section_number

This 8-bit binary number has a value of 00_H.

Last_section_number

This 8-bit binary number has a value of 00_H.

Reserved_bits

These three bits have a value of "111."

PCR_PID

This 13-bit binary number indicates the PID of the transport stream packets which contain PCR fields valid for the program specified by *program_number*. The value of this field may have a value of 0010_H–$1FFE_H$. If no PCR is associated with a program definition for private streams, the value is $1FFF_H$.

Reserved_bits

These four bits have a value of "1111."

Program_info_length

This 12-bit binary number specifies the total length in bytes of the descriptors immediately following this field.

Descriptor_loop

[n] descriptors may be present in this *descriptor_loop*.

Note: The following six fields are repeated for each stream type present.

Stream_type

This 8-bit codeword specifies the type of program element carried within the packets with the PID specified by *elementary_PID*. The values of *stream_type* are specified in Table 13.46.

Reserved_bits

These three bits have a value of "111."

Elementary_PID

This 13-bit binary number specifies the PID of the transport stream packets that carry the associated program element.

Reserved_bits

These four bits have a value of "1111."

ES_info_length

This 12-bit binary number specifies the total length in bytes of the descriptors immediately following this field.

Descriptor_loop

[n] descriptors may be present in this *descriptor_loop*.

CRC_32

32-bit CRC value.

Transport Stream Description Table (TSDT)

The optional transport stream description table (TSDT) can be used to include descriptors that apply to an entire transport stream.

The TSDT may be segmented into one or sections before insertion into transport stream packets, using the following syntax:

Table_ID

This 8-bit codeword identifies the type of content and has a value of 03_H as shown in Table 13.50. The Code Point Registry at www.atsc.org provides a complete listing of *table_ID* codes.

Section_syntax_indicator

This 1-bit flag is always "1."

Reserved_bit

This bit has a value of "0."

Reserved_bits

These two bits have a value of "11."

Section_length

This 12-bit binary number specifies the number of bytes of this section, starting immediately following this field and including the CRC. The value in this field may not exceed 1021_D or $3FD_H$.

Reserved_bits

These 18 bits have a value of "11 1111 1111 1111 1111."

Version_number

This 5-bit binary number is the version number of the program map section. The version number is incremented by 1 (modulo 32) when a change in the information carried within the section occurs.

When *current_next_indicator* is "1," *version_number* is the currently applicable program map section. When *current_next_indicator* is "0," *version_number* is the next applicable program map section.

Current_next_indicator

When this 1-bit flag is "1," the program map section sent is currently applicable. When it is "0," the program map section sent is not yet applicable and is the next program map section to become valid.

Section_number

This 8-bit binary number specifies the number of this section. The *section_number* of the first section must be 00_H. It is incremented by 1 with each additional section.

Last_section_number

This 8-bit binary number specifies the number of the last section (that is, the section with the highest *section_number*).

Descriptor_loop

[n] descriptors may be present in this *descriptor_loop*.

CRC_32

32-bit CRC value.

Conditional Access Table (CAT)

The conditional access table (CAT) provides association between one or more conditional access (CA) systems, their entitlement management messages (EMM) and any special parameters associated with them. The PIDs for entitlement control messages (ECM) and entitlement management messages (EMM) are included in the CAT.

The CAT may be segmented into one or sections before insertion into transport stream packets, using the following syntax:

Table_ID

This 8-bit codeword identifies the type of content and has a value of 01_H as shown in Table 13.50. The Code Point Registry at www.atsc.org provides a complete listing of *table_ID* codes.

Section_syntax_indicator

This 1-bit flag is always "1."

Reserved_bits

These three bits have a value of "011."

Section_length

This 12-bit binary number specifies the number of bytes of this section, starting immediately following this field and including the CRC. The value in this field may not exceed 1021_D or $3FD_H$.

Reserved_bits

These 18 bits have a value of "11 1111 1111 1111 1111."

Version_number

This 5-bit binary number is the version number of the whole CAT. The version number is incremented by 1 (modulo 32) when a change in the information carried within the CAT changes.

When *current_next_indicator* is "1," *version_number* is the currently applicable CAT. When *current_next_indicator* is "0," *version_number* is the next applicable CAT.

Current_next_indicator

When this 1-bit flag is "1," the CAT sent is currently applicable. When it is "0," the CAT sent is not yet applicable and is the next CAT to become valid.

Section_number

This 8-bit binary number indicates the number of this section. The *section_number* of the first section in the CAT must be 00_H. It is incremented by 1 with each additional section in the CAT.

Last_section_number

This 8-bit binary number specifies the number of the last section (that is, the section with the highest *section_number*) of the complete CAT.

Descriptor_loop

[n] descriptors may be present in this *descriptor_loop*.

CRC_32

32-bit CRC value.

Network Information Table (NIT)

The first entry in the PAT (program 0) is reserved for network data and contains the PID of the network information table (NIT). The NIT includes information of other transport streams that may be available, for example, by tuning to a different RF channel or satellite. Each transport stream may include a descriptor that specifies the radio frequency, satellite orbital position, etc. In MPEG-2, only the NIT is mandatory for this purpose. In DVB, additional metadata, known as DVB-SI, is used, and NIT is considered to be part of DVB-SI. Therefore, the term PSI/SI is a commonly used term.

IPMP Control Information Table (ICIT)

The IPMP (Intellectual Property Management and Protection) control information table contains IPMP (also known as DRM or Digital Rights Management) related information including tool list, rights container and tool container.

The *tool list* identifies, and enables selection of, IPMP tools required to process content. The *tool container* enables the carriage of binary tools in content streams. The *rights container* may contain rights description that describes usage rules associated with the IPMP protected content.

The IPMP stream carries all types of IPMP data (including key, ECM, EMM) to be delivered to the tools.

MPEG-2 PMT/PSM Descriptors

These MPEG-2 descriptors are used to identify commonly used private (non-MPEG-2) information that is present in the MPEG-2 transport or program stream.

A descriptor is typically contained within a *descriptor_loop* in the MPEG-2 PMT or PSM. The general format of a descriptor is:

    descriptor_tag (8 bits)
    descriptor_length (8 bits)
    data

Descriptor tag values of 0, 1, and 42–63 are reserved. Values of 19–26 are reserved for MPEG-2 Part 6 data. Unless otherwise indicated, descriptors may be present in both transport and program streams.

Audio Stream Descriptor

This MPEG-2 descriptor provides basic information which identifies the coding version of an audio elementary stream.

Descriptor_tag

This 8-bit field has a value of "0000 0011."

Descriptor_length

This 8-bit binary number specifies the number of bytes following this field. It has a value of "0000 0001."

Free_format_flag

A "1" for this bit indicates the *bitrate_index* field in the audio stream is "0000."

ID

This bit is set to the same value as the ID field in the audio stream.

Layer

This 2-bit binary number is set to the same or higher value as the highest layer in any audio stream.

Variable_rate_audio_indicator

A "0" for this bit indicates that the bit rate of the audio stream does not vary between audio frames.

Reserved_bits

These three bits are always "111."

AVC Video Descriptor

For individual MPEG-4 Part 10 (H.264) streams carried in PES packets, this MPEG-4 descriptor provides basic information for identifying the coding parameters.

Descriptor_tag

This 8-bit field has a value of "0011 0110."

Descriptor_length

This 8-bit binary number specifies the number of bytes following this field. It has a value of "0000 0001."

AVC_profile_and_level

This 8-bit field identifies the profile and level of the MPEG-4 Part 10 (H.264) video stream.

CA Descriptor

This MPEG-2 descriptor indicates the PIDs of transport stream packets which contain ECM or EMM information. If present in CAT, then a system-wide conditional access management system exists. If present in PMT, *CA_PID* points to packets containing program related access control information (ECM).

Descriptor_tag

This 8-bit field has a value of "0000 1001."

Descriptor_length

This 8-bit binary number specifies the number of bytes following this field.

CA_system_ID

This 16-bit binary number specifies the type of conditional access used. The coding of this field is privately defined.

Reserved_bits

These three bits are always "111."

CA_PID

This 13-bit binary number specifies the PID of the transport stream packets which contain either ECM or EMM information for the conditional access system specified by *CA_system_ID*.

For transport streams, a value of 0003_H indicates that there is IPMP used by components in the transport stream. For program streams, the presence of *stream_ID_extension* value 00_H indicates that IPMP is used by components in the program stream.

Private_data_byte

These optional [n] bytes of private data are defined by the CA owner.

Caption Service Descriptor

For OpenCable™ and generic MPEG-2 decoders, this EIA–708 descriptor must be present for each program that has closed captioning. For ATSC and OpenCable™, it must also present in the *descriptor_loop* of the event information table (EIT).

Up to sixteen individual service descriptions are supported, with each service description being six bytes in length.

Descriptor_tag

This 8-bit field has a value of "1000 0110."

Descriptor_length

This 8-bit binary number specifies the number of bytes following this field.

Reserved_bits

These three bits are always "111."

Number_of_services

This 5-bit field has a range of 1–16 to indicate the number of captioning services present.

Note: [number_of_services] specifies how many times the following nine fields are repeated.

Language

This 3-byte code specifies the language associated with the caption service, per ISO 639.2/B.

CC_type

A "1" for this 1-bit flag indicates EIA–708 captioning service is present. A "0" indicates EIA–608 captioning is present.

Reserved_bit

This bit is always "1."

Reserved_bits

These five optional bits are always "1 1111." They are present only if *caption_type* = "0."

Line21_field

A "1" for this optional 1-bit flag indicates that EIA–608 captioning for field 2 is present. A "0" indicates EIA–608 captioning for field 1 is present. This bit is present only if *caption_type* = "0."

Caption_service_number

This optional 6-bit field has a range of 1–63 to indicate the service number of the captioning stream. These bits are present only if *caption_type* = "1."

Easy_reader

A "1" for this 1-bit flag indicates that the caption service contains text formatted for beginning readers. A "0" indicates that the caption service is not tailored for this.

Wide_aspect_ratio

A "1" for this 1-bit flag indicates that the caption service is formatted for 16:9 displays. A "0" indicates that the caption service is formatted for 4:3 displays, and may be optionally displayed centered on 16:9 displays.

Reserved_bits

These fourteen bits are always "11 1111 1111 1111."

Copyright Descriptor

This MPEG-2 descriptor provides a method to enable audio-visual works identification. For the DVB standard, this descriptor is optional, and may be ignored by the decoder if present.

Descriptor_tag

This 8-bit binary number has a value of "0000 1101."

Descriptor_length

This 8-bit binary number specifies the number of bytes following this field.

Copyright_ID

This 32-bit value is obtained from the Registration Authority.

Additional_copyright_info

These optional [n] bytes of data are defined by the copyright owner and are never changed.

Data Stream Alignment Descriptor

This MPEG-2 descriptor describes the alignment of video stream syntax with respect to the start of the PES packet payload. ATSC requires this descriptor to be present in the program element loop of the PMT section that describes the video elementary stream. For the DVB standard, this descriptor is optional, and may be ignored by the decoder if present.

Descriptor_tag

This 8-bit binary number has a value of "0000 0110."

Descriptor_length

This 8-bit binary number specifies the number of bytes following this field. It has a value of "0000 0001."

Alignment_type

This 8-bit codeword specifies the audio or video alignment type as shown in Table 13.51. For the ATSC standard, the value must be "0000 0010."

Video Stream Alignment Type	Audio Stream Alignment Type	Code
reserved	reserved	0000 0000
slice, picture, GOP, or sequence	sync word	0000 0001
picture, GOP, or sequence	reserved	0000 0010
GOP or sequence		0000 0011
sequence		0000 0100
		0000 0101
reserved		:
		1111 1111

Table 13.51. *alignment_type* **Codewords.**

DTCP Descriptor

This descriptor is used to control HDCP- and DTCP-protected digital outputs, such as IEEE 1394, USB, DVI, HDMI and Ethernet (IP).

Descriptor_tag
This 8-bit binary number has a value of "1000 1000."

Descriptor_length
This 8-bit binary number specifies the number of bytes following this field.

CA_system_ID
This 16-bit binary number specifies the type of conditional access used. It has a value of $0FFE_H$ (DTLA).

Reserved_bits
These five bits are always "1 1111."

EPN
A "1" for this 1-bit Encryption Plus Non-Assertion flag specifies that content not otherwise copy controlled is not to be retransmitted over the Internet.

DTCP_CCI
This 2-bit codeword specifies the digital copy generation management:

00 = copy free
01 = no more copies
10 = copy one generation
11 = copy never

Reserved_bits
These five bits are always "1 1111."

Image_constrain_token
A "1" for this 1-bit flag specifies that high-definition content must be constrained to 520,000 total samples or less (960 × 540p for example) when output onto unprotected high-definition analog video outputs.

APS

This 2-bit codeword specifies the analog copy generation management.

> 00 = no Macrovision pseudo-sync pulse
> 01 = Macrovision pseudo-sync pulse on;
>   color striping off
> 10 = Macrovision pseudo-sync pulse on;
>   2-line color striping on
> 11 = Macrovision pseudo-sync pulse on;
>   4-line color striping on

DTS Audio Descriptor

PES packets containing DTS audio may be included in a MPEG-2 program or transport stream in the same way as MPEG or Dolby Digital audio can be included.

MPEG-2 does not explicitly support a DTS bitstream. Also, the MPEG-2 audio stream descriptor is inadequate to describe the contents of the DTS bitstream in the PSI tables.

Therefore, PES packets containing DTS audio data are sent using private stream 1. A *DTS Registration Descriptor* (*descriptor_tag* = "0000 0101") and *DTS Audio Descriptor* (*descriptor_tag* = "1001 0001") are also required.

Hierarchy Descriptor

This MPEG-2 descriptor provides information to identify the program elements containing components of hierarchically-coded video, audio and private streams.

Descriptor_tag

This 8-bit field has a value of "0000 0100."

Descriptor_length

This 8-bit binary number specifies the number of bytes following this field. It has a value of "0000 0100."

Reserved_bits

These four bits are always "1111."

Hierarchy_type

This 4-bit codeword indicates the relationship between the hierarchy layer and its embedded layer as shown in Table 13.52.

Hierarchy Type	Code
reserved	0000
spatial scalability	0001
SNR scalability	0010
temporal scalability	0011
data partitioning	0100
extension bitstream	0101
private bitstream	0110
multi-view profile	0111
reserved	1000–1110
base layer	1111

Table 13.52. *hierarchy_type* **Codewords.**

Reserved_bits

These two bits are always "11."

Hierarchy_layer_index

This 6-bit binary number indicates a unique index of the stream.

Reserved_bits

These two bits are always "11."

Hierarchy_embedded_layer_index

This 6-bit binary number defines the hierarchical table index of the stream that must be accessed prior to decoding. This parameter is undefined for a *hierarchy_type* value of "1111."

Reserved_bits

These two bits are always "11."

Hierarchy_channel

This 6-bit binary number indicates the intended channel number. The most robust channel has the lowest value.

IBP Descriptor

This MPEG-2 descriptor provides information about some characteristics of the sequence of frame types in the video sequence. For the DVB standard, this descriptor is optional, and may be ignored by the decoder if present.

Descriptor_tag

This 8-bit field has a value of "0001 0010."

Descriptor_length

This 8-bit binary number specifies the number of bytes following this field. It has a value of "0000 0010."

Closed_gop_flag

A "1" for this 1-bit flag indicates that a group of pictures header is encoded before every I-frame and that the *closed_gop* flag is set to "1" in all group of pictures headers in the video sequence.

Identical_gop_flag

A "1" for this 1-bit flag indicates the number of P-frames and B-frames between I-frames, and the picture coding types and sequence of picture types between I-pictures, is the same throughout the sequence, except possibly for pictures up to the second I-picture.

Max_gop_length

This 14-bit binary number indicates the maximum number of the coded pictures between any two consecutive I-pictures in the sequence. The value zero may not be used.

IPMP Descriptor

This MPEG-2 descriptor signals IPMP tool protection, associating IPMP tools with each protected program and indicating the control point at which a specific IPMP tool should be running.

The MPEG-2 IPMP descriptor has a *descriptor_tag* value of "0010 1001."

ISO 639 Language Descriptor

This MPEG-2 descriptor provides a method to indicate the language(s) of each audio elementary stream. If present, ATSC requires this descriptor to be in *descriptor_loop* following *ES_info_length* in the PMT for each Dolby® Digital or Dolby® Digital Plus audio elementary stream. For the DVB standard, this descriptor must be present and decoded if more than one audio (or video) stream with different languages is used for a program.

Descriptor_tag

This 8-bit field has a value of "0000 1010."

Descriptor_length

This 8-bit binary number specifies the number of bytes following this field.

Note: The following two fields are present for each language.

ISO_639_language_code
This 24-bit field contains a 3-character language code.

Audio_type
This 8-bit codeword identifies the audio type:

00_H = reserved
01_H = clean effects (no language)
02_H = hearing impaired
03_H = visual impaired commentary
04_H–FF_H = reserved

Maximum Bitrate Descriptor

This MPEG-2 descriptor provides a method to indicate information about the maximum bit rate present. It only applies to transport streams, not program streams. For the DVB standard, this descriptor is optional, and may be ignored by the decoder if present.

Descriptor_tag
This 8-bit field has a value of "0000 1110."

Descriptor_length
This 8-bit binary number specifies the number of bytes following this field. It has a value of "0000 0011."

Reserved_bits
These two bits are always "11."

Maximum_bitrate
This 22-bit binary number indicates the maximum bit rate present, in units of 50 bytes per second.

Metadata Descriptors

Various MPEG-2 descriptors are available to enable including metadata information that describes audiovisual content and data essence.

MPEG-2 defines two tools for synchronous delivery of the metadata:

PES packet payload

DSM-CC synchronized download protocol

In addition, MPEG-2 defines three tools for asynchronous delivery of metadata:

Carriage in metadata sections

DSM-CC data carousels

DSM-CC object carousels

Content Labelling Descriptor
This MPEG-2 descriptor assigns a label to content which can be used by metadata to reference the associated content.

This descriptor has a *descriptor_tag* value of "0010 0100."

Metadata Pointer Descriptor
This MPEG-2 descriptor points to one metadata service and associates its service with audiovisual content in a MPEG-2 stream.

This descriptor has a *descriptor_tag* value of "0010 0101."

Metadata Descriptor

This MPEG-2 descriptor specifies the format of the associated metadata carried in program or transport stream. It can also convey information to identify the metadata service from a collection of metadata transmitted in a DSM-CC carousel.

This descriptor has a *descriptor_tag* value of "0010 0110."

Metadata STD Descriptor

This MPEG-2 descriptor defines parameters of the standard model for the processing of the metadata stream.

This descriptor has a *descriptor_tag* value of "0010 0111."

Multiplex Buffer Utilization Descriptor

This MPEG-2 descriptor provides bounds on the occupancy of the STD multiplex buffer. For the DVB standard, this descriptor is optional, and may be ignored by the decoder if present.

Descriptor_tag

This 8-bit field has a value of "0000 1100."

Descriptor_length

This 8-bit binary number specifies the number of bytes following this field. It has a value of "0000 0100."

Bound_valid_flag

A "1" for this bit indicates *LTW_offset_lower_bound* and *LTW_offset_upper_bound* are valid.

LTW_offset_lower_bound

This 15-bit binary number is in units of (27 MHz / 300) clock periods. It specifies the lowest value any *LTW_offset* field will have.

Reserved_bit

This bit is always a "1."

LTW_offset_upper_bound

This 15-bit binary number is in units of (27 MHz / 300) clock periods. It specifies the upper value any *LTW_offset* field will have.

Private Data Indicator Descriptor

This MPEG-2 descriptor provides a method for carrying private information. For the DVB standard, this MPEG-2 descriptor is optional, and may be ignored by the decoder if present.

Descriptor_tag

This 8-bit field has a value of "0000 1111."

Descriptor_length

This 8-bit binary number specifies the number of bytes following this field. It has a value of "0000 0100."

Private_data_indicator

These 32 bits are private data and are not defined by the MPEG-2 specification.

Registration Descriptor

This MPEG-2 descriptor provides a method to uniquely identify formats of private data.

Programs that conform to ATSC are identified by this descriptor in *descriptor_loop* after *program_info_length* of the PMT.

This descriptor is also placed in *descriptor_loop* after *ES_info_length* of the PMT for each program element having a *stream_type* value in the ATSC-user private range ($C4_H$ to FF_H), to establish the private entity associated with that program element.

For the DVB standard, this descriptor is optional, and may be ignored by the decoder if present.

Descriptor_tag
This 8-bit field has a value of "0000 0101."

Descriptor_length
This 8-bit binary number specifies the number of bytes following this field.

Format_identifier
This 32-bit value is obtained from the Registration Authority (SMPTE), as illustrated in Table 13.53.

Format	Code
ATSC	47413934_H
Dolby® Digital audio	$41432D33_H$
DTS audio (512 frame size)	44545331_H
DTS audio (1024 frame size)	44545332_H
DTS audio (2048 frame size)	44545333_H

Table 13.53. *format_identifier* **Codewords.**

Additional_identification_info
These optional [n] bytes of data are defined by the registration owner and are never changed.

Smoothing Buffer Descriptor

This MPEG-2 descriptor conveys information about the size of a smoothing buffer associated with this descriptor and the associated leak rate out of that buffer. ATSC requires the PMT to have a *Smoothing Buffer Descriptor* pertaining to that program. For the DVB standard, this descriptor is recommended, but may be ignored by the decoder if present.

Descriptor_tag
This 8-bit field has a value of "0001 0000."

Descriptor_length
This 8-bit binary number specifies the number of bytes following this field. It has a value of "0000 0110."

Reserved_bits
These two bits are always "11."

Sb_leak_rate
This 22-bit binary number specifies the value of the leak rate out of the buffer for the associated elementary stream or other data in units of 400 bps.

Reserved_bits
These two bits are always "11."

Sb_size
This 22-bit binary number specifies the smoothing buffer size for the associated elementary stream or other data in 1-byte units. For ATSC and OpenCable™, this field has a value ≤ 2048.

STD Descriptor

This MPEG-2 descriptor only applies to transport streams, not program streams. For the DVB standard, this descriptor may be ignored by the decoder.

Descriptor_tag
This 8-bit field has a value of "0001 0001."

Descriptor_length

This 8-bit binary number specifies the number of bytes following this field. It has a value of "0000 0001."

Reserved_bits

These seven bits are always "111 1111."

Leak_valid_flag

This 1-bit flag specifies the technique used to transfer data between memory buffers.

System Clock Descriptor

This MPEG-2 descriptor provides a method to indicate information about the system clock used to generate timestamps. For the DVB standard, this descriptor is recommended, but may be ignored by the decoder if present.

Descriptor_tag

This 8-bit field has a value of "0000 1011."

Descriptor_length

This 8-bit binary number specifies the number of bytes following this field. It has a value of "0000 0010."

External_clock_reference_indicator

A "1" for this bit indicates that the system clock was derived from a reference that may be available at the decoder.

Reserved_bit

This bit is always a "1."

Clock_accuracy_integer

Combined with *clock_accuracy_exponent*, this 6-bit binary number provides the fractional frequency accuracy of the system clock.

Clock_accuracy_exponent

Combined with *clock_accuracy_integer*, this 3-bit binary number provides the fractional frequency accuracy of the system clock.

Reserved_bits

These five bits are always "1 1111."

Target Background Grid Descriptor

This MPEG-2 descriptor provides displaying the video within a specified location of the display. It is useful when the video is not intended to use the full area of the display. For the DVB standard, this descriptor is required when the resolution is greater than 720×576 (25 Hz bitstream) or 720×480 (30 Hz bitstream).

Descriptor_tag

This 8-bit field has a value of "0000 0111."

Descriptor_length

This 8-bit binary number specifies the number of bytes following this field. It has a value of "0000 0100."

Horizontal_size

This 14-bit binary number specifies the horizontal size of the target background grid in samples.

Vertical_size

This 14-bit binary number specifies the vertical size of the target background grid in lines.

Aspect_ratio_information

This 4-bit codeword specifies the aspect ratio as defined in the video sequence header.

Video Stream Descriptor

This MPEG-2 descriptor provides basic information which identifies the coding parameters of a video elementary stream.

Descriptor_tag

This 8-bit field has a value of "0000 0010."

Descriptor_length

This 8-bit binary number specifies the number of bytes following this field.

Multiple_frame_rate_flag

A "1" for this bit indicates multiple frame rates may be present in the video stream.

Frame_rate_code

This 4-bit codeword indicates the video frame rate, as shown in Table 13.54. When *multiple_frame_rate_flag* is a "1," the indication of a specific frame rate also allows other frame rates to be present in the video stream.

MPEG_1_only_flag

A "1" for this bit indicates the video stream contains only MPEG-1 video data.

Constrained_parameter_flag

If *MPEG_1_only_flag* is a "0," this bit must be a "1." If *MPEG_1_only_flag* is a "1," this bit reflects the value of the *constrained_parameter_flag* in the MPEG-1 video stream.

Indicated Frame Rate	May Also Include These Frame Rates	Code
forbidden		0000
23.976		0001
24.0	23.976	0010
25.0		0011
29.97	23.976	0100
30.0	23.976, 24.0, 29.97	0101
50.0	25.0	0110
59.94	23.976, 29.97	0111
60.0	23.976, 24.0, 29.97, 30.0, 59.94	1000
reserved		1001
:		:
reserved		1111

Table 13.54. *frame_rate_code* **Codewords.**

Still_picture_flag

A "1" for this bit indicates the video stream contains only still pictures. A "0" indicates the video stream may have either still or moving pictures.

Profile_and_level_indication

This optional 8-bit codeword reflects the same or higher profile and level as indicated by the *profile_and_level_indication* field in the MPEG-2 video stream. This field is present only if *MPEG_1_only_flag* = "0."

Chroma_format

This optional 2-bit codeword reflects the same or higher chroma format as indicated by the *chroma_format* field in the MPEG-2 video stream. This field is present only if *MPEG_1_only_flag* = "0."

Frame_rate_extension_flag

A "1" for this optional bit indicates that either or both of the *frame_rate_extension_n* and *frame_rate_extension_d* fields in any MPEG-2 video stream are non-zero. This field is present only if *MPEG_1_only_flag* = "0."

Reserved_bits

These five optional bits are always "1 1111." This field is present only if *MPEG_1_only_flag* = "0."

Video Window Descriptor

This MPEG-2 descriptor provides a method to indicate information about the associated video elementary stream. For the DVB standard, if this field is present, the decoder must process the data.

Descriptor_tag

This 8-bit field has a value of "0000 1000."

Descriptor_length

This 8-bit binary number specifies the number of bytes following this field. It has a value of "0000 0100."

Horizontal_offset

This 14-bit binary number indicates the horizontal position of the top left pixel of the video window on the target background grid.

Vertical_offset

This 14-bit binary number indicates the vertical position of the top left pixel of the video window on the target background grid.

Window_priority

This 4-bit binary number indicates how video windows overlap. A value of "0000" is lowest priority and "1111" is highest priority. Higher priority windows are visible over lower priority windows.

MPEG-4 PMT/PSM Descriptors

These MPEG-4 descriptors are used to identify MPEG-4-specific information that is present in the MPEG-2 transport or program stream. They are typically contained within a *descriptor_loop* in the MPEG-2 PMT or PSM, and may also be present in other MPEG-4-specific tables.

Audio Descriptor

For individual MPEG-4 Part 3 streams carried in PES packets, this MPEG-4 descriptor provides basic information for identifying the coding parameters. It does not apply to MPEG-4 Part 3 streams encapsulated in MPEG-4 SL-packets or FlexMux packets.

Descriptor_tag

This 8-bit field has a value of "0001 1100."

Descriptor_length

This 8-bit binary number specifies the number of bytes following this field. It has a value of "0000 0001."

MPEG-4_audio_profile_and_level

This 8-bit field identifies the profile and level of the MPEG-4 Part 3 audio stream.

External ES ID Descriptor

This MPEG-4 descriptor assigns an *ES_ID*, defined in MPEG-4 Part 1, to a program element which has no *ES_ID* value. *ES_ID* allows reference to a non-MPEG-4 component in the scene description or to associate a non-MPEG-4 component within an IPMP stream.

For a transport stream, this descriptor is in *descriptor_loop* after *ES_info_length* within the PMT. For a program stream, within the PSM, this descriptor is in *descriptor_loop* after *elementary_stream_info_length*.

Descriptor_tag

This 8-bit field has a value of "0010 0000."

Descriptor_length

This 8-bit binary number specifies the number of bytes following this field. It has a value of "0000 0010."

External_ES_ID

This 16-bit field assigns *ES_ID*, as defined in MPEG-4 Part 1, to a component of a program.

FMC Descriptor

This MPEG-4 descriptor indicates that the MPEG-4 FlexMux tool has been used to multiplex MPEG-4 SL-packetized streams into a FlexMux stream before encapsulation in MPEG-2 PES packets or MPEG-4 sections. It associates FlexMux channels to the *ES_ID* values of the SL-packetized streams in the FlexMux stream. A FMC Descriptor is required for each program element referenced by an *elementary_PID* value in a transport stream and for each *elementary_stream_ID* in a program stream that conveys a FlexMux stream.

For a transport stream, this descriptor is in the *descriptor_loop* after *ES_info_length* within the PMT. For a program stream, within the PSM, this descriptor is in *descriptor_loop* after *elementary_stream_info_length*.

Descriptor_tag

This 8-bit field has a value of "0001 1111."

Descriptor_length

This 8-bit binary number specifies the number of bytes following this field.

ES_ID

This 16-bit field specifies the identifier of a SL-packetized stream.

FlexMuxChannel

This 8-bit field specifies the number of the FlexMux channel used for this SL-packetized stream.

FmxBufferSize Descriptor

This MPEG-4 descriptor conveys the size of the FlexMux buffer for each MPEG-4 SL-packetized stream multiplexed in a MPEG-4 FlexMux stream. One FmxBufferSize Descriptor is associated with each *elementary_PID* or *elementary_stream_ID* conveying a FlexMux stream.

For a transport stream, this descriptor is in *descriptor_loop* after *ES_info_length* within the PMT. For a program stream, within the PSM, this descriptor is in *descriptor_loop* after *elementary_stream_info_length*.

Descriptor_tag

This 8-bit field has a value of "0010 0010."

Descriptor_length

This 8-bit binary number specifies the number of bytes following this field.

DefaultFlexMuxBufferDescriptor()

This descriptor specifies the default FlexMux buffer size for this FlexMux stream. It is defined in MPEG-4 Part 1.

FlexMux Buffer Descriptor()

This descriptor specifies the FlexMux buffer size for one SL-packetized stream carried within the FlexMux stream. It is defined in MPEG-4 Part 1.

IOD Descriptor

This MPEG-4 descriptor encapsulates the MPEG-4 Part 1 InitialObjectDescriptor structure. It allows access to MPEG-4 streams by identifying the *ES_ID* values of the MPEG-4 Part 1 scene description and object descriptor streams. Both contain further information about the MPEG-4 streams that are part of the scene.

For a transport stream, this descriptor is in *descriptor_loop* after *program_info_length* within the PMT. For a program stream, within the PSM, this descriptor is in *descriptor_loop* after *program_stream_info_length*.

Descriptor_tag

This 8-bit field has a value of "0001 1101."

Descriptor_length

This 8-bit binary number specifies the number of bytes following this field.

Scope_of_IOD_label

This 8-bit field specifies the scope of the *IOD_label* field. A value of 10_H indicates that the *IOD_label* is unique within the program stream or program in a transport stream. A value of 11_H indicates that the *IOD_label* is unique within the transport stream in which the IOD descriptor is carried. All other values are reserved.

IOD_label

This 8-bit field specifies the label of the IOD descriptor.

Initial Object Descriptor()

This structure is defined in MPEG-4 Part 1.

MultiplexBuffer Descriptor

This MPEG-4 descriptor is associated to each *elementary_PID* that contains an MPEG-4 FlexMux or SL-packetized stream, including those containing MPEG-4 sections.

This descriptor only applies to transport streams, not program streams. For a transport stream, this descriptor is in *descriptor_loop* after *ES_info_length* within the PMT.

Descriptor_tag

This 8-bit field has a value of "0010 0011."

Descriptor_length

This 8-bit binary number specifies the number of bytes following this field. It has a value of "0000 0110."

MB_buffer_size

This 24-bit field specifies the size (in bytes) of buffer [MB] of the elementary stream.

TB_leak_rate

This 24-bit field specifies the rate (in units of 400 bps) at which data is transferred from transport buffer [TB] to multiplex buffer [MB] for the elementary stream.

Muxcode Descriptor

This MPEG-4 descriptor conveys MuxCodeTableEntry structures as defined in MPEG-4 Part 1. MuxCodeTableEntries configure the MuxCode mode of FlexMux. One or more Muxcode Descriptors may be associated with each *elementary_PID* or *elementary_stream_ID* conveying an MPEG-4 FlexMux stream that utilizes the MuxCode mode.

Within a transport stream, this descriptor is in *descriptor_loop* after *ES_info_length* in the PMT. Within a program stream, within the PSM, this descriptor is in *descriptor_loop* after *elementary_stream_info_length*.

Descriptor_tag

This 8-bit field has a value of "0010 0001."

Descriptor_length

This 8-bit binary number specifies the number of bytes following this field.

MuxCodeTableEntry()

This structure is defined in MPEG-4 Part 1.

SL Descriptor

This MPEG-4 descriptor is used when a single MPEG-4 SL-packetized stream is encapsulated in MPEG-2 PES packets. It associates the *ES_ID* of the SL-packetized stream to an *elementary_PID* or *elementary_stream_ID*.

This descriptor only applies to transport streams, not program streams. Within a transport stream, this descriptor is in *descriptor_loop* after *ES_info_length* within the PMT.

Descriptor_tag

This 8-bit field has a value of "0001 1110."

Descriptor_length

This 8-bit binary number specifies the number of bytes following this field. It has a value of "0000 0010."

ES_ID

This 16-bit field specifies the identifier of a SL-packetized stream.

Video Descriptor

For MPEG-4 Part 2 streams carried in PES packets, this MPEG-4 video descriptor provides basic information for identifying the coding parameters. It does not apply to MPEG-4 Part 2 streams encapsulated in SL-packets or FlexMux packets.

Descriptor_tag

This 8-bit field has a value of "0001 1011."

Descriptor_length

This 8-bit binary number specifies the number of bytes following this field. It has a value of "0000 0001."

MPEG-4_visual_profile_and_level

This 8-bit field identifies the profile and level of the MPEG-4 Part 2 video stream. It has the same value as *profile_and_level_indication* in the Visual Object sequence header in the associated MPEG-4 Part 2 stream.

ARIB PMT Descriptors

These ARIB descriptors are used to identify ARIB-specific information that is present in the MPEG-2 transport stream. They are typically contained within a *descriptor_loop* in the MPEG-2 PMT, and may also be present in other ARIB-specific tables. ARIB descriptors not associated with the PMT are discussed in Chapter 18.

AVC Timing and HRD Descriptor

This ARIB descriptor describes the video stream time information and the reference decoder information for H.264 (MPEG-4 Part 10).

AVC Video Descriptor

This ARIB descriptor describes basic coding parameters of the H.264 (MPEG-4 Part 10) video stream.

Carousel Compatible Composite Descriptor

This ARIB descriptor uses descriptors defined in the data carousel transmission specification (ARIB STD-B24 Part 3) as sub-descriptors, and describes accumulation control by applying the functions of the sub-descriptors.

Component Descriptor

This ARIB descriptor is the same as the one discussed in the DVB Descriptors section.

Conditional Playback Descriptor

This ARIB descriptor conveys the description of conditional playback and the PID that transmits the ECM and EMM.

Content Availability Descriptor

This ARIB descriptor describes information to control the recording and output of content by receivers. The *encryption_mode* flag indicates whether or not to encrypt the digital video outputs. It is used in combination with the *Digital Copy Control Descriptor*.

Country Availability Descriptor

This ARIB descriptor is the same as the one discussed in the DVB Descriptors section.

Data Component Descriptor

This ARIB descriptor identifies the data coding system standard. The syntax is the same as for the DVB "*Data Broadcast ID Descriptor*" except for different field names.

Descriptor_tag

This 8-bit field has a value of "1111 1101."

Descriptor_length

This 8-bit binary number specifies the number of bytes following this field.

Data_component_ID

This 16-bit field identifies the data broadcast specification that is used to broadcast the data in the broadcast network.

Note: The following field may be repeated [n] times.

Additional_identifier_information

The definition of this data depends on *data_component_ID*.

Digital Copy Control Descriptor

This ARIB descriptor contains information to control copy generation. For digital recording, broadcasting service provider uses it to inform digital recording equipment about event recording and copyright information. It has a *descriptor_tag* value of "1100 0001."

Emergency Information Descriptor

This ARIB descriptor is used to transmit an emergency alarm. It may only be used with terrestrial digital audio, terrestrial digital television, BS digital or broadband CS broadcasting. It is also present in the *descriptor_loop* of the ARIB network information table (NIT).

Descriptor_tag

This 8-bit field has a value of "1111 1100."

Descriptor_length

This 8-bit binary number specifies the number of bytes following this field.

Note: The following seven fields may be repeated [n] times.

Service_ID

This 16-bit field identifies the broadcast program number.

Start/end_flag

This 1-bit flag has a value of "1" at the start of emergency information and a value of "0" when transmission ends.

Signal_type

This 1-bit flag has a value of "0" and "1", respectively, when Class 1 and 2 start signals are transmitted.

Reserved_bits

These six bits are always "11 1111."

Area_code_length

This 8-bit binary number indicates the number of bytes following this field.

Note: The following two fields may be repeated [n] times.

Area_code

This 12-bit field indicates the area code as defined in Notification No. 405.

Reserved_bits

These four bits are always "1111."

Hierarchical Transmission Descriptor

This ARIB descriptor indicates the relationship between hierarchical streams when transmitting events hierarchically. etc.). It has a *descriptor_tag* value of "1100 0000."

Linkage Descriptor

This ARIB descriptor provides a link to another service, transport stream, program guide, service information, software upgrade, etc.

Mosaic Descriptor

This ARIB descriptor is the same as the one discussed in the DVB Descriptors section.

Parental Rating Descriptor

This ARIB descriptor is the same as the one discussed in the DVB Descriptors section.

Stream Identifier Descriptor

This ARIB descriptor is the same as the one discussed in the DVB Descriptors section.

System Management Descriptor

This ARIB descriptor identifies the type of broadcasting. It is also present in the *descriptor_loop* of the ARIB network information table (NIT).

Descriptor_tag
This 8-bit field has a value of "1111 1110."

Descriptor_length
This 8-bit binary number specifies the number of bytes following this field.

System_management_ID
This 16-bit field identifies the type of broadcasting:

b0–b1:
    00 = broadcasting
    01 = non-broadcasting
    10 = non-broadcasting
    11 = reserved

b2–b7:
    000000 = reserved
    000001 = CS digital broadcast
    000010 = BS digital broadcast
    000011 = terrestrial digital TV broadcast
    000100 = broadband CS digital broadcast
    000101 = terrestrial digital audio broadcast
    000110 - 111111 = reserved

The remaining eight bits (b8–b15) make up the *additional_identifier_information* field, used to extend the broadcasting signal standard.

Target Region Descriptor

This ARIB descriptor describes the target region of an event or a part of the stream comprising an event.

Video Decode Control Descriptor

This ARIB descriptor is used to control video decoding when receiving MPEG-based still pictures and to smoothly display when encode format changes at a video splice point.

Descriptor_tag

This 8-bit field has a value of "1100 1000."

Descriptor_length

This 8-bit binary number specifies the number of bytes following this field.

Still_picture_flag

A "1" for this 1-bit flag indicates a still (MPEG) picture; a "0" indicates animation.

Sequence_end_code_flag

A "1" for this 1-bit flag indicates that the video stream has sequence end code at the end of sequence. A "0" indicates that it does not have a sequence end code.

Video_encode_format

This is a 4-bit codeword indicate the video encode format:

```
0000 = 1080p
0001 = 1080i
0010 = 720p
0011 = 480p
0100 = 480i
0101 = 240p
0110 = 120p
0111 - 1111 = reserved
```

ATSC PMT Descriptors

These ATSC descriptors are used to identify ATSC-specific information that is present in the MPEG-2 transport stream. They are typically contained within a *descriptor_loop* in the MPEG-2 PMT, and may also be present in other ATSC-specific tables. ATSC descriptors not associated with the PMT are discussed in Chapter 15.

AC-3 Audio Stream Descriptor

A Dolby® Digital (AC-3) audio elementary bitstream may be included within a MPEG-2 bitstream in much the same way a standard MPEG audio stream is included. Like the MPEG audio bitstream, the Dolby® Digital bitstream is packetized into PES packets.

MPEG-2 does not explicitly support a Dolby® Digital bitstream. Also, the MPEG-2 audio stream descriptor is inadequate to describe the contents of the Dolby® Digital bitstream in the PSI tables.

Therefore, PES packets containing Dolby® Digital or Dolby® Digital Plus audio data are sent using private stream 1. A *Registration Descriptor* (not required for DVB systems) and *AC-3 Audio Stream Descriptor* with a *descriptor_tag* value of "1000 0001" are also required.

Note that for ATSC and OpenCable™, the AC-3 audio descriptor is titled "*AC-3 Audio Stream Descriptor*" while for DVB, the AC-3 audio descriptor is titled "*AC-3 Descriptor*". The syntax of these descriptors differs significantly between the two systems.

ATSC Private Information Descriptor

This ATSC descriptor provides a way to carry private information. More than one descriptor may appear within a single *descriptor_loop*.

Descriptor_tag

This 8-bit field has a value of "1010 1101."

Descriptor_length

This 8-bit binary number specifies the number of bytes following this field.

Format_identifier

This 32-bit binary number specifies the owner of the following private information, registered with the SMPTE Registration Authority, as illustrated in Table 13.53.

Private_data_byte

These optional [n] bytes of private data are defined by the *format_identifier* owner.

Component Name Descriptor

This ATSC and OpenCable™ descriptor defines an optional textual name tag for any component of the service.

Descriptor_tag

This 8-bit field has a value of "1010 0011."

Descriptor_length

This 8-bit binary number specifies the number of bytes following this field.

Component_name_string()

Name string, based on ATSC's Multiple String Structure.

Content Advisory Descriptor

This ATSC and OpenCable™ descriptor defines the ratings for a given program. It is also present in the *descriptor_loop* of the ATSC and OpenCable™ event information table (EIT).

Descriptor_tag

This 8-bit field has a value of "1000 0111."

Descriptor_length

This 8-bit binary number specifies the number of bytes following this field.

Rating_region_count

This 6-bit binary number indicates the number (1–8) of rating region specifications that follow.

Note: [rating_region_count] specifies how many times the following seven fields are repeated.

Rating_region

This 8-bit binary number specifies the rating region for which the following data is defined.

Rated_dimensions

This 8-bit binary number specifies the number of rating dimensions for which content advisories are specified for this program.

Note: [rated_dimensions] specifies how many times the following three fields are repeated.

Rating_dimension_j

This 8-bit binary number specifies the dimension index into the ATSC RRT instance for the region specified by the field *rating_region*.

Reserved_bits

These four bits are always "1111."

Rating_value

This 4-bit binary number represents the rating value of the dimension specified by the field *rating_dimension_j* for the region given by *rating_region*.

Rating_description_length

This 8-bit binary number specifying the length (0–80) of the *rating_description_text* field that follows.

Rating_description_text()

Rating description string, based on ATSC's Multiple String Structure.

Enhanced Signaling Descriptor

This ATSC descriptor identifies the terrestrial broadcast transmission method of a program element. If the program element is an alternative to another, it is linked to the alternative one and the broadcaster's preferences are specified.

Descriptor_tag

This 8-bit field has a value of "1011 0010."

Descriptor_length

This 8-bit binary number specifies the number of bytes following this field.

Linkage_preference

This 2-bit codeword indicates if the program element is linked to another program element. If linked, it also identifies the broadcaster's preference.

00 = not linked
01 = linked, no preference
10 = linked, preferred
11 = linked, not preferred

Tx_method

This 2-bit codeword specifies the VSB transmission method used to transmit the associated program element.

00 = main: main coding
01 = half-rate: rate-1/2 enhanced coding
10 = quarter-rate: rate-1/4 enhanced coding
11 = reserved

Linked_component_tag

An optional 4-bit value that links the program element to an alternative. The alternative is the program element with the same *linked_component_tag* value in the transport stream PMT labeled with an equivalent value of *program_number* as the transport stream PMT that carries this descriptor.

Reserved_bits

These optional four bits are always "1111." This field is present only when *linkage_preference* = "00."

MAC Address List Descriptor

This ATSC and OpenCable™ descriptor is used when implementing IP (Internet Protocol) multicasting over MPEG-2 transport streams. Streams carrying IP data are identified as containing DSM-CC sections by assigning *stream_type* = $0D_H$ within the PMT.

This descriptor is used to identify the data, by multicast MAC group addresses, being carried by each data elementary stream. It has a *descriptor_tag* value of "1010 1100."

Redistribution Control Descriptor

This ATSC and OpenCable™ descriptor (also known as the "broadcast flag") conveys any redistribution control information held by the program rights holder for the content. It is also present in the *descriptor_loop* of the ATSC and OpenCable™ event information table (EIT).

Descriptor_tag
This 8-bit field has a value of "1010 1010."

Descriptor_length
This 8-bit binary number specifies the number of bytes following this field.

RC_information()
[n] bytes of optional additional redistribution control information that may be defined in the future.

DVB PMT Descriptors

These DVB descriptors are used to identify DVB-specific information that is present in the MPEG-2 transport stream. They are typically contained within a *descriptor_loop* in the MPEG-2 PMT, and may also be present in other DVB-specific tables. DVB descriptors not associated with the PMT are discussed in Chapter 17.

AC-3 Descriptor

A Dolby® Digital (AC-3) audio elementary bitstream may be included within a MPEG-2 bitstream in much the same way a standard MPEG audio stream is included. Like the MPEG audio bitstream, the Dolby® Digital bitstream is packetized into PES packets.

MPEG-2 does not explicitly support a Dolby® Digital bitstream. Also, the MPEG-2 audio stream descriptor is inadequate to describe the contents of the Dolby® Digital bitstream in the PSI tables.

Therefore, PES packets containing Dolby® Digital or Dolby® Digital Plus audio data are sent using private stream 1. An *AC-3 Descriptor* with a *descriptor_tag* value of "1000 0001" is also required.

Note that for ATSC and OpenCable™, the AC-3 audio descriptor is titled *"AC-3 Audio Stream Descriptor"* while for DVB, the AC-3 audio descriptor is titled *"AC-3 Descriptor"*. The syntax of these descriptors differs significantly between the two systems.

Adaptation Field Data Descriptor

This DVB descriptor is used to indicate the type of data fields supported in the private data field of the adaptation field.

Descriptor_tag
This 8-bit field has a value of "0111 0000."

Descriptor_length
This 8-bit binary number specifies the number of bytes following this field, and as a value of 01_H.

Adaptation_field_data_identifier
This 8-bit field identifies data fields transmitted in the private data field of the adaptation field. If a bit is set to "1" it indicates that the corresponding data field is supported.

b0 = announcement switching data field
b1-b7 = reserved

Ancillary Data Descriptor

This DVB descriptor is used to indicate the presence and type of ancillary data in MPEG audio elementary streams.

Descriptor_tag
This 8-bit field has a value of "0110 1011."

Descriptor_length

This 8-bit binary number specifies the number of bytes following this field, and as a value of 01_H.

Ancillary_data_identifier

This 8-bit field identifies data fields transmitted in the private data field of the adaptation field. If a bit is set to "1" it indicates that the corresponding data field is supported.

    b0 = DVD-Video ancillary data
    b1 = extended ancillary data
    b2 = announcement switching data
    b3 = DAB ancillary data
    b4 = scale factor error check
    b5 = reserved
    b6 = reserved
    b7 = reserved

Component Descriptor

This ARIB and DVB descriptor indicates the type of stream and may be used to provide a text description of the stream. For DVB, it is only present in the EIT and SIT.

Descriptor_tag

This 8-bit field has a value of "0101 0000."

Descriptor_length

This 8-bit binary number specifies the number of bytes following this field.

Reserved_bits

These four bits are always "0000."

Stream_content

This 4-bit codeword indicates the type of content in the stream (audio, video or data).

Component_type

This 8-bit codeword indicates the type of audio, video or data.

Component_tag

This 8-bit field has the same value as *component_tag* in the *Stream Identifier Descriptor* for the component stream.

ISO_639_language_code

This 24-bit field contains a 3-character language code.

Text_char

[n] bytes that specify a text description of the stream.

Country Availability Descriptor

This ARIB and DVB descriptor identifies countries that are either allowed or not allowed to receive the service. The descriptor may appear twice for each service, once for listing countries allowed to receive the service, and a second time for listing countries not allowed to receive the service. The latter list overrides the former list.

Descriptor_tag

This 8-bit field has a value of "0100 1001."

Descriptor_length

This 8-bit binary number specifies the number of bytes following this field.

Country_availability_flag

A "1" for this 1-bit flag indicates the country codes specify countries that may receive the service. A "0" indicates the country codes specify the countries that may not receive the service.

Country_code

[n] 24-bit fields that identify countries, using the 3-character code as specified in ISO 3166.

Data Broadcast ID Descriptor

This DVB descriptor identifies the data coding system standard.

Descriptor_tag

This 8-bit field has a value of "0110 0100."

Descriptor_length

This 8-bit binary number specifies the number of bytes following this field.

Data_broadcast_ID

This 16-bit field identifies the data broadcast specification that is used to broadcast the data in the broadcast network. Allocations of the value of this field are found in ETR 162.

Note: The following field may be repeated [n] times.

ID_selector_byte

The definition of this data depends on *data_broadcast_ID*.

Mosaic Descriptor

A mosaic component is a collection of different video images to form a coded video component. The information is organized so that each specific information when displayed appears on a small area of a screen.

This ARIB and DVB descriptor partitions a digital video component into elementary cells, the allocation of elementary cells to logical cells, and links the content of the logical cell and the corresponding information (e.g. bouquet, service, event etc.). It has a *descriptor_tag* value of "0101 0001."

Parental Rating Descriptor

This ARIB and DVB descriptor gives a rating based on age and offers extensions to be able to use other rating criteria. For DVB, it is only present in the EIT and SIT.

Descriptor_tag

This 8-bit field has a value of "0101 0101."

Descriptor_length

This 8-bit binary number specifies the number of bytes following this field.

Note: The following two fields are repeated [n] times.

Country_code

This 24-bit field identifies a country using the 3-character code, as specified in ISO 3166.

Rating

This 8-bit field indicates the recommended minimum age in years of the viewer.

Private Data Specifier Descriptor

This DVB descriptor is used identify the specifier of any private descriptors or private fields within descriptors.

Descriptor_tag

This 8-bit field has a value of "0101 1111."

Descriptor_length

This 8-bit binary number specifies the number of bytes following this field, and as a value of 04_H.

Private_data_specifier

The assignment of values for this field is given in ETR 162.

Service Move Descriptor

This DVB descriptor enables a decoder to track a service when it is moved from one transport stream to another.

Descriptor_tag

This 8-bit field has a value of "0110 0000."

Descriptor_length

This 8-bit binary number specifies the number of bytes following this field.

New_original_network_ID

This 16-bit field specifies the *original_network_ID* of the transport stream in which the service is found after the move.

New_transport_stream_ID

This 16-bit field specifies the *transport_stream_ID* of the transport stream in which the service is found after the move.

New_service_ID

This 16-bit field specifies the *service_ID* of the service after the move.

Stream Identifier Descriptor

This ARIB and DVB descriptor enables specific streams to be associated with a description in the EIT. This is used where there is more than one stream of the same type within a service.

Descriptor_tag

This 8-bit field has a value of "0101 0010."

Descriptor_length

This 8-bit binary number specifies the number of bytes following this field, and has a value of 01_H.

Component_tag

This 8-bit field identifies the component stream associated with a component descriptor. Within the PMT, each *Stream Identifier Descriptor* has a different value for this field.

Subtitling Descriptor

This DVB descriptor identifies ETSI EN 300 743 subtitle data.

Descriptor_tag

This 8-bit field has a value of "0101 1001."

Descriptor_length

This 8-bit binary number specifies the number of bytes following this field.

Note: The following four fields may be repeated [n] times to allow identifying multiple data services using a single descriptor.

ISO_639_language_code

This 24-bit field contains a 3-character language code.

Subtitling_type

This 8-bit field provides information on the content of the subtitle and the intended display.

Composition_page_ID

This 16-bit field identifies the composition page.

Ancillary_page_ID

This 16-bit field identifies the (optional) ancillary page.

Teletext Descriptor

This DVB descriptor is used to identify elementary streams which carry EBU Teletext data.

Descriptor_tag

This 8-bit field has a value of "0101 0110."

Descriptor_length

This 8-bit binary number specifies the number of bytes following this field.

Note: The following four fields may be repeated [n] times to allow identifying multiple data services using a single descriptor.

ISO_639_language_code

This 24-bit field contains a 3-character language code.

Teletext_type

This 5-bit codeword specifies the type of teletext page:

01_H = initial teletext page
02_H = teletext subtitle page
03_H = additional information page
04_H = programme schedule page
05_H = teletext subtitle page for hearing impaired

Teletext_magazine_number

This 3-bit binary number identifies the magazine number.

Teletext_page_number

This 8-bit field specifies the teletext page number as two 4-bit hex digits.

VBI Data Descriptor

This DVB descriptor defines the VBI service type in the associated elementary stream.

Descriptor_tag

This 8-bit field has a value of "0100 0101."

Descriptor_length

This 8-bit binary number specifies the number of bytes following this field.

Note: The following fields may be repeated [n] times to allow identifying multiple data services using a single descriptor.

Data_service_ID

This 8-bit binary number identifies the type of VBI data present in the associated elementary stream. It has a value of:

01_H = EBU teletext
02_H = EBU teletext with inverted framing code
04_H = video program system (VPS)
05_H = widescreen signaling (WSS)
06_H = closed captioning
07_H = monochrome 4:2:2 samples

Data_service_description_length

This 8-bit binary number indicates the number of bytes following this field.

Note: The following fields are present, and may be repeated [n] times, when data_service_ID = 01_H, 02_H, 04_H, 05_H, 06_H, or 07_H.

Reserved_bits

These two bits always have a value of "11."

Field_parity

A "1" for this bit indicates field 1 data; a "0" indicates field 2 data.

Line_offset

This 5-bit binary number specifies the line number the teletext data is to be inserted on for a 576i video signal. Only values of "0 0111" to "1 0110" are valid. When *field_parity* = "0", a value of 313_D is added to the *line_offset* value to obtain the line number.

Note: The following field is present, and may be repeated [n] times, when data_service_ID ≠ 01_H, 02_H, 04_H, 05_H, 06_H or 07_H.

Reserved_bits

These eight bits always have a value of "1111 1111."

VBI Teletext Descriptor

The syntax for this descriptor is the same as for *Teletext Descriptor.* The only difference is that it is not used to associate *stream_type* 06_H with either the VBI or EBU teletext standard. Decoders use the languages in this descriptor to select magazines and subtitles. It has a *descriptor_tag* value of "0100 0110."

OpenCable PMT Descriptors

These additional descriptors are used to carry OpenCable™-related information. They are contained within a *descriptor_loop* in the MPEG-2 PMT and may also be present in other OpenCable™-specific tables. OpenCable™ descriptors not associated with the PMT are discussed in Chapter 16.

AC-3 Audio Stream Descriptor

This OpenCable™ descriptor is the same as the one discussed in the ATSC Descriptors section.

Component Name Descriptor

This OpenCable™ descriptor is the same as the one discussed in the ATSC Descriptors section.

Content Advisory Descriptor

This OpenCable™ descriptor is the same as the one discussed in the ATSC Descriptors section.

MAC Address List Descriptor

This OpenCable™ descriptor is the same as the one discussed in the ATSC Descriptors section.

Redistribution Control Descriptor

This OpenCable™ descriptor is the same as the one discussed in the ATSC Descriptors section.

Closed Captioning

EIA–608 and EIA–708 are the primary closed captioning standards. While EIA–608 (discussed in Chapter 8) was originally designed for use with NTSC broadcasts, EIA–708 was designed for use with digital TV broadcasts.

EIA–708

The EIA–708 DTV closed captioning standard makes a number of changes to the NTSC-based EIA–608 closed captioning standard. The focus is on giving viewers better looking information, and giving them more control over it.

Most importantly, more information can now be included. EIA–608 supports up to 960 bps for captioning information, while EIA–708 reserves a constant 9600 bps for captioning (including the 960 bps used for EIA–608 captioning).

Viewers can control the size of the captioning text. Those with poor vision can make it larger, those who prefer captions not cover so much of the picture can make it smaller, and everyone else can leave them as they are.

EIA–708 also offers more letters and symbols, supporting multilingual captioning. While the EIA–608 character set doesn't have all of the letters and accent marks needed for proper captioning in languages like French, Spanish, German, Italian, or Portuguese, EIA–708 provides all of these and more.

Support for multiple fonts and more colors eliminates the familiar clunky monospaced white-on-black look. Eight fonts (including proportional spaced, casual, and script fonts) and up to 64 text and background colors are specified, although caption decoders aren't required to support all the fonts and colors. This allows captioners to improve the look of the captions, however, they will have to take into consideration how the captions will appear on televisions without the multiple font support.

The additional color support means the traditional black box background can be replaced by a colored box, done away with entirely in favor of edged or drop-shadowed text. The caption box can also be made translucent (see-through).

EIA–708 allows adding closed captioning data to a MPEG-2 transport stream using *user_data* at the sequence, GOP, or picture layer of the video bitstream; ATSC specifies that it be present only at the picture layer. Figure 13.25 illustrates the DTV closed captioning protocol model.

EIA–608 captioning data is not embedded within the DTV protocol stack. This allows it to be extracted at the transport layer, enabling simpler captioning decoder designs since the entire DTV closed caption channel bitstream does not have to be parsed to recover the two bytes of EIA–608 data.

MPEG coded pictures are transmitted in a different order than they are displayed. Captioning data is similarly reordered, so must be reordered (by the decoder), along with the pictures to which they correspond, prior to packet location and extraction.

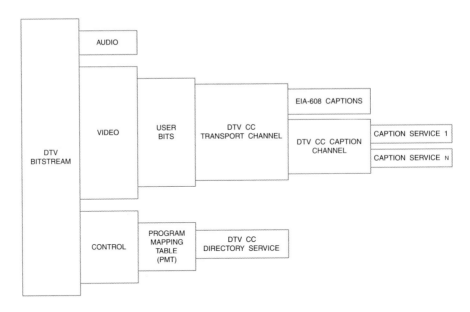

Figure 13.25. DTV Closed Captioning Data in MPEG-2 Bitstream.

CC Type	Code	ATSC	OpenCable	SCTE 21
closed captions	03_H	yes	yes	yes
additional_EIA_608_data	04_H	-	yes	yes
luma_PAM_data	05_H	-	yes	yes
bar information	06_H	yes	yes	yes

Table 13.55. *user_data_type_code* Codewords.

ATSC

ATSC closed captioning uses a continuous 9600 bps stream allocated from the ATSC signal capacity. The captioning data is allocated on a frame-by-frame basis so that 1200 bytes of data are transported per second. This enables up to 20 bytes of caption data per frame for a 480i60 or 1080i60 ATSC signal. On average, EIA–608 captions are allocated 960 bps, and EIA–708 captions are allocated 8640 bps.

Captioning may only be present at the picture layer. The bitstream syntax is:

User_data_start_code

This 32-bit string has a value of $000001B2_H$.

ATSC_identifier

This 32-bit value of 47413934_H indicates that the user data conforms to the ATSC A/53 specification.

User_data_type_code

The value of this 8-bit codeword specifies the type of information that follows, as indicated in Table 13.55. Other values are either in use in other standards or are reserved for future ATSC use.

Note: The following twelve fields are present when user_data_type_code = 03_H.

Reserved_bit

This bit is always "1."

Process_cc_data_flag

If this bit is a "1," the *cc_data* must be processed. If it is a "0," the *cc_data* can be discarded.

Additional_data_flag

This flag is set to a "1" to indicate the presence of additional user data.

Cc_count

This 5-bit binary number, with a range of 0–31, indicates the number of closed caption constructs following this field. The value is set such that a fixed bandwidth of 9600 bps is maintained for the closed caption data.

Reserved_bits

These eight bits are always "1111 1111."

Note: [cc_count] specifies how many times the following five fields are repeated.

Marker_bits

These five bits are always "1 1111."

Cc_valid

If this bit is a "1," the two closed caption data bytes are valid. If it is a "0," the two data bytes are invalid. If invalid, EIA–608 clock run-in and start bits are not generated.

Cc_type

These two bits specify the type of closed caption data that follows, as shown in Table 13.56.

CC Type	Code
EIA–608 line 21 field 1 captions	00
EIA–608 line 21 field 2 captions	01
EIA–708 channel packet data	10
EIA–708 channel packet start	11

Table 13.56. *cc_type* Codewords.

Cc_data_1

The first eight bits of closed caption data. They are processed only if *process_cc_data_flag* is a "1."

Cc_data_2

The second eight bits of closed caption data. They are processed only if *process_cc_data_flag* is a "1."

Marker_bits

These eight bits are always "1111 1111."

Additional_cc_data

These optional eight bits are present if *additional_data_flag* = "1" and are reserved for future ATSC definition.

Note: The following seven fields are present when user_data_type_code = 04$_H$.

Marker_bits

These three bits are always "111."

Additional_cc_count

This 5-bit binary specifies the number of lines of EIA–608 defined following the field.

Note: [additional_cc_count] specifies how many times the following five fields are repeated.

Additional_cc_valid

If this bit is a "1," the two closed caption data bytes are valid. If it is a "0," the two data bytes are invalid.

Additional_cc_line_offset

This 5-bit binary number specifies the offset in lines from which the EIA–608 closed caption data originated relative lines 9 and 272 for 480i systems or lines 5 and 318 for 576i systems.

Additional_cc_field_number

This 2-bit codeword indicates the number of the field, in display order, from which the EIA–608 data originated.

```
00 = forbidden
01 = 1st field
10 = 2nd field
11 = 3rd display field (repeated field in film mode)
```

Additional_cc_data_1

The first eight bits of EIA–608 closed caption data. They are processed only if *additional_cc_valid* is a "1."

Additional_cc_data_2

The second eight bits of EIA–608 closed caption data. They are processed only if *additional_cc_valid* is a "1."

Note: The following field is present when a user_data_type_code is used that is not listed in Table 13.56.

ATSC_reserved_user_data

These optional eight bits are reserved for use by ATSC or used by other standards.

Digital Cable - SCTE 21

The SCTE 21 specification forms the basis for digital cable closed captioning. OpenCable™ references the SCTE 21 specification.

As illustrated in Table 13.54, SCTE 21 extends ATSC closed captioning, as discussed in the ATSC closed captioning section, to better support EIA–608 captions. This is because some cable systems use the EIA–608 closed caption format to carry non-captioning data on other VBI lines.

A pulse-amplitude modulation (PAM) scheme is also available to transfer basic VBI waveforms, such as:

EIA–608-compliant closed captioning for one or more VBI lines other than line 21

Nielsen Source Identification (SID)/Automated Measurement Of Lineups (AMOL) signals

North American Basic Teletext per the EIA–516 NABTS Specification

World System Teletext (WST)

Vertical Interval Time Code (VITC)

Although most standards use two-level luminance encoding, multi-level PAM coding is also supported.

Digital Cable - SCTE 20

The SCTE 20 specification is another standard for digital cable closed captioning. OpenCable™ also references the SCTE 20 specification.

SCTE 20 is an earlier captioning specification, which may be present at the picture layer as *user_data*. Both SCTE 20 and 21 must be supported by OpenCable™ and other digital cable decoders to ensure backwards compatibility with current systems.

User_data_start_code

This 32-bit string has a value of $000001B2_H$.

User_data_type_code

The value of this 8-bit codeword is 03_H, indicating closed captioning information.

Reserved_bits

These seven bits are always "100 0000." However, some early cable systems instead use "000 0000," so the value of this field should be ignored by decoders.

VBI_data_flag

If this bit is a "1," one or more VBI data constructs follow.

Cc_count

This 5-bit binary number, with a range of 0–31, indicates the number of closed caption constructs following this field.

Note: [cc_count] specifies how many times the following six fields are repeated.

Cc_priority

This 2-bit codeword indicates the priority of constructs in picture reconstruction. "00" is highest priority and "11" is lowest priority.

Field_number

This 2-bit codeword indicates the number of the field, in display order, from which the EIA–608 data originated.

00 = forbidden
01 = 1st field
10 = 2nd field
11 = 3rd display field (repeated field in film mode)

Note that the MPEG-2 defined association of top field with field 1 or odd field and bottom field with field 2 or even field is used.

Line_offset

This 5-bit binary number specifies the offset in lines from which the EIA–608 closed caption data originated relative to lines 10 and 273 for 480i systems or lines 6 and 319 for 576i systems.

Cc_data_1

The first eight bits of closed caption data.

Cc_data_2

The second eight bits of closed caption data.

Marker_bit

This bits is always "1."

Non_real_time_video_count

This 4-bit binary number has a value of 0–15, and indicates the number of non-real-time video constructs that follow.

Note: [non_real_time_video_count] specifies how many times the following eight fields are repeated.

Non_real_time_video_priority

This 2-bit codeword indicates the priority of constructs in non-real-time VBI data. "00" is highest priority and "11" is lowest priority.

Sequence_number

This 2-bit binary number increments by one between sequences. A value of "00" indicates the non-real- time-sampled video line is not to be reconstructed (is inactive) until a segment is received with a non-zero *sequence_number.*

Non_real_time_video_field_number

This one-bit flag indicates whether to reconstruct the data into the odd field ("0") or even field ("1").

Segment_number

This 5-bit binary number specifies the number of the non-real-time sampled video segment, starting with "0 0001."

Non-real-time sampled video is segmented into 64-byte segments and transport each as an array of 32 luminance (Y) samples followed by an array of 16 chrominance sample pairs (Cb, Cr), starting with the most significant bit of the left-most sample. All segments of the sequence shall be transmitted in order before any segment of a new sample of the same non-real-time video line.

Note: The following field is repeated thirty-two times.

Non_real_time_video_Y_data

Eight bits of non-real-time Y data for this segment.

Note: The following two fields are repeated sixteen times.

Non_real_time_video_Cb_data

Eight bits of non-real-time Cb data for this segment.

Non_real_time_video_Cr_data

Eight bits of non-real-time Cr data for this segment.

DVB

The ETSI EN 301 775 standard defines how to add closed captioning data to a MPEG-2 transport stream, as discussed in the teletext section.

Teletext

Two standards for teletext transmission over MPEG-2 are available, the "VBI standard" and the newer "EBU Teletext" standard. The "EBU Teletext" standard only supports the transmission of teletext data. The "VBI standard" allows transmission of VBI and EBU Teletext data using the same elementary stream. This technique is useful when minimizing the number of synchronized PID streams is important.

Some systems may support only the "EBU Teletext" standard, others just the "VBI standard," and others both standards. In the case where both standards must be supported, separate PIDs are used; teletext data is broadcast on both PIDs.

DVB VBI Standard

The ETSI EN 301 775 standard defines how to add closed captioning, teletext, video program system (VPS), and widescreen signalling (WSS) data to a MPEG-2 transport stream for DVB applications. The data is carried in MPEG-2 PES packets as private stream 1 which are in turn carried by transport packets. Although designed for the DVB standard, it is applicable to any MPEG-2 bitstream. Use of the *DVB VBI Teletext Descriptor* is required.

The syntax for the PES data field is:

Data_identifier

This 8-bit binary number identifies the type of data carried in the PES packet. It has a value of 10_H to $1F_H$ and 99_H to $9B_H$.

Note: The following fields may be repeated [n] times to allow transmission of multiple types of data within a single stream.

Data_unit_ID

This 8-bit binary number identifies the type of data present. It has a value of:

02_H = EBU teletext non-subtitle data
03_H = EBU teletext subtitle data
$C0_H$ = EBU teletext with inverted framing code
$C3_H$ = video program system (VPS)
$C4_H$ = widescreen signaling (WSS)
$C5_H$ = closed captioning
$C6_H$ = monochrome 4:2:2 samples
FF_H = stuffing

Data_unit_length

This 8-bit binary number indicates the number of bytes following this field. If *data_identifier* has a value between 10_H and $1F_H$ inclusive, this field must be set to $2C_H$.

Note: The following fields are present when data_unit_ID = 02_H, 03_H or $C0_H$. This teletext packet can be used to transfer 43 bytes of teletext information (ETSI EN 300 706).

Reserved_bits

These two bits always have a value of "11."

Field_parity

A "1" for this bit indicates field 1 data; a "0" indicates field 2 data.

Line_offset

This 5-bit binary number specifies the line number the teletext data is to be inserted on for a 576i video signal. Only values of "0 0111" to "1 0110" are valid. When *field_parity* = "0", a value of 313_D is added to the *line_offset* value to obtain the line number.

Framing_code

This 8-bit field specifies the framing code to be used. For EBU teletext, it has a value of "1110 0100." For inverted teletext, it has a value of "0001 1011."

Inverted teletext is used to carry signals that are not intended for public reception, such as down-stream controls for cable head-end equipment, schedules, etc. The use of inverted teletext is on the decline with many broadcasters now instead using teletext packet 31.

Txt_data_block

This 336-bit field corresponds to the 336 bits (42 bytes) of teletext data.

Note: The following fields are present when data_unit_ID = C3$_H$. This video program system (VPS) packet can be used to transfer 13 bits of VPS information (ETSI EN 300 231).

Reserved_bits

These two bits always have a value of "11."

Field_parity

Decoders need only implement *field_parity* when it is a "1." They may ignore packets when this bit is a "0."

Line_offset

This 5-bit binary number specifies the line number the VPS data is to be inserted on for a 576i video signal. Decoders need only imple-ment *line_offset* when it is "1 0000," they may ignore other lines.

VPS_data_block

This 104-bit field corresponds to the thirteen data bytes of VPS data, excluding run-in and start code byte.

Note: The following fields are present when data_unit_ID = C4$_H$. This widescreen signaling (WSS) packet can be used to transfer 14 bits of WSS information (ETSI EN 300 294).

Reserved_bits

These two bits always have a value of "11."

Field_parity

Decoders need only implement *field_parity* when it is a "1." They may ignore packets when this bit is a "0."

Line_offset

This 5-bit binary number specifies the line number the WSS data is to be inserted on for a 576i video signal. Decoders need only imple-ment *line_offset* when it is "1 0111," they may ignore other lines.

WSS_data_block

This 14-bit field corresponds to the 14 bits of WSS data.

Reserved_bits

These two bits always have a value of "11."

Note: The following fields are present when data_unit_ID = C5$_H$. This closed captioning packet can be used to transfer 16 bits of EIA–608 closed captioning information.

Reserved_bits

These two bits always have a value of "11."

Field_parity

A "1" for this bit indicates field 1 (line 21) data; a "0" indicates field 2 (line 284) data.

Line_offset

This 5-bit binary number specifies the line number the caption data is to be inserted on for a 480i video signal. Decoders need only implement *line_offset* when it is "1 0101," they may ignore other lines.

Closed_captioning_data_block

This 16-bit field corresponds to the 16 bits of EIA–608 closed captioning data.

Note: The following fields are present when data_unit_ID = FF$_H$.

Stuffing_byte

These [n] bytes have a value of "1111 1111." Any number of stuffing bytes may present and they are ignored by the decoder.

DVB EBU Teletext Standard

The ETSI EN 300 472 standard defines how to add EBU teletext to a MPEG-2 transport stream for DVB applications. The data is carried in MPEG-2 PES packets as private stream 1 which are in turn carried by transport packets. Although designed for the DVB standard, it is applicable to any MPEG-2 bitstream. Use of the *DVB Teletext Descriptor* is required.

The syntax for the PES data field is:

Data_identifier

This 8-bit binary number identifies the type of data carried in the PES packet. It has a value of 10_H to $1F_H$.

Note: The following fields may be repeated [n] times.

Data_unit_ID

This 8-bit binary number identifies the type of data present. It has a value of:

02_H = EBU teletext non-subtitle data
03_H = EBU teletext subtitle data

Data_unit_length

This 8-bit binary number indicates the number of bytes following this field, and has a value of $2C_H$.

Reserved_bits

These two bits always have a value of "11."

Field_parity

A "1" for this bit indicates field 1 data; a "0" indicates field 2 data.

Line_offset

This 5-bit binary number specifies the line number the teletext data is to be inserted on for a 576i video signal. Only values of "0 0111" to "1 0110" are valid. When *field_parity* = "0", a value of 313_D is added to the *line_offset* value to obtain the line number.

Framing_code

This 8-bit field specifies the framing code to be used. It has a value of "1110 0100."

Magazine_and_packet_address

This 16-bit field corresponds to the magazine and packet address.

Data_block

This 320-bit field corresponds to the remaining 40 bytes of teletext data.

Widescreen Signaling (WSS)

The ETSI EN 301 775 standard defines how to add widescreen signaling (WSS) data to a MPEG-2 transport stream, as discussed in the teletext section.

Subtitles

Subtitles consist of one or more compressed bitmap images, along with optional rectangular backgrounds for each. They are positioned at a defined location, being displayed at a defined start time and for a defined number of video frames.

The bitmap technique enables support for any language, rather than just those supported by the decoder, and enables the subtitle author complete control over the appearance of the characters, including font size and kerning. In addition, characters and symbols that are not a part of any standard character set can easily be used, such as those characters in ideographic languages which represent proper names.

Digital Cable Subtitles

The *subtitle_message()* in the SCTE 27 specification defines subtitle bitmaps associated with a program. Timing for the display of subtitle text is given as a Presentation Time Stamp (PTS) referenced to the program's program clock (PCR).

The *subtitle_message()* is carried in transport stream packets with PID = $1FFB_H$. The syntax is:

Table_ID

This 8-bit codeword has a value of $C6_H$.

Reserved_bits

These two bits are always "00."

Reserved_bits

These two bits are always "11."

Section_length

This 12-bit binary number specifies the number of bytes after this field, up to and including *CRC_32*.

Reserved_bit

This bit is always "0."

Segmentation_overlay_included

A "1" for this 1-bit flag indicates the message includes the segmentation definition.

Protocol_version

This 6-bit binary number allows, in the future, this message type to carry parameters that may be structured fundamentally differently than those defined in the current protocol. At present, the subtitle message is defined for *protocol_version* "00 0000" only.

Table_extension

This 16-bit binary number is used to differentiate between various segmented *message_body()s* that are present simultaneously on the Transport Stream, all delivered using *subtitle_message()s*. This field is present if *segmentation_overlay_included* = "1."

Last_segment_number

This 12-bit binary number indicates the segment number of the last segment needed to recover the full message. This field is present if *segmentation_overlay_included* = "1."

Segment_number

This 12-bit binary number indicates which part of a (perhaps) multi-part message is present. This field is present if *segmentation_overlay_included* = "1."

ISO_639_language_code

This 24-bit field contains a 3-character language code.

Pre_clear_display

A "1" for this flag indicates that the entire display is to be made transparent prior to the display of the subtitle text. Otherwise, the subtitle text is to be added to the text already on the screen (cumulative display).

Immediate

A "1" for this 1-bit flag indicates that the subtitle is to be displayed immediately upon receipt. Otherwise, it should be cued for display at the *display_in_PTS* time.

Reserved_bit

This bit is always "0."

Display_standard

This 5-bit codeword specifies the display format for which the subtitle was prepared.

00000 = 720×480, 29.97 or 30 frames per second
00001 = 720×576, 25 frames per second
00010 = 1280×720, 29.97 or 30 frames per second
00011 = 1920×1080, 29.97 or 30 frames per second

Display_in_PTS

When this 32-bit value matches the 32 least significant bits of the 33-bit MPEG program clock (90KHz portion), the subtitle is to be displayed.

Subtitle_type

This 4-bit codeword indicates the format of the subtitle data block. Currently, only a value of "0001" is defined.

Reserved_bit

This bit is always "0."

Display_duration

This 11-bit binary number indicates the number of video frames, from 1–2000, for which the subtitle data is to be displayed.

Block_length

This 16-bit binary number indicates the number of bytes that follow, excluding the CRC and any descriptors.

Note: The following fields are present when subtitle_type = "0001."

Reserved_bits

These five bits are always "0 0000."

Background_style

This 1-bit flag specifies the background style:

0 = transparent
1 = framed

Outline_style

This 2-bit codeword specifies the text outline style:

00 = none
01 = outline
10 = drop shadow
11 = reserved

Bitmap_Y_component

This 5-bit binary number specifies the value of Y, with a range 0–31, for the text color.

Bitmap_opaque_enable

A "1" for this 1-bit flag indicates that the text color shall be opaque (no video blend). A "0" indicates a 50% mix with the video is to be done.

Bitmap_Cr_component

This 5-bit binary number specifies the value of Cr, with a range 0–31, for the text color.

Bitmap_Cb_component

This 5-bit binary number specifies the value of Cb, with a range 0–31, for the text color.

Bitmap_top_H_coordinate

This 12-bit binary number, with a range of 0–1919, specifies the horizontal coordinate of the left-most pixel of the decompressed bitmap.

Bitmap_top_V_coordinate

This 12-bit binary number, with a range of 0–1079, specifies the vertical coordinate of the top line of the decompressed bitmap.

Bitmap_bottom_H_coordinate

This 12-bit binary number, with a range of 0–1919, specifies the horizontal coordinate of the right-most pixel of the decompressed bitmap.

Bitmap_bottom_V_coordinate

This 12-bit binary number, with a range of 0–1079, specifies the vertical coordinate of the bottom line of the decompressed bitmap.

Frame_top_H_coordinate

This 12-bit binary number, with a range of 0–1919, specifies the horizontal coordinate of the left-most pixel of the frame. This field is present if *background_style* = "1."

Frame_top_V_coordinate

This 12-bit binary number, with a range of 0–1079, specifies the vertical coordinate of the top line of the frame. This field is present if *background_style* = "1."

Frame_bottom_H_coordinate

This 12-bit binary number, with a range of 0–1919, specifies the horizontal coordinate of the right-most pixel of the frame. This field is present if *background_style* = "1."

Frame_bottom_V_coordinate

This 12-bit binary number, with a range of 0–1079, specifies the vertical coordinate of the bottom line of the frame. This field is present if *background_style* = "1."

Frame_Y_component

This 5-bit binary number specifies the value of Y, with a range 0–31, for the frame color. This field is present if *background_style* = "1."

Frame_opaque_enable

A "1" for this 1-bit flag indicates that the frame color shall be opaque (no video blend). A "0" indicates a 50% mix with the video is to be done. This field is present if *background_style* = "1."

Frame_Cr_component

This 5-bit binary number specifies the value of Cr, with a range 0–31, for the frame color. This field is present if *background_style* = "1."

Frame_Cb_component

This 5-bit binary number specifies the value of Cb, with a range 0–31, for the frame color. This field is present if *background_style* = "1."

Reserved_bits

These four bits are always "0000." This field is present if *outline_style* = "01."

Outline_thickness

This 4-bit binary number, with a range 0–15, specifies the text outline thickness. This field is present if *outline_style* = "01."

Outline_Y_component

This 5-bit binary number specifies the value of Y, with a range 0–31, for the text outline color. This field is present if *outline_style* = "01."

Outline_opaque_enable

A "1" for this 1-bit flag indicates that the text outline color shall be opaque (no video blend). A "0" indicates a 50% mix with the video is to be done. This field is present if *outline_style* = "01."

Outline_Cr_component

This 5-bit binary number specifies the value of Cr, with a range 0–31, for the text outline color. This field is present if *outline_style* = "01."

Outline_Cb_component

This 5-bit binary number specifies the value of Cb, with a range 0–31, for the text outline color. This field is present if *outline_style* = "01."

Shadow_right

This 4-bit binary number, with a range 0–15, specifies the text right shadow thickness. This field is present if *outline_style* = "10."

Shadow_bottom

This 4-bit binary number, with a range 0–15, specifies the text bottom shadow thickness. This field is present if *outline_style* = "10."

Shadow_Y_component

This 5-bit binary number specifies the value of Y, with a range 0–31, for the text shadow color. This field is present if *outline_style* = "10."

Shadow_opaque_enable

A "1" for this 1-bit flag indicates that the text shadow color shall be opaque (no video blend). A "0" indicates a 50% mix with the video is to be done. This field is present if *outline_style* = "10."

Shadow_Cr_component

This 5-bit binary number specifies the value of Cr, with a range 0–31, for the text shadow color. This field is present if *outline_style* = "10."

Shadow_Cb_component

This 5-bit binary number specifies the value of Cb, with a range 0–31, for the text shadow color. This field is present if *outline_style* = "10."

Reserved_bits

Each of these three bytes have a value of "0000 0000." This field is present if *outline_style* = "11."

Bitmap_length

This 16-bit binary number specifying the number of bytes in the following compressed bitmap.

Compressed_bitmap()

Reserved_bits

Each of these [n] bytes have a value of "0000 0000." They are only is present when *subtitle_type* ≠ "0001."

Descriptor_loop

[n] descriptors may be present in this *descriptor_loop*.

CRC_32

32-bit CRC value.

DVB Subtitles

DVB subtitles (ETSI EN 300 743) are much more complex than digital cable subtitles.

Subtitle streams are carried in PES packets with *stream_id* = private stream 1. The timing of their display is given as a presentation time stamp (PTS) referenced to the program's pro-gram clock (PCR). A subtitle stream conveys one or more *subtitle services*.

Each *subtitle service* contains text or graphics needed to provide subtitles for a particular purpose; separate subtitle services may be used, for example, to convey subtitles in several languages. Each subtitle service displays its information in a sequence of *subtitle pages*.

Subtitle pages are overlaid on the video image. A subtitle page contains one or more *subtitle regions*.

Each *subtitle region* is a rectangular area with attributes such as position, size, pixel depth and background color. A subtitle region is used as the background structure into which one or more *subtitle objects* are placed.

An *subtitle object* represents a character, word, line of text, entire sentence, logo or icon.

The PES data field syntax is:

Data_identifier

The value for this 8-bit field is 20_H, indicating DVB subtitle stream.

Subtitle_stream_ID

The value for this 8-bit field is 00_H, indicating DVB subtitle stream.

Sync_byte

This 8-bit field has a value of "0000 1111."

Segment_type

This 8-bit field indicates the type of data contained in *segment_data_field*. The following *segment_type* values are defined for subtitling:

00_H = page composition segment
01_H = region composition segment
02_H = CLUT definition segment
03_H = object data segment
03_H = end of display set segment
FF_H = stuffing

Page_ID

This 16-bit binary number identifies the subtitle service. Segments with a value matching *composition_page_ID* in the DVB *Subtitling Descriptor* carry data for one subtitle service. Segments with a value matching *ancillary_page_ID* carry data that may be shared by multiple subtitle services.

A frequent, and often preferred method, is to convey distinct services by using different streams on separate PIDs.

Segment_length

This 16-bit binary number indicates the number of bytes contained in *segment_data_field*.

Segment_data_field

This is the payload of the segment. Several segment types are defined:

> *Page composition segment*: carries information on the page composition, such as list of included regions, each region's position and any time-out information for the page.

> *Region composition segment*: carries information on the region composition and attributes, such as the size, background color, the pixel depth, which color lookup table (CLUT) to use and a list of included objects with their position within the region.

> *CLUT definition segment*: contains information on a specific CLUT, such as the colors used for a CLUT entries.

> *Object data segment*: carries information on a specific text or graphical object. Object data segments for graphical objects contain run-length encoded bitmap colors; for text objects, a string of character codes is carried.

> *End of display set segment*: used to signal that no more segments need to be received before the decoding of the current display set can begin.

End_of_PES_data_field_marker

An 8-bit field with a value of "1111 1111."

Enhanced Television Programming

As discussed in Chapter 8, SMPTE 363M Transport Type B broadcast data using IP multicast binding is delivered as three components: announcements, triggers and resources. Announcements are delivered on a known multicast IP address and UDP port, and point to triggers and resources that are available on specified multicast IP addresses and UDP ports.

The SCTE 42 specification defines how these announcements, triggers and resources can be included as part of a MPEG-2 stream, as shown in Figure 13.26. The data is associated to a program by identifying it within the program's PMT through the use of the *MAC Address List Descriptor*.

In some cases, it is desirable to carry announcements in their own streams with unique PIDs; triggers and resources are carried in separate a stream with a unique PID, as illustrated in Figure 13.27. The advantage of this technique is that the decoder does not have to process IP datagrams, associated with triggers and resources, if the application is not enabled to receive them. Announcements are always processed, but are small in comparison. Triggers and resources may also be carried on separate streams based on their characteris-

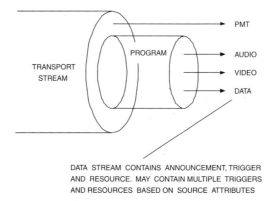

DATA STREAM CONTAINS ANNOUNCEMENT, TRIGGER
AND RESOURCE. MAY CONTAIN MULTIPLE TRIGGERS
AND RESOURCES BASED ON SOURCE ATTRIBUTES

Figure 13.26. Announcement, Trigger and Resource Data Carried on a Single PID Stream.

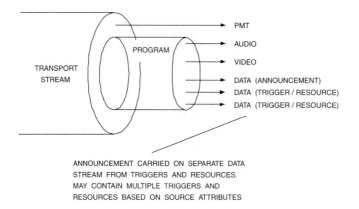

ANNOUNCEMENT CARRIED ON SEPARATE DATA
STREAM FROM TRIGGERS AND RESOURCES.
MAY CONTAIN MULTIPLE TRIGGERS AND
RESOURCES BASED ON SOURCE ATTRIBUTES

Figure 13.27. Announcement Carried Separately from Triggers and Resources.

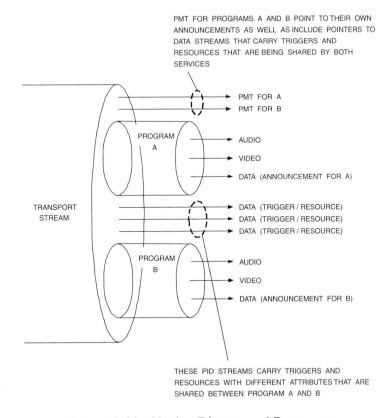

PMT FOR PROGRAMS A AND B POINT TO THEIR OWN ANNOUNCEMENTS AS WELL AS INCLUDE POINTERS TO DATA STREAMS THAT CARRY TRIGGERS AND RESOURCES THAT ARE BEING SHARED BY BOTH SERVICES

PMT FOR A
PMT FOR B

PROGRAM A
AUDIO
VIDEO
DATA (ANNOUNCEMENT FOR A)

TRANSPORT STREAM
DATA (TRIGGER / RESOURCE)
DATA (TRIGGER / RESOURCE)
DATA (TRIGGER / RESOURCE)

PROGRAM B
AUDIO
VIDEO
DATA (ANNOUNCEMENT FOR B)

THESE PID STREAMS CARRY TRIGGERS AND RESOURCES WITH DIFFERENT ATTRIBUTES THAT ARE SHARED BETWEEN PROGRAM A AND B

Figure 13.28. Sharing Triggers and Resources.

tics, such as language type, target audience, size of resource, maximum bit rate, etc.

Figure 13.28 illustrates an extension to Figure 13.27. The trigger and resource streams are shared between multiple programs. For example, two programs may wish to provide data from the same weather service.

Intellectual Property Management and Protection (IPMP)

IPMP, also called digital rights management (DRM), provides an interface and tools,

rather than a complete system, for implementing intellectual property rights management.

The level and type of management and protection provided is dependent on the value of the content and the business model. For this reason, the complete design of the IPMP system is left to application developers.

The architecture enables both open and proprietary solutions to be used, while enabling interoperability, supporting the use of more than one type of protection (i.e., decryption, watermarking, rights management, and so on) and supporting the transferring of content between devices using a defined inter-device message (reflecting the issue of content distribution over home networks).

For protected content, the IPMP tool requirements are communicated to the decoder before the presentation starts. Tool configuration and initialization information are conveyed by the IPMP Descriptor or IPMP elementary stream. Needed tools can be embedded, downloaded or acquired by other means.

Control point and ordering sequence information in the IPMP Descriptor allow different tools to function at different places in the system. IPMP data, carried in either an IPMP Descriptor or IPMP elementary stream, includes rights containers, key containers and tool initialization data.

MPEG-4 over MPEG-2

Instead of MPEG-2 video, the MPEG-2 transport or program stream can carry MPEG-4 Part 2 video. This enables existing infrastructures and equipment to accommodate the MPEG-4 Part 2 video codec easily.

The PES packet *stream_id* = "1110 xxxx" for MPEG-4 Part 2 video. *Stream_type* = 10_H within the PMT or PSM. Carriage of the MPEG-4 Part 2 stream must also be signaled by using the MPEG-4 Video Descriptor.

MPEG-4 Part 3 audio, MPEG-4 SL-packetized streams and MPEG-4 FlexMux streams can also be carried by a MPEG-2 transport or program stream.

H.264 over MPEG-2

Instead of MPEG-2 video, MPEG-2 transport and program streams can carry H.264 (MPEG-4 Part 10) video in PES packets. This enables existing infrastructures and equipment to accommodate the H.264 video codec easily.

The PES packet *stream_id* = "1110 xxxx" for MPEG-4 Part 10 (H.264) video. *Stream_type* = $1B_H$ within the PMT or PSM. Carriage of the H.264 stream must also be signaled by using the MPEG-2 AVC Video Descriptor.

SMPTE VC-9 over MPEG-2

Instead of MPEG-2 video, MPEG-2 transport and program streams can carry SMPTE VC-9 video in PES packets. This enables existing infrastructures and equipment to accommodate the VC-9 video codec easily.

The PES packet *stream_ID* = FD_H and *stream_ID_extension* = "101 0101" for VC-9 video. *Stream_type* = 88_H within the PMT or PSM.

Data Broadcasting

MPEG-2 supports a variety of content distribution tools and protocols via the Digital Storage Media Command and Control (DSM-CC) specification (MPEG-2 Part 6). Applications that can take advantage of the DSM-CC tools include:

- Video-on-Demand
- Data broadcasting
- Internet access
- IP multicasting

At first, DSM-CC simply offered VCR-like functions (fast-forward, rewind, pause, etc.) as an annex to MPEG-2 Part 1. It was later expanded into MPEG-2 Part 6 to handle the selection, access and control of distributed content. As a result DSM-CC now encompasses a larger set of tools:

- Network session and resource control
- Client configuration
- Downloading of a client
- Stream control, file access
- Interactive and broadcast download
- Data and object carousels
- Switched digital broadcast channel change protocol

Figure 13.29 illustrates how ATSC and DVB data broadcasting use DSM-CC to implement a variety of data broadcast features.

Carousels

Carousels cyclically repeat their content. If a receiver wants to access particular data from a carousel, it simply waits for the next time that the data is broadcast.

Data Carousels contain unspecified data, so a receiver has to know what to do with it when it is received. Data Carousels are often used for downloading new system software.

Object Carousels contain identifiable data such as data streams, pictures, trigger events, executable applications, etc. along with a directory listing all objects in the carousel. Object Carousels are often used for shopping services, electronic program guides (EPG), advertisements and other interactive functions. Unlike Data Carousels, Object Carousels can vary the repetition rate of individual objects.

IP Multicasting over MPEG-2 Transport

IP multicasting conveys IP datagrams over MPEG-2 transport streams based on DMS-CC. ATSC (A/92) and DVB (ETSI EN 301 192) use slightly different forms of LAN emulation to convey packet data. SCTE 42 for digital cable systems requires support for both techniques.

There are two primary protocols: DVB MultiProtocol Encapsulation (MPE) and ATSC DSM-CC addressable sections. Digital cable systems commonly use one or the other, but not both simultaneously. However, encapsulation changes may occur at any time within a program.

The DVB implementation is compliant with DSM-CC Sections containing private data. *Table_id* = $3E_H$, indicating a DSM-CC Section containing private data (see Table 13.50), in this case, MPE Datagram Sections.

The ATSC implementation is also compliant with DSM-CC Sections containing private data. The *table_id* field is $3F_H$, a DSM-CC Section containing addressable sections (see Table 13.50).

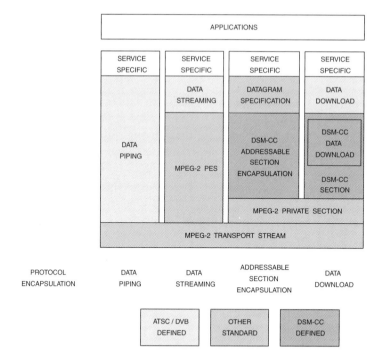

Figure 13.29. Encapsulation Overview.

Each stream that carries IP multicasting data has a *stream_type* = $0D_H$ associated with it within the PMT, indicating that it carries DSM-CC Sections.

Formulation of the MAC Address

IETF RFC 1112 specifies the mapping of an IP multicast address to an Ethernet MAC address (ATSC calls it *deviceId*). The IP multicast address is mapped into the corresponding hardware multicast address by placing the low-order 23 bits of the IP multicast address into the lower order 23 bits of the MAC address 01:00:5E:xx:xx:xx (base 16). Bit 23 of the MAC address is always "0" per IETF RFC 1700.

Transporting Over MPEG-2

Figure 13.30 illustrates how the IP datagrams are encapsulated and segmented into MPEG-2 transport packets. IP datagrams are fragmented at the IP layer so that they do not exceed the specified Maximum Transfer Unit (MTU) size, typically 4080 bytes.

A single Datagram Section may span multiple MPEG-2 packets of the same PID. Also, messages may be placed back-to-back in the MPEG packet payload. This requires the use of *pointer_field* (PF) to point to the location of the beginning of the next message.

The *MAC Address List Descriptor* is used to identify data, by multicast MAC group addresses, being carried by each stream.

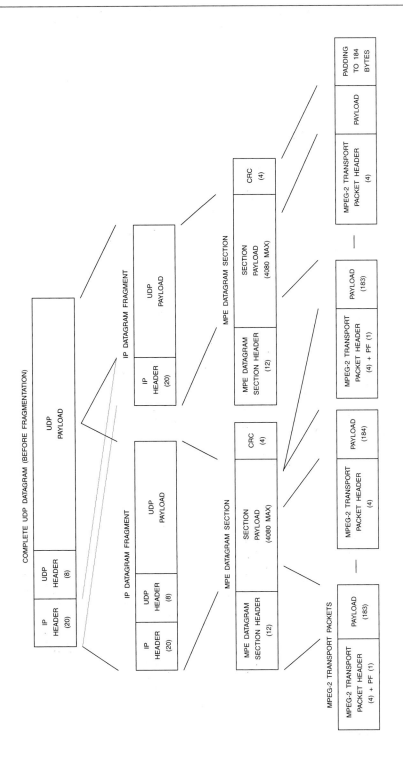

Figure 13.30. Carrying IP Multicast Datagrams over MPEG-2 Transport Streams.

Data Broadcasting Mechanisms

There are a wide variety of encapsulation protocols used to transport data within a MPEG-2 transport stream. IP multicasting has been previously discussed. Other common techniques include *asynchronous data streaming*, *synchronous data streaming* and *synchronized data streaming*.

Asynchronous Data Streaming

Asynchronous data streams, carried in DSM-CC Sections (ATSC) or PES packets (ARIB and DVB), are used for applications where the delivery of data is not subject to any timing constraints.

For ARIB and DVB, the PES packet *stream_id* = BF_H for private stream 2. Since the maximum size of a PES packet is 64 KB, it is segmented into as many 184-byte units as needed, to match the transport stream requirements.

Stream_type = $0D_H$ (ARIB and ATSC) or 06_H (DVB) within the PMT.

Synchronous Data Streaming

Synchronous data streams, carried as PES packet payloads, are used for applications requiring continuous streaming of data to a receiver at a regular and constant data rate. To achieve this, timing information is included within the stream.

Since the maximum size of a PES packet is 64 KB, it is segmented into as many 184-byte units as needed, to match the transport stream requirements.

The PES packet *stream_id* = BD_H for private stream 1, allowing the use of PES header fields, including the Presentation Time Stamp (PTS). However, the resolution of the PTS is extended from 11.1 us to 74 ns. *Stream_type* = $0D_H$ (ARIB), $C2_H$ (ATSC) or 06_H (DVB) within the PMT.

Synchronized Data Streaming

Synchronized data streams, carried as PES packet payloads, are used for applications requiring presentation of data at precise but not necessarily regular times. The presentation times are usually associated with a video, audio or data stream.

Since the maximum size of a PES packet is 64 KB, it is usually segmented into as many 184-byte units as needed, to match the transport stream requirements.

The PES packet *stream_id* = BD_H for private stream 1, allowing the use of PES header fields, including the Presentation Time Stamp (PTS). *Stream_type* = $0D_H$ (ARIB) or 06_H (ATSC and DVB) within the PMT.

Data Piping

Data Piping is a basic asynchronous transportation mechanism for data over MPEG-2 transport streams – data is inserted directly in the payload of MPEG-2 transport stream packets. Sections, tables and PES data structures are not used. There is no standardized way for the splitting and reassembly of the datagrams; this is defined by the application.

Decoder Considerations

The video decoder essentially performs the inverse function of the encoder. From the coded bitstream, it reconstructs the I frames. Using I frames, additional coded data, and motion vectors, the P and B frames are generated. Finally, the frames are output in the proper order.

Figure 13.31 illustrates the block diagram of a basic MPEG-2 video decoder. Figure 13.32 illustrates the block diagram of a MPEG-2 video decoder that supports SNR scalability. Figure 13.33 illustrates the block diagram of a MPEG-2 video decoder that supports temporal scalability.

Audio and Video Synchronization

A MPEG-2 encoder produces PES packets having a different PID (packet identification) for each program. The MPEG-2 decoder recognizes only packets with the PID for the selected program and ignores the others.

The MPEG-2 encoder contains a 27 MHz oscillator and 33-bit counter, called the STC (system time clock). STC is a 33-bit value driven by a 90 kHz clock, obtained by dividing the 27 MHz clock by 300. It belongs to a particular program and is the master clock of the video and audio encoders for that program.

Time Stamps

After compression, pictures may be sent out of sequence due to any bi-directional coding that may be present. Each picture has a variable amount of data and may have a variable delay due to multiplexing and transmission. In order to keep the audio and video synchronized, time stamps are periodically sent with a picture.

The MPEG-2 encoder notes the time of occurrence of an input picture or audio block (and of the appearance of its coded output) by sampling the STC. A constant value equal to the sum of encoder and decoder buffer delays is added, creating a 33-bit presentation time stamp (PTS). A 33-bit decode time stamp (DTS) may also be added, indicating the time at which the data should be taken from the MPEG-2 decoder's buffer and decoded. DTS and PTS are identical, except in the case of picture reordering for bi-directional (B) pictures.

Since presentation times are evenly spaced, it is not always necessary to include a time stamp (they can be interpolated by the decoder), but they must not be more than 700 ms apart. Lip sync is obtained by incorporating time stamps into the headers of both video and audio PES packets.

PTS and DTS

When B pictures are present, a picture may have to be decoded before it is presented, so that it can act as a reference for a B picture. Although, for example, pictures can be presented in the order IBBP, they are transmitted in the order IPBB. Consequently, two types of time stamps exist.

DTS indicates the time when a picture must be decoded, whereas PTS indicates when it must be present at the output of the MPEG-2 decoder. B pictures are decoded and presented simultaneously so they only contain PTS. When an IPBB sequence is received, both I and P must be decoded before the first B picture. A MPEG-2 decoder can only decode one picture at a time; therefore the I picture is decoded first and stored. While the P picture is being decoded, the decoded I picture is output so that it can be followed by the B pictures.

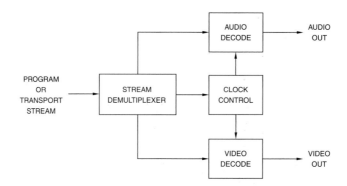

Figure 13.31. Simplified MPEG-2 Decoder Block Diagram.

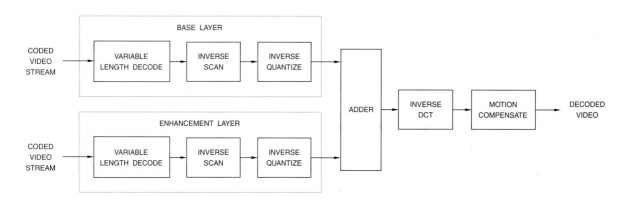

Figure 13.32. Simplified MPEG-2 SNR Scalability Decoder Block Diagram.

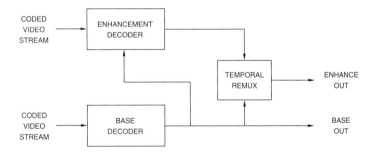

Figure 13.33. Simplified MPEG-2 Temporal Scalability Decoder Block Diagram.

The PTS and DTS flags in the PES packet header are set to indicate the presence of PTS alone or both PTS and DTS. Audio PES packet headers contain only a PTS. Since audio PES packets are never transmitted out of sequence, they do not contain DTS fields.

PTS (or DTS) is included in the bitstream at intervals not exceeding 700 ms. ATSC further constrains PTS or DTS to be inserted at the beginning of each coded picture.

PCR

The output of the MPEG-2 encoder is also time stamped with STC values, called PCR (program clock reference) or SCR (system clock reference), used to synchronize the MPEG-2 decoder's STC with the MPEG-2 encoder's STC. In a program stream, the clock reference is called SCR; in a transport stream, the clock reference is called PCR.

The adaptation field in the transport stream packet header is periodically used to include the PCR information. MPEG-2 requires a minimum of ten PCRs per second be sent, while DVB specifies a minimum of 25 PCRs per second. SCR is required to occur at least once every 700 ms.

Synchronization

Synchronization may be achieved by locking the MPEG-2 decoder's 27 MHz clock to the received PCR using a VCXO and PLL, as shown in Figure 13.34. This technique ensures that the decoder's receive buffer does not overflow or underflow during long periods of continuous operation as a result of the source clock being slightly faster or slower than the decoder clock. An adjustment range of ±100 ppm is typically required for streaming video applications.

At the MPEG-2 decoder, the VCXO generates a nominal 27 MHz clock that drives the local PCR counter. This local PCR is compared with the PES packet header PCR, resulting in a PCR phase error. The phase error is filtered to control the VCXO that will bring the local PCR count into step with the PES packet header PCRs. The discontinuity indicator may reset the local PCR count and may optionally be used to reduce the filtering to help the MPEG-2 decoder lock more quickly to the new timing.

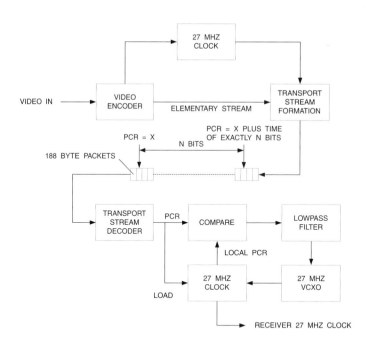

Figure 13.34. 27 MHz Clock Generation by MPEG-2 Decoder.

Lip Sync Issues

Lip sync is an implementation issue that has nothing to do with the MPEG-2 standard. Assuming that the audio and video are in synchronization at the MPEG-2 encoder input, support of the PCR and PTS provides the tools needed to maintain the audio and video timing relationship. MPEG-2 decoders, and possibly MPEG-2 encoders, with incorrect implementations of these tools have been manufactured.

A representative example of an incorrect implementation in MPEG-2 decoders has to do with the reading of the PTS. Some MPEG-2 decoders simply read the PTS when the station is first tuned in, then incorrectly "freewheel" afterwards on the basis of temporal reference values, never checking the PTS again until the channel is changed or turned on again. Others completely ignore the temporal reference values and instead look at the time stamps. As time goes by, early implementation mistakes are being corrected, reducing the occurrence of lip sync problems.

Digital Television Issues

Set-top boxes and digital televisions are incorporating more and more sophisticated video processing. As a result, audio may be delayed relative to the video enough that lip sync becomes an issue. A good decoder design will allow the audio to be delayed a programmable amount to compensate for any video delays that may occur in system design.

Testing Issues

Since they are so complex, MPEG decoders have special testing challenges. Unlike linear systems, the output from a MPEG decoder is not only dependent on the input at any given point in time, but also the previous state of the system before the input was received. Since the output is the result of many combinations of inputs, the number of output states to be tested rapidly grows beyond reason.

The question is how to measure the output of the MPEG decoder. Since the original image has been encoded and decoded, there will be differences between the original and decoded images. Some high-frequency information may be lost, and there may be compression artifacts in the output. For this reason, direct numerical comparison between an original and decoded image is not practical.

Alternately, the output of the decoder could be compared to the output of a reference decoder. However, the inverse discrete cosine transfer (IDCT) accuracy must be taken into account. Different IDCT implementations may have slightly different outputs from the same input. Also, additional differences may occur if the encoder and decoder use different IDCT implementations.

The predictive nature of MPEG can be used to test the proper operation of the decoder. This technique does not require internal access to the decoder or numerical comparisons of the output to a reference. Instead, the output image is designed in such a way that it will tend to highlight decoder errors. If the decoder is working properly, the result is an image that you can inspect visually.

The best approach is to test the range of MPEG features one function at a time, processing a number of test bitstreams one after another. Limiting the scope of each test makes it easier to pinpoint problems. The tester can also quickly determine which (if any) MPEG syntax the decoder fails to process correctly.

Encoder Bitstreams Not Adequate

Decoder testing should include the use of "typical" video sequences produced by one or more production encoders. These sequences give an indication of how the decoder will look in normal use. They are not, however, sufficient to ensure confidence that it will work with all bitstreams.

Syntax Testing

MPEG is a standard that defines the syntax that a decoder must recognize and process, but it does not define what syntax an encoder must use to generate a bitstream.

Most encoders use only a part of the MPEG syntax. As a result, a decoder that works with one encoder may not work with another one that uses different MPEG features. Decoders must be adequately tested to ensure that they will work with future encoders. The easiest way to ensure this is to test the decoder using all the various syntax elements. All conditions that are permitted by a given syntax should be tested.

The development of test bitstreams should be separated from the development of the decoder. If the team that develops the decoder also develops the test bitstreams, misunderstandings about the MPEG specification details could be built into the tests.

The test bitstreams should also make it obvious if there are display problems. This includes distinctive marks at each corner of the image to ensure that the entire image is displayed. Smooth horizontal and vertical motion helps verify field order and alignment. Each frame should also be unique so that it is visually obvious if a frame has been skipped.

More than Just Video

So far, only video has been discussed. However, video is multiplexed with audio, data, and timing information into a program or transport stream. Testing at these levels is also needed.

Testing the audio/video synchronization requires a bitstream with known relationships between unique audio and video events. The test should be designed to facilitate both instrument measurement and visual inspection. A shaped audio pulse correlated to a specific video event works well.

Testing at the program or transport layer must also include checking the ability to receive and utilize the system information that is multiplexed into the bitstream. This includes the program association table and program map table, necessary for the decoder to find the individual bitstreams that comprise a program. There may also be program guides, rating tables, conditional access tables, and other information that should be tested.

Pushing the Limits

So far, we have discussed testing one or a two things at a time. This may help isolate problems and decrease debugging time, but will not answer how a decoder will work under real-world conditions.

Since MPEG is statistical in nature, the information sent to the decoder is not at a constant rate. What happens when many DCT coefficients are received in a short period of time, many MPEG features are enabled simultaneously, and a large percentage of the bits in a frame occur in a single slice? A thorough testing strategy must check a decoder's ability to handle bitstreams that push multiple elements to the limit.

It is also important to see how a decoder recovers from errors, since it is likely to receive noncompliant bitstreams occasionally. Will it freeze or recover gracefully? How long does it take to recover?

If error concealment has been designed into the system, it should be tested by bitstreams with known errors. The output can then be checked to see how effectively the concealment works.

References

1. *A Guide to MPEG Fundamentals and Protocol Analysis (Including DVB and ATSC)*, Tektronix, 1997.
2. ATSC A/53b, *ATSC Standard: Digital Television Standard, Revision B with Amendments 1 and 2*, May 2003.
3. ATSC A/54a, *Recommended Practice: Guide to the Use of the ATSC Digital Television Standard*, December 2003.
4. ATSC A/65b, *ATSC Standard: Program and System Information Protocol for Terrestrial Broadcast and Cable (Revision B)*, March 2003.
5. ATSC A/90, *ATSC Data Broadcast Standard*, July 2000.
6. ATSC A/91, *Recommended Practice: Implementation Guidelines for the ATSC Data Broadcast Standard (Doc. A/90)*, June 2001.
7. ATSC A/92, *ATSC Standard: Delivery of IP Multicast Sessions over ATSC Data Broadcast*, January 2002.
8. Digital Video Magazine, *"Not All MPEGs Are Created Equal,"* by John Toebes, Doug Walker, and Paul Kaiser, August 1995.

9. Digital Video Magazine, "*Squeeze the Most From MPEG,*" by Mark Magel, August 1995.

10. EIA-708-B, *Digital Television (DTV) Closed Captioning*, December 1999.

11. ETSI EN 300 468, *Digital Video Broadcasting (DVB): Specification for Service Information (SI) in DVB Systems*, January 2003.

12. ETSI EN 300 472, *Digital Video Broadcasting (DVB): Specification for Conveying ITU-R System B Teletext in DVB Bitstreams*, January 2003.

13. ETSI EN 300 743, *Digital Video Broadcasting (DVB): Subtitling Systems*, October 2002.

14. ETSI EN 301 192, *Digital Video Broadcasting (DVB): DVB Specification for Data Broadcasting*, January 2003.

15. ETSI EN 301 775, *Digital Video Broadcasting (DVB): Specification for the Carriage of Vertical Blanking Information (VBI) Data in DVB Bitstreams*, January 2003.

16. ETSI TR 101 154, *Digital Video Broadcasting (DVB): Implementation Guidelines for the use of MPEG-2 Systems, Video and Audio in Satellite, Cable and Terrestrial Broadcasting Applications*, July 2000.

17. ISO/IEC 13818–1, Generic coding of moving pictures and associated audio information, Part 1: Systems.

18. ISO/IEC 13818–2, Generic coding of moving pictures and associated audio information, Part 2: Video.

19. ISO/IEC 13818–3, Generic coding of moving pictures and associated audio information, Part 3: Audio.

20. *OpenCable™ Host Device Core Functional Requirements*, April 2004.

21. SCTE 20 2001, *Methods for Carriage of Closed Captions and Non-real Time Sampled Video*.

22. SCTE 21 2001, *Standard for Carriage of NTSC VBI Data in Cable Digital Transport Streams*.

23. SCTE 27 2003, *Subtitling Methods for Broadcast Cable*.

24. SCTE 42 2002, *IP Multicast for Digital MPEG Networks*.

25. Watkinson, John, *The Engineer's Guide to Compression*, Snell and Wilcox Handbook Series.

Chapter 14

MPEG-4 and H.264

MPEG-4 builds upon the success and experience of MPEG-2. It is best known for:

- Lower bit rates than MPEG-2 (for the same quality of video)

- Use of natural or synthetic "objects" that can be rendered together to make a "scene"

- Support for interactivity

For *authors*, MPEG-4 enables creating content that is more reusable and flexible, with better content protection capabilities.

For *consumers*, MPEG-4 can offer more interactivity and, due to the lower bit rate over MPEG-2, the ability to enjoy content over new networks (such as DSL) and mobile products.

MPEG-4 is an ISO standard (ISO/IEC 14496), and currently consists of 19 parts:

systems	ISO/IEC 14496–1
visual	ISO/IEC 14496–2
audio	ISO/IEC 14496–3
conformance testing	ISO/IEC 14496–4
reference software	ISO/IEC 14496–5
DMIF	ISO/IEC 14496–6
reference software	ISO/IEC 14496–7
carriage over IP networks	ISO/IEC 14496–8
reference hardware	ISO/IEC 14496–9
advanced video (H.264)	ISO/IEC 14496–10
scene description	ISO/IEC 14496–11
ISO file format	ISO/IEC 14496–12
IPMP extensions	ISO/IEC 14496–13
MP4 file format	ISO/IEC 14496–14
H.264 file format	ISO/IEC 14496–15
animation extension	ISO/IEC 14496–16
streaming text format	ISO/IEC 14496–17
font compression	ISO/IEC 14496–18
synthesize texture stream	ISO/IEC 14496–19

MPEG-4 provides a standardized way to represent audio, video or still image "media objects" using descriptive elements (instead of actual bits of an image, for example). A media object can be natural or synthetic (computer-generated) and can be represented independent of its surroundings or background.

It also describes how to merge multiple media objects to create a scene. Rather than sending bits of picture, the media objects are sent, and the receiver composes the picture. This allows:

- An object to be placed anywhere

- To do geometric transformations on an object

- To group objects

- To modify attributes and transform data

- To change the view of a scene dynamically

Audio Overview

MPEG-4 audio supports a wide variety of applications, from simple speech to multi-channel high-quality audio.

Audio objects (audio codecs) use specific combinations of *tools* to efficiently represent different types of audio objects. *Profiles* use specific combinations of audio object types to efficiently service a specific market segment. *Levels* specify size, rate and complexity limitations within a profile to ensure interoperability.

Currently, most solutions support a few of the most popular audio codecs (usually AAC-LC and HE-AAC) rather than one or more profiles/levels.

General Audio Object Types

This category supports a wide range of quality, bit rates and number of channels. For natural audio, MPEG-4 supports the AAC (Advanced Audio Coding), BSAC (Bit Sliced Arithmetic Coding) and TwinVQ (Transform domain Weighted Interleave Vector Quantization) algorithms. The following audio objects are available:

AAC-Main Objects

AAC-Main objects add the Perceptual Noise Shaping (PNS) tool to MPEG-2 AAC-Main.

AAC-LC Objects

AAC-LC (Low Complexity) objects add the PNS tool to MPEG-2 AAC-LC. There is also an Error Resilient version, ER AAC-LC.

AAC-SSR Objects

AAC-SSR (Scalable Sampling Rate) objects add the PNS tool to MPEG-2 AAC-SSR.

AAC-LTP Objects

AAC-LTP (Long Term Predictor) objects are similar to AAC-LC objects, with the long term predictor replacing the AAC-LC predictor. This gives the same efficiency with significantly lower implementation cost. There is also an Error Resilient version, ER AAC-LTP.

AAC-Scalable Objects

AAC-Scalable objects allow a large number of scalable combinations. They support only mono or 2-channel stereo sound. There is also an Error Resilient version, ER AAC-Scalable.

ER AAC-LD Objects

Error Resilient AAC-LD (Low Delay) is derived from AAC and all the capabilities for coding of two or more sound channels are supported. coder. They support sample rates up 48 kHz and use frame lengths of 512 or 480 samples (compared to 1024 or 960 samples used by AAC) to enable a maximum algorithmic delay of 20 ms.

ER BSAC Objects

Error Resilient BSAC objects replace the noiseless coding of AAC quantized spectral data and the scale factors. One base layer bitstream and many small enhancement layer bitstreams are used, enabling real-time adjustments to the quality of service.

HE-AAC Objects

HE-AAC (High Efficiency), a combination of AAC and Spectral Band Replication (SBR) technology, is designed for ultra low bit rate coding, as low as 32 kbps for stereo.

TwinVQ Objects

TwinVQ objects are based on fixed rate vector quantization instead of the Huffman coding used in AAC. They operate at lower bit rates than AAC, supporting mono and stereo sound. There is also an Error Resilient version, ER TwinVQ.

Speech Object Types

Speech coding can be done using bit rates from 2–24 kbps. Lower bit rates, such as an average of 1.2 kbps, are possible when variable rate coding is used. The following audio objects are available:

CELP Objects

CELP (Code Excited Linear Prediction) objects support 8 and 16 kHz sampling rates at bit rates of 4–24 kbps. There is also an Error Resilient version, ER CELP.

HVXC Objects

HVXC (Harmonic Vector eXcitation Coding) objects support 8 kHz mono speech at fixed bit rates of 2–4 kbps (below 2 kbps using a variable bitrate mode), along with the ability to change the pitch and speed during decoding. There is also an Error Resilient version, ER HVXC.

Synthesized Speech Object Types

Scalable TTS (Text-to-Speech) objects offer a low bit rate (200–1.2 kbps) phonemic representations of speech. Content with narration can be easily created without recording natural speech. The TTS Interface, allows speech information to be transmitted in the International Phonetic Alphabet (IPA) or in a textual (written) form of any language. The synthesized speech can also be synchronized with a facial animation object.

Synthesized Audio Object Types

Synthetic Audio support is provided by a Structured Audio Decoder implementation that allows the application of score-based control information to musical instruments described in a special language. The following audio objects are available:

Main Synthetic Objects

Main Synthetic objects allow the use of the all MPEG-4 Structured Audio tools. They support synthesis using the Structured Audio Orchestra Language (SAOL) music-synthesis language and wavetable synthesis using Structured Audio Sample-Bank Format (SASBF).

Wavetable Synthesis Objects

Wavetable Synthesis objects are a subset of Main Synthetic, making use of SASBF and MIDI (Musical Instrument Digital Interface) tools. They provide relatively simple sampling synthesis.

General MIDI Objects

General MIDI objects provide interoperability with existing content.

Visual Overview

MPEG-4 visual is divided into two sections. MPEG-4 Part 2 includes the original MPEG-4 video codecs discussed in this section. MPEG-4 Part 10 specifies the "advanced video codec," also known as H.264, and is discussed at the end of this chapter.

The visual specifications are optimized for three primary bit rate ranges:

- less than 64 kbps

- 64–384 kbps

- 0.384–4 Mbps

For high-quality applications, higher bit rates are possible, using the same tools and bit-stream syntax as those used for lower bit rates.

With MPEG-4, *visual objects* (video codecs) use specific combinations of *tools* to efficiently represent different types of visual objects. *Profiles* use specific combinations of visual object types to efficiently service a specific market segment. *Levels* specify size, rate and complexity limitations within a profile to ensure interoperability.

Currently, most solutions support only a couple of the MPEG-4 Part 2 video codecs (usually Simple and Advanced Simple) due to silicon cost issues. Interest in Part 2 video codecs also dropped dramatically with the introduction of the Part 10 (H.264) and SMPTE VC-9 video codecs, which offer about $2\times$ better performance.

YCbCr Color Space

The 4:2:0 YCbCr color space is used for most objects. Each component can be represented by a number of bits ranging from four to twelve bits, with eight bits being the most commonly used.

MPEG-4 Part 2 Simple Studio and Core Studio objects may use 4:2:2, 4:4:4, 4:2:2:4 and 4:4:4:4:4 YCbCr or RGB sampling options, to support the higher picture quality required during the editing process.

Like H.263 and MPEG-2, the MPEG-4 Part 2 video codecs are also macroblock, block and DCT-based.

Visual Objects

Instead of the video "frames" or "pictures" used in earlier MPEG specifications, MPEG-4 uses natural and synthetic *visual objects*. Instances of video objects at a given time are called *visual object planes* (VOPs).

Much like MPEG-2, there are I (intra), P (predicted) and B (bi-directional) VOPs. The S-VOP is a VOP for a sprite object. The S(GMC)-VOP is coded using prediction based on global motion compensation from a past reference VOP.

Arbitrarily shaped video objects, as well as rectangular objects, may be used. A MPEG-2 video stream can be a rectangular video object, for example.

Objects may also be scalable, enabling the reconstruction of useful video from pieces of a total bitstream. This is done by using a base layer and one or more enhancement layers.

Only *natural visual object types* are discussed since they are currently of the most interest in the marketplace.

MPEG-4 Part 2 Natural Visual Object Types

MPEG-4 Part 2 supports many natural visual object types (video codecs), with several interesting ones shown in Table 14.1. The more common object types are:

Tools	Object Type						
	Main	**Core**	**Simple**	**Advanced Simple**	**Advanced Real Time Simple**	**Advanced Coding Efficiency**	**Fine Granularity Scalable**
VOP types	I, P, B	I, P, B	I, P	I, P, B	I, P	I, P, B	I, P, B
chroma format	4:2:0	4:2:0	4:2:0	4:2:0	4:2:0	4:2:0	4:2:0
interlace	×	–	–	×	–	×	×
global motion compensation (GMC)	–	–	–	×	–	×	–
quarter-pel motion compensation (QPEL)	–	–	–	×	–	×	–
slice resynchronization	×	×	×	×	×	×	×
data partitioning	×	×	×	×	×	×	×
reversible VLC	×	×	×	×	×	×	×
short header	×	×	×	×	×	×	×
method 1 and 2 quantization	×	×	–	×	–	×	×
shape adaptive DCT	–	–	–	–	–	×	–
dynamic resolution conversion	–	–	–	–	×	×	–
NEWPRED	–	–	–	–	×	×	–
binary shape	×	×	–	–	–	×	–
grey shape	×	–	–	–	–	×	–
sprite	×	–	–	–	–	–	–
fine granularity scalability (FGS)	–	–	–	–	–	–	×
FGS temporal scalability	–	–	–	–	–	–	×

Table 14.1. Available Tools for Common MPEG-4 Part 2 Natural Visual Object Types.

Main Objects

Main objects provide the highest video quality. Compared to Core objects, they also support grey-scale shapes, sprites, and both interlaced and progressive content.

Core Objects

Core objects use a subset of the tools used by Main objects, although B-VOPs are still supported. They also support scalability by sending extra P-VOPs. Binary shapes can include a constant transparency but cannot do the variable transparency offered by greyscale shape coding.

Simple Objects

Simple objects are low bit rate, error resilient, rectangular natural video objects of arbitrary aspect ratio. Simple objects use a subset of the tools by Core objects.

Advanced Simple Objects

Advanced Simple objects looks much like Simple objects in that only rectangular objects are supported, but adds a few tools to make it more efficient: B-frames, ¼-pixel motion compensation (QPEL) and global motion compensation (GMC).

Fine Granularity Scalable Objects

Fine Granularity Scalable objects can use up to eight scalable layers so delivery quality can easily adapt to transmission and decoding circumstances.

MPEG-4 Part 2 Natural Visual Profiles

MPEG-4 Part 2 supports many visual profiles and levels. Only *natural visual profiles* (Tables 14.2 and 4.3) are discussed since they are currently of the most interest in the marketplace. The more common profiles are:

Main Profile

Main profile was created for broadcast applications, supporting both progressive and interlaced content. It combines highest quality video with arbitrarily shaped objects.

Core Profile

Core profile is useful for higher quality interactive services, combining good quality with limited complexity and supporting arbitrary shape objects. Mobile broadcast services can also be supported by this profile.

Simple Profile

Simple profile was created with low complexity applications in mind. Primary applications are mobile and the Internet.

Advanced Simple Profile

Advanced Simple profile provides the ability to distribute single-layer frame-based video at a wide range of bit rates.

Fine Granularity Scalable Profile

Fine Granularity Scalable profile was created with Internet streaming and wireless multimedia in mind.

Graphics Overview

Graphics profiles specify which graphics elements of the BIFS tool can be used to build a scene. Although it is defined in the Systems specification, graphics is really just another media profile like audio and video, so it is discussed here.

Four hierarchical graphics profiles are defined: Simple 2D, Complete 2D, Complete and 3D Audio Graphics. They differ in the graphics elements of the BIFS tool to be supported by the decoder, as shown in Table 14.4.

MPEG-4 Part 2 Profile	Supported Shapes	Notes	Level	Typical Resolution	Maximum Number of Objects	Maximum Bit Rate
Main	arbitrary	additional tools and functionality	L4 L3 L2	BT.709 BT.601 CIF	32 32 16	38.4 Mbps 15 Mbps 2 Mbps
Core	arbitrary	additional tools and functionality	L2 L1	CIF QCIF	16 4	2 Mbps 384 kbps
Advanced Core	arbitrary	higher coding efficiency	L2 L1	CIF QCIF	16 4	2 Mbps 384 kbps
N-Bit	arbitrary		L2	CIF	16	2 Mbps
Simple	rectangular		L3 L2 L1	CIF CIF QCIF	4 4 4	384 kbps 128 kbps 64 kbps
Advanced Simple	rectangular	higher coding efficiency	L5 L4 L3b L3 L2 L1 L0	BT.601 352×576 CIF CIF CIF QCIF QCIF	4 4 4 4 4 4 1	8 Mbps 3 Mbps 1.5 Mbps 768 kbps 384 kbps 128 kbps 128 kbps
Advanced Real Time Simple	rectangular	higher error resilience	L4 L3 L2 L1	CIF CIF CIF QCIF	16 4 4 4	2 Mbps 384 kbps 128 kbps 64 kbps
Core Scalable	arbitrary	spatial and temporal scalability	L3 L2 L1	BT.601 CIF CIF	16 8 4	4 Mbps 1.5 Mbps 768 kbps
Simple Scalable	rectangular	spatial and temporal scalability	L2 L1 L0	CIF CIF QCIF	4 4 1	256 kbps 128 kbps 128 kbps
Fine Granularity Scalable	rectangular	SNR and temporal scalability	L5 L4 L3 L2 L1 L0	BT.601 352×576 CIF CIF QCIF QCIF	4 4 4 4 4 1	8 Mbps 3 Mbps 768 kbps 384 kbps 128 kbps 128 kbps
Advanced Coding Efficiency	arbitrary	higher coding efficiency	L4 L3 L2 L1	BT.709 BT.601 CIF CIF	32 32 16 4	38.4 Mbps 15 Mbps 2 Mbps 384 kbps

Table 14.2a. MPEG-4 Part 2 Natural Vision Profiles and Levels.

MPEG-4 Part 2 Profile	Supported Shapes	Notes	Level	Typical Resolution	Maximum Number of Objects	Maximum Bit Rate
Core Studio (uses 10-bit pixel data)	arbitrary	additional tools and functionality	L4	BT.709, 60P, 4:4:4 BT.709, 30I, 4:4:4:4:4	16	900 Mbps
			L3	BT.709, 30I, 4:4:4 BT.601, 4:2:2:4	8	450 Mbps
			L2	BT.709, 30I, 4:2:2 BT.601 4:4:4:4:4	4	300 Mbps
			L1	BT.601, 4:2:2:4 BT.601, 4:4:4	4	90 Mbps
Simple Studio (uses 10- or 12-bit pixel data)	arbitrary		L4	BT.709, 60P, 4:4:4 BT.709, 30I, 4:4:4:4:4	1	1800 Mbps
			L3	BT.709, 30I, 4:4:4 BT.709, 30I, 4:2:2:4	1	900 Mbps
			L2	BT.709, 30I, 4:2:2 BT.601, 4:4:4:4:4	1	600 Mbps
			L1	BT.601, 4:2:4 BT.601, 4:4:4	1	180 Mbps

Table 14.2b. MPEG-4 Part 2 Natural Vision Profiles and Levels. 4:4:4:4:4:4 means 4:4:4 RGB + 3 auxiliary channels. 4:2:2:4 means 4:2:2 YCbCr + 1 auxiliary channel.

MPEG-4 Part 2 Object Type	MPEG-4 Part 2 Profile						
	Main	Core	Simple	Advanced Simple	Advanced Real Time Simple	Advanced Coding Efficiency	Fine Granularity Scalable
Main	×	–	–	–	–	–	–
Core	×	×	–	–	–	×	–
N-Bit	–	–	–	–	–	–	–
Simple	×	×	×	×	×	×	×
Advanced Simple	–	–	–	×	–	–	×
Advanced Real Time Simple	–	–	–	–	×	–	–
Advanced Coding Efficiency	–	–	–	–	–	×	–
Fine Granularity Scalable	–	–	–	–	–	–	×

Table 14.3. Objects Supported by Common MPEG-4 Part 2 Profiles.

Graphics Element of BIFS Tool	Graphics Profile			Graphics Tool (BIFS node)	Graphics Profile		
	Simple 2D	Complete 2D	Complete		Simple 2D	Complete 2D	Complete
appearance	×	×	×	fog	–	–	×
box	–	–	×	font style	–	×	×
bitmap	×	×	×	indexed face set	–	–	×
background	–	–	×	indexed face set 2D	–	×	×
background 2D	–	×	×	indexed line set	–	–	×
circle	–	×	×	indexed line set 2D	–	×	×
color	–	×	×	line properties	–	×	×
cone	–	–	×	material	–	–	×
coordinate	–	–	×	material 2D	–	×	×
coordinate 2D	–	×	×	normal	–	–	×
curve 2D	–	×	×	pixel texture	–	×	×
cylinder	–	–	×	point light	–	–	×
directional light	–	–	×	point set	–	–	×
elevation grid	–	–	×	point set 2D	–	×	×
expression	–	–	×	rectangle	–	×	×
extrusion	–	–	×	shape	×	×	×
face	–	–	×	sphere	–	–	×
face def mesh	–	–	×	spot light	–	–	×
face def table	–	–	×	text	–	×	×
face def transform	–	–	×	texture coordinate	–	×	×
FAP	–	–	×	texture transform	–	×	×
FDP	–	–	×	viseme	–	–	×
FIT	–	–	×				

Table 14.4. Graphics Elements (BIFS Tools) Supported by MPEG-4 Graphics Profiles.

Simple 2D profile provides the basic features needed to place one or more visual objects in a scene.

Complete 2D profile provides 2D graphics functions and supports features such as arbitrary 2D graphics and text, possibly in conjunction with visual objects.

Complete profile provides advanced capabilities such as elevation grids, extrusions and sophisticated lighting. It enables complex virtual worlds to exhibit a high degree of realism.

3D Audio Graphics profile may be used to define the acoustical properties of the scene (geometry, acoustics absorption, diffusion, material transparency). This profile is useful for applications that do environmental equalization of the audio signals.

Visual Layers

A MPEG-4 visual scene consists of one or more video objects. Currently, the most common video object is a simple rectangular frame of video.

Each video object may have one or more layers to support temporal or spatial scalable coding. This enables the reconstruction of video in a layered manner, starting with a base layer and adding a number of enhancement layers. Where a high degree of scalability is needed, such as when an image is mapped onto a 2D or 3D object, a wavelet transform is available.

The visual bitstream provides a hierarchical description of the scene. Each level of hierarchy can be accessed through the use of unique start codes in the bitstream.

Visual Object Sequence (VS)

This is the complete scene which contains all the 2D or 3D, natural or synthetic, objects and any enhancement layers.

Video Object (VO)

A video object corresponds to a particular object in the scene. In the most simple case this can be a rectangular frame, or it can be an arbitrarily shaped object corresponding to an object or background of the scene.

Video Object Layer (VOL)

Each video object can be encoded in scalable (multi-layer) or nonscalable form (single layer), depending on the application, represented by the video object layer (VOL). The VOL provides support for scalable coding. A video object can be encoded using spatial or temporal scalability, going from coarse to fine resolution. Depending on parameters such as available bandwidth, computational power, and user preferences, the desired resolution can be made available to the decoder.

There are two types of video object layers, the video object layer that provides full MPEG-4 functionality, and a reduced functionality video object layer, the video object layer with short headers. The latter provides bitstream compatibility with base-line H.263.

Group of Video Object Plane (GOV)

Each video object is sampled in time, each time sample of a video object is a video object plane. Video object planes can be grouped together to form a group of video object planes.

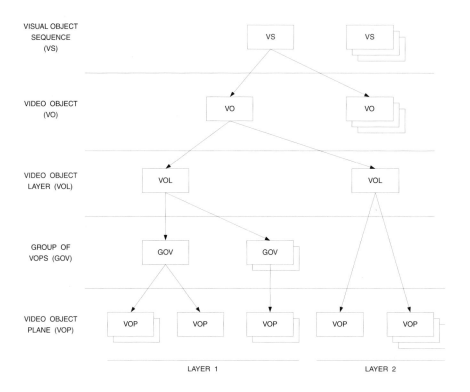

Figure 14.1. Example MPEG-4 Video Bitstream Logical Structure.

The GOV groups together video object planes. GOVs can provide points in the bitstream where video object planes are encoded independently from each other, and can thus provide random access points into the bitstream. GOVs are optional.

Video Object Plane (VOP)

A VOP is a time sample of a video object. VOPs can be encoded independently of each other, or dependent on each other by using motion compensation. A conventional video frame can be represented by a VOP with rectangular shape.

Object Description Framework

Unlike MPEG-2, MPEG-4 does not multiplex multiple elementary streams together into a single transport or program stream.

Data for each object (audio, one layer of one visual object, etc.), scene description information (to declare the spatial-temporal relationship of objects) and object control information are carried in separate *elementary streams*. Synthetic objects may be generated using BIFS to provide the graphics and audio. BIFS is more than a scene description language – it integrates natural and synthetic objects into the same composition space.

The *object description framework* is a set of *object descriptors* used to identify, describe and associate elementary streams to each other, and to objects used in the scene description, as illustrated in Figure 14.2.

An *initial object descriptor*, a derivative of the object descriptor, contains two descriptors. One descriptor points to the *scene description [elementary] stream*; the other points to the corresponding *object descriptor [elementary] stream*.

Object Descriptor (OD) Stream

The object descriptors are transported in a dedicated elementary stream, called the *object descriptor stream*.

The object descriptor effectively associates sets of related elementary streams so they are seen as a single entity by the decoder. Each object descriptor contains other descriptors that typically point to one or more elementary streams associated to a single node and a single audio or visual object. This allows support for multiple alternative streams, such as different languages.

In addition, an object descriptor can point to auxiliary data such as *object content information* (OCI) and *intellectual property rights management and protection* (IPMP).

Object descriptors are not simply present in an object descriptor stream one after the other. Rather, they are encapsulated in *object descriptor commands*. These commands enable object descriptors to be dynamically conveyed, updated or removed at a specific point in time. This allows new elementary streams for an object to be advertised as they become available, or to remove references to elementary streams that are no longer available. Updates are time stamped to indicate when they are to take effect. The time stamp is placed on the sync layer, as with any other elementary stream.

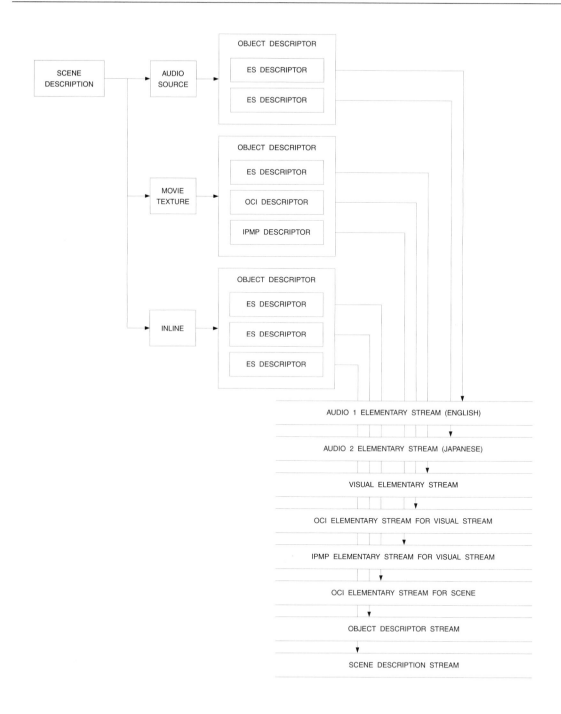

Figure 14.2. Linking Elementary Streams to a Scene.

Object Content Information (OCI)

The *OCI elementary stream* conveys *OCI events*. Each OCI event consists of *OCI descriptors*.

OCI descriptors communicate a number of features of the associated object, such as keywords, text description of the content, language, parental rating, creation date, authors etc.

If the OCI information will never change, it may instead be conveyed using CCI descriptors within the object descriptor stream.

Intellectual Property Management and Protection (IPMP)

The *IPMP elementary stream* conveys IPMP messages to one or more IPMP systems. The IPMP system provides intellectual property management and content protection functions in the receiver.

If the IPMP information will rarely change, it may instead be conveyed using IPMP descriptors within the object descriptor stream.

Scene Description

To assemble a multimedia scene at the receiver, it is not sufficient to simply send just the multiple streams of data. For example, objects may be located in 2D or 3D space, and each has its local coordinate system. Objects are positioned within a scene by transforming each of them to the scene's coordinate system. Therefore, additional data is required for the receiver to assemble a meaningful scene for the user. This additional data is called *scene description*.

Scene graph elements (which are BIFS tools) describe audio-visual primitives and attributes. These elements, and any relationship between them, form a hierarchical *scene graph*, as illustrated in Figure 14.3. The scene graph is not necessarily static; elements may be added, deleted or modified as needed.

The *scene graph profile* defines the allowable set of scene graph elements that may be used.

BIFS

BIFS (BInary Format for Scenes) is used to not only describe the scene composition information, but also graphical elements. A fundamental difference between the BIFS and VRML is that BIFS is a *binary* format, whereas VRML is a *textual* format. BIFS supports the elements used by VRML and several that VRML does not, including compressed binary format, streaming, streamed animation, 2D primitives, enhanced audio and facial animation.

Compressed Binary Format

BIFS supports an efficient binary representation of the scene graph information. The coding may be either lossless or lossy. Lossy compression is possible due to context knowledge: if some scene graph data has been received, it is possible to anticipate the type and format of subsequent data.

Streaming

BIFS is designed so that a scene may be transmitted as an initial scene, followed by modifications to the scene.

Streamed Animation

BIFS includes a low-overhead method for the continuous animation of changes to numerical values of the elements in a scene. This provides an alternative to the interpolator elements supported in BIFS and VRML.

2D Primitives

BIFS has native support for 2D scenes to support low-complexity, low-cost solutions such as traditional television. Rather than partitioning the world into 2D vs. 3D, BIFS allows both 2D and 3D elements in a single scene.

Enhanced Audio

BIFS improves audio support through the use an *audio scene graph*, enabling audio sources to be mixed or the generation of sound effects.

Facial Animation

BIFS exposes the animated face properties to the scene level. This enables it to be a full member of a scene that can be integrated with any other BIFS functionality, similar to other audio-visual objects.

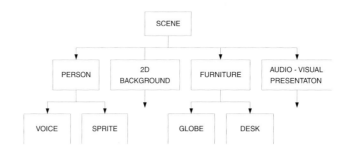

Figure 14.3. Example MPEG-4 Hierarchical Scene Graph.

Synchronization of Elementary Streams

Sync Layer

The *sync layer* (Figure 14.4) partitions each elementary stream into a sequence of *access units,* the smallest entity to which timing information can be associated. It then encodes all the relevant properties using a flexible syntax, generating *SL packets*. These SL packets are then passed on to a delivery (transport) layer. A sequence of SL packets from a single elementary stream is called a *SL-packetized stream.*

Unlike the MPEG-2 PES, the sync layer is not a self-contained stream. Instead, it is an intermediate format that is mapped to a specific delivery layer, such as IP, MPEG-2 transport stream, etc. For this reason, there is no need to include start codes, SL packet lengths, etc., in the sync layer—these are already included in the various delivery layer protocols.

SL packets serve two purposes. First, access units can be fragmented in any way during adaptation to a specific delivery layer. Second, it makes sense to have the encoder guide the fragmentation when it knows about delivery layer characteristics, such as the size of the maximum transfer unit (MTU).

Synchronization of multiple elementary streams is done by conveying object clock reference (OCR), decoding time stamps (DTS), composition time stamps (CTS) and clock references within the sync layer.

The sync layer syntax is flexible in that it can be configured individually for each elementary stream. For example, low bit rate audio streams may desire time stamps with minimum overhead; high bit rate video streams may need very precise time stamps.

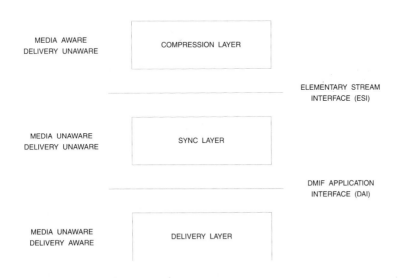

Figure 14.4. Relationship Between the MPEG-4 Compression, Sync and Delivery Layers.

DMIF Application Interface

Unlike MPEG-2, MPEG-4 supports multiple simultaneous usage scenarios (local retrieval, remote interaction, broadcast, multicast, etc.) and multiple simultaneous delivery technologies.

The DMIF (Delivery Multimedia Integration Framework) Application Interface, or DAI, is the interface that controls the data exchanged between a sync layer and a delivery layer (Figure 14.4) during both transmission and reception. It enables accessing, presenting and synchronizing MPEG-4 content transmitted or received using different technologies, such as MPEG-2 transport stream, IP multicast, RTP, etc., even simultaneously.

As a result, a specification for control and data mapping to a specific delivery or storage protocol (also called a *payload format specification*) has to be done jointly with the organization that manages the delivery layer specification. In the example of MPEG-4 transport over IP, development work was done jointly with the Internet Engineering Task Force (IETF).

Multiplexing of Elementary Streams

Delivery of MPEG-4 content is a task that is usually dealt with outside the specification.

However, an analysis of existing delivery layers indicated a need for an additional layer of multiplexing. The occasionally bursty and low bitrate MPEG-4 streams sometimes have to map to a delivery layer that uses fixed packet sizes (such as MPEG-2 transport streams) or high multiplexing overhead (such as RTP/UDP/IP). The potentially large number of delivery streams may also have a burden in terms of management and cost.

To address this situation, FlexMux, a very simple multiplex packet syntax, was defined. It allows multiplexing a number of SL-packetized streams into a self-contained FlexMux stream with low overhead.

In addition, the specifications for the encapsulation of SL-packetized streams into common delivery layer protocols, including MPEG-2 transport and program streams, IP (see Chapter 19) and MPEG-4 file format, have already been done.

FlexMux

FlexMux multiplexes one or more SL-packetized streams with varying instantaneous bit rates into *FlexMux packets* that have variable lengths.

Identification of SL packets originating from different elementary streams is through FlexMux channel numbers. Each SL-packetized stream is mapped into one FlexMux channel. FlexMux packets with data from different SL-packetized streams can therefore be arbitrarily interleaved. The sequence of FlexMux packets that are interleaved into one stream is called a *FlexMux stream*.

MPEG-4 Over MPEG-2

The MPEG-2 PES is the common denominator for encapsulating content. MPEG-4 defines the encapsulation of SL-packetized and FlexMux streams within PES packets.

One SL-packetized stream is mapped to one PID or *stream_id* of the MPEG-2 multiplex. Each SL packet is mapped to one PES packet. PES and SL packet header redundancy is reduced by conveying information only in the PES header, removing duplicate data from the SL packet header.

An integer number of FlexMux packets can also be conveyed in a PES packet to further reduce multiplex overhead. Several SL-packetized streams can then be mapped to one MPEG-2 PID or *stream_id*. The PES header is not used at all, since synchronization can be done with the time stamp information conveyed in the SL packet headers.

MP4 File Format

A file format for the exchange of MPEG-4 content has also been defined. The file format supports metadata in order to allow indexing, fast searches and random access.

Intellectual Property Management and Protection (IPMP)

IPMP, also called digital rights management (DRM), provides an interface and tools, rather than a complete system, for implementing intellectual property rights management.

The level and type of management and protection provided is dependent on the value of the content and the business model. For this reason, the complete design of the IPMP system is left to application developers.

The architecture enables both open and proprietary solutions to be used, while enabling interoperability, supporting the use of more than one type of protection (i.e., decryption, watermarking, rights management, etc.) and supporting the transferring of content between devices using a defined inter-device message (reflecting the issue of content distribution over home networks).

For protected content, the IPMP tool requirements are communicated to the decoder before the presentation starts. Tool configuration and initialization information are conveyed by the IPMP Descriptor or IPMP elementary stream. Needed tools can be embedded, downloaded or acquired by other means.

Control point and ordering sequence information in the IPMP Descriptor allow different tools to function at different places in the system. IPMP data, carried in either an IPMP Descriptor or IPMP elementary stream, includes rights containers, key containers and tool initialization data.

MPEG-4 Part 10 (H.264) Video

Previously known as "H.26L", "JVT", "JVT codec", "AVC" and "Advanced Video codec", ITU-T H.264 is one of two new video codecs, the other being SMPTE "VC-9" which is based on Microsoft Windows Media Video 9 codec. H.264 is incorporated into the MPEG-4 specifications as Part 10.

Rather than a single major advancement, H.264 employs many new tools designed to improve performance. These include:

- Support for 8-, 10- and 12-bit 4:2:2 and 4:4:4 YCbCr

- Integer transform

- UVLC, CAVLC and CABAC entropy coding

- Multiple reference frames

- Intra prediction

- In-loop de-blocking filter

- SP and SI slices

- Many new error resilience tools

Profiles and Levels

Similar to other video codecs, profiles specify the syntax (i.e., algorithms) and levels specify various parameters (resolution, frame rate, bit rate, etc.).

At the time of printing, H.264 supported three profiles. The various levels are described in Table 14.5. Additional profiles and levels are in development to add support for DCT, etc.

Baseline Profile

Baseline profile is designed for progressive video such as video conferencing, video-over-IP and mobile applications. Tools used by baseline profile include:

- I and P slice types

- ¼-pixel motion compensation

- UVLC and CAVLC entropy coding

- Arbitrary slice ordering

- Flexible macroblock ordering

- Redundant slices

- 4:2:0 YCbCr format

Main Profile

Main profile is designed for a wide range of broadcast applications. Additional tools over baseline profile include:

- Interlaced pictures

- B slice type

- CABAC entropy coding

- Weighted prediction

- 4:2:2 and 4:4:4 YCbCr, 10- and 12-bit formats

- Arbitrary slice ordering not supported

- Flexible macroblock ordering not supported

- Redundant slices not supported

Extended Profile

Extended profile is designed for mobile and Internet streaming applications. Additional tools over baseline profile include:

- B, SP and SI slice types

- Slice data partitioning

- Weighted prediction

Video Coding Layer

YCbCr Color Space

H.264 uses the YCbCr color space, supporting 4:2:0, 4:2:2 and 4:4:4 sampling. The 4:2:2 and 4:4:4 sampling options increase the chroma resolution over 4:2:0, resulting in better picture quality. In addition to 8-bit YCbCr data, H.264 supports 10- and 12-bit YCbCr data to further improve picture quality.

The 4:2:0 sampling structure for H.264 is shown in Figures 3.8 through 3.10. The 4:2:2 and 4:4:4 sampling structures are shown in Figures 3.2 and 3.3.

Macroblocks

With H.264, the partitioning of the 16×16 macroblocks as been extended, as illustrated in Figure 14.5.

Such fine granularity leads to a potentially large number of motion vectors per macroblock (up to 32) and number of blocks that must be interpolated (up to 96). To constrain encoder/decoder complexity, there are limits on the number of motion vectors used for two consecutive macroblocks.

Error concealment is improved with Flexible Macroblock Ordering (FMO), which assigns macroblocks to another slice so they are transmitted in a non-scanning sequence. This reduces the chance that an error will affect a large spatial region, and improves

Level	Maximum MB per Second	Maximum Frame Size (MB)	Typical Frame Resolution	Typical Frames per Second	Maximum MVs per Two Consecutive MBs	Maximum Reference Frames	Maximum Bit Rate
1	1,485	99	176 × 144	15	–	4	64 kbps
1.1	3,000	396	176 × 144 320 × 240 352 × 288	30 10 7.5	–	9 3 3	192 kbps
1.2	6,000	396	352 × 288	15	–	6	384 kbps
1.3	11,880	396	352 × 288	30	–	6	768 kbps
2	11,880	396	352 × 288	30	–	6	2 Mbps
2.1	19,800	792	352 × 480 352 × 576	30 25	–	6	4 Mbps
2.2	20,250	1,620	720 × 480 720 × 576	15 12.5	–	5	4 Mbps
3	40,500	1,620	720 × 480 720 × 576	30 25	32	5	10 Mbps
3.1	108,000	3,600	1280 × 720	30	16	5	14 Mbps
3.2	216,000	5,120	1280 × 720	60	16	4	20 Mbps
4	245,760	8,192	1920 × 1080 1280 × 720	30 60	16	4	20 Mbps
4.1	245,760	8,192	1920 × 1080 1280 × 720	30 60	16	4	50 Mbps
4.2	491,520	8,192	1920 × 1080	60	16	4	50 Mbps
5	589,824	22,080	2048 × 1024	72	16	5	135 Mbps
5.1	983,040	36,864	2048 × 1024 4096 × 2048	120 30	16	5	240 Mbps

Table 14.5. MPEG-4 Part 10 (H.264) Levels. "MB" = macroblock, "MV" = motion vector.

error concealment by being able to use neighboring macroblocks for prediction of a missing macroblock.

In-loop De-blocking Filter

H.264 adds an in-loop de-blocking filter. It removes artifacts resulting from adjacent macroblocks having different estimation types and/or different quantizer scales. The filter also removes artifacts resulting from adjacent blocks having different transform/quantization and motion vectors.

The loop filter uses a content adaptive non-linear filter to modify the two samples on either side of a block or macroblock boundary.

Slices

The slice has greater importance in H.264 since it is now the basic independent spatial element. This prevents an error in one slice from affecting other slices. This flexibility allows extending the I-, P- and B-picture types down to the slice level, resulting in I-, P- and B-slice types.

Arbitrary Slice Ordering (ASO) enables slices to be transmitted and received out of order. This can improve low-delay performance in video conferencing and Internet applications.

Redundant slices are also allowed for additional error resilience. This alternative data can be used to recover any corrupted macroblocks.

SP and SI Slices

In addition to I-, P- and B-slices, H.264 adds support for SP-slices (Switching P) and SI-slices (Switching I). SP-slices use motion compensated prediction, taking advantage of temporal redundancy to enable reconstruction of a slice even when different reference slices are used. SI-slices take advantage of spatial prediction to enable identically reconstructing a corresponding SP-slice.

Use of S-slices enable efficient bit-stream switching, random access, fast-forward and error resilience/recovery, as illustrated in Figures 14.6 and 14.7.

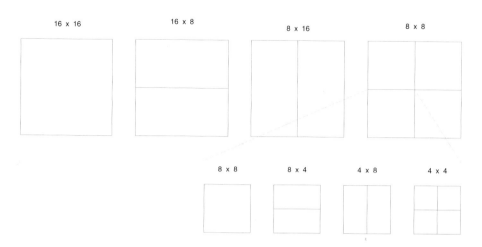

Figure 14.5. Segmentations of H.264 Macroblocks for Motion Compensation. Top: segmentation of macroblocks. Bottom: segmentation of 8 × 8 partitions.

Intra Prediction

When motion estimation is not efficient, intra prediction can be used to eliminate spatial redundancies. This technique attempts to predict the current block based on adjacent blocks. The difference between the predicted block and the actual block is then coded. This tool is very useful in flat backgrounds where spatial redundancies often exist.

Motion Compensation

¼-pixel Motion Compensation

Motion compensation accuracy is improved from the ½-pixel accuracy used by most earlier video codecs. H.264 supports the same ¼-pixel accuracy that is used on the latest MPEG-4 video codec.

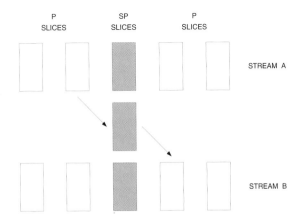

Figure 14.6. Using SP Slices to Switch to Another Stream.

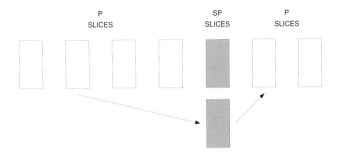

Figure 14.7. Using SP Slices to Fast Forward.

Multiple Reference Frames

H.264 adds supports for multiple reference frames. This increases compression by improving the prediction process and increases error resilience by being able to use another reference frame in the event that one was lost.

A single macroblock can use up to eight reference frames (up to three for HDTV), with a total limit of 16 reference frames used within a frame.

To compensate for the different temporal distances between current and reference frames, predicted blocks are averaged with configurable weighting parameters. These parameters can either be embedded within the bitstream or the decoder may implicitly derive them from temporal references.

Unrestricted Motion Search

This allows for reference frames that are outside the picture. Missing data is spatially predicted from boundary data.

Transform, Scaling and Quantization

H.264 uses a simple 4×4 integer transform. In contrast, older video codecs use an 8×8 DCT that operates on floating-point coefficients. An additional 2×2 transform is applied to the four CbCr DC coefficients. Intra-16×16 macroblocks have an additional 4×4 transform performed for the sixteen Y DC coefficients.

Blocking and ringing artifacts are reduced as a result of the smaller block size used by H.264. The use of integer coefficients eliminates rounding errors that cause drifting artifacts common with DCT-based video codecs.

For quantization, H.264 uses a set of 52 uniform scalar quantizers, with a step increment of about 12.5% between each.

The quantized coefficients are then scanned, from low frequency to high frequency, using one of two scan orders.

Entropy Coding

After quantization and zig-zag scanning, H.264 uses two types of entropy encoding: variable-length coding (VLC) and Context Adaptive Binary Arithmetic Coding (CABAC).

For everything but the transform coefficients, H.264 uses a single Universal VLC (UVLC) table that uses an infinite-extend codeword set (Exponential Golomb). Instead of multiple VLC tables as used by other video codecs, only the mapping to the single UVLC table is customized according to statistics.

For transform coefficients, which consume most of the bandwidth, H.264 uses Context-Adaptive Variable Length Coding (CAVLC). Based upon previously processed data, the best VLC table is selected.

Additional efficiency (5–10%) may be achieved by using Context Adaptive Binary Arithmetic Coding (CABAC). CABAC continually updates the statistics of incoming data and real-time adaptively adjusts the algorithm using a process called context modeling.

Network Abstraction Layer (NAL)

The NAL facilitates mapping H.264 data to a variety of transport layers including:

- RTP/IP for wired and wireless Internet services

- File formats such as MP4

- H.32X for conferencing

- MPEG-2 systems

The data is organized into NAL units, packets that contain an integer number of bytes. The first byte of each NAL unit indicates the payload data type and the remaining bytes contain the payload data. The payload data may be interleaved with additional data to prevent a start code prefix from being accidentally generated.

When data partitioning is used, each slice is divided into three separate partitions, with each partition using a specific NAL unit type. This enables data partitioning to be used as an efficient layering method that separates the data different levels of importance. By partitioning data into different NAL units, it is much easier to use different error protection for various parts of the data.

References

1. *H.264 Advanced Video Coding: A Whirlwind Tour,* by PixelTools, 2003.
2. *H.264 Coding Efficiency Has a Price*, by Eric Barrau, Philippe Durieuz and Stephane Muta, Sophia Antipolis Micro-Electronics Forum, January 2003.
3. ISO/IEC 14996–1, *Information Technology—Coding of Audio-Visual Objects—Part 1: Systems.*
4. ISO/IEC 14996–2, *Information Technology—Coding of Audio-Visual Objects—Part 2: Visual.*
5. ISO/IEC 14996–3, *Information Technology—Coding of Audio-Visual Objects—Part 3: Audio.*
6. ITU-T H.264, *Advanced Video Coding for Generic Audiovisual Services*, May 2003.
7. *Technical Overview of H.264/AVC*, by R. Schafer, T. Wiegand and H. Schwarz, EBU Technical Review, January 2003.
8. *The H.264/AVC Video Coding Standard for the Next Generation Multimedia Communication*, by M. Mahdi Ghandi and Mohammad Ghanbari, IAEEE invited paper.

Chapter 15

ATSC
Digital Television

The ATSC (Advanced Television Systems Committee) digital television (DTV) broadcast standard is used in the United States, Canada, South Korea, Taiwan and Argentina.

The three other primary DTV standards are DVB (Digital Video Broadcast), ISDB (Integrated Services Digital Broadcasting) and OpenCable™. A comparison between the standards is shown in Table 15.1. The basic audio and video capabilities are very similar. The major differences are the RF modulation schemes and the level of definition for non-audio/video services.

The ATSC standard is actually a group of standards:

A/52   Digital Audio Compression (AC-3) Standard

A/53   ATSC Digital Television Standard

A/54   Guide to the Use of the ATSC Digital Television Standard

A/57   Content Identification and Labeling for ATSC Transport

A/64   Transmission Measurement and Compliance for Digital Television

A/65   Program and System Information Protocol for Terrestrial Broadcast and Cable

A/69   Program and System Information Protocol Implementation Guidelines for Broadcasters

A/70   Conditional Access System for Terrestrial Broadcast

A/80   Modulation and Coding Requirements for Digital TV (DTV) Applications Over Satellite

A/81   Direct-to-Home Satellite Broadcast Standard

A/90   ATSC Data Broadcast Standard

A/91   Implementation Guidelines for the Data Broadcast Standard

A/92   Delivery of IP Multicast Sessions over Data Broadcast Standard

A/93   Synchronized/Asynchronous Trigger Standard

A/94   ATSC Data Application Reference Model

A/95   Transport Stream File System Standard

A/96   ATSC Interaction Channel Protocols

A/100 DTV Application Software Environment - Level 1 (DASE-1)

The ATSC standard uses an MPEG-2 transport stream to convey compressed digital video, compressed digital audio and ancillary data over a single 6 MHz channel. Within this transport stream can be a single HDTV program, multiple SDTV programs, data or a combination of these.

The MPEG-2 transport stream has a maximum bit rate of ~19.4 Mbps (6 MHz over-the-air channel) or ~38.8 Mbps (6 MHz digital cable channel).

The 19.4 Mbps bit rate can be used in a very flexible manner, trading-off the number of programs offered versus video quality and resolution. For example,

(1) HDTV program

(1) HDTV program + (1) SDTV program + data

(4) SDTV programs

Parameter	US (ATSC)	Europe, Asia (DVB)	Japan (ISDB)	OpenCable™
video compression	MPEG-2			
audio compression	Dolby® Digital	MPEG-2, Dolby® Digital, DTS®	MPEG-2 AAC-LC	Dolby® Digital
multiplexing	MPEG-2 transport stream			
terrestrial modulation	8-VSB	COFDM	BST-OFDM[2]	–
channel bandwidth	6 MHz	6, 7, or 8 MHz	6, 7, or 8 MHz	6 MHz
cable modulation	16-VSB[3]	QAM	QAM	QAM
channel bandwidth	6 MHz	6, 7, or 8 MHz	6, 7, or 8 MHz	6 MHz
satellite modulation	QPSK, 8PSK	QPSK	PSK	–

Notes:
1. ISDB = Integrated Services Digital Broadcasting.
2. BST-OFDM = Bandwidth Segmented Transmission of OFDM.
3. Most digital cable systems use QAM instead of 16-VSB.

Table 15.1. Comparison of the Various DTV Standards.

Video Capability

Although any resolution may be used as long as the maximum bit rate is not exceeded, there are several standardized resolutions, indicated in Table 15.2. Both interlaced and progressive pictures are permitted for most of the resolutions.

Video compression is based on MPEG-2. However, there are some minor constraints on some of the MPEG-2 parameters, as discussed within the MPEG-2 chapter.

Audio Capability

Audio compression is implemented using Dolby® Digital and supports 1–5.1 channels.

The main audio, or associated audio which is a complete service (containing all necessary program elements), has a bit rate ≤448 kbps (384 kbps is typically used). A single channel associated service containing a single program element has a bit rate ≤128 kbps. A two channel associated service containing only dialogue has a bit rate ≤192 kbps. The combined bit rate of a main and associated service which are intended to be decoded simultaneously must be ≤576 kbps.

There are several types of audio service defined.

Main Audio Service: Complete Main (CM)

This type of audio service contains a complete audio program (dialogue, music and effects). This is the type of audio service normally provided, and may contain 1–5.1 audio channels.

The CM service may be further enhanced by using the VI, HI, C or VO associated services. Audio in multiple languages may be provided by supplying multiple CM services, each in a different language.

Main Audio Service: Music and Effects (ME)

This type of audio service contains the music and effects of an audio program, but not the dialogue. It may contain 1–5.1 audio channels. The primary program dialogue (if any exists) is supplied by a D service.

Associated Service: Visually Impaired (VI)

This service typically contains a narrative description of the program content. The VI service uses a single audio channel. The simultaneous decoding of both the VI and CM allows the visually impaired to enjoy the program.

Besides providing VI as a single narrative channel, it may be provided as a complete program mix containing music, effects, dialogue and narration. In this case, the service may use up to 5.1 channels.

Associated Service: Hearing Impaired (HI)

This service typically contains only dialogue which is intended to be reproduced simultaneously with the CM service. In this case, HI is a single audio channel.

Besides providing HI as a single dialogue channel, it may be provided as a complete program mix containing music, effects and dialogue with enhanced intelligibility. In this case, the service may use up to 5.1 channels.

Active Resolution (Y)	SDTV or HDTV	Frame Rate (p = progressive, i = interlaced)			
		23.976p 24p	29.97i 30i	29.97p 30p	59.94p 60p
480 × 480	SDTV	×	×	×	×
528 × 480		×	×	×	×
544 × 480		×	×	×	×
640 × 480		×	×	×	×
704 × 480		×	×	×	×
1280 × 720	HDTV	×		×	×
1280 × 1080		×	×	×	
1440 × 1080		×	×	×	
1920 × 1080		×	×	×	

Table 15.2. Common Active Resolutions for ATSC Content.

Associated Service: Dialogue (D)

This service contains program dialogue intended for use with the ME service.

A complete audio program is formed by simultaneously decoding both the D and ME services and mixing the D service into the center channel of the ME service.

If the ME service contains more than two audio channels, the D service is monophonic. If the ME service contains two channels, the D service may also contain two channels. In this case, a complete audio program is formed by simultaneously decoding the D and ME services, mixing the left channels of the ME and D service, and mixing the right channels of the D and ME service. The result will be a two-channel stereo signal containing music, effects and dialogue.

Audio in multiple languages may be provided by supplying multiple D services (each in a different language) along with a single ME

service. This is more efficient than providing multiple CM services. However, in the case of more than two audio channels in the ME service, this requires that the dialogue be restricted to the center channel.

Associated Service: Commentary (C)

This service is similar to the D service, except that instead of conveying essential program dialogue, it conveys an optional program commentary using a single audio channel.

In addition, it may be provided as a complete program mix containing music, effects, dialogue and commentary. In this case, the service may use up to 5.1 channels.

Associated Service: Voice-Over (VO)

This service is a single-channel service intended to be decoded and mixed with the ME service.

Program and System Information Protocol (PSIP)

Enough bandwidth is available within the MPEG-2 transport stream to support several low-bandwidth non-television services such as program guide, closed captioning, weather reports, stock indices, headline news, software downloads, pay-per-view information, etc. The number of additional non-television services (virtual channels) may easily reach ten or more. In addition, the number and type of services will constantly be changing.

To support these non-television services in a flexible, yet consistent, manner, the *Program and System Information Protocol* (PSIP) was developed. PSIP is a small collection of hierarchically associated tables (see Figure 15.1 and Table 15.3) designed to extend the MPEG-2 PSI tables. It describes the information for all virtual channels carried in a particular MPEG-2 transport stream. Additionally, information for analog broadcast channels may be incorporated.

Required Tables

Event Information Table (EIT)

There are up to 128 EITs, EIT-0 through EIT-127, each of which describes the events or TV programs associated with each virtual channel listed in the VCT. Each EIT is valid for three hours. Since there are up to 128 EITs, up to 16 days of programming may be advertised in advance. The first four EITs are required (the first 24 are recommended) to be present.

Information provided by the EIT includes start time, duration, title, pointer to optional descriptive text for the event, advisory data, caption service data, audio service descriptor, and so on.

Master Guide Table (MGT)

This table provides general information about the other tables. It defines table sizes, version numbers and packet identifiers (PIDs).

Rating Region Table (RRT)

This table transmits the rating system, commonly referred to as the "V-chip."

System Time Table (STT)

This table serves as a reference for the time of day. Receivers use it to maintain the correct local time.

Terrestrial Virtual Channel Table (TVCT)

This table, also referred to as the VCT although there is also a Cable VCT (CVCT) and Satellite VCT (SVCT), contains a list of all the channels in the transport stream that are or will be available, plus their attributes. It may also include the broadcaster's analog channel and digital channels in other transport streams.

Attributes for each channel include major/minor channel number, short name, Transport/Transmission System ID (TSID) that uniquely identifies each station, etc. The *Service Location Descriptor* is used to list the PIDs for the video, audio, data, and other related elementary streams.

Optional Tables

Extended Text Table (ETT)

For text messages, there can be several ETTs, each having its PID defined by the MGT. Messages can describe channel information, coming attractions, movie descriptions, and so on.

Descriptor	Descriptor Tag	Terrestrial Broadcast Tables									
		PMT	MGT	TVCT	RRT	EIT	ETT	STT	DCCT	DCCSCT	CAT
PID		per PAT	$1FFB_H$	$1FFB_H$	$1FFB_H$	per MGT	per MGT	$1FFB_H$	$1FFB_H$	$1FFB_H$	0001_H
Table_ID		02_H	$C7_H$	$C8_H$	CA_H	CB_H	CC_H	CD_H	$D3_H$	$D4_H$	80_H, 81_H (ECM) 82_H - $8F_H$ (EMM)
repetition rate		400 ms	150 ms	400 ms	1 min	0.5 sec	1 min	1 sec	400 ms	1 hour	
AC-3 audio stream	1000 0001	M				M					
ATSC CA	1000 1000			O		O					
ATSC private information*	1010 1101										
CA	0000 1001	M									M
caption service	1000 0110	M				M					
component name	1010 0011	M									
content advisory	1000 0111	M				M					
content identifier	1011 0110					O					
DCC arriving request	1010 1001								M		
DCC departing request	1010 1000								M		
enhanced signaling	1011 0010										
extended channel name	1010 0000			M							
MAC address list	1010 1100	M									
redistribution control	1010 1010	M				M					
service location	1010 0001			S							
stuffing*	1000 0000										
time-shifted service	1010 0010			M							

Notes:
1. PMT: MPEG-2 Program Map Table. CAT: MPEG-2 Conditional Access Table.
2. M = when present, required in this table. S = always present in this table for terrestrial broadcast.
   O = may be present in this table also. * = no restrictions.

Table 15.3. List of ATSC PSIP Tables, Descriptors and Descriptor Locations.

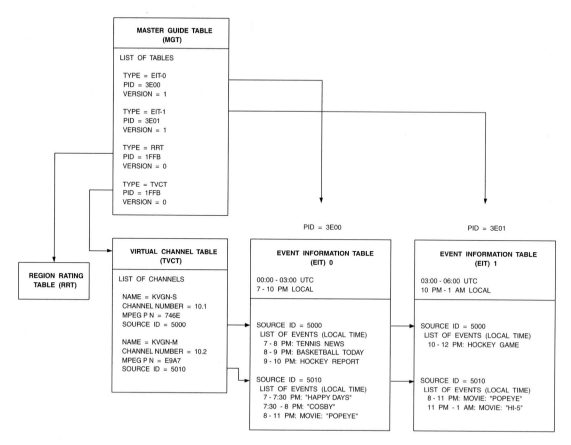

Figure 15.1. ATSC PSIP Table Relationships.

Directed Channel Change Table (DCCT)

The DCCT contains information needed for a channel change to be done at a broadcaster-specified time. The requested channel change may be unconditional or may be based upon criteria specified by the viewer.

Directed Channel Change Selection Code Table (DCCSCT)

The DCCSCT permits a broadcast program categorical classification table to be downloaded for use by some Directed Channel Change requests.

Descriptors

Much like MPEG-2, ATSC uses descriptors to add new functionality. In addition to various MPEG-2 descriptors, one or more of these ATSC-specific descriptors may be included within the PMT or one or more PSIP tables (see Table 15.3) to extend data within the tables. A descriptor not recognized by a decoder must be ignored by that decoder. This enables new descriptors to be implemented without affecting receivers that cannot recognize and process the descriptors.

AC-3 Audio Stream Descriptor

This ATSC descriptor indicates Dolby® Digital or Dolby® Digital Plus audio is present, and is discussed in Chapter 13.

ATSC CA Descriptor

This ATSC descriptor has a syntax almost the same as the MPEG-2 CA descriptor.

ATSC Private Information Descriptor

This ATSC descriptor provides a way to carry private information, and is discussed in Chapter 13. More than one descriptor may appear within a single descriptor.

Component Name Descriptor

This ATSC descriptor defines a variable-length text-based name for any component of the service, and is discussed in Chapter 13.

Content Advisory Descriptor

This ATSC descriptor defines the ratings for a given program, and is discussed in Chapter 13.

Content Identifier Descriptor

This ATSC descriptor is used to uniquely identify content with the ATSC transport.

DCC Arriving Request Descriptor

This ATSC descriptor provides instructions for the actions to be performed by a receiver upon arrival to a newly changed channel:

> Display text for at least 10 seconds, or for a less amount of time if the viewer issues a "continue", "OK" or equivalent command.

> Display text indefinitely, or until the viewer issues a "continue", "OK" or equivalent command.

DCC Departing Request Descriptor

This ATSC descriptor defines provides instructions for the actions to be performed by a receiver prior to leaving a channel:

> Cancel any outstanding things and immediately perform the channel change.

> Display text for at least 10 seconds, or for a less amount of time if the viewer issues a "continue", "OK" or equivalent command.

> Display text indefinitely, or until the viewer issues a "continue", "OK" or equivalent command.

Enhanced Signaling Descriptor

This ATSC descriptor identifies the terrestrial broadcast transmission method of a program element, and is discussed in Chapter 13.

Extended Channel Name Descriptor

This ATSC descriptor provides a variable-length channel name for the virtual channel.

MAC Address List Descriptor

This ATSC descriptor is used when implementing IP (Internet Protocol) multicasting over MPEG-2 transport streams, and is discussed in Chapter 13.

Redistribution Control Descriptor

This ATSC descriptor conveys any redistribution control information held by the program rights holder for the content, and is discussed in Chapter 13.

Service Location Descriptor

This ATSC descriptor specifies the stream type, PID and language code for each elementary stream. It is present in the TVCT for each active channel.

Time-shifted Service Descriptor

This ATSC descriptor links one virtual channel with up to 20 other virtual channels carrying the same programming, but time-shifted. A typical application is for Near Video On Demand (NVOD) services.

Data Broadcasting

The ATSC data broadcast standard describes various ways to transport data within the MPEG-2 transport stream. It can be used for a number of applications, such as:

Delivering declarative data (HTML code)

Delivering procedural data (Java code)

Delivering software and images

Delivering MPEG-4, H.264 or VC-9 video streams

Carouseling MPEG-2 video or MP3 audio files

The key elements defined within the standard are:

Data services announcement

Data delivery models such as data piping, data streaming, addressable sections and data download

Application signaling

MPEG-2 systems tools

Protocols

A data service must be contained in a virtual channel, and each virtual channel may have at most one data service. One data service may consist of multiple applications, and each application may contain multiple data elements.

Data broadcasting is also discussed in Chapter 13.

Data Service Announcements

Data broadcasting utilizes and extends PSIP to announce and find data services in the broadcast stream. Data services are announced by an event in either the PSIP EIT or Data Event Tables (DET).

Additional tables and descriptors for data service announcements include:

Data Event Table (DET)

There are up to 128 DETs, DET-0 through DET-127, each of which describes information (titles, start times, etc.) for data services on virtual channels. Each DET is valid for three hours. A minimum of four DETs (DET-0 through DET-3) are required. Any change in a DET shall trigger a change in version of the MGT.

Extended Text Table (ETT)

The ETT is used to provide detailed descriptions of a data event. The syntax is mostly identical to ETT used for AV services.

Long Term Service Table (LTST)

The LTST is used to pre-announce data events that will occur on a time scale outside what the EITs/DETs can support.

Data Service Descriptor

This ATSC descriptor indicates the maximum transmission bandwidth and buffer model requirements of the data service.

PID Count Descriptor

This ATSC descriptor provides a total count of PIDs used in the service. It may also specify the minimum number of PID values a receiver must be able to monitor simultaneously to provide a meaningful rendition of the data service.

Service Description Framework (SDF)

Due to the wide range of protocol encapsulations possible, there is a need to signal which encapsulation is used in each data stream. While it is possible to use the PMT for signaling, this approach is not very scalable, working only for simple cases. The *Service Description Framework* (SDF) was developed to provide a scalable framework.

Additional tables and descriptors for the service description framework include:

Data Service Table (DST)

The DST provides the description of a data service comprised of one or more receiver applications. It also provides information to allow data receivers to associate applications with references to data consumed by them.

Network Resources Table (NRT)

The NRT provides a list of all network resources outside those in the current MPEG-2 program or transport stream.

A data service may use the NRT to get data packets or datagrams other than the ones published in the *Service Location Descriptor* of the TVCT. This includes data elementary streams in another program within the same transport stream, data elementary streams in other transport streams, and bi-directional communication channels using other protocols such as IP.

Association Tag Descriptor

This MPEG-2 DSM-CC descriptor binds data resource references (taps) to specific data elementary streams in the MPEG-2 transport stream.

Download Descriptor

This MPEG-2 DSM-CC descriptor includes information specific to the DSM-CC download protocol.

DSM-CC Compatibility Descriptor

This MPEG-2 DSM-CC descriptor indicates receiver hardware and/or software requirements for proper acquisition and processing of a data service.

DSM-CC Resource Descriptor

This MPEG-2 DSM-CC descriptor provides a list of DSM-CC resource descriptors used in the NRT. Resource descriptors contain information required for the network to allocate a resource, track the resource once it has been allocated, and de-allocate the resource once it is no longer needed. Supported descriptors include:

Deferred MPEG Program Element Descriptor: This descriptor is used to identify a MPEG-2 resource in another virtual channel (in the same or a different transport stream).

IP Resource Descriptor: This descriptor is used to identity a resource by its IP address and port.

IPV6 Resource Descriptor: This descriptor is used to identity a resource by its IPv6 address and port.

URL Resource Descriptor: This descriptor is used to identity a resource by its Uniform Resource Locator (URL).

Multiprotocol Encapsulation Descriptor

This MPEG-2 DSM-CC descriptor provides information defining the mapping of a receiver's MAC address to a specific addressing scheme.

Terrestrial Transmission Format

The VSB (vestigial sideband) modulation system offers two modes: a terrestrial broadcast mode and a high data rate mode. The terrestrial broadcast mode provides maximum coverage, supporting one HDTV signal or multiple SDTV signals in a 6 MHz channel. The high data rate mode supports two HDTV signals or multiple SDTV signals in a 6 MHz channel, trading off robustness for a higher data rate. The two modes share the same pilot, symbol rate, data frame structure, interleaving, Reed-Solomon coding, and synchronization pulses.

VSB modulation requires only processing the in-phase (I) channel signal, sampled at the symbol rate. The decoder only requires one ADC and a real (not complex) equalizer operating at the symbol rate of 10.76M samples per second.

8-VSB Overview

Figure 15.2 illustrates a simplified block diagram of the 8-VSB modulation process.

Data Synchronization

Initially, an 8-VSB (8-level vestigial sideband) modulator must synchronize to the incoming MPEG-2 transport bitstream packets. Before any processing can be done, it must correctly identify the start and end points of each MPEG-2 packet by using its sync byte. This sync byte is then discarded and replaced with an ATSC segment sync later in the processing stage.

Data Randomizer

In general, the 8-VSB signal must have a random, noise-like nature. This results in a flat, noise-like spectrum that maximizes channel bandwidth efficiency and reduces possible interference with other DTV and NTSC channels.

To accomplish this, the data randomizer uses a pseudo-random number generator to scramble the bitstream. The receiver reverses this process to recover the original data.

Reed-Solomon Encoding

Reed-Solomon (RS) encoding provides forward error correction (FEC) for the data stream.

The RS encoder takes the 187 bytes of an incoming MPEG-2 packet (remember, the packet sync byte has been removed), calculates 20 parity bytes (known as Reed-Solomon parity bytes), and adds them to the end of the packet.

The receiver compares the received 187 bytes to the 20 parity bytes in order to determine the validity of the recovered data. If errors are detected, the receiver can use the parity bytes to locate and correct the errors.

Data Interleaver

The data interleaver scrambles the sequential order of the data stream in order to minimize sensitivity to burst-type interference. It does this by assembling new data packets that incorporate portions of many different MPEG-2 packets. These new data packets are the same length as the original MPEG-2 packets: 207 bytes after RS coding.

Data interleaving is done according to a known pattern. The process is reversed in the receiver in order to recover the proper data.

Trellis Encoder

Trellis coding is another form of forward error correction, known as convolutional coding, that tracks the data stream over time. Twelve different Trellis encoders operate in parallel, providing another form of interleaving to provide further protection against burst-type interference.

In general, each 8-bit data word is split into a stream of four, 2-bit words. A 3-bit binary code is mathematically generated to describe the transition from previous 2-bit words to the current one. For every two bits that go into the trellis coder, three bits come out. For this reason, the trellis coder in the 8-VSB system is said to be a 2/3 rate coder.

These 3-bit codes are substituted for the original 2-bit words and transmitted as one of the eight level symbols of 8-VSB.

The receiver's trellis decoder uses the 3-bit transition codes to reconstruct the evolution of the data stream from one 2-bit word to the next.

Sync and Pilot Insertion

The next step is the insertion of various "helper" signals (ATSC pilot, segment sync, and field sync) that aid the 8-VSB receiver to locate and demodulate the transmitted signal. These signals are inserted after the randomization and error coding stages so as not to destroy the relationships that these signals must have in order to be effective.

The first "helper" signal is the ATSC *pilot*, which gives the 8-VSB receiver something to lock on to independent of the transmitted data. Just before modulation, a small DC offset is added to the 8-VSB baseband signal. This causes a small residual carrier to appear at the zero frequency point of the resulting modulated spectrum. This is the ATSC pilot.

Another "helper" signal is the ATSC *segment sync*. A data segment is composed of the 207 bytes of an interleaved data packet. After trellis coding, the 207-byte segment has been converted into a baseband stream of 828 eight-level symbols. The ATSC segment sync is a four-symbol pulse that is added to the front of each data segment to replace the missing sync byte of the original MPEG-2 data packet. The segment sync thus appears once every 832 symbols. 8-VSB receivers can detect the repetitive nature of the ATSC segment sync, recover it, and use it to help regenerate the sample clock.

The last "helper" signal is the ATSC *field sync*. 313 consecutive data segments comprise a data field. The ATSC field sync is one data segment that is added at the beginning of each data field. It has a known symbol pattern and is used by the 8-VSB receiver to eliminate signal ghosts caused by reflections. This can be done by comparing the received ATSC field sync against the known field sync sequence. The difference is used to adjust the taps of the receiver's ghost-canceling equalizer.

AM Modulation

The eight-level baseband signal is then amplitude modulated onto an intermediate frequency (IF) carrier. This creates a large, double-sideband spectrum about the carrier frequency that is far too wide to be transmitted in a 6 MHz channel. A Nyquist VSB filter trims the bandwidth down to fit within the channel by removing redundant information.

As a result of the overhead added to the data stream due to forward error correction coding and sync insertion, the bit rate has gone from 19.39 Mbps to 32.28 Mbps at the output of the trellis coder. Since three bits are transmitted in each symbol, the resulting symbol rate is 10.76 million symbols per second. This can be transmitted with a minimum frequency bandwidth of 5.38 MHz, which fits nicely within a 6 Mhz channel bandwidth.

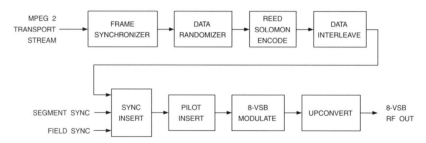

Figure 15.2. Simplified Block Diagram of a 8-VSB Modulator.

After the Nyquist VSB filter, the 8-VSB intermediate frequency (IF) signal is up-converted to the assigned UHF or VHF channel frequency.

E-VSB

E-VSB, also referred to as "Enhanced VSB" and "Enhanced 8-VSB", enables ATSC broadcasters to include a secondary lower bit-rate program that is more robust than typical HDTV programs in low-signal conditions. If interference degrades the primary HDTV signal, the receiver switches to a more robust SDTV version of the same program that has been multiplexed into that transport stream. It also enables the use of secondary, low-cost SDTVs using an indoor antenna.

For example, within the 19 Mbps channel, 14 Mbps could be used for the HD program, 4 Mbps for the robust SD program and 1 Mbps for management overhead.

To use minimal bandwidth for the robust SD program, the Dolby® Digital Plus audio codec and the H.264 and/or SMPTE VC-9 video codec are used.

Audio Capability

Audio compression for the enhanced service will likely be implemented using Dolby® Digital Plus and supports 1–5.1 channels. Dolby® Digital Plus offers new coding tools to improve performance, and new features to allow operation over a wider range of bit-rates (96 kbps to 6 Mbps) and number of channels. 192 kbps, 5.1-channel audio and 96 kbps stereo audio is targeted.

Converting to Dolby® Digital is possible, ensuring compatibility with consumer's existing surround sound receivers via the S/PDIF input.

There are four types of audio service defined:

Main audio service: complete main (CM)

Associated service: visually impaired (VI)

Associated service: hearing impaired (HI)

Associated service: commentary (C)

Video Capability

Video compression for the enhanced service will likely be implemented using H.264 or SMPTE VC-9.

Modulation System

Head-end Reference Model

As seen in Figure 15.3, the E-VSB exciter has three separate inputs, one each for transport stream packets to be sent via 1/2 rate (IS-Ea), 1/4 rate (IS-Eb) and normal rate (IS-N) modes. Only one of the three inputs has a transport stream packet at any one time.

The decisions about whether a packet is 1/4 rate enhanced, 1/2 rate enhanced or not enhanced are made on based on the PID value in each transport stream packet.

Receiver Reference Model

The receiver model (Figure 15.4) has a packet selector and has additional deinterleaving for the enhanced data.

TS-N: The portion of TS-R that contains only transport stream packets transmitted by the main mode.

TS-R: The recombined transport stream containing all transport packets delivered by all transmission modes (main, 1/2 rate and 1/4 rate).

TS-E: The portion of TS-R that contains only transport stream packets transmitted by 1/2 rate and/or 1/4 rate modes.

TS-Ea: The portion of TS-E that contains only transport stream packets transmitted by 1/2 rate mode.

TS-Eb: The portion of TS-E that contains only transport stream packets transmitted by 1/4 rate mode.

Program and System Information Protocol (PSIP-E)

PSIP-E is Program and System Information carried on the TS-E. Similar to the PSIP, the PSIP-E consists of the following tables.

EIT-E: Event Information Table in the TS-E. Syntax is the same as for the EIT.

ETT-E: Event Text Table in the TS-E. Syntax is the same as for the ETT.

MGT-E: Master Guide Table in the TS-E. Syntax is the same as for the MGT. The PID value for the table sections that make up the MGT-E is $1FF9_H$.

RRT-E: Region Rating Table in the TS-E. Syntax is the same as for the RRT. The PID value for the table sections that make up the RRT-E is $1FF9_H$.

STT-E System Time Table in the TS-E. Syntax is the same as for the STT. The PID value for STT-E is $1FF8_H$.

TVCT-E: Terrestrial Virtual Channel Table in the TS-E. Syntax is the same as for the TVCT. The PID value for the table sections that make up the TVCT-E is $1FF9_H$.

Application Block Diagrams

Figure 15.5 illustrates a typical ATSC receiver set-top box block diagram. A common requirement is the ability to output both high-definition and standard-definition versions of a program simultaneously.

Figure 15.6 illustrates a typical ATSC digital television block diagram. A common requirement is the ability to decode two programs simultaneously to support Picture-in-Picture (PIP).

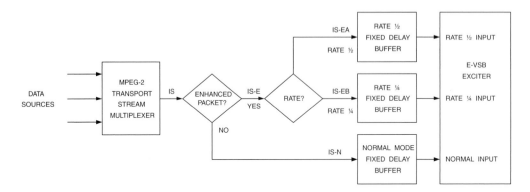

Figure 15.3. ATSC E-VSB Head-end Model.

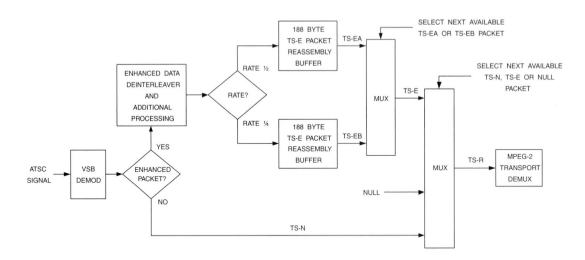

Figure 15.4. ATSC E-VSB Receiver Model.

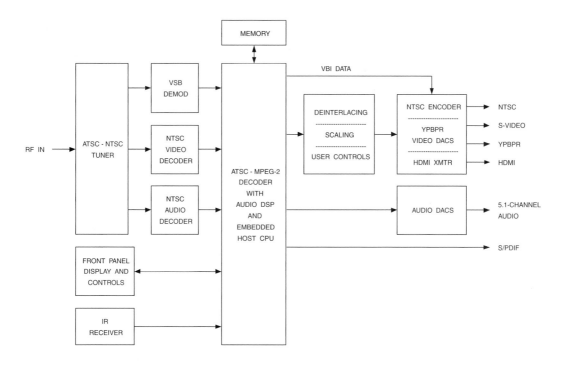

Figure 15.5. Typical ATSC Receiver Set-top Box Block Diagram.

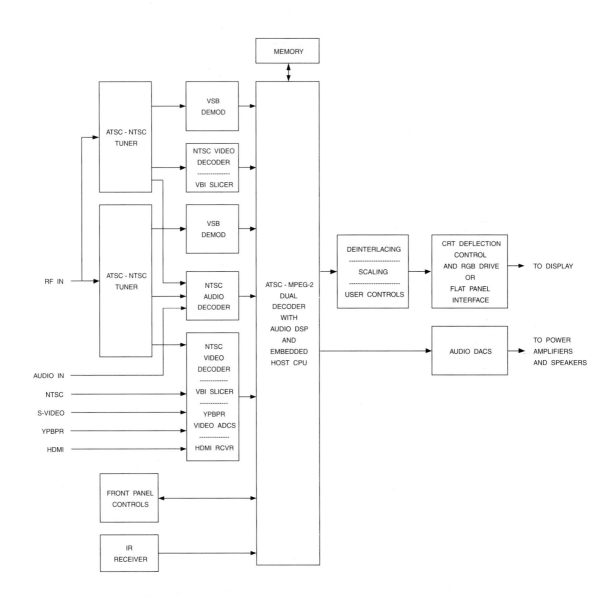

Figure 15.6. Typical ATSC Digital Television Block Diagram.

References

1. ATSC A/52a, *Digital Audio Compression Standard (AC-3)*, August 2001.
2. ATSC A/53b, *ATSC Standard: Digital Television Standard, Revision B with Amendments 1 and 2*, May 2003.
3. ATSC A/54a, *Recommended Practice: Guide to the Use of the ATSC Digital Television Standard*, December 2003.
4. ATSC A/57a, *ATSC Standard: Content Identification and Labeling for ATSC Transport*, July 2003.
5. ATSC A/58, *Harmonization With DVB SI in the Use of the ATSC Digital Television Standard*, September 1996.
6. ATSC A/64, *Transmission Measurement and Compliance for Digital Television*, May 2000.
7. ATSC A/65b, *ATSC Standard: Program and System Information Protocol for Terrestrial Broadcast and Cable (Revision B)*, March 2003.
8. ATSC A/69, *ATSC Recommended Practice: Program and System Information Protocol Implementation Guidelines for Broadcasters*, June 2002.
9. ATSC A/70, *Conditional Access System for Terrestrial Broadcast*, May 2000.
10. ATSC A/80, *Modulation and Coding Requirements for Digital TV (DTV) Applications Over Satellite*, July 1999.
11. ATSC A/81, *ATSC Direct-to-Home Satellite Broadcast Standard*, July 2003.
12. ATSC A/90, *ATSC Data Broadcast Standard*, July 2000.
13. ATSC A/91, *Recommended Practice: Implementation Guidelines for the ATSC Data Broadcast Standard*, June 2001.
14. ATSC A/92, *ATSC Standard: Delivery of IP Multicast Sessions over ATSC Data Broadcast*, January 2002.
15. ATSC A/93, *ATSC Standard: Synchronized/Asynchronous Trigger*, April 2002.
16. ATSC A/95, *ATSC Standard: Transport Stream File System Standard*, February 2003.

Chapter 16

OpenCable™
Digital Television

OpenCable™ is a digital cable standard for the United States, designed to offer interoperability between different hardware and software suppliers. A subset of the standard is being incorporated inside digital televisions.

The three other primary DTV standards are ATSC (Advanced Television Systems Committee), DVB (Digital Video Broadcast) and ISDB (Integrated Services Digital Broadcasting). A comparison between the standards is shown in Table 16.1. The basic audio and video capabilities are very similar. The major differences are the RF modulation schemes and the level of definition for non-audio/video services.

OpenCable™ is based on the following ATSC and SCTE standards:

A/52   Digital Audio Compression (AC-3) Standard

A/53   ATSC Digital Television Standard

A/65   Program and System Information Protocol for Terrestrial Broadcast and Cable

A/90   ATSC Data Broadcast Standard

SCTE 07   Digital Video Transmission Standard for Cable Television

SCTE 18   Emergency Alert Message for Cable

SCTE 20   Standard Methods for Carriage of Closed Captions and Non-Real Time Sampled Video

SCTE 26   Home Digital Network Interface Specification with Copy Protection

SCTE 40   Digital Cable Network Interface Standard

SCTE 43   Digital Video Systems Characteristics Standard for Cable Television

SCTE 54   Digital Video Service Multiplex and Transport System Standard for Cable Television

SCTE 55   Digital Broadband Delivery System: Out-of-Band Transport

SCTE 65   Service Information Delivered Out-of-Band for Digital Cable Television

SCTE 80   In-Band Data Broadcast Standard including Out-of-Band Announcements

OpenCable™ receivers use the following four communications channels over the digital cable network:

6 MHz NTSC analog channels. They are typically located in the 54–450 MHz range. Each channel carries one program.

6 MHz Forward Application Transport (FAT) channels, which carry content via MPEG-2 transport streams. They use QAM encoding and are typically located in the 450–864 MHz range. Each channel can carry multiple programs.

Out-of-Band (OOB) Forward Data Channels (FDC). They use QPSK modulation and are typically located in the 70–130 MHz range, spaced between the 6-MHz NTSC analog and/or FAT channels. SCTE 55-1 and SCTE 55-2 are two alternative implementations.

Out-of-Band (OOB) Reverse Data Channels (RDC). They use QPSK modulation and are typically located in the 5–42 MHz range. SCTE 55-1 and SCTE 55-2 provide two alternative implementations.

OpenCable™ receivers obtain content by tuning to one of many 6 MHz channels available via the cable TV connection. When the selected channel is a legacy analog channel, the signal is processed using a NTSC audio/video/VBI decoder.

When the selected channel is a digital channel, it is processed by a QAM demodulator and then a CableCARD™ for content descrambling (conditional access descrambling). The conditional access descrambling is specific to a given cable system and is usually proprietary. The CableCARD™ then rescram-

Parameter	US (ATSC)	Europe, Asia (DVB)	Japan (ISDB)	OpenCable™
video compression	MPEG-2			
audio compression	Dolby® Digital	MPEG-2, Dolby® Digital, DTS®	MPEG-2 AAC-LC	Dolby® Digital
multiplexing	MPEG-2 transport stream			
terrestrial modulation	8-VSB	COFDM	BST-OFDM[2]	–
channel bandwidth	6 MHz	6, 7, or 8 MHz	6, 7, or 8 MHz	6 MHz
cable modulation	16-VSB[3]	QAM	QAM	QAM
channel bandwidth	6 MHz	6, 7, or 8 MHz	6, 7, or 8 MHz	6 MHz
satellite modulation	QPSK, 8PSK	QPSK	PSK	–

Notes:
1. ISDB = Integrated Services Digital Broadcasting.
2. BST-OFDM = Bandwidth Segmented Transmission of OFDM.
3. Most digital cable systems use QAM instead of 16-VSB.

Table 16.1. Comparison of the Various DTV Standards.

bles the content to a common algorithm and passes it on to the MPEG-2 decoder. The multi-stream CableCARD™ is capable of handling up to six different channels simultaneously, enabling picture-in-picture and DVR (digital video recording) capabilities.

When the CableCARD™ is not inserted, the output of the QAM demodulator is routed directly to the MPEG-2 decoder. However, encrypted content will not be viewable.

OpenCable™ receivers also obtain control information and other data by tuning to the OOB FDC channel. Using a dedicated tuner, the receiver remains tuned to the OOB FDC to receive information continuously. This information is also passed to the CableCARD™ and MPEG-2 decoder for processing.

The bi-directional OpenCable™ receiver can also transmit data using the OOB RDC.

The OpenCable™ standard uses a MPEG-2 transport stream to convey compressed digital video, compressed digital audio and ancillary data over a single 6 MHz FAT channel. Within this transport stream can be a single HDTV program, multiple SDTV programs, data or a combination of these.

The MPEG-2 transport stream has a constant bit rate of ~27 Mbps (64-QAM modulation), ~38.8 Mbps (256-QAM) or ~44.3 Mbps (1024-QAM).

The available bit rate can be used in a very flexible manner, trading-off the number of programs offered versus video quality and resolution. For example, if statistical multiplexing and 256-QAM are used,

(4) HDTV programs

(2) HDTV programs + (6) SDTV programs + data

(18) SDTV programs

Video Capability

Digital video compression is implemented using MPEG-2 and has the same requirements as ATSC. There are some minor constraints on some of the MPEG-2 parameters, as discussed within the MPEG-2 chapter.

Although any resolution may be used as long as the maximum bit rate is not exceeded, there are several standardized resolutions, indicated in Table 16.2. Both interlaced and progressive pictures are permitted for most of the resolutions.

Compliant receivers must also be capable of tuning to and decoding analog NTSC signals, discussed in Chapter 8.

Audio Capability

Digital audio compression is implemented using Dolby® Digital and has the same requirements as ATSC.

Compliant receivers must also be capable of decoding the audio portion of analog NTSC signals. NTSC audio standards are discussed in Chapter 8.

In-Band System Information (SI)

Enough bandwidth is available within the MPEG-2 transport stream to support several low-bandwidth non-television services such as program guide, closed captioning, weather reports, stock indices, headline news, software downloads, pay-per-view information, etc. The number of additional non-television services (virtual channels) may easily reach ten or more. In addition, the number and type of services will be constantly changing.

Active Resolution (Y)	SDTV or HDTV	Frame Rate (p = progressive, i = interlaced)			
		23.976p 24p	29.97i 30i	29.97p 30p	59.94p 60p
480 × 480	SDTV	×	×	×	×
528 × 480		×	×	×	×
544 × 480		×	×	×	×
640 × 480		×	×	×	×
704 × 480		×	×	×	×
1280 × 720	HDTV	×		×	×
1280 × 1080		×	×	×	
1440 × 1080		×	×	×	
1920 × 1080		×	×	×	

Table 16.2. Common Active Resolutions for OpenCable™ Content.

To support these non-television services in a flexible, yet consistent, manner, *System Information* (SI) was developed. SI is a small collection of hierarchically associated tables (see Figure 16.1 and Table 16.3) designed to extend the MPEG-2 PSI tables. It describes the information for all virtual channels carried in a particular MPEG-2 transport stream. Additionally, information for analog broadcast channels may be incorporated.

For in-band SI, OpenCable™ pretty much follows the ATSC PSIP standard, with some extensions.

Required Tables

Cable Virtual Channel Table (CVCT)

This table contains a list of all the channels in the transport stream that are or will be available plus their attributes. It may also include the broadcaster's analog channel and digital channels in other transport streams.

Attributes for each channel include major/minor channel number, short name, Transport/Transmission System ID (TSID) that uniquely identifies each station, carrier frequency, modulation mode, etc. The *Service Location Descriptor* is used to list the PIDs for the video, audio, data, and other related elementary streams.

ATSC also uses a version of this table called the Terrestrial Virtual Channel Table (TVCT).

Emergency Alert (EA) Table

This table provides a signaling method that enables cable TV operators to send emergency messages to digital set-top boxes, digital televisions receivers, digital VCR's., etc. These devices must be able to store any EA events for later use; the start time and duration information are used to delete expired events.

Typically, transport streams originating from terrestrial digital broadcast sources located in the same geographic region as the cable hub also provide any Emergency Alert information within their broadcast.

Event Information Table (EIT)

There are up to 128 EITs, EIT-0 through EIT-127, each of which describes the events or TV programs associated with each virtual channel listed in the CVCT. Each EIT is valid for three hours. Since there are up to 128 EITs, up to 16 days of programming may be advertised in advance. The first four EITs are required (the first 24 are recommended) to be present.

Information provided by the EIT includes start time, duration, title, pointer to optional descriptive text for the event, advisory data, caption service data, audio service descriptor, etc. ATSC also uses this table.

Master Guide Table (MGT)

This table provides general information about the other tables. It defines table sizes, version numbers and packet identifiers (PIDs). ATSC also uses this table.

Rating Region Table (RRT)

This table transmits the rating system, commonly referred to as the "V-chip." ATSC also uses this table.

System Time Table (STT)

This table serves as a reference for the time of day. Receivers use it to maintain the correct local time. ATSC also uses this table.

Optional Tables

Directed Channel Change Table (DCCT)

The DCCT contains information needed for a channel change to be done at a broadcaster-specified time. The requested channel change may be unconditional or may be based upon criteria specified by the viewer. ATSC also optionally uses this table.

Directed Channel Change Selection Code Table (DCCSCT)

The DCCSCT permits a broadcast program categorical classification table to be downloaded for use by some Directed Channel Change requests. ATSC also optionally uses this table.

Extended Text Table (ETT)

For text messages, there can be several ETTs, each having its PID defined by the MGT. Messages can describe channel information, coming attractions, movie descriptions, etc. ATSC also optionally uses this table.

Descriptor	Descriptor Tag	Tables									
		PMT	MGT	CVCT	RRT	EIT	ETT	STT	DCCT	DCCSCT	CAT
PID		per PAT	$1FFB_H$	$1FFB_H$	$1FFB_H$	per MGT	per MGT	$1FFB_H$	$1FFB_H$	$1FFB_H$	0001_H
Table_ID		02_H	$C7_H$	$C9_H$	CA_H	CB_H	CC_H	CD_H	$D3_H$	$D4_H$	80_H, 81_H (ECM) 82_H - $8F_H$ (EMM)
repetition rate		400 ms	150 ms	400 ms	1 min	0.5 sec	1 min	10 sec	400 ms	1 hour	
AC-3 audio stream	1000 0001	M									
ATSC CA	1000 1000			O		O					
ATSC private information*	1010 1101										
CA	0000 1001	M									M
caption service	1000 0110	M				M					
component name	1010 0011	M									
content advisory	1000 0111	M				M					
content identifier	1011 0110					O					
DCC arriving request	1010 1001								M		
DCC departing request	1010 1000								M		
enhanced signaling	1011 0010										
extended channel name	1010 0000			M							
MAC address list	1010 1100	M									
redistribution control	1010 1010	M				M					
service location	1010 0001			M							
stuffing*	1000 0000										
time-shifted service	1010 0010			M							

Notes:
1. PMT: MPEG-2 Program Map Table. CAT: MPEG-2 Conditional Access Table.
2. M = when present, required in this table. O = may be present in this table also. * = no restrictions.

Table 16.3. List of OpenCable™ In-Band SI Tables, Descriptors and Descriptor Locations.

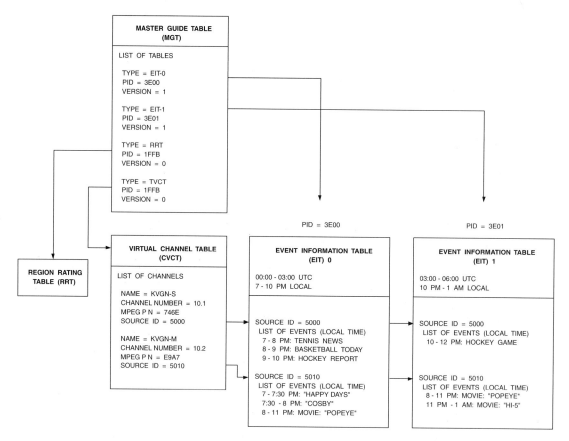

Figure 16.1. OpenCable™ In-Band SI Table Relationships.

Descriptors

Much like MPEG-2, OpenCable™ uses descriptors to add new functionality. In addition to various MPEG-2 descriptors, one or more of these OpenCable™-specific descriptors may be included within the PMT or one or more SI tables (see Table 16.3) to extend data within the tables. A descriptor not recognized by a decoder must be ignored by that decoder. This enables new descriptors to be implemented without affecting receivers that cannot recognize and process the descriptors.

AC-3 Audio Stream Descriptor

This OpenCable™ descriptor indicates Dolby® Digital audio is present, and is discussed in Chapter 13. ATSC also uses this descriptor.

ATSC Private Information Descriptor

This OpenCable™ descriptor provides a way to carry private information, and is discussed in Chapter 13. More than one descriptor may appear within a single descriptor. ATSC also uses this descriptor.

Component Name Descriptor

This OpenCable™ descriptor defines a variable-length text-based name for any component of the service, and is discussed in Chapter 13. ATSC also uses this descriptor.

Content Advisory Descriptor

This OpenCable™ descriptor defines the ratings for a given program, and is discussed in Chapter 13. ATSC also uses this descriptor.

DCC Arriving Request Descriptor

This OpenCable™ descriptor provides instructions for the actions to be performed by a receiver upon arrival to a newly changed channel:

> Display text for at least 10 seconds, or for a less amount of time if the viewer issues a "continue", "OK" or equivalent command.

> Display text indefinitely, or until the viewer issues a "continue", "OK" or equivalent command.

ATSC also uses this descriptor.

DCC Departing Request Descriptor

This OpenCable™ descriptor provides instructions for the actions to be performed by a receiver prior to leaving a channel:

> Cancel any outstanding things and immediately perform the channel change.

> Display text for at least 10 seconds, or for a less amount of time if the viewer issues a "continue", "OK" or equivalent command.

> Display text indefinitely, or until the viewer issues a "continue", "OK" or equivalent command.

ATSC also uses this descriptor.

Extended Channel Name Descriptor

This OpenCable™ descriptor provides a variable-length channel name for the virtual channel. ATSC also uses this descriptor.

MAC Address List Descriptor

This OpenCable™ descriptor is used when implementing IP (Internet Protocol) multicasting over MPEG-2 transport streams, and is discussed in Chapter 13. ATSC also uses this descriptor.

Redistribution Control Descriptor

This OpenCable™ descriptor conveys any redistribution control information held by the content rights holder, and is discussed in Chapter 13. ATSC also uses this descriptor.

Service Location Descriptor

This OpenCable™ descriptor specifies the stream type, PID and language code for each elementary stream. It is present in the CVCT for each active channel. ATSC also uses this descriptor.

Time-shifted Service Descriptor

This OpenCable™ descriptor links one virtual channel with up to 20 other virtual channels carrying the same programming, but time-shifted. A typical application is for Near Video On Demand (NVOD) services. ATSC also uses this descriptor.

Out-of-Band System Information (SI)

SI data may also be conveyed out-of-band (OOB). The CableCARD™ converts the OOB SI data, which may or may not be within a compliant MPEG-2 transport stream, into compliant table sections, each associated with an appropriate PID value.

Based on the network configuration, OOB messaging is implemented either over the OOB FDC and OOB RDC channels or over the DOCSIS® channel. Which system is to be used by the receiver is communicated by the Cable-CARD™ to the receiver.

Tables

Six profiles are defined (Table 16.4) that indicate required and optional tables. Adherence to these profiles is required for compliance.

Aggregate Event Information Table (AEIT)

This table delivers event title and schedule information for supporting an EPG. To reduce the total number of PID values used for SI data, the format allows instances of table sections for different time periods to be associated with common PID values.

Aggregate Extended Text Table (AETT)

This table contains Extended Text Messages (ETM), which may be used to convey detailed event descriptions. An ETM is a multiple string data structure, and is therefore capable of conveying a description in several different languages.

Emergency Alert (EA) Table

This table provides a signaling method that enables cable TV operators to send emergency messages to digital set-top boxes, digital televisions receivers, digital VCR's., and so on. These devices must be able to store any EA events for later use; the start time and duration information are used to delete expired events.

Long-form Virtual Channel Table

This table is the CVCT transmitted using MPEG-2 private sections.

Master Guide Table (MGT)

This table provides general information about the other tables. It defines table sizes, version numbers and packet identifiers (PIDs). ATSC also uses this table.

Network Information Table (NIT)

This table groups a number of transport streams together, providing tuning information the receiver.

Network Text Table (NTT)

This table delivers system-wide multilingual text strings.

Rating Region Table (RRT)

This table transmits the rating system, commonly referred to as the "V-chip." ATSC also uses this table.

Short-form Virtual Channel Table

This table delivers portions of the Virtual Channel Map (VCM), Defined Channels Map (DCM) and the Inverse Channel Map (ICM).

Table	Table ID	Profile 1 Baseline	Profile 2 Revision Detection	Profile 3 Parental Advisory	Profile 4 Standard EPG Data	Profile 5 Combination	Profile 6 SI Only
NIT	$C2_H$						
carrier definition subtable		M	M	M	M	M	
modulation mode subtable		M	M	M	M	M	
NTT	$C3_H$						
source name subtable		O	O	O	M	M	
Short-form VCT	$C4_H$						
virtual channel map		M	M	M	M	M	
defined channels map		M	M	M	M	M	
inverse channel map		O	O	O	O	O	
STT	$C5_H$	M	M	M	M	M	M
MGT	$C7_H$			M	M	M	M
RRT	CA_H			M	M	M	M
Long-form VCT	$C9_H$					M	M
AEIT	$D6_H$				M	M	M
AETT	$D7_H$				O	O	O

Notes:
1. M = when present, required in this table. O = may be present in this table also.

Table 16.4. Usage of Tables in Various Profiles.

Descriptor	Tag	Profile 1 Baseline	Profile 2 Revision Detection	Profile 3 Parental Advisory	Profile 4 Standard EPG Data	Profile 5 Combination	Profile 6 SI Only
AC-3 audio	81_H				O	O	O
caption service	86_H				O	O	O
channel properties	95_H				O	O	
component name	$A3_H$				O	O	O
content advisory	87_H			M	M	M	M
daylight savings time	96_H			O	M	M	M
extended channel name	$A0_H$					O	O
revision detection	93_H		M	M	M	M	
time shifted service	$A2_H$					O	O
two part channel number	94_H				O	O	

Notes:
1. M = required in this profile. O = may be present in this profile.

Table 16.5. Usage of Descriptors in Various Profiles.

System Time Table (STT)

This table serves as a reference for the time of day. Receivers use it to maintain the correct local time. ATSC also uses this table.

Descriptors

Tables 16.5 and 16.6 illustrate the usage of descriptors in the profiles and tables, respectively.

AC-3 Audio Stream Descriptor

This descriptor indicates Dolby® Digital audio is present, and is discussed in Chapter 13. ATSC also uses this descriptor.

Channel Properties Descriptor

This descriptor enables receivers to become aware of various channel aspects. Otherwise, the receiver must tune the channel and self-discover the channel's aspects.

Component Name Descriptor

This descriptor defines a variable-length text-based name for any component of the service, and is discussed in Chapter 13. ATSC also uses this descriptor.

Content Advisory Descriptor

This descriptor defines the ratings for a given program, and is discussed in Chapter 13. ATSC also uses this descriptor.

Descriptor	Tag	Table								
		PMT	**NIT**	**NTT**	**S-VCT**	**STT**	**MGT**	**L-VCT**	**RRT**	**AEIT**
PID		per PAT	1FFC$_H$	1FFC$_H$	1FFC$_H$	1FFC$_H$	1FFC$_H$	1FFC$_H$	1FFC$_H$	per MGT
Table_ID		02$_H$	C2$_H$	C3$_H$	C4$_H$	C5$_H$	C7$_H$	C9$_H$	CA$_H$	D6$_H$
AC-3 audio	81$_H$	×								×
caption service	86$_H$	×								×
channel properties	95$_H$				×					
component name	A3$_H$	×								
content advisory	87$_H$	×								×
daylight savings time	96$_H$					×				
extended channel name	A0$_H$							×		
revision detection	93$_H$		×	×	×					
time shifted service	A2$_H$							×		
two part channel number	94$_H$				×					

Table 16.6. Usage of Descriptors in Various Tables.

Daylight Savings Time Descriptor

This descriptor indicates whether or not daylight saving time is currently being observed and the time/day on which the daylight savings time transition occurs.

The receiver must not assume that the lack of this descriptor means that daylight savings time is not currently in effect.

Extended Channel Name Descriptor

This descriptor provides a variable-length channel name for the virtual channel. ATSC also uses this descriptor.

Revision Detection Descriptor

This descriptor indicates if new information is contained in the table section in which it appears. It should be the first descriptor in the list to minimize processing overhead.

Time-shifted Service Descriptor

This descriptor links one virtual channel with up to 20 other virtual channels carrying the same programming, but time-shifted. A typical application is for Near Video On Demand (NVOD) services. ATSC also uses this descriptor.

Two Part Channel Number Descriptor

This descriptor may be used to associate a two-part channel number (i.e., 10-2) with any virtual channel.

In-Band Data Broadcasting

The OpenCable™ in-band data broadcast standard describes various ways to transport data within the MPEG-2 transport stream. It can be used for a number of applications, such as:

Delivering declarative data (HTML code)

Delivering procedural data (Java code)

Delivering software and images

Delivering MPEG-4, H.264 or VC-9 video streams

Carouseling MPEG-2 video or MP3 audio files

The key elements defined within the standard are:

Data services announcement

Data delivery models such as data piping, data streaming, addressable sections and data download

Application signaling

MPEG-2 systems tools

Protocols

A data service must be contained in a virtual channel, and each virtual channel may have at most one data service. One data service may consist of multiple applications, and each application may contain multiple data elements.

For in-band data broadcasting, OpenCable™ pretty much follows the ATSC data broadcast standard. The major difference is the addition of out-of-band announcements.

Data broadcasting is also discussed in Chapter 13.

Data Service Announcements

Data broadcasting utilizes and extends SI to announce and find data services in the broadcast stream. Data services are announced by an event in either the EIT or Data Event Table (DET).

Additional tables and descriptors for data service announcements include:

Data Event Table (DET)

There are up to 128 DETs, DET-0 through DET-127, each of which describes information (titles, start times, etc.) for data services on virtual channels. Each DET is valid for three hours. A minimum of four DETs (DET-0 through DET-3) are required. Any change in a DET shall trigger a change in version of the MGT. ATSC also supports this table.

Extended Text Table (ETT)

The ETT is used to provide detailed descriptions of a data event. The syntax is mostly identical to the ETT used for audio and video services. ATSC also supports this table.

Long Term Service Table (LTST)

The LTST is used to pre-announce data events that will occur on a time scale outside what the EITs/DETs can support. ATSC also supports this table.

Data Service Descriptor

This OpenCable™ descriptor (also called the *Data Broadcast Descriptor* in older specifications) indicates the maximum transmission bandwidth and buffer model requirements of the data service. ATSC also uses this descriptor.

PID Count Descriptor

This OpenCable™ descriptor provides a total count of PIDs used in the service. It may also specify the minimum number of PID values a receiver must be able to monitor simultaneously to provide a meaningful rendition of the data service. ATSC also uses this descriptor.

Service Description Framework (SDF)

Due to the wide range of protocol encapsulations possible, there is a need to signal which encapsulation is used in each data stream. While it is possible to use the PMT for signaling, this approach is not very scalable, working only for simple cases. The *Service Description Framework* (SDF) was developed to provide a scalable framework.

Additional tables and descriptors for the service description framework include:

Aggregate Data Event Table (ADET)

There are up to 128 ADETs, ADET-0 through ADET-127, They deliver event title and schedule information for implementing an Electronic Program Guide (EPG). The purpose of the ADET is:

To announce a data service in a virtual channel which does not include any audio-visual event.

To allow separate announcement of the data service portion of an audio/video/data service or audio/data service in a virtual channel.

The transmission format allows ADET table sections to share a common PID value since the CableCARD™ can support only a small number of concurrent data flows (each associated with one PID value). Each ADET is valid for three hours. A minimum of four ADETs (ADET-0 through ADET-3) are required. Any change in a ADET triggers a change in version of the MGT.

ADET section tables may be delivered to the receiver either in-band (via MPEG-2 transport streams) or out-of-band.

Data Service Table (DST)

The DST provides the description of a data service comprised of one or more receiver applications. It also provides information to allow data receivers to associate applications with references to data consumed by them. ATSC also supports this table.

Network Resources Table (NRT)

The NRT provides a list of all network resources outside those in the current MPEG-2 program or transport stream.

A data service may use the NRT to get data packets or datagrams other than the ones published in the *Service Location Descriptor* of the CVCT. This includes data elementary streams in another program within the same transport stream, data elementary streams in other transport streams, and bi-directional communication channels using other protocols such as IP. ATSC also supports this table.

Association Tag Descriptor

This MPEG-2 DSM-CC descriptor binds data resource references (taps) to specific data elementary streams in the MPEG-2 transport stream. ATSC also uses this descriptor.

Download Descriptor

This MPEG-2 DSM-CC descriptor includes information specific to the DSM-CC download protocol. ATSC also uses this descriptor.

DSM-CC Compatibility Descriptor

This MPEG-2 DSM-CC descriptor indicates receiver hardware and/or software requirements for proper acquisition and processing of a data service. ATSC also uses this descriptor.

DSM-CC Resource Descriptor

This MPEG-2 DSM-CC descriptor provides a list of DSM-CC resource descriptors used in the NRT. Resource descriptors contain information required for the network to allocate a resource, track the resource once it has been allocated, and de-allocate the resource once it is no longer needed. Supported descriptors include:

Deferred MPEG Program Element Descriptor: This descriptor is used to identify a MPEG-2 resource in another virtual channel (in the same or a different transport stream).

IP Resource Descriptor: This descriptor is used to identify a resource by its IP address and port.

IPV6 Resource Descriptor: This descriptor is used to identify a resource by its IPv6 address and port.

URL Resource Descriptor: This descriptor is used to identify a resource by its Uniform Resource Locator (URL).

ATSC also uses this descriptor.

Multiprotocol Encapsulation Descriptor

This MPEG-2 DSM-CC descriptor provides information defining the mapping of a receiver's MAC address to a specific addressing scheme. ATSC also uses this descriptor.

Conditional Access

The conditional access (CA) scheme used by OpenCable™ is similar to Multicrypt and the DVB Common Interface discussed in Chapter 17. OpenCable™ calls their CA module "Cable-CARD™" and it is also based on the EIA-679 NRSS-B interface (PCMCIA or PC Card form factor). Two major additions over DVB's solution are the ability to support up to six simultaneous streams (requires the multi-stream CableCARD™) and a DFAST-encrypted interface between the CableCARD™ output and the MPEG-2 decoder input.

Related Technologies

In addition to OpenCable™, CableLabs® has developed, and is continually developing, a wide variety of cable-related standards.

DOCSIS®

DOCSIS® (Data Over Cable Service Interface Specification) defines interface requirements for cable modems.

PacketCable™

PacketCable™ defines a common platform to deliver real-time multimedia services. Built on top of the DOCSIS®, PacketCable™ uses Internet Protocol (IP) technology.

Application Block Diagrams

Figure 16.2 illustrates an OpenCable™ set-top box. Part of the requirements is the ability to output both high-definition and standard-definition versions of HD content simultaneously.

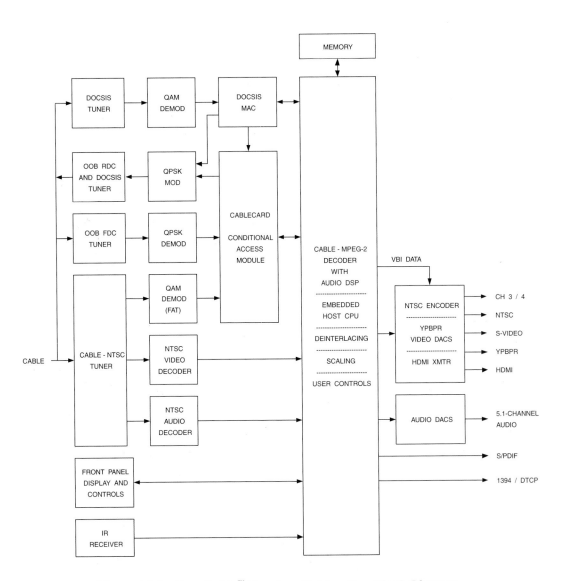

Figure 16.2. OpenCable™ Receiver Set-top Box Block Diagram.

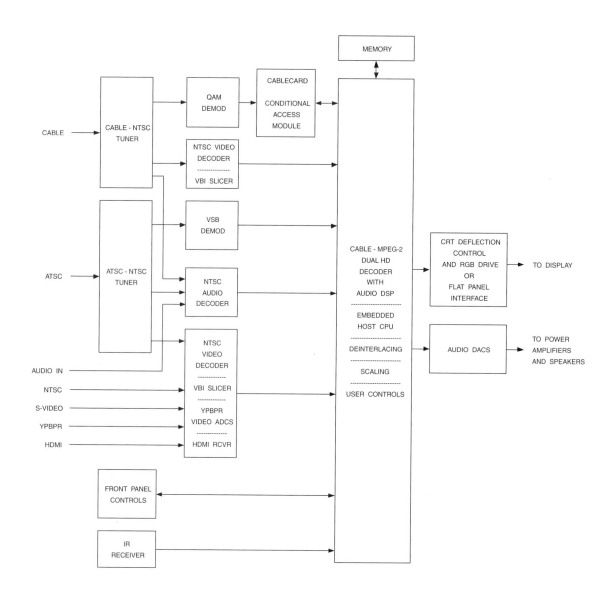

Figure 16.3. OpenCable™ Receiver Inside a Digital Television

Figure 16.3 illustrates the incorporation of an one-way OpenCable™ receiver into a digital television.

References

1. ATSC A/53b, *ATSC Standard: Digital Television Standard, Revision B with Amendments 1 and 2*, May 2003.
2. ATSC A/65b, *ATSC Standard: Program and System Information Protocol for Terrestrial Broadcast and Cable (Revision B)*, March 2003.
3. ATSC A/90, *ATSC Data Broadcast Standard*, July 2000.
4. SCTE 07, *Digital Video Transmission Standard for Cable Television*, 2000.
5. SCTE 28, *HOST-POD Interface Standard*, 2003.
6. SCTE 40, *Digital Cable Network Interface Standard*, 2004.
7. SCTE 54, *Digital Video Service Multiplex and Transport System Standard for Cable Television*, 2004.
8. SCTE 65, *Service Information Delivered Out-of-Band for Digital Cable Television*, 2002.
9. SCTE 80, *In-Band Data Broadcast Standard including Out-of-Band Announcements*, 2002.

Chapter 17

DVB
Digital Television

The DVB (Digital Video Broadcast) digital television (DTV) broadcast standard is used in most regions except the United States, Canada, South Korea, Taiwan and Argentina.

The three other primary DTV standards are ISDB (Integrated Services Digital Broadcasting), ATSC (Advanced Television Systems Committee) and OpenCable™. A comparison between the standards is shown in Table 17.1. The basic audio and video capabilities are very similar. The major differences are the RF modulation schemes and the level of definition for non-audio/video services.

The DVB standard is actually a group of ETSI standards:

EN 300 421   DVB-S: Framing Structure, Channel Coding and Modulation for 11/12 GHz Satellite Services

EN 300 429   DVB-C: Framing Structure, Channel Coding and Modulation for Cable Systems

EN 300 468   Specification for Service Information (SI) in DVB Systems

EN 300 472   Specification for Conveying ITU-R System B Teletext in DVB Bitstreams

EN 300 743   Subtitling Systems

EN 300 744   DVB-T: Framing Structure, Channel Coding and Modulation for Digital Terrestrial Television

EN 301 192   DVB Specification for Data Broadcasting

EN 301 775   Specification for the Carriage of Vertical Blanking Information (VBI) data in DVB Bitstreams

ES 200 800   DVB Interaction Channel for Cable TV Distribution Systems (CATV)

ETS 300 801   Interaction Channel through Public Switched Telecommunications Network (PSTN) / Integrated Services Digital Networks (ISDN)

ETS 300 802   Network-independent Protocols for DVB Interactive Services

TR 101 154	Implementation Guidelines for the Use of MPEG-2 Systems, Video and Audio in Satellite, Cable and Terrestrial Broadcasting Applications
TR 101 190	Implementation Guidelines for DVB Terrestrial Services; Transmission Aspects
TR 101 194	Guidelines for Implementation and Usage of the Specification of Network Independent Protocols for DVB Interactive Services
TR 101 200	A Guideline for the Use of DVB Specifications and Standards
TR 101 202	Implementation Guidelines for Data Broadcasting

TR 101 211	Guidelines on Implementation and Usage of Service Information (SI)
EN 50221	Common Interface Specification for Conditional Access and other Digital Video Broadcasting Decoder Applications
ETR 289	Support for Use of Scrambling and Conditional Access (CA) within Digital Broadcasting Systems
TS 101 699	Extensions to the Common Interface Specification

Parameter	US (ATSC)	Europe, Asia (DVB)	Japan (ISDB)	OpenCable™
video compression	MPEG-2			
audio compression	Dolby® Digital	MPEG-2, Dolby® Digital, DTS®	MPEG-2 AAC-LC	Dolby® Digital
multiplexing	MPEG-2 transport stream			
terrestrial modulation	8-VSB	COFDM	BST-OFDM[2]	–
channel bandwidth	6 MHz	6, 7, or 8 MHz	6, 7, or 8 MHz	6 MHz
cable modulation	16-VSB[3]	QAM	QAM	QAM
channel bandwidth	6 MHz	6, 7, or 8 MHz	6, 7, or 8 MHz	6 MHz
satellite modulation	QPSK, 8PSK	QPSK	PSK	–

Notes:
1. ISDB = Integrated Services Digital Broadcasting.
2. BST-OFDM = Bandwidth Segmented Transmission of OFDM.
3. Most digital cable systems use QAM instead of 16-VSB.

Table 17.1. Comparison of the Various DTV Standards.

DVB uses a MPEG-2 transport stream to convey compressed digital video, compressed digital audio and ancillary data over a 6, 7 or 8 MHz channel. This enables addressing the needs of different countries that may adopt the standard.

The MPEG-2 transport stream has a maximum bit rate of ~24.1 Mbps (8 MHz DVB-T) or ~51 Mbps (8 MHz 256-QAM DVB-C). DVB-S bit rates are dependent on the transponder bandwidth and code rates used, and can approach 54 Mbps. The bit rate can be used in a very flexible manner, trading-off the number of programs offered versus video quality and resolution.

Video Capability

Although any resolution may be used as long as the maximum bit rate is not exceeded, there are several standardized resolutions, indicated in Table 17.2. Both interlaced and progressive pictures are permitted for most of the resolutions.

Video compression is based on MPEG-2. However, there are some minor constraints on some of the MPEG-2 parameters, as discussed within the MPEG-2 chapter. Support for H.264 is in the process of being added to the specifications.

Audio Capability

Audio compression is implemented using MPEG-1 Layer II, MPEG-2 BC multi-channel Layer II, Dolby® Digital or DTS®. Support for MPEG-4 HE-AAC is in the process of being added to the specifications.

System Information (SI)

ETSI EN 300 468 specifies the Service Information (SI) data which forms a part of DVB bitstreams. SI is a small collection of hierarchically associated tables (see Figure 17.1 and Table 17.3) designed to extend the MPEG-2 PSI tables. It provides information on what is available on other transport streams and even other networks. The method of information presentation to the user is not specified, allowing receiver manufacturers to choose appropriate presentation methods.

Required Tables

Event Information Table (EIT)
There are up to 128 EITs, EIT-0 through EIT-127, each of which describes the events or TV programs associated with each channel. Each EIT is valid for three hours. Since there are up to 128 EITs, up to 16 days of programming may be advertised in advance. The first four EITs are required (the first 24 are recommended) to be present.

Information provided by the IET includes start time, duration, title, pointer to optional descriptive text for the event, advisory data, caption service data, audio service descriptor, etc.

Network Information Table (NIT)
The NIT provides information about the physical network, including any grouping of transport streams and the relevant tuning information. It can be used during receiver set-up and the relevant tuning information stored in non-volatile memory. The NIT can also be used to signal changes of tuning information.

Active Resolution (Y)	SDTV or HDTV	Frame Rate (p = progressive, i = interlaced)						
		23.976p 24p	25i	29.97i 30i	25p	29.97p 30p	50p	59.94p 60p
480 × 480	SDTV	×		×		×		×
480 × 576		×	×		×		×	
544 × 480		×		×		×		×
544 × 576		×	×		×		×	
704 × 480		×		×		×		×
704 × 576		×	×		×		×	
1280 × 720	HDTV	×			×	×	×	×
1280 × 1080		×	×	×	×	×		
1440 × 1080		×	×	×	×	×		
1920 × 1080		×	×	×	×	×		

Table 17.2. Common Active Resolutions for DVB Digital Television.

Service Description Table (SDT)

The SDT describes the available services, such as the service names, the service providers, etc.

Time and Date Table (TDT)

The TDT contains the actual UTC-time coded as Modified Julian Date (MJD). Receivers can use it to maintain the correct local time.

Optional Tables

Bouquet Association Table (BAT)

The BAT provides information regarding bouquets (groups of services that may traverse the network boundary). Along with the name of the bouquet, it provides a list of services for each bouquet.

Discontinuity Information Table (DIT)

The DIT is present at transition points where the SI information is discontinuous. The use of this table is restricted to partial transport streams, they are not used in broadcasts.

Running Status Table (RST)

The RST updates the running status of one or more events. These are sent out only once, at the time of an event status change, unlike other tables which are usually transmitted repeatedly.

Selection Information Table (SIT)

The SIT describes services and events carried by a partial transport stream. The use of this table is restricted to partial transport streams, they are not used in broadcasts.

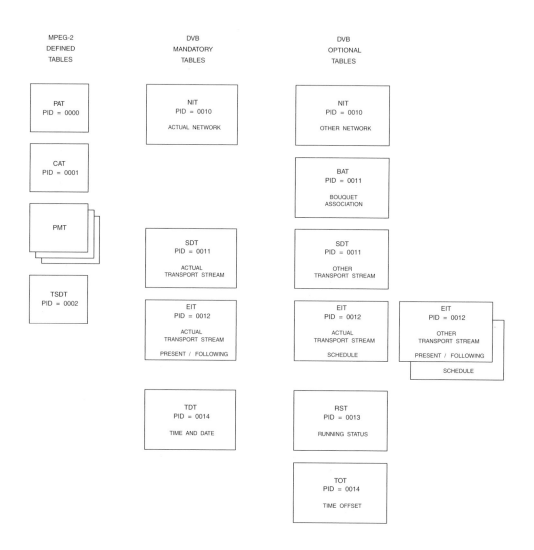

Figure 17.1. DVB SI Table Relationships.

Descriptor	Descriptor Tag	Tables						
		PMT	**NIT**	**BAT**	**SDT**	**EIT**	**TOT**	**SIT**
PID		per PAT	0010_H	0011_H	0011_H	0012_H	0014_H	$001F_H$
Table_ID		02_H	40_H 41_H	$4A_H$	42_H 46_H	$4E_H$ - $6F_H$	73_H	$7F_H$
repetition rate		100 ms	10 sec	10 sec	2 - 10 sec	2 - 10 sec	30 sec	30 sec
AC-3	0110 1010	×						
adaptation field data	0111 0000	×						
application signaling	0110 1111	×						
ancillary data	0110 1011	×						
announcement support	0110 1110				×			
bouquet name	0100 0111			×	×			×
cable delivery system	0100 0100		×					
CA identifier	0101 0011			×	×	×		×
cell frequency link	0110 1101		×					
cell list	0110 1100		×					
component	0101 0000					×		×
content	0101 0100					×		×
country availability	0100 1001			×	×			×
data broadcast	0110 0100				×	×		×
data broadcast ID	0110 0110	×						
DSNG	0110 1000							
DTS audio	0110 1000	×						×
DTS registration	0110 1000	×						×

Notes:
1. PMT: MPEG-2 Program Map Table.
2. SIT only present in partial transport streams.

Table 17.3a. List of DVB SI Tables, Descriptors and Descriptor Locations.

Descriptor	Descriptor Tag	Tables						
		PMT	**NIT**	**BAT**	**SDT**	**EIT**	**TOT**	**SIT**
extended event	0100 1110					×		×
frequency list	0110 0010		×					
linkage	0100 1010		×	×	×	×		×
local time offset	0101 1000						×	
mosaic	0101 0001	×			×			×
multilingual bouquet name	0101 1100			×				
multilingual component	0101 1110					×		×
multilingual network name	0101 1011		×					
multilingual service name	0101 1101				×			×
NVOD reference	0100 1011				×			×
network name	0100 0000		×					
parental rating	0101 0101					×		×
partial transport stream	0110 0011							×
PDC	0110 1001					×		
private data specifier	0101 1111	×	×	×	×	×		×

Notes:
1. PMT: MPEG-2 Program Map Table.
2. SIT only present in partial transport streams.

Table 17.3b. List of DVB SI Tables, Descriptors and Descriptor Locations.

Descriptor	Descriptor Tag	Tables						
		PMT	**NIT**	**BAT**	**SDT**	**EIT**	**TOT**	**SIT**
satellite delivery system	0100 0011		×					
short smoothing buffer	0110 0001					×		×
service	0100 1000				×			×
service availability	0111 0010				×			
service list	0100 0001		×	×				
service move	0110 0000	×						
short event	0100 1101					×		×
stream identifier	0101 0010	×						
stuffing	0100 0010		×	×	×	×		×
subtitling	0101 1001	×						
telephone	0101 0111				×	×		×
teletext	0101 0110	×						
terrestrial delivery system	0101 1010		×					
time shifted event	0100 1111					×		×
time shifted service	0100 1100				×			×
transport stream	0110 0111							
VBI data	0100 0101	×						
VBI teletext	0100 0110	×						

Notes:
1. PMT: MPEG-2 Program Map Table.
2. SIT only present in partial transport streams.

Table 17.3c. List of DVB SI Tables, Descriptors and Descriptor Locations.

Stuffing Table (ST)

The ST is used to replace or invalidate sub-tables or complete SI tables.

Time Offset Table (TOT)

The TOT is the same as the TDT, except that it includes local time offset information.

Descriptors

Much like MPEG-2, DVB uses descriptors to add new functionality. In addition to various MPEG-2 descriptors, one or more of these DVB-specific descriptors may be included within the PMT or one or more SI tables (see Table 17.3) to extend data within the tables. A descriptor not recognized by a decoder must be ignored by that decoder. This enables new descriptors to be implemented without affecting receivers that cannot recognize and process the descriptors.

AC-3 Descriptor

This DVB descriptor indicates Dolby® Digital audio is present, and is discussed in Chapter 13.

Adaptation Field Data Descriptor

This DVB descriptor, discussed in Chapter 13, indicates the type of data fields within the private data field of the MPEG-2 adaptation field.

Ancillary Data Descriptor

The DVB descriptor, discussed in Chapter 13, indicates the presence and type of ancillary data in MPEG audio elementary streams.

Announcement Support Descriptor

This DVB descriptor identifies the type of announcements that are supported by the service. It also indicates the announcement transport method and gives linkage information so that the announcement stream can be monitored.

Bouquet Name Descriptor

This DVB descriptor provides the bouquet name as variable-length text, such as "Max Movie Channels."

CA Identifier Descriptor

This DVB descriptor indicates whether a bouquet, service or event is associated with a conditional access system and if so, identifies conditional access used.

Cable Delivery System Descriptor

This DVB descriptor used to transmit the physical parameters of the cable network, including frequency, modulation and symbol rate.

Cell Frequency Link Descriptor

This DVB descriptor is used in the NIT to describe the terrestrial network. It provides links between a cell and the frequencies that are used in the cell for the transport stream.

Cell List Descriptor

This DVB descriptor provides a list of all network cells about which the NIT informs and describes their coverage areas.

Component Descriptor

This DVB descriptor, discussed in Chapter 13, indicates the type of stream and may be used to provide a text description of the stream.

Content Descriptor

This DVB descriptor is used to identify the type of content (comedy, talk show, etc.).

Country Availability Descriptor

This DVB descriptor, discussed in Chapter 13, identifies countries that are either allowed or not allowed to receive the service. The descriptor may appear twice for each service, once for listing countries allowed to receive the service, and a second time for listing countries not allowed to receive the service. The latter list overrides the former list.

Data Broadcast Descriptor

This DVB descriptor identifies within the SI available data broadcast services.

Data Broadcast ID Descriptor

This DVB descriptor, discussed in Chapter 13, identifies the data coding system standard. It is a short form of the *Data Broadcast Descriptor* and may be present the PMT.

DSNG Descriptor

This DVB descriptor is present only in DSNG (Digital Satellite News Gathering) transmissions.

DTS Audio Descriptor

When a DTS® audio stream is included in a DVB transport stream, this descriptor must also be included. Either this descriptor or the *DTS Registration Descriptor* must also be located in the PMT and SIT to identify a DTS® stream.

DTS Registration Descriptor

Either this descriptor or the *DTS Audio Descriptor* must be located in the PMT and SIT to identify a DTS® audio stream.

Extended Event Descriptor

This DVB descriptor provides a text description of an event, which may be used in addition to the *Short Event Descriptor*. More than one descriptor can be used to convey more than 256 bytes of information.

Frequency List Descriptor

This DVB descriptor may be present in the NIT. It conveys the additional frequencies when content is transmitted on other frequencies.

Linkage Descriptor

This DVB descriptor provides a link to another service, transport stream, program guide, service information, software upgrade, etc.

Local Time Offset Descriptor

This DVB descriptor may be present in the TOT to describe country-specific dynamic changes of the local time offset relative to UTC. This enables a receiver to automatically adjust between summer and winter times.

Mosaic Descriptor

This DVB descriptor, discussed in Chapter 13, partitions a digital video component into elementary cells, controls the allocation of elementary cells to logical cells, and links the content of the logical cell and the corresponding information (e.g. bouquet, service, event etc.).

Multilingual Bouquet Name Descriptor

This DVB descriptor provides a bouquet name in text form in one or more languages.

Multilingual Component Descriptor

This DVB descriptor provides a component name in text form in one or more languages. The component is identified by its component tag value.

Multilingual Network Name Descriptor

This DVB descriptor provides a network name in text form in one or more languages.

Multilingual Service Name Descriptor

This DVB descriptor provides service provider names and offered services in text form in one or more languages.

Network Name Descriptor

This DVB descriptor conveys the network name in text form, such as "Munich Cable."

NVOD (Near Video On Demand) Reference Descriptor

This DVB descriptor, in conjunction with the *Time Shifted Service Descriptor* and the time *Time Shifted Event Descriptor*, provides an efficient way of describing a number of services which carry the same sequence of events, but with the start times offset from one another.

Parental Rating Descriptor

This DVB descriptor, discussed in Chapter 13, gives a rating based on age and offers extensions to be able to use other rating criteria.

Partial Transport Stream Descriptor

The SIT contains all the information needed to control, play and copy partial transport streams. This DVB descriptor describes this information.

PDC Descriptor

This DVB descriptor extends the DVB system with the functionality of PDC (Program Delivery Control), defined by ETSI EN 300 231 and ITU-R BT.809, and discussed in Chapter 8.

Private Data Specifier Descriptor

This DVB descriptor, discussed in Chapter 13, is used to identify the source of any private descriptors or private fields within descriptors.

Satellite Delivery System Descriptor

This DVB descriptor is used to transmit the physical parameters of the satellite network, including frequency, orbital position, west-east flag, polarization, modulation and symbol rate.

Service Descriptor

This DVB descriptor provides the name of the service and the service provider in text form.

Service Availability Descriptor

This DVB descriptor is present in the SDT of a terrestrial network. It indicates whether or not service is available for the identified cells.

Service List Descriptor

This DVB descriptor provides a list of the services and service types for each transport stream.

Service Move Descriptor

This DVB descriptor, discussed in Chapter 13, provides a way for a receiver to follow a service as it moves from one transport stream to another. Some disturbance in the video and audio will occur at such a transition.

Short Event Descriptor

This DVB descriptor provides the name and a short description of an event.

Short Smoothing Buffer Descriptor

This MPEG-2 descriptor enables the bitrate of a service to be indicated in the PSI.

Stream Identifier Descriptor

This DVB descriptor, discussed in Chapter 13, enables streams to be associated with a description in the EIT, useful when there is more than one stream of the same type within a service.

Stuffing Descriptor

This DVB descriptor is used to stuff tables for any reason or to disable descriptors that are no longer valid.

Subtitling Descriptor

This DVB descriptor, discussed in Chapter 13, is used to identify ETSI EN 300 743 subtitle data.

Telephone Descriptor

This DVB descriptor indicates a telephone number, which may be used in conjunction with a PSTN or cable modem, to support narrowband interactive channels.

Teletext Descriptor

This DVB descriptor, discussed in Chapter 13, is used to identify elementary streams which carry EBU Teletext data.

Terrestrial Delivery System Descriptor

This DVB descriptor used to transmit the physical parameters of the terrestrial network, including center frequency, bandwidth, constellation, hierarchy information, code rate, guard interval and transmission mode.

Time-shifted Event Descriptor

This DVB descriptor indicates that an event is the time-shifted copy of another event.

Time-shifted Service Descriptor

This DVB descriptor links one service with up to 20 other services carrying the same programming, but time-shifted. A typical application is for Near Video On Demand (NVOD) services.

Transport Stream Descriptor

This DVB descriptor, transmitted only in the TSDT, is used to indicate that the MPEG transport stream is DVB or DSNG complaint.

VBI Data Descriptor

This DVB descriptor, discussed in Chapter 13, defines the VBI service type in the associated elementary stream.

VBI Teletext Descriptor

The syntax for this descriptor is the same as for *Teletext Descriptor*. The only difference is that it is not used to associate *stream_type* = 06_H with either the VBI or EBU teletext standard. Decoders use the languages in this descriptor to select magazines and subtitles.

Teletext

Two standards for teletext transmission over MPEG-2 are available, the "VBI standard" and the newer "EBU Teletext" standard. Teletext is discussed in detail in Chapters 8 and 13.

Subtitles

Subtitles consist of one or more compressed bitmap images, along with optional rectangular backgrounds for each. They are positioned at a defined location, being displayed at a defined start time and for a defined number of video frames. Subtitles are discussed in detail in Chapter 13.

Widescreen Signaling (WSS)

The ETSI EN 301 775 standard defines how to add widescreen signalling (WSS) data to a MPEG-2 transport stream. WSS is discussed in detail in Chapters 8 and 13.

Data Broadcasting

The DVB data broadcast standard describes the available encapsulation protocols used to transport data within a DVB stream. Based on MPEG-2 DSM-CC, it specifies DVB data piping, DVB data streaming, DVB multiprotocol encapsulation, DVB data carousels and DVB object carousels. DVB has added specific information to get the DSM-CC framework working in the DVB environment, particularly with the DVB SI.

Five different application areas with different broadcast requirements have been identified. For each application area, a profile is defined and additional descriptors are used to support the application. Additional data broadcasting information is available in Chapter 13.

Application Areas and Profiles

Data Piping
This profile supports data broadcast services that use a simple, asynchronous, end-to-end delivery of data. Data is carried directly in the payloads of MPEG-2 transport stream packets.

Data Streaming
This profile supports data broadcast services that use a streaming-oriented, end-to-end delivery of data in either an asynchronous, synchronous, or synchronized way. Data is carried in MPEG-2 PES packets.

Multiprotocol Encapsulation
This profile supports data broadcast services that use the transmission of datagrams of communication protocols, such as IP multicasting. The transmission of datagrams is done by encapsulating the datagrams in MPEG-2 DSM-CC sections.

Data Carousels
This profile supports data broadcast services that use the periodic transmission of data modules. These modules are of known sizes and may be updated, added to, or removed from the data carousel in time. Data is broadcast using a MPEG-2 DSM-CC Data Carousel.

Object Carousels

This profile supports data broadcast services that use the periodic broadcasting of DSM-CC User-User (U-U) Objects. Data is broadcast using the MPEG-2 DSM-CC Object Carousel and DSM-CC Data Carousel.

Transmission Format: Terrestrial (DVB-T)

The COFDM (Coded Orthogonal Frequency Division Multiplexing) system is used for terrestrial broadcast. Note that COFDM is not a modulation method; it is a multiplexing method.

Frequency division multiplexing (the "FDM" in COFDM) means that the data is distributed over many carriers, as opposed to using a single carrier. As a result, the data rate transmitted on each COFDM carrier is much lower than the data rate required of a single high-speed carrier. All the COFDM carriers are orthogonal (the "O" in COFDM), or mutually perpendicular. Coded (the "C" in COFDM) means that forward error correction is incorporated.

Standard modulation methods, such as QPSK, 16-QAM or 64-QAM, are used to modulate the COFDM carriers.

Two modes of operation are defined. A "2K mode" is available for single transmitter operation and for small SFN (Single Frequency Network) networks with limited transmitter distances. An "8K mode" can be used for both single transmitter operation and small or large SFN networks.

The system also allows different levels of QAM modulation and different inner code rates that can be used to trade off bit rate versus ruggedness.

Figure 17.2 illustrates a simplified block diagram of the COFDM process.

COFDM Overview

Basic OFDM spreads the transmitted data over thousands of carriers. The "2K mode" generates 2048 carriers, resulting in 1705 active carriers remaining after removal of carriers that might be subject to co-channel and adjacent-channel interference. The "8K mode" generates 8192 carriers, with 6817 active carriers. Data symbols on each carrier are arranged to occur simultaneously. The number of bits carried by each modulation symbol is dependent on the choice of modulation: for example, 2 bits per symbol for QPSK, 4 bits per symbol for 16-QAM, etc.

The carriers have a common, precise frequency spacing to ensure orthogonality of the carriers. Thus, the OFDM demodulator for one carrier doesn't "see" the modulation of the others, reducing crosstalk.

Another refinement is the *guard interval* to reduce inter-symbol and inter-carrier interference caused by echoes. Naturally, using a guard interval reduces the data capacity slightly.

The process of simultaneously modulating or demodulating thousands of carriers is equivalent to Discrete Fourier Transform operations, for which efficient Fast Fourier Transform (FFT) algorithms exist. Thus, the implementation is not as difficult as it might first appear.

Adaptation

Initially, the COFDM system must synchronize to the incoming MPEG-2 transport bitstream packets. Before any processing can be done, it must correctly identify the start and

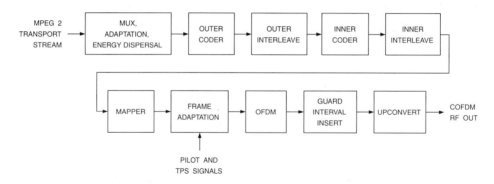

Figure 17.2. Simplified Block Diagram of a COFDM Modulator.

end points of each MPEG-2 packet by using its sync byte.

Energy Dispersal

In general, the COFDM signal must have a random, noise-like nature. This results in a flat, noise-like spectrum that maximizes channel bandwidth efficiency and reduces possible interference with other DTV and NTSC/PAL channels.

To accomplish this, the data randomizer uses a pseudo-random number generator to scramble the bitstream. This receiver reverses this process to recover the original data.

Outer Coder

Like the ATSC standard, DVB uses Reed-Solomon (RS) encoding to provide forward error correction (FEC) for the data stream.

The RS encoder takes the 188 bytes of an incoming MPEG-2 packet, calculates 16 parity bytes (known as Reed-Solomon parity bytes), and adds them to the end of the packet. This results in an error-protected packet length of 204 bytes.

The receiver compares the received 188 bytes to the 16 parity bytes in order to determine the validity of the recovered data. If errors are detected, the receiver can use the parity bytes to locate and correct the errors.

Outer Interleave

Convolutional byte-wise interleaving with a depth (I) of 12 is then applied to the error-protected packets. The interleaved data bytes are then delimited by normal or inverted MPEG-2 sync bytes. The periodicity of the 204 bytes is preserved.

Inner Coder

A range of *punctured convolutional codes* may be used, based on a mother convolutional code of rate one-half with 64 states. This part of the coding process can operate at many rates: 1/2, 2/3, 3/4, or 7/8.

Inner Interleave

This process consists of bit-wise interleaving followed by symbol interleaving. Both are block-based.

The *bit interleaver* scrambles the sequential order of the data stream to minimize sensitivity to burst-type interference. The bitstream is demultiplexed into [n] sub-streams, where n = 2 for QPSK, n = 4 for 16-QAM, and n = 6 for 64-QAM. Each sub-stream is processed by a separate bit interleaver, with a bit interleaving block size of 126 bits. The block interleaving process is therefore repeated twelve times per OFDM symbol in the 2K mode and forty-eight times per symbol in the 8K mode.

The *symbol interleaver* maps [n] bit words onto the 1512 (2K mode) or 6048 (8K mode) active carriers per OFDM symbol. In the 2K mode, twelve groups of 126 data words are used; for the 8K mode, 48 groups of 126 data words are used.

Mapper

All data carriers in one OFDM frame are either QPSK, 16-QAM, 64-QAM, non-uniform-16-QAM, or non-uniform-64-QAM using Gray mapping.

The transmitted signal is then organized into frames, with each frame consisting of 68 OFDM symbols. Four frames constitute a super-frame. Since the OFDM signal uses many separately modulated carriers, each symbol can be further divided into cells.

Frame Adaptation

In addition to the audio, video, and ancillary data, an OFDM frame contains:

Scattered pilot cells

Continual pilot carriers

TPS carriers

The pilots are used for frame synchronization, frequency synchronization, time synchronization, channel estimation, transmission mode identification, and to follow phase noise.

Various cells within a OFDM frame are also modulated with reference data whose value is known to the receiver. Cells containing this reference data are transmitted at an increased power level.

The TPS (Transmission Parameter Signalling) carriers are used to convey information related to the transmission scheme, such as channel coding and modulation. The TPS is transmitted over 17 carriers for the 2K mode and over 68 carriers for the 8K mode (at the rate of 1 bit per carrier). The TPS carriers convey information on:

Modulation including the value of the QAM constellation pattern

Hierarchy information

Guard interval

Inner code rates

Transmission mode

Frame number in a super-frame

Transmission Format: Cable (DVB-C)

QAM (Quadrature Amplitude Modulation) modulation is used for cable broadcast. 16-QAM, 32-QAM, 64-QAM, 128-QAM and 256-QAM are supported. Receivers must support at least 64-QAM modulation.

Interactivity can be supported through the use of in-band or out-of-band forward (downstream) channels and out-of-band reverse (upstream) channels.

Transmission Format: Satellite (DVB-S)

QPSK (Quaternary Phase Shift Keying) modulation system is used for satellite broadcast.

Transmission Format: Satellite (DVB-S2)

DVB-S2 is the next-generation satellite broadcast standard, supporting the QPSK, 8PSK, 16APSK and 32APSK modulation systems. It may transport single or multiple streams in a variety of formats (MPEG-2 transport stream, IP, etc.), and each stream may be protected in a different manner. DVB-S2 offers a 25-35% bit-rate capacity gain over DVB-S.

Combining DVB-S2 with H.264 or VC-9 video compression, a conventional 36 MHz transponder could carry 20–25 SDTV or 5–6 HDTV programs.

Transmission Format: Mobile (DVB-H)

DVB-H (DVB-Handheld), based on DVB-T, is designed to deliver multimedia and other data (up to 15 Mbps assuming an 8 MHz channel) to a mobile device. Like DVB-T, DVB-H can also use 6, 7 or 8 MHz channels. A 5 MHz channel is also supported for use in non-broadcast environments. A significant feature of DVB-H is the ability to co-exist with DVB-T in the same multiplex. This enables a service provider to have 2 DVB-T services and one DVB-H service within a single DVB-T multiplex.

The payload used by DVB-H is either IP datagrams or other network layer datagrams encapsulated into multiprotocol encapsulated (MPE) sections.

To reduce power consumption, and enable seamless service handovers, time-slicing is used. Time-slicing consists of sending data in bursts at a substantially higher instantaneous bit rate compared to normal streaming rates. Forward error correction for the MPE data (MPE-FEC) improves C/N, Doppler and impulse noise tolerance.

Conditional Access

Conditional Access (CA) is the encryption of the content prior to transmission so that only authorized users may enjoy it. In order to decrypt the protected content, the receiver uses a CA module. The CA module enables decrypting only those programs that have been authorized.

There are two basic techniques to implement DVB conditional access: *Simulcrypt* and *Multicrypt*.

Simulcrypt

Simulcrypt relies on the DVB *Common Scrambling Algorithm* (CSA), a tool for the secure scrambling and descrambling of transport streams or program elementary streams. The CSA descrambling circuitry is embedded inside the MPEG-2 decoder rather than inside the CA module. The CA module does not process the encrypted data directly; it simply provides decryption key information to the CSA descrambling circuitry so the stream can be decrypted.

Since Simulcrypt-based receivers do not need to use the *DVB Common Interface*, the CA module is either embedded inside the receiver or a detachable CA module is used based on the NRSS-A interface. An embedded CA module improves security since access to it is more difficult; however, the receiver is not easily upgraded or adapted for use with other CA systems and therefore may quickly become obsolete.

Using a detachable CA module based the EIA-679 NRSS-A interface (ISO 7816 form factor) allows changing CA systems as easily as changing a low-cost CA card. CA cards are supplied by the service providers to each subscriber.

Simulcrypt also enables the use of more than one CA system in a broadcast or receiver. Each CA system's ECMs (Entitlement Control Messages) and EMMs (Entitlement Management Messages) are transmitted in the stream. Receivers recognize and use the appropriate ECM and EMM needed for decrypting. Thus, a broadcast containing data for multiple CA systems can be viewed on receivers that support any of these CA systems. It also enables a new CA solution to be deployed while maintaining compatibility with a legacy CA system.

Multicrypt

Multicrypt is an open system based a detachable CA module, supplied by the service provider to each subscriber. Encrypted streams are sent to the CA module. The CA module finds and extracts needed data, such as ECM and EMM, directly from the streams. Decrypted streams are then output to the MPEG-2 decoder.

The CA module is plugged into the receiver via the *DVB Common Interface*. Multicrypt's advantage is that a receiver can be easily configured to receive services from different service providers using different and incompatible CA systems. As a result, the receiver is less likely to become obsolete.

A receiver can support multiple CA systems by using multiple DVB Common Interfaces to support multiple CA modules. Encrypted streams are passed sequentially through the different CA modules, with each of them extracting from the stream its own ECMs and EMMs.

DVB Common Interface

The *DVB Common Interface* provides a physical separation between receiver and CA functions. Based on the EIA-679 NRSS-B interface (PCMCIA or PC Card form factor), it is the key to the Multicrypt system.

The transport stream interface consists of an 8-bit parallel input, 8-bit parallel output, control signals and clocks. The command (host) interface consists of an 8-bit bi-directional data bus, address and control signals.

The interface can also used to add new features to a receiver, such as supporting a new audio codec or adding visually impaired audio capabilities.

Application Block Diagrams

Figures 17.3 and 17.4 illustrate a typical DVB-S set-top box.

References

1. ETSI EN 300 421, *Digital Video Broadcasting (DVB): Framing Structure, Channel Coding and Modulation for 11/12 GHz Satellite Services*, August 1997.
2. ETSI EN 300 429, *Digital Video Broadcasting (DVB): Framing Structure, Channel Coding and Modulation for Cable Systems*, April 1998.
3. ETSI EN 300 468, *Digital Video Broadcasting (DVB): Specification for Service Information (SI) in DVB Systems*, January 2003.
4. ETSI EN 300 472, *Digital Video Broadcasting (DVB): Specification for Conveying ITU-R System B Teletext in DVB Bitstreams*, January 2003.
5. ETSI EN 300 743, *Digital Video Broadcasting (DVB): Subtitling Systems*, October 2002.
6. ETSI EN 300 744, *Digital Video Broadcasting (DVB): Framing Structure, Channel Coding and Modulation for Digital Terrestrial Television*, January 2001.
7. ETSI EN 301 192, *Digital Video Broadcasting (DVB): DVB Specification for Data Broadcasting*, January 2003.
8. ETSI EN 301 775, *Digital Video Broadcasting (DVB): Specification for the Carriage of Vertical Blanking Information (VBI) Data in DVB Bitstreams*, January 2003.
9. ETSI ES 200 800, *Digital Video Broadcasting (DVB): DVB Interaction Channel for Cable TV Distribution Systems (CATV)*, October 2001.
10. ETSI ETS 300 801, *Digital Video Broadcasting (DVB): Interaction Channel through Public Switched Telecommunications Network (PSTN) / Integrated Services Digital Networks (ISDN)*, August 1997.
11. ETSI ETS 300 802, *Digital Video Broadcasting (DVB): Network-independent Protocols for DVB Interactive Services*, November 1997.
12. ETSI TR 101 154, *Digital Video Broadcasting (DVB): Implementation Guidelines for the use of MPEG-2 Systems, Video and Audio in Satellite, Cable and Terrestrial Broadcasting Applications*, July 2000.
13. ETSI TR 101 190, *Digital Video Broadcasting (DVB): Implementation Guidelines for DVB Terrestrial Services; Transmission Aspects*, July 2000.
14. ETSI TR 101 190, *Digital Video Broadcasting (DVB): Implementation Guidelines for DVB Terrestrial Services; Transmission Aspects*, December 1997.
15. ETSI TR 101 194, *Digital Video Broadcasting (DVB): Guidelines for Implementation and Usage of the Specification of Network Independent Protocols for DVB Interactive Services*, June 1997.
16. ETSI TR 101 200, *Digital Video Broadcasting (DVB): A Guideline for the Use of DVB Specifications and Standards*, September 1997.
17. ETSI TR 101 200, *Digital Video Broadcasting (DVB): Implementation Guidelines for Data Broadcasting*, January 2003.
18. ETSI TR 101 211, *Digital Video Broadcasting (DVB): Guidelines on Implementation and Usage of Service Information (SI)*, January 2003.
19. ETSI TS 101 699, *Digital Video Broadcasting (DVB): Extensions to the Common Interface Specification*, November 1999.

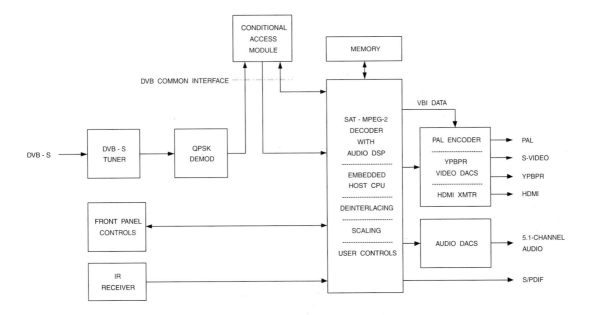

Figure 17.3. DVB Receiver Set-top Box Block Diagram (Multicrypt).

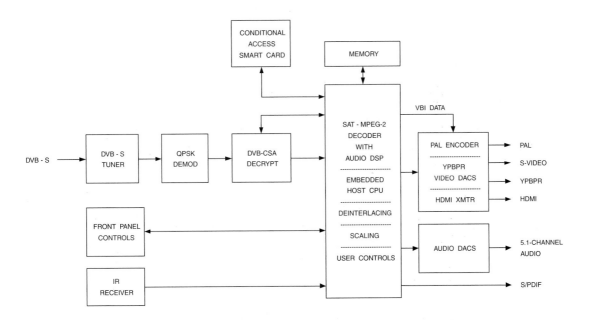

Figure 17.4. DVB Receiver Set-top Box Block Diagram (Simulcrypt).

Chapter 18

ISDB

Digital Television

The ISDB (Integrated Services Digital Broadcasting) digital television (DTV) broadcast standard is used in Japan.

The three other primary DTV standards are ATSC (Advanced Television Systems Committee), DVB (Digital Video Broadcast) and OpenCable™. A comparison between the standards is shown in Table 18.1. The major differences are the RF modulation schemes and the level of definition for non-audio/video services. ISDB builds on DVB, adding additional services required for Japan.

The ISDB standard is actually a group of ARIB standards:

STD-B10   Service Information For Digital Broadcasting System

STD-B16   Standard Digital Receiver Commonly Used For Digital Satellite Broadcasting Services Using Communication Satellite

STD-B20   ISDB-S: Transmission System For Digital Satellite Broadcasting

STD-B21   Receiver For Digital Broadcasting (Desirable Specifications)

STD-B23   Application Execution Engine Platform for Digital Broadcasting

STD-B24   Data Coding And Transmission Specification For Digital Broadcasting

STD-B25   Conditional Access System Specifications for Digital Broadcasting

STD-B31   ISDB-T: Transmission System For Digital Terrestrial Television Broadcasting

STD-B32   Video Coding, Audio Coding And Multiplexing Specifications For Digital Broadcasting

STD-B40   PES Packet Transport Mechanism for Ancillary Data

ISDB uses an MPEG-2 transport stream to convey compressed digital video, compressed digital audio and ancillary data. Like DVB, this transport stream is then transmitted either via terrestrial, cable or satellite. Interactive applications are based on BML (Broadcast Mark-up Language).

ISDB-S (Satellite)

Two satellite standards exist, ISDB-S, also known as the BS (broadcast satellite) system, and DVB-S, also known as the CS (communications satellite) system.

ISDB-S (BS) is also specified by ITU-R BO.1408. It has a maximum bit rate of ~52.2 Mbps using TC8PSK modulation and a 34.5 MHz transponder.

CS supports only one transport stream per transport channel, and supports up to ~34 Mbps on a 27 MHz channel. The modulation scheme is QPSK, just like DVB-S. Unlike the other variations of ISDB, CS uses MPEG-2 MP@ML video (480i or 480p) and MPEG-2 BC audio.

ISDB-C (Cable)

ISDB-C uses 64-QAM modulation, with two versions: one that supports only a single transport stream per transmission channel and one that supports multiple transport streams per transmission channel. On a 6 MHz channel, ISDB-C can transmit up to ~29.16 Mbps. As the bit rate on an ISDB-S satellite channel is 2× that, two cable channels can be used to rebroadcast satellite information.

Parameter	US (ATSC)	Europe, Asia (DVB)	Japan (ISDB)	OpenCable™
video compression	MPEG-2			
audio compression	Dolby® Digital	MPEG-2, Dolby® Digital, DTS®	MPEG-2 AAC-LC	Dolby® Digital
multiplexing	MPEG-2 transport stream			
terrestrial modulation	8-VSB	COFDM	BST-OFDM[2]	–
channel bandwidth	6 MHz	6, 7, or 8 MHz	6, 7, or 8 MHz	6 MHz
cable modulation	16-VSB[3]	QAM	QAM	QAM
channel bandwidth	6 MHz	6, 7, or 8 MHz	6, 7, or 8 MHz	6 MHz
satellite modulation	QPSK, 8PSK	QPSK	PSK	–

Notes:
1. ISDB = Integrated Services Digital Broadcasting.
2. BST-OFDM = Bandwidth Segmented Transmission of OFDM.
3. Most digital cable systems use QAM instead of 16-VSB.

Table 18.1. Comparison of the Various DTV Standards.

Active Resolution (Y)	SDTV or HDTV	Frame Rate (p = progressive, i = interlaced)			
		23.976p 24p	29.97i 30i	29.97p 30p	59.94p 60p
720 × 480	SDTV	×	×	×	×
1280 × 720	HDTV	×		×	×
1920 × 1080		×	×	×	×

Table 18.2. Common Active Resolutions for ISDB Digital Television.

ISDB-C also allows passing through OFDM-based ISDB-T signals since the channel bandwidths are the same.

ISDB-S (BS) signals can also be passed through by downconverting the satellite signals at the cable head-end, then up-converting them at the receiver. This technique is suitable only for cable systems that have many (up to 29) unused channels.

ISBD-T (Terrestrial)

ISDB-T, the terrestrial broadcast standard, is also specified by ITU-R BT.1306. It has a maximum bit rate of ~23.2 Mbps using 5.6 MHz of bandwidth. ISDB-T also supports bandwidths of 6, 7 and 8 MHz.

The bandwidth is divided into 13 OFDM segments; each segment can be divided in up to three segment groups (hierarchical layers) having different transmission parameters such as the carrier modulation scheme, inner-code coding rate, and time interleaving length. This enables the same program to be broadcast in different resolutions, allowing a mobile receiver to show a standard-definition picture while a stationary receiver shows a high-definition picture.

Video Capability

There are several standardized resolutions, indicated in Table 18.2. Both interlaced and progressive pictures are permitted for most of the resolutions.

Primary video compression is based on MPEG-2. However, there are some minor constraints on some of the MPEG-2 parameters, as discussed within the MPEG-2 chapter.

ISDB also supports MPEG-4 video using CIF, SIF, QCIF, QSIF and QVGA resolutions.

Audio Capability

Primary audio compression is implemented using MPEG-2 AAC-LC. Up to 5.1 channels are supported. ISDB also supports MPEG-4 audio.

Still Picture Capability

Still pictures are supported using JPEG (ISO/IEC 10918-1), PNG (Portable Network Graphics), MNG (Multiple-image Network Graphics), MPEG-2 I-frame and MPEG-4 I-VOP formats.

Graphics Capability

Graphics commands include *Domain, Texture* (fill, vertical hatch, horizontal hatch, cross hatch), *Set Color* (foreground, background), *Select Color, Blink, Set Pattern, Point, Line* (solid, dotted, broken, dotted-broken), *Arc* (outlined, filled), *Rectangle* (outlined, filled) and *Polygon* (outlined, filled).

Figure 18.1 illustrates the 5-plane video/graphics architecture used by ISDB.

System Information (SI)

ARIB STD-B10 specifies the Service Information (SI) data which forms a part of ISDB bitstreams. SI is a small collection of hierarchically associated tables (see Table 18.3) designed to extend the MPEG-2 PSI tables. It provides information on what is available on other transport streams and even other networks. The method of information presentation to the user is not specified, allowing receiver manufacturers to choose appropriate presentation methods.

Tables

Application Information Table (AIT)
The AIT transmits dynamic control information concerning ARIB-J application and information for execution.

Bouquet Association Table (BAT)
The BAT provides information regarding bouquets (groups of services that may traverse the network boundary). Along with the name of the bouquet, it provides a list of services for each bouquet.

Broadcaster Information Table (BIT)
The BIT is used to submit broadcaster information on network.

Common Data Table (CDT)
The CDT transmits data which is required for all receivers and is to be stored in non-volatile memory.

Discontinuity Information Table (DIT)
The DIT is present at transition points where the SI information is discontinuous. The use of this table is restricted to partial transport streams, they are not used in broadcasts.

Download Table (DLT)
The DLT is used to transmit software for downloading.

Download Control Table (DCT)
The DCT is used to transmit of information to indicate how to process the DLT.

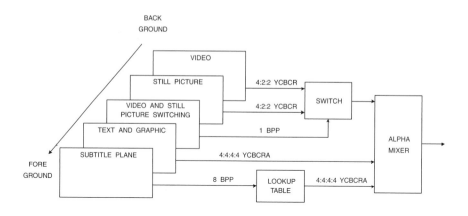

Figure 18.1. ISDB 5-Plane Video/Graphics Standard.

Event Information Table (EIT)

There are up to 128 EITs, EIT-0 through EIT-127, each of which describes the events or TV programs associated with each channel. Each EIT is valid for three hours. Since there are up to 128 EITs, up to 16 days of programming may be advertised in advance. The first four EITs are required (the first 24 are recommended) to be present.

Information provided by the IET includes start time, duration, title, pointer to optional descriptive text for the event, advisory data, caption service data, audio service descriptor, etc.

Event Relation Table (ERT)

The ERT indicates relationships between programs or events and their attributes.

Index Transmission Table (ITT)

The ITT is used to convey program index information with a program.

Linked Description Table (LDT)

The LDT is used to link various descriptions from other tables.

Local Event Information Table (LIT)

The LIT conveys information in a program that relates to a local event, such as time, name and explanation of the local event.

Network Board Information Table (NBIT)

The NBIT transmits board information on the network, e.g. a guide.

Network Information Table (NIT)

The NIT provides information about the physical network, including any grouping of transport streams and the relevant tuning information. It can be used during receiver set-up and the relevant tuning information stored in non-volatile memory. The NIT can also be used to signal changes of tuning information.

Partial Content Announcement Table (PCAT)

The PCAT conveys partial content announcement for data broadcasting.

Running Status Table (RST)

The RST updates the running status of one or more events. These are sent out only once, at the time of an event status change, unlike other tables which are usually transmitted repeatedly.

Selection Information Table (SIT)

The SIT describes services and events carried by a partial transport stream. The use of this table is restricted to partial transport streams, they are not used in broadcasts.

Service Description Table (SDT)

The SDT conveys information related to the channel, such as channel name and broadcasting company name.

Software Download Trigger Table (SDTT)

The SDTT conveys notification information, such as download service ID, schedule information and receiver types for revision.

Stuffing Table (ST)

The ST is used to replace or invalidate sub-tables or complete SI tables.

Time and Date Table (TDT)

The TDT contains the actual UTC-time coded as Modified Julian Date (MJD). Receivers can use it to maintain the correct local time.

Time Offset Table (TOT)

The TOT is the same as the TDT, except it includes local time offset information.

Descriptors

Much like MPEG-2, ISDB uses descriptors to add new functionality. In addition to various MPEG-2 descriptors, one or more of these ISDB-specific descriptors may be included within the PMT or one or more SI tables (see Table 18.3) to extend data within the tables. A descriptor not recognized by a decoder must be ignored by that decoder. This enables new descriptors to be implemented without affecting receivers that cannot recognize and process the descriptors.

Audio Component Descriptor

This ARIB descriptor indicates the parameters of an audio elementary stream.

AVC Timing and HRD Descriptor

This ARIB descriptor, also discussed in Chapter 13, describes the video stream time information and the reference decoder information for H.264 (MPEG-4 Part 10).

Descriptor	Descriptor Tag	Tables								
		PMT	NIT	BAT	SDT	EIT	TOT	BIT	NBIT	LDT
PID		per PAT	0010_H	0011_H	0011_H	0012_H	0014_H	0024_H	0025_H	0025_H
Table_ID		02_H	40_H 41_H	$4A_H$	42_H 46_H	$4E_H$ - $6F_H$	73_H	$C4_H$	$C5_H$ $C6_H$	$C7_H$
repetition rate		100 ms	10 sec	10 sec	2 - 10 sec	2 - 10 sec	30 sec	20 sec	20 sec	20 sec
audio component	1100 0100					×				
AVC timing and HRD	0010 1010	×								
AVC video	0010 1000	×								
basic local event	1101 0000									
board information	1101 1011								×	
bouquet name	0100 0111			×	×					
broadcaster name	1101 1000							×		
CA contract information	1100 1011									
CA EMM TS	1100 1010									
CA identifier	0101 0011			×	×	×				
CA service	1100 1100									
cable TS division system	1111 1001									
cable distribution system	0100 0100									
carousel compatible composite descriptor	1111 0111	×				×				
component	0101 0000	×				×				
component group	1101 1001					×				
conditional playback	1111 1000	×								
connected transmission	1101 1101									

Notes:
1. PMT: MPEG-2 Program Map Table.

Table 18.3a. List of ISDB SI Tables, Descriptors and Descriptor Locations.

Descriptor	Descriptor Tag	Tables								
		PMT	NIT	BAT	SDT	EIT	TOT	BIT	NBIT	LDT
content	0101 0100					×				
content availability	1101 1110	×		×	×					
country availability	0100 1001	×		×	×					
data component	1111 1101	×								
data content	1100 0111					×				
digital copy control	1100 0001	×			×	×				
download content	1100 1001									
emergency information	1111 1100	×	×			×				
event group	1101 0110					×				
extended broadcaster	1100 1110							×		
extended event	0100 1110					×				×
hierarchical transmission	1100 0000	×								
hyperlink	1100 0101					×				
LDT linkage	1101 1100					×				
linkage	0100 1010	×	×	×	×	×				
local time offset	0101 1000						×			
logo transmission	1100 1111				×					
mosaic	0101 0001	×			×					
network identification	1100 0010									
network name	0100 0000		×							
node relation	1101 0010									
NVOD reference	0100 1011				×					
parental rating	0101 0101	×				×				
partial reception	1111 1011		×							

Notes:
1. PMT: MPEG-2 Program Map Table.

Table 18.3b. List of ISDB SI Tables, Descriptors and Descriptor Locations.

Descriptor	Descriptor Tag	Tables								
		PMT	NIT	BAT	SDT	EIT	TOT	BIT	NBIT	LDT
partial transport stream	0110 0011									
partial transport stream time	1100 0011									
reference	1101 0001									
satellite delivery system	0100 0011		×							
series	1101 0101					×				
service	0100 1000				×					
service list	0100 0001		×	×				×		
short event	0100 1101					×				×
short node information	1101 0011									
SI parameter	1101 0111							×		
SI Prime_TS	1101 1010							×		
STC reference	1101 0100									
stream identifier	0101 0010	×								
stuffing	0100 0010		×	×	×	×			×	×
system management	1111 1110	×	×							
target region	1100 0110	×								
terrestrial delivery system	1111 1010		×							
time shifted event	0100 1111					×				
time shifted service	0100 1100				×					
TS information	1100 1101									
video decode control	1100 1000	×								

Notes:

1. PMT: MPEG-2 Program Map Table.

Table 18.3c. List of ISDB SI Tables, Descriptors and Descriptor Locations.

AVC Video Descriptor

This ARIB descriptor, also discussed in Chapter 13, describes basic coding parameters of the H.264 (MPEG-4 Part 10) video stream.

Basic Local Event Descriptor

This ARIB descriptor indicates the local event identifier information.

Board Information Descriptor

This ARIB descriptor indicates the title and content of the board information in text format.

Bouquet Name Descriptor

This ARIB descriptor provides the bouquet name as variable-length text, such as "Max Movie Channels." DVB also uses this descriptor.

Broadcaster Name Descriptor

The ARIB descriptor indicates the name of the broadcaster.

CA Contract Information Descriptor

This ARIB descriptor describes the conditional access service type for the scheduled program.

CA EMM TS Descriptor

This ARIB descriptor indicates the special trap-on when the EMM transmission is made by the special trap-on method.

CA Identifier Descriptor

This ARIB descriptor indicates whether a bouquet, service or event is associated with a conditional access system and if so, identifies the conditional access used. DVB also uses this descriptor.

CA Service Descriptor

This ARIB descriptor conveys the broadcast service provider servicing the automatic indication message indication.

Carousel Compatible Composite Descriptor

This ARIB descriptor, also discussed in Chapter 13, uses descriptors defined in the data carousel transmission specification (ARIB STD-B24 Part 3) as sub-descriptors, and describes accumulation control by applying the functions of the sub-descriptors.

Component Descriptor

This ARIB descriptor, also discussed in Chapter 13, indicates the type of stream and may be used to provide a text description of the stream. DVB also uses this descriptor.

Component Group Descriptor

This ARIB descriptor defines and identifies component grouping in an event.

Conditional Playback Descriptor

This ARIB descriptor, also discussed in Chapter 13, conveys the description of conditional playback and the PID that transmits the ECM and EMM.

Connected Transmission Descriptor

This ARIB descriptor indicates the physical condition when connected to a transmission in the terrestrial audio transmission path.

Content Descriptor

This ARIB descriptor is used to identify the type of content (comedy, talk show, etc.). DVB also uses this descriptor.

Content Availability Descriptor

This ARIB descriptor, also discussed in Chapter 13, describes information to control the recording and output of content by receivers. The *encryption_mode* flag indicates whether or not to encrypt the digital video outputs. It is used in combination with the *Digital Copy Control Descriptor.*

Country Availability Descriptor

This ARIB descriptor, also discussed in Chapter 13, identifies countries that are either allowed or not allowed to receive the service. The descriptor may appear twice for each service, once for listing countries allowed to receive the service, and a second time for listing countries not allowed to receive the service. The latter list overrides the former list. DVB also uses this descriptor.

Data Component Descriptor

This ARIB descriptor, also discussed in Chapter 13, identifies data components.

Data Content Descriptor

This ARIB descriptor describes the detailed information relating to individual contents of a data broadcasting event.

Digital Copy Control Descriptor

This ARIB descriptor, also discussed in Chapter 13, signals copy generation information, including *copy-free*, *copy-one-generation* and *copy-never*.

For content which is either copy-restricted by *digital_recording_control_data* in the *Digital Copy Control Descriptor*, or copy-protected by *encryption_mode* in the *Content Availability Descriptor*, receivers are prohibited from transferring the content to any output that potentially allows redistribution of it over the Internet.

Download Content Descriptor

This ARIB descriptor conveys download attribute information such as size, type and ID.

Emergency Information Descriptor

This ARIB descriptor, also discussed in Chapter 13, is used to broadcast an emergency message.

Event Group Descriptor

This ARIB descriptor, when there is a relationship between multiple events, indicates that these events are in a group.

Extended Broadcaster Descriptor

This ARIB descriptor specifies the extended broadcaster identification information and defines the relationships with other extended broadcasters and broadcasters of other networks.

Extended Event Descriptor

This ARIB descriptor provides a text description of an event, which may be used in addition to the *Short Event Descriptor*. More than one descriptor can be used to convey more than 256 bytes of information. DVB also uses this descriptor.

Hierarchical Transmission Descriptor

This ARIB descriptor, also discussed in Chapter 13, indicates the relationship between hierarchical streams.

Hyperlink Descriptor

This ARIB descriptor describes the linkage to other events, event contents and information events.

LDT Linkage Descriptor

This ARIB descriptor describes the linkage of the information collected in the LDT.

Linkage Descriptor

This ARIB descriptor, also discussed in Chapter 13, provides a link to another service, transport stream, program guide, service information, software upgrade, etc. DVB also uses this descriptor.

Local Time Offset Descriptor

This ARIB descriptor may be present in the TOT to describe country-specific dynamic changes of the local time offset relative to UTC. This enables a receiver to adjust automatically between summer and winter times. DVB also uses this descriptor.

Logo Transmission Descriptor

This ARIB descriptor describes service logo information, such as pointing to PNG logo data transmitted by ARIB STD-B21, logo identifier, logo version and the 8-unit code alphanumeric character string for a simple logo.

Mosaic Descriptor

This ARIB descriptor, also discussed in Chapter 13, partitions a digital video component into elementary cells, the allocation of elementary cells to logical cells, and links the content of the logical cell and the corresponding information (e.g. bouquet, service, event etc.). DVB also uses this descriptor.

Network Identification Descriptor

This ARIB descriptor identifies the network.

Network Name Descriptor

This ARIB descriptor conveys the network name in text form, such as "Tokyo Cable." DVB also uses this descriptor.

Node Relation Descriptor

This ARIB descriptor describes the relationship between two nodes.

NVOD (Near Video On Demand) Reference Descriptor

This ARIB descriptor, in conjunction with the *Time Shifted Service Descriptor* and the time *Time Shifted Event Descriptor*, provides an efficient way of describing a number of services which carry the same sequence of events, but with the start times offset from one another. DVB also uses this descriptor.

Parental Rating Descriptor

This ARIB descriptor, also discussed in Chapter 13, gives a rating based on age and offers extensions to be able to use other rating criteria. DVB also uses this descriptor.

Partial Reception Descriptor

This ARIB descriptor indicates the *service_id* transmitted by the partial reception hierarchy of the terrestrial transmission path.

Partial Transport Stream Descriptor

The SIT contains all the information needed to control, play and copy partial transport streams. This ARIB descriptor describes this information. DVB also uses this descriptor.

Partial Transport Stream Time Descriptor

This ARIB descriptor describes partial transport stream time information.

Reference Descriptor

This ARIB descriptor indicates the node reference from programs and local event.

Satellite Delivery System Descriptor

This ARIB descriptor conveys the physical parameters of the satellite network, including frequency, orbital position, west-east flag, polarization, modulation and symbol rate. DVB also uses this descriptor.

Series Descriptor

This ARIB descriptor identifies a series event.

Service Descriptor

This ARIB descriptor provides the name of the service and the service provider in text form. DVB also uses this descriptor, although the *service_type_id* types are different between ARIB and DVB.

Service List Descriptor

This ARIB descriptor provides a list of the services and service types for each transport stream. DVB also uses this descriptor.

Short Event Descriptor

This ARIB descriptor provides the name and a short description of an event. DVB also uses this descriptor.

Short Node Information Descriptor

This ARIB descriptor indicates the node name and simple explanation.

SI Parameter Descriptor

This ARIB descriptor indicates the SI parameter.

SI Prime_TS Descriptor

This ARIB descriptor indicates the identifier information of the SI prime TS and its transmission parameter.

STC Reference Descriptor

This ARIB descriptor indicates the relationship between the identification time of local event and the STC.

Stream Identifier Descriptor

This ARIB descriptor, also discussed in Chapter 13, enables streams to be associated with a description in the EIT, useful when there is more than one stream of the same type within a service. DVB also uses this descriptor.

Stuffing Descriptor

This ARIB descriptor is used to stuff tables for any reason or to disable descriptors that are no longer valid. DVB also uses this descriptor.

System Management Descriptor

This ARIB descriptor, also discussed in Chapter 13, identifies broadcasting and non-broadcasting format used.

Target Region Descriptor

This ARIB descriptor, also discussed in Chapter 13, describes the target region of an event or a part of the stream comprising an event.

Terrestrial Delivery System Descriptor

This ARIB descriptor is used to transmit the physical parameters of the terrestrial network, including center frequency, bandwidth, constellation, hierarchy information, code rate, guard interval and transmission mode.

Time-shifted Event Descriptor

This ARIB descriptor indicates that an event is the time-shifted copy of another event. DVB also uses this descriptor.

Time-shifted Service Descriptor

This ARIB descriptor links one service with up to 20 other services carrying the same programming, but time-shifted. A typical application is for Near Video On Demand (NVOD) services. DVB also uses this descriptor.

TS Information Descriptor

This ARIB descriptor specifies the remote control key identifier assigned to the applicable transport stream and indicates the relationship between the service identifier and the transmission layer during hierarchical transmission.

Video Decode Control Descriptor

This ARIB descriptor, also discussed in Chapter 13, controls the decoding of MPEG-based still pictures transmitted at low transmission speed and to achieve smooth decoding at video splice points when changing video coding method.

Captioning

Japanese captioning data (ARIB STD-B24 Part 3) can be present in video PES, audio PES or independent PES (preferred). Captioning not related to video content is called "superimpose" (ARIB STD-B5).

Both the horizontal and vertical writing formats may be used. Supported character sets include Mosaic, Chinese, Kanji, Hiragana, Katakana, Symbol, and alpha-numeric. Attributes include reverse polarity, flash, underline, hem, shade, bold, italic and bold-italic. Bitmap graphics are also supported.

Display control includes display timing, erase timing, cut, dissolve, wipe, slide and roll. It also supports flexible viewing, recording and playback options.

Data Broadcasting

The ARIB data broadcast standard describes the available encapsulation protocols used to transport data within a ARIB stream. Based on MPEG-2 DSM-CC, it also supports a XML-based multimedia coding scheme.

Five different data broadcast specifications have been identified. Most of the specifications have additional descriptors used to support the specification.

Data Carousel Transmission

This specification transmits general synchronous and asynchronous data, allowing a receiver to obtain data during its transmission period. Used for download and multimedia services.

Data Piping

If required, this specification may be used to deliver data to a receiver. Data is carried directly in the payloads of MPEG-2 transport stream packets.

Event Message Transmission

This specification is used for synchronous and asynchronous message notification (either immediately or at a specified time) to an application in the receiver. Used for multimedia services.

Independent PES Transmission

This specification supports data broadcast services that use a streaming-oriented delivery of data in either an asynchronous or synchronous way. Data is carried in MPEG-2 PES packets. Also used for subtitles and superimposed characters.

Interaction Channel Protocols

This specification provides the transmission protocols used over public networks including PSTN, ISDN, and mobile networks for bi-directional interactive services.

Application Block Diagrams

Figure 18.2 illustrates a typical ISDB-S set-top box.

References

1 STD-B10, *Service Information for Digital Broadcasting System*, version 3.8.
2. STD-B16, *Standard Digital Receiver Commonly Used for Digital Satellite Broadcasting Services Using Communication Satellite.*
3. STD-B20, *ISDB-S: Transmission System for Digital Satellite Broadcasting.*
4. STD-B21, *Receiver for Digital Broadcasting (Desirable Specifications)*, version 4.2.
5. STD-B23, *Application Execution Engine Platform for Digital Broadcasting.*
6. STD-B24, *Data Coding and Transmission Specification for Digital Broadcasting*, version 3.2.
7. STD-B25, *Conditional Access System Specifications for Digital Broadcasting.*
8. STD-B31, *ISDB-T: Transmission System for Digital Terrestrial Television Broadcasting*, version 1.5.
9. STD-B32, *Video Coding, Audio Coding and Multiplexing Specifications for Digital Broadcasting*, version 1.5.
10. STD-B40, *PES Packet Transport Mechanism for Ancillary Data.*

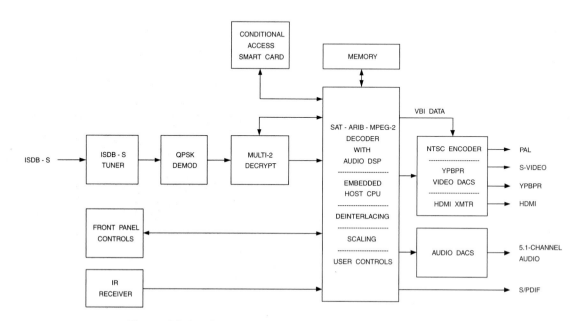

Figure 18.2. ISDB Receiver Set-top Box Block Diagram.

Chapter 19

IPTV

With the increased use of digital video and high-speed broadband networks, transferring real-time audio and video over a broadband network has become popular. The technology is known by several names, including IPTV (Internet Protocol TV), streaming video, video over IP and IP video.

Rather than downloading and storing large audio and video files, then playing them back, data is sent across the network in "streams." Streaming breaks the audio and video data into small packets suitable for transmission. The real-time audio and video data flows from a video server or real-time video encoder, through a network, and is decoded and played by the receiver (or "client") in real time. Thus, the user can start viewing a video without waiting until the end of the download process.

Telcos are adopting IPTV over DSL and FTTH as a way of offering video services to compete with cable and satellite TV. They are now able to offer VoIP (Voice over IP), video-on-demand (VOD), gaming, music, interactive television and local, national, and premium television programming.

Considerations

Streaming video over a network is not a trivial task. First, even compressed video data requires relatively high bandwidth. Limiting the bit rate to about 700 kbps is desirable to support streaming two standard-definition video streams over a single 1.5 Mbps DSL connection. For this reason, using the new H.264 and SMPTE VC-9 video codecs are highly desirable. The lower bit rates achievable with H.264 and VC-9 also enable a larger area to be serviced since DSL bit rate decreases with distance.

Second, streaming video requires "real-time" transfers to avoid interruptions in the playback process. This requires the video servers and real-time encoders to be able to stream the video continuously and avoid network congestion. To address this issue, standards are available to reserve bandwidth resources along the network. Multicasting is also being used to reduce network bandwidth requirements further.

Third, streaming video is usually "bursty". Streaming video clients have a receive buffer of limited size. If measures are not taken to smooth the transmitted bit rate, the receive buffer may overflow or underflow. To address this issue, additional protocols are used to manage the timing issues.

Multicasting

There are three common techniques of streaming real-time audio and video over a network:

> *Unicast*, where a server sends data to one receiver, as shown in Figure 19.1. The port number is chosen by the receiver.
>
> *Broadcast*, where data is sent from one server to all receivers, as illustrated in Figure 19.2.
>
> *Multicast*, where data is sent from one server to a group of receivers, as shown in Figure 19.3. The server picks the multicast IP address and port. This is a typical case for live and near-video-on-demand (NVoD) applications.

The recent support for multicasting is a result of needing real-time distribution of large amounts of data, such as audio and video, combined with an increasing number of users. In this environment, multicasting is an excellent way to save network and server capacity.

RTSP-Based Solutions

With proprietary solutions, each video server vendor has their own unique streaming protocols and file formats, requiring a client to support multiple protocols or be tailored to a specific vendor. In an effort to develop an open, standards-based solution, the Internet Engineering Task Force (IETF) developed several protocols to enable cross-platform connectivity and communications between clients and servers.

> *RTSP* is the control protocol for initiating and directing delivery of streaming data from video servers, implementing a "remote control" capability. RTSP does not deliver the multimedia data, though the RTSP connection may be used to tunnel RTP data for ease of use with firewalls and other network devices.
>
> *RTP* is the transport protocol for the delivery of real-time data, including streaming audio and video. RTP and RTSP are usually used together, but either protocol can be used without the other.
>
> *RTCP* is a part of RTP and helps with lip synchronization and Quality-of-Service (QoS) management.
>
> *RSVP* is the protocol for establishing and maintaining desired QoS levels, ensuring adequate network resources (such as bandwidth) are available.

RTSP

The Real-Time Streaming Protocol (RTSP) establishes and controls one or more time-synchronized streams of audio and video data between a server (source) and client (receiver). The server provides playback or recording services for the streams while the client requests continuous data from the server.

RTSP provides "VCR-style" control functionality for the audio and video streams, including play, pause, fast forward, and reverse. It also provides:

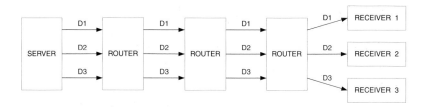

Figure 19.1. Unicast Example. Three copies of the same data are sent point-to-point as streams D1, D2 and D3 to receivers 1, 2, and 3.

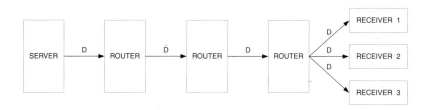

Figure 19.2. Broadcast Example. One copy of the same data (D) is sent to all receivers.

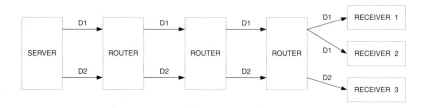

Figure 19.3. Multicast Example. One copy of the same data (D1) is multicast to receivers 1 and 2. Note the bandwidth savings locally and across the networks as the number of receivers increases.

Retrieval of program information from a server. The client can request a list of available programs and their description ("program description") via a web browser (HTTP) or other technique. If a program is being multicast, the program description also contains the multicast addresses and ports used. If a program is to be sent to only one client (unicast), the client provides the destination address.

Invitation of a server to a conference. A server can be "invited" to join an existing conference, either to provide or record data.

Adding media to an existing presentation. Particularly useful for live presentations, this enables a server to inform a client if additional data is available.

RTSP versus HTTP

RTSP provides the same services for streaming audio and video as HTTP does for text and graphics when browsing the web. It is designed to have a similar syntax and operation, enabling most HTTP extensions to be easily adopted to RTSP. For example, the RTSP URL

rtsp://media.example.com:554/twister

identifies the presentation "twister", which may be composed of audio and video streams. The RTSP URL

rtsp://media.example.com:554/twister/audio

identifies the audio stream within the presentation "twister", which can be controlled via RTSP requests to port 554 of server media.example.com.

There is some overlap in functionality between RTSP and HTTP since the user interface is often implemented using web pages. For this reason, RTSP supports different hand-off points between a web and video server. For example, the presentation description can be retrieved using HTTP or RTSP, allowing standalone RTSP servers and clients which do not support HTTP. Figure 19.4 illustrates using a web server for the presentation and a separate video server for the content.

RTSP differs from HTTP in two major areas. First, unlike HTTP, a RTSP compatible video server has to maintain "session states" in order to correlate RTSP requests with a stream. Second, while HTTP is basically an asymmetric protocol (the client issues requests and the server responds), both the video server and client can issue requests with RTSP. For example, the video server can issue a request to set the playback parameters of a stream.

Stream Properties

The properties of a stream are defined in a *presentation description file*, which may include the encoding format, language, RTSP URLs, destination address, port, and other parameters. The presentation description file is obtained by the client using HTTP or other means. RTSP requests are usually sent on a channel independent of the data channel.

RTP

The Real-Time Transport Protocol (RTP) is a packet-based protocol for the transfer of real-time data, such as audio and video streams. Designed primarily for multicast, RTP can be also used for unicast and video-on-demand.

Packets sent over a network have unpredictable delay and jitter, complicating the streaming of real-time video. To overcome these issues, the RTP packet header includes timestamping, loss detection, payload identification, source identification and security. This information is used at the applications level to implement lost packet recovery, congestion control, etc.

RTP is typically run on top of UDP to make use of its multiplexing and checksum functions. While TCP provides a connection-oriented and reliable flow between two hosts, UDP provides a connectionless (but unreliable) datagram service over the network. UDP was chosen as the target transport protocol for RTP for two reasons. First, RTP is primarily designed for multicast; the connection-oriented TCP does not scale well and therefore is not suitable. Second, for real-time data, reliability is not as important as timely delivery; the higher reliability provided by TCP using retransmission is not desirable. For example, in network congestion, some packets might

get lost and the application would result in lower but acceptable quality. If the protocol insists on a reliable transmission, the retransmitted packets could possibly increase the delay, jam the network, and eventually starve the receiving application. Figure 19.5 illustrates a RTP packet encapsulated within a UDP/IP packet.

RTP itself does not provide mechanisms to ensure timely delivery. It requires support from lower layers that control resources in switches and routers. RSVP may be used to reserve such resources and to provide the requested QoS. RTP is also designed to work in conjunction with RTCP to get feedback on quality of data transmission and information about participants in the session.

RTP is also designed to work in conjunction with RTCP to get feedback on quality of data transmission and information about participants in the session.

RTP is a protocol framework that is deliberately not complete. It is open to new payload formats and new multimedia software. By add-

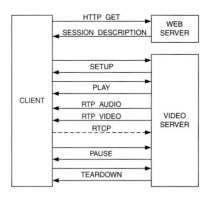

Figure 19.4. Client, Web Server and Video Server Communications.

ing new profile and payload specifications, RTP can easily be tailored to new data formats and new applications.

RTP Sessions

To set up a RTP session, the application defines a pair of destination addresses (one network address plus two ports for RTP and RTCP). In a multimedia session, each medium is usually carried using its own RTP session, with corresponding RTCP packets reporting the reception quality for that session. For example, audio and video typically use separate RTP sessions, enabling a receiver to select whether or not to receive a particular medium.

Timestamps

Timestamping is important for real-time applications. The receiver uses timestamps to reconstruct the original timing in order to play the data at the correct rate. Timestamps are also used to synchronize different streams, such as audio and video data. However, RTP itself is not responsible for the synchronization; this is done at the application level.

In addition, UDP does not deliver packets in a timely order. Therefore, sequence numbers are used to place incoming data packets in the correct order and for packet loss detection. When a video frame is split into several RTP packets, some video formats allow all of them to have the same timestamp. Thus, timestamps are not enough to ensure packets can be put back into the correct sequence.

Payload Identification

A payload identifier specifies the type of content and compression format. This enables the receiver to know how to interpret and present the content. Several types of payloads are supported:

Various audio formats, including CELP, linear PCM, ADPCM, G.711, G.721, G.722, and so on

MPEG-1 audio elementary streams

MPEG-1 video elementary streams

MPEG-1 system streams

MPEG-2 audio elementary streams

MPEG-2 video elementary streams

MPEG-2 program streams

MPEG-2 transport streams

MPEG-4 audio streams

MPEG-4 visual streams

MPEG-4 OD, BIFS, OCI and IPMP streams

M-JPEG video streams

H.261 and H.263 streams

ASF

IP HEADER	UDP HEADER	RTP HEADER	RTP PAYLOAD

Figure 19.5. RTP Packet Encapsulated within a UDP/IP Packet.

Additional payload types may be added by providing a profile and payload format specification. At any given time of transmission, an RTP sender can only send one type of payload, although the payload type may change during transmission, for example, to adjust to network congestion.

RTCP

The Real-Time Control Protocol (RTCP) is a control protocol designed to work in conjunction with RTP. In a RTP session, clients periodically send RTCP packets to the server to convey feedback on quality of data delivery and information of membership.

Five types of RTCP packets to convey control information are defined:

Receiver report. Receiver reports contain information about data delivery, including the highest packet number received, number of packets lost, inter-arrival jitter, and timestamps to calculate the round-trip delay between the server and the client.

Sender report. Sender reports contain the receiver report information and information on inter-media synchronization, cumulative packet counters, and number of bytes sent.

Source description items. They contain information to describe the sources.

Bye: Indicates end of participation.

Application specific functions. Intended for experimental use as new applications and new features are developed.

Through these control information packets, RTCP provides:

QoS monitoring and congestion control. Servers can adjust transmission based on the client feedback. Clients can determine whether a congestion is local, regional or global. Network performance can be evaluated during multicast distribution.

Source identification. In RTP data packets, sources are identified by randomly generated 32-bit identifiers, not convenient for users. Source description packets contain textual information such as user's name, telephone number, e-mail address, etc.

Inter-media synchronization. Used in inter-media synchronization, such as lip synchronization for audio and video.

Control information scaling. When the number of participants increases, steps must be taken to prevent the control traffic from overwhelming network resources. RTP limits the control traffic to 5% of the overall session traffic. This is enforced by adjusting the RTCP generating rate according to the number of participants.

Combined, RTP and RTCP provide the necessary functionality and control mechanisms for transmitting real-time content. However, RTP and RTCP themselves are not responsible for higher-level tasks such as assembly and synchronization. These are done at the application level.

RSVP

The Resource Reservation Protocol (RSVP) enhances the network with support for QoS.

RSVP is used to set up reservations for network resources, such as bandwidth. When a client requests a specific QoS for its data stream, it delivers its request to nodes (or routers) along the network path using RSVP. At each node, RSVP attempts to make a resource reservation for the stream. Once a reservation is setup, RSVP is also responsible for maintaining the requested level of service.

Reservation requests do not need to travel all the way to the server. Instead, each reservation request travels upstream until it meets another reservation request for the same data stream, then merges with that reservation. This reservation merging is the primary advantage of RSVP: scalability—a large number of clients can be added to a multicast without increasing the data traffic significantly. RSVP easily scales to large multicast groups; the average protocol overhead decreases as the number of participants increases.

RSVP supports both multicast and unicast, and adapts to changing memberships and routes. Designed to utilize the robustness of current Internet routing algorithms, RSVP uses underlying routing protocols to determine where it should carry reservation requests. As routing changes paths to adapt to network changes, RSVP adapts its reservation to the new paths.

ISMA

The Internet Streaming Media Alliance (ISMA) is a nonprofit industry alliance founded by Apple Computer, Cisco Systems, IBM, Kasenna, Philips and Sun Microsystems. Since its inception, it has received wide industry support. The mission is to facilitate and promote the adoption of an open architecture for streaming audio and video over the Internet.

The ISMA v1.0 specification provides tools to stream audio and video over networks at up to 1.5 Mbps. It uses MPEG-4 audio/video compression and IETF protocols (RTP, RTSP and SDP) for content transport and control.

ISMA v1.0 defines two hierarchical profiles: Profile 0 and Profile 1. Profile 1 supports all the tools supported by Profile 0, along with some additional tools.

Profile 0

Profile 0 is aimed at streaming audio and video over wireless and narrowband networks to devices with limited audio and video capabilities.

Video uses MPEG-4 Part 2 SP@L1 (QCIF, 176×144. Audio uses MPEG-4 HQ@L2. Up to two channels of audio are supported, with a sampling rate up to 48 kHz. It has a maximum total bit rate of 128 kbps.

Profile 1

Profile 1 is aimed at streaming audio and video over broadband networks to provide the user with a better viewing experience.

Video uses MPEG-4 Part 2 ASP@L3 (CIF, 352 × 288). Audio uses MPEG-4 HQ@2. Up to two channels of audio are supported, with a sampling rate up to 48 kHz. It has a maximum total bit rate of 1.5 Mbps.

Broadcast over IP

ARIB and DVB transport streams may also be transmitted over a broadband IP network. The transport stream packets are encapsulated in RTP packets and sent via IP multicast to receivers.

For DVB, this is called DVB-IPI or Digital Video Broadcasting - Internet Protocol Infrastructure. Don't confuse DVD-IPI with DVB-IP, which enables IP services over DVB.

Conditional Access (DRM)

For broadband IP networks, conditional access is commonly called DRM (Digital Rights Management).

DRM solutions used in early IPTV deployments are similar in principle to DVB Simulcrypt. The MPEG decoder chip contained an embedded (and usually modified) AES or 3DES decryption block; an ISO 7816 "smartcard" provided decryption key information to the AES/3DES descrambling circuitry.

Newer DRM solutions do not use a "smartcard." Software securely running inside the MPEG decoder chip replaces the "smartcard," lowering cost and providing a more secure solution. The software DRM solutions also typically include the ability to control the usage and re-distribution of the content after it has been initially received and decrypted. Capabilities of the DRM can include:

Turning analog and/or digital video copy protection on/off

Limiting the resolution of the analog video outputs (constrained image)

Limiting the sample rate and size of the digital audio outputs

Disabling analog and/or digital audio and/or video outputs

References

1. ETSI TR 102 033, *Digital Video Broadcasting (DVB); Architectural Framework for the Delivery of DVB-services Over IP-based Networks*, April 2002.
2. ETSI TS 102 813, *Digital Video Broadcasting (DVB); IEEE 1394 Home Network Segment*, November 2002.
3. ETSI TS 102 814, *Digital Video Broadcasting (DVB); Ethernet Home Network Segment*, April 2003.
4. IETF RFC 2205, *Resource ReSerVation Protocol (RSVP)—Version 1 Functional Specification*, September 1997.
5. IETF RFC 2206, *RSVP Management Information Base using SMIv2*, September 1997.
6. IETF RFC 2207, *RSVP Extensions for IPSEC Data Flows*, September 1997.
7. IETF RFC 2208, *Resource ReSerVation Protocol (RSVP) Version 1 Applicability Statement Some Guidelines on Deployment*, September 1997.
8. IETF RFC 2209, *Resource ReSerVation Protocol (RSVP)—Version 1 Message Processing Rules*, September 1997.
9. IETF RFC 2250, *RTP Payload Format for MPEG1/MPEG2 Video*, January 1998.

10. IETF RFC 2326, *Real Time Streaming Protocol (RTSP)*, April 1998.
11. IETF RFC 2327, *SDP: Session Description Protocol*, April 1998.
12. IETF RFC 2343, *RTP Payload Format for Bundled MPEG*, May 1998.
13. IETF RFC 2750, *RSVP Extensions for Policy Control*, January 2000.
14. IETF RFC 3016, *RTP Payload Format for MPEG-4 Audio/Visual Streams*, November 2000.
15. *RTP Payload Format for MPEG-4 Streams*, IETF Internet Draft, April 2001.
16. IETF RFC 3266, Support for IPv6 in Session Description Protocol (SDP), June 2002.
17. IETF RFC 3550, *RTP: A Transport Protocol for Real-Time Applications*, July 2003.
18. IETF RFC 3551, *RTP Profile for Audio and Video Conferences with Minimal Control*, July 2003.
19. IETF RFC 3640, *RTP Payload Format for Transport of MPEG-4 Elementary Streams*, November 2003.

Chapter 20

Glossary

8-VSB	See Vestigial Sideband.
AC-3	An early name for Dolby® Digital.
AC'97, AC'98	These are definitions by Intel for the audio I/O implementation for PCs. Two chips are defined: an analog audio I/O chip and a digital controller chip. The digital chip will eventually be replaced by a software solution. The goal is to increase the audio performance of PCs and lower cost.
AC Coupled	AC coupling passes a signal through a capacitor to remove any DC offset, or the overall voltage level that the video signal "rides". One way to find the signal is to remove the DC offset by AC coupling, and then do DC restoration to add a known DC offset (one that we selected). Another reason AC coupling is important is that it can remove large (and harmful) DC offsets.
Active Video	The part of the video waveform that contains picture information. Most of the active video, if not all of it, is visible on the display.
A/D, ADC	Analog-to-Digital Converter. This device is used to digitize audio and video. An ADC for digitizing video must be capable of sampling at 10 to 150 million samples per second (MSPS).
AFC	See Automatic Frequency Control.
AGC	See Automatic Gain Control.
Alpha	See Alpha Channel and Alpha Mix.

Alpha Channel	The alpha channel is used to specify an alpha value for each video sample. The alpha value is used to control the blending, on a sample-by-sample basis, of two images.

$$\text{new sample} = (\text{alpha})\,(\text{sample A color}) + (1 - \text{alpha})\,(\text{sample B color})$$

	Alpha typically has a normalized value of 0 to 1. In a graphics environment, the alpha values can be stored in additional memory. When you hear about 32-bit frame buffers, what this really means is that there are 24 bits of color, eight each for red, green, and blue, along with an 8-bit alpha channel. Also see Alpha Mix.
Alpha Mix	This is a way of combining two images. How the mixing is performed is specified by the alpha channel. The little box that appears over the left-hand shoulder of a news anchor is put there by an alpha mixer. Wherever the little box is to appear, a "1" is put in the alpha channel. Wherever it doesn't appear, a "0" is used. When the alpha mixer sees a "1" coming from the alpha channel, it displays the little box. Whenever it sees a "0," it displays the news anchor. Of course, it doesn't matter if a "1" or a "0" is used, but you get the point.
AM	See Amplitude Modulation.
Amplitude Modulation	A method of encoding data onto a carrier, such that the amplitude of the carrier is proportional to the data value.
Ancillary Timecode	ITU-R BT.1366 defines how to transfer VITC and LTC as ancillary data in digital component interfaces.
Anti-Alias Filter	A lowpass filter used to bandwidth-limit a signal to less than one-half the sampling rate.
Aperture Delay	Aperture delay is the time from an edge of the input clock of the ADC until the time the ADC actually takes the sample. The smaller this number, the better.
Aperture Jitter	The uncertainty in the aperture delay. This means the aperture delay time changes a little bit each time, and that little bit of change is the aperture jitter.
ARIB	Short for Association of Radio Industries and Businesses, a standards organization in Japan. The ARIB provides several specifications that form the core of the ISDB digital television system used in Japan.
ARIB STD-B10	Japan ISDB-S and ISDB-T digital television service information specification.

ARIB STD-B20	Japan ISDB-S (satellite) digital television system specification.
ARIB STD-B21	Japan ISDB-S and ISDB-T digital television receiver specification.
ARIB STD-B24	Japan ISDB-S and ISDB-T digital television data broadcasting specification.
ARIB STD-B25	Japan ISDB-S and ISDB-T digital television access control specification.
ARIB STD-B31	Japan ISDB-T (terrestrial) digital television system specification.
ARIB STD-B32	Japan ISDB-S and ISDB-T digital television video coding, audio coding and multiplexing specification.
ARIB STD-B38	Japan ISDB-S and ISDB-T digital television home server specification.
Artifacts	In the video domain, artifacts are blemishes, noise, snow, spots, whatever. When you have an image artifact, something is wrong with the picture from a visual standpoint. Don't confuse this term with not having the display properly adjusted. For example, if the hue control is set wrong, the picture will look bad, but this is not an artifact. An artifact is some physical disruption of the image.
Aspect Ratio	The ratio of the width of the picture to the height. Displays commonly have a 4:3 or 16:9 aspect ratio. Program material may have other aspect ratios (such as 2.35:1), resulting in it being "letterboxed" on the display.
Asynchronous	Refers to circuitry without a common clock or timing signal.
ATC	See Ancillary Timecode.
ATSC	Advanced Television Systems Committee. They defined the HDTV standards for the United States. Other countries are also adopting the ATSC HDTV standard.
ATSC A/49	Defines the ghost cancellation reference signal for NTSC.
ATSC A/52	Defines the Dolby® Digital audio compression for ATSC HDTV.

ATSC A/53, ATSC A/54	Defines ATSC HDTV for the United States.
ATSC A/57	Defines a means to uniquely identify content for ATSC HDTV.
ATSC A/64	Defines the transmission measurement and compliance for ATSC HDTV.
ATSC A/65, ATSC A/69	Defines the program and system information protocol (PSIP) for ATSC HDTV.
ATSC A/70	Defines a standard for the conditional access system for ATSC HDTV.
ATSC A/80	Defines a standard for modulation and coding of ATSC data delivered over satellite for digital television contribution and distribution applications.
ATSC A/81	Describes the transmission system for ATSC Direct-to-Home (DTH) satellite broadcast system.
ATSC A/90, ATSC A/91	Defines the data broadcast standard for ATSC.
ATSC A/92	Defines the delivery of Internet Protocol (IP) multicast sessions and usage of the ATSC A/90 data broadcast standard for IP multicast.
ATSC A/93	Defines the transmission of synchronized data elements, and synchronized and asynchronous events.
ATSC A/94	Defines an Application Reference Model (ARM) including a binding of application environment facilities onto the ATSC A/90 data broadcast standard.
ATSC A/95	Defines the ATSC Transport Stream File System (TSFS) standard for delivery of hierarchical name-spaces, directories and files. It builds on the ATSC A/90 data service delivery scheme.
ATSC A/96	Defines a core suite of protocols to enable remote interactivity in ATSC television environments.
ATSC A/100	This DTV Application Software Environment (DASE) defines a software layer (middleware) that allows programming content and applications to run on a common ATSC receiver.

Audio Modulation	Refers to modifying an audio subcarrier with audio information so that it may be mixed with the video information and transmitted.
Audio Subcarrier	A specific frequency that is modulated with audio data.
Automatic Frequency Control (AFC)	A technique to lock onto and track a desired frequency.
Automatic Gain Control (AGC)	A circuit that has a constant output amplitude, regardless of the input amplitude.
AVC	Short for Advanced Video Codec, an early name for the H.264 video codec.
Back Porch	The portion of the video waveform between the trailing edge of the horizontal sync and the start of active video.
Bandpass Filter	A circuit that allows only a selected range of frequencies to pass through.
Bandwidth (BW)	The range of frequencies a circuit will respond to or pass through. It may also be the difference between the highest and lowest frequencies of a signal.
Bandwidth Segmented Orthogonal Frequency Division Multiplexing	BST-OFDM attempts to improve on COFDM by modulating some OFDM carriers differently from others within the same multiplex. A given transmission channel may therefore be "segmented," with different segments being modulated differently.
Baseband	When applied to audio and video, baseband means an audio or video signal that is not modulated onto another carrier (such as RF modulated to channel 3 or 4 for example). In DTV, baseband also may refer to the basic (unmodulated) MPEG-2 program or system stream.
BBC	British Broadcasting Corporation.
BITC	Burned-In Time Code. The timecode information is displayed within a portion of the picture, and may be viewed on any monitor or TV.

Black Burst   Black burst is a composite video signal with a totally black picture. It is used to synchronize video equipment so the video outputs are aligned. Black burst tells the video equipment the vertical sync, horizontal sync, and the chroma burst timing.

Black Level   This level represents the darkest an image can get, defining what black is for a particular video system. If for some reason the video goes below this level, it is referred to as blacker-than-black. You could say that sync is blacker-than-black.

Blanking   On a CRT display, the scan line moves from the left edge to the right edge, jumps back to the left edge, and starts out all over again, on down the screen. When the scan line hits the right side and is about to be brought back to the left side, the video signal is blanked so that you can't "see" the return path of the scan beam from the right to the left-hand edge. To blank the video signal, the video level is brought down to the blanking level, which is below the black level if a pedestal is used.

Blanking Level   That level of the video waveform defined by the system to be where blanking occurs. This could be the black level if a pedestal is not used or below the black level if a pedestal is used.

Blooming   This is an effect, sometimes caused when video becomes whiter-than-white, in which a line that is supposed to be nice and thin becomes fat and fuzzy on the screen.

Breezeway   That portion of the video waveform between the trailing edge of horizontal sync and the start of color burst.

Brightness   This refers to how much light is emitted by the display, and is controlled by the intensity of the video level.

BS.707   This ITU recommendation specifies the stereo audio specifications (Zweiton and NICAM 728) for the PAL and SECAM video standards.

BST-OFDM   Short for Bandwidth Segmented Orthogonal Frequency Division Multiplexing.

BT.470   This ITU recommendation specifies the various NTSC, PAL, and SECAM video standards used around the world. SMPTE 170M also specifies the (M) NTSC video standard used in the United States. BT.470 has replaced BT.624.

BT.601 This ITU recommendation specifies the 720×480 (59.94 Hz), 960×480 (59.94 Hz), 720×576 (50 Hz), and 960×576 (50 Hz) 4:2:2 YCbCr interlaced standards.

BT.653 This ITU recommendation defines the various teletext standards used around the world. Systems A, B, C, and D for both 525-line and 625-line TV systems are defined.

BT.656 This ITU recommendation defines a parallel interface (8-bit or 10-bit, 27 MHz) and a serial interface (270 Mbps) for the transmission of 4:3 BT.601 4:2:2 YCbCr digital video between pro-video equipment. Also see SMPTE 125M.

BT.709 This ITU recommendation specifies the 1920×1080 RGB and 4:2:2 YCbCr interlaced and progressive 16:9 digital video standards. Frame refresh rates of 60, 59.94, 50, 30, 29.97, 25, 24, and 23.976 Hz are supported.

BT.799 This ITU recommendation defines the transmission of 4:3 BT.601 4:4:4:4 YCbCrK and RGBK digital video between pro-video equipment. Two parallel interfaces (8-bit or 10-bit, 27 MHz) or two serial interfaces (270 Mbps) are used.

BT.809 This ITU recommendation defines Programme Delivery Control (PDC) system for video recording of PAL broadcasts.

BT.1119 This ITU recommendation defines the widescreen signaling (WSS) information for NTSC and PAL video signals. For (B, D, G, H, I) PAL systems, WSS may be present on line 23, and on lines 22 and 285 for (M) NTSC.

BT.1124 This ITU recommendation defines the ghost cancellation reference (GCR) signal for NTSC and PAL.

BT.1197 This ITU recommendation defines the PALplus standard, allowing the transmission of 16:9 programs over normal PAL transmission systems.

BT.1302 This ITU recommendation defines the transmission of 16:9 BT.601 4:2:2 YCbCr digital video between pro-video equipment. It defines a parallel interface (8-bit or 10-bit, 36 MHz) and a serial interface (360 Mbps).

BT.1303 This ITU recommendation defines the transmission of 16:9 BT.601 4:4:4:4 YCbCrK and RGBK digital video between pro-video equipment. Two parallel interfaces (8-bit or 10-bit, 36 MHz) or two serial interfaces (360 Mbps) are used.

BT.1304   This ITU recommendation specifies the checksum for error detection and status for pro-video digital interfaces.

BT.1305   This ITU recommendation specifies the digital audio format for ancillary data for pro-video digital interfaces. Also see SMPTE 272M.

BT.1358   This ITU recommendation defines the 720 × 480 (59.94 Hz) and 720 × 576 (50 Hz) 4:2:2 YCbCr pro-video progressive standards. Also see SMPTE 293M.

BT.1362   This ITU recommendation defines the pro-video serial interface for the transmission of BT.1358 digital video between equipment. Two 270 Mbps serial interfaces are used.

BT.1364   This ITU recommendation specifies the ancillary data packet format for pro-video digital interfaces. Also see SMPTE 291M.

BT.1365   This ITU recommendation specifies the 24-bit digital audio format for pro-video HDTV serial interfaces. Also see SMPTE 299M.

BT.1366   This ITU recommendation specifies the transmission of timecode as ancillary data for pro-video digital interfaces. Also see SMPTE 266M.

BT.1381   This ITU recommendation specifies a serial digital interface (SDI) based transport interface for compressed television signals in networked television production based on BT.656 and BT.1302.

BT.1618   This ITU recommendation specifies a data structure for DV-based audio, data and compressed video at data rates of 25 and 50 Mbps. Also see SMPTE 314M.

BT.1620   This ITU recommendation specifies a data structure for DV-based audio, data and compressed video at data rates of 100 Mbps. Also see SMPTE 370M.

BTSC   This EIA TVSB5 standard defines a technique of implementing stereo audio for NTSC video. One FM subcarrier transmits an L+R signal, and an AM subcarrier transmits an L–R signal.

Burst   See Color Burst.

Burst Gate	This is a signal that tells a video decoder where the color burst is located within the scan line.
B–Y	The blue-minus-luma signal, also called a color difference signal. When added to the luma (Y) signal, it produces the blue video signal.
Carrier	A frequency that is modulated with data to be transmitted.
CATV	Community antenna television, now generally meaning cable TV.
CBC	Canadian Broadcasting Corporation.
CBR	Abbreviation for constant bit rate.
CCIR	Comite Consultatif International des Radiocommunications or International Radio Consultative Committee. The CCIR no longer exists—it has been absorbed into the parent body, the ITU. For a given "CCIR xxx" specification, see "BT.xxx."
CGMS-A	Copy Generation Management System - Analog. See EIA-608.
Chaoji VideoCD	Another name for Super VideoCD.
Checksum	An error-detecting scheme which is the sum of the data values transmitted. The receiver computes the sum of the received data values and compares it to the transmitted sum. If they are equal, the transmission was error-free.
Chroma	The NTSC, PAL, or SECAM video signal contains two parts that make up what you see on the display: the intensity part, and the color part. Chroma is the color part.
Chroma Bandpass	In a NTSC or PAL video signal, the luma (black and white) and the chroma (color) information are combined together. If you want to decode an NTSC or PAL video signal, the luma and chroma must be separated. A chroma bandpass filter removes the luma from the video signal, leaving the chroma relatively intact. This works reasonably well except in images where the luma and chroma information overlap, meaning that we have luma and chroma stuff at the same frequency. The filter can't tell the difference between the two and passes everything. This can make for a funny-looking picture. Next time you're watching TV and someone is wearing a herringbone jacket or a shirt with thin, closely spaced stripes, take a good look. You may see a rainbow color effect moving through that area. What's happening is that the video

decoder thinks that the luma is chroma. Since the luma isn't chroma, the video decoder can't figure out what color it is and it shows up as a rainbow pattern. This problem can be overcome by using a comb filter.

Chroma Burst

See Color Burst.

Chroma Demodulator

After the NTSC or PAL video signal makes its way through the Y/C separator, the colors must be decoded. That's what a chroma demodulator does. It takes the chroma output of the Y/C separator and recovers two color difference signals (typically I and Q or U and V). Now, with the luma information and two color difference signals, the video system can figure out what colors to display.

Chroma Key

This is a method of combining two video images. An example of chroma keying in action is the nightly news person standing in front of a giant weather map. In actuality, the person is standing in front of a blue or green background and their image is mixed with a computer-generated weather map. This is how it works: a TV camera is pointed at the person and fed along with the image of the weather map into a box. Inside the box, a decision is made. Wherever it sees the blue or green background, it displays the weather map. Otherwise, it shows the person. So, whenever the person moves around, the box figures out where he is, and displays the appropriate image.

Chroma Trap

In a NTSC or PAL video signal, the luma (black and white) and the chroma (color) information are combined together. If you want to decode the video signal, the luma and chroma must be separated. The chroma trap is one method for separating the chroma from the luma, leaving the luma relatively intact. How does it work? The NTSC or PAL signal is fed to a trap filter. For all practical purposes, a trap filter allows certain frequencies to pass through, but not others. The trap filter is designed with a response to remove the chroma so that the output of the filter only contains the luma. Since this trap stops chroma, it's called a chroma trap. The sad part about all of this is that not only does the filter remove chroma, it removes luma as well if it exists within the frequencies where the trap exists. The filter only knows ranges and, depending on the image, the luma information may overlap the chroma information. The filter can't tell the difference between the luma and chroma, so it traps both when they are in the same range. What's the big deal? Well, you lose luma and this means that the picture is degraded somewhat. Using a comb filter for a Y/C separator is better than a chroma trap or chroma bandpass.

Chrominance

In video, the terms chrominance and chroma are commonly (and incorrectly) interchanged. See the definition of Chroma.

CIF
Common Interface Format or Common Image Format. The Common Interface Format was developed to support video conferencing. It has an active resolution of 352 × 288 and a refresh rate of 29.97 frames per second. The High-Definition Common Image Format (HD-CIF) is used for HDTV production and distribution, having an active resolution of 1920 × 1080 with a frame refresh rate of 23.976, 24, 29.97, 30, 50, 59.94, or 60 Hz.

Clamp
This is basically another name for the DC-restoration circuit. It can also refer to a switch used within the DC-restoration circuit. When it means DC restoration, then it's usually used as "clamping." When it's the switch, then it's just "clamp."

Clipping Logic
A circuit used to prevent illegal conversion. Some colors can exist in one color space but not in another. Right after the conversion from one color space to another, a color space converter might check for illegal colors. If any appear, the clipping logic is used to limit, or clip, part of the information until a legal color can be represented. Since this circuit clips off some information and is built using logic, it's not too hard to see how the name "clipping logic" was developed.

Closed Captioning
A service which decodes text information transmitted with the video signal and displays it on the display. For NTSC, the caption signal may be present on lines 21 and 284. For PAL, the caption signal may be present on lines 22 and 334. See the EIA-608 specification for (M) NTSC usage of closed captioning and the EIA-708 specification for DTV support.

For MPEG-2 video, including ATSC and DVB, the closed caption data are multiplexed as a separate data stream within the MPEG-2 bitstream. It may use the picture layer user_data bits as specified by EIA-708, or in PES packets (private_stream_1) as specified by ETSI EN 301 775.

For DVD, caption data may be 8-bit user_data in the group_of_pictures header (525/60 systems), a digitized caption signal (quantized to 16 levels) that is processed as normal video data (625/50 systems), or a subpicture that is simply decoded and mixed with the decoded video.

Closed Subtitles
See subtitles.

CMYK
This is a color space primarily used in color printing. CMYK is an acronym for Cyan, Magenta, Yellow, and blacK. The CMYK color space is subtractive, meaning that cyan, magenta, yellow and black pigments or inks are applied to a white surface to remove color information from the white surface to create the final color. The reason black is used is because even if a printer could print hues of cyan, magenta, and yellow inks perfectly enough to make black

(which it can't for large areas), it would be too expensive since colored inks cost more than black inks. So, when black is used, instead of putting down a lot of CMY, they just use black.

Coded Orthogonal Frequency Division Multiplexing	Coded orthogonal frequency division multiplexing, or COFDM, transmits digital data differently than 8-VSB or other single-carrier approaches. *Frequency division multiplexing* means that the data to be transmitted is distributed over many carriers (1705 or 6817 for DVB-T), as opposed to modulating a single carrier. Thus, the data rate on each COFDM carrier is much lower than that required of a single carrier. The COFDM carriers are *orthogonal,* or mutually perpendicular, and forward error correction ("*coded*") is used.

COFDM is a multiplexing technique rather than a modulation technique. One of any of the common modulation methods, such as QPSK, 16-QAM or 64-QAM, is used to modulate the COFDM carriers.

COFDM	See Coded Orthogonal Frequency Division Multiplexing.
Color Bars	This is a test pattern used to check whether a video system is calibrated correctly. A video system is calibrated correctly if the colors are the correct brightness, hue, and saturation. This can be checked with a vectorscope.
Color Burst	A waveform of a specific frequency and amplitude that is positioned between the trailing edge of horizontal sync and the start of active video. The color burst tells the color decoder how to decode the color information contained in that line of active video. By looking at the color burst, the decoder can determine what's blue, orange, or magenta. Essentially, the decoder figures out what the correct color is.
Color Decoder	See Chroma Demodulator.
Color Demodulator	See Chroma Demodulator.
Color Difference	All of the color spaces used in color video require three components. These might be RGB, YIQ, YUV or Y(R–Y)(B–Y). In the Y(R–Y)(B–Y) color space, the R–Y and B–Y components are often referred to as color difference signals for obvious reasons. They are made by subtracting the luma (Y) from the red and blue components. I and Q and U and V are also color difference signals since they are scaled versions of R–Y and B–Y. All the Ys in each of the YIQ, YUV and Y(R–Y)(B–Y) are basically the same, although they are slightly different between SDTV and HDTV.

Color Edging	Extraneous colors that appear along the edges of objects, but don't have a color relationship to those areas.
Color Encoder	The color encoder does the exact opposite of the color decoder. It takes two color difference signals, such as I and Q or U and V, and combines them into a chroma signal.
Color Key	This is essentially the same thing as Chroma Key.
Color Killer	A color killer is a circuit that shuts off the color decoding if the incoming video does not contain color information. How does this work? The color killer looks for the color burst and if it can't find it, it shuts off the color decoding. For example, let's say that a color TV is going to receive material recorded in black and white. Since the black and white signal does not contain a color burst, the color decoding is shut off. Why is a color killer used? Well, in the old days, the color decoder would still generate a tiny little bit of color if a black and white transmission was received, due to small errors in the color decoder, causing a black and white program to have faint color spots throughout the picture.
Color Modulator	See Color Encoder.
Color Purity	This term is used to describe how close a color is to the theoretical. For example, in the Y'UV color space, color purity is specified as a percentage of saturation and $\pm q$, where q is an angle in degrees, and both quantities are referenced to the color of interest. The smaller the numbers, the closer the actual color is to the color that it's really supposed to be. For a studio-grade device, the saturation is $\pm 2\%$ and the hue is ± 2 degrees. On a vectorscope, if you're in that range, you're studio quality.
Color Space	A color space is a mathematical representation for a color. No matter what color space is used—RGB, YIQ, YUV, etc.—orange is still orange. What changes is how you represent orange. For example, the RGB color space is based on a Cartesian coordinate system and the HSI color space is based on a polar coordinate system.
ColorStream, ColorStream Pro, ColorStreamHD	The name Toshiba uses for the analog YPbPr video interface on their consumer equipment. If the interface supports progressive SDTV resolutions, it is called ColorStream Pro. If the interface supports HDTV resolutions, it is called ColorStream HD.

Color Subcarrier	The color subcarrier is a signal used to control the color encoder or color decoder. For (M) NTSC, the frequency of the color subcarrier is about 3.58 MHz and for (B, D, G, H, I) PAL it's about 4.43 MHz. In the color encoder, a portion of the color subcarrier is used to create the color burst, while in the color decoder, the color burst is used to reconstruct a color subcarrier.
Color Temperature	Color temperature is measured in degrees Kelvin. If a TV has a color temperature of 8,000 degrees Kelvin, that means the whites have the same shade as a piece of pure carbon heated to that temperature. Low color temperatures have a shift towards red; high color temperatures have a shift towards blue.
	The standard for video is 6,500 degrees Kelvin. Thus, professional TV monitors use a 6,500-degree color temperature. However, most consumer TVs have a color temperature of 8,000 degrees Kelvin or higher, resulting in a bluish cast. By adjusting the color temperature of the TV, more accurate colors are produced, at the expense of picture brightness.
Comb Filter	This is another method of performing Y/C separation. A comb filter is used in place of a chroma bandpass or chroma trap. The comb filter provides better video quality since it does a better job of separating the luma from chroma. It reduces the amount of creepy-crawlies or zipper artifacts. It's called a comb filter because the frequency response looks like a comb. The important thing to remember is that the comb filter is a better method for Y/C separation than chroma bandpass or chroma trap.
Common Image Format	See CIF.
Common Interface Format	See CIF.
Component Video	Video using three separate color components, such as digital Y'CbCr, analog Y'PbPr, or R'G'B'.
Composite Video	A single video signal that contains brightness, color, and timing information. If a video system is to receive video correctly, it must have several pieces of the puzzle in place. It must have the picture that is to be displayed on the screen, and it must be displayed with the correct colors. This piece is called the active video. The video system also needs information that tells it where to put each pixel. This is called sync. The display needs to know when to shut off the electron beam so the viewer can't see the spot retrace across the CRT display.

This piece of the video puzzle is called blanking. Now, each piece could be sent in parallel over three separate connections, and it would still be called video and would still look good on the screen. This is a waste, though, because all three pieces can be combined together so that only one connection is needed. Composite video is a video stream that combines all of the pieces required for displaying an image into one signal, thus requiring only one connection. NTSC and PAL are examples of composite video. Both are made up of active video, horizontal sync, horizontal blanking, vertical sync, vertical blanking, and color burst. RGB is not an example of composite video, even though each red, green, and blue signal may contain sync and blanking information, because all three signals are required to display the picture with the right colors.

Compression Ratio

Compression ratio is a number used to tell how much information is squeezed out of an image when it has been compressed. For example, suppose we start with a 1 MB image and compress it down to 128 kB. The compression ratio would be:

$$1,048,576 / 131,072 = 8$$

This represents a compression ratio of 8:1; 1/8 of the original amount of storage is now required. For a given compression technique—MPEG, for example—the higher the compression ratio, the worse the image looks. This has nothing to do with which compression method is better, for example JPEG vs. MPEG. A video stream that is compressed using MPEG at 100:1 may look better than the same video stream compressed to 100:1 using JPEG.

Conditional Access

This is a technology by which service providers enable subscribers to decode and view content. It consists of key decryption (using a key obtained from changing coded keys periodically sent with the content) and descrambling. The decryption may be proprietary (such as Canal+, DigiCipher, Irdeto Access, Nagravision, NDS, Viaccess, etc.) or standardized, such as the DVB common scrambling algorithm and OpenCable™. Conditional access may be thought of as a simple form of digital rights management.

Two common DVB conditional access (CA) techniques are SimulCrypt and MultiCrypt. With SimulCrypt, a single transport stream can contain several CA systems. This enables receivers with different CA systems to receive and correctly decode the same video and audio streams. With MultiCrypt, a receiver permits the user to manually switch between CA systems. Thus, when the viewer is presented with a CA system which is not installed in his receiver, they simply switch CA cards.

Constant Bit Rate	Constant bit rate (CBR) means that a bitstream (compressed or uncompressed) has the same number of bits each second.
Contouring	This is an image artifact caused by not having enough bits to represent the image. The reason the effect is called "contouring" is because the image develops vertical bands of brightness.
Contrast	A video term referring to how far the whitest whites are from the blackest blacks in a video waveform. If the peak white is far away from the peak black, the image is said to have high contrast. With high contrast, the image is very stark and very "contrasty," like a black-and-white tile floor. If the two are very close to each other, the image is said to have poor, or low, contrast. With poor contrast, an image may be referred to as being "washed out"—you can't tell the difference between white and black, and the image looks gray.
Creepy-crawlies	Yes, this is a real video term! Creepy-crawlies refers to a specific image artifact that is a result of the NTSC system. When the nightly news is on, and a little box containing a picture appears over the anchorperson's shoulder, or when some computer-generated text shows up on top of the video clip being shown, get up close to the TV and check it out. Along the edges of the box, or along the edges of the text, you'll notice some jaggies "rolling" up (or down) the picture. That's the creepy-crawlies. Some people refer to this as zipper because it looks like one.
Cross Color	This occurs when the video decoder incorrectly interprets high-frequency luma information (brightness) to be chroma information (color), resulting in color being displayed where it shouldn't.
Cross Luma	This occurs when the video decoder incorrectly interprets chroma information (color) to be high-frequency luma information (brightness).
Cross Modulation	A condition when one signal erroneously modulates another signal.
Crosstalk	Interference from one signal that is detected on another.
CVBS	Abbreviation for "Composite Video Baseband Signal" or "Composite Video, Blanking, and Synchronization".
D/A, DAC	These are short for Digital-to-Analog Converter.

DAVIC

Abbreviation for Digital Audio Visual Council. Its goal was to create an industry standard for the end-to-end interoperability of broadcast and interactive digital audio-visual information, and of multimedia communication. The specification is now ISO/IEC 16500 (normative part) and ITR 16501 (informative part).

dB

Abbreviation for decibels, a standard unit for expressing relative power, voltage, or current.

dBm

Measure of power in communications. 0 dBm = 1 mW, with a logarithmic relationship as the values increase or decrease. In a 50-ohm system, 0 dBm = 0.223 volts.

dBw

Decibels referenced to 1 watt.

DC Restoration

DC restoration is what you have to do to a video signal after it has been AC-coupled and has to be digitized. Since the video waveform has been AC-coupled, we no longer know absolutely where it is. For example, is the bottom of the sync tip at –5v or at 1v? In fact, not only don't we know where it is, it also changes over time, since the average voltage level of the active video changes over time. Since the ADC requires a known input level and range to work properly, the video signal needs to be referenced to a known DC level. DC restoration essentially adds a known DC level to an AC-coupled signal. In decoding video, the DC level used for DC restoration is usually such that when the sync tip is digitized, it will be generate the number 0.

DCT

This is short for Discrete Cosine Transform, used in the MPEG, H.261, and H.263 video compression algorithms.

Decibel

One-tenth of a Bel, used to define the ratio of two powers, voltages, or currents, in terms of gains or losses. It is 10× the log of the power ratio and 20× the voltage or current ratio.

Decimation

When a video signal is digitized so that 100 samples are produced, but only every other one is stored or used, the signal is decimated by a factor of 2:1. The image is now 1/4 of its original size, since 3/4 of the data is missing. If only one out of five samples were used, then the image would be decimated by a factor of 5:1, and the image would be 1/25 its original size. Decimation, then, is a quick-and-easy method for image scaling.

Decimation can be performed in several ways. One way is the method just described, where data is literally thrown away. Even though this technique is easy to implement and cheap, it introduces aliasing artifacts. Another method is to use a decimation filter, which reduces the aliasing artifacts, but is more costly to implement.

Decimation Filter

A decimation filter is a lowpass filter designed to provide decimation without the aliasing artifacts associated with simply throwing data away.

De-emphasis

Also referred to as post-emphasis and post-equalization. De-emphasis performs a frequency-response characteristic that is complementary to that introduced by pre-emphasis.

De-emphasis Network

A circuit used to restore a frequency response to its original form.

Demodulation

The process of recovering an original signal from a modulated carrier.

Demodulator

In NTSC and PAL video, demodulation is the technique used to recover the color difference signals. See the definitions for Chroma Demodulator and Color Decoder; these are two other names for the demodulator used in NTSC/PAL video applications. Demodulation is also used after DTV tuners to convert the transmitted DTV signal to a baseband MPEG-2 stream.

Differential Gain

Differential gain is how much the color saturation changes when the luma level changes (it isn't supposed to). For a video system, the better the differential gain—that is, the smaller the number specified—the better the system is at figuring out the correct color.

Differential Phase

Differential phase is how much the hue changes when the luma level changes (it isn't supposed to). For a video system, the better the differential phase—that is, the smaller the number specified—the better the system is at figuring out the correct color.

Digital 8

Digital 8 compresses video using standard DV compression, but records it in a manner that allows it to use standard Hi-8 tape. The result is a DV "box" that can also play standard Hi-8 and 8 mm tapes. On playback, analog tapes are converted to a 25 Mbps compressed signal available via the i-Link digital output interface. Playback from analog tapes has limited video quality. New recordings are digital and identical in performance to DV; audio specs and other data also are the same.

Digital Component Video	Digital video using three separate color components, such as YCbCr or RGB.
Digital Composite Video	Digital video that is essentially the digitized waveform of NTSC or PAL video signals, with specific digital values assigned to the sync, blank, and white levels.
Digital Rights Management (DRM)	Digital Rights Management (DRM) is a generic term for a number of capabilities that allow a content producer or distributor to determine under what conditions their product can be acquired, stored, viewed, copied, loaned, and so on.
Digital Transmission Content Protection (DTCP)	An encryption method (also known as "5C" and "DTCP") developed by Sony, Hitachi, Intel, Matsushita and Toshiba for IP, USB and IEEE 1394 interfaces.
Digital VCR	Digital VCRs are similar to analog VCRs in that tape is still used for storage. Instead of recording an analog audio/video signal, digital VCRs record digital signals, usually using compressed audio/video.
Digital Versatile Disc (DVD)	See DVD–Video and DVD–Audio.
Digital Vertical Interval Timecode	DVITC digitizes the analog VITC waveform to generate 8-bit values. This allows the VITC to be used with digital video systems. For 525-line video systems, it is defined by SMPTE 266M. BT.1366 defines how to transfer VITC and LTC as ancillary data in digital component interfaces.
Digital Video Recorder (DVR)	DVRs can be thought of a digital version of the VCR, with several enhancements. Instead of a tape, the DVR uses an internal hard disk to store compressed audio/video, and has the ability to record and playback at the same time. The main advantage that DVRs have over VCRs is their ability to time shift viewing the program as it is being recorded. This is accomplished by continuing to record the incoming live program, while retrieving the earlier part of the program that was just recorded. The DVR also offers pause, rewind, slow motion, and fast forward control, just as with a VCR.

Discrete Cosine Transform (DCT)	A DCT is just another way to represent an image. Instead of looking at it in the time domain—which, by the way, is how we normally do it—it is viewed in the frequency domain. It's analogous to color spaces, where the color is still the color but is represented differently. Same thing applies here—the image is still the image, but it is represented in a different way. Why do JPEG, MPEG, H.261, and H.263 base part of their compression schemes on the DCT? Because it is more efficient to represent an image that way. In the same way that the YCbCr color space is more efficient than RGB in representing an image, the DCT is even more efficient at image representation.
Discrete Time Oscillator (DTO)	A discrete time oscillator is a digital version of the voltage-controlled oscillator.
dNTSC™	Technique developed by Dotcast for data broadcasting within the NTSC video signal. It supports up to 4.5Mbps per analog TV channel.
Dolby® Digital	An audio compression technique developed by Dolby®. It is a multi-channel surround sound format used in DVD and HDTV.
Dot Pitch	The distance between screen pixels measured in millimeters. The smaller the number, the better the horizontal resolution.
Double Buffering	As the name implies, you are using two buffers—for video, this means two frame buffers. While buffer 1 is being read, buffer 2 is being written to. When finished, buffer 2 is read out while buffer 1 is being written to.
Downconverter	A circuit used to change a high-frequency signal to a lower frequency.
Downlink	The frequency satellites use to transmit data to Earth stations.
DRM	See Digital Rights Management.
Drop Field Scrambling	This method is identical to the sync suppression technique for scrambling analog TV channels, except there is no suppression of the horizontal blanking intervals. Sync pulse suppression only takes place during the vertical blanking interval. The descrambling pulses still go out for the horizontal blanking intervals (to fool unauthorized descrambling devices). If a descrambling device is triggering on descrambling pulses only, and does not know that the scrambler is using the drop field scrambling technique, it will try to reinsert the horizontal intervals (which were never suppressed). This is known as double reinsertion, which causes compression of the active video signal. An unauthorized

descrambling device creates a washed-out picture and loss of neutral sync during drop field scrambling.

DTCP　　　　　Short for Digital Transmission Content Protection.

DTS®　　　　　DTS® stands for Digital Theater Systems. It is a multi-channel surround sound format, similar to Dolby® Digital. For DVDs that use DTS® audio, the DVD–Video specification still requires that PCM or Dolby® Digital audio still be present. In this situation, only two channels of Dolby® Digital audio may be present (due to bandwidth limitations).

DTV　　　　　Short for digital television, including SDTV, EDTV, and HDTV.

DV　　　　　Short for Digital Video, the standard used for digital camcorders that record on tape. It is defined by the BT.1618, BT.1620, IEC 61834, SMPTE 314M and 370M specifications.

DVB　　　　　Short for digital video broadcast, a method of transmitting digital audio and video (SDTV or HDTV resolution), based on MPEG-2. There are several variations: DVB-T for terrestrial broadcasting (ETSI EN 300 744), DVB-S for satellite broadcasting (ETSI EN 300 421), and DVB-C for cable broadcasting (ETSI EN 300 429). MPEG, Dolby® Digital and DTS® compressed audio are supported.

DVB-S uses the QPSK modulation system to guard against errors in satellite transmissions caused by reduced signal-to-noise ratio, with channel coding optimized to the error characteristics of the channel. A typical set of parameter values and a 36 MHz transponder gives a useful data rate of around 38 Mbps.

DVB-C uses Quadrature Amplitude Modulation (QAM), which is optimized for maximum data rate since the cable environment is less prone to interference than satellite or terrestrial. Systems from 16-QAM up to 256-QAM can be used, but the system centers on 64-QAM, in which an 8 MHz channel can accommodate a physical payload of about 38 Mbps. The cable return path uses Quadrature Phase Shift Keying (QPSK) modulation in a 200 kHz, 1 MHz, or 2 MHz channel to provide a return path of up to about 3 Mbps. The path to the user may be either in-band (embedded in the MPEG-2 Transport Stream in the DVB-C channel) or out-of-band (on a separate 1 or 2 MHz frequency band).

DVD–Audio　　　　　DVDs that contain linear PCM audio data in any combination of 44.1, 48.0, 88.2, 96.0, 176.4, or 192 kHz sample rates, 16, 20, or 24 bits per sample, and 1 to 6 channels, subject to a maximum bit rate of 9.6 Mbps. With a 176.4 or 192 kHz sample rate, only two channels are allowed.

Meridian Lossless Packing (MLP) is a lossless compression method that has an approximate 2:1 compression ratio. The use of MLP is optional, but the decoding capability is mandatory on all DVD–Audio players.

Dolby® Digital compressed audio is required for any video portion of a DVD–Audio disc.

DVD–Interactive

DVD-Interactive is under development (due summer 2002), and is intended to provide additional capability for users to do interactive operations with content on DVDs or at Web sites on the Internet. It will probably be based on one of three technologies: MPEG-4, Java/HTML, or software from InterActual.

DVD–Video

DVDs that contain about two hours of digital audio, video, and data. The video is compressed and stored using MPEG-2 MP@ML. A variable bit rate is used, with an average of about 4 Mbps (video only), and a peak of 10 Mbps (audio and video). The audio is either linear PCM or Dolby® Digital compressed audio. DTS® compressed audio may also be used as an option.

Linear PCM audio can be sampled at 48 or 96 kHz, 16, 20, or 24 bits per sample, and 1 to 8 channels. The maximum bit rate is 6.144 Mbps, which limits sample rates and bit sizes in some cases.

For Dolby® Digital audio, the bit rate is 64 to 448 kbps, with 384 kbps being the normal rate for 5.1 channels and 192 kbps being the normal rate for stereo. The channel combinations are (front/surround): 1/0, 1+1/0 (dual mono), 2/0, 3/0, 2/1, 3/1, 2/2, and 3/2. The LFE channel (0.1) is optional with all 8 combinations.

For DTS® audio, the bit rate is 64 to 1,536 kbps. The channel combinations are (front/surround): 1/0, 2/0, 3/0, 2/1, 2/2, 3/2. The LFE channel (0.1) is optional with all 6 combinations.

Columbia Tristar Home Entertainment has introduced a Superbit DVD that has an average bit rate of about 7 Mbps (video only) for improved video quality. This is achieved by having minimal "extras" on the DVD.

DVI

Abbreviation for Digital Visual Interface. This is a digital video interface to a display, designed to replace the analog Y'PbPr or R'G'B' interface. For analog displays, the D/A conversion resides in the display. The EIA-861 standard specifies how to include data such as aspect ratio and format information. The VESA EEDID and DI-EXT standards document data structures and mechanisms to communicate data across DVI.

DVI-D is a digital-only connector; a DVI-I connector handles both analog and digital. DVI-A is available as a plug (male) connector only and mates to the analog-only pins of a DVI-I connector. DVI-A is only used in adapter cables, where there is the need to convert to or from a traditional analog VGA signal.

DVITC

See Digital Vertical Interval Timecode.

DVR	See Digital Video Recorder.
Dynamic Range	The weakest to the strongest signal a circuit will accept as input or generate as an output.
EDTV	See Enhanced Definition Television.
EIA	Electronics Industries Alliance.
EIA-516	United States teletext standard, also called NABTS.
EIA-608	United States closed captioning and extended data services (XDS) standard. Revision B adds Copy Generation Management System - Analog (CGMS-A), content advisory (V-chip), Internet Uniform Resource Locators (URLs) using Text-2 (T-2) service, 16-bit Transmission Signal Identifier and transmission of DTV PSIP data.
EIA-708	United States DTV closed captioning standard. EIA CEB-8 also provides guidance on the use and processing of EIA-608 data streams embedded within the ATSC MPEG-2 video elementary transport stream, and augments EIA-708.
EIA-744	NTSC "V-chip" operation. This standard added content advisory filtering capabilities to NTSC video by extending the EIA-608 standard. It is now included in the latest EIA-608 standard, and has been withdrawn.
EIA-761	Specifies how to convert QAM to 8-VSB, with support for OSD (on screen displays).
EIA-762	Specifies how to convert QAM to 8-VSB, with no support for OSD (on screen displays).
EIA-766	United States HDTV content advisory standard.
EIA-770	This specification consists of three parts (EIA-770.1, EIA-770.2, and EIA-770.3). EIA-770.1 and EIA-770.2 define the analog YPbPr video interface for 525-line interlaced and progressive SDTV systems. EIA-770.3 defines the analog YPbPr video interface for interlaced and progressive HDTV systems. EIA-805 defines how transfer VBI data over these YPbPr video interfaces.
EIA-775	EIA-775 defines a specification for a baseband digital interface to a DTV using IEEE 1394 and provides a level of functionality that is similar to the analog system. It is designed to enable interoperability between a DTV and various types

of consumer digital audio/video sources, including settop boxes and DVRs or VCRs.

EIA-775.1 adds mechanisms to allow a source of MPEG services to utilize the MPEG decoding and display capabilities in a DTV.

EIA-775.2 adds information on how a digital storage device, such as a D-VHS or hard disk digital recorder, may be used by the DTV or by another source device such as a cable set-top box to record or time-shift digital television signals. This standard supports the use of such storage devices by defining Service Selection Information (SSI), methods for managing discontinuities that occur during recording and playback, and rules for management of partial transport streams.

EIA-849 specifies profiles for various applications of the EIA-775 standard, including digital streams compliant with ATSC terrestrial broadcast, direct-broadcast satellite (DBS), OpenCable™, and standard definition Digital Video (DV) camcorders.

EIA-805

This standard specifies how VBI data are carried on component video interfaces, as described in EIA-770.1 (for 480p signals only), EIA-770.2 (for 480p signals only) and EIA-770.3. This standard does not apply to signals which originate in 480i, as defined in EIA-770.1 and EIA-770.2. The first VBI service defined is Copy Generation Management System (CGMS) information, including signal format and data structure when carried by the VBI of standard definition progressive and high definition YPbPr type component video signals. It is also intended to be usable when the YPbPr signal is converted into other component video interfaces including RGB and VGA.

EIA-861

The EIA-861 standard specifies how to include data, such as aspect ratio and format information, on DVI and HDMI.

**EIA-J
CPR-1204**

This EIA-J recommendation specifies another widescreen signaling (WSS) standard for NTSC video signals. WSS may be present on 20 and 283.

**Enhanced
Definition
Television
(EDTV)**

EDTV is content or a display capable of displaying a maximum of 576 progressive active scan lines. No aspect ratio is specified.

**Equalization
Pulses**

These are two groups of pulses, one that occurs before the serrated vertical sync and another group that occurs after. These pulses happen at twice the normal horizontal scan rate. They exist to ensure correct 2:1 interlacing in early televisions.

Error Concealment	The ability to hide transmission errors that corrupt the content beyond the ability of the receiver to properly display it. Techniques for video include replacing the corrupt region with either earlier video data, interpolated video data from previous and next frames, or interpolated data from neighboring areas within the current frame. Decoded MPEG video may also be processed using deblocking filters to reduce blocking artifacts. Techniques for audio include replacing the corrupt region with interpolated audio data.
Error Resilience	The ability to handle transmission errors without corrupting the content beyond the ability of the receiver to properly display it. MPEG-4 supports error resilience through the use of resynchronization markers, extended header code, data partitioning, and reversible VLCs.
ETSI EN 300 163	This specification defines NICAM 728 digital audio for PAL.
ETSI EN 300 231	This specification defines information sent during the vertical blanking interval using PAL teletext (ETSI EN 300 706) to control VCRs in Europe (PDC).
ETSI EN 300 294	Defines the widescreen signaling (WSS) information for PAL video signals. For (B, D, G, H, I) PAL systems, WSS may be present on line 23.
ETSI EN 300 421	This is the DVB-S specification.
ETSI EN 300 429	This is the DVB-C specification.
ETSI EN 300 468	This is the DVB SI (service information) specification.
ETSI EN 300 472	This is the specification for the carriage of teletext data (ETSI EN 300 706) in DVB bitstreams.
ETSI EN 300 706	This is the enhanced PAL teletext specification.
ETSI EN 300 708	This specification defines data transmission using PAL teletext (ETSI EN 300 706).
ETSI EN 300 743	This is the DVB subtitling specification.

ETSI EN 300 744	This is the DVB-T specification.
ETSI EN 301 775	This is the specification for the carriage of Vertical Blanking Information (VBI) data in DVB bitstreams.
ETSI ETS 300 731	Defines the PALplus standard, allowing the transmission of 16:9 programs over normal PAL transmission systems.
ETSI ETS 300 732	Defines the ghost cancellation reference (GCR) signal for PAL.
Fade	Fading is a method of switching from one video source to another. Next time you watch a TV program (or a movie), pay extra attention when the scene is about to end and go on to another. The scene fades to black, then a fade from black to another scene occurs. Fading between scenes without going to black is called a dissolve. One way to do a fade is to use an Alpha Mixer.
Field	An interlaced display is made using two fields, each one containing one-half of the scan lines needed to make up one frame of video. Each field is displayed in its entirety—therefore, the odd field is displayed, then the even, then the odd, and so on. Fields only exist for interlaced scanning systems. So for (M) NTSC, which has 525 lines per frame, a field has 262.5 lines, and two fields make up a 525-line frame.
Firewire	When Apple Computer initially developed IEEE 1394, they called it Firewire.
Flicker	Flicker occurs when the frame rate of the video is too low. It's the same effect produced by an old fluorescent light fixture. The two problems with flicker are that it's distracting and tiring to the eyes.
FM	See Frequency Modulation.
Frame	A frame of video is essentially one picture or "still" out of a video stream. By playing these individual frames fast enough, it looks like people are "moving" on the screen. It's the same principle as flip cards, cartoons, and movies.
Frame Buffer	A frame buffer is a memory used to hold an image for display. How much memory are we talking about? Well, let's assume a horizontal resolution of 640 pixels and 480 scan lines, and we'll use the RGB color space. This works out to be:

$$640 \times 480 \times 3 = 921{,}600 \text{ bytes or } 900 \text{ kB}$$

So, 900 kB are needed to store one frame of video at that resolution.

Frame Rate   The frame rate of a video source is how fast a new still image is available. For example, with the NTSC system, the entire display is repainted about once every 30th of a second, for a frame rate of about 30 frames per second. For PAL, the frame rate is 25 frames per second. For computer displays, the frame rate is usually about 75 frames per second.

Frame Rate Conversion   Frame rate conversion is the act of converting one frame rate to another.

Front Porch   This is the area of the video waveform that sits between the start of horizontal blanking and the start of horizontal sync.

FVD   FVD (Finalized Versatile Disc) is an increased-capacity red-laser DVD disc and player specification from Taiwan. It uses Microsoft's Windows Media Video 9 (WMV9) and Windows Media Audio 9 (WMA9) codecs. The 5.4GB/9.8GB FVD-1 disc supports up to 135 minutes of WMV9 720p content. The 6GB/11GB FVD-2 disc will support up to 1080i content.

Gamma   The transfer characteristics of most cameras and displays are nonlinear. For a display, a small change in amplitude when the signal level is small produces a change in the display brightness level, but the same change in amplitude at a high level will not produce the same magnitude of brightness change. This nonlinearity is known as gamma.

Gamma Correction   Before being displayed, linear RGB data must be processed (gamma corrected) to compensate for the nonlinearity of the display.

GCR   See Ghost Cancellation Reference Signal.

Genlock   A video signal provides all of the information necessary for a video decoder to reconstruct the picture. This includes brightness, color, and timing information. To properly decode the video signal, the video decoder must lock to all the timing information embedded within the video signal, including the color burst, horizontal sync, and vertical sync. The decoder looks at the color burst of the video signal and reconstructs the original color subcarrier that was used by the encoder. This is needed to decode the color information properly. It also generates a sample clock (done by looking at the sync information within the video signal), used to clock pixel data out of the decoder into a memory or another circuit for processing. The circuitry within the decoder that does all of this work is called the genlock circuit. Although it sounds simple, the genlock circuit must be able to handle very bad video sources, such as the output of

VCRs, cameras, and toys. In reality, the genlock circuit is the most complex section of a video decoder.

Ghost Cancellation Reference

A reference signal on (M) NTSC scan lines 19 and 282 and (B, D, G, H, I) PAL scan line 318 that allows the removal of ghosting from TVs. Filtering is employed to process the transmitted GCR signal and determine how to filter the entire video signal to remove the ghosting. ITU-R BT.1124 and ETSI ETS 300 732 define the standard each country uses. ATSC A/49 also defines the standard for NTSC.

Gray Scale

The term gray scale has several meanings. It some cases, it means the luma component of color video signals. In other cases, it means a black-and-white video signal.

H.261, H.263

The ITU-T H.261 and H.263 video compression standards were developed to implement video conferencing over ISDN, LANs, regular phone lines, etc. H.261 supports video resolutions of 352×288 and 176×144 at up to 29.97 frames per second. H.263 supports video resolutions of 1408×1152, 704×576, 352×288, 176×144, and 128×96 at up to 29.97 frames per second.

H.264

The "next-generation" video codec. Previously known as "H.26L", "JVT", and "AVC" (advanced video codec), it is now also a MPEG-4 Part 10 standard.

ITU-T H.264 offers bit rates up to 50% less than the MPEG-4 advanced simple profile (ASP) video codec for the same video quality. It is designed to compete with the SMPTE VC-9 video codec in bit rate and quality.

H.26L

Early name for the H.264 video codec.

HD-CIF

See CIF.

HD-SDTI

High Data-Rate Serial Data Transport Interface, defined by SMPTE 348M.

HDMI

Abbreviation for High Definition Multimedia Interface, a single-cable digital audio/video interface for consumer equipment. It is designed to replace DVI in a backwards compatible fashion and supports EIA-861 and HDCP.

Digital RGB or YCbCr data at rates up to 5 Gbps are supported (HDTV requires 2.2 Gbps). Up to 8 channels of 32-192 kHz digital audio are also supported, along with AV.link (remote control) capability and a smaller 15mm 19-pin connector.

HDTV

See High Definition Television.

HDV

High Definition DV.

High Definition Television (HDTV)	HDTV is capable of displaying at least 720 progressive or 1080 interlaced active scan lines. It must be capable of displaying a 16:9 image using at least 540 progressive or 810 interlaced active scan lines.
Highpass Filter	A circuit that passes frequencies above a specific frequency (the cutoff frequency). Frequencies below the cutoff frequency are reduced in amplitude to eliminate them.
Horizontal Blanking	During the horizontal blanking interval, the video signal is at the blank level so as not to display the electron beam when it sweeps back from the right to the left side of the CRT screen.
Horizontal Resolution	See Resolution.
Horizontal Scan Rate	This is how fast the scanning beam in a display sweeps from side to side. In the NTSC system, this rate is 63.556 ms, or 15.734 kHz. That means the scanning beam moves from side to side 15,734 times a second.
Horizontal Sync	This is the portion of the video signal that tells the display where to place the image in the left-to-right dimension. The horizontal sync pulse tells the receiving system where the beginning of the new scan line is.
House Sync	This is another name for black burst.
HSI	HSI stands for Hue, Saturation and Intensity. HSI is based on polar coordinates, while the RGB color space is based on a three-dimensional Cartesian coordinate system. The intensity, analogous to luma, is the vertical axis of the polar system. The hue is the angle and the saturation is the distance out from the axis. HSI is more intuitive to manipulate colors as opposed to the RGB space. For example, in the HSI space, if you want to change red to pink, you decrease the saturation. In the RGB space, what would you do? My point exactly. In the HSI space, if you wanted to change the color from purple to green, you would adjust the hue. Take a guess what you would have to do in the RGB space. However, the key thing to remember, as with all color spaces, is that it's just a way to represent a color—nothing more, nothing less.
HSL	This is similar to HSI, except that HSL stands for Hue, Saturation and Lightness.
HSV	This is similar to HSI, except that HSV stands for Hue, Saturation and Value.

HSYNC Check out the Horizontal Sync definition.

Hue In technical terms, hue refers to the wavelength of the color. That means that hue is the term used for the base color—red, green, yellow, etc. Hue is completely separate from the intensity or the saturation of the color. For example, a red hue could look brown at low saturation, bright red at a higher level of saturation, or pink at a high brightness level. All three "colors" have the same hue.

Huffman Coding Huffman coding is a method of data compression. It doesn't matter what the data is—it could be image data, audio data, or whatever. It just so happens that Huffman coding is one of the techniques used in JPEG, MPEG, H.261, and H.263 to help with the compression. This is how it works. First, take a look at the data that needs to be compressed and create a table that lists how many times each piece of unique data occurs. Now assign a very small code word to the piece of data that occurs most frequently. The next largest code word is assigned to the piece of data that occurs next most frequently. This continues until all of the unique pieces of data are assigned unique code words of varying lengths. The idea is that data that occurs most frequently is assigned a small code word, and data that rarely occurs is assigned a long code word, resulting in space savings.

HVD HVD (High-definition Versatile Disc) is a red-laser DVD disc and player specification from China. It supports up to 150 minutes of 720p MPEG-2 content. The player supports 1080i and 720p video outputs.

IDTV See Improved Definition Television.

IEC 60461 Defines the longitudinal (LTC) and vertical interval (VITC) timecode for NTSC and PAL video systems. LTC requires an entire field time to transfer timecode information, using a separate track. VITC uses one scan line each field during the vertical blanking interval. Also see SMPTE 12M.

IEC 60958 Defines a serial digital audio interface for consumer (SPDIF) and professional applications.

IEC 61834 Defines the DV (originally the "Blue Book") standard. Also see BT.1618 and SMPTE 314M.

IEC 61880 Defines the widescreen signaling (WSS) information for NTSC video signals. WSS may be present on lines 20 and 283.

IEC 61883   Defines the methods for transferring data, audio, DV (IEC 61834), and MPEG-2 data over IEEE 1394.

IEC 62107   Defines the Super VideoCD standard.

IEEE 1394   A high-speed "daisy-chained" serial interface. Digital audio, video, and data can be transferred with either a guaranteed bandwidth or a guaranteed latency. It is hot-pluggable, and uses a small 6-pin or 4-pin connector, with the 6-pin connector providing power.

iLink   Sony's name for their IEEE 1394 interface.

Illegal Video   Some colors that exist in the RGB color space can't be represented in the NTSC and PAL video domain. For example, 100% saturated red in the RGB space (which is the red color on full strength and the blue and green colors turned off) can't exist in the NTSC video signal, due to color bandwidth limitations. The NTSC encoder must be able to determine that an illegal color is being generated and stop that from occurring, since it may cause over-saturation and blooming.

Improved Definition Television (IDTV)   IDTV is different from HDTV. IDTV is a system that improves the display on TVs by adding processing in the TV; standard NTSC or PAL signals are transmitted.

Intensity   This is the same thing as Brightness.

Interlaced   An interlaced video system is one where two interleaved fields are used to generate one video frame. Therefore, the number of lines in a field is one-half of the number of lines in a frame. In NTSC, there are 262.5 lines per field (525 lines per frame), while there are 312.5 lines per field (625 lines per frame) in PAL. Each field is drawn on the screen consecutively—first one field, then the other.

Interpolation   Interpolation is a mathematical way of generating additional information. Let's say that an image needs to be scaled up by a factor of two, from 100 samples to 200 samples. The "missing" samples are generated by calculating (interpolating) new samples between two existing samples. After all of the "missing" samples have been generated—presto!—200 samples exist where only 100 existed before, and the image is twice as big as it used to be.

IRE Unit   An arbitrary unit used to describe the amplitude characteristics of a video signal. White is defined to be 100 IRE and the blanking level is defined to be 0 IRE.

ISMA   Abbreviation for the Internet Streaming Media Alliance. ISMA is a group of industry leaders in content management, distribution infrastructure and media streaming working together to promote open standards for developing end-to-end media streaming solutions. The ISMA specification defines the exact features of the MPEG-4 standard that have to be implemented on the server, client and intermediate components to ensure interoperability between the entire streaming workflow. Similarly, it also defines the exact features and the selected formats of the RTP, RTSP, and SDP standards that have to be implemented.

The ISMA v1.0 specification defines two hierarchical profiles. Profile 0 is aimed to stream audio/video content on wireless and narrowband networks to low-complexity devices, such as cell phones or PDAs, that have limited viewing and audio capabilities. Profile 1 is aimed to stream content over broadband-quality networks to provide the end user with a richer viewing experience. Profile 1 is targeted to more powerful devices, such as set-top boxes and personal computers

ITU-R BT.xxx   See BT.xxx.

Jitter   Short-term variations in the characteristics (such as frequency, amplitude, etc.) of a signal.

JPEG   JPEG stands for Joint Photographic Experts Group. However, what people usually mean when they use the term "JPEG" is the image compression standard they developed. JPEG was developed to compress still images, such as photographs, a single video frame, something scanned into the computer, and so forth. You can run JPEG at any speed that the application requires. For a still picture database, the algorithm doesn't have to be very fast. If you run JPEG fast enough, you can compress motion video—which means that JPEG would have to run at 50 or 60 fields per second. This is called motion JPEG or M-JPEG. You might want to do this if you were designing a video editing system. Now, M-JPEG running at 60 fields per second is not as efficient as MPEG-2 running at 60 fields per second because MPEG was designed to take advantage of certain aspects of motion video.

JVT   Abbreviation for Joint Video Team, a group of video experts that work on new video codecs. It is a collaborative effort between the International Telecommunication Union (ITU), the International Electrotechnical Commission (IEC)

and the International Organization for Standardization (ISO). "JVT" was also an early name for the H.264 video codec.

kbps
Abbreviation for kilobits per second.

kBps
Abbreviation for kilobytes per second.

Line-Locked Clock
A design that ensures that there is always a constant number of samples per scan line, even if the timing of the line changes.

Line Store
A line store is a memory used to hold one scan line of video. If the horizontal resolution of the active display is 640 samples and RGB is used as the color space, the line store would have to be 640 locations long by 3 bytes wide. This amounts to one location for each sample and each color. Line stores are typically used in filtering algorithms. For example, a comb filter is made up of one or more line stores.

Linearity
Linearity is a basic measurement of how well an ADC or DAC is performing. Linearity is typically measured by making the ADC or DAC attempt to generate a linearly increasing signal. The actual output is compared to the ideal of the output. The difference is a measure of the linearity. The smaller the number, the better. Linearity is typically specified as a range or percentage of LSBs (Least Significant Bits).

Locked
When a PLL is accurately producing timing that is precisely lined up with the timing of the incoming video source, the PLL is said to be "locked." When a PLL is locked, the PLL is stable and there is minimum jitter in the generated sample clock.

Longitudinal Timecode
Timecode information is stored on a separate track from the video, requiring an entire field time to store or read it.

Lossless
Lossless is a term used with compression. Lossless compression is when the decompressed data is exactly the same as the original data. It's lossless because you haven't lost anything.

Lossy
Lossy compression is the exact opposite of lossless. The regenerated data is different from the original data. The differences may or may not be noticeable, but if the two images are not identical, the compression was lossy.

Lowpass Filter
A circuit that passes frequencies below a specific frequency (the cutoff frequency). Frequencies above the cutoff frequency are reduced in amplitude to eliminate them.

LTC	See Longitudinal Timecode.
Luma	As mentioned in the definition of chroma, the NTSC and PAL video systems use a signal that has two pieces: the black and white part, and the color part. The black and white part is the luma. It was the luma component that allowed color TV broadcasts to be received by black and white TVs and still remain viewable.
Luminance	In video, the terms luminance and luma are commonly (and incorrectly) interchanged. See the definition of Luma.
Mbps	Abbreviation for megabits per second.
MBps	Abbreviation for megabytes per second.
MESECAM	A technique of recording SECAM video. Instead of dividing the FM color subcarrier by four and then multiplying back up on playback, MESECAM uses the same heterodyne conversion as PAL.
MHP	See Multimedia Home Platform.
M-JPEG	See Motion JPEG.
Modulator	A modulator is basically a circuit that combines two different signals in such a way that they can be pulled apart later. What does this have to do with video? Let's take the NTSC system as an example, although the example applies equally as well to PAL. The NTSC system may use the YIQ or YUV color space, with the I and Q or U and V signals containing all of the color information for the picture. Two 3.58-MHz color subcarriers (90 degrees out of phase) are modulated by the I and Q or U and V components and added together to create the chroma part of the NTSC video.
Moiré	This is a type of image artifact. A moiré effect occurs when a pattern is created on the display where there really shouldn't be one. A moiré pattern is typically generated when two different frequencies beat together to create a new, unwanted frequency.
Monochrome	A monochrome signal is a video source having only one component. Although usually meant to be the luma (or black-and-white) video signal, the red video signal coming into the back of a computer display is monochrome because it only has one component.

Monotonic	This is a term that is used to describe ADCs and DACs. An ADC or DAC is said to be monotonic if for every increase in input signal, the output increases. Any ADC or DAC that is nonmonotonic—meaning that the output decreases for an increase in input—is bad! Nobody wants a nonmonotonic ADC or DAC.
Motion Estimation	Motion estimation is trying to figure out where an object has moved to from one video frame to the other. Why would you want to do that? Well, let's take an example of a video source showing a ball flying through the air. The background is a solid color that is different from the color of the ball. In one video frame the ball is at one location and in the next video frame the ball has moved up and to the right by some amount. Now let's assume that the video camera has just sent the first video frame of the series. Now, instead of sending the second frame, wouldn't it be more efficient to send only the position of the ball? Nothing else moves, so only two little numbers would have to be sent. This is the essence of motion estimation. By the way, motion estimation is an integral part of MPEG, H.261, and H.263.
Motion JPEG	JPEG compression or decompression that is applied real-time to video. Each field or frame of video is individually processed.
MPEG	MPEG stands for Moving Picture Experts Group. This is an ISO/IEC (International Standards Organization) body that is developing various compression algorithms. MPEG differs from JPEG in that MPEG takes advantage of the redundancy on a frame-to-frame basis of a motion video sequence, whereas JPEG does not.
MPEG-1	MPEG-1 was the first MPEG standard defining the compression format for real-time audio and video. The video resolution is typically 352×240 or 352×288, although higher resolutions are supported. The maximum bit rate is about 1.5 Mbps. MPEG-1 is used for the Video CD format.
MPEG-2	MPEG-2 extends the MPEG-1 standard to cover a wider range of applications. Higher video resolutions are supported to allow for HDTV applications, and both progressive and interlaced video are supported. MPEG-2 is used for DVD–Video and the ARIB, ATSC, DVB and OpenCable™ digital television standards.
MPEG-3	MPEG-3 was originally targeted for HDTV applications. This was incorporated into MPEG-2, so there is no MPEG-3 standard.

MPEG-4

MPEG-4 (ISO/IEC 14496) supports an object-based approach, where scenes are modeled as compositions of objects, both natural and synthetic, with which the user may interact. Visual objects in a scene can be described mathematically and given a position in a two- or three-dimensional space. Similarly, audio objects can be placed in a sound space. Thus, the video or audio object need only be defined once; the viewer can change his viewing position, and the calculations to update the audio and video are done locally. Classical "rectangular" video, as from a camera, is one of the visual objects supported. In addition, there is the ability to map images onto computer-generated shapes, and a text-to-speech interface.

H.264, a "next-generation" video codec, has been included in the MPEG-4 standard as Part 10.

MPEG-7

MPEG-7 standardizes the description of multimedia material (referred to as metadata), such as still pictures, audio, and video, regardless if locally stored, in a remote database, or broadcast. Examples are finding a scene in a movie, finding a song in a database, or selecting a broadcast channel. The searcher for an image can use a sketch or a general description. Music can be found using a "query by humming" format.

MTS

Multichannel Television Sound. A generic name for various stereo audio implementations, such as BTSC and Zweiton.

NABTS

North American Broadcast Teletext Specification (EIA-516). This is also ITU-R BT.653 525-line system C teletext. However, the NABTS specification goes into much more detail.

NexTView

An electronic program guide (EPG) based on ETSI ETS 300 707.

NICAM 728

A technique of implementing digital stereo audio for PAL video using another audio subcarrier. The bit rate is 728 kbps. It is discussed in BS.707 and ETSI EN 300 163. NICAM 728 is also used to transmit non-audio digital data in China.

Noninterlaced

This is a method of scanning out a video display that is the total opposite of interlaced. All of the lines in the frame are scanned out sequentially, one right after the other. The term "field" does not apply in a noninterlaced system. Another term for a noninterlaced system is progressive scan.

NTSC Never Twice the Same Color, Never The Same Color, or National Television Standards Committee, depending on who you're talking to. Technically, NTSC is just a color modulation scheme. To specify the color video signal fully, it should be referred to as (M) NTSC. "NTSC" is also commonly (though incorrectly) used to refer to any 525/59.94 or 525/60 video system. See also NTSC 4.43.

NTSC 4.43 This is a NTSC video signal that uses the PAL color subcarrier frequency (about 4.43 MHz). It was developed by Sony in the 1970s to more easily adapt European receivers to accept NTSC signals.

nVOD Abbreviation for near-video-on-demand. See Video-on-Demand.

OIRT Organisation Internationale de Radiodiffusion et Television.

Open Subtitles See Subtitles.

Oversampled VBI Data See Raw VBI Data.

Overscan When an image is displayed, it is "overscanned" if a small portion of the image extends beyond the edges of the screen. Overscan is common in TVs that use CRTs to allow for aging and variations in components, temperature and power supply.

PAL PAL stands for Phase Alternation Line, Picture Always Lousy, or Perfect At Last depending on your viewpoint. Technically, PAL is just a color modulation scheme. To fully specify the color video signal it should be referred to as (B, D, G, H, I, M, N, or N_C) PAL. (B, D, G, H, I) PAL is the color video standard used in Europe and many other countries. (M, N, N_C) PAL is also used in a few places, but is not as popular. "PAL" is also commonly (though incorrectly) used to refer to any 625/50 video system. See also PAL 60.

PAL 60 This is a NTSC video signal that uses the PAL color subcarrier frequency (about 4.43 MHz) and PAL-type color modulation. It is a further adaptation of NTSC 4.43, modifying the color modulation in addition to changing the color subcarrier frequency. It was developed by JVC in the 1980s for use with their video disc players, hence the early name of "Disk-PAL."

There is a little-used variation, also called PAL 60, which is a PAL video signal that uses the NTSC color subcarrier frequency (about 3.58 MHz), and PAL-type color modulation.

PALplus	PALplus is 16:9 aspect ratio version of PAL, designed to be transmitted using normal PAL systems. 16:9 TVs without the PALplus decoder, and standard 4:3 TVs, show a standard picture. It is defined by BT.1197 and ETSI ETS 300 731.
PDC	See Program Delivery Control.
Pedestal	Pedestal is an offset used to separate the black level from the blanking level by a small amount. When a video system doesn't use a pedestal, the black and blanking levels are the same. (M) NTSC uses a pedestal, (B, D, G, H, I) PAL does not. (M) NTSC-J used in Japan also does not use a pedestal.
Phase Adjust	This is a term used to describe a method of adjusting the hue in a NTSC video signal. The phase of the color subcarrier is moved, or adjusted, relative to the color burst. PAL and SECAM systems do not usually have a phase (or hue) adjust control.
Pixel	A pixel, which is short for picture element, is the smallest sample that makes up a scan line. For example, when the horizontal resolution is defined as 640 pixels, that means that there are 640 individual locations, or samples, that make up the visible portion of each horizontal scan line. Pixels may be square or rectangular.
Pixel Clock	The pixel clock is used to divide the horizontal line of video into samples. The pixel clock has to be stable (a very small amount of jitter) relative to the video or the image will not be stored correctly. The higher the frequency of the pixel clock, the more samples per line there are.
Pixel Drop Out	This can be a real troublemaker, since it can cause artifacts. In some instances, a pixel drop out looks like black spots on the screen, either stationary or moving around. Several things can cause pixel drop out, such as the ADC not digitizing the video correctly. Also, the timing between the ADC and the frame buffer might not be correct, causing the wrong number to be stored in memory. For that matter, the timing anywhere in the video stream might cause a pixel drop out.
Primary Colors	A set of colors that can be combined to produce any desired set of intermediate colors, within a limitation call the "gamut." The primary colors for color television are red, green, and blue. The exact red, green, and blue colors used are dependent on the television standard.
Program Delivery Control	Information sent during the vertical blanking interval using teletext to control VCRs in Europe. The specification is ETSI ETS 300 231.

Progressive Scan	See Noninterlaced.
Pseudo Color	Pseudo color is a term used to describe a technique that applies color, or shows color, where it does not really exist. We are all familiar with the satellite photos that show temperature differences across a continent or the multicolored cloud motion sequences on the nightly weather report. These are real-world examples of pseudo color. The color does not really exist. A computer uses a lookup table memory to add the color so information, such as temperature or cloud height, is viewable.
Px64	This is basically the same as H.261. The term is starting to fade away since H.261 is used in applications other than ISDN video conferencing.
QAM	See Quadrature Amplitude Modulation.
QCIF	Quarter Common Interface Format. This video format was developed to allow the implementation of cheaper video phones. The QCIF format has a resolution of 176 × 144 active pixels and a refresh rate of 29.97 frames per second.
QSIF	Quarter Standard Interface Format. The computer industry, which uses square pixels, has defined QSIF to be 160 × 120 active pixels, with a refresh rate of whatever the computer is capable of supporting.
Quad Chroma	Quad chroma refers to a technique where the sample clock is four times the frequency of the color burst. For NTSC this means that the sample clock is about 14.32 MHz (4× 3.579545 MHz), while for PAL the sample clock is about 17.73 MHz (4× 4.43361875 MHz). The reason these are popular sample clock frequencies is that, depending on the method chosen, they make the chrominance (color) decoding and encoding easier.
Quadrature Amplitude Modulation	A method of encoding digital data onto a carrier for RF transmission. QAM is typically used for cable transmission of digital SDTV and HDTV signals. DVB-C supports 16-QAM, 32-QAM, 64-QAM, 128-QAM, and 256-QAM, although receivers need only support up to 64-QAM.
Quadrature Modulation	The modulation of two carrier components which are 90 degrees apart in phase.
Quantization	The process of converting a continuous analog signal into a set of discrete levels (digitizing).

Quantization Noise	This is the inherent uncertainty introduced during quantization since only discrete, rather than continuous, levels are generated. Also called quantization distortion.
Raster	Essentially, a raster is the series of scan lines that make up a picture. You may from time to time hear the term raster line—it's the same as scan line. All of the scan lines that make up a frame of video form a raster.
Raw VBI Data	A technique where VBI data (such as teletext and captioning data) is sampled by a fast sample clock (i.e. 27 MHz), and output. This technique allows software decoding of the VBI data to be done.
RC Time Code	Rewritable time code, used in consumer video products.
Real Time Control Protocol	See RTCP.
Real Time Streaming Protocol	See RTSP.
Real Time Transport Protocol	See RTP.
Rectangular Pixels	Pixels that are not "square pixels" are "rectangular pixels."
Residual Subcarrier	This is the amount of color subcarrier information present in white, gray, or black areas of a composite color video signal (ideally, there is none present). The number usually appears as –n dB. The larger "n" is, the better.
Resolution	This is the basic measurement of how much information is visible for an image. It is usually described as "h" × "v." The "h" is the horizontal resolution (across the display) and the "v" is the vertical resolution (down the display). The higher the numbers, the better, since that means there is more detail to see. If only one number is specified, it is the horizontal resolution. Displays specify the maximum resolution they can handle, determined by the display technology and the electronics used. The actual resolution will be the resolution of either the source or the display, whichever is lower.

Vertical resolution is the number of white-to-black and black-to-white transitions that can be seen from the top to the bottom of the picture. The maximum number is the number of active scan lines used by the image. The actual vertical resolution may be less due to processing, interlacing, overscanning, or limited by the source.

Horizontal resolution is the number of white-to-black and black-to-white transitions that can be seen from the left to the right of the picture. For digital displays, the maximum number is the number of active pixels used by a scan line. For both analog and digital displays, the actual horizontal resolution may be less due to processing, overscanning, or limited by the source.

Resource Reservation Protocol

See RSVP.

Retrace

Retrace is what the electron beam does when it gets to the right-hand edge of the CRT display to get back to the left-hand edge. Retrace happens during the horizontal blanking time.

RGB

Abbreviation for red, green, blue.

RS-170, RS-170A

RS-170 is the United States standard that was used for black-and-white TV, and defines voltage levels, blanking times, the width of the sync pulses, and so forth. The specification spells out everything required for a receiver to display a monochrome picture. Now, SMPTE 170M is essentially the same specification, modified for color TV by adding the color components. They modified RS-170 just a tiny little bit so that color could be added (RS-170A), with the final result being SMPTE 170M for NTSC. This tiny little change was so small that the existing black-and-white TVs didn't even notice it.

RS-343

RS-343 does the same thing as RS-170, defining a specification for transferring analog video, but the difference is that RS-343 is for high-resolution computer graphics analog video, while RS-170 is for TV-resolution NTSC analog video.

RSDL

RSDL stands for Reverse Spiral Dual Layer. It is a storage method that uses two layers of information on one side of a DVD. For movies that are longer than can be recorded on one layer, the disc stops spinning, reverses direction, and begins playing from the next layer.

RSVP

RSVP (Resource Reservation Protocol) is a control protocol that allows a receiver to request a specific quality of service level over an IP network. Real-time applications, such as streaming video, use RSVP to reserve necessary resources at routers along the transmission paths so that the requested bandwidth can be available when the transmission actually occurs.

RTCP

RTCP (Real-Time Control Protocol) is a control protocol designed to work in conjunction with RTP. During a RTP session, participants periodically send RTCP packets to convey status on quality of service and membership management. RTCP also uses RSVP to reserve resources to guarantee a given quality of service.

RTP

RTP (Real-Time Transport Protocol) is a packet format and protocol for the transport of real-time audio and video data over an IP network. The data may be any file format, including MPEG-2, MPEG-4, ASF, QuickTime, etc. Implementing time reconstruction, loss detection, security and content identification, it also supports multicasting (one source to many receivers) and unicasting (one source to one receiver) of real-time audio and video. One-way transport (such as video-on-demand) as well as interactive services (such as Internet telephony) are supported. RTP is designed to work in conjunction with RTCP.

RTSP

RTSP (Real-Time Streaming Protocol) is a client-server protocol to enable controlled delivery of streaming audio and video over an IP network. It provides "VCR-style" remote control capabilities such as play, pause, fast forward, and reverse. The actual data delivery is done using RTP.

Run Length Coding

Run length coding is a type of data compression. Let's say that this page is wide enough to hold a line of 80 characters. Now, imagine a line that is almost blank except for a few words. It's 80 characters long, but it's just about all blanks—let's say 50 blanks between the words "coding" and "medium." These 50 blanks could be stored as 50 individual codes, but that would take up 50 bytes of storage. An alternative would be to define a special code that said a string of blanks is coming and the next number is the amount of blanks in the string. So, using our example, we would need only 2 bytes to store the string of 50 blanks, the first special code byte followed by the number 50. We compressed the data; 50 bytes down to 2. This is a compression ration of 25:1. Not bad, except that we only compressed one line out of the entire document, so we should expect that the total compression ratio would be much less.

Run length coding all by itself as applied to images is not as efficient as using a DCT for compression, since long runs of the same "number" rarely exist in real-world images. The only advantage of run length coding over the DCT is that it is easier to implement. Even though run length coding by itself is not efficient for compressing images, it is still used as part of the JPEG, MPEG, H.261, and H.263 compression schemes.

R–Y

In video, the red-minus-luma signal, also called a color difference signal. When added to the luma (Y) signal, it produces the red video signal.

SABC

South Africa Broadcasting Corporation.

Sample

To obtain values of a signal at periodic intervals. Also the value of a signal at a given moment in time.

Sample and Hold

A circuit that samples a signal and holds the value until the next sample is taken.

Sample Rate

Sample rate is how often a sample of a signal is taken. The sample rate is determined by the sample clock.

SAP

See Secondary Audio Program.

Saturation

Saturation is the amount of color present. For example, a lightly saturated red looks pink, while a fully saturated red looks like the color of a red crayon. Saturation does not mean the brightness of the color, just how much "pigment" is used to make the color. The less "pigment," the less saturated the color is, effectively adding white to the pure color.

Scaling

Scaling is the act of changing the resolution of an image. For example, scaling a 640 × 480 image by one-half results in a 320 × 240 image. Scaling by 2× results in an image that is 1280 × 960. There are many different methods for image scaling, and some "look" better than others. In general, though, the better the algorithm "looks," the more expensive it is to implement.

Scan Line

A scan line is an individual line across the display. It takes 525 of these scan lines to make up a NTSC TV picture and 625 scan lines to make up a PAL TV picture.

Scan Velocity Modulation

See Velocity Scan Modulation.

SCART

Syndicat des Constructeurs d'Appareils Radio Recepteurs et Televiseurs. This is a 21-pin connector supported by many consumer video components in Europe. It allows mono or stereo audio and composite, s-video, or RGB video to be transmitted between equipment.

The IEC 60933-1 and 60933-2 standards specify the basic SCART connector, including signal levels.

SDI

Serial Digital I/O. Another name for the 270 Mbps or 360 Mbps serial interface defined by BT.656. It is used primarily on professional and studio video equipment.

SDTI

Serial Data Transport Interface, defined by SMPTE 305M.

SDTV

See Standard Definition Television.

SECAM

This is another color video format similar to PAL. The major differences between the two are that in SECAM the chroma is FM modulated and the R–Y and B–Y signals are transmitted line sequentially. SECAM stands for Sequentiel Couleur Avec Memoire or Sequential Color with Memory.

Secondary Audio Program

Generally used to transmit audio in a second language. May also be used to transmit the [aural] description of key visual elements of a program, inserted into the natural pauses in the audio of the programming.

Serration Pulses

These are pulses that occur during the vertical sync interval of NTSC, PAL, and SECAM, at twice the normal horizontal scan rate. The reason these exist was to ensure correct 2:1 interlacing in early televisions and eliminate DC offset buildup.

Setup

Setup is the same thing as Pedestal.

SIF

Standard (or Source) Input Format. This video format was developed to allow the storage and transmission of digital video. The 625/50 SIF format has a resolution of 352 × 288 active pixels and a refresh rate of 25 frames per second. The 525/60 SIF format has a resolution of 352 × 240 active pixels and a refresh rate of 30 frames per second. Note that MPEG-1 allows resolutions up to 4095 × 4095 active pixels; however, there is a "constrained subset" of parameters defined as SIF. The computer industry, which uses square pixels, has defined SIF to be 320 × 240 active pixels, with a refresh rate of whatever the computer is capable of supporting.

Signal-to-Noise Ratio (SNR)	Signal-to-noise ratio is the magnitude of the signal divided by the amount of unwanted stuff that is interfering with the signal (the noise). SNR is usually described in decibels, or "dB," for short; the bigger the number, the better looking the picture.
Silent Radio	Silent Radio is a service that feeds data that is often seen in hotels and nightclubs. It's usually a large red sign that shows current news, events, scores, etc. It is present on NTSC lines 10–11 and 273–274, and uses encoding similar to EIA-608.
Sliced VBI Data	A technique where a VBI decoder samples the VBI data (such as teletext and captioning data), locks to the timing information, and converts it to binary 0's and 1's. DC offsets, amplitude variations, and ghosting must be compensated for by the VBI decoder to accurately recover the data.
SMPTE 12M	Defines the longitudinal (LTC) and vertical interval (VITC) timecode for NTSC and PAL video systems. LTC requires an entire field time to store timecode information, using a separate track. VITC uses one scan line each field during the vertical blanking interval.
SMPTE 125M	720×480 pro-video interlaced standard (29.97 Hz). Covers the digital representation and the digital parallel interface. Also see BT.601 and BT.656.
SMPTE 170M	NTSC video specification for the United States. See RS-170A and BT.470.
SMPTE 240M	1920×1035 pro-video interlaced standard (29.97 or 30 Hz). Covers the analog RGB and YPbPr representation. The digital parallel interface is defined by SMPTE 260M. The digital serial interface is defined by SMPTE 292M.
SMPTE 244M	768×486 pro-video interlaced standard (29.97 Hz). Covers the digital representation (composite NTSC video sampled at $4\times$ Fsc) and the digital parallel interface. The digital serial interface is defined by SMPTE 259M.
SMPTE 253M	Analog RGB video interface specification for pro-video SDTV systems.
SMPTE 259M	Pro-video serial digital interface for SMPTE 244M.
SMPTE 260M	Digital representation and parallel interface for SMPTE 240M video.
SMPTE 266M	Defines the digital vertical interval timecode (DVITC). Also see BT.1366.

SMPTE 267M	960 × 480 pro-video interlaced standard (29.97 Hz). Covers the digital representation and the digital parallel interface. Also see BT.601 and BT.1302.
SMPTE 272M	Formatting AES/EBU digital audio and auxiliary data into the digital blanking intervals. Also see BT.1305.
SMPTE 274M	1920 × 1080 pro-video interlaced and progressive standards (29.97, 30, 59.94 and 60 Hz). Covers the digital representation, the analog RGB and YPbPr interfaces, and the digital parallel interface. The digital serial interface is defined by SMPTE 292M.
SMPTE 276M	Transmission of AES/EBU digital audio and auxiliary data over coaxial cable.
SMPTE 291M	Ancillary data packet and space formatting for pro-video digital interfaces. Also see BT.1364.
SMPTE 292M	1.485 Gbps pro-video HDTV serial interfaces.
SMPTE 293M	720 × 480 pro-video progressive standards (59.94 Hz). Covers the digital representation, the analog RGB and YPbPr interfaces, and the digital parallel interface. The digital serial interface is defined by SMPTE 294M. Also see BT.1358 and BT.1362.
SMPTE 294M	Pro-video serial digital interface for SMPTE 293M.
SMPTE 296M	1280 × 720 pro-video progressive standards. Covers the digital representation and the analog RGB and YPbPr interfaces. The digital parallel interface uses SMPTE 274M. The digital serial interface is defined by SMPTE 292M.
SMPTE 299M	24-bit digital audio format for pro-video HDTV serial interfaces. Also see BT.1365.
SMPTE 305M	Serial data transport interface (SDTI). This is a 270 or 360 Mbps serial interface based on BT.656 that can be used to transfer almost any type of digital data, including MPEG-2 program and transport bitstreams, DV bitstreams, etc. You cannot exchange material between devices that use different data types. Material that is created in one data type can only be transported to other devices that support the same data type. There are separate map documents that format each data type into the 305M transport.
SMPTE 308M	MPEG-2 4:2:2 profile at high level.

SMPTE 314M	Defines the data structure for DV audio, data and compressed video at 25 and 50 Mbps. Also see BT.1618 and IEC 61834.
SMPTE 322M	Data stream format for the exchange of DV audio, data and compressed video over the Serial Data Transport Interface (SDTI or SMPTE 305M).
SMTPE 344M	Defines a 540 Mbps serial digital interface for pro-video applications.
SMPTE 348M	High data-rate serial data transport interface (HD-SDTI). This is a 1.485 Gbps serial interface based on SMPTE 292M that can be used to transfer almost any type of digital data, including MPEG-2 program and transport bitstreams, DV bitstreams, etc. You cannot exchange material between devices that use different data types. Material that is created in one data type can only be transported to other devices that support the same data type. There are separate map documents that format each data type into the 348M transport.
SMPTE 370M	This SMPTE standard specifies a data structure for DV-based audio, data and compressed video at data rates of 100 Mbps. Also see BT.1620.
SMPTE RP160	Analog RGB and YPbPr video interface specification for pro-video HDTV systems.
SPDIF	Short for Sony/Philips Digital InterFace. This is a consumer interface used to transfer digital audio. A serial, self-clocking scheme is used, based on a coax or fiber interconnect. The audio samples may be 16–24 bits each. 16 different sampling rates are supported, with 32, 44.1, and 48 kHz being the most common. IEC 60958 now fully defines this interface for consumer and professional applications.
Split Sync Scrambling	Split sync is a video scrambling technique, usually used with either horizontal blanking inversion, active video inversion, or both. In split sync, the horizontal sync pulse is "split," with the second half of the pulse at +100 IRE instead of the standard –40 IRE. Depending on the scrambling mode, either the entire horizontal blanking interval is inverted about the +30 IRE axis, the active video is inverted about the +30 IRE axis, both are inverted, or neither is inverted. By splitting the horizontal sync pulse, a reference of both –40 IRE and +100 IRE is available to the descrambler. Since a portion of the horizontal sync is still at –40 IRE, some sync separators may still lock on the shortened horizontal sync pulses. However, the timing circuits that look for color burst a fixed interval after the end of horizontal sync may be confused. In addition, if the active video is inverted, some video information may fall below 0 IRE, possibly confusing sync detector circuits.

The burst is always present at the correct frequency and timing; however, the phase is shifted 180 degrees when the horizontal blanking interval is inverted.

Square Pixels

When the ratio of active pixels per line to active lines per frame is the same as the display aspect ratio. This is the same as the sampling lattice having equal spatial horizontal and vertical spacing of the sampling points.

Standard Definition Television (SDTV)

SDTV is content or a display capable of displaying a maximum of 576 interlaced active scan lines. No aspect ratio is specified.

Streaming Video

Compressed audio and video that is transmitted over the Internet or other network in real time. It usually offers "VCR-style" remote control capabilities such as play, pause, fast forward, and reverse.

Subcarrier

A secondary signal containing additional information that is added to a main signal.

Subsampled

Subsampled means that a signal has been sampled at a lower rate than some other signal in the system. A prime example of this is the 4:2:2 YCbCr color space used in ITU-R BT.601. For every two luma (Y) samples, only one Cb and Cr sample is present. This means that the Cb and Cr signals are subsampled.

Subtitles

Text that is added below or over a picture that usually reflects what is being said, possibly in another language. Open subtitles are transmitted as video that already has the subtitles present. Closed subtitles are transmitted during the VBI, and relies on the TV to decode it and position it below or over the picture. Closed captioning is a form of subtitling. Subtitling for DVB is specified in ETSI ETS 300 743.

Superblack

A keying signal that is embedded within the composite video signal as a level between black and sync. It is usually used to improve luma self-keying because the video signal contains black, making a good luma self-key hard to implement. When a downstream keyer detects the super black level, it inserts the second composite video signal.

Super VideoCD (Super VCD, SVCD)

Defined by the China National Technical Committee of Standards on Recording, this CD standard holds 35–70 minutes of digital audio and video information. MPEG-2 video is used, with a resolution of 480 × 480 (29.97 Hz frame rate) or 480 × 576 (25 Hz frame rate). Audio uses MPEG layer 2 at a bit rate of 32–384 kbps, and supports four mono, two stereo, or 5.1 channels. Subtitles use overlays rather than subpictures (DVD–Video) or being encoded as video (VideoCD). Variable bit-rate encoding is used, with a maximum bit rate of 2.6 Mbps. IEC 62107 defines the Super VideoCD standard.

XSVCD, although not an industry standard, increases the video resolution and bit rate to improve the video quality over SVCD. MPEG-2 video is still used, with a resolution of 720×480 (29.97 Hz frame rate) or 720×576 (25 Hz frame rate). Variable bit-rate encoding is still used, with a maximum bit rate of 9.8 Mbps.

Superbit DVD   See DVD-Video.

S-VHS   S-VHS is an enhancement to regular VHS video tape decks. S-VHS provides better resolution and less noise than VHS. S-VHS video tape decks support s-video inputs and outputs, although this is not required. It does, however, improve the quality by not having to separate and then merge the luma and chroma signals.

S-Video   Separate video, also called Y/C video. Separate luma (Y) and chroma (C) video signals are used, rather than a single composite video signal. By simply adding together the Y and C signals, you generate a composite video signal.

A DC offset of +2.3v may be present on the C signal when a letterbox picture format is present. A DC offset of +5v may be present to indicate when a 16:9 anamorphic picture format is present. A standard 4:3 receiver ignores all DC offsets, thus displaying a typical letterboxed picture.

The IEC 60933-5 standard specifies the s-video connector, including signal levels.

SVM   See Velocity Scan Modulation.

Sync   Sync is a fundamental, you gotta have it, piece of information for displaying any type of video. Essentially, the sync signal tells the display where to put the picture. The horizontal sync, or HSYNC for short, tells the display where to put the picture in the left-to-right dimension. The vertical sync, or VSYNC for short, tells the display where to put the picture from top-to-bottom.

Analog SDTV and EDTV signals use a bi-level sync, where the sync level is a known value below the blanking level. Analog HDTV signals use a tri-level sync, where the sync levels are known values above and below the blanking level.

The reason analog HDTV signals use a tri-level sync signal is timing accuracy. The horizontal timing reference point for a bi-level sync signal is defined as the 50% point of the leading edge of the horizontal sync pulse. In order to ascertain this point precisely, it is necessary to determine both the blanking level and sync-tip level and determine the mid-point value. If the signal is in any way distorted, this will reduce the timing accuracy.

With a tri-level sync signal, the timing reference point is the rising edge of the sync signal as it passes through the blacking level. This point is much easier to accurately determine, and can be implemented relatively easily. It is also more immune to signal distortion.

Sync Generator

A sync generator is a circuit that provides sync signals. A sync generator may have genlock capability.

Sync Noise Gate

A sync noise gate is used to define an area within the video waveform where the video decoder is to look for the sync pulse. Anything outside of this defined window will be rejected. The main purpose of the sync noise gate is to make sure that the output of the video decoder is nice, clean, and correct.

Sync Stripper

A video signal contains video information, which is the picture to be displayed, and timing (sync) information that tells the receiver where to put this video information on the display. A sync stripper pulls out the sync information from the video signal and throws the rest away.

Synchronous

Refers to two or more events that happen in a system or circuit at the same time.

SVCD

See Super VideoCD.

TDF

Telediffusion de France.

Teletext

A method of transmitting data with a video signal. ITU-R BT.653 lists the major teletext systems used around the world, while ETSI ETS 300 706 defines in detail the teletext standard for PAL. North American Broadcast Teletext Specification (NABTS) is 525-line system C.

For digital transmissions such as HDTV and SDTV, the teletext characters are multiplexed as a separate stream along with the video and audio data. It is common practice actually to embed this stream in the MPEG video bitstream itself, rather than at the transport layer. Unfortunately there is no wide-spread standard for this teletext stream—each system (DSS, DVB, ATSC, DVD) has its own solution.

Tessellated Sync

This is what the Europeans call serrated sync. See Serration Pulses and Sync.

Timebase Corrector	Certain video sources have their sync signals screwed up. The most common of these sources is the VCR. A timebase corrector "fixes" a video signal that has bad sync timing.
Tri-Level Sync	A sync signal that has three levels, and is commonly used for analog HDTV signals. See the definition for Sync.
True Color	True color means that each sample of an image is individually represented using three color components, such as RGB or YCbCr.
Underscan	When an image is displayed, it is "underscanned" if all of the image, including the top, bottom, and side edges, are visible on the display. Underscan is common in computer displays.
Uplink	The carrier used by Earth stations to transmit information to a satellite.
V-chip	See EIA-608.
Variable Bit Rate	Variable bit rate (VBR) means that a bitstream (compressed or uncompressed) has a changing number of bits each second. Simple scenes can be assigned a low bit rate, with complex scenes using a higher bit rate. This enables maintaining the audio and video quality at a more consistent level.
VBI	See Vertical Blanking Interval.
VBR	Abbreviation for Variable Bit Rate.
VCD	Abbreviation for VideoCD.
Velocity Scan Modulation	Commonly used in TVs to increase the apparent sharpness of a picture. At horizontal dark-to-light transitions, the beam scanning speed is momentarily increased approaching the transition, making the display relatively darker just before the transition. Upon passing into the lighter area, the beam speed is momentarily decreased, making the display relatively brighter just after the transition. The reverse occurs in passing from light to dark.
Vertical Blanking Interval (VBI)	During the vertical blanking interval, the video signal is at the blanking level so as not to display the electron beam when it sweeps back from the bottom to the top side of the CRT screen.
Vertical Interval Timecode	Timecode information is stored on a scan line during each vertical blanking interval.

Vertical Resolution	See Resolution.
Vertical Scan Rate	For noninterlaced video, this is the same as the frame rate. For interlaced video, it is usually one-half the field rate.
Vertical Sync	This is the portion of the video signal that tells the video decoder where the top of the picture is.
Vestigial Sideband	A method of encoding digital data onto a carrier for RF transmission. 8-VSB is used for over-the-air broadcasting of ATSC HDTV in the United States.
Video Carrier	A specific frequency that is modulated with video data before being mixed with the audio data and transmitted.
Video Interface Port	A digital video interface designed to simplify interfacing video ICs together. One portion is a digital video interface (based on BT.656) designed to simplify interfacing video ICs together. A second portion is a host processor interface. VIP is a VESA specification.
Video Mixing	Video mixing is taking two independent video sources (they must be gen-locked) and merging them together. See Alpha Mix.
Video Modulation	Converting a baseband video signal to an RF signal.
Video Module Interface (VMI)	A digital video interface designed to simplify interfacing video ICs together. It is being replaced by VIP.
Video-on-Demand	Video-on-demand, or VOD, allows a user to select which program to view at their convenience and playing starts almost immediately. When used over the Internet or other network, it is commonly called "streaming video". For broadcast, satellite and cable networks, it is commonly called "pay-per-view" and is usually confined to specific start times. For this reason, it may also be referred to as "near video-on-demand" or nVOD.
Video Program System (VPS)	VPS is used in some countries instead of PDC to control VCRs. The data format is the same as for PDC, except that it is transmitted on a dedicated line during the vertical blanking interval, usually line 16.

VideoCD	Compact discs that hold up to about an hour of digital audio and video information. MPEG-1 video is used, with a resolution of 352×240 (29.97 Hz frame rate) or 352×288 (25 Hz frame rate). Audio uses MPEG layer 2 at a fixed bit rate of 224 kbps, and supports two mono or one stereo channels (with optional Dolby® pro-logic). Fixed bit-rate encoding is used, with a bit rate of 1.15 Mbps. The next generation, defined for the Chinese market, is Super VideoCD.

XVCD, although not an industry standard, increases the video resolution and bit rate to improve the video quality over VCD. MPEG-1 video is still used, with a resolution of up to 720×480 (29.97 Hz frame rate) or 720×576 (25 Hz frame rate). Fixed bit-rate encoding is still used, with a bit rate of 3.5 Mbps.

VIP	See Video Interface Port.
VITC	See Vertical Interval Timecode.
VMI	See Video Module Interface.
VOB	DVD-Video movies are stored on the DVD using VOB files. They usually contain multiplexed Dolby Digital audio and MPEG-2 video. VOB Files are named as follows: vts_XX_Y.vob where XX represents the title and Y the part of the title. There can be 99 titles and 10 parts, although vts_XX_0.vob never contains video, usually just menu or navigational information.
VOD	See Video-on-Demand.
VPS	See Video Program System.
VSB	See Vestigial Sideband.
VSM	See Velocity Scan Modulation.
VSYNC	Check out the Vertical Sync definition.
White Level	This level defines what white is for the particular video system.
Wide Screen Signalling (WSS)	WSS may be used on (B, D, G, H, I) PAL line 23 and (M) NTSC lines 20 and 283 to specify the aspect ratio of the program and other information. 16:9 TVs may use this information to allow displaying of the program in the correct aspect ratio. ITU-R BT.1119 and ETSI EN 300 294 specify the WSS signal for PAL and NTSC systems. EIA-J CPR-1204 and IEC 61880 also specify another WSS signal for NTSC systems.

World System Teletext (WST)	BT.653 525-line and 625-line system B teletext.
WSS	See Wide Screen Signaling.
WST	See World System Teletext.
XSVCD	Abbreviation for eXtended Super VideoCD. See Super VideoCD.
XVCD	Abbreviation for eXtended VideoCD. See VideoCD.
Y/C Video	See S-video.
Y/C Separator	A Y/C separator is what's used in a video decoder to separate the luma and chroma in a NTSC or PAL system. This is the first thing that any video decoder must do. The composite video signal is fed to a Y/C separator so that the chroma can then be decoded further.
YCbCr	YCbCr is the color space originally defined by BT.601, and now used for all digital component video formats. Y is the luma component and the Cb and Cr components are color difference signals. The technically correct notation is Y'Cb'Cr' since all three components are derived from R'G'B'. Many people use the YCbCr notation rather than Y'CbCr or Y'Cb'Cr'.
	4:4:4 Y'CbCr means that for every Y sample, there is one sample each of Cb and Cr.
	4:2:2 Y'CbCr means that for every two horizontal Y samples, there is one sample each of Cb and Cr.
	4:1:1 Y'CbCr means that for every four horizontal Y samples, there is one sample each of Cb and Cr.
	4:2:0 Y'CbCr means that for every block of 2×2 Y samples, there is one sample each of Cb and Cr. There are three variations of 4:2:0 YCbCr, with the difference being the position of Cb and Cr sampling relative to Y.
	Note that the coefficients to convert R'G'B' to YCbCr are different for SDTV and HDTV applications.
YIQ	YIQ is a color space optionally used by the NTSC video system. The Y component is the black-and-white portion of the image. The I and Q parts are the color difference components; these are effectively nothing more than color placed over the black and white, or luma, component. Many people use the YIQ notation rather than Y'IQ or Y'I'Q'. The technically correct notation is Y'I'Q' since all three components are derived from R'G'B'.

YPbPr   YPbPr is a scaled version of the YUV color space, with specific levels and timing signals, designed to interface equipment together. Consumer video standards are defined by EIA-770; the professional video standards are defined by numerous SMPTE standards. VBI data formats for EIA-770 are defined by EIA-805. Many people use the YPbPr notation rather than Y'PbPr or Y'Pb'Pr'. The technically correct notation is Y'Pb'Pr' since all three components are derived from R'G'B'.

YUV   YUV is the color space used by the NTSC and PAL video systems. As with the YIQ color space, the Y is the luma component while the U and V are the color difference components. Many people use the YUV notation when they actually mean YCbCr data. Most use the YUV notation rather than Y'UV or Y'U'V'. The technically correct notation is Y'U'V' since all three components are derived from R'G'B'.

YUV is also the name for some component analog interfaces on consumer equipment. Some manufacturers incorrectly label it YCbCr. THX certification requires it to be labeled YPbPr.

YUV9   Intel's 4:1:0 YCbCr format. The picture is divided into blocks, with each block comprising 4×4 samples. For each block, sixteen 8-bit values of Y, one 8-bit value of Cb, and one 8-bit value of Cr are assigned. The result is an average of 9 bits per pixel.

YUV12   Intel's notation for MPEG-1 4:2:0 YCbCr stored in memory in a planar format. The picture is divided into blocks, with each block comprising 2×2 samples. For each block, four 8-bit values of Y, one 8-bit value of Cb, and one 8-bit value of Cr are assigned. The result is an average of 12 bits per pixel.

YUY2   Intel's notation for 4:2:2 YCbCr format.

Zipper   See the definition for Creepy-Crawlies.

Zoom   Zoom is a type of image scaling. Zooming is making the picture larger so that you can see more detail. The examples described in the definition of scaling are also examples that could be used here.

Zweiton   A technique of implementing stereo or dual-mono audio for NTSC and PAL video. One FM subcarrier transmits an L+R signal, and a second FM subcarrier transmits a R signal (for stereo) or a second L+R signal. It is discussed in BS.707, and is similar to the BTSC technique.

Index